Studies in Systems, Decision and Control

Volume 558

The series "Studies in Systems, Decision and Control" (SSDC) covers both new developments and advances, as well as the state of the art, in the various areas of broadly perceived systems, decision making and control–quickly, up to date and with a high quality. The intent is to cover the theory, applications, and perspectives on the state of the art and future developments relevant to systems, decision making, control, complex processes and related areas, as embedded in the fields of engineering, computer science, physics, economics, social and life sciences, as well as the paradigms and methodologies behind them. The series contains monographs, textbooks, lecture notes and edited volumes in systems, decision making and control spanning the areas of Cyber-Physical Systems, Autonomous Systems, Sensor Networks, Control Systems, Energy Systems, Automotive Systems, Biological Systems, Vehicular Networking and Connected Vehicles, Aerospace Systems, Automation, Manufacturing, Smart Grids, Nonlinear Systems, Power Systems, Robotics, Social Systems, Economic Systems and other. Of particular value to both the contributors and the readership are the short publication timeframe and the worldwide distribution and exposure which enable both a wide and rapid dissemination of research output.

Indexed by SCOPUS, DBLP, WTI Frankfurt eG, zbMATH, SCImago.

All books published in the series are submitted for consideration in Web of Science.

Laxminarayan Sahoo · Tapan Senapati ·
Madhumangal Pal · Ronald R. Yager
Editors

Decision Making Under Uncertainty Via Optimization, Modelling, and Analysis

Springer

Editors
Laxminarayan Sahoo
Department of Computer and Information Science
Raiganj University
Raiganj, West Bengal, India

Madhumangal Pal
Department of Applied Mathematics
Vidyasagar University
Midnapore, West Bengal, India

Tapan Senapati
School of Mathematics and Statistics
Southwest University
Chongqing, China

Ronald R. Yager
Iona College
Machine Intelligence Institute
New Rochelle, NY, USA

ISSN 2198-4182 ISSN 2198-4190 (electronic)
Studies in Systems, Decision and Control
ISBN 978-981-96-0084-7 ISBN 978-981-96-0085-4 (eBook)
https://doi.org/10.1007/978-981-96-0085-4

This Springer imprint is published by the registered company Springer Nature Singapore Pte Ltd.
The registered company address is: 152 Beach Road, #21-01/04 Gateway East, Singapore 189721, Singapore

I hereby dedicate this edited book to my esteemed teacher, Shri Sujit Kumar Kar, who imparted the invaluable lessons of education, knowledge, and wisdom.

—Laxminarayan Sahoo

I dedicate this edited book to my parents, my better half Smt. Swagata Dutta Senapati; and son, Ayush.

—Tapan Senapati

I dedicate this edited book to my parents, who impart the initial lessons of education.

—Madhumangal Pal

This edited book is dedicated to my wife, Rachel.

—Ronald R. Yager

Preface

With the use of the optimization technique, decision-makers may state that the best choice under consideration should be optimal. This optimization technique has an affinity to the scientific method or approaches that makes optimal to the decision-makers. The challenges posed by contemporary technology, which is also supported by cutting-edge scientific methodologies, call for a significant number of novel procedures and sophisticated analytical tools. It causes modeling and optimization to develop as a method for quickly selecting the best course of action from a wide range of scenarios and possibilities. Even if there are multiple divides of possible activities for which the decision-makers must make decisions in all aspects of daily life and they have come to realize that various issues can arise in various situations. Modeling and optimization are thus used everywhere, but both have a lot of same distinctive features. In addition, a common analytical technique can be utilized to identify the best answer to issues ranging under the identical umbrella to resolve real-world decision-making issues. As usual, in today's world, the demands of users and decision-makers, as well as the quantity of businesses, are what keep any industry competitive. Several businesses concentrate on creating and adding to highly reliable and dependable solutions. Additionally, it has become an impossible task for many manufacturing companies in the present day to determine the best strategies to manage corporate policies. Further, because of ongoing monetary liberalization, improved safety, rising consumer demands, rising globalization, sophisticated modern technical systems, and a constantly evolving business model, the manager of the company faces regular bouts of stress for sustaining the business systems. Moreover, due to uncontrollable factors of dynamic economic environment, the manager or decision-makers of the manufacturing company faces a lot of difficulties. Consequently, parameters relating to the decision-making may arising in terms of uncertainty or lack of proper information and uncertainty occurs at each step at the point of suitable decisions. During the last decades, various solution methodologies have been proposed to deal with the challenges about uncertainties. Decision-making under uncertainty via optimization, modeling, and analysis have provided their effectiveness to manage the difficulties and solving the real-world problems with imprecise parameters. In recent years, the landscape of medical treatment has undergone a profound transformation, driven by advances in machine learning and the rising importance of addressing uncertainty within the healthcare system. This edited book also explores the integration of smart city planning, waste management system, strategic decision-making in different industries, based on utility theory, game theory, heuristic approaches offering innovative solutions to modern challenges. Through meticulous modeling and optimization, it elucidates the applications of Markovian processes and their critical role in business sectors. The convergence of game theory and inventory management systems is examined, highlighting how strategic interactions can optimize supply chains and enhance efficiency. By leveraging these theoretical frameworks, the complexities of urban infrastructures are addressed, presenting new avenues for research and implementation. This comprehensive guide is intended to provide a robust foundation for professionals and researchers alike, seeking to navigate the intricacies of modern managerial decision-making through modeling and optimization. Emphasis is placed on the practical applications of these theories, ensuring relevance to real-world scenarios. The present edited book has been divided into 34 chapters. "Linguistic Z Numbers-Based FMEA of the Delivery of Stereotactic Body Radiation Therapy for Lung Cancer Treatment," has been highlighted in the first chapter. Through a Linguistic Z numbers-based Failure Mode and Effects Analysis (FMEA), this chapter investigates potential failure modes and their effects in delivering stereotactic body radiation therapy for lung cancer treatment. It offers insights into enhancing safety protocols and minimizing risks associated with this critical medical procedure. In the second chapter, entitled "A Novel Triangular Divergence Distance-Based TOPSIS Method for Interval-Valued Fermatean Fuzzy Group Decision-Making," an innovative approach to group decision-making is presented. The use of triangular divergence distance in the TOPSIS method for interval-valued Fermatean fuzzy data is explored. Advanced methods for making decisions are discussed in the third chapter via titled, "Archimedean Averaging Operator with Linear Programming Model for Generalized Orthopair Fuzzy Decision-Making." An in-depth analysis of the combination of the Archimedean averaging operator and a linear programming model is presented for managing fuzzy generalized orthopair data, offering a thorough method for addressing decisions in difficult situations. In the fourth chapter, titled "Mean Squared Error Utility for Fuzzy Preference Relations," the application of mean squared error utility in analyzing fuzzy preference relations is examined. The effectiveness of this method in enhancing decision-making accuracy within fuzzy systems is explored in depth. The fifth chapter deals with "MCDA/F-DEMATEL/ICTs Method Under Uncertainty in Mathematics Education: How to Make a Decision with Flipped, Gamified, and Sustainable Criteria." In this chapter, decision-making strategies under uncertainty in educational settings are examined. The integration of modern teaching approaches and sustainability is explored using advanced MCDA and F-DEMATEL methods. "Uncertainty Approaches for Spatial Data Management," has been addressed in the sixth chapter. Here, the handling of spatial data under uncertainty is explored. Various techniques for managing and mitigating uncertainties in spatial data are examined, providing insights into effective spatial data management practices. In the seventh chapter,

"Evaluating Sustainable Production Barriers and Efficiency-Enhancing Techniques in Manufacturing Companies Using Bipolar Neutrosophic REF," the challenges to sustainable production have been examined. Efficiency-enhancing techniques are analyzed within the context of manufacturing companies. The application of Bipolar Neutrosophic REF in identifying and addressing these barriers is explored, providing valuable insights into achieving sustainability and efficiency in the manufacturing sector. In the eighth chapter, titled "Assessment of Smart City Performance Using Multi-Criteria Decision-Making Methods," the evaluation of smart city performance is examined. Through the utilization of various multi-criteria decision-making methods, this chapter investigates the effectiveness of smart city initiatives. By analyzing multiple criteria, it provides valuable insights into assessing and improving the performance of smart cities, contributing to the advancement of urban development strategies in the digital age. "Impact of Technological Innovation on Employment Under Intuitionistic Fuzzy Einstein Aggregation Information with Z-Numbers," has been addressed in the ninth chapter. Through the utilization of intuitionistic fuzzy Einstein aggregation information with Z-numbers, this chapter examines the complex dynamics of technological advancement and its implications for the workforce. It provides valuable insights into the evolving relationship between technology and employment, informing strategies for navigating the future of work. In the tenth chapter, entitled "Cubic Analytic Hierarchy Process with Application in Decision-Making," innovative decision-making methodologies are explored. The application of the Cubic Analytic Hierarchy Process (CAHP) is examined, offering a robust framework for making complex decisions. This chapter delves into the practical applications of CAHP across various domains, providing valuable insights into its effectiveness in facilitating informed and structured decision-making processes. In the eleventh chapter, entitled "Decision-Making Process Under Uncertain Domain of Pythagorean Fuzzy Sets Based on an Enhanced Similarity Operator," the complexities of decision-making under uncertainty have been covered. Enhanced similarity operators are applied to navigate the uncertain domain of Pythagorean fuzzy sets, enabling a nuanced approach to decision-making. This chapter gives significant insights into addressing uncertainty, boosting the effectiveness of decision-making processes in varied contexts. "Bipolar Fuzzy Generalized Dombi Aggregation Operators for Group Decision-Making" has been addressed in the twelfth chapter. In this chapter, innovative aggregation techniques are investigated for group decision-making scenarios. The utilization of bipolar fuzzy sets in generalized Dombi aggregation operators is explored, offering a robust framework for reconciling diverse viewpoints within group deliberations. This chapter sheds light on advanced methodologies to streamline the decision-making process in complex group settings. In the thirteenth chapter, provides "Utility-Theory-Based Product Selection: Strategic Decision-Making in Automotive Industry," where, strategic decision-making processes within the automotive sector are examined. The application of utility theory for product selection is explored, providing a framework for making informed choices. This chapter highlights the importance of aligning product selection with strategic goals to enhance competitiveness and sustainability in the automotive industry. The issues of managing inventory within multi-production periods are addressed in the fourteenth chapter, "Imperfect Production Inventory Under Multi-Production Cycle for Non-deteriorating Items with Carbon Tax and Green Investment." With the carbon tax and green investment methods, sustainability is emphasized. In the fifteenth chapter, headed "An EOQ Model of a Fresh Product Having Variable Demand Under Trade Credit Policy," the dynamics of inventory management for perishable goods are examined. The impact of variable demand on the Economic Order Quantity (EOQ) model is explored within the context of trade credit policies. This chapter offers insights into optimizing inventory levels to meet fluctuating demands effectively. "Inventory Model with Negative Exponential Demand and Instantaneous Replenishment Under Preservation Technology," has been discussed in the sixteenth chapter. The chapter explores how preservation technology influences inventory models with negative exponential demand and instantaneous replenishment. Through rigorous analysis, it offers insights into optimizing inventory levels while accounting for preservation requirements. This chapter contributes to enhancing efficiency in inventory management systems, particularly in industries where preservation of goods is paramount. "A Real Coded Heuristic Technique in Solving a Plant Location Optimization Problem with Interval-Valued Transportation Costs and Quantities," has been highlighted in the seventeenth chapter. In this chapter, advanced methodologies for plant location optimization are examined. The application of real coded heuristic techniques is explored to address challenges posed by interval-valued transportation costs and quantities. This chapter provides valuable insights into optimizing plant locations under uncertain transportation parameters. In the eighteenth chapter, outlined "Interval Quadratic Programming Problem with Interval-Valued Decision Variables," the complexities of quadratic programming are explored. The focus is on problems where decision variables are interval-valued. This chapter delves into techniques for solving such problems, offering insights into optimizing decision-making processes when faced with uncertainty in variable values. In the nineteenth chapter, entitled "A Case Study on Medical Company Selection in the Health Sector by Using an Integrated Fuzzy AHP-MOORA Method," the process of allocating reliability resources is investigated. The chapter explores methodologies for allocating reliability in systems under uncertainty, considering both optimistic and pessimistic scenarios. Through rigorous analysis, it aims to optimize reliability allocation strategies, ensuring robust performance in systems where uncertainty is prevalent. This chapter contributes to enhancing reliability engineering practices, particularly in contexts where uncertainty poses significant challenges. In the twentieth chapter, headlined "Novel Pythagorean Fuzzy Hamacher Aggregation Operator and Its Application to Green Supplier Selection in Pharmaceutical Industry," novel approaches to game theory are discussed. The use of heuristic strategies is examined for solving games with interval-valued payoffs and objectives. This chapter presents a novel perspective on solving difficult decision-making situations providing useful insights into the application of heuristics in game theory. The twenty-first chapter, provides "On the Maintenance Oversight of the Healthcare Sector Based on Artificial Intelligence." Through the utilization of an integrated fuzzy AHP-MOORA method, this chapter examines the systematic approach to evaluating and selecting medical companies, providing valuable insights into decision-making processes within the healthcare industry. In the twenty-second chapter, entitled "Early Diagnosis of Medical Images in Healthcare Management by Artificial Intelligence," the development and application of a novel aggregation operator are explored. The chapter investigates how this operator enhances the process of green supplier selection within the pharmaceutical industry. Through rigorous analysis, it offers insights into improving sustainability practices and supplier evaluation methodologies, contributing to greener supply chain management in the pharmaceutical sector. The twenty-third chapter, discusses "Comparing Metrics of Classification Algorithms in Sentiment Analysis: A Comparative Study of Logistic Regression and KNN Using Count Vectorizer." The utilization of artificial intelligence for maintenance oversight in healthcare is investigated in this chapter. Advanced AI techniques are examined for optimizing maintenance operations, ensuring the efficient functioning of healthcare facilities. This chapter offers insights into leveraging AI to enhance the reliability and performance of maintenance processes within the healthcare sector. "Healthcare Waste Disposal Location Selection Using q-Rung Orthopair Fuzzy MEREC—CoCoSo Technique—A Case Study," has been explored in the twenty-fourth chapter. The chapter investigates the potential of AI to facilitate early diagnosis through the analysis of medical images. By employing advanced algorithms and deep learning techniques, it aims to enhance medical decision-making processes, leading to more timely and accurate diagnoses for improved patient care and outcomes. In the twenty-fifth chapter, titled "Optimal Plastic Waste Recycling Technology Selection Using Picture Fuzzy SWARA–MARCOS Technique," the effectiveness of classification algorithms in sentiment analysis is examined. A comparative study between Logistic Regression and KNN is conducted, utilizing Count Vectorizer. This chapter offers valuable insights into the performance metrics of different algorithms, enhancing our understanding of their applicability in sentiment analysis tasks. The twenty-sixth chapter presents "MCDM PROMETHEE Method in Identifying Best Teacher Awardee." In this chapter, the selection process for healthcare waste disposal locations is examined. Utilizing the q-Rung Orthopair Fuzzy MEREC—CoCoSo technique, this chapter investigates optimal site selection methodologies. Through a detailed case study, it provides valuable insights into efficient healthcare waste management strategies, ensuring environmentally sound and sustainable disposal practices. In the twenty-seventh chapter, entitled "Evaluating Challenges in Smart Transportation with Grey Theory-Based Multi-criteria Decision-Making," the process of selecting optimal recycling technologies for plastic waste is investigated. The Picture Fuzzy SWARA—MARCOS technique is utilized to analyze and prioritize various recycling methods. This chapter offers insights into the systematic approach of choosing the most effective recycling technologies, contributing to sustainable waste management practices and environmental conservation efforts. The twenty-eighth chapter, provides "T-spherical Fuzzy Group Decision-Making Using Subjective and Objective Weights of Experts and Copula Aggregation Operators." Here, the process of identifying outstanding teachers is explored. The Multi-Criteria Decision-Making (MCDM) Promethee method is employed to assess and rank teacher candidates based on various criteria. This chapter provides insights into the systematic approach of selecting the best teacher awardees, contributing to the recognition and appreciation of exemplary educators within educational institutions. In the twenty-ninth chapter, titled "Chap "Effect of Internal Delay on the Dynamics of a Mean-Field Diffusive Coupled Oscillating System"," the complexities of smart transportation are analyzed. Utilizing a Grey Theory-based Multi-Criteria Decision-Making approach, this chapter delves into the assessment of challenges within smart transportation systems. Through rigorous analysis, it offers insights into addressing multifaceted issues such as traffic congestion, environmental sustainability, and technological integration. This chapter serves as a valuable resource for policymakers, urban planners, and transportation stakeholders seeking to enhance smart transportation initiatives. "Markovian Brand Switching Model for Long-Term Steady-State Market Shares: A Study on Toothpaste Market," has been investigated in the thirtieth chapter. The chapter investigates the integration of subjective and objective weights of experts in T-spherical fuzzy decision-making processes. Through the utilization of Copula aggregation operators, it aims to provide a comprehensive framework for effectively synthesizing diverse expert opinions in decision-making scenarios. The thirty-first chapter deals with "Root Hair Algorithm: A Swarm Intelligence Algorithm." Through rigorous analysis, the chapter investigates how delays within coupled oscillating systems influence their behavior. By examining the dynamics of mean field diffusive coupling, it sheds light on the intricate interplay between internal delays and system oscillations, contributing to a deeper understanding of complex dynamical systems. "Pythagorean Fuzzy Ordered Weighted Averaging Aggregation Operator Based on Appropriate Score Function and Their Application to Multi-criteria Decision-Making in IT Project Management," has been provided in the thirty-second chapter. In this chapter, the dynamics of market share in the toothpaste industry are investigated. Utilizing a Markovian approach, long-term steady-state market shares are analyzed. This chapter provides valuable insights into the factors influencing brand switching behavior among consumers, offering a comprehensive study of market dynamics within the toothpaste market. In the thirty-third chapter, entitled "33 Root Hair Algorithm: A Swarm Intelligence Algorithm," the exploration of swarm intelligence is undertaken. This chapter delves into the principles and applications of the Root Hair Algorithm, elucidating its effectiveness in solving optimization problems inspired by the collective behavior of root hairs. Through a comprehensive analysis, it offers insights into harnessing natural phenomena to design efficient optimization algorithms. Finally, "Pythagorean fuzzy ordered weighted averaging aggregation operator based on appropriate score function and their application to multi-criteria decision-making in IT Project Management" has been highlighted in Chapter 34. In this chapter, the Pythagorean fuzzy ordered weighted averaging (PFOWA) aggregation operator, utilizing an appropriate score function, is examined. Its application to multi-criteria decision-making in IT project management is explored. Emphasis is placed on the operator's effectiveness in handling uncertainty and improving decision-making quality in IT projects.

Researchers, students, engineers, managers of manufacturing firms, decision-makers, and anyone else interested in decision-making under uncertainty through modeling, analysis, and optimization can all benefit from this book. The editors extend their utmost gratitude to the authors from all over the world who contributed to this edited work under a variety of conditions. Special thanks go out to all the reviewers whose comments and suggestions have helped improve the quality of the chapters. Mr. Aninda Bose, Executive Editor (Interdisciplinary Applied Sciences, Computational Intelligence, Energy) at Springer Nature, London, is also thanked for his kind consideration and endless support for this edited book. It is hoped that the excitement felt in presenting this edited book will be shared by the reader and that it will be found beneficial.

Raiganj, India
April 2024

Laxminarayan Sahoo
Tapan Senapati
Madhumangal Pal
Ronald R. Yager

Acknowledgments

The completion of this edited book was made possible through the invaluable contributions of chapter contributors and reviewers, whose names may not all be enumerated. Their contributions are deeply appreciated and sincerely acknowledged. Gratitude and indebtedness are extended to the following individuals and organizations.

Mr. Aninda Bose, Executive Editor (Interdisciplinary Applied Sciences, Computational Intelligence, Energy) at Springer Nature, London, is especially acknowledged for his kind consideration and unwavering support for this edited book.

Ramamoorthy Rajangam, Project Coordinator at Springer Nature, is recognized for his immeasurable assistance throughout the process of this edited book's completion.

Springer Nature Singapore Pte Ltd., located at 152 Beach Road, #21-01/04 Gateway East, Singapore 189721, Singapore, along with the publishing teams of Springer Nature, provided the opportunity and professional support essential for producing this edited book in its present form. Their contributions were instrumental in the successful publication of this work.

The efforts of all who contributed, whether explicitly mentioned or not, were vital in bringing this book to completion.

Raiganj, India
April 2024

Laxminarayan Sahoo
Tapan Senapati
Madhumangal Pal
Ronald R. Yager

Contents

Editors and Contributors

About the Editors

Laxminarayan Sahoo is currently an associate professor of Computer and Information Science, Raiganj University, Raiganj, India. He obtained his M.Sc. from Vidyasagar University, India, and his Ph.D. from the University of Burdwan, India. He has received MHRD fellowship from Government of India during his M.Tech. course at ISM, Dhanbad, India, and received Prof. M. N. Gopalan Award for Best Ph.D. thesis in Operations Research from Operational Research Society of India (ORSI). Dr. Sahoo has successfully guided four research scholars for Ph.D. degree and four students continuing Ph.D. degree and has published more than 80 articles in international and national journals. He has also successfully completed one UGC minor research project. He is a reviewer of several international journals and an academic editor of International Journal "Mathematical Problems in Engineering," Hindawi Publication.

Tapan Senapati is a postdoctoral fellow at Southwest University, School of Mathematics and Statistics, 400715 Chongqing, China. He received the B.Sc., M.Sc., and Ph.D. degrees in Mathematics from the Vidyasagar University, India, in 2006, 2008, and 2013, respectively. He has published three books and more than 160 articles in international journals. He is a reviewer of several international journals and a member of the editorial boards of many journals. His name has recently been enlisted in the World's Top 2% Scientists in the field of Artificial Intelligence (2022, 2023), identified by Elsevier BV, Stanford University. He is also the editor of the book *Real Life Applications of Multiple Criteria Decision-Making Techniques in Fuzzy Domain* published by Springer. His main scientific interests are fuzzy sets, fuzzy optimization, soft computing, multi-attribute decision-making, and aggregation operators.

Prof. Madhumangal Pal is currently a professor and head of Applied Mathematics and Director of IQAC, Vidyasagar University. He has received gold and silver medals from Vidyasagar University for ranking first and second in the M.Sc. and B.Sc.

examinations, respectively. Also, he received the "Computer Division Medal" from the Institute of Engineers (India) in 1996 for best research work. In 2013, he received Bharat Jyoti Award for his significant contribution in academics. In 2023, Prof. Pal received the Vidyasagar Memorial Research Award from Vidyasagar University for best research work during the last ten years. Professor Pal has successfully guided 43 research scholars for Ph.D. degrees and has published more than 404 articles in international and national journals. His specializations include algorithms and fuzzy graph theory, fuzzy matrix, genetic algorithms, and parallel algorithms.

Prof. Ronald R. Yager has worked in the area of machine intelligence for over 25 years. He has published near about 1000 papers and more than 30 books in areas related to artificial intelligence, fuzzy sets, decision-making under uncertainty, and the fusion of information. He is among the world's top 1% most highly cited researchers with over 95237 citations. He was the recipient of the IEEE Computational Intelligence Society's highly prestigious Frank Rosenblatt Award in 2016. He was the recipient of the IEEE Systems, Man and Cybernetics Society 2018 Lotfi Zadeh Pioneer Award. He was also the recipient of the IEEE Computational Intelligence Society Pioneer Award in Fuzzy Systems. He received honorary doctorates from the Azerbaijan Technical University, the State University of Information Technologies, Sofia Bulgaria, and the Rostov on the Don University, Russia.

Contributors

Mohini Agarwal College of Humanities and Social Sciences, Grand Canyon University, Phoenix, USA

Deepti Aggrawal USME, Delhi Technological University, Delhi, India

Maria Akram Institute of Mathematics, Khwaja Fareed University of Engineering and Information Technology, Rahim Yar Khan, Pakistan

Adarsh Anand Department of Operational Research, University of Delhi, Delhi, India

Shahzaib Ashraf Institute of Mathematics, Khwaja Fareed University of Engineering and Information Technology, Rahim Yar Khan, Pakistan

Muhammad Asif Department of Mathematics and Statistics, Hazara University Mansehra, Mansehra, KP, Pakistan

Ahmet Aytekin Department of Business Administration, Hopa Faculty of Economics and Administrative Sciences, Artvin Çoruh University, Hopa, Artvin, Türkiye

Chandan Bandyopadhyay Department of Computer Science and Engineering (Data Science), Dr. B. C. Roy Engineering College, Durgapur, India;
Department of Computer Science and Engineering, University of Bremen, Bremen, Germany

Sudatta Banerjee Department of Business Administration, Burdwan Raj College, Burdwan, WB, India

Ismat Beg Department of Mathematics and Statistical Sciences, Lahore School of Economics, Lahore, Pakistan

Nabajyoti Bhattacharjee Department of Mathematics, Assam University, Silchar, India;
Department of Mathematics, Pandit Deendayal Upadhyaya Govt. Model College, Katlicherra, India

Baby Bhattacharya Department of Mathematics, National Institute of Technology Agartala, Tripura, India

Sovan Bhattacharya Department of Computer Science and Engineering (Data Science), Dr. B. C. Roy Engineering College, Durgapur, India

Abhijit Biswas Department of Management Science, Maulana Abul Kalam Azad University of Technology (Formerly WBUT), Kolkata, West Bengal, India;
Department of PGDM-Business Analytics, Globsyn Business School, Kolkata, West Bengal, India

Arindam Biswas Center for IOT, AI Integration with Education-Industry-Agriculture, Kazi Nazrul University, Asansol, India;
School of Mines and Metallurgy, Kazi Nazrul University, Asansol, India

Jogjiban Chakraborty Department of Mathematics, Bankura University, Bankura, India

Saumen Chakraborty Department of Physics, Bidhan Chandra College, Asansol, West Bengal, India

Sayanta Chakraborty Department of Mathematics, National Institute of Technology Agartala, Tripura, India

Chanchal Department of Operational Research, University of Delhi, Delhi, India

Prayosi Chatterjee Department of Mathematics, Kazi Nazrul University, Asansol, India

Titas Roy Chowdhury Department of PGDM-Information Systems, Globsyn Business School, Kolkata, West Bengal, India

Amrit Das Department of Mathematics, School of Advanced Sciences, Vellore Institute of Technology, Vellore, Tamil Nadu, India

Surajit Das Department of Artificial Intelligence and Data Science, B V Raju Institute of Technology, Narsapur, Telangana, India

Saumendra Sankar De Sarkar Department of Physics, Raniganj Girls' College, Searsole Rajbari, West Bengal, India

Saima Debbarma Department of Mathematics, National Institute of Technology Agartala, Tripura, India

Bijoy Krishna Debnath Department of Applied Sciences, School of Engineering, Tezpur University, Sonitpur, Assam, India

Bishnupada Debnath Department of Mathematics, Rajiv Gandhi University, Doimukh, Papumpare, Arunachal Pradesh, India

Gülay Demir High School of Health Vocational Services, Sivas Cumhuriyet University, Sivas, Turkey

R. Narmada Devi Department of Mathematics, Vel Tech Rangarajan Dr, Sagunthala R&D Institute of Science and Technology, Chennai, Tamil Nadu, India

Paul Augustine Ejegwa Department of Mathematics, Joseph Sarwuan Tarka University, Makurdi, Nigeria

Diego García-Zamora Department of Mathematics, Universidad De Jaén, Jaén, Spain

Meghdoot Ghosh Department of Hospital Administration, Post Graduate Institute of Hospital Administration (PGIHA), Peerless Hospital Campus, Kolkata, West Bengal, India

Lavanya Golipally Department of Artificial Intelligence and Data Science, B V Raju Institute of Technology, Narsapur, Telangana, India

David González-Gómez Department of Science and Mathematics Education, Teacher Training College, University of Extremadura, Cáceres, Spain

Muhammad Gulistan Department of Mathematics and Statistics, Hazara University Mansehra, Mansehra, KP, Pakistan

Ranjan Kumar Gupta Department of Management and Marketing, West Bengal State University, Barasat, WB, India

Rakib Hasan Department of Mechanical, Aerospace and Industrial Engineering, University of Texas, San Antonio, TX, USA

Chiranjibe Jana Saveetha School of Engineering, Saveetha Institute of Medical and Technical Sciences (SIMATS), Chennai, Tamil Nadu, India

Jin Su Jeong Department of Science and Mathematics Education, Teacher Training College, University of Extremadura, Cáceres, Spain

S. Kalaiselvi Department of Computer Science and Engineering, National Engineering College, KR Nagar, Kovilpatti, Tamil Nadu, India

Jewel Karmakar Department of Mathematics, Bankura Zilla Saradamani Mahila Mahavidyapith, Bankura, West Bengal, India;
Department of Mathematics, Sidho Kanho Birsha University, Purulia, West Bengal, India

Samiran Karmakar Department of Mathematics, Bankura Sammilani College, Bankura, West Bengal, India

Debdip Khan Department of Business Administration, Burdwan Raj College, Burdwan, WB, India

Komal Department of Mathematics, School of Physical Sciences, Doon University, Dehradun, India

Selçuk Korucuk Department of Logistics Management, Bulancak Kadir Karabaş Vocational School, Giresun University, Giresun, Türkiye

Abhay Kumar Department of Artificial Intelligence and Data Science, B V Raju Institute of Technology, Narsapur, Telangana, India

Jambi Ratna Raja Kumar Computer Engineering Department, Genba Sopanrao Moze College of Engineering, Pune, India

Ranjan Kumar Jana Department of Mathematics, Sardar Vallabhbhai National Institute of Technology, Surat, Gujarat, India

Sathish Kumar Kumaravel Department of Mathematics, Vel Tech Rangarajan Dr, Sagunthala R&D Institute of Science and Technology, Chennai, Tamil Nadu, India

Sanat Kumar Mahato Department of Mathematics, Sidho Kanho Birsha University, Purulia, West Bengal, India

Saibal Majumder Department of Computer Science and Engineering (Data Science), Dr. B. C. Roy Engineering College, Durgapur, India

Sanjukta Malakar Assam University Silchar, Assam, India

Prasenjit Mandal Department of Applied Mathematics with Oceanology and Computer Programming, Vidyasagar University, Midnapur, W.B., India

Luis Martínez Department of Computer Sciences, Universidad De Jaén, Jaén, Spain

Kala Raja Mohan Department of Mathematics, Panimalar Engineering College, Chennai, India

Rituparna Mondal Department of Mathematics, Sardar Vallabhbhai National Institute of Technology, Surat, Gujarat, India;
Department of Applied Science, Haldia Institute of Technology, Haldia, West Bengal, India

Sarbast Moslem School of Architecture Planning and Environmental Policy, University College Dublin, Belfield, Ireland

Sathi Mukherjee Department of Mathematics, Bankura, India

Regan Murugesan Department of Mathematics, Vel Tech Rangarajan Dr, Sagunthala R&D Institute of Science and Technology, Chennai, Tamil Nadu, India

Usha Rani Naathi Department of Artificial Intelligence and Data Science, B V Raju Institute of Technology, Narsapur, Telangana, India

Prodosh Kiran Nath Assam University, Silchar, India

Moddassir Khan Nayeem Department of Mechanical, Aerospace and Industrial Engineering, University of Texas, San Antonio, TX, USA

Madhumangal Pal Department of Applied Mathematics with Oceanology and Computer Programming, Vidyasagar University, Midnapore, W.B., India;
Department of Mathematics and Innovation, Saveetha School of Engineering, Saveetha Institute of Medical and Technical Sciences, Chennai, Tamilnadu, India

Tapas Kumar Paul Department of Applied Mathematics, Vidyasagar University, Midnapore, India

Frederick E. Petry Naval Research Laboratory, Stennis Space Center, Hancock, MS, USA

Sukhwinder Singh Rawat Department of Mathematics, School of Physical Sciences, Doon University, Dehradun, India

Abhijit Saha Department of Computing Technologies, SRM Institute of Science and Technology (SRMIST), Kattankulathur, Tamil Nadu, India

Apu Kumar Saha Department of Mathematics, National Institute of Technology Agartala, Tripura, India

Brajamohan Sahoo Department of Applied Sciences, School of Engineering, Tezpur University, Sonitpur, Assam, India

Laxminarayan Sahoo Department of Computer and Information Science, Raiganj University, Raiganj, India

Sovan Samanta Department of Mathematics, Tamralipta Mahavidyalaya, Tamluk, India;
Department of Technical Sciences, Western Caspian University, Baku, Azerbaijan;
Research Center of Performance and Productivity Analysis, Istinye University, Istanbul, Turkey

Sovan Samanta Department of Technical Sciences, Algebra University, Zagreb, Croatia

P. Santhiya Department of Mathematics, School of Advanced Sciences, Vellore Institute of Technology, Vellore, Tamil Nadu, India

Mijanur Rahaman Seikh Department of Mathematics, Kazi Nazrul University, Asansol, India

Nabendu Sen Department of Mathematics, Assam University, Silchar, Assam, India

Dola Sinha Department of Electrical Engineering, Dr. B.C. Roy Engineering College, Durgapur, India

Chayel Tripura Department of Mathematics, National Institute of Technology Agartala, Tripura, India

Linguistic Z Numbers-Based FMEA of the Delivery of Stereotactic Body Radiation Therapy for Lung Cancer Treatment

Prasenjit Mandal, Sovan Samanta, Madhumangal Pal, and Jambi Ratna Raja Kumar

Abstract Failure mode and effect analysis (FMEA) is a common management tool for analysing the causes and consequences of possible failures in different cancer treatments, goods, processes, systems, and services. As a result of individual differences in reason and cognition as well as the impact of social ties, the experts involved in an FMEA with many participants may have conflicting effects on the decision-making process. Incomplete weights of risk variables are also essential for reflecting the uncertainty with which experts see a situation. As a result, a consensus-based FMEA approach is proposed to investigate stereotactic body radiation therapy (SBRT) for lung cancer. First, we identified the failure modes of SBRT of lung cancer and then analysed their effect on the treatment process. Generally, the choice of lung cancer treatment is usually made by oncology experts, who select the most suitable option for the patient. However, the oncology expert's bounded personality and rationality characteristics can influence the patient's treatment choice. Moreover, the American Cancer Society states that the patient's treatment choice may vary depending on the oncology expert. Therefore, we propose a group decision-making model (GDM) for FMEA of lung cancer treatment in a linguistic Z numbers environment to enhance the survival rate of lung cancer patients.

Keywords Linguistic Z number · Group decision-making · Failure mode and effect analysis

P. Mandal · M. Pal
Department of Applied Mathematics with Oceanology and Computer Programming, Vidyasagar University, Midnapur, W.B. 721102, India

S. Samanta (✉)
Department of Mathematics, Tamralipta Mahavidyalaya, Tamluk 721636, India
e-mail: ssamantavu@gmail.com

Department of Technical Sciences, Western Caspian University, 1001 Baku, Azerbaijan

Research Center of Performance and Productivity Analysis, Istinye University, Istanbul, Turkey

Department of Technical Sciences, Algebra University, 10000 Zagreb, Croatia

J. R. R. Kumar
Computer Engineering Department, Genba Sopanrao Moze College of Engineering, Pune, India

L. Sahoo et al. (eds.), *Decision Making Under Uncertainty Via Optimization, Modelling, and Analysis*, Studies in Systems, Decision and Control 558,
https://doi.org/10.1007/978-981-96-0085-4_1

1 Introduction

Risk analysis (RA) in a problem, failure It has been commonly used in various sectors, such as safety [54], treatment [24], and construction [4], among others. However, it is not flawless. It has its own weaknesses that sometimes lead to failures [7, 60]. In the classical FMEA model [54], three types of risk factors (RFs) named occurrence, detection, and severity are considered. However, some limitations of the classic FMEA model are pointed out in [50]. Multiattribute decision-making [11–19] is a useful tool for solving the limitations of the classical FMEA model. In this connection, a fuzzy environment with multiattribute group decision-making (MAGDM) is considered for analysing different RFs [10, 21] of the FMEA model. However, linguistic terms (LTs) can convey the expert opinion more clearly than fuzzy statistics in [30–35, 58]. In this connection, various proposals are introduced for analysing different RFs of the FMEA model under the linguistic environment in [4, 24, 55]. Many interesting works have been made related to fuzzy and other sets as they have many applications [6, 8, 20, 26–29, 38–46, 48–50, 54, 58].

Z number is a suitable tool for dealing with relative information type uncertainty, and it has received much attention in various fields recently [59]. Then, with the advantages of LTs, linguistic Z number (LZN) is proposed in [53]. After that, many works are done in LZN [3, 5, 22, 36, 52]. However, we find three major drawbacks in [3, 5, 22, 25, 52, 53]. There are the following: (1) They do not include the consensus-reaching process. (2) They assume equal or predetermined experts' weight. (3) They assume equal or predetermined attribute weight.

Stereotactic body radiation therapy (SBRT) is a common way to destroy cancer cells in lung cancer [47, 51]. To prevent radiation from harming their heart, they may need to hold their breath deeply while receiving the therapy [57]. SBRT enhances the precision and safety of radiation therapy. However, it also adds complexity and difficulty to the procedure. FMEA is used by Perks et al. [2] in SBRT to improve the SBRT process for patients undergoing radiation therapy. The FMEA model works on a problem to identify and solve the potential causes. In view of this point, hare we proposed an FMEA GDM model for identifying and ranking the RFs of SBRT for lung cancer treatment under the LZN environment, which is our primary aim.

Liu et al. [23] emphasized the importance of reaching consensus among group members for making collective decisions. However, many FMEA studies [10, 60] neglect this aspect and focus little on reaching a consensus among group member-based-RA in FMEA [4, 21, 24]. These models do not account for the semantic meaning of aggregations, and they simply combine individual preferences to form the matrix with consensus among group members, which may result that most experts do not agree with the decision. In this paper, we address these limitations by proposing a three-level consensus measurement degree and a minimum adjustment sum based on the relativity among experts and the group consensus opinion for identifying and ranking the RFs of SBRT for lung cancer treatment under the LZN environment.

Risk factors' weight is crucial for ranking failure modes, as shown by the studies in [4, 24, 50]. However, experts may be unable to provide weight on risk factors due to the problem's complexity and limited capacity. In this connection, various methods were proposed to find the weight of RFs. Some of these methods are found in [4, 10, 24]. These methods have their own pros and cons. Hence, this paper aims to combine subjective and objective weights to obtain the weight of RFs, which is the second aim. FMEA aims to rank the FMs according to multiple risk factors. Several methods have been suggested for this purpose in [4, 10, 22, 24, 50, 54, 60]. The gained and lost dominance score (GLDS)-based ranking method performed better in dealing with the object [24]. Therefore, for LZN information-based failure mode ranking of SBRT for lung cancer treatment, we use the GLDS method in this paper.

This study consists of the following sections: The concept of LZNs is reviewed in Sect. 2. Section 3 proposes LZN-based FMEA of the delivery of SBRT for lung cancer treatment. This section covers consensus measures with three levels, inconsistent judgement element identification, the FAM, the aggregation tools, RFs weight computation and the FMs ranking. Section 4 presents a real-world case study with sensitivity and comparative analysis. Section 6 summarizes the study and suggests directions for future research.

2 LZNs

Let $\mathscr{X} = \{x_g \mid g = -\tau_1, \ldots, 1, 0, 1, \ldots, \tau_1\}$ and $\mathscr{Y} = \{y_\hbar \mid \hbar = -\tau_2, \ldots, 1, 0, 1, \ldots, \tau_2\}$ be two distinct linguistic term sets (LTS) with cardinality $2\iota_1 + 1$ and $2\tau_2 + 1$, $(\tau_1, \tau_2 = \{1, 2, 3, \ldots\})$, and the following properties: (1) $x_{g_1} > x_{g_2}$ (and $y_{\hbar_1} > y_{\hbar_2}$) if and only if $g_1 > g_2$ $(\hbar_1 > \hbar_2)$, (2) $neg(x_g) = x_{-g}$ (and $neg(y_\hbar) = y_{-\hbar}$), (3) $\max(x_{g_1}, x_{g_2}) = x_{g_1}$ if $x_{g_1} \geq x_{g_2}$ (and $\max(y_{\hbar_1}, y_{\hbar_2}) = y_{\hbar_1}$ if $y_{\hbar_1} \geq y_{\hbar_2}$), and (4) $\min(x_{g_1}, x_{g_2}) = x_{g_1}$ if $x_{g_1} \leq x_{g_2}$ (and $\min(y_{\hbar_1}, y_{\hbar_2}) = y_{\hbar_1}$ if $y_{\hbar_1} \leq y_{\hbar_2}$).

According to studies Xu [56], $\tilde{\mathscr{X}} = \{x_g \mid x_{-\tau_1} \leq x_g \leq x_{\tau_1}, g \in [-\tau_1, \tau_1]\}$ and $\tilde{\mathscr{Y}} = \{y_\hbar \mid y_{-\tau_2} \leq y_\hbar \leq y_{\tau_2}, \hbar \in [-\tau_2, \tau_2]\}$ are continuous LTSs.

Definition 1 [59]. An LZN is an ordered pair of information that denotes $\mathscr{A} \in \mathscr{X}$ and relative information of $\mathscr{A}$ denotes $\mathscr{B} \in \mathscr{Y}$, which is represented by $z = (\mathscr{A}, \mathscr{B}) = (x_g, y_\hbar)$. Especially, $\tilde{1} = (x_{\tau_1}, y_{\tau_2})$ and $\tilde{0} = (x_{-\tau_1}, y_{-\tau_2})$.

To avoid information losses, we use the linguistic scale function Δ and ∇ of Wang et al. [53] for dealing with any LZN $z = (\mathscr{A}, \mathscr{B})$, where $\Delta : [-\tau_1, \tau_1] \to [0, 1]$, $\Delta(\mathscr{A}) = \{\Delta(x_g) = \frac{g}{2\tau_1} + \frac{1}{2} \mid g \in [-\tau_1, \tau_1]\} = \hbar$ and $\nabla : [-\tau_2, \tau_2] \to [0, 1]$, $\nabla(\mathscr{B}) = \{\nabla(y_\hbar) = \frac{\hbar}{2\tau_2} + \frac{1}{2} \mid \hbar \in [-\tau_2, \tau_2]\} = \wp$. The inverse functions of Δ and ∇ are denoted by Δ^{-1} and ∇^{-1}, where $\Delta^{-1} : [0, 1] \to [-\tau_1, \tau_1]$, $\Delta^{-1}(\hbar) = \{x_{(2\hbar-1)\tau_1} \mid \hbar \in [0, 1]\} = x_g$ and $\nabla^{-1} : [0, 1] \to [-\tau_2, \tau_2]$, $\nabla^{-1}(\wp) = \{y_{(2\wp-1)\tau_2} \mid \wp \in [0, 1]\} = y_\hbar$.

Definition 2 [3]. For any two LZNs $z_1 = (\mathscr{A}_1, \mathscr{B}_1) = (x_{g_1}, y_{\hbar_1})$ and $z_2 = (\mathscr{A}_2, \mathscr{B}_2) = (x_{g_2}, y_{\hbar_2})$, and $\lambda \in [0, 1]$ we have

(1) $\mathscr{z}_1 \oplus \mathscr{z}_2 = \left(\max\{x_{-\tau_1}, \min\{x_{g_1+g_2}, x_{\tau_1}\}\}, y_{\frac{h_1+h_2}{2}}\right)$.
(2) $\lambda \mathscr{z}_1 = (x_{\lambda g_1}, y_{h_1})$.
(3) $neg(\mathscr{z}_1) = (x_{-g_1}, y_{h_1})$.

Definition 3 [3]. The distance between two LZNs $\mathscr{z}_1 = (\mathscr{A}_1, \mathscr{B}_1) = (x_{g_1}, y_{h_1})$ and $\mathscr{z}_2 = (\mathscr{A}_2, \mathscr{B}_2) = (x_{g_2}, y_{h_2})$ is

$$\mathscr{D}(\mathscr{z}_1, \mathscr{z}_2) = \sqrt{\varsigma|\Delta(\mathscr{A}_1) - \Delta(\mathscr{A}_2)|^2 + \varrho|\nabla(\mathscr{B}_1) - \nabla(\mathscr{B}_2)|^2}, \tag{1}$$

where $\varsigma, \varrho \in [0, 1]$, $\varsigma + \varrho = 1$ and $\varsigma \geq \varrho$.

Definition 4 [3]. The expected value of any LZN $\mathscr{z} = (\mathscr{A}, \mathscr{B})$ is $\mathscr{EXP}(\mathscr{z}) = \Delta(\mathscr{A}) \times \nabla(\mathscr{B})$.

3 LZN-Based FMEA GDM Model

3.1 Process of Consensus Measure

- Object set: $\mathscr{O} = \{o_i \mid i = 1, 2, \ldots, n\}$ $(n \geq 2)$.
- Attribute set: $\mathscr{C} = \{c_j \mid j = 1, 2, \ldots, m\}$ $(m \geq 2)$.
- Expert set: $\mathscr{E} = \{e^p \mid p = 1, 2, \ldots, k\}$ $(k \geq 2)$.
- Opinion of expert e^p on o_i with respect to c_j: $\mathscr{Z}^p = (\mathscr{z}_{ij}^p)_{n\times m}$, where $\mathscr{z}_{ij}^p = (\mathscr{A}_{ij}^p, \mathscr{B}_{ij}^p)$.

3.1.1 Three-Level Degree of Consensus Measure

In the following, we develop an LZNs-based three-level degree of consensus measure. (1) Attribute level: The consensus degree of the opinions of experts e^p and e^q on o_i $(i = 1, 2, \ldots, n)$ for c_j $(j = 1, 2, \ldots, m)$ is

$$\mathscr{AL}_{ij}(e^p, e^q) = 1 - \mathscr{D}(e^p, e^q) = 1 - (\mathscr{z}_{ij}^p, \mathscr{z}_{ij}^q). \tag{2}$$

The weight consensus degree of the expert e^p with remaining experts in $\mathscr{E}$ on o_i $(i = 1, 2, \ldots, n)$ for c_j $(j = 1, 2, \ldots, m)$ is

$$\mathscr{WAL}_{ij}^p = \sum_{q=1,q\neq p}^{k} \varpi'_{pq}\mathscr{AL}_{ij}(e^p, e^q), \tag{3}$$

where ϖ'_{pq} denotes normalized weight of (e^p, e^q). For computation of ϖ'_{pq} we use the following formulae:

$$\varpi'_{pq} = \frac{\varpi_{pq}}{\sum_{q=1,q\neq p}^{k-1} \varpi_{pq}}, \ \varpi_{pq} = \frac{\varpi_{e^p}\varpi_{e^q}}{\sum_{p=1}^{k-1}\sum_{q=p+1}^{k} \varpi_{e^p}\varpi_{e^q}}. \tag{4}$$

For computation of ϖ_{e^p} $(p = 1, 2, \ldots, k)$ of e^p, we use the following formulae:

$$\varpi_{e^p} = \frac{\psi(\mathscr{Z}^p)}{\sum_{p=1}^{k} \psi(\mathscr{Z}^p)}, \ \psi(\mathscr{Z}^p) = \alpha \frac{\sum_{j=1}^{m}\sum_{i=1}^{n} \Delta(\mathscr{A}_{ij}^p)}{mn} + (1-\alpha)\frac{\sum_{j=1}^{m}\sum_{i=1}^{n} \nabla(\mathscr{B}_{ij}^p)}{mn}, \tag{5}$$

where $\alpha \in [0, 1]$.
(2) Object level: The consensus degree of the opinions of experts e^p and e^q on o_i $(i = 1, 2, \ldots, n)$ is

$$\mathscr{OL}_i(e^p, e^q) = \sum_{j=1}^{m} \omega'_{j,pq} \mathscr{AL}_{ij}(e^p, e^q). \tag{6}$$

Here, $\omega'_{j,pq}$ denotes normalized weight of (ω_j^p, ω_j^q) for (e^p, e^q). For computation of $\omega'_{j,pq}$ we use the following formulae:

$$\omega'_{j,pq} = \frac{\omega_j^{pq}}{\sum_{j=1}^{m} \omega_j^{pq}}, \ \omega_j^{pq} = \frac{\omega_j^p \omega_j^q}{\sum_{j=1}^{m}\sum_{p=1}^{k-1}\sum_{q=p+1}^{k} w_j^p \omega_j^q}, \tag{7}$$

where ω_j^p denotes c_j $(j = 1, 2, \ldots, m)$ attribute weight of e^p with judgement matrix $\mathscr{Z}^p$. For computation of ω_j^p we use the following formulae:

$$\omega_j^p = \frac{\sum_{i=1}^{n}\sum_{l=1}^{n} \mathscr{D}(t_{ij}^p, t_{lj}^p)}{\sum_{j=1}^{m}\sum_{i=1}^{n}\sum_{l=1}^{n} \mathscr{D}(t_{ij}^p, t_{lj}^p)}, \tag{8}$$

where t_{ij}^p is a (i, j) entity of $\mathscr{T}^p = (t_{ij}^p)_{n\times m}$ with $t_{ij}^p = (\Delta^{-1}(\mathscr{U}_{ij}^p), \nabla^{-1}(\mathscr{V}_{ij}^p))$ and

$$\mathscr{U}_{ij}^p = \frac{\Delta(\mathscr{A}_{ij}^p)}{\sqrt{\sum_{i=1}^{n}(\Delta(\mathscr{A}_{ij}^p))^2}}, \ \mathscr{V}_{ij}^p = \frac{\nabla(\mathscr{B}_{ij}^p)}{\sqrt{\sum_{i=1}^{n}(\nabla(\mathscr{B}_{ij}^p))^2}}, \ i = 1, 2, \ldots, n; \ j = 1, 2, \ldots, m. \tag{9}$$

Now, in the following we give the weight consensus degree of e^p with remaining experts in $\mathscr{E}$ on o_i $(i = 1, 2, \ldots, n)$ is

$$\mathscr{WOL}_i^p = \sum_{q=1,q\neq p}^{k} \varpi'_{pq} \mathscr{OL}_i(e^p, e^q). \tag{10}$$

(3) Expert level: The consensus degree between e^p and e^q is

$$\mathscr{EL}(e^p, e^q) = \frac{1}{n}\sum_{i=1}^{n}\sum_{j=1}^{m}\omega'_{j,pq}\mathscr{AL}_{ij}(e^p, e^q). \tag{11}$$

For expert e^p, the weight consensus degree of $\mathscr{EL}(e^p, e^q)$ is

$$\mathscr{WEL}^p = \sum_{q=1,q\neq p}^{k} \varpi'_{pq}\mathscr{EL}(e^p, e^q). \tag{12}$$

Clearly, $0 \leq \mathscr{WEL}^p \leq 1$.

Consensus measures are used for accepting the final decision of all experts. In this connection, we needed pre-defined thresholds for comparing $\mathscr{WEL}^p$ of expert e^q. Let ξ be a pre-defined thresholds such that $\xi \in [0, 1]$. Therefore, to accept the final decision of all experts if $\mathscr{WEL}^p \geq \xi$, otherwise we apply the identification rule (IR), which is discussed in the following section.

3.1.2 IR

Here, we discussed three types of IRs: expert, object, and attribute levels.
(1) Expert level: To find the set of inconsistent experts, we use the following IR:

$$\mathscr{IAL} = \{p \mid \mathscr{WEL}^p < \xi\},\ p = 1, 2, \ldots, k. \tag{13}$$

(2) Object level: To find object level inconsistent position of the set of inconsistent experts, we use the following IR:

$$\mathscr{IOL} = \{(p, i) \mid e^p \in \mathscr{IAL} \wedge \mathscr{WOL}_i^p < \xi\}, \tag{14}$$

$(i = 1, 2, \ldots, n;\ p = 1, 2, \ldots, k)$.
(3) Attribute level: To find attribute level inconsistent position, we use the following IR:

$$\mathscr{IEL} = \{(p, i, j) \mid (p, i) \in \mathscr{IOL} \wedge \mathscr{WAL}_{ij}^p < \xi\}, \tag{15}$$

$(i = 1, 2, \ldots, n;\ j = 1, 2, \ldots, m;\ p = 1, 2, \ldots, k)$.

After finding the inconsistent position, we needed the method to improve the consistent degree of $\mathscr{WEL}^p$ with respect to ξ. This method is called the feedback advice mechanism (FAM), discussed in the following section.

3.1.3 FAM

We use the following FAM to improve the position $(p, i, j) \in \mathscr{I}\mathscr{E}\mathscr{L}$:

$$\varkappa_{ij,\ell+1}^{p} = \begin{cases} (1-\gamma)\varkappa_{ij,\ell}^{p} \oplus \gamma\overline{\varkappa}_{ij}, \ (p,i,j) \in \mathscr{I}\mathscr{E}\mathscr{L} \\ \varkappa_{ij,\ell}^{p}, \ (p,i,j) \notin \mathscr{I}\mathscr{E}\mathscr{L}, \end{cases} \tag{16}$$

where $\gamma \in [0, 1]$ and ℓ indicates the number of iteration. $\overline{\varkappa}_{ij} = (\overline{\mathscr{A}}_{ij}, \overline{\mathscr{B}}_{ij})$, $\overline{\mathscr{Z}} = (\overline{\varkappa}_{ij})_{n\times m}$ is the adjustment decision matrix called minimum adjustment sum obtained by the group consensus in the following way:

$$\begin{aligned} &\mathscr{M} = \min \sum_{p=1}^{k} \mathscr{D}(\varkappa_{ij}^{p}, \overline{\varkappa}_{ij}) \\ &\text{s.t.} \quad \begin{cases} x_{-\tau_1} \le \overline{\mathscr{A}}_{ij} \le x_{\tau_1}, \ i = 1,2,\ldots,n; \ j = 1,2,\ldots,m \\ y_{-\tau_2} \le \overline{\mathscr{B}}_{ij} \le y_{\tau_2}, \ i = 1,2,\ldots,n; \ j = 1,2,\ldots,m \end{cases} \end{aligned} \tag{17}$$

After solving the model (17), we find out $\overline{\mathscr{Z}}$, i.e., $\overline{\mathscr{A}}_{ij}, \overline{\mathscr{B}}_{ij}$.

Theorem 1 *The iterations of FAM (16) are convergent.*

Proof Let $(p, i, j) \in \mathscr{I}\mathscr{E}\mathscr{L}$; $\mathscr{Z}_{\ell}^{p} = (\varkappa_{ij,\ell}^{p})_{n\times m}$, $\varkappa_{ij,\ell}^{p} = (\mathscr{A}_{ij,\ell}^{p}, \mathscr{B}_{ij,\ell}^{p})$ and $\mathscr{Z}_{\ell+1}^{p} = (\varkappa_{ij,\ell+1}^{p})_{n\times m}$, $\varkappa_{ij,\ell+1}^{p} = (\mathscr{A}_{ij,\ell+1}^{p}, \mathscr{B}_{ij,\ell+1}^{p})$ the ℓ-th and $\ell+1$-th iteration evaluation matrix of expert e^{p}; $\overline{\mathscr{Z}} = (\overline{\varkappa}_{ij})_{n\times m}$, $\overline{\varkappa}_{ij} = (\overline{\mathscr{A}}_{ij}, \overline{\mathscr{B}}_{ij})$ is an adjustment matrix found by (17); $\beta = \mathscr{D}(\mathscr{Z}_{\ell+1}^{p}, \overline{\mathscr{Z}}) - \mathscr{D}(\mathscr{Z}_{\ell}^{p}, \overline{\mathscr{Z}})$. By Definition 3, $\beta = \sqrt{\frac{2}{nm}\sum_{i=1}^{n}\sum_{j=1}^{m}(\mathscr{D}(\varkappa_{ij,\ell+1}^{p}, \overline{\varkappa}_{ij}))^2} - \sqrt{\frac{2}{nm}\sum_{i=1}^{n}\sum_{j=1}^{m}(\mathscr{D}(\varkappa_{ij,\ell}^{p}, \overline{\varkappa}_{ij}))^2}$. Therefore, for $\gamma \in [0, 1]$, we have

$$\begin{aligned}
&\mathscr{D}(\varkappa_{ij,\ell+1}^{p}, \overline{\varkappa}_{ij}) \\
&= \left(\varsigma\left|\Delta\left(\mathscr{A}_{ij,\ell+1}^{p}\right) - \Delta(\overline{\mathscr{A}}_{ij})\right|^2 + \varrho\left|\nabla\left(\mathscr{B}_{ij,t+1}^{p}\right) - \nabla(\overline{\mathscr{B}}_{ij})\right|^2\right)^{\frac{1}{2}} \\
&= \left(\varsigma\left|\Delta\left((1-\gamma)\mathscr{A}_{ij,\ell}^{p} + \gamma\overline{\mathscr{A}}_{ij}\right) - \Delta\left(\overline{\mathscr{A}}_{ij}\right)\right|^2 + \varrho\left|\nabla\left(\frac{\mathscr{B}_{ij,\ell}^{p} + \overline{\mathscr{B}}_{ij}}{2}\right) - \nabla\left(\overline{\mathscr{B}}_{ij}\right)\right|^2\right)^{\frac{1}{2}} \\
&= \left(\varsigma\left|\Delta\left((1-\gamma)\mathscr{A}_{ij,\ell}^{p}\right) + \Delta\left(\gamma\overline{\mathscr{A}}_{ij}\right) - \frac{1}{2} - \Delta\left(\overline{\mathscr{A}}_{ij}\right)\right|^2 + \varrho\left|\nabla\left(\frac{\mathscr{B}_{ij,\ell}^{p}}{2}\right) + \nabla\left(\frac{\overline{\mathscr{B}}_{ij}}{2}\right) - \frac{1}{2} - \nabla\left(\overline{\mathscr{B}}_{ij}\right)\right|^2\right)^{\frac{1}{2}} \\
&= \left(\Delta\left|(1-\gamma)\Delta\left(\mathscr{A}_{ij,\ell}^{p}\right) + \frac{\gamma}{2} + \gamma\Delta\left(\overline{\mathscr{A}}_{ij}\right) + \frac{1-\gamma}{2} - \frac{1}{2} - \Delta\left(\overline{\mathscr{A}}_{ij}\right)\right|^2\right. \\
&\qquad \left. + \varrho\left|\frac{1}{2}\nabla\left(\mathscr{B}_{ij,\ell}^{p}\right) + \frac{1}{4} + \frac{1}{2}\nabla\left(\overline{\mathscr{B}}_{ij}\right) + \frac{1}{4} - \frac{1}{2} - \nabla\left(\overline{\mathscr{B}}_{ij}\right)\right|^2\right)^{\frac{1}{2}} \\
&= \left(\varsigma\left|(1-\gamma)\Delta\left(\mathscr{A}_{ij,\ell}^{p}\right) - (1-\gamma)\Delta\left(\overline{\mathscr{A}}_{ij}\right)\right|^2 + \varrho\left|\frac{1}{2}\left(\nabla\left(\mathscr{B}_{ij,\ell}^{p}\right) - \nabla\left(\overline{\mathscr{B}}_{ij}\right)\right)\right|^2\right)^{\frac{1}{2}}
\end{aligned}$$

$$= \left(\varsigma \left| (1-\gamma) \left(\triangle \left(\mathscr{A}^{p}_{ij,\ell} \right) - \triangle \left(\overline{\mathscr{A}}_{ij} \right) \right) \right|^2 + \varrho \left| \frac{1}{2} \left(\triangledown \left(\mathscr{B}^{p}_{ij,\ell} \right) - \triangledown \left(\overline{\mathscr{B}}_{ij} \right) \right) \right|^2 \right)^{\frac{1}{2}}$$

$$< \left(\varsigma \left| \left(\triangle \left(\mathscr{A}^{p}_{ij,\ell} \right) - \triangle \left(\overline{\mathscr{A}}_{ij} \right) \right) \right|^2 + \varrho \left| \left(\triangledown \left(\mathscr{B}^{p}_{ij,\ell} \right) - \triangledown \left(\overline{\mathscr{B}}_{ij} \right) \right) \right|^2 \right)^{\frac{1}{2}} = \mathscr{D}(\varkappa^{p}_{ij,\ell}, \overline{\varkappa}_{ij}). \tag{18}$$

Thus $\beta = \mathscr{D}(\mathscr{Z}^{p}_{\ell+1}, \overline{\mathscr{Z}}) - \mathscr{D}(\mathscr{Z}^{p}_{\ell}, \overline{\mathscr{Z}}) < 0 \Rightarrow \mathscr{D}(\mathscr{Z}^{p}_{\ell+1}, \overline{\mathscr{Z}}) < \mathscr{D}(\mathscr{Z}^{p}_{\ell}, \overline{\mathscr{Z}})$.

In Algorithm 1, we summarized our proposed model.

Algorithm 1 Acceptable judgement matrix of all experts in $\mathscr{E}$ and their corresponding weights.

Input: (1) $\mathscr{Z}^p = (\varkappa^p_{ij})_{n\times m}$, where $\varkappa^p_{ij} = (\mathscr{A}^p_{ij}, \mathscr{B}^p_{ij})$, $p = 1, 2, \ldots, k$. (2) $\varsigma, \varrho, \alpha, \gamma \in [0, 1]$. (3) $\xi \in [0, 1]$.
Output: (1) Acceptable judgement matrix $\tilde{\mathscr{Z}}^p = (\tilde{\varkappa}^p_{ij})_{n\times m}$, where $\tilde{\varkappa}^p_{ij} = (\tilde{\mathscr{A}}^p_{ij}, \tilde{\mathscr{B}}^p_{ij})$ and $p = 1, 2, \ldots, k$. (2) Acceptable weight vector of experts $\tilde{\varpi} = (\tilde{\varpi}_{e^1}, \tilde{\varpi}_{e^2}, \ldots, \tilde{\varpi}_{e^k})^T$, $\tilde{\varpi}_{e^p} \geq 0$. (3) Acceptable criterion weights $\tilde{\omega}^p_j$, for each $p = 1, 2, \ldots, k$ and $j = 1, 2, \ldots, m$.
Step 1: Let ℓ be the number of iteration. Assume $\ell = 0$ ans calculate $\mathscr{W}\mathscr{E}\mathscr{L}^p_\ell$ for each p, where $p = 1, 2, \ldots, k$, by formulas (2)–(12).
Step 2: If $\mathscr{W}\mathscr{E}\mathscr{L}^p_\ell < \xi$, then follow Step 3. Otherwise, follow Step 6.
Step 3: Obtained $\mathscr{I}\mathscr{E}\mathscr{L}$, by formulas (13)–(15).
Step 4: Find out adjustment decision matrix $\overline{\mathscr{Z}} = (\overline{\varkappa}_{ij})_{n\times m}$, where $\overline{\varkappa}_{ij} = (\overline{\mathscr{A}}_{ij}, \overline{\mathscr{B}}_{ij})$ using model (17).
Step 5: Improve consistency degree of the position in $\mathscr{I}\mathscr{E}\mathscr{L}$, by FAM technique given in formula (16). Then follow Step 2.
Step 6: Acceptable out put.

3.2 Process of Aggregation

For aggregating acceptable expert judgements, we adopt NLZWA operator from literature [3].

Definition 5 [3]. The aggregation of k number acceptable expert judgement matrix $\tilde{\mathscr{Z}}^p = (\tilde{\varkappa}^p_{ij})_{n\times m}$, where $\tilde{\varkappa}^p_{ij} = (\tilde{\mathscr{A}}^p_{ij}, \tilde{\mathscr{B}}^p_{ij})$ and $p = 1, 2, \ldots, k$ with their corresponding weight vector $\tilde{\varpi} = (\tilde{\varpi}_{e^1}, \tilde{\varpi}_{e^2}, \ldots, \tilde{\varpi}_{e^k})^T$ such that $\sum_{p=1}^{k} \tilde{\varpi}_{e^p} = 1, \tilde{\varpi}_{e^p} \geq 0$ is denoted by $\mathscr{Z}^c = (\varkappa^c_{ij})_{n\times m}$, where $\varkappa^c_{ij} = (\mathscr{A}^c_{ij}, \mathscr{B}^c_{ij})$ and defined in the following way:

$$\Omega(\mathscr{A}^c_{ij}) = \max\left\{ -\tau_1, \min\left\{ \sum_{p=1}^{k} \tilde{\varpi}_{e^p} \Omega(\tilde{\mathscr{A}}^p_{ij}), \tau_1 \right\} \right\}, \quad \Omega(\mathscr{B}^c_{ij}) = \frac{1}{k} \sum_{p=1}^{k} \Omega(\tilde{\mathscr{B}}^p_{ij}), \tag{19}$$

where Ω is subscripts of linguistic terms.

3.3 Determination of Risk Factor Weight

In MAGDM, there are two types of criterion weights: subjective and objective weights. But each of these has some advantages and some disadvantages. In view of this, we proposed a combined weight determination method here. The detailed process is discussed in the following.

Step 1: Determination of the risk factor subjective weight

The determination of the risk factor subjective weight formula is

$$\omega_{sj} = \frac{\sum_{p=1}^{k} \tilde{\varpi}_{e^p} \tilde{\omega}_j^p}{\sum_{j=1}^{m} \sum_{p=1}^{k} \tilde{\varpi}_{e^p} \tilde{\omega}_j^p}, \tag{20}$$

where $p = 1, 2, \ldots, k$ and $j = 1, 2, \ldots, m$. According to weight vector, it is represent by $\mathscr{w}_s = (\omega_{s1}, \omega_{s2}, \ldots, \omega_{sm})$.

Step 2: Determination of the risk factor objective weight according to entropy method

(1) Create the normalized judgement aggregative matrix $\mathscr{NEXP}(\mathscr{Z}^c) = (\mathscr{NEXP}(z_{ij}^c))_{n\times m}$, where

$$\mathscr{NEXP}(z_{ij}^c) = \frac{\mathscr{EXP}(z_{ij}^c)}{\sum_{i=1}^{n} \mathscr{EXP}(z_{ij}^c)}, i = 1, 2, \ldots, n;\ j = 1, 2, \ldots, m. \tag{21}$$

(2) Calculate entropy:

$$E_j = -\frac{1}{ln(n)} \sum_{i=1}^{n} \mathscr{NEXP}(z_{ij}^c) ln(\mathscr{NEXP}(z_{ij}^c)), \ j = 1, 2, \ldots, m. \tag{22}$$

(3) The objective weight of c_j is

$$\omega_{oej} = \frac{1 - E_j}{\sum_{j=1}^{m} (1 - E_j)}, \ j = 1, 2, \ldots, m. \tag{23}$$

According to weight vector, it is represent by $\mathscr{w}_o e = (\omega_{oe1}, \omega_{oe2}, \ldots, \omega_{oem})$.

Step 3: Determination of the risk factor objective weight according to division method

The determination of the risk factor objective weight formula by division method is

$$\omega_{odj} = \frac{\sum_{i=1}^{n} \sum_{l=1}^{n} \mathscr{D}(t_{ij}, t_{lj})}{\sum_{j=1}^{m} \sum_{i=1}^{n} \sum_{l=1}^{n} \mathscr{D}(t_{ij}, t_{lj})}, \tag{24}$$

where t_{ij} is a (i, j) entity of $\mathscr{T} = (t_{ij})_{n\times m}$ with $t_{ij} = (\Delta^{-1}(\mathscr{U}_{ij}), \nabla^{-1}(\mathscr{V}_{ij}))$ and

$$\mathscr{U}_{ij} = \frac{\Delta(\mathscr{A}^c_{ij})}{\sqrt{\sum_{i=1}^{n}(\Delta(\mathscr{A}^c_{ij}))^2}},\ \mathscr{V}^c_{ij} = \frac{\nabla(\mathscr{B}^c_{ij})}{\sqrt{\sum_{i=1}^{n}(\nabla(\mathscr{B}^c_{ij}))^2}},\ i = 1, 2, \ldots, n;\ j = 1, 2, \ldots, m. \tag{25}$$

According to weight vector, it is represent by $w_{od} = (\omega_{od1}, \omega_{od2}, \ldots, \omega_{odm})$.

Step 4: Determination of the risk factor weight according to combined method

The determination of the risk factor weight formula by combined method is

$$\omega_j = \frac{\omega_{oj} \times \omega_{oej} \times \omega_{odj}}{\sum_{j=1}^{m}(\omega_{oj} \times \omega_{oej} \times \omega_{odj})},\ j = 1, 2, \ldots, m. \tag{26}$$

According to weight vector, it is represent by $w = (\omega_1, \omega_2, \ldots, \omega_m)$.

3.4 Ranking Method

Here we discuss GLDS-based ranking method [55].

Definition 6 For k number acceptable expert judgement matrix $\tilde{\mathscr{Z}}^p = (\tilde{z}^p_{ij})_{n\times m}$, where $\tilde{z}^p_{ij} = (\tilde{\mathscr{A}}^p_{ij}, \tilde{\mathscr{B}}^p_{ij})$ and $p = 1, 2, \ldots, k$, a similar proximity degree of $\tilde{z}^p_{ij}$ is

$$\mathscr{CV}(\tilde{z}^p_{ij}) = \frac{\mathscr{D}^-_{ij}}{\mathscr{D}^-_{ij} + \mathscr{D}^+_{ij}} \tag{27}$$

where $\mathscr{D}^+_{ij} = \begin{cases} \mathscr{D}(\tilde{z}^p_{ij}, \tilde{1}), & \text{for profit criterion } c_j \\ \mathscr{D}(\tilde{z}^p_{ij}, \tilde{0}), & \text{for loss criterion } c_j \end{cases}$, $\mathscr{D}^-_{ij} = \begin{cases} \mathscr{D}(\tilde{z}^p_{ij}, \tilde{0}), & \text{for profit criterion } c_j \\ \mathscr{D}(\tilde{z}^p_{ij}, \tilde{1}), & \text{for loss criterion } c_j \end{cases}$.

Clearly, $\mathscr{CV}(\tilde{z}^p_{ij}) \in [0, 1]$, for $i = 1, 2, \ldots, n$ and $j = 1, 2, \ldots, m$.

Therefore, the dominance degree for each criterion c_j $(j = 1, 2, \ldots, m)$ of o_i and o_l $(i, l = 1, 2, \ldots, n)$ is

$$\mathscr{DD}^p_j(o_i, o_l) = \left|\mathscr{CV}(\tilde{z}^p_{ij}) - \mathscr{CV}(\tilde{z}^p_{lj})\right|. \tag{28}$$

Clearly, $\mathscr{DD}^p_j(o_i, o_l) \in [0, 1]$ and it holds properties of ideal distance.

By expression (28), the dominance flow for each criterion c_j $(j = 1, 2, \ldots, m)$ of o_i and o_l $(i, l = 1, 2, \ldots, n)$ is

$$\mathscr{DF}_j(o_i, o_l) = \sum_{p=1}^{k} \tilde{\varpi}_{e^p} \mathscr{DD}^p_j(o_i, o_l). \tag{29}$$

The normalized dominance flow of $\mathscr{DF}_j(o_i, o_l)$ is

$$\mathscr{NDF}_j(o_i, o_l) = \frac{\mathscr{DF}_j(o_i, o_l)}{\sqrt{\sum_{l=1}^{n}\sum_{i=1}^{n}\left(\mathscr{DF}_j(o_i, o_l)\right)^2}}. \tag{30}$$

The gained and lost dominance scores of o_i are

$$\mathscr{GDS}(o_i) = \sum_{j=1}^{m}\sum_{l=1}^{n} w_j \mathscr{NDF}_j(o_i, o_l), \tag{31}$$

$$\mathscr{LDS}(o_i) = \max_j \omega_j \left(\max_l \mathscr{NDF}_j(o_i, o_l)\right). \tag{32}$$

The final, i.e., global score is

$$\mathscr{GS}(o_i) = \frac{\mathscr{GDS}(o_i)}{\sqrt{\sum_{i=1}^{n}(\mathscr{GDS}(o_i))^2}} \frac{n - r_1(o_i) + 1}{n(n+1)/2} - \frac{\mathscr{LDS}(o_i)}{\sqrt{\sum_{i=1}^{n}(\mathscr{GDS}(o_i))^2}} \frac{r_2(o_i)}{n(n+1)/2}, \tag{33}$$

where $r_1(o_i)$ is the descending order object o_i ranking of $\mathscr{GDS}(o_i)$ for $i = 1, 2, \ldots, n$ and $r_2(o_i)$ is the ascending order object o_i ranking of $\mathscr{LDS}(o_i)$, for $i = 1, 2, \ldots, n$. According to value of $\mathscr{GS}(o_i)$, we obtain final ranking of object in object set.

4 A Case Study: Ranking of FMs of SBRT for Lung Cancer Treatment

Lung cancer is a fast-growing and lethal malignant tumour that endangers human health [1]. Most primary lung cancer cases (85%) are non-small-cell lung cancers, which have poor survival rates and high malignancy [37]. Chemotherapy, radiotherapy, and surgery are the main treatments for non-small-cell lung cancer [9]. SBRT, a widely used radiotherapy technique, has been used to treat lung cancer in clinical settings for over 10 years. The treatment steps and the main process of LSBRT are described in [2]. The FMEA method is necessary to analyse the complex process, which requires experts from different disciplines to improve the safety and reliability of SBRT for lung cancer treatment.

4.1 Expert Evaluation and FM Identification

According to studies in [2], six experts are involved in analyses of FMs of SBRT for lung cancer treatment and identified the FMs, which are included in Table 1. The roles of experts are as follows: e^1, e^2: Radiation oncologists; e^3, e^4, e^5: Medical physicists; e^6: Radiation therapist.

After identifying FMs, six experts gave their judgement in Table 2 concerning RFs O (the occurrence of FMs), S (the severity of FMs), and D (the likelihood of FMs not being detected) for each FM by using the LTSs $\mathscr{X}$ and $\mathscr{Y}$ given in (34):

Table 1 FMs of SBRT for lung cancer treatment

	Description of FMs	Process steps
$\mathscr{FM}_1$	Patient alignment checked to ceiling laser when couch rotated	Improper laser marking of patient, reassessed values after procedural changes
$\mathscr{FM}_2$	Incorrect calibration of couch motors	Automatic setup failure
$\mathscr{FM}_3$	Transcription error in coordinates	SBRT frame at isocenter
$\mathscr{FM}_4$	Loss of beam	Machine faults leading to interlocks
$\mathscr{FM}_5$	Difference with beam isocenter	CBCT scan isocenter misaligned
$\mathscr{FM}_6$	Shifts should all be less than 1 cm	Overly large CBCT shifts
$\mathscr{FM}_7$	Transcription error	Documentation of couch coordinates

Table 2 The judgement of the experts on FMs

Experts	RFs	$\mathscr{FM}_1$	$\mathscr{FM}_2$	$\mathscr{FM}_3$	$\mathscr{FM}_4$	$\mathscr{FM}_5$	$\mathscr{FM}_6$	$\mathscr{FM}_7$
e^1	O	(x_0, y_3)	(x_{-2}, y_2)	(x_1, y_3)	(x_2, y_{-1})	(x_1, y_2)	(x_1, y_1)	(x_{-1}, y_1)
	S	(x_0, y_1)	(x_{-1}, y_1)	(x_0, y_3)	(x_{-1}, y_{-2})	(x_{-1}, y_{-3})	(x_0, y_2)	(x_{-1}, y_1)
	D	(x_{-1}, y_3)	(x_{-1}, y_3)	(x_0, y_3)	(x_1, y_2)	(x_1, y_3)	(x_1, y_{-2})	(x_2, y_{-1})
e^2	O	(x_2, y_3)	(x_0, y_{-2})	(x_0, y_2)	(x_{-2}, y_{-2})	(x_0, y_{-2})	(x_0, y_{-2})	(x_1, y_3)
	S	(x_{-2}, y_3)	(x_0, y_1)	(x_2, y_3)	(x_0, y_{-2})	(x_2, y_3)	(x_1, y_{-2})	(x_1, y_{-1})
	D	(x_0, y_2)	(x_{-2}, y_3)	(x_2, y_3)	(x_1, y_2)	(x_0, y_1)	(x_1, y_2)	(x_0, y_3)
e^3	O	(x_1, y_3)	(x_1, y_2)	(x_0, y_3)	(x_{-1}, y_{-2})	(x_2, y_3)	(x_{-1}, y_1)	(x_2, y_3)
	S	(x_{-1}, y_1)	(x_1, y_3)	(x_0, y_3)	(x_{-1}, y_3)	(x_0, y_{-1})	(x_{-1}, y_3)	(x_2, y_3)
	D	(x_{-2}, y_3)	(x_0, y_2)	(x_{-1}, y_{-1})	(x_{-1}, y_3)	(x_{-1}, y_{-3})	(x_2, y_3)	(x_0, y_3)
e^4	O	(x_1, y_{-1})	(x_{-2}, y_1)	(x_1, y_3)	(x_{-2}, y_1)	(x_1, y_1)	(x_{-1}, y_3)	(x_2, y_3)
	S	(x_2, y_3)	(x_{-2}, y_3)	(x_1, y_1)	(x_0, y_2)	(x_0, y_3)	(x_1, y_3)	(x_{-1}, y_{-3})
	D	(x_0, y_3)	(x_{-1}, y_1)	(x_{-1}, y_2)	(x_{-1}, y_{-1})	(x_1, y_1)	(x_2, y_3)	(x_{-2}, y_2)
e^5	O	(x_0, y_3)	(x_{-1}, y_1)	(x_1, y_3)	(x_{-2}, y_3)	(x_{-2}, y_1)	(x_0, y_1)	(x_{-1}, y_3)
	S	(x_1, y_3)	(x_{-2}, y_3)	(x_1, y_{-1})	(x_0, y_3)	(x_{-1}, y_2)	(x_{-1}, y_2)	(x_1, y_{-3})
	D	(x_1, y_{-2})	(x_0, y_{-1})	(x_{-1}, y_3)	(x_{-2}, y_3)	(x_{-2}, y_3)	(x_2, y_3)	(x_1, y_1)
e^6	O	(x_0, y_2)	(x_1, y_3)	(x_{-1}, y_{-2})	(x_1, y_{-1})	(x_1, y_3)	(x_0, y_3)	(x_{-2}, y_0)
	S	(x_{-2}, y_3)	(x_1, y_3)	(x_1, y_1)	(x_0, y_3)	(x_1, y_0)	(x_0, y_1)	(x_1, y_{-2})
	D	(x_1, y_3)	(x_{-2}, y_3)	(x_2, y_3)	(x_1, y_{-1})	(x_1, y_3)	(x_2, y_3)	(x_1, y_{-1})

$$\tilde{\mathscr{X}} = \begin{cases} \mathscr{x}_{-2} = \text{very low}, \\ \mathscr{x}_{-1} = \text{low}, \\ \mathscr{x}_{0} = \text{normal}, \\ \mathscr{x}_{1} = \text{high}, \\ \mathscr{x}_{2} = \text{very high} \end{cases} \quad \text{and} \quad \tilde{\mathscr{Y}} = \begin{cases} \mathscr{y}_{-3} = \text{not all at sure}, \\ \mathscr{y}_{-2} = \text{not sure}, \\ \mathscr{y}_{-1} = \text{slightly not sure}, \\ \mathscr{y}_{0} = \text{fair}, \\ \mathscr{y}_{1} = \text{slightly sure}, \\ \mathscr{y}_{2} = \text{sure}, \\ \mathscr{y}_{3} = \text{very sure} \end{cases}. \tag{34}$$

4.2 *Apply the Proposed LZN-Based FMEA GDM Model to Solve the Case*

Let $\varsigma = 0.6$, $\alpha = 0.5$, $\xi = 0.9$, and $\gamma = 0.15$; by Algorithm 1, we have found (1) Acceptable judgement matrix $\tilde{\mathscr{Z}}^p = (\tilde{\mathscr{z}}^p_{ij})_{6\times3}$, where $\tilde{\mathscr{z}}^p_{ij} = (\tilde{\mathscr{A}}^p_{ij}, \tilde{\mathscr{B}}^p_{ij})$, and $p = 1, 2, \ldots, 6$ provided in Tables 3 and 4.

Table 3 Acceptable judgement matrix of all experts

e^p	$\tilde{\mathscr{Z}}^p = (\tilde{\mathscr{z}}^p_{ij})_{6\times3}, \tilde{\mathscr{z}}^p_{ij} = (\tilde{\mathscr{A}}^p_{ij}, \tilde{\mathscr{B}}^p_{ij})$
e^1	$\begin{bmatrix} (\mathscr{x}_{0.151}, \mathscr{y}_{2.536}) & (\mathscr{x}_{-0.297}, \mathscr{y}_{1.809}) & (\mathscr{x}_{-0.389}, \mathscr{y}_{2.200}) \\ (\mathscr{x}_{-1.377}, \mathscr{y}_{1.016}) & (\mathscr{x}_{-0.689}, \mathscr{y}_{1.874}) & (\mathscr{x}_{-1.014}, \mathscr{y}_{2.014}) \\ (\mathscr{x}_{0.997}, \mathscr{y}_{2.999}) & (\mathscr{x}_{0.623}, \mathscr{y}_{1.031}) & (\mathscr{x}_{-0.046}, \mathscr{y}_{2.691}) \\ (\mathscr{x}_{-0.133}, \mathscr{y}_{-0.840}) & (\mathscr{x}_{-0.416}, \mathscr{y}_{2.058}) & (\mathscr{x}_{0.403}, \mathscr{y}_{1.361}) \\ (\mathscr{x}_{1}, \mathscr{y}_{2}) & (\mathscr{x}_{-0.333}, \mathscr{y}_{0.358}) & (\mathscr{x}_{0.607}, \mathscr{y}_{1.529}) \\ (\mathscr{x}_{0.376}, \mathscr{y}_{1.002}) & (\mathscr{x}_{-0.007}, \mathscr{y}_{1.815}) & (\mathscr{x}_{1.623}, \mathscr{y}_{2.922}) \\ (\mathscr{x}_{0.229}, \mathscr{y}_{2.729}) & (\mathscr{x}_{0.313}, \mathscr{y}_{-1.528}) & (\mathscr{x}_{0.980}, \mathscr{y}_{1.276}) \end{bmatrix}$
e^2	$\begin{bmatrix} (\mathscr{x}_{0.905}, \mathscr{y}_{2.536}) & (\mathscr{x}_{-1.051}, \mathscr{y}_{1.840}) & (\mathscr{x}_{-0.012}, \mathscr{y}_{2.184}) \\ (\mathscr{x}_{-0.623}, \mathscr{y}_{0.953}) & (\mathscr{x}_{-0.311}, \mathscr{y}_{1.874}) & (\mathscr{x}_{-1.391}, \mathscr{y}_{2.014}) \\ (\mathscr{x}_{0.620}, \mathscr{y}_{2.983}) & (\mathscr{x}_{1.377}, \mathscr{y}_{1.031}) & (\mathscr{x}_{0.709}, \mathscr{y}_{2.691}) \\ (\mathscr{x}_{-1.460}, \mathscr{y}_{-0.857}) & (\mathscr{x}_{-0.038}, \mathscr{y}_{2.058}) & (\mathscr{x}_{0.454}, \mathscr{y}_{1.367}) \\ (\mathscr{x}_{0.623}, \mathscr{y}_{1.937}) & (\mathscr{x}_{0.799}, \mathscr{y}_{0.452}) & (\mathscr{x}_{0.230}, \mathscr{y}_{1.498}) \\ (\mathscr{x}_{-0.001}, \mathscr{y}_{0.955}) & (\mathscr{x}_{0.370}, \mathscr{y}_{1.752}) & (\mathscr{x}_{1.623}, \mathscr{y}_{2.984}) \\ (\mathscr{x}_{0.881}, \mathscr{y}_{2.747}) & (\mathscr{x}_{0.954}, \mathscr{y}_{-1.536}) & (\mathscr{x}_{0.315}, \mathscr{y}_{1.319}) \end{bmatrix}$
e^3	$\begin{bmatrix} (\mathscr{x}_{0.528}, \mathscr{y}_{2.536}) & (\mathscr{x}_{-0.674}, \mathscr{y}_{1.809}) & (\mathscr{x}_{-0.766}, \mathscr{y}_{2.200}) \\ (\mathscr{x}_{-0.246}, \mathscr{y}_{1.016}) & (\mathscr{x}_{0.066}, \mathscr{y}_{1.905}) & (\mathscr{x}_{-0.637}, \mathscr{y}_{1.999}) \\ (\mathscr{x}_{0.620}, \mathscr{y}_{2.999}) & (\mathscr{x}_{0.623}, \mathscr{y}_{1.031}) & (\mathscr{x}_{-0.423}, \mathscr{y}_{2.629}) \\ (\mathscr{x}_{-1.083}, \mathscr{y}_{-0.857}) & (\mathscr{x}_{-0.416}, \mathscr{y}_{2.137}) & (\mathscr{x}_{-0.300}, \mathscr{y}_{1.383}) \\ (\mathscr{x}_{1.377}, \mathscr{y}_{2.015}) & (\mathscr{x}_{0.045}, \mathscr{y}_{0.390}) & (\mathscr{x}_{-0.147}, \mathscr{y}_{1.435}) \\ (\mathscr{x}_{-0.378}, \mathscr{y}_{1.002}) & (\mathscr{x}_{-0.384}, \mathscr{y}_{1.831}) & (\mathscr{x}_{2}, \mathscr{y}_{3}) \\ (\mathscr{x}_{1.259}, \mathscr{y}_{2.747}) & (\mathscr{x}_{1.331}, \mathscr{y}_{-1.473}) & (\mathscr{x}_{0.315}, \mathscr{y}_{1.319}) \end{bmatrix}$

Table 4 Continued of Table 3

e^p	$\tilde{\mathscr{Z}}^p = (\tilde{z}_{ij}^p)_{6\times 3}, \tilde{z}_{ij}^p = (\tilde{\mathscr{A}}_{ij}^p, \tilde{\mathscr{B}}_{ij}^p)$
e^4	$\begin{bmatrix} (x_{0.528}, y_{2.474}) & (x_{0.457}, y_{1.840}) & (x_{-0.012}, y_{2.200}) \\ (x_{-1.377}, y_{1}) & (x_{-1.066}, y_{1.905}) & (x_{-1.014}, y_{1.983}) \\ (x_{0.997}, y_{2.999}) & (x_{1}, y_{1}) & (x_{-0.423}, y_{2.675}) \\ (x_{-1.460}, y_{-0.810}) & (x_{-0.038}, y_{2.121}) & (x_{-0.300}, y_{1.321}) \\ (x_{1}, y_{1.984}) & (x_{0.045}, y_{0.452}) & (x_{0.607}, y_{1.498}) \\ (x_{-0.378}, y_{1.033}) & (x_{0.370}, y_{1.831}) & (x_{2}, y_{3}) \\ (x_{1.259}, y_{2.747}) & (x_{0.200}, y_{-1.567}) & (x_{-0.440}, y_{1.303}) \end{bmatrix}$
e^5	$\begin{bmatrix} (x_{0.151}, y_{2.536}) & (x_{-0.003}, y_{1.830}) & (x_{0.309}, y_{2.153}) \\ (x_{-1}, y_{1}) & (x_{-0.981}, y_{1.895}) & (x_{-0.637}, y_{1.952}) \\ (x_{0.997}, y_{2.999}) & (x_{1}, y_{0.969}) & (x_{-0.423}, y_{2.691}) \\ (x_{-1.416}, y_{-0.809}) & (x_{-0.038}, y_{2.137}) & (x_{-0.559}, y_{1.369}) \\ (x_{0.038}, y_{1.992}) & (x_{-0.276}, y_{0.422}) & (x_{-0.388}, y_{1.517}) \\ (x_{-0.001}, y_{1.002}) & (x_{-0.384}, y_{1.815}) & (x_{2}, y_{3}) \\ (x_{0.127}, y_{2.747}) & (x_{0.954}, y_{-1.567}) & (x_{0.692}, y_{1.287}) \end{bmatrix}$
e^6	$\begin{bmatrix} (x_{0.151}, y_{2.521}) & (x_{-1.051}, y_{1.840}) & (x_{0.366}, y_{2.200}) \\ (x_{-0.246}, y_{1.031}) & (x_{0.066}, y_{1.905}) & (x_{-1.391}, y_{2.014}) \\ (x_{0.243}, y_{2.920}) & (x_{1}, y_{1}) & (x_{0.709}, y_{2.691}) \\ (x_{-0.329}, y_{-0.842}) & (x_{-0.038}, y_{2.137}) & (x_{0.454}, y_{1.321}) \\ (x_{1}, y_{2.015}) & (x_{0.422}, y_{0.405}) & (x_{0.607}, y_{1.529}) \\ (x_{-0.001}, y_{1.033}) & (x_{-0.007}, y_{1.799}) & (x_{2}, y_{3}) \\ (x_{-0.250}, y_{2.700}) & (x_{0.954}, y_{-1.551}) & (x_{0.692}, y_{1.256}) \end{bmatrix}$

Table 5 Acceptable weight vector of experts and criterion weights for each experts

e^p	$\tilde{\varpi}_{e^p}$	$\tilde{\omega}_j^p$		
		O	S	D
e^1	0.1663	0.3681	0.3009	0.3310
	0.1697	0.3490	0.3530	0.2980
	0.1665	0.3564	0.3195	0.3241
	0.1653	0.3945	0.2912	0.3143
	0.1618	0.3516	0.3319	0.3165
	0.1704	0.2994	0.3659	0.3347

(2) Acceptable weight vector of experts $\tilde{\varpi} = (\tilde{\varpi}_{e^1}, \tilde{\varpi}_{e^2}, \ldots, \tilde{\varpi}_{e^6})^T$, $\tilde{\varpi}_{e^p} \geq 0$; and criterion weights $\tilde{\omega}_j^p$, for each $p = 1, 2, \ldots, 6$ and $j = 1, 2, 3$ are provided in Table 5.

Using the results shown in Tables 3 and 5, we can aggregate the expert acceptable judgement matrix by formula (19). The result is display in (35):

Table 6 RF combined weights

RFs	w_s	w_{oe}	w_{od}	ω
O	0.3426	0.5176	0.3469	5609
S	0.3366	0.1339	0.3201	1297
D	0.3208	0.3485	0.3330	3094

Table 7 Ranking of FMs

Failure mode	$\mathscr{GDS}(o_i)$	$r_1(o_i)$	$\mathscr{LDS}(o_i)$	$r_2(o_i)$	$\mathscr{GS}(o_i)$	Final position of ranking
$\mathscr{FM}_1$	0.2477	7	0.0520	3	−0.0282	5
$\mathscr{FM}_2$	0.3317	3	0.0430	1	0.0610	2
$\mathscr{FM}_3$	0.2958	4	0.0608	6	−0.0399	6
$\mathscr{FM}_4$	0.3812	1	0.0608	6	0.0243	3
$\mathscr{FM}_5$	0.2590	6	0.0548	5	−0.0505	7
$\mathscr{FM}_6$	0.3750	2	0.0430	1	0.0866	1
$\mathscr{FM}_7$	0.2639	5	0.0571	4	−0.0228	4

$$\mathscr{Z}^c = \begin{bmatrix} (x_{0.404}, y_{2.523}) & (x_{-0.444}, y_{1.828}) & (x_{-0.083}, y_{2.19}) \\ (x_{-0.807}, y_{1.003}) & (x_{-0.48}, y_{1.893}) & (x_{-1.018}, y_{1.996}) \\ (x_{0.742}, y_{2.983}) & (x_{0.939}, y_{1.01}) & (x_{0.024}, y_{2.678}) \\ (x_{-0.977}, y_{-0.836}) & (x_{-0.164}, y_{2.108}) & (x_{0.31}, y_{1.354}) \\ (x_{0.843}, y_{1.991}) & (x_{0.122}, y_{0.413}) & (x_{0.256}, y_{1.501}) \\ (x_{-0.063}, y_{1.005}) & (x_{-0.005}, y_{1.807}) & (x_{1.873}, y_{2.984}) \\ (x_{0.583}, y_{2.736}) & (x_{0.786}, y_{-1.537}) & (x_{0.426}, y_{1.293}) \end{bmatrix}. \tag{35}$$

According to weight computation process of RFs recommended in Sect. 3.3, the weight of RFs are given in Table 6.

According to ranking process recommended in Sect. 3.4, the result of ranking value with final ranking of each FM shown in Table 7. From Table 7 we get $\mathscr{FM}_6 \succ \mathscr{FM}_2 \succ \mathscr{FM}_4 \succ \mathscr{FM}_7 \succ \mathscr{FM}_1 \succ \mathscr{FM}_3 \succ \mathscr{FM}_5$.

5 Analysis

Here, we mainly perform sensitive and comparative analyses.

5.1 Sensitive Analysis

Our proposed method included four parameters ς, ϱ, α, and γ. In this section, we show the effect of these parameters on ranking result.

Table 8 The effect of ς and ϱ on ranking result

ς	ϱ	Ranking
1	0	$\mathscr{FM}_4 \succ \mathscr{FM}_6 \succ \mathscr{FM}_7 \succ \mathscr{FM}_3 \succ \mathscr{FM}_5 \succ \mathscr{FM}_2 \succ \mathscr{FM}_1$
0.9	0.1	$\mathscr{FM}_6 \succ \mathscr{FM}_2 \succ \mathscr{FM}_1 \succ \mathscr{FM}_4 \succ \mathscr{FM}_3 \succ \mathscr{FM}_7 \succ \mathscr{FM}_5$
0.8	0.2	$\mathscr{FM}_6 \succ \mathscr{FM}_2 \succ \mathscr{FM}_4 \succ \mathscr{FM}_1 \succ \mathscr{FM}_7 \succ \mathscr{FM}_3 \succ \mathscr{FM}_5$
0.7	0.3	$\mathscr{FM}_6 \succ \mathscr{FM}_2 \succ \mathscr{FM}_4 \succ \mathscr{FM}_1 \succ \mathscr{FM}_7 \succ \mathscr{FM}_3 \succ \mathscr{FM}_5$
0.6	0.4	$\mathscr{FM}6 \succ \mathscr{FM}2 \succ \mathscr{FM}4 \succ \mathscr{FM}7 \succ \mathscr{FM}1 \succ \mathscr{FM}3 \succ \mathscr{FM}5$
0.5	0.5	$\mathscr{FM}_6 \succ \mathscr{FM}_2 \succ \mathscr{FM}_4 \succ \mathscr{FM}_1 \succ \mathscr{FM}_5 \succ \mathscr{FM}_7 \succ \mathscr{FM}_3$
0.4	0.6	$\mathscr{FM}6 \succ \mathscr{FM}2 \succ \mathscr{FM}4 \succ \mathscr{FM}7 \succ \mathscr{FM}1 \succ \mathscr{FM}5 \succ \mathscr{FM}3$
0.3	0.7	$\mathscr{FM}_6 \succ \mathscr{FM}_4 \succ \mathscr{FM}_2 \succ \mathscr{FM}_7 \succ \mathscr{FM}_5 \succ \mathscr{FM}_1 \succ \mathscr{FM}_3$
0.2	0.8	$\mathscr{FM}_6 \succ \mathscr{FM}_2 \succ \mathscr{FM}_4 \succ \mathscr{FM}_7 \succ \mathscr{FM}_5 \succ \mathscr{FM}_1 \succ \mathscr{FM}_3$
0.1	0.9	$\mathscr{FM}_6 \succ \mathscr{FM}_4 \succ \mathscr{FM}_7 \succ \mathscr{FM}_2 \succ \mathscr{FM}_5 \succ \mathscr{FM}_1 \succ \mathscr{FM}_3$

5.1.1 The Effect of ς and ϱ on Ranking Result

To change ς and ϱ for fixed $\alpha = 0.5$, and $\gamma = 0.15$, we get different ranking results of FMs, which is shown in Table 8. As Table 8 shows, changing ς and ϱ affects the ranking results. ς goes up, but ϱ goes down. This implies that decision-making depends a lot on opinions and their associated reliable information. But when we look at all the ranking results in Table 8, we see that the FM with the most risk stays the same, except for the first one.

5.1.2 The Effect of α on Ranking Result

To change α for fixed $\varsigma = 0.6$, $\varrho = 0.4$, and $\gamma = 0.15$, we get the ranking results of FMs, shown in Table 9. We can see in Table 9 that the ranking results are unchanged. However, without consensus, α moderately affects the ranking result, as shown in Table 10.

5.1.3 The Effect of γ on Ranking Result

To change γ for fix $\varsigma = 0.6$, $\varrho = 0.4$, and $\alpha = 0.5$, we get the ranking results of FMs, which is shown in Table 11. We can see in Table 11 that the ranking results are changed, but the highest FM in most of the cases is unchanged.

Table 9 The effect of α on ranking results

α	Ranking
0	$\mathscr{FM}_6 \succ \mathscr{FM}_2 \succ \mathscr{FM}_4 \succ \mathscr{FM}_7 \succ \mathscr{FM}_1 \succ \mathscr{FM}_3 \succ \mathscr{FM}_5$
0.1	$\mathscr{FM}_6 \succ \mathscr{FM}_2 \succ \mathscr{FM}_4 \succ \mathscr{FM}_7 \succ \mathscr{FM}_1 \succ \mathscr{FM}_3 \succ \mathscr{FM}_5$
0.2	$\mathscr{FM}_6 \succ \mathscr{FM}_2 \succ \mathscr{FM}_4 \succ \mathscr{FM}_7 \succ \mathscr{FM}_1 \succ \mathscr{FM}_3 \succ \mathscr{FM}_5$
0.3	$\mathscr{FM}_6 \succ \mathscr{FM}_2 \succ \mathscr{FM}_4 \succ \mathscr{FM}_7 \succ \mathscr{FM}_1 \succ \mathscr{FM}_3 \succ \mathscr{FM}_5$
0.4	$\mathscr{FM}_6 \succ \mathscr{FM}_2 \succ \mathscr{FM}_4 \succ \mathscr{FM}_7 \succ \mathscr{FM}_1 \succ \mathscr{FM}_3 \succ \mathscr{FM}_5$
0.5	$\mathscr{FM}_6 \succ \mathscr{FM}_2 \succ \mathscr{FM}_4 \succ \mathscr{FM}_7 \succ \mathscr{FM}_1 \succ \mathscr{FM}_3 \succ \mathscr{FM}_5$
0.6	$\mathscr{FM}_6 \succ \mathscr{FM}_2 \succ \mathscr{FM}_4 \succ \mathscr{FM}_7 \succ \mathscr{FM}_1 \succ \mathscr{FM}_3 \succ \mathscr{FM}_5$
0.7	$\mathscr{FM}_6 \succ \mathscr{FM}_2 \succ \mathscr{FM}_4 \succ \mathscr{FM}_7 \succ \mathscr{FM}_1 \succ \mathscr{FM}_3 \succ \mathscr{FM}_5$
0.8	$\mathscr{FM}_6 \succ \mathscr{FM}_2 \succ \mathscr{FM}_4 \succ \mathscr{FM}_7 \succ \mathscr{FM}_1 \succ \mathscr{FM}_3 \succ \mathscr{FM}_5$
0.9	$\mathscr{FM}_6 \succ \mathscr{FM}_2 \succ \mathscr{FM}_4 \succ \mathscr{FM}_7 \succ \mathscr{FM}_1 \succ \mathscr{FM}_3 \succ \mathscr{FM}_5$
1	$\mathscr{FM}_6 \succ \mathscr{FM}_2 \succ \mathscr{FM}_4 \succ \mathscr{FM}_7 \succ \mathscr{FM}_1 \succ \mathscr{FM}_3 \succ \mathscr{FM}_5$

Table 10 The effect of α on ranking results without consensus

κ	Ranking
0	$\mathscr{FM}_7 \succ \mathscr{FM}_3 \succ \mathscr{FM}_6 \succ \mathscr{FM}_1 \succ \mathscr{FM}_4 \succ \mathscr{FM}_2 \succ \mathscr{FM}_5$
0.1	$\mathscr{FM}_7 \succ \mathscr{FM}_3 \succ \mathscr{FM}_6 \succ \mathscr{FM}_1 \succ \mathscr{FM}_4 \succ \mathscr{FM}_2 \succ \mathscr{FM}_5$
0.2	$\mathscr{FM}_7 \succ \mathscr{FM}_3 \succ \mathscr{FM}_6 \succ \mathscr{FM}_1 \succ \mathscr{FM}_4 \succ \mathscr{FM}_2 \succ \mathscr{FM}_5$
0.3	$\mathscr{FM}_7 \succ \mathscr{FM}_3 \succ \mathscr{FM}_6 \succ \mathscr{FM}_1 \succ \mathscr{FM}_2 \succ \mathscr{FM}_4 \succ \mathscr{FM}_5$
0.4	$\mathscr{FM}_7 \succ \mathscr{FM}_3 \succ \mathscr{FM}_6 \succ \mathscr{FM}_1 \succ \mathscr{FM}_2 \succ \mathscr{FM}_4 \succ \mathscr{FM}_5$
0.5	$\mathscr{FM}_7 \succ \mathscr{FM}_3 \succ \mathscr{FM}_6 \succ \mathscr{FM}_1 \succ \mathscr{FM}_2 \succ \mathscr{FM}_4 \succ \mathscr{FM}_5$
0.6	$\mathscr{FM}_3 \succ \mathscr{FM}_7 \succ \mathscr{FM}_6 \succ \mathscr{FM}_1 \succ \mathscr{FM}_4 \succ \mathscr{FM}_5 \succ \mathscr{FM}_2$
0.7	$\mathscr{FM}_3 \succ \mathscr{FM}_7 \succ \mathscr{FM}_6 \succ \mathscr{FM}_1 \succ \mathscr{FM}_4 \succ \mathscr{FM}_5 \succ \mathscr{FM}_2$
0.8	$\mathscr{FM}_3 \succ \mathscr{FM}_7 \succ \mathscr{FM}_6 \succ \mathscr{FM}_1 \succ \mathscr{FM}_4 \succ \mathscr{FM}_5 \succ \mathscr{FM}_2$
0.9	$\mathscr{FM}_3 \succ \mathscr{FM}_7 \succ \mathscr{FM}_6 \succ \mathscr{FM}_1 \succ \mathscr{FM}_4 \succ \mathscr{FM}_5 \succ \mathscr{FM}_2$
1	$\mathscr{FM}_3 \succ \mathscr{FM}_7 \succ \mathscr{FM}_6 \succ \mathscr{FM}_1 \succ \mathscr{FM}_4 \succ \mathscr{FM}_5 \succ \mathscr{FM}_2$

5.2 Comparative Analysis

This section conducts a comparative study to demonstrate the benefits of our method with different methods for FMEA of SBRT for lung cancer treatment. The studies in [5, 22, 25, 52, 53] propose different types of MAGDM methods based on LZNs. Here, we compare these methods versus our method for ranking the FMs of SBRT for lung cancer treatment. The detailed ranking results are shown in Table 12.

To sum up, our model has the following general benefits compared with [3, 5, 22, 25, 52, 53]:

(1) LZNs can capture more information from experts in a situation of uncertainty compared to relative information. Unlike the data used in [3, 5, 22, 25, 52, 53], our model includes LZNs that reflect the experts' full views of the options.

Table 11 The effect of γ on ranking results

γ	Ranking
0.05	$\mathcal{FM}_4 \succ \mathcal{FM}_2 \succ \mathcal{FM}_6 \succ \mathcal{FM}_3 \succ \mathcal{FM}_7 \succ \mathcal{FM}_1 \succ \mathcal{FM}_5$
0.1	$\mathcal{FM}_5 \succ \mathcal{FM}_2 \succ \mathcal{FM}_6 \succ \mathcal{FM}_3 \succ \mathcal{FM}_7 \succ \mathcal{FM}_1 \succ \mathcal{FM}_4$
0.15	$\mathcal{FM}_5 \succ \mathcal{FM}_2 \succ \mathcal{FM}_6 \succ \mathcal{FM}_3 \succ \mathcal{FM}_7 \succ \mathcal{FM}_1 \succ \mathcal{FM}_4$
0.2	$\mathcal{FM}_5 \succ \mathcal{FM}_2 \succ \mathcal{FM}_6 \succ \mathcal{FM}_3 \succ \mathcal{FM}_7 \succ \mathcal{FM}_1 \succ \mathcal{FM}_4$
0.25	$\mathcal{FM}_5 \succ \mathcal{FM}_2 \succ \mathcal{FM}_6 \succ \mathcal{FM}_3 \succ \mathcal{FM}_7 \succ \mathcal{FM}_1 \succ \mathcal{FM}_4$
0.3	$\mathcal{FM}_5 \succ \mathcal{FM}_2 \succ \mathcal{FM}_6 \succ \mathcal{FM}_3 \succ \mathcal{FM}_7 \succ \mathcal{FM}_1 \succ \mathcal{FM}_4$
0.35	$\mathcal{FM}_5 \succ \mathcal{FM}_2 \succ \mathcal{FM}_6 \succ \mathcal{FM}_3 \succ \mathcal{FM}_7 \succ \mathcal{FM}_1 \succ \mathcal{FM}_4$
0.4	$\mathcal{FM}_5 \succ \mathcal{FM}_2 \succ \mathcal{FM}_6 \succ \mathcal{FM}_3 \succ \mathcal{FM}_7 \succ \mathcal{FM}_1 \succ \mathcal{FM}_4$
0.45	$\mathcal{FM}_5 \succ \mathcal{FM}_2 \succ \mathcal{FM}_6 \succ \mathcal{FM}_3 \succ \mathcal{FM}_7 \succ \mathcal{FM}_1 \succ \mathcal{FM}_4$
0.5	$\mathcal{FM}_4 \succ \mathcal{FM}_2 \succ \mathcal{FM}_7 \succ \mathcal{FM}_3 \succ \mathcal{FM}_6 \succ \mathcal{FM}_1 \succ \mathcal{FM}_5$

Table 12 The ranking results of our method versus different methods

Studies in	Ranking
[53]	$\mathcal{FM}_4 \succ \mathcal{FM}_3 \succ \mathcal{FM}_1 \succ \mathcal{FM}_6 \succ \mathcal{FM}_5 \succ \mathcal{FM}_7 \succ \mathcal{FM}_2$
[52]	$\mathcal{FM}_3 \succ \mathcal{FM}_7 \succ \mathcal{FM}_6 \succ \mathcal{FM}_5 \succ \mathcal{FM}_2 \succ \mathcal{FM}_1 \succ \mathcal{FM}_4$
[22]	$\mathcal{FM}_4 \succ \mathcal{FM}_1 \succ \mathcal{FM}_3 \succ \mathcal{FM}_7 \succ \mathcal{FM}_6 \succ \mathcal{FM}_2 \succ \mathcal{FM}_5$
[25]	$\mathcal{FM}_2 \succ \mathcal{FM}_5 \succ \mathcal{FM}_6 \succ \mathcal{FM}_4 \succ \mathcal{FM}_7 \succ \mathcal{FM}_3 \succ \mathcal{FM}_1$
[5]	$\mathcal{FM}_3 \succ \mathcal{FM}_7 \succ \mathcal{FM}_2 \succ \mathcal{FM}_6 \succ \mathcal{FM}_4 \succ \mathcal{FM}_1 \succ \mathcal{FM}_5$
Our method	$\mathcal{FM}_6 \succ \mathcal{FM}_2 \succ \mathcal{FM}_4 \succ \mathcal{FM}_7 \succ \mathcal{FM}_1 \succ \mathcal{FM}_3 \succ \mathcal{FM}_5$

(2) We develop the MAGDM model based on consensus measures for handling LZNs. The proposed model considers the relative importance of the experts' weights and the attributes' weights according to the evaluation information given by each expert. Therefore, we provide three levels of consensus degree measurement, and we compute the weighted consensus degree at each level.

(3) The aggregation operator is used to combine individual assessments to get the collective consensus opinion. The weights of experts can have a significant impact on the group's consensus opinion. Our model produces the proposed consensus opinion, which allows all experts to reach a balanced state between their personal goals (reducing changes) and the group goal (achieving consensus). This is done by minimizing the total changes made by the experts during the consensus process.

6 Conclusion

In our proposed approach for FMEA under LZN-based information with a case study, we have found the following:

(1) Establishing relativity among expert-based three-level consensus measurement degree and minimum adjustment sum with group consensus opinion-based FAM for delivering SBRT of lung cancer treatment under LZN-based information.
(2) The weight of experts is constructed by consensus-driven techniques.
(3) Creating the GLDS-based ranking approach for ranking FMs under LZN-based information.
(4) Validating our method by applying a case study of the delivery of SBRT for lung cancer treatment under LZN-based information.

Two shortcomings in our paper are as follows:

(1) Here we consider three types of risk factors for our FMEA method. However, it is essential to consider more risk factors for FMEA for better results.
(2) In FAM, experts' psychological behaviour is not taken into account in this paper. Our future attention is on these shortcomings.

Conflict of interest The authors declare that they have no conflicts of interest.

References

1. Bray, F., Ferlay, J., Soerjomataram, I., Siegel, R.L., Torre, L.A., Jemal, A.: Global cancer statistics: GLOBOCAM estimates of incidence and mortality worldwide for 36 cancers in 185 countries. CA Cancer J. Clin. **68**(6), 394–424 (2018)
2. Perks, J.R., Stanic, S., Stern, R.L., Henk, B., Nelson, M.S., Harse, R.D., Mathai, M., Purdy, J.A., Valicenti, R.K., Siefkin, A.D., Chen, A.M.: Failure mode and effect analysis for delivery of lung stereotactic body radiation therapy. Int. J. Radiat. Oncol. Biol. Phys. **83**(4), 1324–1329 (2012)
3. Chai, J., Su, Y., Lu, S.: Linguistic Z-number preference relation for group decision making and its application in digital transformation assessment of SMEs. Expert Syst. Appl. **213**, 118749 (2023)
4. Chen, Z., Chen, J.Y., Chen, Y.H., Yang, Y., Jin, L., Herrera-Viedma, E., Pedrycz, W.: Large-group failure mode and effects analysis for risk management of angle grinders in the construction industry. Inf. Fusion **97**, 101803 (2023)
5. Chen, B., Cai, Q., Wei, G., Mo, Z.: Novel Aczel-Alsina operations-based linguistic Z-number aggregation operators and their applications in multi-attribute group decision-making process. Eng. Appl. Artif. Intell. **124**, 106541 (2023)
6. Das, A., Pal, M.: A mathematical study of an imprecise SIR epidemic model with treatment control. J. Appl. Math. Comput. **56**, 477–500 (2018)
7. Du, J., Liu, S., Tao, L., Dong, W.: Three-way failure mode and effect analysis approach for reliability management in multigranular unbalanced linguistic contexts. Comput. Ind. Eng. **175**, 108909 (2023)
8. Ghorai, G., Pal, M.: Certain types of product bipolar fuzzy graphs. Int. J. Appl. Comput. Math. **3**, 605–619 (2017)
9. Goldstraw, P., Ball, D., Jett, J.R., Le Chevalier, T., Lim, E., Nicholson, A.G., Shepherd, F.A.: Nonsmall-cell lung cancer. Lancet **378**(9804), 1727–1740 (2011)
10. Hua, Z., Jing, X., Martínez, L.: An ELICIT information-based ORESTE method for failure mode and effect analysis considering risk correlation with GRA-DEMATEL. Inf. Fusion **93**, 396–411 (2023)

11. Jana, C., Pal, M., Wang, J.: A robust aggregation operator for multi-criteria decision-making method with bipolar fuzzy soft environment. Iran. J. Fuzzy Syst. **16**(6), 1–16 (2019)
12. Jana, C., Pal, M.: Assessment of enterprise performance based on picture fuzzy Hamacher aggregation operators. Symmetry **11**(1), 75 (2019)
13. Jana, C., Muhiuddin, G., Pal, M.: Multiple-attribute decision making problems based on SVTNH methods. J. Ambient Intell. Hum. Comput. **11**(9), 3717–3733 (2020)
14. Jana, C., Pal, M.: Multi-criteria decision making process based on some single-valued neutrosophic Dombi power aggregation operators. Soft Comput. **25**(7), 5055–5072 (2021)
15. Jana, C., Pal, M.: A dynamical hybrid method to design decision making process based on GRA approach for multiple attributes problem. Eng. Appl. Artif. Intell. **100**, 104203 (2021)
16. Jana, C., Muhiuddin, G., Pal, M.: Multi-criteria decision making approach based on SVTrN Dombi aggregation functions. Artif. Intell. Rev. **54**(5), 3685–3723 (2021)
17. Jana, C., Pal, M.: Extended bipolar fuzzy EDAS approach for multi-criteria group decision-making process. Comput. Appl. Math. **40**(1), 9 (2021)
18. Jana, C., Muhiuddin, G., Pal, M.: Intuitionistic fuzzy Dombi hybrid decision-making method and their applications to enterprise financial performance evaluation. Math. Probl. Eng. **2021**, 3218133 (2021)
19. Jana, C., Garg, H., Pal, M.: Multi-attribute decision making for power Dombi operators under Pythagorean fuzzy information with MABAC method. J. Ambient Intell. Hum. Comput. **14**(8), 10761–10778 (2023)
20. Jana, C., Pal, M.: Extended bipolar fuzzy EDAS approach for multi-criteria group decision-making process. Comput. Appl. Math. **40**, 9 (2021)
21. Jin, C., Ran, Y., Zhang, G.: Interval-valued q-rung orthopair fuzzy FMEA application to improve risk evaluation process of tool changing manipulator. Appl. Soft Comput. **104**, 107192 (2021)
22. Liu, Z., Bi, Y., Wang, X., Sha, L., Liu, P.: A generalized TODIM-ELECTRE II method based on linguistic Z-numbers and Dempster-Shafer evidence theory with unknown weight information. Complex Intell. Syst. **8**, 949–971 (2021)
23. Liu, X., Xu, Y., Gong, Z., Herrera, F.: Democratic consensus reaching process for multi-person multi-criteria large scale decision making considering participants individual attributes and concerns. Inf. Fusion **77**, 220–232 (2022)
24. Liu, F., Liao, H., Wu, X., Al-Barakati, A.: Consensus-based failure mode and effect analysis in group decision-making with incomplete weights of risk factors: case study of stereotactic body radiation therapy for lung cancer. Appl. Soft Comput. **142**, 110322 (2023)
25. Mao, L.X., Liu, R., Mou, X., Liu, H.C.: New approach for quality function deployment using linguistic Z-numbers and EDAS method. Informatica **32**(3), 365–582 (2021)
26. Mahapatra, R., Samanta, S., Pal, M., Xin, Q.: RSM index: a new way of link prediction in social networks. J. Intell. Fuzzy Syst. **37**(2), 2137–2151 (2019)
27. Mahapatra, R., Samanta, S., Pal, M.: Applications of edge colouring of fuzzy graphs. Informatica **31**(3), 1–18 (2020)
28. Mahapatra, R., Samanta, S., Pal, M.: Generalized neutrosophic planar graphs and its application. J. Appl. Math. Comput. **65**, 693–712 (2021)
29. Mandal, S., Sahoo, S., Ghorai, G., Pal, M.: Genus value of m -polar fuzzy graphs. J. Intell. Fuzzy Syst. **34**(3), 1947–1957 (2018)
30. Mandal, P., Samanta, S., Pal, M., Ranadive, A.S.: Pythagorean linguistic preference relations and their applications to group decision making using group recommendations based on consistency matrices and feedback mechanism. Int. J. Intell. Syst. **35**(5), 826–849 (2020)
31. Mandal, P., Samanta, S., Pal, M.: Multiplicative consistency analysis of linguistic preference relation with self-confidence level and self-doubting level and its application in a group decision making. Int. J. Intell. Syst. **36**(10), 5389–5418 (2021)
32. Mandal, P., Samanta, S., Pal, M., Ranadive, A.S.: Three-way decision model under a large-scale group decision-making environment with detecting and managing non-cooperative behaviors in consensus reaching process. Artif. Intell. Rev. **55**, 5517–5542 (2022)

33. Mandal, P., Samanta, S., Pal, M.: Large-scale group decision-making based on Pythagorean linguistic preference relations using experts clustering and consensus measure with non-cooperative behavior analysis of clusters. Complex Intell. Syst. **8**, 819–833 (2022)
34. Mandal, P., Samanta, S., Pal, M., Ranadive, A.S.: Social network trust relationship environment based advanced ovarian cancer treatment decision-making model: an approach based on linguistic information with experts' multiple confidence levels. Exp. Syst. Appl. **229**(Part A), 120407 (2023)
35. Mandal, P., Samanta, S., Pal, M.: Large-scale alternative processing group decision- making under Pythagorean linguistic preference environment. Soft Comput. (2023)
36. Mandal, P., Samanta, S., Pal, M.: Failure mode and effects analysis in consensus-based GDM for surface-guided deep inspiration breath-hold breast radiotherapy for breast cancer under the framework of linguistic Z-number. Inf. Sci. **658**, 120016 (2024)
37. Moitra, D., Mandal, R.K.: Classification of non-small cell lung cancer using one-dimensional conventional neural network. Rev. Syst. Appl. **159**, 113–564 (2020)
38. Mondal, U., Mahapatra, T., Xin, Q., Pal, M.: Solution of road network problem with the help of m-polar fuzzy graph using isometric and antipodal concept. Sci. Rep. **13**(1), 6452 (2023)
39. Mondal, U., Mahapatra, T., Xin, Q., Pal, M.: Generalized m-polar fuzzy planar graph and its application. IEEE Access **11**, 138399–138413 (2023)
40. Mondal, U., Pal, M., Das, K.: Interval-valued intuitionistic fuzzy soft graphs with application. TWMS J. Appl. Eng. Math. **14**(1), 1–18 (2024)
41. Nayak, P.K., Pal, M.: Linear programming technique to solve two person matrix games with interval pay-offs. Asia-Pac. J. Oper. Res. **26**(2), 285–305 (2009)
42. Rashmanlou, H., Muhiuddin, G., Amanathulla, S.K., Mofidnakhaei, F., Pal, M.: A study on cubic graphs with novel application. J. Intell. Fuzzy Syst. **40**(1), 89–101 (2021)
43. Saha, A., Pal, M., Pal, T.K.: Selection of programme slots of television channels for giving advertisement: a graph theoretic approach. Inf. Sci. **177**(12), 2480–2492 (2007)
44. Sahoo, S., Pal, M., Rashmanlou, H., Borzooei, R.A.: Covering and paired domination in intuitionistic fuzzy graphs. J. Intell. Fuzzy Syst. **33**(6), 4007–4015 (2017)
45. Samanta, S., Sarkar, B., Shin, D., Pal, M.: Completeness and Regularity of Generalized Fuzzy Graphs, vol. 5. Springer Plus (2016)
46. Senapati, T., Kim, C.S., Bhowmik, M., Pal, M.: Cubic subalgebras and cubic closed ideals of B-algebras. Fuzzy Inf. Eng. **7**(2), 129–149 (2015)
47. Siegel, R.L., Miller, K.D., Jemal, A.: Cancer statistics, 2020. CA Cancer J. Clin. **70**(1), 7–30 (2020)
48. Shyamal, A.K., Pal, M.: Triangular fuzzy matrices. Iran. J. Fuzzy Syst. **4**(1), 75–87 (2007)
49. Shyamal, A.K., Pal, M.: Two new operators on fuzzy matrices. J. Appl. Math. Comput. **15**, 91–107 (2004)
50. Sun, J., Liu, Y., Xu, J., Wang, N., Zhu, F.: A probabilistic uncertain linguistic FMEA model based on the extended ORESTE and regret theory. Comput. Ind. Eng. **180**, 109251 (2023)
51. Sung, H., Ferlay, J., Siegel, R.L., Laversanne, M., Soerjomataram, I., Bray, A.F.: Global cancer statistics 2020: GLOBOCAN estimates of incidence and mortality worldwide for 36 cancers in 185 countries. CA Cancer J. Clin. **71**, 209–249 (2021)
52. Tao, R., Xiao, F.: A GMCDM approach with linguistic Z-numbers based on TOPSIS and Choquet integral considering risk preference. J. Intell. Fuzzy Syst. **39**, 4285–4298 (2020)
53. Wang, J.Q., Cao, Y.X., Zhang, H.Y.: Multi-criteria decision-making method based on distance measure and Choquet integral for linguistic Z-numbers. Cognit. Comput. **9**(6), 827–842 (2017)
54. Wang, W., Ding, L., Liu, X., Liu, S.: An interval 2-Tuple linguistic Fine-Kinney model for risk analysis based on extended ORESTE method with cumulative prospect theory. Inf. Fusion **78**, 40–56 (2022)
55. Wang, Y., Zhan, J., Zhang, C., Xu, Z.: A group consensus model with prospect theory under probabilistic linguistic term sets. Inf. Sci. **635**, 119800 (2024)
56. Xu, Z.S.: Deviation measures of linguistic preference relations in group decision making. Omega **33**, 249–254 (2005)

57. Zagar, T.M., Kaidar-Person, O., Tang, X., Jones, E.E., Matney, J., Das, S.K., Green, R.L., Sheikh, A., Khandani, A.H., McCartney, W.H., Oldan, J.D., Wong, T.Z., Marks, L.B.: Utility of deep inspiration breath hold for left-sided breast radiation therapy in preventing early cardiac perfusion defects: a prospective study. Int. J. Radiat. Oncol. Biol. Phys. **97**(5), 903–909 (2017)
58. Zhang, H., Dong, Y., Xiao, J., Chiclana, F., Herrera-Viedma, E.: Consensus and opinion evolution-based failure mode and effect analysis approach for reliability management in social network and uncertainty contexts. Reliab. Eng. Syst. Saf. **208**, 107425 (2021)
59. Zhang, J., Li, M.: Group decision-making method based on expert credibility with multi-granularity probabilistic linguistic Z-number preference relation. Inf. Sci. **650**, 119664 (2023)
60. Zhu, G.N., Ma, J., Hu, J.: A fuzzy rough number extended AHP and VIKOR for failure mode and effects analysis under uncertainty. Adv. Eng. Inform. **51**, 101454 (2022)

A Novel Triangular Divergence Distance-Based TOPSIS Method for Interval-Valued Fermatean Fuzzy Group Decision-Making

Prayosi Chatterjee and Mijanur Rahaman Seikh

Abstract The Technique for Order of Preference by Similarity to Ideal Solution (TOPSIS) method is widely recognized and proven as an effective approach for ranking alternatives within a multi-attribute group decision-making (MAGDM) scenario. In such contexts, the selection of appropriate distance metrics holds paramount importance. Interval-valued Fermatean fuzzy numbers are renowned for their effectiveness in encapsulating fuzzy information better than intuitionistic and Fermatean fuzzy numbers within MAGDM problems. Despite the availability of distance metrics for interval-valued Fermatean fuzzy numbers, certain scenarios exist where these metrics are inadequate in discerning the differences between two such fuzzy numbers, consequently leading to conflicting results. Addressing this issue is the primary objective of the present research endeavour. First, the triangular divergence measure has been utilized to propose a modified distance metric for interval-valued Fermatean fuzzy sets and its properties are investigated. Through a comparative analysis with existing metrics, the superiority of the proposed distance measure is demonstrated. Subsequently, the traditional TOPSIS method undergoes modification by incorporating the newly proposed triangular divergence-based distance metric in the interval-valued Fermatean fuzzy environment. To illustrate the proposed method's practicality, a numerical example is presented. The efficacy of the proposed modified TOPSIS method is evaluated through comparison with existing approaches within the context of interval-valued Fermatean fuzzy sets. The results of the comparative analysis illustrate that the proposed method effectively distinguishes fine differences in interval-valued Fermatean fuzzy sets. The results of this study show that the proposed triangular divergence-based distance metric not only modifies the existing TOPSIS approach but also improves its overall effectiveness and precision across various MAGDM techniques.

Keywords Distance measure · Triangular divergence measure · Interval-valued Fermatean fuzzy sets · TOPSIS · MAGDM

P. Chatterjee · M. R. Seikh (✉)
Department of Mathematics, Kazi Nazrul University, Asansol 713340, India
e-mail: mrseikh@ymail.com

L. Sahoo et al. (eds.), *Decision Making Under Uncertainty Via Optimization, Modelling, and Analysis*, Studies in Systems, Decision and Control 558,
https://doi.org/10.1007/978-981-96-0085-4_2

1 Introduction

Distance measures play an important role in handling fuzzy information. Several distance measures have been developed for different fuzzy environments over the years. Recently, Kirişci et al. [7] developed cosine similarity and distance measures for Fermatean fuzzy sets and applied the proposed distance measure to the Technique for Order of Preference by Similarity to Ideal Solution (TOPSIS) method. Zeng et al. [25] proposed the Fermatean fuzzy hybrid weighted distance measure and used it for a comprehensive evaluation of low-carbon cities using TOPSIS. Yang et al. [23] developed Fermatean fuzzy integrated weighted distance and applied them to evaluate green low-carbon ports. Onyeke and Ejegwa [12] modified the distance measure given by Senapati and Yager for Fermatean fuzzy set (FFS) and applied it to address students' course placement in tertiary institutions.

Jeevaraj [5] presented interval-valued Fermatean fuzzy numbers (IVFFNs) in 2021, modifying the theory of Fermatean fuzzy numbers [19]. The advantage of IVFFNs lies in the representation of membership degree and non-membership degree in intervals, where the cube of the upper bounds of membership degree and non-membership degree adds up to less than one. It facilitates a better representation of uncertain information in decision-making problems. There are a few distance measures for interval-valued Fermatean fuzzy sets (IVFFSs) in the literature. Jeevaraj [5] proposed the general-Euclidean distance measure (GEDM) for IVFFNs and used TOPSIS to solve some illustrative multi-attribute decision-making problems. Mathew et al. [11] used the Hamming distance measure (HDM) for IVFFNs to solve multi-attribute decision-making problems. Mandal et al. [10] proposed a new distance measure (NDM) for IVFFNs and applied it to a multi-attribute decision-making problem. However, a significant research gap arises here. These distances are not entirely competent for calculating accurate distances between IVFFNs. IVFFSs are more advanced and efficient in depicting fuzzy information and automatically call for the development of modified distance measures for better decision-making methods.

Recently, the triangular divergence measure, a method generally used in probability distributions and proposed by Yehudayoff et al. [24], has been researched to develop distance measures based on it. Qin et al. [13] used the triangular divergence measure to form a new distance measure for interval-valued intuitionistic fuzzy sets. Furthermore, TOPSIS method was modified using the proposed distance measure and used to evaluate teaching satisfaction. Deng and Wang [3] proposed a distance measure based on triangular divergence for FFSs along with Hellinger distance. Very recently, Liu [9] modified the triangular divergence-based distance measure for FFSs proposed by Deng and Wang [3]. Triangular divergence has proved to form successful distance measures for interval-valued intuitionistic fuzzy sets and FFSs. However, this distance measure has not been extended to any other form of fuzzy sets. In this study, the triangular divergence has been used to form a distance measure for an extension of FFS, namely the IVFFSs.

Moreover, none of the studies on triangular divergence-based distance measures address the hesitancy degree of fuzzy numbers. Non-involvement of hesitancy degree results in leaving out a significant portion of fuzzy information and leads to discrepancies in results. To address this research gap, this study includes the hesitancy degree of IVFFNs in the proposed distance measure. Consequently, this makes the proposed distance better and more accurate.

Various studies use fuzzy data to address real-life multi-attribute group decision-making (MAGDM) problems. Sahoo et al. [15] studied the applications of different fuzzy numbers to real-life decision-making. Out of these popular MAGDM methods, many use distance measures to identify the best alternative. One such method is TOPSIS. TOPSIS method was first proposed by Hwang and Yoon [4]. The fundamental premise underlying TOPSIS is the identification of optimal schemes by minimizing the distance to the positive ideal solution (PIS) and maximizing the distance to the negative ideal solution (NIS) [20]. TOPSIS method has been modified in several studies The TOPSIS method has been utilized by Mandal et al. [10] in the IVFFN environment to identify sustainable development programs. The TOPSIS method was also used by Jeevaraj [5] to solve some illustrative multi-attribute decision-making problems. Further, Ilieva et al. [6] employed the TOPSIS method for the evaluation and benchmarking of COVID-19 vaccines. To assess the causes of accidents at construction sites, Bouraima et al. [1] used the TOPSIS method. Kirişci [8] applied TOPSIS method to IVFFNs for assessing risk of self-driving vehicles. Most of these studies utilized the GEDM or HDM in the TOPSIS method, which often fails to provide a clear distinction between two fuzzy numbers. Addressing this research gap, the study modifies the TOPSIS method by introducing the proposed triangular divergence-based distance measure for IVFFNs.

The following research gaps motivate the study:

- IVFFSs have two popular distance measures, but they are not entirely competent in calculating distances between all IVFFNs. It leads to the requirement of a new and better distance measure for achieving more accurate results in decision-making problems.
- Triangular divergence measure is a popular classical method, mostly used in probability distributions. It has been utilized for distance measures for interval-valued intuitionistic fuzzy sets and Fermatean fuzzy sets. IVFFNs are better at expressing fuzzy information than interval-valued intuitionistic fuzzy numbers and Fermatean fuzzy sets. Hence extending the triangular divergence-based distance measure to IVFFNs will be more beneficial and realistic.
- The previous application of triangular divergence-based measure does not include the hesitancy degree. It leaves out certain fuzzy information from the calculated data and thus may cause discrepancies in results. Hence including the hesitancy degree in the distance measure will make it more precise.

Motivated by the identified research gaps, this study makes the following contributions:

- A new triangular divergence-based distance measure for IVFFSs is proposed and its properties are discussed.
- The hesitancy degree is also included in the triangular divergence-based distance measure for IVFFSs to overcome the loss of information.

- The proposed distance measure is utilized to modify the TOPSIS method, eventually modifying the method to give better results to decision-making problems.

The study has been organized in the following way: Sect. 2 consists of the preliminaries. Section 3 has the newly proposed distance measure with its properties, and we establish the superiority of the proposed triangular divergence-based distance measure. In Sect. 4, we iterate the modified-TOPSIS method used to solve an illustrative MAGDM problem. In Sect. 5, we use an example to apply our proposed method and solve it. A comparative analysis of the proposed method is done in Sect. 5.1. Lastly, Sect. 6 has the conclusion, research implications, limitations and future research scopes.

2 Preliminaries

In this section, some basic definitions and preliminaries are recalled. Throughout this paper, the universal set is referred to as $\overline{\Lambda}$.

Definition 2.1 [5] Let the universal set be $\overline{\Lambda}$. Let $\bar{F}$ be an IVFFS over $\overline{\Lambda}$ and be defined as follows:

$$\bar{F} = \{\langle \mathfrak{f}_i, [\sigma^l_{\bar{F}}(\mathfrak{f}_i), \sigma^u_{\bar{F}}(x_i)], [\varsigma^l_{\bar{F}}(\mathfrak{f}_i), \varsigma^u_{\bar{F}}(\mathfrak{f}_i)]\rangle | \mathfrak{f}_i \in \overline{\Lambda}\}$$
$$\text{where } 0 \le \sigma^l_{\bar{F}}(\mathfrak{f}_i) \le \sigma^u_{\bar{F}}(\mathfrak{f}_i) \le 1, \ \ 0 \le \varsigma^l_{\bar{F}}(\mathfrak{f}_i) \le \varsigma^u_{\bar{F}}(\mathfrak{f}_i) \le 1$$
$$\text{and } (\sigma^u_{\bar{F}}(\mathfrak{f}_i))^3 + (\varsigma^u_{\bar{F}}(\mathfrak{f}_i))^3 \le 1.$$

Here, $\sigma_{\bar{F}}(\mathfrak{f}_i) = [\sigma^l_{\bar{F}}(\mathfrak{f}_i), \sigma^u_{\bar{F}}(\mathfrak{f}_i)]$ is the membership degree, and $\varsigma_{\bar{F}}(\mathfrak{f}_i) = [\varsigma^l_{\bar{F}}(\mathfrak{f}_i), \varsigma^u_{\bar{F}}(\mathfrak{f}_i)]$ is the non-membership degree of $\mathfrak{f}_i$. Furthermore, the hesitancy degree is $\delta_{\bar{F}}(\mathfrak{f}_i) = [\delta^l_{\bar{F}}(\mathfrak{f}_i), \delta^u_{\bar{F}}(\mathfrak{f}_i)]$, where

$$\delta^l_{\bar{F}}(\mathfrak{f}_i) = \sqrt[3]{1 - (\sigma^l_{\bar{F}}(\mathfrak{f}_i))^3 - (\varsigma^l_{\bar{F}}(\mathfrak{f}_i))^3}$$
$$\text{and } \delta^u_{\bar{F}}(\mathfrak{f}_i) = \sqrt[3]{1 - (\sigma^u_{\bar{F}}(\mathfrak{f}_i))^3 - (\varsigma^u_{\bar{F}}(\mathfrak{f}_i))^3}.$$

Definition 2.2 [14] Let $\mathfrak{f}_1$ and $\mathfrak{f}_2$ be two IVFFNs. Let $\mathfrak{f}_1 = ([\sigma^l_{\mathfrak{f}_1}, \sigma^u_{\mathfrak{f}_1}], [\varsigma^l_{\mathfrak{f}_1}, \varsigma^u_{\mathfrak{f}_1}])$ and $\mathfrak{f}_2 = ([\sigma^l_{\mathfrak{f}_2}, \sigma^u_{\mathfrak{f}_2}], [\varsigma^l_{\mathfrak{f}_2}, \varsigma^u_{\mathfrak{f}_2}])$. Then,

- $\mathfrak{f}_1 = \mathfrak{f}_2$ iff $\sigma^l_{\mathfrak{f}_1} = \sigma^l_{\mathfrak{f}_2}, \sigma^u_{\mathfrak{f}_1} = \sigma^u_{\mathfrak{f}_2}, \varsigma^l_{\mathfrak{f}_1} = \varsigma^l_{\mathfrak{f}_2}$ and $\varsigma^u_{\mathfrak{f}_1} = \varsigma^u_{\mathfrak{f}_2}$.
- $\mathfrak{f}_1 < \mathfrak{f}_2$ iff $\sigma^l_{\mathfrak{f}_1} \le \sigma^l_{\mathfrak{f}_2}, \sigma^u_{\mathfrak{f}_1} \le \sigma^u_{\mathfrak{f}_2}, \varsigma^l_{\mathfrak{f}_1} \ge \varsigma^l_{\mathfrak{f}_2}$ and $\varsigma^u_{\mathfrak{f}_1} \ge \varsigma^u_{\mathfrak{f}_2}$.

Definition 2.3 [14] The definition of score function of $\mathfrak{f} = ([\sigma_{\mathfrak{f}}^{l}, \sigma_{\mathfrak{f}}^{u}], [\varsigma_{\mathfrak{f}}^{l}, \varsigma_{\mathfrak{f}}^{u}])$ is as follows:

$$\mathcal{SI}(\mathfrak{f}) = \frac{1}{2}\left((\sigma_{\mathfrak{f}}^{l})^3 + (\sigma_{\mathfrak{f}}^{u})^3 - (\varsigma_{\mathfrak{f}}^{l})^3 - (\varsigma_{\mathfrak{f}}^{u})^3\right)$$

and the accuracy function of $\mathfrak{f}$ is

$$\mathcal{AI}(\mathfrak{f}) = \frac{1}{2}\left((\sigma_{\mathfrak{f}}^{l})^3 + (\sigma_{\mathfrak{f}}^{u})^3 + (\varsigma_{\mathfrak{f}}^{l})^3 + (\varsigma_{\mathfrak{f}}^{u})^3\right).$$

Here $\mathcal{SI}(\mathfrak{f}) \in [-1, 1]$ and $\mathcal{AI}(\mathfrak{f}) \in [0, 1]$.

For comparing two IVFFNs, we use the score and accuracy functions. For $\mathfrak{f}_1 = ([\sigma_{\mathfrak{f}_1}^{l}, \sigma_{\mathfrak{f}_1}^{u}], [\varsigma_{\mathfrak{f}_1}^{l}, \varsigma_{\mathfrak{f}_1}^{u}])$ and $\mathfrak{f}_2 = ([\sigma_{\mathfrak{f}_2}^{l}, \sigma_{\mathfrak{f}_2}^{u}], [\varsigma_{\mathfrak{f}_2}^{l}, \varsigma_{\mathfrak{f}_2}^{u}])$, two IVFFNs,

- If $\mathcal{SI}(\mathfrak{f}_1) > \mathcal{SI}(\mathfrak{f}_2)$ then $\mathfrak{f}_1 > \mathfrak{f}_2$.
- If $\mathcal{SI}(\mathfrak{f}_1) < \mathcal{SI}(\mathfrak{f}_2)$ then $\mathfrak{f}_1 < \mathfrak{f}_2$.
- If $\mathcal{SI}(\mathfrak{f}_1) = \mathcal{SI}(\mathfrak{f}_2)$ and
  - if $\mathcal{AI}(\mathfrak{f}_1) > \mathcal{AI}(\mathfrak{f}_2)$ then $\mathfrak{f}_1 > \mathfrak{f}_2$,
  - if $\mathcal{AI}(\mathfrak{f}_1) < \mathcal{AI}(\mathfrak{f}_2)$ then $\mathfrak{f}_1 < \mathfrak{f}_2$ and
  - if $\mathcal{AI}(\mathfrak{f}_1) = \mathcal{AI}(\mathfrak{f}_2)$ then $\mathfrak{f}_1 = \mathfrak{f}_2$.

Definition 2.4 [14] Let $\mathfrak{f} = ([\sigma_{\mathfrak{f}}^{l}, \sigma_{\mathfrak{f}}^{u}], [\varsigma_{\mathfrak{f}}^{l}, \varsigma_{\mathfrak{f}}^{u}])$ be an IVFFN. The normalized score function is

$$\widehat{\mathcal{SI}}(\mathfrak{f}) = \frac{1}{2}(\mathcal{SI}(\mathfrak{f}) + 1). \quad (1)$$

Here, $\widehat{\mathcal{SI}}(\mathfrak{f}) \in [0, 1]$.

Definition 2.5 [5] A mapping $d : IVFFS(\overline{\Lambda}) \times IVFFS(\overline{\Lambda}) \rightarrow [0, 1]$ denotes the distance measure between the IVFFSs $\bar{F}_1$ and $\bar{F}_2$ if the following conditions are satisfied:

I. $d(\bar{F}_1, \bar{F}_2) = 0 \Leftrightarrow \bar{F}_1 = \bar{F}_2$;
II. $d(\bar{F}_1, \bar{F}_2) = d(\bar{F}_2, \bar{F}_1)$;
III. $0 \leq d(\bar{F}_1, \bar{F}_2) \leq 1$;
IV. If $\bar{F}_1 \leq \bar{F}_2 \leq \bar{F}_3$, then $d(\bar{F}_1, \bar{F}_2) \leq d(\bar{F}_1, \bar{F}_3)$ and $d(\bar{F}_2, \bar{F}_3) \leq d(\bar{F}_1, \bar{F}_3)$.

Suppose $\mathfrak{f}_1$ and $\mathfrak{f}_2$ be two IVFFNs, where $\mathfrak{f}_1 = ([\sigma_{\mathfrak{f}_1}^{l}, \sigma_{\mathfrak{f}_1}^{u}], [\varsigma_{\mathfrak{f}_1}^{l}, \varsigma_{\mathfrak{f}_1}^{u}])$ and $\mathfrak{f}_2 = ([\sigma_{\mathfrak{f}_2}^{l}, \sigma_{\mathfrak{f}_2}^{u}], [\varsigma_{\mathfrak{f}_2}^{l}, \varsigma_{\mathfrak{f}_2}^{u}])$. Then, the general-Euclidean distance measure ($d_{GE}(\mathfrak{f}_1, \mathfrak{f}_2)$) [5], the Hamming distance measure ($d_H(\mathfrak{f}_1, \mathfrak{f}_2)$) [11] and the modified new distance measure ($d_N(\mathfrak{f}_1, \mathfrak{f}_2)$) [10] for two IVFFNs $\mathfrak{f}_1$ and $\mathfrak{f}_2$ are defined as follows:

$$d_{GE}(\mathfrak{f}_1,\mathfrak{f}_2)=\Big[\frac{1}{6}\Big\{((\sigma^l_{\mathfrak{f}_1})^3-(\sigma^l_{\mathfrak{f}_2})^3)^2+((\sigma^u_{\mathfrak{f}_1})^3-(\sigma^u_{\mathfrak{f}_2})^3)^2+((\varsigma^l_{\mathfrak{f}_1})^3-(\varsigma^l_{\mathfrak{f}_2})^3)^2$$
$$+((\varsigma^u_{\mathfrak{f}_1})^3-(\varsigma^u_{\mathfrak{f}_2})^3)^2+((\delta^l_{\mathfrak{f}_1})^3-(\delta^l_{\mathfrak{f}_2})^3)^2+((\delta^u_{\mathfrak{f}_1})^3-(\delta^u_{\mathfrak{f}_2})^3)^2\Big\}\Big]^{\frac{1}{2}} \quad (2)$$
$$d_H(\mathfrak{f}_1,\mathfrak{f}_2)=\frac{1}{4}\Big(|(\sigma^l_{\mathfrak{f}_1})^3-(\sigma^l_{\mathfrak{f}_2})^3|+|(\sigma^u_{\mathfrak{f}_1})^3-(\sigma^u_{\mathfrak{f}_2})^3|+|(\varsigma^l_{\mathfrak{f}_1})^3-(\varsigma^l_{\mathfrak{f}_2})^3|$$
$$+|(\varsigma^u_{\mathfrak{f}_1})^3-(\varsigma^u_{\mathfrak{f}_2})^3|+|(\delta^l_{\mathfrak{f}_1})^3-(\delta^l_{\mathfrak{f}_2})^3|+|(\delta^u_{\mathfrak{f}_1})^3-(\delta^u_{\mathfrak{f}_2})^3|\Big) \quad (3)$$
$$d_N(\mathfrak{f}_1,\mathfrak{f}_2)=\frac{1}{4}\Big(|(\sigma^l_{\mathfrak{f}_1})^3-(\sigma^l_{\mathfrak{f}_2})^3|+|(\sigma^u_{\mathfrak{f}_1})^3-(\sigma^u_{\mathfrak{f}_2})^3|+|(\varsigma^l_{\mathfrak{f}_1})^3-(\varsigma^l_{\mathfrak{f}_2})^3|$$
$$+|(\varsigma^u_{\mathfrak{f}_1})^3-(\varsigma^u_{\mathfrak{f}_2})^3|\Big). \quad (4)$$

Definition 2.6 [14] Let $\mathfrak{f}_s$ be IVFFNs where $s=1,2,\ldots k$ and $IVFFN: F^k \rightarrow F$. Then interval-valued Fermatean fuzzy weighted averaging (IVFFWA) aggregation operator is formulated as

$$IVFFWA(\mathfrak{f}_1,\mathfrak{f}_2,\ldots,\mathfrak{f}_k)=\bigoplus_{s=1}^{k}\varpi_s\mathfrak{f}_s$$
$$=\Big(\Big[(1-\prod_{s=1}^{k}(1-(\sigma^l_{\mathfrak{f}_s})^3)^{\varpi_s})^{\frac{1}{3}},(1-\prod_{s=1}^{k}(1-(\sigma^u_{\mathfrak{f}_s})^3)^{\varpi_s})^{\frac{1}{3}}\Big],$$
$$\Big[\prod_{s=1}^{k}(\varsigma^l_{\mathfrak{f}_s})^{\varpi_s},\prod_{s=1}^{k}(\varsigma^u_{\mathfrak{f}_s})^{\varpi_s}\Big]\Big) \quad (5)$$

where ϖ_s are weights of $\mathfrak{f}_s$ for $s=1,2,\ldots k$ such that $\varpi_s\in[0,1]$ and $\sum_{s=1}^{k}\varpi_s=1$.

Definition 2.7 [24] Let $\Theta_n=\{S=(s_1,s_2,\ldots,s_n)|s_i>0,\sum_{i=1}^{n}s_i=1\}$, $n\geq 2$ be a set of finite discrete probability distributions. For $\forall\, S,T\in\Theta_n$, the classical triangular divergence measure between S and T is defined as

$$\Delta(S,T)=\sum_{i=1}^{n}\frac{(s_i-t_i)^2}{s_i+t_i}.$$

Greater triangular divergence indicates greater difference between the probability distributions S and T. Using the above-mentioned equation, the square root of the triangular divergence is given as follows:

$$d(S,T)=\sqrt{\sum_{i=1}^{n}\frac{(s_i-t_i)^2}{s_i+t_i}}$$

where, by convention, $0/0=0$.

3 A Triangular Divergence-Based Modified Distance Measure for IVFFSs

In this section, we propose the new distance measure for IVFFSs based on triangular divergence measure.

Definition 3.1 Let $\bar{F}_p = \langle \mathfrak{f}_q, [\sigma_p^l(\mathfrak{f}_q), \sigma_p^u(\mathfrak{f}_q)], [\varsigma_p^l(\mathfrak{f}_q), \varsigma_p^u(\mathfrak{f}_q)] \rangle$ for $p = 1, 2$ be two IVFFSs in $\overline{\Lambda} = \{\mathfrak{f}_1, \mathfrak{f}_2, \ldots, \mathfrak{f}_r\}$, then the triangular divergence-based distance measure (TDDM) between IVFFSs $\bar{F}_1$ and $\bar{F}_2$, denoted by d_{TV}, is given by

$$d_{TV}(\bar{F}_1, \bar{F}_2) = \left[\frac{1}{8r}\sum_{q=1}^{r}\left\{\frac{((\sigma_1^l(\mathfrak{f}_q))^3 - (\sigma_2^l(\mathfrak{f}_q))^3)^2}{(\sigma_1^l(\mathfrak{f}_q))^3 + (\sigma_2^l(\mathfrak{f}_q))^3} + \frac{((\sigma_1^u(\mathfrak{f}_q))^3 - (\sigma_2^u(\mathfrak{f}_q))^3)^2}{(\sigma_1^u(\mathfrak{f}_q))^3 + (\sigma_2^u(\mathfrak{f}_q))^3} + \frac{((\varsigma_1^l(\mathfrak{f}_q))^3 - (\varsigma_2^l(\mathfrak{f}_q))^3)^2}{(\varsigma_1^l(\mathfrak{f}_q))^3 + (\varsigma_2^l(\mathfrak{f}_q))^3} + \frac{((\varsigma_1^u(\mathfrak{f}_q))^3 - (\varsigma_2^u(\mathfrak{f}_q))^3)^2}{(\varsigma_1^u(\mathfrak{f}_q))^3 + (\varsigma_2^u(\mathfrak{f}_q))^3} + \frac{((\delta_1^l(\mathfrak{f}_q))^3 - (\delta_2^l(\mathfrak{f}_q))^3)^2}{(\delta_1^l(\mathfrak{f}_q))^3 + (\delta_2^l(\mathfrak{f}_q))^3} + \frac{((\delta_1^u(\mathfrak{f}_q))^3 - (\delta_2^u(\mathfrak{f}_q))^3)^2}{(\delta_1^u(\mathfrak{f}_q))^3 + (\delta_2^u(\mathfrak{f}_q))^3}\right\}\right]^{\frac{1}{2}}. \quad (6)$$

Theorem 3.1 *The distance measure $d_{TV}(\bar{F}_1, \bar{F}_2)$, between the two IVFFSs $\bar{F}_1$ and $\bar{F}_2$, follows the following properties. Here $\bar{F}_1$, $\bar{F}_2$ and $\bar{F}_3$ are IVFFSs:*

I. $d_{TV}(\bar{F}_1, \bar{F}_2) = 0 \Leftrightarrow \bar{F}_1 = \bar{F}_2$;
II. $d_{TV}(\bar{F}_1, \bar{F}_2) = d_{TV}(\bar{F}_2, \bar{F}_1)$;
III. $0 \le d_{TV}(\bar{F}_1, \bar{F}_2) \le 1$;
IV. *If* $\bar{F}_1 \le \bar{F}_2 \le \bar{F}_3$, *then* $d_{TV}(\bar{F}_1, \bar{F}_2) \le d_{TV}(\bar{F}_1, \bar{F}_3)$ *and* $d_{TV}(\bar{F}_2, \bar{F}_3) \le d_{TV}(\bar{F}_1, \bar{F}_3)$.

Proof I. Let $d_{TV}(\bar{F}_1, \bar{F}_2) = 0$ for any $\mathfrak{f}_q \in \overline{\Lambda}$. Then we can say that

$$d_{TV}(\bar{F}_1, \bar{F}_2) = \left[\frac{1}{8r}\sum_{q=1}^{r}\left\{\frac{((\sigma_1^l(\mathfrak{f}_q))^3 - (\sigma_2^l(\mathfrak{f}_q))^3)^2}{(\sigma_1^l(\mathfrak{f}_q))^3 + (\sigma_2^l(\mathfrak{f}_q))^3} + \frac{((\sigma_1^u(\mathfrak{f}_q))^3 - (\sigma_2^u(\mathfrak{f}_q))^3)^2}{(\sigma_1^u(\mathfrak{f}_q))^3 + (\sigma_2^u(\mathfrak{f}_q))^3} + \frac{((\varsigma_1^l(\mathfrak{f}_q))^3 - (\varsigma_2^l(\mathfrak{f}_q))^3)^2}{(\varsigma_1^l(\mathfrak{f}_q))^3 + (\varsigma_2^l(\mathfrak{f}_q))^3} + \frac{((\varsigma_1^u(\mathfrak{f}_q))^3 - (\varsigma_2^u(\mathfrak{f}_q))^3)^2}{(\varsigma_1^u(\mathfrak{f}_q))^3 + (\varsigma_2^u(\mathfrak{f}_q))^3} + \frac{((\delta_1^l(\mathfrak{f}_q))^3 - (\delta_2^l(\mathfrak{f}_q))^3)^2}{(\delta_1^l(\mathfrak{f}_q))^3 + (\delta_2^l(\mathfrak{f}_q))^3} + \frac{((\delta_1^u(\mathfrak{f}_q))^3 - (\delta_2^u(\mathfrak{f}_q))^3)^2}{(\delta_1^u(\mathfrak{f}_q))^3 + (\delta_2^u(\mathfrak{f}_q))^3}\right\}\right]^{\frac{1}{2}} = 0.$$

Then we have

$$\frac{((\sigma_1^l(\mathfrak{f}_q))^3 - (\sigma_2^l(\mathfrak{f}_q))^3)^2}{(\sigma_1^l(\mathfrak{f}_q))^3 + (\sigma_2^l(\mathfrak{f}_q))^3} = \frac{((\sigma_1^u(\mathfrak{f}_q))^3 - (\sigma_2^u(\mathfrak{f}_q))^3)^2}{(\sigma_1^u(\mathfrak{f}_q))^3 + (\sigma_2^u(\mathfrak{f}_q))^3} = \frac{((\varsigma_1^l(\mathfrak{f}_q))^3 - (\varsigma_2^l(\mathfrak{f}_q))^3)^2}{(\varsigma_1^l(\mathfrak{f}_q))^3 + (\varsigma_2^l(\mathfrak{f}_q))^3} =$$

$$\frac{((\varsigma_1^u(\mathfrak{f}_q))^3-(\varsigma_2^u(\mathfrak{f}_q))^3)^2}{(\varsigma_1^u(\mathfrak{f}_q))^3+(\varsigma_2^u(\mathfrak{f}_q))^3}=\frac{((\delta_1^l(\mathfrak{f}_q))^3-(\delta_2^l(\mathfrak{f}_q))^3)^2}{(\delta_1^l(\mathfrak{f}_q))^3+(\delta_2^l(\mathfrak{f}_q))^3}=\frac{((\delta_1^u(\mathfrak{f}_q))^3-(\delta_2^u(\mathfrak{f}_q))^3)^2}{(\delta_1^u(\mathfrak{f}_q))^3+(\delta_2^u(\mathfrak{f}_q))^3}=0,$$
$$\Rightarrow ((\sigma_1^l(\mathfrak{f}_q))^3-(\sigma_2^l(\mathfrak{f}_q))^3)^2=((\sigma_1^u(\mathfrak{f}_q))^3-(\sigma_2^u(\mathfrak{f}_q))^3)^2=((\varsigma_1^l(\mathfrak{f}_q))^3-(\varsigma_2^l(\mathfrak{f}_q))^3)^2=$$
$$((\varsigma_1^u(\mathfrak{f}_q))^3-(\varsigma_2^u(\mathfrak{f}_q))^3)^2=((\delta_1^l(\mathfrak{f}_q))^3-(\delta_2^l(\mathfrak{f}_q))^3)^2=((\delta_1^u(\mathfrak{f}_q))^3-(\delta_2^u(\mathfrak{f}_q))^3)^2=0.$$

From Definition 2.1, we can say that

$$\sigma_1^l(\mathfrak{f}_q)=\sigma_2^l(\mathfrak{f}_q),\ \sigma_1^u(\mathfrak{f}_q)=\sigma_2^u(\mathfrak{f}_q),\ \varsigma_1^l(\mathfrak{f}_q)=\varsigma_2^l(\mathfrak{f}_q),\ \varsigma_1^u(\mathfrak{f}_q)=\varsigma_2^u(\mathfrak{f}_q),$$
$$\delta_1^l(\mathfrak{f}_q)=\delta_2^l(\mathfrak{f}_q),\ \delta_1^u(\mathfrak{f}_q)=\delta_2^u(\mathfrak{f}_q).$$

Thus we can say that $\bar{F}_1=\bar{F}_2$.
Now, conversely let $\bar{F}_1=\bar{F}_2$. This implies that

$$\sigma_1^l(\mathfrak{f}_q)=\sigma_2^l(\mathfrak{f}_q),\ \sigma_1^u(\mathfrak{f}_q)=\sigma_2^u(\mathfrak{f}_q),\ \varsigma_1^l(\mathfrak{f}_q)=\varsigma_2^l(\mathfrak{f}_q),\ \varsigma_1^u(\mathfrak{f}_q)=\varsigma_2^u(\mathfrak{f}_q),$$
$$\delta_1^l(\mathfrak{f}_q)=\delta_2^l(\mathfrak{f}_q),\ \delta_1^u(\mathfrak{f}_q)=\delta_2^u(\mathfrak{f}_q).$$

Then we can easily show that $d_{TV}(\bar{F}_1,\bar{F}_2)=0$.
II. Next we prove that $d_{TV}(\bar{F}_1,\bar{F}_2)=d_{TV}(\bar{F}_2,\bar{F}_1)$. We know that

$$\begin{aligned}
d_{TV}(\bar{F}_1,\bar{F}_2)&=\Bigg[\frac{1}{8r}\sum_{q=1}^{r}\Bigg\{\frac{((\sigma_1^l(\mathfrak{f}_q))^3-(\sigma_2^l(\mathfrak{f}_q))^3)^2}{(\sigma_1^l(\mathfrak{f}_q))^3+(\sigma_2^l(\mathfrak{f}_q))^3}+\frac{((\sigma_1^u(\mathfrak{f}_q))^3-(\sigma_2^u(\mathfrak{f}_q))^3)^2}{(\sigma_1^u(\mathfrak{f}_q))^3+(\sigma_2^u(\mathfrak{f}_q))^3}\\
&\quad+\frac{((\varsigma_1^l(\mathfrak{f}_q))^3-(\varsigma_2^l(\mathfrak{f}_q))^3)^2}{(\varsigma_1^l(\mathfrak{f}_q))^3+(\varsigma_2^l(\mathfrak{f}_q))^3}+\frac{((\varsigma_1^u(\mathfrak{f}_q))^3-(\varsigma_2^u(\mathfrak{f}_q))^3)^2}{(\varsigma_1^u(\mathfrak{f}_q))^3+(\varsigma_2^u(\mathfrak{f}_q))^3}\\
&\quad+\frac{((\delta_1^l(\mathfrak{f}_q))^3-(\delta_2^l(\mathfrak{f}_q))^3)^2}{(\delta_1^l(\mathfrak{f}_q))^3+(\delta_2^l(\mathfrak{f}_q))^3}+\frac{((\delta_1^u(\mathfrak{f}_q))^3-(\delta_2^u(\mathfrak{f}_q))^3)^2}{(\delta_1^u(\mathfrak{f}_q))^3+(\delta_2^u(\mathfrak{f}_q))^3}\Bigg\}\Bigg]^{\frac{1}{2}}\\
&=\Bigg[\frac{1}{8r}\sum_{q=1}^{r}\Bigg\{\frac{((\sigma_2^l(\mathfrak{f}_q))^3-(\sigma_1^l(\mathfrak{f}_q))^3)^2}{(\sigma_2^l(\mathfrak{f}_q))^3+(\sigma_1^l(\mathfrak{f}_q))^3}+\frac{((\sigma_2^u(\mathfrak{f}_q))^3-(\sigma_1^u(\mathfrak{f}_q))^3)^2}{(\sigma_2^u(\mathfrak{f}_q))^3+(\sigma_1^u(\mathfrak{f}_q))^3}\\
&\quad+\frac{((\varsigma_2^l(\mathfrak{f}_q))^3-(\varsigma_1^l(\mathfrak{f}_q))^3)^2}{(\varsigma_2^l(\mathfrak{f}_q))^3+(\varsigma_1^l(\mathfrak{f}_q))^3}+\frac{((\varsigma_2^u(\mathfrak{f}_q))^3-(\varsigma_1^u(\mathfrak{f}_q))^3)^2}{(\varsigma_2^u(\mathfrak{f}_q))^3+(\varsigma_1^u(\mathfrak{f}_q))^3}\\
&\quad+\frac{((\delta_2^l(\mathfrak{f}_q))^3-(\delta_1^l(\mathfrak{f}_q))^3)^2}{(\delta_2^l(\mathfrak{f}_q))^3+(\delta_1^l(\mathfrak{f}_q))^3}+\frac{((\delta_2^u(\mathfrak{f}_q))^3-(\delta_1^u(\mathfrak{f}_q))^3)^2}{(\delta_2^u(\mathfrak{f}_q))^3+(\delta_1^u(\mathfrak{f}_q))^3}\Bigg\}\Bigg]^{\frac{1}{2}}\\
&=d_{TV}(\bar{F}_2,\bar{F}_1).
\end{aligned}$$

Hence, property II holds.

III. We then prove that $0\leq d_{TV}(\bar{F}_1,\bar{F}_2)\leq 1$. From Definition 2.1, it is obvious that $0\leq d_{TV}(\bar{F}_1,\bar{F}_2)$ and we observe that

$$0 \le (\sigma_1^l(\mathfrak{f}_q))^3 + (\varsigma_1^l(\mathfrak{f}_q))^3 \le (\sigma_1^u(\mathfrak{f}_q))^3 + (\varsigma_1^u(\mathfrak{f}_q))^3 \le 1,$$
$$0 \le (\sigma_2^l(\mathfrak{f}_q))^3 + (\varsigma_2^l(\mathfrak{f}_q))^3 \le (\sigma_2^u(\mathfrak{f}_q))^3 + (\varsigma_2^u(\mathfrak{f}_q))^3 \le 1$$
$$\text{and } 0 \le \delta_1^u \le \delta_1^l \le 1, 0 \le \delta_2^u \le \delta_2^l \le 1.$$

Then we have

$$\begin{aligned}
d_{TV}(\bar{F}_1, \bar{F}_2) &= \Bigg[\frac{1}{8r}\sum_{q=1}^{r}\Bigg\{\frac{((\sigma_1^l(\mathfrak{f}_q))^3 - (\sigma_2^l(\mathfrak{f}_q))^3)^2}{(\sigma_1^l(\mathfrak{f}_q))^3 + (\sigma_2^l(\mathfrak{f}_q))^3} + \frac{((\sigma_1^u(\mathfrak{f}_q))^3 - (\sigma_2^u(\mathfrak{f}_q))^3)^2}{(\sigma_1^u(\mathfrak{f}_q))^3 + (\sigma_2^u(\mathfrak{f}_q))^3} \\
&\quad + \frac{((\varsigma_1^l(\mathfrak{f}_q))^3 - (\varsigma_2^l(\mathfrak{f}_q))^3)^2}{(\varsigma_1^l(\mathfrak{f}_q))^3 + (\varsigma_2^l(\mathfrak{f}_q))^3} + \frac{((\varsigma_1^u(\mathfrak{f}_q))^3 - (\varsigma_2^u(\mathfrak{f}_q))^3)^2}{(\varsigma_1^u(\mathfrak{f}_q))^3 + (\varsigma_2^u(\mathfrak{f}_q))^3} \\
&\quad + \frac{((\delta_1^l(\mathfrak{f}_q))^3 - (\delta_2^l(\mathfrak{f}_q))^3)^2}{(\delta_1^l(\mathfrak{f}_q))^3 + (\delta_2^l(\mathfrak{f}_q))^3} + \frac{((\delta_1^u(\mathfrak{f}_q))^3 - (\delta_2^u(\mathfrak{f}_q))^3)^2}{(\delta_1^u(\mathfrak{f}_q))^3 + (\delta_2^u(\mathfrak{f}_q))^3}\Bigg\}\Bigg]^{\frac{1}{2}} \\
&\le \Bigg[\frac{1}{8r}\sum_{q=1}^{r}\Bigg\{\frac{((\sigma_1^l(\mathfrak{f}_q))^3 + (\sigma_2^l(\mathfrak{f}_q)^3)^2}{(\sigma_1^l(\mathfrak{f}_q))^3 + (\sigma_2^l(\mathfrak{f}_q))^3} + \frac{((\sigma_1^u(\mathfrak{f}_q))^3 + (\sigma_2^u(\mathfrak{f}_q))^3)^2}{(\sigma_1^u(\mathfrak{f}_q))^3 + (\sigma_2^u(\mathfrak{f}_q))^3} \\
&\quad + \frac{((\varsigma_1^l(\mathfrak{f}_q))^3 + (\varsigma_2^l(\mathfrak{f}_q))^3)^2}{(\varsigma_1^l(\mathfrak{f}_q))^3 + (\varsigma_2^l(\mathfrak{f}_q))^3} + \frac{((\varsigma_1^u(\mathfrak{f}_q))^3 + (\varsigma_2^u(\mathfrak{f}_q))^3)^2}{(\varsigma_1^u(\mathfrak{f}_q))^3 + (\varsigma_2^u(\mathfrak{f}_q))^3} \\
&\quad + \frac{((\delta_1^l(\mathfrak{f}_q))^3 + (\delta_2^l(\mathfrak{f}_q))^3)^2}{(\delta_1^l(\mathfrak{f}_q))^3 + (\delta_2^l(\mathfrak{f}_q))^3} + \frac{((\delta_1^u(\mathfrak{f}_q))^3 + (\delta_2^u(\mathfrak{f}_q))^3)^2}{(\delta_1^u(\mathfrak{f}_q))^3 + (\delta_2^u(\mathfrak{f}_q))^3}\Bigg\}\Bigg]^{\frac{1}{2}} \\
&= \Bigg[\frac{1}{8r}\sum_{q=1}^{r}\Bigg\{(\sigma_1^l(\mathfrak{f}_q))^3 + (\sigma_2^l(\mathfrak{f}_q))^3 + (\sigma_1^u(\mathfrak{f}_q))^3 + (\sigma_2^u(\mathfrak{f}_q))^3 \\
&\quad + (\varsigma_1^l(\mathfrak{f}_q))^3 + (\varsigma_2^l(\mathfrak{f}_q))^3 + (\varsigma_1^u(\mathfrak{f}_q))^3 + (\varsigma_2^u(\mathfrak{f}_q))^3 + (\delta_1^l(\mathfrak{f}_q))^3 \\
&\quad + (\delta_2^l(\mathfrak{f}_q))^3 + (\delta_1^u(\mathfrak{f}_q))^3 + (\delta_2^u(\mathfrak{f}_q))^3\Bigg\}\Bigg]^{\frac{1}{2}} \\
&\le \Bigg[\frac{1}{8r}\sum_{q=1}^{r}\Big\{2((\sigma_1^u(\mathfrak{f}_q))^3 + (\sigma_2^u(\mathfrak{f}_q))^3) + 2((\varsigma_1^u(\mathfrak{f}_q))^3 + (\varsigma_2^u(\mathfrak{f}_q))^3) \\
&\quad + 2(\delta_1^l(\mathfrak{f}_q))^3 + 2(\delta_2^l(\mathfrak{f}_q))^3\Big\}\Bigg]^{\frac{1}{2}} \\
&\le \left[\frac{1}{8r}\sum_{q=1}^{r} 8\right]^{\frac{1}{2}} \\
&= 1.
\end{aligned}$$

Hence, property III holds.

IV. Lastly, we prove that if $\bar{F}_1 \le \bar{F}_2 \le \bar{F}_3$, then $d_{TV}(\bar{F}_1, \bar{F}_2) \le d_{TV}(\bar{F}_1, \bar{F}_3)$ and $d_{TV}(\bar{F}_2, \bar{F}_3) \le d_{TV}(\bar{F}_1, \bar{F}_3)$. Following the ordering properties of IVFFSs in Rani et al. [14] and Jeevaraj's work [5], if $\bar{F}_1 \le \bar{F}_2 \le \bar{F}_3$ then, $\sigma_{\bar{F}_1}^l \le \sigma_{\bar{F}_2}^l \le \sigma_{\bar{F}_3}^l, \sigma_{\bar{F}_1}^u \le \sigma_{\bar{F}_2}^u \le$

$\sigma^u_{\bar{F}_3}$, $\varsigma^l_{\bar{F}_1} \geq \varsigma^l_{\bar{F}_2} \geq \varsigma^l_{\bar{F}_3}$ and $\varsigma^u_{\bar{F}_1} \geq \varsigma^u_{\bar{F}_2} \geq \varsigma^u_{\bar{F}_3}$. For $0 \leq \alpha_1 + \alpha_3 \leq 1$, $0 \leq \alpha_2 + \alpha_4 \leq 1$, where $0 \leq \alpha_k \leq 1$, and α_k are real numbers $\forall k$, we write a function $w(y_1, y_2, y_3, y_4)$ as

$$w(y_1, y_2, y_3, y_4) = \sum_{k=1}^{4} \frac{(y_k - \alpha_k)^2}{y_k + \alpha_k}, \quad y_k \in [0, 1],$$

then the partial derivatives of w with respect to y_k is as follows:

$$\frac{\partial w}{\partial y_k} = \frac{(y_k - \alpha_k)(y_k + 3\alpha_k)}{(y_k + \alpha_k)^2}, \quad k = 1, 2, 3, 4.$$

Thus we get

$$\begin{cases} \frac{\partial w}{\partial y_k} \geq 0, & 0 \leq \alpha_k \leq y_k \leq 1, \\ \frac{\partial w}{\partial y_k} < 0, & 0 \leq y_k \leq \alpha_k \leq 1. \end{cases} \tag{7}$$

Therefore, for $y_k \geq \alpha_k$ and $y_k \leq \alpha_k$, w is a monotonically increasing and monotonically decreasing function for y_k respectively.

Let $\alpha_1 = (\sigma^l_{\bar{F}_1})^3$, $\alpha_2 = (\sigma^u_{\bar{F}_1})^3$, $\alpha_3 = (\varsigma^l_{\bar{F}_1})^3$ and $\alpha_4 = (\varsigma^u_{\bar{F}_1})^3$. w is monotonically increasing when $\alpha_1 \leq y_1$, if $\sigma^l_{\bar{F}_3} \geq \sigma^l_{\bar{F}_2}$, then

$$w(\sigma^l_{\bar{F}_3}, \sigma^u_{\bar{F}_3}, \varsigma^l_{\bar{F}_3}, \varsigma^u_{\bar{F}_3}) \geq w(\sigma^l_{\bar{F}_2}, \sigma^u_{\bar{F}_3}, \varsigma^l_{\bar{F}_3}, \varsigma^u_{\bar{F}_3}).$$

w is monotonically increasing when $\alpha_2 \leq y_2$, if $\sigma^u_{\bar{F}_3} \geq \sigma^u_{\bar{F}_2}$, then

$$w(\sigma^u_{\bar{F}_3}, \sigma^u_{\bar{F}_3}, \varsigma^l_{\bar{F}_3}, \varsigma^l_{\bar{F}_3}) \geq w(\sigma^l_{\bar{F}_2}, \sigma^u_{\bar{F}_2}, \varsigma^l_{\bar{F}_3}, \varsigma^u_{\bar{F}_3}),$$

whereas w is monotonically decreasing when $\alpha_3 \geq y_3$, if $\varsigma^l_{\bar{F}_3} \leq \varsigma^l_{\bar{F}_2}$, then

$$w(\sigma^l_{\bar{F}_2}, \sigma^u_{\bar{F}_2}, \varsigma^l_{\bar{F}_3}, \varsigma^u_{\bar{F}_3}) \geq w(\sigma^l_{\bar{F}_2}, \sigma^u_{\bar{F}_2}, \varsigma^l_{\bar{F}_2}, \varsigma^u_{\bar{F}_3})$$

and w is monotonically decreasing when $\alpha_4 \geq y_4$, if $\varsigma^u_{\bar{F}_3} \leq \varsigma^u_{\bar{F}_2}$, then

$$w(\sigma^l_{\bar{F}_2}, \sigma^u_{\bar{F}_2}, \varsigma^l_{\bar{F}_2}, \varsigma^u_{\bar{F}_3}) \geq w(\sigma^l_{\bar{F}_2}, \sigma^u_{\bar{F}_2}, \varsigma^l_{\bar{F}_2}, \varsigma^u_{\bar{F}_2}).$$

Using the above four inequalities, we get

$$w(\sigma^l_{\bar{F}_3}, \sigma^u_{\bar{F}_3}, \varsigma^l_{\bar{F}_3}, \varsigma^u_{\bar{F}_3}) \geq w(\sigma^l_{\bar{F}_2}, \sigma^u_{\bar{F}_2}, \varsigma^l_{\bar{F}_2}, \varsigma^u_{\bar{F}_2}).$$

Since the hesitancy degree is dependent on the membership and non-membership degree, we can say that $d_{TV}(\bar{F}_1, \bar{F}_2) \leq d_{TV}(\bar{F}_1, \bar{F}_3)$. Similarly, we can prove that

Table 1 Comparison of distance measure for IVFFNs

IVFFN pair	GEDM [5]	HDM [11]	NDM [10]	TDDM (proposed)
$(\bar{F}_1, \bar{F}_2)$	0.166	0.203	0.117	0.178
$(\bar{F}_1, \bar{F}_3)$	0.166	0.203	0.117	0.188

$d_{TV}(\bar{F}_2, \bar{F}_3) \leq d_{TV}(\bar{F}_1, \bar{F}_3)$. Thus property IV holds as well. Hence, the theorem is proved. □

Next, we utilize the following example to establish the superiority of the proposed triangular divergence-based distance measure for IVFFNs.

Example 1 Let there be three IVFFNs, $\bar{F}_1 = \langle [0.5, 0.5], [0.5, 0.5] \rangle$, $\bar{F}_2 = \langle [0.4, 0.6], [0.606, 0.7] \rangle$ and $\bar{F}_3 = \langle [0.3, 0.6], [0.57, 0.7] \rangle$. Clearly $\bar{F}_2 \neq \bar{F}_3$, so distance between $(\bar{F}_1, \bar{F}_2)$ and $\bar{F}_1, \bar{F}_3)$ should not be equal. We now calculate the GEDM (proposed by Jeevaraj [5]), HDM (proposed by Mathew et al. [11]), NDM (proposed by Mandal et al. [10]) and TDDM (proposed) between $(\bar{F}_1, \bar{F}_2)$ and $(\bar{F}_1, \bar{F}_3)$ using Eqs. (2), (3), (4) and (6) respectively, to establish superiority of the proposed distance measure. Table 1 gives the values of the calculated distances.

Thus we see that even though the GEDM and HDM give equal distances for the pairs, the proposed TDDM gives different distances for the given pair, thus establishing the superiority of the proposed distance measure.

From Fig. 1, we notice that there is a significant difference in the distance measures between the given pairs in case of TDDM, whereas the existing distance measures fail to distinguish between them.

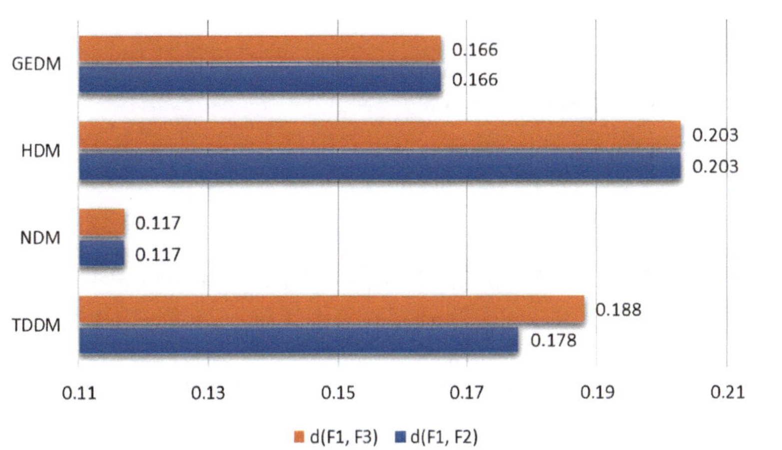

Fig. 1 Superiority of the proposed distance measure

4 A MAGDM Process Using Modified Distance-Based IVFF-TOPSIS Method

In this section, we propose a MAGDM process utilizing a modified version of the TOPSIS method by implementing the proposed distance measure for IVFFNs. The entropy method is used to calculate the weights of decision experts (DEs) and criteria while the IVFF-based modified-TOPSIS method is employed to rank the alternatives.

Let the set of alternatives be $E = \{E_1, E_2, \ldots, E_a\}$ such that there are a alternatives and C criteria such that $C = \{C_1, C_2, \ldots, C_c\}$ is the set of criteria where their weights are $\varpi_1, \varpi_2, \ldots, \varpi_c$ respectively. Here $0 \leq \varpi_\upsilon \leq 1$ and $\sum\limits_{\upsilon=1}^{c} \varpi_\upsilon = 1$. Let there be h decision experts $DE_1, DE_2, \ldots, DE_h$ with weights $\pi_1, \pi_2, \ldots, \pi_h$ respectively such that $0 \leq \pi_t \leq 1$ and $\sum\limits_{t=1}^{h} \pi_t = 1$. The steps of the MAGDM method are covered extensively in the following segment.

First, we calculate the weights of DEs. We convert the importance of DEs in linguistic variables (LVs) to respective IVFFNs. Then, to calculate the weights of DEs, we utilize the entropy method as proposed by Seikh et al. [18]. The following equation gives the entropy formula:

$$\pi_t = \frac{\sqrt{\frac{1}{2}[(1-\delta_t^l)^2 + (1-\delta_t^u)^2]}}{\sum\limits_{t=1}^{h} \sqrt{\frac{1}{2}[(1-\delta_t^l)^2 + (1-\delta_t^u)^2]}}. \tag{8}$$

To evaluate the weights of the criteria, we ask the DEs to assess the importance of criteria using LVs. The LVs are then converted to respective IVFFNs and aggregated for each criteria using Eq. (5). Then, the entropy method (Eq. 8) is used to determine the weights of each criteria.

The TDDM-based TOPSIS method is used for ranking the alternatives. The algorithm of the method is as follows:

Step 1: Obtain the decision matrices (DMs) in LVs from each DE. Then, convert the LVs into IVFFNs. Hence we get h DMs of order $a \times c$. We represent them as

$$M_{(t)} = \left[\mathfrak{f}_{y\upsilon(t)} = ([\sigma^l_{y\upsilon(t)}, \sigma^u_{y\upsilon(t)}], [\varsigma^l_{y\upsilon(t)}, \varsigma^u_{y\upsilon(t)}])\right]_{a\times c}, \quad \text{for } t = 1, 2, \ldots, h$$
$$\text{where } y = 1, 2, \ldots, a \text{ and } \upsilon = 1, 2, \ldots, c.$$

Here, $\mathfrak{f}_{y\upsilon(t)}$ denotes the performance of yth alternative with respect to υth attribute as rated by tth DE.

Step 2: Aggregate the h DMs using IVFFWA aggregation operator (Eq. 5). The aggregated matrix is denoted as $\overline{M} = [\overline{\mathfrak{f}_{y\upsilon}}]_{a\times c}$.

Step 3: Calculate the weighted matrix using the formula

$$M' = [\mathfrak{f}'_{yv}]_{a\times c} = [\varpi_v \overline{\mathfrak{f}_{yv}}]_{a\times c} \text{ where } y = 1, 2, \ldots, a \text{ and } v = 1, 2, \ldots, c. \tag{9}$$

Step 4: Calculate the positive ideal solution (PIS) denoted by I_v^+ and negative ideal solution (NIS) denoted by I_v^-, using the following formula:

$$([\max_y \sigma_{yv}^l, \max_y \sigma_{yv}^u], [\min_y \varsigma_{yv}^l, \min_y \varsigma_{yv}^u]) = \begin{cases} I_v^+ & \text{when } C_v \text{ is benefit criteria} \\ I_v^- & \text{when } C_v \text{ is cost criteria} \end{cases}$$

and

$$([\min_y \sigma_{yv}^l, \min_y \sigma_{yv}^u], [\max_y \varsigma_{yv}^l, \max_y \varsigma_{yv}^u]) = \begin{cases} I_v^- & \text{when } C_v \text{ is benefit criteria} \\ I_v^+ & \text{when } C_v \text{ is cost criteria} \end{cases}. \tag{10}$$

Step 5: Evaluate the distance between aggregated performance of alternatives and PIS and NIS using the proposed TDDM as given in Eq. (6), as follows:

$$D_y^+(E_y, PIS) = \sum_{v=1}^{c} \left(d_{TV}(\mathfrak{f}'_{yv}, I_v^+)\right)$$

$$\text{and } D_y^-(E_y, NIS) = \sum_{v=1}^{c} \left(d_{TV}(\mathfrak{f}'_{yv}, I_v^-)\right). \tag{11}$$

Step 6: Calculate the closeness coefficient of alternatives following the equation

$$CC(E_y) = \frac{D_y^-(E_y, NIS)}{D_y^+(E_y, PIS) + D_y^-(E_y, NIS)}. \tag{12}$$

Step 7: Rank the alternatives based on the closeness coefficient. The highest closeness coefficient gives the best alternative.

Figure 2 shows the MAGDM process utilizing the modified distance-based IVFF-TOPSIS method.

5 Illustrative Example

This section illustrates the proposed MAGDM technique by solving a numerical problem. Suppose an automobile company wishes to invest in the manufacturing and promotion of electric cars in the state of West Bengal, India. The company has four alternative designs or models: E_1, E_2, E_3 and E_4. There are three decision experts available, say, D_1, D_2 and D_3. These three decision experts will assess the alternative designs or models based on the following five criteria: C_1 is the driving range of the car; C_2 is its energy consumption; C_3 is the investment cost; C_4 is the

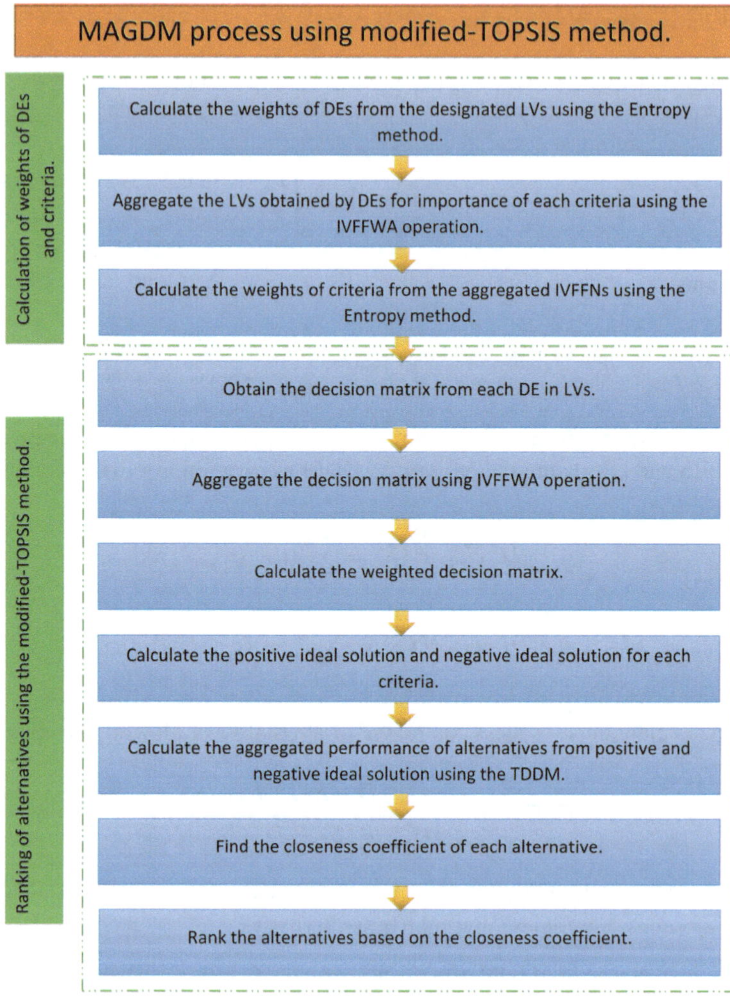

Fig. 2 MAGDM process using the TDDM-based modified-TOPSIS method on IVFFN environment

cost of the car; and C_5 is the market potential of the car. Among these criteria, C_2, C_3 and C_4 are "cost criteria" whereas C_1 and C_5 are "benefit criteria".

The LVs used to rate the importance of decision experts, criteria and evaluate the alternatives are given in Table 2. The importance of the DEs in LVs are EH for D_1, H for D_2 and L for D_3. Then, using Eq. (8), we find the weights of DEs to be $(\pi_1, \pi_2, \pi_3) = (0.57, 0.226, 0.204)$ respectively, for D_1, D_2 and D_3.

The LVs denoting the importance of criteria and for assessing alternatives are given in Table 3. We then convert the LVs into IVFFNs and aggregate them for each criterion using Eq. (5). The aggregated value for each criterion is given in Table 3.

Table 2 LVs for assessing importance of DEs, criteria and alternatives

LVs	IVFFN
Extremely high (EH)	([0.9, 0.95], [0.05, 0.1])
Very high (VH)	([0.8, 0.9], [0.2, 0.35])
High (H)	([0.7, 0.75], [0.3, 0.35])
Medium (M)	([0.5, 0.65], [0.5, 0.6])
Low (L)	([0.4, 0.45], [0.65, 0.7])
Very low (VL)	([0.2, 0.25], [0.6, 0.65])
Extremely low (EL)	([0.02, 0.05], [0.6, 0.65])

Table 3 The importance of criteria in LVs

Criteria	D_1	D_2	D_3	Aggregated value
C_1	EH	VH	EH	([0.884, 0.942], [0.068, 0.133])
C_2	M	M	M	([0.5, 0.65], [0.5, 0.6])
C_3	L	L	VL	([0.376, 0.424], [0.639, 0.69])
C_4	H	M	H	([0.669, 0.731], [0.336, 0.395])
C_5	VH	VH	M	([0.767, 0.875], [0.241, 0.391])

Table 4 Decision matrices for D_1, D_2 and D_3

Alternative	C_1	C_2	C_3	C_4	C_5
E_1	(L, M, VL)	(H, VH, VH)	(M, L, H)	(H, VH, H)	(H, M, M)
E_2	(VH, H, H)	(H, M, M)	(EH, VH, EH)	(VH, EH, H)	(VL, L, EL)
E_3	(EH, VH, EH)	(L, L, VL)	(H, M, M)	(M, M, L)	(VH, H, EH)
E_4	(VH, H, M)	(L, H, M)	(H, H, H)	(H, M, VH)	(VH, H, H)

We then calculate the weights of the criteria using Eq. (8). The weights of the criteria are found to be $\varpi_1 = 0.346$, $\varpi_2 = 0.142$, $\varpi_3 = 0.122$, $\varpi_4 = 0.137$ and $\varpi_5 = 0.253$ for C_1 to C_5 respectively.

The next and last step is the application of the proposed TDDM-based TOPSIS method for ranking of the alternatives.

Step 1 The decision matrices for the three decision makers are combined and written in Table 4.

Step 2 We first aggregate the decision matrices for each decision maker using IVF-FWA aggregation operator given in Eq. (5). The aggregated decision matrix is given in Table 5.

Table 5 Aggregated decision matrix

Alternative	C_1	C_2	C_3	C_4	C_5
E_1	([0.408, 0.496], [0.602, 0.666])	([0.75, 0.835], [0.252, 0.35])	([0.545, 0.647] [0.478, 0.556])	([0.728, 0.801], [0.273, 0.35])	([0.636, 0.713], [0.373, 0.442])
E_2	([0.764, 0.855], [0.238, 0.35])	([0.636, 0.713], [0.373, 0.442])	([0.884, 0.942], [0.068, 0.133])	([0.817, 0.899], [0.159, 0.264])	([0.271, 0.311], [0.61, 0.661])
E_3	([0.884, 0.942], [0.068, 0.133])	([0.376, 0.424], [0.639, 0.69])	([0.636, 0.713], [0.373, 0.442])	([0.482, 0.621], [0.528, 0.619])	([0.813, 0.895], [0.166, 0.271])
E_4	([0.745, 0.847], [0.265, 0.391])	([0.529, 0.6], [0.517, 0.58])	([0.700, 0.750], [0.3, 0.35])	([0.698, 0.782], [0.31, 0.395])	([0.764, 0.855], [0.238, 0.35])

Table 6 Weighted decision matrix

Alternative	C_1	C_2	C_3	C_4	C_5
E_1	([0.289, 0.353], [0.839, 0.869])	([0.421, 0.488], [0.822, 0.862])	([0.277, 0.336], [0.914, 0.931])	([0.401, 0.455], [0.837, 0.866])	([0.417, 0.476], [0.779, 0.813])
E_2	([0.57, 0.66], [0.609, 0.695])	([0.346, 0.396], [0.869, 0.891])	([0.511, 0.583], [0.72, 0.782])	([0.468, 0.546], [0.777, 0.833])	([0.172, 0.197], [0.882, 0.901])
E_3	([0.694, 0.775], [0.395, 0.498])	([0.198, 0.224], [0.938, 0.949])	([0.329, 0.377], [0.887, 0.905])	([0.253, 0.333], [0.916, 0.936])	([0.562, 0.649], [0.635, 0.719])
E_4	([0.552, 0.652], [0.632, 0.723])	([0.282, 0.324], [0.911, 0.926])	([0.368, 0.401], [0.863, 0.88])	([0.381, 0.44], [0.852, 0.881])	([0.518, 0.603], [0.695, 0.767])

Step 3 We calculate the weighted decision matrix using Eq. (9). The weighted decision matrix is given in Table 6.

Step 4 Then we find out the PIS and NIS. We use Eq. (10) to calculate the PIS and NIS (for criteria C_1 to C_5 respectively) as

$$\begin{aligned} PIS = \{&([0.694, 0.775], [0.395, 0.498]), ([0.198, 0.224], [0.938, 0.949]),\\ &([0.277, 0.336], [0.914, 0.931]), ([0.253, 0.333], [0.916, 0.936]),\\ &([0.562, 0.649], [0.635, 0.719])\} \text{ and}\\ NIS = \{&([0.289, 0.353], [0.839, 0.869]), ([0.421, 0.488], [0.822, 0.862]),\\ &([0.511, 0.583], [0.72, 0.782]), ([0.468, 0.546], [0.777, 0.833]),\\ &([0.172, 0.197], [0.882, 0.901])\}. \end{aligned}$$

Table 7 PDM and NDM

Alternative	C_1		C_2		C_3		C_4		C_5	
	D^+	D^-	D^+	D^-	D^+	D^-	D^+	D^-	D^+	D^-
E_1	0.429	0	0.204	0	0	0.269	0.143	0.081	0.167	0.179
E_2	0.186	0.278	0.14	0.075	0.269	0	0.213	0	0.319	0
E_3	0	0.429	0	0.204	0.056	0.22	0	0.213	0	0.319
E_4	0.21	0.258	0.067	0.146	0.1	0.182	0.119	0.102	0.063	0.267

Table 8 Closeness coefficient of alternatives and rank

Alternative	D_y^+	D_y^-	CC_y	Rank
E_1	0.943	0.529	0.359	3
E_2	1.127	0.353	0.239	4
E_3	0.056	1.385	0.961	1
E_4	0.559	0.955	0.631	2

Step 5 Next, we find the distance of each alternative from the PIS and NIS using the proposed TDDM given in Eq. (6), following Eq. (11). Given below is the positive distance matrix (PDM) and negative distance matrix (NDM) in Table 7.

Step 6 The last step is the calculation of the closeness coefficient of each alternative. We calculate the closeness coefficient using Eq. (12). The values of closeness coefficient and the final ranking is given in Table 8.

Step 7 We rank the alternatives to find that the 3rd alternative E_3 is the best choice. It is followed by E_4 and E_1. E_2 holds the last position.

5.1 Comparative Analysis

In this section, the proposed modified TOPSIS method is compared with the existing methods and distance measures. First, we compare the TOPSIS method proposed by Jeevaraj [5] in the IVFF environment, which uses the GEDM given in Eq. (2), to evaluate the distance between the alternatives and the PIS and NIS. The HDM was proposed by Mathew et al. [11] and utilized in the multi-attribute border approximation area comparison (MABAC) method in the IVFFN environment. The comparison is performed by using the HDM in place of the proposed TDDM in the proposed modified TOPSIS method. Lastly, a comparison is done with the NDM proposed by Mandal et al. [10]. A novel TOPSIS method was proposed by Mandal et al., utilizing the NDM. For comparison, the NDM is used in place of the proposed TDDM in the proposed modified TOPSIS method. Given in Table 9 is the distance of aggregated alternatives from PIS and NIS, the closeness coefficient of each alternative, and their

Table 9 Comparison of different methods

Distance measure used in TOPSIS	E_1			E_2			E_3			E_4			Ranking order of alternatives
	D_1^+	D_1^-	CC_1	D_2^+	D_2^-	CC_2	D_3^+	D_3^-	CC_3	D_4^+	D_4^-	CC_4	
GEDM [5]	0.837	0.478	0.363	1.004	0.31	0.236	0.049	1.258	0.963	0.469	0.851	0.645	$E_3 > E_4 > E_1 > E_2$
HDM [11]	1.141	0.658	0.366	1.373	0.427	0.237	0.066	1.733	0.963	0.635	1.166	0.647	$E_3 > E_4 > E_1 > E_2$
NDM [10]	0.89	0.466	0.344	1.02	0.336	0.248	0.041	1.315	0.97	0.475	0.883	0.65	
TDDM (proposed)	0.943	0.529	0.359	1.127	0.353	0.239	0.056	1.385	0.961	0.559	0.955	0.631	$E_3 > E_4 > E_1 > E_2$

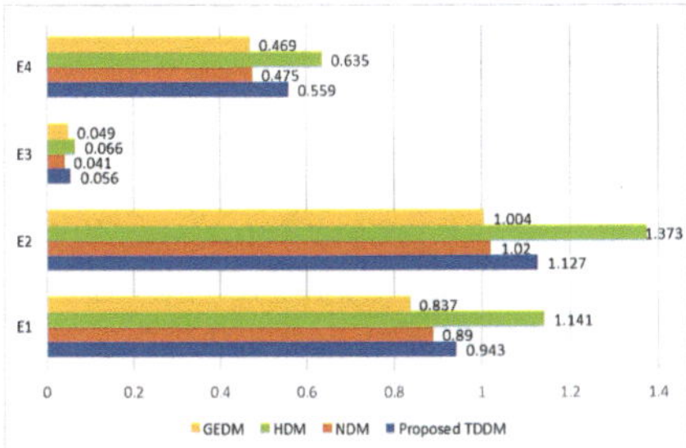

(a) Distance of aggregated performance of alternatives from PIS.

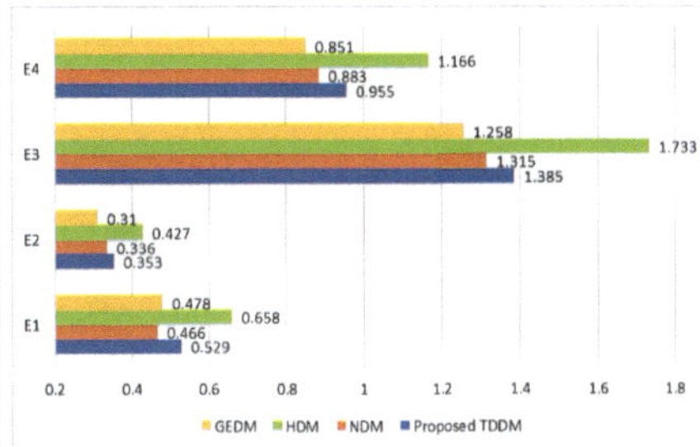

(b) Distance of aggregated performance of alternatives from NIS.

Fig. 3 Distance of aggregated performance of alternatives from PIS and NIS

ranking orders using different distance measures. Thus, we can say that the proposed TDDM method is superior as well as reliable.

Thus, we see that the ranking order is the same in all the cases, but the value of the closeness coefficient of alternatives differs slightly in all three cases. Thus, we can say that our proposed modified TOPSIS method is practically applicable.

Figure 3 shows the comparative analysis concerning distance. The graphical representations of the distance of aggregated performance of alternatives from PIS and NIS are shown in Fig. 3a and b respectively. From Fig. 3a, we see that the $D_y^+(E_y, PIS)$ for GEDM and NDM are very close to each other, but for the proposed TDDM, it is more than GEDM and NDM but less than that of HDM. The $D_y^+(E_y, PIS)$ using the proposed TDDM is very close to the average value of $D_y^+(E_y, PIS)$ using GEDM, HDM and NDM. In the case of Fig. 3b also, we observe a similar trend where the $D_y^-(E_y, NIS)$ for TDDM is equal (up to 2 decimal places) to the average of $D_y^-(E_y, NIS)$ for GEDM, HDM and NDM for E_4, E_2 and E_1. There is a slight variation in the case of E_3.

Figure 4 depicts the comparison of closeness coefficients of alternatives in Fig. 4a and ranking orders of the alternatives in Fig. 4b. From Fig. 4a, we observe that for TDDM, the CC_y is equal (up to 2 decimal places) to the average of CC_y for GEDM, HDM and NDM for E_4, E_3 and E_2. There is a slight variation in the case of E_1. Figure 4b shows the ranking order of alternatives, where we observe that despite the differences in $D_y^+(E_y, PIS)$, $D_y^-(E_y, NIS)$ and CC_y, the ranking order using the different distance measures remains the same.

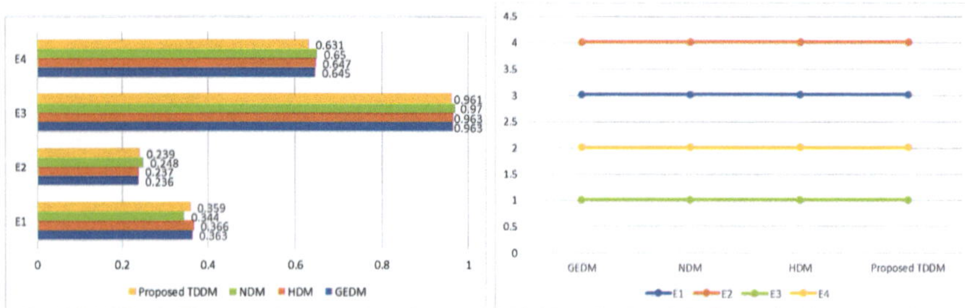

(a) Closeness coefficient of each alternative using different distance measure.

(b) Rank of each alternative using different distance measure.

Fig. 4 Comparative analysis of closeness coefficient and ranking order

6 Conclusion

The study's main goal is to address the constraints inherent in current distance metrics for IVFFNs and, consequently, to suggest a new distance metric based on triangular divergence customized for interval-valued Fermatean fuzzy sets. This innovative metric aims to reduce any information loss by integrating the inherent hesitancy degree in IVFFNs. Moreover, the study proposes a modified TOPSIS method based on the proposed TDDM in the IVFFN environment. The effectiveness of the proposed TDDM is validated through an illustrative example where it accurately distinguishes the distances between two specified pairs of IVFFNs while the existing distance measures fail to do so. Additionally, the practical usability of the proposed adapted TOPSIS technique is confirmed through a numerical illustration. The proposed adapted TOPSIS technique is compared with existing methods and distance metrics. The results of the comparative analysis demonstrate that the proposed TDDM closely aligns with the average distances from other methods, showcasing its efficacy in capturing relative distances from PIS and NIS. Moreover, it highlights consistent closeness coefficients across alternatives, with TDDM yielding values akin to the average of closeness coefficients from other methods. Despite minor variations, the ranking orders of alternatives remain consistent across different distance measures, underscoring the stability and significance of the proposed approach in the decision-making scenario. The results contribute to enhancing MAGDM in situations of uncertainty by providing a more robust structure for managing imprecise data.

Every research endeavour inevitably encounters certain constraints. In this paper, only the TOPSIS method has been modified using the proposed TDDM. Utilizing the proposed TDDM in other distance-based MAGDM methods can establish the practicality of the proposed distance measure. Another limitation is that we have addressed only one numerical problem using the proposed modified TOPSIS method.

There are numerous avenues for future investigation. The proposed distance measure can be extended to other fuzzy environments like 3,4-quasi-rung fuzzy sets [17], quasi-rung orthopair fuzzy sets [16], hesitant fuzzy sets [21], Pythagorean fuzzy sets

[22] and neutrosophic fuzzy sets [2]. Also, the proposed TDDM can be applied to other distance-based MAGDM methods. The proposed TDDM-based modified TOPSIS can be applied to other real-life MAGDM problems to establish further applicability of the method.

Conflict of interest The authors declare that they have no conflict of interest.

References

1. Bouraima, M.B., Gore, A., Ayyildiz, E. et al.: (2023), Assessing of causes of accidents based on a novel integrated interval-valued Fermatean fuzzy methodology: towards a sustainable construction site. Neural Comput. Appl. https://doi.org/10.1007/s00521-023-08948-5
2. Das, S., Roy, B.K., Kar, M.B., Kar, S., Pamucar, D.: Neutrosophic fuzzy set and its application in decision making. J. Ambient Intell. Hum. Comput. **11**, 5017–5029 (2020)
3. Deng, Z., Wang, J.: New distance measure for Fermatean fuzzy sets and its application. Int. J. Intell. Syst. **37**(3), 1903–1930 (2022)
4. Hwang, C. L., Yoon, K.: Multiple Attribute Decision Making: Methods and Applications a State-of-the-Art Survey, vol. 186. Springer Science & Business Media (2012)
5. Jeevaraj, S.: Ordering of interval-valued Fermatean fuzzy sets and its applications. Expert Syst. Appl. **185**, 115613 (2021)
6. Ilieva, G., Yankova, T.: Extension of interval-valued Fermatean fuzzy TOPSIS for evaluating and benchmarking COVID-19 vaccines. Mathematics **10**(19), 3514 (2022)
7. Kirişci, M.: New cosine similarity and distance measures for Fermatean fuzzy sets and TOPSIS approach. Knowl. Inf. Syst. **65**(2), 855–868 (2023)
8. Kirişci, M.: Interval-valued Fermatean fuzzy based risk assessment for self-driving vehicles. Appl. Soft Comput. **152**, 111265 (2024)
9. Liu, Z.: A distance measure of Fermatean fuzzy sets based on triangular divergence and its application in medical diagnosis. J. Oper. Intell. **2**(1), 167–178 (2024)
10. Mandal, U., Seikh, M.R.: Interval-valued Fermatean fuzzy TOPSIS method and its application to sustainable development program. In: Congress on Intelligent Systems: Proceedings of CIS 2021, vol. 2, pp. 783–796. Springer Nature Singapore, Singapore (2022)
11. Mathew, M., Chakrabortty, R.K., Ryan, M.J., Ljaz, M.F., Khan, S.A.R.: The Multi-Attributive Border Approximation Area Comparison (Mabac) Method for Decision Making under Interval-Valued Fermatean Fuzzy Environment for Green Supplier Selection. (2021) (Preprint)
12. Onyeke, I.C., Ejegwa, P.A.: Modified Senapati and Yager's Fermatean fuzzy distance and its application in students' course placement in tertiary institution. In: Real Life Applications of Multiple Criteria Decision Making Techniques in Fuzzy Domain, pp. 237–253. Singapore, Springer Nature Singapore (2022)
13. Qin, Y., Hashim, S.R.M., Sulaiman, J.: A new distance measure and corresponding TOPSIS method for interval-valued intuitionistic fuzzy sets in multi-attribute decision-making. AIMS Math. **8**(11), 26459–26483 (2023)
14. Rani, P., Mishra, A.R.: Interval-valued Fermatean fuzzy sets with multi-criteria weighted aggregated sum product assessment-based decision analysis framework. Neural Comput. Appl. **34**(10), 8051–8067 (2022)
15. Sahoo, L., Senapati, T., Yager, R.R. (eds.): Real life applications of multiple criteria decision making techniques in fuzzy domain. In: Studies in Fuzziness and Soft Computing (STUDFUZZ), vol. 420. Springer Nature (2023)
16. Seikh, M.R., Mandal, U.: Multiple attribute group decision making based on quasirung orthopair fuzzy sets: application to electric vehicle charging station site selection problem. Eng. Appl. Artif. Intell. **115**, 105299 (2022)

17. Seikh, M.R., Mandal, U.: Multiple attribute decision-making based on 3,4-quasirung fuzzy sets. Granular Comput. **7**, 965–978 (2022)
18. Seikh, M.R., Mandal, U.: Interval-valued Fermatean fuzzy Dombi aggregation operators and SWARA based PROMETHEE II method to bio-medical waste management. Expert Syst. Appl. **226**, 120082 (2023)
19. Senapati, T., Yager, R.R.: Fermatean fuzzy sets. J. Ambient Intell. Hum. Comput. **11**, 663–674 (2020)
20. Solangi, Y.A., Longsheng, C., Shah, S.A.A.: Assessing and overcoming the renewable energy barriers for sustainable development in Pakistan: an integrated AHP and fuzzy TOPSIS approach. Renew. Energy **173**, 209–222 (2021)
21. Torra, V.: Hesitant fuzzy sets. Int. J. Intell. Syst. **25**, 529–539 (2010)
22. Yager, R.R.: Pythagorean membership grades in multicriteria decision making. IEEE Trans. Fuzzy Syst. **22**(4), 958–965 (2013)
23. Yang, S., Pan, Y., Zeng, S.: Decision making framework based Fermatean fuzzy integrated weighted distance and TOPSIS for green low-carbon port evaluation. Eng. Appl. Artif. Intell. **114**, 105048 (2022)
24. Yehudayoff, A.: Pointer chasing via triangular discrimination. Combin. Prob. Comput. **29**(4), 485–494 (2020)
25. Zeng, S., Gu, J., Peng, X.: Low-carbon cities comprehensive evaluation method based on Fermatean fuzzy hybrid distance measure and TOPSIS. Artif. Intell. Rev. 1–17 (2023)

Archimedean Averaging Operators with Linear Programming Model for Generalized Orthopair Fuzzy Decision-Making

Sukhwinder Singh Rawat and Komal

Abstract The averaging operator stands as a fundamental and widely employed aggregation tool in decision-making processes. This chapter delves into the realm of generalized orthopair fuzzy numbers, a domain typically governed by a t-norms (TNs) and their corresponding t-conorms (TCNs). Archimedean TN and TCN establish a versatile foundation for intersection and union operations within the context of fuzzy sets. In this context, we discuss the q-rung orthopair fuzzy Archimedean weighted averaging (q-ROFAWA) and q-rung orthopair fuzzy Archimedean ordered weighted averaging (q-ROFOWA) operators, extending the applicability of these operators. Additionally, we propose a q-rung orthopair fuzzy Archimedean ordinal priority approach (q-ROFAOPA) for the systematic evaluation of attribute weights. The culmination of this work manifests in a multi-attribute group decision-making (MAGDM) approach, along with its practical implementation, offering a comprehensive solution for decision-making scenarios.

Keywords Archimedean t-norm · MAGDM · Ordinal priority approach · q-ROFS

1 Introduction

MAGDM is a complex and critical process essential for addressing intricate decision-making scenarios where multiple criteria contribute to the final choice [25]. In such contexts, decision-makers (DMs) are often confronted with the challenge of evaluating and prioritizing alternatives based on a set of diverse attributes. MAGDM provides a systematic framework to facilitate collaborative decision-making by integrating the preferences, opinions, and expertise of multiple experts or DMs. The significance of MAGDM arises in various fields, including business, engineering, environmental management, etc., where decisions involve a multiple factors that need to be considered simultaneously [18]. One of the primary objectives of MAGDM is to

S. S. Rawat (✉) · Komal
Department of Mathematics, School of Physical Sciences, Doon University, Dehradun, India
e-mail: sukhwinderawat@gmail.com

L. Sahoo et al. (eds.), *Decision Making Under Uncertainty Via Optimization, Modelling, and Analysis*, Studies in Systems, Decision and Control 558,
https://doi.org/10.1007/978-981-96-0085-4_3

achieve a comprehensive and well-informed decision that accommodates the diverse perspectives and priorities of the experts group. This involves not only aggregating individual preferences but also addressing conflicts and uncertainties inherent in the decision-making process [14, 17]. Consequently, MAGDM methodologies often incorporate mathematical models and optimization techniques to identifying a satisfactory attribute importance.

The integration of fuzzy sets (FSs) into MAGDM represents a significant advancement that addresses the inherent uncertainties present in real-world decision-making scenarios. FS [28] establish a mathematical framework for dealing with imprecision and ambiguity, making it particularly well-suited for applications in decision science. An extension of FS named intuitionistic fuzzy set (IFS) [2] emerged as a more effective way to handle the uncertainty. It encompasses not only membership and non-membership degrees but also a hesitation degree, providing DMs a more nuanced representation of uncertainty. Building upon these advancements, Pythagorean fuzzy set (PFS) [26] and then Fermatean fuzzy set (FFS) [22] offer a more comprehensive framework. This extension allows DMs to express their preferences and uncertainties more explicitly, contributing to a more accurate representation of decision contexts. In recent years, q-rung orthopair fuzzy set (q-ROFS) [27] has emerged as a versatile extension of orthopair family of fuzzy sets. The q-ROFS incorporates a parameter q to adjust the level of indeterminacy, providing DMs with a flexible tool to model and manage uncertainties in decision-making processes. Such a generalized orthopair fuzzy sets allow DMs to model and represent the inherent vagueness in human perceptions and linguistic expressions associated with evaluating attributes.

Archimedean TNs and TCNs provide a versatile mathematical framework for modeling the intersection and union operations in fuzzy set context. The Archimedean class of TNs is not necessarily continuous. But if an Archimedean TN is continuous, then it is strictly increasing on the subset where its value is greater than zero [4]. The continuous Archimedean TNs provide an advantage of being represented by an additive generator (AG), which reduces the calculation process from a multivariate function to its univariate generator [10]. Over the years, researchers have extended and adapted these Archimedean TNs and TCNs to generate aggregation operators (AOs) for decision-making approaches [11, 23, 24]. These AOs play a crucial role in consolidating diverse sources of information, preferences, and attributes in decision processes. Specifically, AOs, such as weighted averaging (WA) and ordered weighted averaging (OWA), have been instrumental in aggregating conflicting criteria and preferences within fuzzy MCGDM contexts [3]. The WA operator for q-rung orthopair fuzzy numbers (q-ROFNs) has been extensively studied by various researchers under different TNs, that is, Liu and Wang [12] used algebraic TN, Jana et al. [9] applied Dombi TN, Darko and Liang [6] utilized Hamacher TN, Seikh and Mandal [19] used Frank TN, and Senapati et al. [21] implemented Aczél-Alsina TN, all of these TNs are Archimedean. Recently, Seikh and Mandal [20] utilized the general framework of Archimedean TNs for developing various AO under q-rung orthopair fuzzy (q-ROF) environment.

One important aspect of weighted AOs is attributes weights, which influences the overall process. Over the years, researchers have explored and developed diverse methodologies to address the challenge of assigning appropriate weights to different attributes based on their relative importance [5]. Attribute weights evaluation techniques aim to provide DMs with systematic and objective approaches to quantify the relative importance of each attribute. One such approach is ordinal priority approach (OPA) in which a linear programming model is used for computing final attribute weights. Deveci et al. [8] developed the q-ROF OPA approach and Pamucar et al. [13] suggested a Schweizer–Sklar norm-based OPA.

In this chapter, we have discussed WA and OWA operators with their special cases and a OPA for q-ROFNs under Archimedean TNs and TCNs. Finally, this chapter introduces a MAGDM approach by using the proposed OPA model.

The chapter is structured as follows: Sect. 2 discuses the definitions of q-ROFS, Archimedean TNs and TCNs, and basic operations for q-ROFNs. In Sect. 3, we have discussed the q-ROF Archimedean WA (q-ROFAWA) operator with some specific cases and q-ROF Archimedean OWA (q-ROFAOWA) operator. In Sect. 4, firstly, we developed a q-ROF Archimedean OPA (q-ROFAOPA), secondly we proposed a q-ROF MAGDM approach, and finally we have discussed the practical implementation of the developed approach. Section 5 discusses an overall conclusion of this chapter.

2 Preliminary

2.1 q-ROFS [27]

A q-ROFS Q on a domain of discourse U is a collection of elements with their orthopair membership degree $(\mu_Q(x), \nu_Q(x))$ and is defined as follows:

$$Q = \{\langle x, (\mu_Q(x), \nu_Q(x))\rangle \mid x \in U\} \tag{1}$$

where $\mu_Q(x) \in [0, 1]$ and $\nu_Q(x) \in [0, 1]$ indicates support for and support against membership of x in Q and satisfy $\mu_Q^q(x) + \nu_Q^q(x) \leq 1$, for at least one $q \geq 1$.

The hesitancy of x in Q is represented by $\pi_Q(x) = \sqrt[q]{1 - (\mu_Q(x))^q - (\nu_Q(x))^q}$. The orthopair $(\mu_Q, \nu_Q) \in \mathcal{Q}$ is known as q-ROFN, where $\mathcal{Q}$ is the collection of all q-ROFNs.

2.2 Ordering Procedure for q-ROFNs

To sort any two q-ROFNs, ℓ_1 and ℓ_2, usually score function (Sc) and accuracy function (Ac) are used and based on their values, the ordering can be done as follows:

(a) If $Sc(\ell_1) < Sc(\ell_2) \implies \ell_1 < \ell_2$.
(b) If $Sc(\ell_1) = Sc(\ell_2)$, then

(i) $Ac(\ell_1) < Ac(\ell_2) \implies \ell_1 < \ell_2$.
(ii) If $Ac(\ell_1) = Ac(\ell_2) \implies \ell_1 = \ell_2$.

The score (Sc) and accuracy (Ac) functions for q-ROFNs are given by Liu and Wang [12].

$$Sc(\ell) = \mu_\ell^q - \nu_\ell^q \text{ and } Ac(\ell) = \mu_\ell^q + \nu_\ell^q. \tag{2}$$

In the literature, numerous score functions assert their capacity to establish a direct ordering of q-ROFNs without necessitating the incorporation of an accuracy function. Adding to this discourse, Rawat et al. [16] have recently introduced a new score function, articulated as follows:

$$Sc(\ell) = \frac{1}{2}\left(\left(\mu_\ell^q - \nu_\ell^q + 1\right) - \frac{1}{\pi^3}\cos\left(\left(1 - \pi_\ell^q\right)\frac{\pi}{2}\right)\right) \tag{3}$$

where π_ℓ is a hesitancy associated with a q-ROFN (μ_ℓ, ν_ℓ).

2.3 *Triangular Norm and Conorm [10]*

For any $u, v, w \in [0, 1]$, if a binary function $T : [0, 1]^2 \to [0, 1]$ holds the conditions (T1)-(T4), then it is called a TN on [0, 1].

(T1) Commutativity: $T(u, v) = T(v, u)$.
(T2) Monotonicity: $T(u, w) \leq T(v, w)$ whenever $u \leq v$.
(T3) Associativity: $T(T(u, v), w) = T(u, T(v, w))$.
(T4) Neutrality: $T(u, 1) = u$.

Similarly, a binary function $S : [0, 1]^2 \to [0, 1]$ that satisfies (T1)-(T3) and has neutral element 0 is known as a TCN.

A TCN $S : [0, 1]^2 \to [0, 1]$ is a dual form of TN, hence they exists iff there exists a TN (T) such that $\forall\ (u, v) \in [0, 1]^2$,

$$S(u, v) = 1 - T(1 - u, 1 - v)$$

2.4 *Archimedean TN and TCN [4]*

A TN T is said to be an Archimedean TN if, for every $(u, v) \in (0, 1)^2$, there exist a $k \in \mathbb{N}$ such that $T(\overbrace{u, \ldots, u}^{k-times}) < v$. This implies $\lim\limits_{k\to\infty} T_k(\overbrace{a, \ldots, a}^{k-times}) = 0\ \forall\ a \in (0, 1)$ and T has only trivial (0 and 1) idempotent elements.

Table 1 Archimedean TNs and AG for q-ROFNs [1]

Name	Archimedean TN (T)	AG (ϕ)
Algebraic	$T^A(u^q, v^q) = u \cdot v$	$\phi^A(t^q) = -\log t^q$
Aczél-Alsina	$T^{AA}_{\gamma \in (0,\infty)}(u^q, v^q) = \left(e^{-((-\log u^q)^\gamma + (-\log v^q)^\gamma)^{\frac{1}{\gamma}}}\right)^{\frac{1}{q}}$	$\phi^{AA}_{\gamma \in (0,\infty)}(t^q) = (-\log t^q)^\gamma$
Dombi	$T^D_{\gamma \in (0,\infty)}(u^q, v^q) = \left(\frac{1}{1 + \left(\left(\frac{1-u^q}{u^q}\right)^\gamma + \left(\frac{1-v^q}{v^q}\right)^\gamma\right)^{\frac{1}{\gamma}}}\right)^{\frac{1}{q}}$	$\phi^D_{\gamma \in (0,\infty)}(t^q) = \left(\frac{1-t^q}{t^q}\right)^\gamma$
Frank	$T^F_{\gamma \in (0,\infty]}(u^q, v^q) = \begin{cases} u \cdot v, & \gamma = 1; \\ \max\left((u^q + v^q - 1)^{\frac{1}{q}}, 0\right), & \gamma = \infty; \\ \left(\log_\gamma\left(1 + \frac{(\gamma^{u^q} - 1)(\gamma^{v^q} - 1)}{(\gamma - 1)}\right)\right)^{\frac{1}{q}}, & \text{else.} \end{cases}$	$\phi^F_{\gamma \in (0,\infty]}(t^q) = \begin{cases} -\log t^q, & \gamma = 1; \\ 1 - t^q, & \gamma = \infty; \\ \log\left(\frac{\gamma - 1}{\gamma^{t^q} - 1}\right), & \text{else.} \end{cases}$
Hamacher	$T^H_{\gamma \in [0,\infty)}(u^q, v^q) = \begin{cases} 0, & \gamma = u = v = 0; \\ \left(\frac{u^q v^q}{\gamma + (1-\gamma)(u^q + v^q - u^q v^q)}\right)^{\frac{1}{q}}, & \text{else.} \end{cases}$	$\phi^H_{\gamma \in [0,\infty)}(t^q) = \begin{cases} \frac{1-t^q}{t^q}, & \gamma = 0; \\ \log\left(\frac{\gamma + (1-\gamma)t^q}{t^q}\right), & \text{else.} \end{cases}$

A continuous Archimedean TN T can be defined by its AG $\phi : [0, 1] \to [0, \infty]$ as follows:

$$T(u, v) = \phi^{-1}(\phi(u) + \phi(v))$$

where ϕ is a strictly decreasing function with $\phi(1) = 0$.

A TCN S is said to be an Archimedean TCN if, for every $(u, v) \in (0, 1)^2$ there exists an $k \in \mathbb{N}$ such that $S(\overbrace{u, \ldots, u}^{k-times}) > v$.

A continuous Archimedean TCN S can be defined by its AG $\psi : [0, 1] \to [0, \infty]$ as follows:

$$S(u, v) = \psi^{-1}(\psi(u) + \psi(v))$$

where ψ is a strictly increasing function with $\psi(0) = 0$.

Furthermore, if T is a TN with AG $\phi : [0, 1] \to [0, \infty]$. Then the AG of TCN S (dual of T) $\psi : [0, 1] \to [0, \infty]$ is given by $\psi(t) = \phi(1 - t)$.

Table 1 presents some continuous q-ROF Archimedean TNs given in [1] from the perspectives of q-ROF representation theorem. These are Algebraic (T^A), Aczél-Alsina $(T^{AA})_{\gamma \in (0,\infty)}$, Dombi $(T^D)_{\gamma \in (0,\infty)}$, Frank $(T^F)_{\gamma \in (0,\infty]}$, and Hamacher $(T^H)_{\gamma \in [0,\infty)}$ TNs with their strictly deceasing AGs, which are also continuous [10].

Table 2 Archimedean TCNs and AG for q-ROFNs [1]

Name	Archimedean TCN (S)	AG (ψ)
Algebraic	$S^A(u^q, v^q) = (u^q + v^q - u^q v^q)^{\frac{1}{q}}$	$\psi^A(t^q) = -\log(1 - t^q)$
Aczél-Alsina	$S^{AA}_{\gamma\in(0,\infty)}(u^q, v^q) = \left(1 - e^{-((-\log(1-u^q))^\gamma + (-\log(1-v^q))^\gamma)^{\frac{1}{\gamma}}}\right)^{\frac{1}{q}}$	$\psi^{AA}_{\gamma\in(0,\infty)}(t^q) = (-\log(1 - t^q))^\gamma$
Dombi	$S^D_{\gamma\in(0,\infty)}(u^q, v^q) = \left(1 - \frac{1}{1 + \left(\left(\frac{u^q}{1-u^q}\right)^\gamma + \left(\frac{v^q}{1-v^q}\right)^\gamma\right)^{\frac{1}{\gamma}}}\right)^{\frac{1}{q}}$	$\psi^D_{\gamma\in(0,\infty)}(t^q) = \left(\frac{t^q}{1-t^q}\right)^\gamma$
Frank	$S^F_{\gamma\in(0,\infty]}(u^q, v^q) = \begin{cases} (u^q + v^q - u^q v^q)^{\frac{1}{q}}, & \gamma = 1; \\ \min\left((u^q + v^q)^{\frac{1}{q}}, 1\right), & \gamma = \infty; \\ \left(1 - \log_\gamma\left(1 + \frac{(\gamma^{1-u^q} - 1)(\gamma^{1-v^q} - 1)}{(\gamma - 1)}\right)\right)^{\frac{1}{q}}, & \text{else.} \end{cases}$	$\psi^F_{\gamma\in(0,\infty]}(t^q) = \begin{cases} -\log(1 - t^q), & \gamma = 1; \\ t^q, & \gamma = \infty; \\ \log\left(\frac{\gamma - 1}{\gamma^{1-t^q} - 1}\right), & \text{else.} \end{cases}$
Hamacher	$S^H_{\gamma\in[0,\infty)}(u^q, v^q) = \begin{cases} 1, & \gamma = 0 \text{ and } u = v = 1; \\ \left(\frac{u^q + v^q - u^q v^q - (1-\gamma)u^q v^q}{1 - (1-\gamma)u^q v^q}\right)^{\frac{1}{q}}, & \text{else.} \end{cases}$	$\psi^H_{\gamma\in[0,\infty)}(t^q) = \begin{cases} \frac{t^q}{1-t^q}, & \gamma = 0; \\ \log\left(\frac{\gamma + (1-\gamma)(1-t^q)}{1-t^q}\right), & \text{else.} \end{cases}$

Similarly, Table 2 contains some continuous q-ROF Archimedean TCNs from the perspectives of q-ROF representation theorem [1]. These TCNs are algebraic (S^A), Aczél-Alsina $(S^{AA})_{\gamma\in(0,\infty)}$, Dombi $(S^D)_{\gamma\in(0,\infty)}$, Frank $(S^F)_{\gamma\in(0,\infty]}$, and Hamacher $(S^H)_{\gamma\in[0,\infty)}$ with their strictly increasing AGs which are continuous [10].

2.5 *Archimedean TN- and TCN-Based Operations for Q-ROFNs [1]*

Let ℓ_1 and ℓ_2 be two q-ROFNs, ϕ and ψ are strictly decreasing and increasing AG of T and S, respectively, and $\lambda > 0$.

(a) $\ell_1 \oplus \ell_2 = \left((\psi^{-1}(\psi(\mu^q_{\ell_1}) + \psi(\mu^q_{\ell_2})))^{\frac{1}{q}}, (\phi^{-1}(\phi(\nu^q_{\ell_1}) + \phi(\nu^q_{\ell_2})))^{\frac{1}{q}}\right)$;

(b) $\ell_1 \otimes \ell_2 = \left((\phi^{-1}(\phi(\mu^q_{\ell_1}) + \phi(\mu^q_{\ell_2})))^{\frac{1}{q}}, (\psi^{-1}(\psi(\nu^q_{\ell_1}) + \psi(\nu^q_{\ell_2})))^{\frac{1}{q}}\right)$;

(c) $\lambda\ell_1 = \left((\psi^{-1}(\lambda\psi(\mu^q_{\ell_1})))^{\frac{1}{q}}, (\phi^{-1}(\lambda\phi(\nu^q_{\ell_1})))^{\frac{1}{q}}\right)$;

(d) $\ell_1^\lambda = \left((\phi^{-1}(\lambda\phi(\mu^q_{\ell_1})))^{\frac{1}{q}}, (\psi^{-1}(\lambda\psi(\nu^q_{\ell_1})))^{\frac{1}{q}}\right)$.

The output values of these operations are q-ROFNs. The operations (a) and (b) can be further extended for any finite numbers of q-ROFNs as follows:

(a) $\bigoplus_{i=1}^{n} \ell_i = \left(\left(\psi^{-1} \left(\sum_{i=1}^{n} \psi(\mu_{\ell_i}^q) \right) \right)^{\frac{1}{q}}, \left(\phi^{-1} \left(\sum_{i=1}^{n} \phi(\nu_{\ell_i}^q) \right) \right)^{\frac{1}{q}} \right)$;

(b) $\bigotimes_{i=1}^{n} \ell_i = \left(\left(\phi^{-1} \left(\sum_{i=1}^{n} \phi(\mu_{\ell_i}^q) \right) \right)^{\frac{1}{q}}, \left(\psi^{-1} \left(\sum_{i=1}^{n} \psi(\nu_{\ell_i}^q) \right) \right)^{\frac{1}{q}} \right)$.

Let ℓ, ℓ_1, and ℓ_2 be three q-ROFNs, and λ, λ_1, and $\lambda_2 > 0$, then some fundamental properties hold by these operations are as follows [11]:

(a) $\ell_1 \oplus \ell_2 = \ell_2 \oplus \ell_1$;
(b) $\ell_1 \otimes \ell_2 = \ell_2 \otimes \ell_1$;
(c) $\lambda(\ell_1 \oplus \ell_2) = \lambda\ell_1 \oplus \lambda\ell_2$;
(d) $\ell_1^\lambda \otimes \ell_2^\lambda = (\ell_1 \otimes \ell_2)^\lambda$;
(e) $\lambda_1\ell \oplus \lambda_2\ell = (\lambda_1 + \lambda_2)\ell$;
(f) $\ell^{\lambda_1} \otimes \ell^{\lambda_2} = \ell^{\lambda_1+\lambda_2}$.

By selecting various types of continuous Archimedean TNs (with corresponding TCNs) and their AGs ϕ and ψ shown in Tables 1 and 2, different rules for operations can be generated.

3 Averaging Operators

3.1 Archimedean TN and TCN based WA Operator [20]

Let $L = \{\ell_1, \ell_2, \ldots, \ell_n\}$ is a finite set of q-ROFNs and the associated weight vector to this set is $w = (w_1, w_2, \ldots, w_n)$. Then the WA operator for these n-inputs under Archimedean TN and TCN and weighting vector w is $\text{q} - \text{ROFAWA} : \mathcal{Q}^n \to \mathcal{Q}$ and define as follows:

$$\text{q} - \text{ROFAWA}(\ell_1, \ell_2, \ldots, \ell_n) = \bigoplus_{j=1}^{n} w_j \ell_j$$
$$= \left(\left(\psi^{-1} \left(\sum_{j=1}^{n} w_j \psi(\mu_j^q) \right) \right)^{\frac{1}{q}}, \left(\phi^{-1} \left(\sum_{j=1}^{n} w_j \phi(\nu_j^q) \right) \right)^{\frac{1}{q}} \right) \tag{4}$$

where $w_j \in [0, 1]$ such that $\sum_{j=1}^{n} w_j = 1$ and ϕ and ψ are AGs of TN and TCN, respectively.

Some desirable properties like boundedness, idempotency, and monotonicity are holds for the q-ROFAWA operator. It is worth noting that Eq. (4) represents a generalized WA operator for q-ROFNs. Thus, any specific AG of discussed Archimedean

TNs will provide a particular TN-based weighted average operator. To give a more insight into this, the generated WA operators are discusses hereafter.

(i) (Algebraic). For $\phi(t^q) = -\log t^q$ and $\psi(t^q) = -\log(1 - t^q)$. The q-ROFAWA operator (4) is converted into the q-ROF-WA operator [12].

$$\mathrm{q-ROFAWA}_{T^A}(\ell_1, \ell_2, \ldots, \ell_n) = \left(\left(1 - \prod_{j=1}^{n}\left(1 - \mu_{\ell_j}^q\right)^{w_j}\right)^{\frac{1}{q}}, \prod_{j=1}^{n} \nu_{\ell_j}^{w_j}\right) \tag{5}$$

(ii) (Aczél-Alsina). For $\phi(t^q) = (-\log t^q)^\gamma$ and $\psi(t^q) = (-\log(1 - t^q))^\gamma$. The q-ROFAWA operator (4) is reduced to the q-ROF Aczél-Alsina WA operator [21].

$$\mathrm{q-ROFAWA}_{T^{AA}}(\ell_1, \ell_2, \ldots, \ell_n)$$

$$= \left(\left(1 - e^{-\left(\sum_{j=1}^{n} w_j \left(-\log(1-\mu_{\ell_j}^q)\right)^\gamma\right)^{\frac{1}{\gamma}}}\right)^{\frac{1}{q}}, \left(e^{-\left(\sum_{j=1}^{n} w_j \left(-\log \nu_{\ell_j}^q\right)^\gamma\right)^{\frac{1}{\gamma}}}\right)^{\frac{1}{q}}\right) \tag{6}$$

(iii) (Dombi). For $\phi(t^q) = \left(\frac{1 - t^q}{t^q}\right)^\gamma$ and $\psi(t^q) = \left(\frac{t^q}{1 - t^q}\right)^\gamma$. The q-ROFAWA operator (4) is converted into the q-ROF Dombi WA operator [9].

$$\mathrm{q-ROFAWA}_{T^D}(\ell_1, \ell_2, \ldots, \ell_n)$$

$$= \left(\left(1 - \frac{1}{1 + \left(\sum_{j=1}^{n} w_j \left(\frac{u_{\ell_j}^q}{1-\mu_{\ell_j}^q}\right)^\gamma\right)^{\frac{1}{\gamma}}}\right)^{\frac{1}{q}}, \left(\frac{1}{1 + \left(\sum_{j=1}^{n} w_j \left(\frac{1-\nu_{\ell_j}^q}{\nu_{\ell_j}^q}\right)^\gamma\right)^{\frac{1}{\gamma}}}\right)^{\frac{1}{q}}\right) \tag{7}$$

(iv) (Frank). For $\phi(t^q) = \log\left(\frac{\gamma - 1}{\gamma^{t^q} - 1}\right)$ and $\psi(t^q) = \log\left(\frac{\gamma - 1}{\gamma^{1-t^q} - 1}\right)$. The q-ROFAWA operator (4) is reduced to the q-ROF Frank WA operator [19].

$$\mathrm{q-ROFAWA}_{T^F}(\ell_1, \ell_2, \ldots, \ell_n)$$

$$= \left(\left(1 - \log_{\gamma} \left(1 + \prod_{j=1}^{n} \left(\gamma^{1-\mu_{\ell_j}^q} - 1 \right)^{w_j} \right) \right)^{\frac{1}{q}}, \right.$$

$$\left. \left(\log_{\gamma} \left(1 + \prod_{j=1}^{n} \left(\gamma^{\nu_{\ell_j}^q} - 1 \right)^{w_j} \right) \right)^{\frac{1}{q}} \right) \tag{8}$$

(v) (Hamacher). For $\phi(t^q) = \log\left(\frac{\gamma + (1-\gamma)t^q}{t^q}\right)$ and $\psi(t^q) = \log\left(\frac{\gamma + (1-\gamma)(1-t^q)}{1-t^q}\right)$. The q-ROFAWA operator (4) is converted into the q-ROF Hamacher WA operator [6].

$$\mathrm{q-ROFAWA}_{TH}(\ell_1, \ell_2, \ldots, \ell_n)$$

$$= \left(\left(\frac{\prod_{j=1}^{n} \left(1 + (\gamma - 1)\mu_{\ell_j}^q\right)^{w_j} - \prod_{j=1}^{n} (1 - \mu_{\ell_j}^q)^{w_j}}{\prod_{j=1}^{n} \left(1 + (\gamma - 1)\mu_{\ell_j}^q\right)^{w_j} + (\gamma - 1) \prod_{j=1}^{n} (1 - \mu_{\ell_j}^q)^{w_j}} \right)^{\frac{1}{q}}, \right.$$

$$\left. \frac{\gamma^{\frac{1}{q}} \prod_{j=1}^{n} (\nu_{\ell_j})^{w_j}}{\left(\prod_{j=1}^{n} (1 + (\gamma - 1)(1 - \nu_{\ell_j}^q))^{w_j} + (\gamma - 1) \prod_{j=1}^{n} (\nu_{\ell_j}^q)^{w_j} \right)^{\frac{1}{q}}} \right) \tag{9}$$

3.2 Archimedean TN and TCN based OWA Operator [20]

Let $L = \{\ell_1, \ell_2, \ldots, \ell_n\}$ be a finite set of q-ROFNs and the associated weight vector to this set is $w = (w_1, w_2, \ldots, w_n)$. Then the OWA operator for these n-inputs under Archimedean TN and TCN and weighting vector w is $\mathrm{q-ROFAOWA} : \mathcal{Q}^n \to \mathcal{Q}$ and defined as

$$\mathrm{q-ROFAOWA}(\ell_1, \ell_2, \ldots, \ell_n) = w_1 \ell_{(1)} \oplus w_2 \ell_{(2)}, \ldots \oplus w_n \ell_{(n)}$$

$$= \left(\left(\psi^{-1} \left(\sum_{j=1}^{n} w_j \psi(\mu_{(j)}^q) \right) \right)^{\frac{1}{q}}, \left(\phi^{-1} \left(\sum_{j=1}^{n} w_j \phi(\nu_{(j)}^q) \right) \right)^{\frac{1}{q}} \right) \tag{10}$$

where $w_j \in [0, 1]$ such that $\sum_{j=1}^{n} w_j = 1$ and $(\cdot)$ represent a permutation on L such that $\ell_{(1)} \geq \ell_{(2)} \geq \cdots \geq \ell_{(n)}$.

Boundedness, idempotency, and monotonicity properties are also satisfied for the q-ROFAOWA operator.

4 q-Rung Orthopair Fuzzy MAGDM Approach

In first part of this section, we have discussed a q-ROFAOPA for calculating the weights of the decision attributes. Subsequently, in second part, the q-ROFAWA or the q-ROFAOWA operator and q-ROFAOPA-based group decision-making technique is discussed.

4.1 q-Rung Orthopair Fuzzy Archimedean Ordinal Priorities Approach

In this section, a brief discussion on the step-wise procedure of q-ROFAOPA is presented. This methodology is used to find the weights of the decision attributes and it has three steps, which are as follows.

Suppose there are n decision attributes and d decision-makers (DMs) in a MAGDM problem. Each DM assigns a relative significance to these n attributes in the form of a linguistic matrix $P^h = \left[l_{1j}^h\right]_{1\times n}$ where $j \in \mathbb{N}_n$, $h \in \mathbb{N}_d$ and l_{1j}^h represent the relative importance of jth attribute provided by the hth DM.

Step 1. Select an appropriate q-ROF linguistic scale and create an attribute by DM q-ROF matrix $R = \left[l_{jh}\right]_{n\times d}$.

Step 2. Utilize the q-ROFAWA operator to aggregate the DMs' assessment values for every attribute. Then utilize the score function defined in Eq. (3) for overall score value $(\mathcal{S}_j)$ of each attribute. Finally rank the attributes based on their final score values.

Step 3. The attribute weights should satisfy the condition $w_j^{(t)} \geq w_j^{(t+1)}$, where $w_j^{(t)}$ represents the weight of the tth rank attribute, which leads to the following condition:

$$\frac{\min\limits_{1\leq j\leq n} \mathcal{S}_j}{\mathcal{S}_j^{(t)}} \left(w_j^{(t)} - w_j^{(t+1)}\right) \geq 0 \tag{11}$$

Based on the above conditions, a linear programming model for computing attributes' weights can be defined as follows:

$$\begin{cases} \text{Max } Z \\ \text{s. t.} \\ \dfrac{\min\limits_{1\leq j\leq n} \mathcal{S}_j}{\mathcal{S}_j^{(t)}}\left(w_j^{(t)} - w_j^{(t+1)}\right) \geq Z; \ \dfrac{\min\limits_{1\leq j\leq n} \mathcal{S}_j}{\mathcal{S}_j^{(n)}} w_j^{(n)} \geq Z; \\ \sum\limits_{j=1}^{n} w_j = 1; \ w_j \geq 0 \ \forall \ j \end{cases} \tag{12}$$

4.2 q-Rung Orthopair Fuzzy Group Decision-Making Approach

Consider a q-ROFMAGDM problem having m alternatives ($\mathcal{A}_i$, where $i \in \mathbb{N}_m$), n attributes ($\mathcal{C}_j$, where $j \in \mathbb{N}_n$), and d DMs ($\mathcal{E}_h$, where $h \in \mathbb{N}_d$). Also, the weights of an attribute and DM are denoted by $w_j \in [0, 1]$ s.t. $\sum_{j=1}^{n} w_j = 1$ and $\omega_h \in [0, 1]$ s.t. $\sum_{h=1}^{d} \omega_h = 1$, respectively. The expert's assessment of an alternative $\mathcal{A}_i$ with respect to an attribute $\mathcal{C}_j$ is given as a ij^{th} entry of a q-ROF matrix $\mathcal{M}^h = \left[\ell_{ij}^h\right]_{m\times n}$, where $\ell_{ij}^h = \left(\mu_{ij}^h, \nu_{ij}^h\right)$ is a q-ROFN provided by the hth DM. Thus, we have d numbers of decision matrices $\mathcal{M}^1, \mathcal{M}^2, \ldots, \mathcal{M}^d$. Further to select a best alternative, the proposed decision procedure is used as follows.

Step 1 (Normalization). The decision-making process may involve both cost and benefit types of attributes. In such scenario, convert the each decision matrix $\mathcal{M}^h = \left[\ell_{ij}^h\right]_{m\times n}$ into normalized decision matrix $\tilde{\mathcal{M}}^h = \left[\tilde{\ell}_{ij}^h\right]_{m\times n}$ by utilizing the following procedure:

$$\tilde{\ell}_{ij}^h = \left(\tilde{\mu}_{ij}^h, \tilde{\nu}_{ij}^h\right) = \begin{cases} \ell_{ij}^h = (\mu_{ij}^h, \nu_{ij}^h), \ \text{for benefit type} \\ \left(\ell_{ij}^h\right)^c = (\nu_{ij}^h, \mu_{ij}^h), \ \text{for cost type} \end{cases} \tag{13}$$

Step 2 (Aggregation of decision matrices). Use the given weights $(\omega_1, \omega_2, \ldots, \omega_d)$ of DMs $(\mathcal{E}_1, \mathcal{E}_2, \ldots, \mathcal{E}_d)$, and the q-ROFAWA operator to aggregate the decision matrices $\tilde{\mathcal{M}}^1$ to $\tilde{\mathcal{M}}^d$ given by d DMs and form a single decision matrix $\tilde{\mathcal{M}} = \left[\tilde{\ell}_{ij}\right]_{m\times n}$ as follows:

$$\tilde{\ell}_{ij} = \text{q-ROFAWA}(\tilde{\ell}_{ij}^1, \tilde{\ell}_{ij}^2, \ldots, \tilde{\ell}_{ij}^d) \tag{14}$$

Step 3 (Attributes' weights evaluation). To obtain the weights of all decision attributes, utilize the discussed q-ROFOPA approach as described in Sect. 4.1. Thus, we have n computed weights $(w_1, w_2, \ldots, w_n)$ for all n attributes $(\mathcal{C}_1, \mathcal{C}_j, \ldots, \mathcal{C}_n)$.

Step 4 (Aggregation of assessment values). Now use the single decision matrix $\tilde{\mathcal{M}}$, calculated attribute weights $(w_1, w_2, \ldots, w_n)$, and the q-ROFAWA operator or the q-ROFAOWA operator for an overall assessment values $(\tilde{\ell}_1, \tilde{\ell}_2, \ldots, \tilde{\ell}_m)$ for all m alternative $(\mathcal{A}_1, \mathcal{A}_2, \ldots, \mathcal{A}_m)$ as follows:

$$\tilde{\ell}_i = \text{q-ROFAWA}(\tilde{\ell}_{i1}, \tilde{\ell}_{i2}, \ldots, \tilde{\ell}_{in}) \tag{15}$$

Step 5 (Score and ranking of alternatives). Finally obtain a score value $(\mathcal{B}_i)$ for each alternative $(\mathcal{A}_i)$ by using Eq. (3) and rank accordingly.

$$\mathcal{B}_i = Sc(\tilde{\ell}_i) \tag{16}$$

4.3 Real-World Implementation of the Proposed Approach

Mining, though often temporary, has lasting implications. Closure and rehabilitation are crucial aspects, with poor rehabilitation leaving lasting issues for society. Sustainable economic benefits can be achieved by diversifying capital investments. Non-renewable resource mining aligns with sustainability by reinvesting revenues in human, social, and physical capital. Depleted resources can contribute to well-being through strategic investments. Corporate social responsibility can involve diverting profits to community purposes, protecting jobs, and creating favorable working conditions. Supporting affected regions aids the transition to agriculture, aligning mining activities with sustainability. After closure, a mining company faces alternatives like rehabilitation $(\mathcal{A}_1)$, rehabilitation with business investment $(\mathcal{A}_2)$, and rehabilitation with social transition subsidy $(\mathcal{A}_3)$. DM interviews focused on mining and sustainability, with critical perspectives. Attributes for socially responsible rehabilitation include 4 main attributes and 12 sub-attributes taken from [7]. The complete decision hierarchy of attributes, alternative assessments with respect to attributes, and attributes' significance evaluated by four DMs $(\mathcal{E}_1, \mathcal{E}_2, \mathcal{E}_3, \mathcal{E}_4)$ is shown in Tables 3, 4, and 5, respectively.

Table 6 can be used for converting the linguistic information of the selected problem into q-ROFNs.

Step 1. The attributes $\mathcal{C}_3, \mathcal{C}_4, \mathcal{C}_{11}$, and $\mathcal{C}_{12}$ are cost type and rest are benefit type. So the normalized decision matrices $\tilde{\mathcal{M}}^1$ to $\tilde{\mathcal{M}}^4$ corresponding to four DMs are shown in Table 7.

Step 2. To aggregate four decision matrices (Table 7), distribute the weight of DMs $(\omega_1, \omega_2, \omega_3, \omega_4)$ equally, i.e., $\omega_h = 0.25 \; \forall \; h$. Then apply the q-ROFAWA operator as suggested in Step 2 of Sect. 4.2. The aggregated decision matrix $\tilde{\mathcal{M}} = \left[\tilde{\ell_{ij}}\right]_{3\times 12}$ under algebraic TN (T^A) is present in Table 8.

Similarly for parametric TNs such as Aczél-Alsina (T_γ^{AA}), Dombi (T_γ^D), Frank (T_γ^F), and Hamacher (T_γ^H), with $\gamma = 2$, the aggregated decision matrix $(\tilde{\mathcal{M}})$ can be generated for final assessment evaluation.

Table 3 List of attributes involved in the problem [7]

Main attributes	Sub-attributes	Type
($\mathcal{G}_1$) Economic aspect	($\mathcal{C}_1$) Income of the residents in the region	Benefit
	($\mathcal{C}_2$) Employment in the region	Benefit
	($\mathcal{C}_3$) Socially responsible activities	Cost
($\mathcal{G}_2$) Social aspect	($\mathcal{C}_4$) Migration to other cities	Cost
	($\mathcal{C}_5$) Social transition after the closure of a mine	Benefit
	($\mathcal{C}_6$) Social justice	Benefit
($\mathcal{G}_3$) Sustainability aspect	($\mathcal{C}_7$) The reputation of the mining company	Benefit
	($\mathcal{C}_8$) Social acceptance	Benefit
	($\mathcal{C}_9$) Providing sustainable land use	Benefit
($\mathcal{G}_4$) Environmental aspect	($\mathcal{C}_{10}$) Biodiversity	Benefit
	($\mathcal{C}_{11}$) GHG emissions	Cost
	($\mathcal{C}_{12}$) Contamination of soil	Cost

Table 4 Linguistic decision matrices provided by various DMs [7]

DM	Alternative	Attributes ($\mathcal{C}_j$)											
$\mathcal{E}_h$	$\mathcal{A}_i$	$\mathcal{C}_1$	$\mathcal{C}_2$	$\mathcal{C}_3$	$\mathcal{C}_4$	$\mathcal{C}_5$	$\mathcal{C}_6$	$\mathcal{C}_7$	$\mathcal{C}_8$	$\mathcal{C}_9$	$\mathcal{C}_{10}$	$\mathcal{C}_{11}$	$\mathcal{C}_{12}$
$\mathcal{E}_1$	$\mathcal{A}_1$	L	ML	EH	EH	L	L	H	M	ML	MH	VL	L
	$\mathcal{A}_2$	MH	H	MH	ML	M	M	EH	MH	H	MH	VL	L
	$\mathcal{A}_3$	ML	L	ML	H	EH	MH	EH	H	EH	EH	VL	VL
$\mathcal{E}_2$	$\mathcal{A}_1$	VL	EL	EH	EH	L	EL	EH	M	M	VL	EL	EH
	$\mathcal{A}_2$	H	EH	ML	EH	M	L	EH	M	EH	VL	EL	L
	$\mathcal{A}_3$	M	ML	EL	EH	MH	L	EH	MH	EH	VL	EL	VL
$\mathcal{E}_3$	$\mathcal{A}_1$	VL	EL	EH	EH	EL	ML	H	MH	EH	EH	EL	EL
	$\mathcal{A}_2$	EH	MH	VL	VL	EH	MH	EH	H	MH	M	ML	M
	$\mathcal{A}_3$	H	MH	ML	VL	MH	EH	EH	EH	EH	EH	EL	VL
$\mathcal{E}_4$	$\mathcal{A}_1$	VL	L	M	H	ML	M	H	M	EH	EH	L	VL
	$\mathcal{A}_2$	EH	H	L	VL	MH	H	EH	EH	M	MH	M	L
	$\mathcal{A}_3$	EH	EH	EL	EL	EH	EH	EH	EH	H	EH	VL	EL

Table 5 Linguistic matrices of attributes' significance provided by various DMs [7]

DMs	Attributes ($\mathcal{G}$ & $\mathcal{C}$)															
$\mathcal{E}_h$	$\mathcal{G}_1$	$\mathcal{G}_2$	$\mathcal{G}_3$	$\mathcal{G}_4$	$\mathcal{C}_1$	$\mathcal{C}_2$	$\mathcal{C}_3$	$\mathcal{C}_4$	$\mathcal{C}_5$	$\mathcal{C}_6$	$\mathcal{C}_7$	$\mathcal{C}_8$	$\mathcal{C}_9$	$\mathcal{C}_{10}$	$\mathcal{C}_{11}$	$\mathcal{C}_{12}$
$\mathcal{E}_1$	EH	MH	M	H	EL	H	MH	EL	M	VH	EH	H	H	VL	VH	H
$\mathcal{E}_2$	EH	ML	M	MH	M	VH	H	ML	MH	EH	EH	MH	VH	L	VH	MH
$\mathcal{E}_3$	VH	M	L	L	ML	MH	VL	VL	MH	ML	EH	L	M	VH	VH	H
$\mathcal{E}_4$	EH	ML	M	MH	MH	H	ML	ML	VH	ML	EH	VH	ML	M	H	VH

Step 3. Apply the q-ROFAOPA approach as suggested in Sect. 4.1.

Step 3.1. Utilize the q-ROF linguistic scaling present in Table 6. The formed attribute by DM q-ROF matrix is shown in Table 9.

Step 3.2. The q-ROFAWA operator is used to aggregate the attribute significance values. Also, the score function (3) for local scores of attributes. The resultant aggregated assessment under algebraic TN (T^A), local scores, and their rank are given in Table 9.

Step 3.3. Create a linear programming model to obtain the weights of the attributes. For that, use the global scores and rank of the attributes in Eqs. (11) and (12), as described. The constructed linear programming model (17) for the selected problem is given below:

$$\begin{cases} \text{Max } Z \\ \text{s. t.} \\ 0.4662\left(w_2^{(1)} - w_1^{(2)}\right) \geq Z;\ 0.5916\left(w_7^{(5)} - w_{10}^{(6)}\right) \geq Z; 0.8555\left(w_9^{(9)} - w_4^{(10)}\right) \geq Z; \\ 0.4789\left(w_1^{(2)} - w_3^{(3)}\right) \geq Z;\ 0.6325\left(w_{10}^{(6)} - w_{12}^{(7)}\right) \geq Z; 0.8750\left(w_4^{(10)} - w_6^{(11)}\right) \geq Z; \\ 0.5050\left(w_3^{(3)} - w_{11}^{(4)}\right) \geq Z;\ 0.6480\left(w_{12}^{(7)} - w_8^{(8)}\right) \geq Z; 0.8750\left(w_6^{(11)} - w_5^{(12)}\right) \geq Z; \\ 0.5858\left(w_{11}^{(4)} - w_7^{(5)}\right) \geq Z;\ 0.7656\left(w_8^{(8)} - w_9^{(9)}\right) \geq Z;\ 1.w_5^{(12)} \geq Z; \\ \sum_{j=1}^{12} w_j = 1;\ w_j \geq 0\ \forall\ j. \end{cases} \tag{17}$$

On solving this model (17), the maximize $Z = 0.0095$ and the obtained weights of the considered attributes under algebraic TN (T^A) are presented in first row of Table 10.

Table 10 also presents the attribute weights from q-ROFAOPA under various parametric TNs including Aczél-Alsina (T_γ^{AA}), Dombi (T_γ^D), Frank (T_γ^F), and Hamacher (T_γ^H), with parameter $\gamma = 2$.

Step 4. With the above aggregated matrix $\tilde{\mathcal{M}}_{3\times 12}$ (Table 8 with T^A) and attribute weights (Table 10 with T^A), use the q-ROFAWA operator for an overall assessment value ($\tilde{\ell}_1, \tilde{\ell}_2, \tilde{\ell}_3$) of three feasible alternative ($\mathcal{A}_1, \mathcal{A}_2, \mathcal{A}_3$). The final assessment values of alternatives under algebraic TN (T^A) are given in the first part of Table 11.

Table 6 q-ROF linguistic scale [15]

Fuzzy degree of q-ROFN (μ & ν)	Linguistic terms								
	Extremely	Very low (VL)	Low (L)	Medium low (ML)	Medium (M)	Medium high (MH)	High (H)	Very high (VH)	Extremely high (EH)
Membership (μ)	0.15	0.25	0.35	0.45	0.55	0.65	0.75	0.85	0.95
Non-membership (ν)	0.95	0.85	0.75	0.65	0.55	0.45	0.35	0.25	0.15

Table 7 Normalized decision matrices $\tilde{\mathcal{M}}^1$ to $\tilde{\mathcal{M}}^4$

DM	Alternative	Attributes ($\mathcal{C}_j$)											
$\mathcal{E}_h$	$\mathcal{A}_i$	$\mathcal{C}_1$	$\mathcal{C}_2$	$\mathcal{C}_3$	$\mathcal{C}_4$	$\mathcal{C}_5$	$\mathcal{C}_6$	$\mathcal{C}_7$	$\mathcal{C}_8$	$\mathcal{C}_9$	$\mathcal{C}_{10}$	$\mathcal{C}_{11}$	$\mathcal{C}_{12}$
$\mathcal{E}_1$	$\mathcal{A}_1$	(0.35, 0.75)	(0.45, 0.65)	(0.15, 0.95)	(0.15, 0.95)	(0.35, 0.75)	(0.35, 0.75)	(0.75, 0.35)	(0.55, 0.55)	(0.45, 0.65)	(0.65, 0.45)	(0.85, 0.25)	(0.75, 0.35)
	$\mathcal{A}_2$	(0.65, 0.45)	(0.75, 0.35)	(0.45, 0.65)	(0.65, 0.45)	(0.55, 0.55)	(0.55, 0.55)	(0.95, 0.15)	(0.65, 0.45)	(0.75, 0.35)	(0.65, 0.45)	(0.85, 0.25)	(0.75, 0.35)
	$\mathcal{A}_3$	(0.45, 0.65)	(0.35, 0.75)	(0.65, 0.45)	(0.35, 0.75)	(0.95, 0.15)	(0.65, 0.45)	(0.95, 0.15)	(0.75, 0.35)	(0.95, 0.15)	(0.95, 0.15)	(0.85, 0.25)	(0.85, 0.25)
$\mathcal{E}_2$	$\mathcal{A}_1$	(0.25, 0.85)	(0.15, 0.95)	(0.15, 0.95)	(0.15, 0.95)	(0.35, 0.75)	(0.15, 0.95)	(0.95, 0.15)	(0.55, 0.55)	(0.55, 0.55)	(0.25, 0.85)	(0.95, 0.15)	(0.15, 0.95)
	$\mathcal{A}_2$	(0.75, 0.35)	(0.95, 0.15)	(0.65, 0.45)	(0.15, 0.95)	(0.55, 0.55)	(0.35, 0.75)	(0.95, 0.15)	(0.55, 0.55)	(0.95, 0.15)	(0.25, 0.85)	(0.95, 0.15)	(0.75, 0.35)
	$\mathcal{A}_3$	(0.55, 0.55)	(0.45, 0.65)	(0.95, 0.15)	(0.15, 0.95)	(0.65, 0.45)	(0.35, 0.75)	(0.95, 0.15)	(0.65, 0.45)	(0.95, 0.15)	(0.25, 0.85)	(0.95, 0.15)	(0.85, 0.25)
$\mathcal{E}_3$	$\mathcal{A}_1$	(0.25, 0.85)	(0.15, 0.95)	(0.15, 0.95)	(0.15, 0.95)	(0.15, 0.95)	(0.45, 0.65)	(0.75, 0.35)	(0.65, 0.45)	(0.95, 0.15)	(0.95, 0.15)	(0.95, 0.15)	(0.95, 0.15)
	$\mathcal{A}_2$	(0.95, 0.15)	(0.65, 0.45)	(0.85, 0.25)	(0.85, 0.25)	(0.95, 0.15)	(0.65, 0.45)	(0.95, 0.15)	(0.75, 0.35)	(0.65, 0.45)	(0.55, 0.55)	(0.65, 0.45)	(0.55, 0.55)
	$\mathcal{A}_3$	(0.75, 0.35)	(0.65, 0.45)	(0.65, 0.45)	(0.85, 0.25)	(0.65, 0.45)	(0.95, 0.15)	(0.95, 0.15)	(0.95, 0.15)	(0.95, 0.15)	(0.95, 0.15)	(0.95, 0.15)	(0.85, 0.25)
$\mathcal{E}_4$	$\mathcal{A}_1$	(0.25, 0.85)	(0.35, 0.75)	(0.55, 0.55)	(0.35, 0.75)	(0.45, 0.65)	(0.55, 0.55)	(0.75, 0.35)	(0.55, 0.55)	(0.95, 0.15)	(0.95, 0.15)	(0.75, 0.35)	(0.85, 0.25)
	$\mathcal{A}_2$	(0.95, 0.15)	(0.75, 0.35)	(0.75, 0.35)	(0.85, 0.25)	(0.65, 0.45)	(0.75, 0.35)	(0.95, 0.15)	(0.95, 0.15)	(0.55, 0.55)	(0.65, 0.45)	(0.55, 0.55)	(0.75, 0.35)
	$\mathcal{A}_3$	(0.95, 0.15)	(0.95, 0.15)	(0.95, 0.15)	(0.95, 0.15)	(0.95, 0.15)	(0.95, 0.15)	(0.95, 0.15)	(0.95, 0.15)	(0.75, 0.35)	(0.95, 0.15)	(0.85, 0.25)	(0.95, 0.15)

Table 8 Aggregated decision matrix ($\tilde{\mathcal{M}}$) under T^A

Alternative	Attributes ($\mathcal{C}_j$)					
$\mathcal{A}_i$	$\mathcal{C}_1$	$\mathcal{C}_2$	$\mathcal{C}_3$	$\mathcal{C}_4$	$\mathcal{C}_5$	$\mathcal{C}_6$
$\mathcal{A}_1$	(0.2791, 0.8238)	(0.3094, 0.8144)	(0.3187, 0.8287)	(0.2203, 0.8955)	(0.3461, 0.7677)	(0.4110, 0.7104)
$\mathcal{A}_2$	(0.8824, 0.2440)	(0.8233, 0.3016)	(0.7167, 0.4000)	(0.7371, 0.4043)	(0.7702, 0.3780)	(0.6107, 0.5049)
$\mathcal{A}_3$	(0.7794, 0.3701)	(0.7446, 0.4259)	(0.8733, 0.2598)	(0.7808, 0.4043)	(0.8733, 0.2598)	(0.8582, 0.2952)
Alternative	$\mathcal{C}_7$	$\mathcal{C}_8$	$\mathcal{C}_9$	$\mathcal{C}_{10}$	$\mathcal{C}_{11}$	$\mathcal{C}_{12}$
$\mathcal{A}_1$	(0.8363, 0.2832)	(0.5785, 0.5231)	(0.8546, 0.2995)	(0.8557, 0.3046)	(0.9032, 0.2106)	(0.8198, 0.3342)
$\mathcal{A}_2$	(0.9500, 0.1500)	(0.7987, 0.3376)	(0.7987, 0.3376)	(0.5627, 0.5547)	(0.8227, 0.3104)	(0.7130, 0.3919)
$\mathcal{A}_3$	(0.9500, 0.1500)	(0.8824, 0.2440)	(0.9263, 0.1854)	(0.9101, 0.2314)	(0.9141, 0.1936)	(0.8868, 0.2200)

Table 9 Significance and rank of attributes

Attributes	Decision-makers ($\mathcal{E}_h$)				Aggregated assessment	Score		
($\mathcal{G}$ & $\mathcal{C}$)	$\mathcal{E}_1$	$\mathcal{E}_2$	$\mathcal{E}_3$	$\mathcal{E}_4$		Local	Global	Rank
$\mathcal{G}_1$	(0.95, 0.15)	(0.95, 0.15)	(0.85, 0.25)	(0.95, 0.15)	(0.9345, 0.1704)	0.9197		
$\mathcal{C}_1$	(0.95, 0.15)	(0.55, 0.55)	(0.45, 0.65)	(0.65, 0.45)	(0.7611, 0.3941)	0.7055	0.6488	2
$\mathcal{C}_2$	(0.75, 0.35)	(0.85, 0.25)	(0.65, 0.45)	(0.75, 0.35)	(0.7626, 0.3426)	0.7247	0.6665	1
$\mathcal{C}_3$	(0.65, 0.45)	(0.75, 0.35)	(0.85, 0.25)	(0.45, 0.65)	(0.7167, 0.4000)	0.6690	0.6153	3
$\mathcal{G}_2$	(0.65, 0.45)	(0.45, 0.65)	(0.55, 0.55)	(0.45, 0.65)	(0.5372, 0.5687)	0.4734		
$\mathcal{C}_4$	(0.95, 0.15)	(0.45, 0.65)	(0.85, 0.25)	(0.45, 0.65)	(0.7986, 0.3548)	0.7501	0.3551	10~11
$\mathcal{C}_5$	(0.55, 0.55)	(0.65, 0.45)	(0.65, 0.45)	(0.85, 0.25)	(0.7042, 0.4085)	0.6564	0.3107	12
$\mathcal{C}_6$	(0.85, 0.25)	(0.95, 0.15)	(0.45, 0.65)	(0.45, 0.65)	(0.7986, 0.3548)	0.7501	0.3551	11~10
$\mathcal{G}_3$	(0.55, 0.55)	(0.55, 0.55)	(0.75, 0.35)	(0.55, 0.55)	(0.6158, 0.4912)	0.5599		
$\mathcal{C}_7$	(0.95, 0.15)	(0.95, 0.15)	(0.95, 0.15)	(0.95, 0.15)	(0.9500, 0.1500)	0.9381	0.5252	5
$\mathcal{C}_8$	(0.75, 0.35)	(0.65, 0.45)	(0.75, 0.35)	(0.85, 0.25)	(0.7626, 0.3426)	0.7247	0.4058	8
$\mathcal{C}_9$	(0.75, 0.35)	(0.85, 0.25)	(0.55, 0.55)	(0.45, 0.65)	(0.7002, 0.4206)	0.6486	0.3632	9
$\mathcal{G}_4$	(0.75, 0.35)	(0.65, 0.45)	(0.75, 0.35)	(0.65, 0.45)	(0.7052, 0.3969)	0.6616		
$\mathcal{C}_{10}$	(0.85, 0.25)	(0.75, 0.35)	(0.85, 0.25)	(0.55, 0.55)	(0.7800, 0.3312)	0.7425	0.4912	6
$\mathcal{C}_{11}$	(0.85, 0.25)	(0.85, 0.25)	(0.85, 0.25)	(0.75, 0.35)	(0.8301, 0.2719)	0.8017	0.5304	4
$\mathcal{C}_{12}$	(0.75, 0.35)	(0.65, 0.45)	(0.75, 0.35)	(0.85, 0.25)	(0.7626, 0.3426)	0.7247	0.4795	7

Table 10 Final weights of attributes under various triangular norms

Triangular norm	Weights of attributes (w_j)											
	w_1	w_2	w_3	w_4	w_5	w_6	w_7	w_8	w_9	w_{10}	w_{11}	w_{12}
Algebraic (T^A)	0.1560	0.1765	0.1361	0.0314	0.0095	0.0204	0.1009	0.0550	0.0425	0.0848	0.1172	0.0697
Aczél-Alsina (T_2^{AA})	0.1752	0.1542	0.1347	0.0441	0.0096	0.0326	0.1161	0.0564	0.0211	0.0850	0.1005	0.0705
Dombi (T_2^D)	0.1741	0.1521	0.1334	0.0456	0.0098	0.0337	0.1151	0.0578	0.0217	0.0854	0.0999	0.0713
Frank (T_2^F)	0.1561	0.1767	0.1363	0.0312	0.0095	0.0204	0.1010	0.0548	0.0423	0.0848	0.1174	0.0696
Hamacher (T_2^H)	0.1562	0.1769	0.1365	0.0310	0.0096	0.0203	0.1010	0.0546	0.0421	0.0847	0.1176	0.0695

Table 11 Final results of alternatives with various triangular norms

Triangular norm	Alternative ($\mathcal{A}_i$)	Collective assessment ($\tilde{\ell}_i$)	Score value ($\mathcal{B}_i$)	Ranking of alternatives
Algebraic (T^A)	$\mathcal{A}_1$	(0.6870, 0.5072)	0.6007	$\mathcal{A}_3 > \mathcal{A}_2 > \mathcal{A}_1$
	$\mathcal{A}_2$	(0.8216, 0.3142)	0.7825	
	$\mathcal{A}_3$	(0.8714, 0.2677)	0.8396	
Aczél-Alsina (T_2^{AA})	$\mathcal{A}_1$	(0.7950, 0.3993)	0.7311	$\mathcal{A}_3 > \mathcal{A}_2 > \mathcal{A}_1$
	$\mathcal{A}_2$	(0.8671, 0.2779)	0.8330	
	$\mathcal{A}_3$	(0.9028, 0.2336)	0.8770	
Dombi (T_2^D)	$\mathcal{A}_1$	(0.8863, 0.2335)	0.8615	$\mathcal{A}_3 > \mathcal{A}_2 > \mathcal{A}_1$
	$\mathcal{A}_2$	(0.9139, 0.1999)	0.8945	
	$\mathcal{A}_3$	(0.9304, 0.1784)	0.9143	
Frank (T_2^F)	$\mathcal{A}_1$	(0.6691, 0.5245)	0.5795	$\mathcal{A}_3 > \mathcal{A}_2 > \mathcal{A}_1$
	$\mathcal{A}_2$	(0.8154, 0.3174)	0.7763	
	$\mathcal{A}_3$	(0.8657, 0.2716)	0.8334	
Hamacher (T_2^H)	$\mathcal{A}_1$	(0.6488, 0.5401)	0.5576	$\mathcal{A}_3 > \mathcal{A}_2 > \mathcal{A}_1$
	$\mathcal{A}_2$	(0.8099, 0.3194)	0.7710	
	$\mathcal{A}_3$	(0.8605, 0.2743)	0.8280	

Step 5. Finally use the score function (3) and get the final score value ($\mathcal{B}_i$) for each alternative. The score values and rank of alternatives with different triangular norms are shown in Table 11.

Similarly for various parametric TNs (T_γ^{AA}, T_γ^D, T_γ^F, and T_γ^H)-based aggregated decision matrix $\tilde{\mathcal{M}}$ and q-ROFAOPA, the corresponding TN-based final assessment values ($\tilde{\ell}_1$, $\tilde{\ell}_2$, $\tilde{\ell}_3$), score values ($\mathcal{B}_1$, $\mathcal{B}_2$, $\mathcal{B}_3$), and ranking order of alternatives are computed and shown in Table 11.

From Table 11, the results are quite clear and consistent. The proposed approach under various Archimedean TNs shows that alternative $\mathcal{A}_3$ (rehabilitation with social transition subsidy) is the best feasible option for a mining company to rehabilitate the people.

5 Conclusion

In conclusion, this chapter has leveraged the versatile framework of Archimedean TNs and TCNs for q-ROFNs. The q-ROFAWA and q-ROFOWA operators along with the q-ROFAOPA for attribute weight evaluation are introduced. These developments lead to a new MAGDM approach, providing a comprehensive solution for decision-making scenarios. The results obtained through the application of the proposed models under various Archimedean TNs demonstrate that the rehabilitation option involving social transition subsidy ($\mathcal{A}_3$) emerging as the most feasible alternative for a mining company. This chapter not only extends the applicability of existing AOs but also addresses the crucial aspect of attribute weight evaluation, thereby enhancing the robustness and effectiveness of decision support systems in real-world contexts.

References

1. Ai, Z., Xu, Z., Yager, R.R., Ye, J.: q-Rung orthopair fuzzy integrals in the frame of continuous Archimedean t-norms and t-conorms and their application. IEEE Trans. Fuzzy Syst. **29**(5), 996–1007 (2020)
2. Atanassov, K.T., Stoeva, S.: Intuitionistic fuzzy sets. Fuzzy Sets Syst. **20**(1), 87–96 (1986)
3. Beliakov, G., Bustince, H., Goswami, D., Mukherjee, U., Pal, N.R.: On averaging operators for Atanassov's intuitionistic fuzzy sets. Inform. Sci. **181**(6), 1116–1124 (2011)
4. Beliakov, G., Pradera, A., Calvo, T.: Aggregation Functions: A Guide for Practitioners, vol. 221. Springer (2007)
5. Boix-Cots, D., Pardo-Bosch, F., Alvarez, P.P.: A systematic review on multi-criteria group decision-making methods based on weights: analysis and classification scheme. Inform. Fusion (2023)
6. Darko, A.P., Liang, D.: Some q-rung orthopair fuzzy Hamacher aggregation operators and their application to multiple attribute group decision making with modified edas method. Eng. Appl. Artif. Intell. **87**, 103259 (2020)
7. Deveci, M., Gokasar, I., Brito-Parada, P.R.: A comprehensive model for socially responsible rehabilitation of mining sites using q-rung orthopair fuzzy sets and combinative distance-based assessment. Expert Syst. Appl. **200**, 117155 (2022)
8. Deveci, M., Pamucar, D., Gokasar, I., Köppen, M., Gupta, B.B.: Personal mobility in metaverse with autonomous vehicles using q-rung orthopair fuzzy sets based opa-rafsi model. IEEE Trans. Intell. Transp. Syst. (2022)
9. Jana, C., Muhiuddin, G., Pal, M.: Some Dombi aggregation of q-rung orthopair fuzzy numbers in multiple-attribute decision making. Int. J. Intell. Syst. **34**(12), 3220–3240 (2019)
10. Klement, E.P., Mesiar, R., Pap, E.: Triangular Norms, vol. 8. Springer Science & Business Media (2013)
11. Liu, P., Wang, P.: Multiple-attribute decision-making based on Archimedean Bonferroni operators of q-rung orthopair fuzzy numbers. IEEE Trans. Fuzzy Syst. **27**(5), 834–848 (2018)
12. Liu, P., Wang, P.: Some q-rung orthopair fuzzy aggregation operators and their applications to multiple-attribute decision making. Int. J. Intell. Syst. **33**(2), 259–280 (2018)
13. Pamucar, D., Deveci, M., Gokasar, I., Delen, D., Köppen, M., Pedrycz, W.: Evaluation of metaverse integration alternatives of sharing economy in transportation using fuzzy schweizer-sklar based ordinal priority approach. Decis. Support Syst. p. 113944 (2023)
14. Petry, F., Yager, R.R.: Information aggregation approaches for multi-criteria applications. In: Real Life Applications of Multiple Criteria Decision Making Techniques in Fuzzy Domain, pp. 195–206. Springer (2022)
15. Pinar, A., Boran, F.E.: A q-rung orthopair fuzzy multi-criteria group decision making method for supplier selection based on a novel distance measure. Int. J. Mach. Learn. Cybern. **11**, 1749–1780 (2020)
16. Rawat, S.S., Komal, Dincer, H., Yüksel, S.: A hybrid weighting method with a new score function for analyzing investment priorities in renewable energy. Comput. Indus. Eng. **185**, 109692 (2023)

17. Rawat, S.S., Komal, Liu, P., Stevic, Z., Senapati, T., Moslem, S.: A novel group decision-making approach based on partitioned hamy mean operators in q-rung orthopair fuzzy context. Complex Intell. Syst. 1–34 (2023)
18. Sahoo, L., Senapati, T., Yager, R.R.: Real Life Applications of Multiple Criteria Decision Making Techniques in Fuzzy Domain, vol. 420. Springer Nature (2022)
19. Seikh, M.R., Mandal, U.: Q-rung orthopair fuzzy frank aggregation operators and its application in multiple attribute decision-making with unknown attribute weights. Granular Comput. 1–22 (2022)
20. Seikh, M.R., Mandal, U.: q-rung orthopair fuzzy Archimedean aggregation operators: application in the site selection for software operating units. Symmetry **15**(9), 1680 (2023)
21. Senapati, T., Martínez, L., Chen, G.: Selection of appropriate global partner for companies using q-rung orthopair fuzzy Aczel-Alsina average aggregation operators. Int. J. Fuzzy Syst. **25**(3), 980–996 (2023)
22. Senapati, T., Yager, R.R.: Fermatean fuzzy sets. J. Ambient Intell. Human. Comput. **11**, 663–674 (2020)
23. Tan, C., Chen, X.: Generalized Archimedean intuitionistic fuzzy averaging aggregation operators and their application to multicriteria decision-making. Int. J. Inform. Technol. Decis. Making **15**(02), 311–352 (2016)
24. Xia, M., Xu, Z., Zhu, B.: Some issues on intuitionistic fuzzy aggregation operators based on Archimedean t-conorm and t-norm. Knowl.-Based Syst. **31**, 78–88 (2012)
25. Xu, Z.: Uncertain Multi-Attribute Decision Making: Methods and Applications. Springer (2015)
26. Yager, R.R.: Pythagorean fuzzy subsets. In: 2013 Joint IFSA World Congress and NAFIPS Annual Meeting (IFSA/NAFIPS), pp. 57–61. IEEE (2013)
27. Yager, R.R.: Generalized orthopair fuzzy sets. IEEE Trans. Fuzzy Syst. **25**(5), 1222–1230 (2016)
28. Zadeh, L.A.: Fuzzy sets. Inform. Control **8**(3), 338–353 (1965)

Mean Squared Error Utility for Fuzzy Preference Relations

Diego García-Zamora and Luis Martínez

Abstract The use of Fuzzy Preference Relations (FPRs) is prevalent in expert-driven decision-making, and thus obtaining the overall performance of each alternative from an FPR is a key topic in the area. This chapter proposes the notion of Mean Squared Error (MSE) utility vector associated with an FPR, which is defined as the utility vector that determines the pairwise comparison matrix that is closest to the original FPR. We demonstrate that this utility vector can be analytically computed and is a valid utility function since it assigns higher utility values to higher rated alternatives. Furthermore, we show that the MSE utilities can be computed although the comparison matrix is elicited using a multiplicative scale, or even when just one row and one column of the matrix are given. The applicability of the MSE utility vectors is then shown in a Large-Scale Group Decision-Making problem, in which we use them to accelerate the computational time necessary to reach a consensus. Additionally, we provide a comparative analysis with other classic methods for deriving priorities from pairwise comparison matrices.

Keywords Converting matrix to vector representation · Utility values for pairwise comparison matrix · Analytical solution for Analytic Hierarchy Process (AHP) · Analytical solution for Best-Worst Method (BWM) · Computational efficiency for large-scale group decision-making (LSGDM)

D. García-Zamora (✉)
Department of Mathematics, Universidad De Jaén, Campus Las Lagunillas, Jaén 23071, Spain
e-mail: dgzamora@ujaen.es

L. Martínez
Department of Computer Sciences, Universidad De Jaén, Campus Las Lagunillas, Jaén 23071, Spain
e-mail: martin@ujaen.es

L. Sahoo et al. (eds.), *Decision Making Under Uncertainty Via Optimization, Modelling, and Analysis*, Studies in Systems, Decision and Control 558,
https://doi.org/10.1007/978-981-96-0085-4_4

1 Introduction

The use of pairwise comparisons to model experts' opinions in decision-making problems is widely extended in the specialized literature [1] because human beings tend to perform better when eliciting preferences by pairwise comparisons among a set of alternatives, rather than comparing each alternative individually against all the others [2, 3]. Essentially, in the decision-making literature, the most widely used type of pairwise comparison matrices are Fuzzy Preference Relations (FPRs), which are given in the [0, 1] interval [4], and Multiplicative Preference relations (MPRs), often provided in a 1–9 Saaty's scale [5].

Pairwise comparisons may provide more precise modeling according to the human psyche than a single preference of one alternative versus all the others [2, 3]. However, some inconsistencies could emerge when they are elicited [6]. These inconsistencies are translated into contradictory evaluations that will negatively impact the decision process if they are not properly managed [6, 7]. When facing inconsistent preference relations, there are two main strategies. The first one is asking the expert to modify their preferences until they are consistent enough [7]. The second one consists of using an automatic mechanism that modifies the matrix until it is consistent [8].

Another aspect to be taken into account when using pairwise comparisons is the fact that ranking the alternatives from pairwise comparisons requires carrying out an additional step. In this regard, some of the most popular approaches are the computation of the dominance of each alternative [9], the eigenvalue algorithm [6], or the Best-Worst Method (BWM) [7]. The first of them provides a simple procedure to compute an overall performance for each alternative but neglects the inconsistencies of the pairwise comparison matrix. On the contrary, the other two approaches consider the consistency of the matrix, but they are computationally more complex: the eigenvalue method requires computing the eigenvalues and eigenvectors of the matrix, whereas BWM needs to solve an optimization problem.

Therefore, several researchers have attempted to solve both problems together by defining methods to derive priorities from pairwise comparison matrices that consider their consistency [10]. In this regard, a common approach is approximating the pairwise comparison matrix by the closest consistent pairwise comparison matrix using an optimization method. Even though such a methodology has led to different results, several aspects remain neglected.

The first one is the relationship between all of these proposals. Different researchers have used such an optimization approach to obtain priority values or weights from pairwise comparison matrices [11, 12]. However, the interaction of such priorities or weights with classical non-optimization-based methods such as AHP has not been studied yet [13].

Second, some methods assume that priority and weights are similar concepts and, consequently, there is no distinction in the methodology to compute them [5].

Third, it should be noticed that obtaining priorities or weights from a pairwise comparison matrix allows for the reduction of the dimensionality from n^2 to n. This fact is especially relevant when managing large volumes of data. Nevertheless, to guarantee that such a dimension reduction is robust, it is necessary to show that after performing computations with the priorities/weights, the resulting priorities/weights are still related to the pairwise comparison matrix that would be obtained if the same computations were performed on the original pairwise comparisons instead.

Therefore, this chapter proposes an alternative method based on the Mean Squared Error (MSE) to obtain an overall performance of each alternative from an FPR by considering its consistency. To do so, we analytically solve an optimization problem based on minimizing the MSE to derive a utility vector that *summarizes* the information within the FPR. Such a utility vector satisfies two main properties. On the one hand, it allows univocally constructing the consistent FPR which is closest to the original preference. On the other hand, such a vector, although it is defined as the solution of an optimization problem, can be computed through an algebraic expression that increasingly depends on the mean dominance of the alternatives. Furthermore, remapping the pairwise comparison matrix into such a utility vector implies a dimensionality reduction, and thus more efficiency in computations with them. Additionally, these utility values can also be obtained when the pairwise comparison matrix is given in a multiplicative scale [6] or BWM preferences [7]. In a parallel way, we also define the MSE weighting vector for FPRs, MPRs, and BWM preferences, which instead of utility values provide weights resembling the importance of the alternatives.

To illustrate the potential of the proposed method, the MSE utility vector is then applied to address to improve the computational time necessary to carry out a Consensus-Reaching Processes (CRPs) [14–16] for Large-Scale Group Decision-Making (LSGDM) [17, 18]. Furthermore, the performance of the MSE weighting vector is later compared with two classic methods to derive weights, namely the eigenvalue method usually applied in Analytic Hierarchy Process (AHP) [6], and classic Best-Worst Method (BWM) [7].

The remainder of this chapter is as follows. Section 2 introduces some basic notions about decision-making and pairwise comparison matrices. Section 3 analyzes first how additive and multiplicative consistency are related through a non-linear rescaling of MPRs and FPRs. Furthermore, it provides a characterization of both additive and multiplicative consistency in terms of utility vectors. Later, we introduce a method to derive such utilities in the case in which the original preference relation is not fully consistent. In Sect. 4 the MSE utility vector is applied to LSGDM. Section 5 compares the MSE weighting vectors with other classic weight determination methods. Finally, Sect. 6 concludes the chapter.

2 Preliminaries: Pairwise Comparisons in Decision-Making

This section introduces the background related to decision-making with pairwise comparison matrices. In a decision-making problem, various alternatives are analyzed to select which one is the most suitable [19]. In this regard, the increasing complexity of decisions nowadays demands the consideration not only of data-driven methods but also of expert knowledge [17].

However, gathering the experts' opinions about the alternatives is not a simple task because when humans are asked to provide their preferences, they do not always reflect what they truly think [2, 3]. Therefore, in the field of decision-making, determining the most appropriate preference structure is a crucial and ongoing area of research. As such, understanding and selecting the most suitable preference structure continues to be a key topic in the decision-making literature [17].

To improve the quality of the decisions, the notion of a pairwise comparison matrix was introduced to better reflect experts' thoughts [4]. In contrast to rating the alternatives separately, when using pairwise comparisons, each expert provides how much he or she prefers an alternative to all the others by comparing each one individually. Once the expert has fulfilled the pairwise comparisons, an auxiliary method is applied to determine the priorities of each alternative. Among other popular methods with this purpose, we can highlight the well-known eigenvalue method [6], the BWM [7], or the dominance computation.

The most widely used types of pairwise comparison matrices are the FPRs and MPRs. Below, their corresponding definitions are introduced.

Definition 1 *(Multiplicative preference relation (MPR)* [6]*)* Let $n \in \mathbb{N}$. An MPR $Y = (y_{ij}) \in \mathcal{M}_{n\times n}(\mathbb{R})$ is a matrix satisfying $y_{ij} \cdot y_{ji} = 1 \ \forall 1 \le i, j \le n$. If $y_{ij} \in [\frac{1}{m}, m] \ \forall 1 \le i, j \le n$, where $m \in \mathbb{N} \setminus \{1\}$, we say that the MPR has been elicited in the scale m. Usually, $m = 9$, but it could be any other natural number greater than one. The set of all the MPRs of dimension n is denoted as

$$\mathbb{M}_n = \{Y \in \mathcal{M}_{n\times n}(\mathbb{R}) \,:\, y_{ij} \cdot y_{ji} = 1, y_{ij} > 0 \,\forall 1 \le i, j \le n\}$$

whereas the MPRs elicited in the scale m will be denoted as

$$\mathbb{M}_n^m = \{Y \in \mathcal{M}_{n\times n}(\mathbb{R}) \,:\, y_{ij} \cdot y_{ji} = 1, y_{ij} \in \left[\frac{1}{m}, m\right] \forall 1 \le i, j \le n\}.$$

Definition 2 *(Additive preference relation (APR)* [4]*)* Let $n \in \mathbb{N}$. An APR $X = (x_{ij}) \in \mathcal{M}_{n\times n}(\mathbb{R})$ is a matrix satisfying $x_{ij} + x_{ji} = 1 \ \forall 1 \le i, j \le n$. If $x_{ij} \in [\frac{1}{2} - m, \frac{1}{2} + m] \ \forall 1 \le i, j \le n$, where $m > 0$, we say that the APR has been elicited in the scale m. If $m = \frac{1}{2}$ the APR is known as an FPR. The set of all the APRs of dimension n is denoted as

$$\mathbb{A}_n = \{X \in \mathcal{M}_{n\times n}(\mathbb{R}) \,:\, x_{ij} + x_{ji} = 1 \,\forall 1 \le i, j \le n\},$$

whereas the APRs elicited in the scale m will be denoted as

$$\mathbb{A}_n^m = \{X \in \mathcal{M}_{n\times n}(\mathbb{R}) \,:\, x_{ij} + x_{ji} = 1, x_{ij} \in \left[\frac{1}{2} - m, \frac{1}{2} + m\right] \forall 1 \le i, j \le n\}.$$

Here, we have introduced the auxiliary notion of APR as a type of pairwise comparison that is slightly more general than an FPR. It should be highlighted that this distinction is only for the sake of simplicity in the notations. These APRs are not used in practice, but such a definition is very convenient to facilitate the reading of this proposal.

When considering pairwise comparisons, it is necessary to take their consistency into account [7, 10]. Intuitively, a pairwise comparison matrix is said to be consistent if it does not contain contradictory evaluations [6].

Definition 3 *(Consistent MPR* [6]*)* An MPR $Y \in \mathbb{M}_n$ is said to be consistent if it satisfies

$$y_{ij} = y_{ik} \cdot y_{kj} \, \forall \, 1 \leq i, j, k \leq n.$$

Definition 4 *(Consistent APR* [9]*)* An APR $X \in \mathbb{A}_n$ is said to be consistent if it satisfies

$$x_{ij} + \frac{1}{2} = x_{ik} + x_{kj} \, \forall \, 1 \leq i, j, k \leq n.$$

3 Mean Squared Error Utility

This section is devoted to introducing the idea of MSE Utility vector as the utility vector that allows reconstructing the consistent APR which is closest to a given FPR. First, we develop some useful results related to the consistency of pairwise comparison matrices. Subsequently, we derive the analytical expression of the MSE Utility vector. Finally, we show how to extend the idea of MSE Utility vector to manage MPRs and BWM preferences, which are a special type of incomplete MPRs.

3.1 On the Consistency of Preference Relations

In the previous section, we have provided definitions for MPRs and APRs which are slightly more general than the classic ones [4, 6]. However, the following result shows that APRs and MPRs are univocally related when their corresponding scales are fixed.

Proposition 1 *Let $n \in \mathbb{N}$ and consider $m_1 \in \mathbb{N} \setminus \{1\}$, $m_2 > 0$. Then the mapping $\Phi : \mathbb{A}_n^{m_2} \to \mathbb{M}_n^{m_1}$ defined by $\Phi(X) = (\phi(x_{ij})) \, \forall \, X \in \mathbb{A}_n^m$, where*

$$\phi : \left[\frac{1}{2} - m_2, \frac{1}{2} + m_2\right] \to \left[\frac{1}{m_1}, m_1\right]$$

is defined as

$$\phi(x) = m_1^{\frac{2x-1}{2m_2}} \, \forall \, x \in \left[\frac{1}{2} - m_2, \frac{1}{2} + m_2\right]$$

is a bijection whose inverse is $\Phi^{-1} : \mathbb{M}_n^{m_1} \to \mathbb{A}_n^{m_1}$ *defined by* $\Phi^{-1}(Y) = (\phi^{-1}(y_{ij}))\,\forall\, Y \in \mathbb{M}_n^{m}$, *where*

$$\phi^{-1} : \left[\frac{1}{m_1}, m_1\right] \to \left[\frac{1}{2} - m_2, \frac{1}{2} + m_2\right]$$

is defined as

$$\phi^{-1}(y) = \frac{1}{2} + m_2 \log_{m_1}(y)\,\forall\, y \in \left[\frac{1}{m_1}, m_1\right].$$

In addition, the mappings ϕ and ϕ^{-1} preserve the consistency.

Proposition 2 *Let* $n \in \mathbb{N}$ *and consider* $m_1 \in \mathbb{N} \setminus \{1\}$, $m_2 > 0$. *An APR* $X \in \mathbb{A}_n^{m_2}$ *is consistent if and only if the MPR* $\Phi(X) \in \mathbb{M}_n^{m_1}$ *is consistent.*

Proof By definition, X is (additively) consistent if, for any $1 \leq i, j, k \leq n$, the following holds:

$$\begin{aligned} x_{ik} + x_{kj} &= x_{ij} + \frac{1}{2} \\ \iff 2x_{ik} + 2x_{kj} - 2 &= 2x_{ij} - 1 \\ \iff m_1^{\frac{2x_{ik}-1}{2m_2}}\, m_1^{\frac{2x_{kj}-1}{2m_2}} &= m_1^{\frac{2x_{ij}-1}{2m_2}}, \end{aligned}$$

which is the (multiplicative) consistency for $\Phi(X)$. □

Proposition 2 guarantees that analyzing the consistency of an FPR is equivalent to studying the consistency of the MPR defined through the bijection Φ and vice versa. Therefore, in the rest of the chapter, we focus on the methods and proofs in the FPR case, i.e., $\mathbb{A}_n^{\frac{1}{2}}$, in order to obtain linearized results. However, all of them could be applied to the MPR scenario by considering the corresponding transformation through Φ^{-1} [12].

The following result shows that consistent FPRs store the same information as any of their rows, i.e., each consistent FPR is univocally determined by any of its rows.

Proposition 3 *Let* $n \in \mathbb{N}$ *and consider a consistent FPR* $X \in \mathbb{A}_n^{\frac{1}{2}}$. *Then, for any* $1 \leq k \leq n$, *the FPR* X *can be reconstructed from the utility vector* $u^k = (1 - x_{k1}, 1 - x_{k2}, ..., 1 - x_{kn})$ *by the equality*

$$x_{ij} := \frac{1}{2} + u_i^k - u_j^k\,\forall\, 1 \leq i, j \leq n.$$

Proof The consistency of X guarantees that

$$\begin{aligned}
& x_{ij} = x_{ik} + x_{kj} - \frac{1}{2} \forall 1 \leq i, j \leq n \\
\iff & x_{ij} = 1 - x_{ki} + x_{kj} - \frac{1}{2} \forall 1 \leq i, j \leq n \\
\iff & x_{ij} = \frac{1}{2} - x_{ki} + x_{kj} \forall 1 \leq i, j \leq n \\
\iff & x_{ij} = \frac{1}{2} + (1 - x_{ki}) - (1 - x_{kj}) \forall 1 \leq i, j \leq n.
\end{aligned}$$

□

Note that, if $x_{ki} > x_{kj}$, the j-th alternative is preferred to the i-th alternative, which motivates the definition of the utility vector u^k. As a consequence, we obtain a characterization for the consistency of FPRs.

Corollary 1 *Let $n \in \mathbb{N}$. An FPR $X \in \mathbb{A}_n^{\frac{1}{2}}$ is consistent if and only if there exists a vector $u \in \mathbb{R}^n$ such that*

$$x_{ij} := \frac{1}{2} + u_i - u_j \forall 1 \leq i, j \leq n.$$

In the same way, an MPR given in scale $m > 1$, $Y \in \mathbb{M}_n^m$ is consistent if and only if there exists a vector $u \in (\mathbb{R}^+)^n$ such that

$$Y_{ij} := \frac{u_i}{u_j} \forall 1 \leq i, j \leq n.$$

In other words, any consistent FPR can be simplified as a utility vector u that contains the same information as the original FPR and satisfies $u_i - u_j \in [-\frac{1}{2}, \frac{1}{2}] \forall 1 \leq i, j \leq n$. Note that the relationship described in Proposition 3 is not a bijection between the utility vectors belonging to $\mathbb{R}^n$ and the set of FPRs. The reason behind that is that given a utility vector $u \in \mathbb{R}^n$, it is not possible to guarantee that $x_{ij} := \frac{1}{2} + u_i - u_j \in [0, 1]$ for all i, j. However, the matrix thus defined is a consistent APR with very convenient properties.

Theorem 1 *Let us consider $n \in \mathbb{N}$ and the mapping $\mu : \mathbb{R}^n \to \mathcal{M}_{n \times n}(\mathbb{R})$ defined as*

$$(\mu(u))_{ij} = \frac{1}{2} + u_i - u_j \forall u \in \mathbb{R}^n.$$

Then, for any $u \in \mathbb{R}^n$, the matrix $\mu(u)$ is a consistent APR. In addition, the mapping μ satisfies the following properties:

- $\mu(u+v) = \mu(u) + \mu(v) - \frac{1}{2}1_{n\times n}$, $\forall\, u, v \in \mathbb{R}^n$,
- $\mu(u + \alpha 1_{1\times n}) = \mu(u)$, $\forall\, u \in \mathbb{R}^n, \alpha \in \mathbb{R}$,
- $\mu(\alpha u) = \alpha\mu(u) + (1-\alpha)\frac{1}{2}_{n\times n}$, $\forall\, u \in \mathbb{R}^n, \alpha \in \mathbb{R}$,
- $\mu(\alpha u + \beta v) = \alpha\mu(u) + \beta\mu(v) + (1-\alpha-\beta)\frac{1}{2}_{n\times n}\, \forall\, u, v \in \mathbb{R}^n, \alpha, \beta \in \mathbb{R}$,
- $\mu(\alpha u + \beta v) = \alpha\mu(u) + \beta\mu(v)\, \forall\, u, v \in \mathbb{R}^n, \alpha, \beta \in \mathbb{R}, \alpha + \beta = 1$.

Proof The additive reciprocity is a consequence of

$$\mu(u)_{ij} + \mu(u)_{ji} = (\frac{1}{2} + u_i - u_j) + (\frac{1}{2} + u_j - u_i) = 1\, \forall\, 1 \leq i, j \leq n$$

and the consistency is given by

$$\begin{aligned}\mu(u)_{ik} + \mu(u)_{kj} - \frac{1}{2} &= (\frac{1}{2} + u_i - u_k) + (\frac{1}{2} + u_k - u_j) - \frac{1}{2} \\ &= \frac{1}{2} + u_i - u_j = \mu(u)_{ij}\, \forall\, 1 \leq i, j, k \leq n.\end{aligned}$$

The rest of the properties are also straightforward from the definition. □

To summarize, given a consistent FPR $X \in \mathbb{A}_n^{\frac{1}{2}}$, it is possible to obtain a utility vector $u \in \mathbb{R}^n$ such that $\mu(u) = X$. Furthermore, if $w_1, \ldots, w_K$ satisfy $w_k \geq 0, \sum_{k=1}^K w_k = 1$ and $X_1, \ldots, X_k$ are consistent FPRs, then it is possible to find $u^1, \ldots, u^K$ verifying $\mu(\sum_{k=1}^K w_k u^k) = \sum_{k=1}^K w_k \mu(u^k) = \sum_{k=1}^K w_k X_k$.

3.2 Mean Squared Error Utility for FPRs

We have shown that given a consistent FPR $X \in \mathbb{A}_n^{\frac{1}{2}}$, it is possible to *summarize* its information in a utility vector $u \in \mathbb{R}^n$ such that $\mu(u) = X$. However, obtaining a consistent FPR is nearly impossible when human DMs are required to provide their opinions. In this section, we develop a method to be able to *summarize* non-consistent FPRs as utility vectors.

To do so, for a given FPR $X \in \mathbb{A}_n^{\frac{1}{2}}$, which is not necessarily consistent, our aim is to find a utility vector $u \in \mathbb{R}^n$ whose associated APR through $\mu(u)$ is close to the original matrix X. In other words, we aim to find the utility vector that makes the differences

$$\left| x_{ij} - \frac{1}{2} - u_i + u_j \right|, 1 \leq i, j \leq n$$

as small as possible. To derive an analytical solution, which is essential to reduce computational costs, we consider an optimization problem consisting of minimizing the corresponding MSE. In addition, since the property $\mu(u + \alpha 1_{1\times n}) = \mu(u)$, $\forall\, u \in \mathbb{R}^n, \alpha \in \mathbb{R}$ in Theorem 1 suggests that such a utility vector is not unique, we include the condition $\sum_{k=1}^n u_k = \sigma$ for some $\sigma \in \mathbb{R}$.

Theorem 2 *Let $X \in \mathbb{A}_n^{\frac{1}{2}}$ be an FPR and consider $\sigma \in \mathbb{R}$. Then, the utility vector $u \in \mathbb{R}^n$ defined as*

$$u_k = \frac{1}{2} + \frac{1}{n}(\sigma - \sum_{i=1}^{n} x_{ik})$$

is the solution to the optimization problem

$$\begin{array}{l} \min\limits_{u \in \mathbb{R}^n} \dfrac{1}{n^2} \xi_X(u) \\ s.t. \left\{ \sum_{k=1}^{n} u_k = \sigma \right. \end{array}$$

where the function $\xi_X : \mathbb{R}^n \to \mathbb{R}$ is defined as

$$\xi_X(u) = \sum_{i=1}^{n} \sum_{j=1}^{n} (x_{ij} - \frac{1}{2} - u_i + u_j)^2 \, \forall u \in \mathbb{R}^n.$$

Proof Let us consider the function $f_{ij} : \mathbb{R}^n \to \mathbb{R}$ defined as $f_{ij}(u) = (x_{ij} - \frac{1}{2} - u_i + u_j)^2 \, \forall u \in \mathbb{R}^n$. Note that, for $1 \le k \le n$, the partial derivative $\partial_k f_{ij} := \frac{\partial f_{ij}}{\partial u_k}$ is given as

$$\partial_k f_{ij}(u) = \begin{cases} 0 & \text{if } i < k, j < k \\ 2(x_{ij} - \frac{1}{2} - u_i + u_j) & \text{if } i < k, j = k \\ 0 & \text{if } i < k, j > k \\ -2(x_{ij} - \frac{1}{2} - u_i + u_j) & \text{if } i = k, j < k \\ 0 & \text{if } i = k, j = k \\ -2(x_{ij} - \frac{1}{2} - u_i + u_j) & \text{if } i = k, j > k \\ 0 & \text{if } i > k, j < k \\ 2(x_{ij} - \frac{1}{2} - u_i + u_j) & \text{if } i > k, j = k \\ 0 & \text{if } i > k, j > k \end{cases}$$

$\forall u \in \mathbb{R}^n$. Consequently, $\partial_k \xi_X := \frac{\partial \xi_X}{\partial u_k}$ can be obtained as follows:

$$\begin{aligned} \partial_k \xi_X(u) &= 2 \sum_{\substack{i=1 \\ i \neq k}}^{n} \left(x_{ik} - \frac{1}{2} - u_i + u_k \right) - 2 \sum_{\substack{j=1 \\ j \neq k}}^{n} \left(x_{kj} - \frac{1}{2} - u_k + u_j \right) \\ &= 2 \sum_{\substack{i=1 \\ i \neq l}}^{n} \left(x_{ik} - \frac{1}{2} - u_i + u_k - x_{ki} + \frac{1}{2} + u_k - u_i \right) \\ &= 2 \sum_{\substack{i=1 \\ i \neq k}}^{n} (2x_{ik} - 1 - 2u_i + 2u_k) = 4 \sum_{i=1}^{n} \left(x_{ik} - \frac{1}{2} - u_i + u_k \right) \end{aligned}$$

$\forall u \in \mathbb{R}^n$. Now, let us consider the Lagrangian function $L : \mathbb{R}^{n+1} \to \mathbb{R}$ defined as

$$L(u, \lambda) = \xi_X(u) + 4\lambda \left(\sum_{i=1}^{n} u_i - \sigma \right) \forall (u, \lambda) \in \mathbb{R}^{n+1}.$$

After computing the corresponding partial derivatives, the resulting equations are as follows:

$$\begin{cases} \sum_{i=1}^{n} x_{ik} - \frac{n}{2} - \sum_{i=1}^{n} u_i + nu_k + \lambda = 0 \,\forall k = 1, 2, \ldots, n \\ \sum_{i=1}^{n} u_i = \sigma. \end{cases}$$

To solve it, note that if the first n equations are added, the following holds:

$$\sum_{k=1}^{n} \left(\sum_{i=1}^{n} x_{ik} - \frac{n}{2} - \sum_{i=1}^{n} u_i + nu_k + \lambda \right) = 0$$
$$\iff \sum_{k=1}^{n} \sum_{i=1}^{n} x_{ik} - \frac{n^2}{2} - n\sigma + n\sigma + n\lambda = 0 \iff \lambda = 0$$

in which the equations $\sum_{i=1}^{n} u_i = \sigma$ and $\sum_{i=1}^{n} \sum_{j=1}^{n} x_{ij} = \frac{n^2}{2}$ have been used. Finally, we can clear the value of each u_k:

$$\sum_{i=1}^{n} x_{ik} - \frac{n}{2} - \sum_{i=1}^{n} u_i + nu_k + \lambda = 0$$
$$\iff \sum_{i=1}^{n} x_{ik} - \frac{n}{2} - \sigma + nu_k = 0$$
$$\iff u_k = \frac{1}{2} - \frac{1}{n} \sum_{i=1}^{n} x_{ik} + \frac{\sigma}{n}.$$

Since ξ_X is a convex continuous non-negative function in $\mathbb{R}^n$, then there exists a global minimum for it. In addition, due to the restriction $\sum_{i=1}^{n} u_i = \sigma$ stands for a hyperplane in $\mathbb{R}^n$, a global minimum of the set $\{\xi_X(u) \,:\, u \in \mathbb{R}^n, \sum_{i=1}^{n} u_i = \sigma\}$ also exists. Consequently, since u is the only solution to the Lagrangian system, it must be the global minimum for the aforementioned optimization problem. □

At this point, we define the MSE utility vector as the vector $u \in \mathbb{R}^n$ obtained in the previous theorem.

Definition 5 *[MSE Utility for FPRs]* Let $X \in \mathbb{A}_n^{\frac{1}{2}}$ be an FPR and consider $\sigma \in \mathbb{R}$. Then, the MSE utility vector for X and σ is the vector $u \in \mathbb{R}^n$ defined as

$$u_k = \frac{1}{2} + \frac{1}{n} \left(\sigma - \sum_{i=1}^{n} x_{ik} \right) \forall k = 1, 2, \ldots, n,$$

and the associated MSE weighting vector $W \in \mathbb{R}^n$ is defined as

$$W_k = \frac{\phi(u_k)}{\sum_{l=1}^n \phi(u_l)} = \frac{m^{\frac{2}{n}\sum_{i=1}^n x_{ki}}}{\sum_{l=1}^n \sum_{l=1}^n m^{\frac{2}{n}\sum_{i=1}^n x_{li}}} \forall k = 1, 2, \ldots, n,$$

where $m \in \mathbb{N} \setminus \{1\}$ stands for the scale.

The MSE utility vector $u \in \mathbb{R}^n$ has the following advantages:

- The APR $\mu(u)$ is the closest consistent APR to the FPR X such that $\sum_{i=1}^n u_i = \sigma$.
- Since $u_k = \frac{1}{n}\sum_{i=1}^n x_{ki} + \frac{\sigma}{n} - \frac{1}{2} \forall k = 1, 2, \ldots, n$, they can be easily computed from the dominance of each alternative, without going through an optimization problem.
- In addition, the higher the dominance of the $k-$th alternative, the higher the value u_k. This motivates the use of the word *utility* to name them [20].
- If positive utilities are required, then it suffices to choose

$$\sigma = n\left(\frac{1}{2} - \min_{k=1,\ldots,n}\{\frac{1}{n}\sum_{i=1}^n x_{ki}\}\right) \geq 0.$$

- The value $\xi_X(u)$ provides a measure of the consistency of the FPR X: if $\xi_X(u) = 0$, then X is fully consistent, and the higher the quantity $\xi_X(u)$, the higher the inconsistency of X.
- The associated MSE weighting vector satisfies $\sum_{k=1}^n W_k = 1$.

Regarding aggregations, the MSE utilities also have interesting behavior.

Corollary 2 *Let $X^1, \ldots, X^m \in \mathbb{A}_n^{\frac{1}{2}}$ be a family of FPRs and consider weights $w_1, \ldots, w_m \geq 0$ such that $\sum_{k=1}^m w_k = 1$. Then, for $\sigma \in \mathbb{R}$, the MSE utility vector u^c associated with the aggregated FPR $X^c := \sum_{k=1}^m w_k X^k$ satisfies*

- $u^c = \sum_{k=1}^m w_k u^k$,
- $\mu(u^c) = \sum_{k=1}^m w_k \mu(u^k)$,
- $\xi_{X^c}(u^c) \leq \sum_{k=1}^m w_k \xi_{X^k}(u^k)$,

where $u^1, \ldots, u^m$ stand for the MSE utilities of $X^1, \ldots, X^m$.

Proof The two first items are straightforward considering Theorems 1 and 2. To show the third one, note that

$$\begin{aligned}\xi_{X^c}(u^c) &= \sum_{i=1}^n \sum_{j=1}^n \left(x_{ij}^c - \frac{1}{2} - u_i^c + u_j^c\right)^2 \\ &= \sum_{i=1}^n \sum_{j=1}^n \left(\sum_{k=1}^m w_k \left(x_{ij}^k - \frac{1}{2} - u_i^k + u_j^k\right)\right)^2\end{aligned}$$

$$\leq \sum_{i=1}^{n}\sum_{j=1}^{n}\sum_{k=1}^{m} w_k \left(x_{ij}^k - \frac{1}{2} - u_i^k + u_j^k\right)^2$$
$$= \sum_{k=1}^{m} w_k \sum_{i=1}^{n}\sum_{j=1}^{n} \left(x_{ij}^k - \frac{1}{2} - u_i^k + u_j^k\right)^2$$
$$= \sum_{k=1}^{m} w_k \xi_{X^k}(u^k)$$

which has been derived from Jensen's inequality. □

This result guarantees that, in GDM problems [21], the collective MSE utility vector and the associated consistent APR can be computed by aggregating the individual MSE utility vectors, which could be very helpful in improving computational efficiency as we show in Sect. 4. Furthermore, the MSE when approximating the collective opinion is bounded according to the MSE associated with the individual FPRs.

In addition, the MSE Utility vector allows recovering the original FPR, provided it is consistent, using the mapping μ as follows.

Corollary 3 *Let $X \in \mathbb{A}_n^{\frac{1}{2}}$ be a consistent FPR and consider its MSE Utility vector $u \in \mathbb{R}^n$ for some $\sigma \in \mathbb{R}$. Then $\mu(u) = X$.*

Proof Since X is consistent, then $x_{ik} + x_{kj} = \frac{1}{2} + x_{ij}\, \forall i, j, k = 1, \ldots, n$. Therefore,

$$\mu(u)_{ij} = \frac{1}{2} + u_i - u_j$$
$$= \frac{1}{2} + \left(\frac{1}{2} + \frac{1}{n}\left(\sigma - \sum_{k=1}^{n} x_{ki}\right)\right) - \left(\frac{1}{2} + \frac{1}{n}\left(\sigma - \sum_{k=1}^{n} x_{kj}\right)\right)$$
$$= \frac{1}{2} + \frac{1}{n}\sum_{k=1}^{n}(x_{kj} - x_{ki}) = \frac{1}{2} + \frac{1}{n}\sum_{k=1}^{n}(x_{ik} + x_{kj} - 1)$$
$$= \frac{1}{2} + \frac{1}{n}\sum_{k=1}^{n}\left(x_{ij} - \frac{1}{2}\right) = x_{ij}\, \forall i, j = 1, \ldots, n.$$

□

Note that this capability of retrieving the original FPR from the MSE utility vector is essential to guarantee that no information is lost during the conversion process. In this sense, other proposals fail to satisfy this property. For instance, let us assume that the utility values for the alternatives are computed as the arithmetic mean of the corresponding row of the FPR. Then, it can be easily checked that it is not possible to recover the original FPR when using mappings such as $\hat{\mu} : \mathbb{R}^n \to \mathcal{M}_{n\times n}(\mathbb{R})$ defined as $(\hat{\mu}(u))_{ij} = \frac{1}{2}(1 + u_i - u_j)$, considered by Herrera-Viedma et al. [9]. In the same way, even though Xu et al. [22] also defined several methods to derive priority values from FPRs, they neglected how to reconstruct the original FPR from the obtained priority values. However, this is a key property of the MSE utility vector because it implies that the information contained in the original FPR can be obtained again, which is very helpful to improve the computational efficiency of processes that simultaneously deal with many FPRs as it is shown in Sect. 4.

3.3 Mean Squared Error Utility for MPRs

Here, the notion of MSE utility vector is adapted for pairwise comparisons given in a multiplicative scale.

Assume $Y \in \mathbb{M}_n^m$ is an MPR for some $m > 1$. Then, we can consider the FPR $X = \Phi^{-1}(Y)$, where $\Phi^{-1} : \mathbb{M}_n^{m_1} \to \mathbb{A}_n$, and compute the corresponding MSE utility vector $u \in \mathbb{R}^n$ as in Theorem 2. Therefore, for the sake of simplicity, we define the multiplicative MSE utilities as follows.

Definition 6 *[MSE utility for MPRs]* Let $Y \in \mathbb{M}_n^m$ be an MPR and consider $\sigma \in \mathbb{R}$. Then the MSE utility vector for Y and σ is the vector $u \in \mathbb{R}^n$ defined as

$$u_k = \frac{1}{2} + \frac{1}{n}\left(\sigma - \sum_{i=1}^{n} \phi^{-1}(y_{ik})\right) \ \forall k = 1, 2, \ldots, n,$$

and the associated MSE weighting vector $W \in \mathbb{R}^n$ is defined as

$$W_k = \frac{\phi(u_k)}{\sum_{l=1}^{n} \phi(u_l)} = \frac{m^{\frac{2}{n}\sum_{i=1}^{n}\phi^{-1}(y_{ki})}}{\sum_{l=1}^{n} m^{\frac{2}{n}\sum_{i=1}^{n}\phi^{-1}(y_{li})}} \ \forall k = 1, 2, \ldots, n.$$

Note that these multiplicative utilities determine the consistent MPR $\tilde{Y} = (\tilde{y}_{ij})$ defined as

$$\begin{aligned}\tilde{y}_{ij} &= \frac{\phi(u_i)}{\phi(u_j)} = \frac{m^{\frac{2}{n}\sum_{k=1}^{n}\phi^{-1}(y_{ik}) + \frac{2\sigma}{n} - 2}}{m^{\frac{2}{n}\sum_{k=1}^{n}\phi^{-1}(y_{jk}) + \frac{2\sigma}{n} - 2}} \\ &= m^{\frac{2}{n}\sum_{k=1}^{n}\phi^{-1}(y_{ik}) - \phi^{-1}(y_{jk})} \ \forall i, j = 1, 2, \ldots, n,\end{aligned}$$

which does not actually depend on the value of σ. Furthermore, note that computing the normalized vector W is equivalent to computing the multiplicative MSE utilities for the value

$$\sigma = n - \frac{n}{2}\log_m\left(\sum_{l=1}^{n} m^{\frac{2}{n}\sum_{i=1}^{n} x_{li}}\right).$$

In addition, the multiplicative utility W constructed through this method is in line with the results obtained in [12] and minimizes the distance function $\Xi_Y : (\mathbb{R}^+)^n \to \mathbb{R}_0^+$ defined as

$$\Xi_Y(W) = \xi_X \circ \Phi^{-1}(W) = \sum_{i=1}^{n}\sum_{j=1}^{n}\left(\frac{1}{2}\log_m\left(y_{ij}\frac{W_j}{W_i}\right)\right)^2 \ \forall\, W_1, W_2, \ldots, W_n > 0,$$

which satisfies

$$\Xi_Y(W) = 0 \iff y_{ij} = \frac{W_i}{W_j}\ \forall i, j = 1, 2, \ldots, n \iff Y \text{ is fully consistent.}$$

Note that the weights computed by this procedure are also compatible with the results given by Xu et al. [8] when they are applied to completely consistent FPRs. Nevertheless, to address the non-consistent case, Xu et al. [8] suggested the use of mathematical optimization models. In this regard, the theoretical discussion developed in this section, which is based on Theorem 2, guarantees that the normalized MSE weighting vector is the analytical solution to the optimization problem described above, even when the FPRs are not consistent. Consequently, by using the MSE weighting vector it is possible to derive the priority weights for FPRs using simple arithmetic operations and skipping the numerical resolution of a complex mathematical programming model.

3.4 Mean Squared Error Utility in Best-Worst Method

The MSE utility vector can also be constructed if the only available pairwise comparisons are those corresponding to the best criteria with all the others, and all the criteria with the worst one. In other words, the MSE can be used to simplify the computations in the BWM because they do not require computationally solving any optimization problem.

If $n \in \mathbb{N}, n > 2$ stands for the number of criteria to compare, let us denote by $O = \{1, 2, 3, \ldots, n\} \setminus \{B, W\}$ the indexes corresponding to the criteria which are neither the best nor the worst. In addition, although it is not necessary, we assume here that the equality $x_{BW} = \frac{1}{2} + u_B - u_W$ holds, where x_{BW} is the pairwise comparison of the best criterion with the worst one and u_B, u_W are, respectively, the MSE utilities corresponding to the best and worst criteria.

Theorem 3 *Let $x_B = (x_{B1}, \ldots, x_{Bn})$ and $x_W = (x_{1W}, \ldots, x_{nW})$ be, respectively, the pairwise comparisons of the best criterion with all the others, and all the criteria with the worst one, and consider $\sigma \in \mathbb{R}$. Then, the utility vector $u \in \mathbb{R}^n$ defined as*

$$u_B = \frac{1}{2}(s + t)$$

$$u_W = \frac{1}{2}(s - t)$$
$$u_i = \frac{1}{2}(s - (x_{Bi} - x_{iW}))\,\forall i \in O$$

where $s := \frac{1}{n}(\sum_{i\in O}(x_{Bi} - x_{iW}) + 2\sigma) = u_B - u_W$ *and* $t := X_{BW} - \frac{1}{2} = u_B - u_W$ *is the solution to the optimization problem*

$$\min_{u\in\mathbb{R}^n} \xi_X(u)$$
$$s.t. \begin{cases} \sum_{i=1}^n u_k = \sigma \\ x_{BW} = \frac{1}{2} + u_B - u_W \end{cases}$$

where the function $\xi_X : \mathbb{R}^n \to \mathbb{R}$ *is defined as*

$$\xi_X(u) = \sum_{i\in O}\left(x_{Bi} - \frac{1}{2} - u_B + u_i\right)^2 + \left(x_{iW} - \frac{1}{2} - u_i + u_W\right)^2 \;\forall u \in \mathbb{R}^n.$$

Proof First, let us define the Lagrangian $L : \mathbb{R}^n \times \mathbb{R}^2 \to \mathbb{R}$ by

$$L(u, \lambda, \mu) = \xi_X(u) + \lambda\left(\sum_{i=1}^n u_i - \sigma\right)$$
$$+ \mu\left(x_{BW} - \frac{1}{2} - u_B + u_W\right) \forall (u, \lambda, \mu) \in \mathbb{R}^{n+2}.$$

The equations obtained from partial derivatives of this function are as follows:

$$\partial_B L(u, \lambda, \mu) = -2\sum_{i\in O}\left(x_{Bi} - \frac{1}{2} - u_B + u_i\right) - \mu + \lambda = 0$$
$$\partial_W L(u, \lambda, \mu) = 2\sum_{i\in O}\left(x_{iW} - \frac{1}{2} - u_i + u_W\right) + \mu + \lambda = 0$$
$$\partial_k L(u, \lambda, \mu) = 2(x_{Bi} - x_{iW} - u_B + 2u_i - u_W) + \lambda,\ \forall k \in O = 0$$
$$\partial_\lambda L(u, \lambda, \mu) = \sum_{i=1}^n u_i - \sigma = 0$$
$$\partial_\mu L(u, \lambda, \mu) = x_{BW} - \frac{1}{2} - u_B + u_W = 0.$$

The addition of the two first equations implies

$$\sum_{i\in O}(x_{Bi} - x_{iW} - u_B + 2u_i - u_W) = \lambda$$

which combined with the third equation guarantees $\lambda = 0$. At this point, the first and second equations can be rewritten as

$$\sum_{i \in O} \left(x_{Bi} - \frac{1}{2} \right) + \sigma - (n-1)u_B - u_W + \frac{\mu}{2}$$
$$\sum_{i \in O} \left(x_{iW} - \frac{1}{2} \right) - \sigma + u_B + (n-1)u_W + \frac{\mu}{2},$$

whose difference is

$$\sum_{i \in O} (x_{Bi} - x_{iW}) + 2\sigma - nu_b - nw_W = 0,$$

and then

$$u_b + u_w = \frac{1}{n} \left(\sum_{i \in O} (x_{Bi} - x_{iW}) + 2\sigma \right) = s.$$

Since $u_B - u_W = t$, the solution of the optimization problem can be computed by combining it with the third Lagrangian equation. □

Note that, since $t \in [0, \frac{1}{2}]$ and $x_{Bk} - x_{kW} \in [-t, t] \, \forall k \in O$, then

$$\begin{aligned} u_B - u_k &= \frac{t}{2} + \frac{1}{2}(x_{Bk} - x_{kW}) \in [0, t] \, \forall k \in O \\ u_k - u_W &= \frac{t}{2} - \frac{1}{2}(x_{Bk} - x_{kW}) \in [0, t] \, \forall k \in O \\ u_k - u_l &= \frac{1}{2}((x_{Bj} - x_{jW}) - (x_{Bi} - x_{iW})) \in [-t, t] \, \forall k, l \in O, \end{aligned}$$

and, consequently, the matrix $\mu(u)$ is an FPR. In addition, it follows that $u_i \in [u_W, u_B]$ and that u_i increases when the quantity $x_{iB} + x_{iW}$ increases $\forall i \in O$.

Note that the MSE utilities obtained in Theorem 3 are given in a linear scale, whereas the weights for BWM are provided for preferences elicited in a multiplicative scale $[1, m]$. Therefore, after using the bijection $\phi^{-1} : [\frac{1}{m}, m] \to [0, 1]$, the weights for the criteria can be obtained as

$$W_i = \frac{m^{2 \cdot u_i - 1}}{\sum_{k=1}^{n} m^{2 \cdot u_k - 1}}, \, i = 1, 2, \ldots, n$$

where $u \in \mathbb{R}^n$ is the MSE utility vector obtained in Theorem 3. Consequently, we can state the following definition.

Definition 7 *[MSE utility in BWM]* Let $x_B = (x_{B1}, \ldots, x_{Bn})$ and $x_W = (x_{1W}, \ldots, x_{nW})$ be, respectively, the pairwise comparisons in a scale where $m \in \mathbb{N} \setminus \{1\}$ of the best criterion with all the others, and all the criteria with the worst one, and consider $\sigma \in \mathbb{R}$. Then, the MSE utility vector for $x_B = (x_{B1}, \ldots, x_{Bn})$ and $x_W = (x_{1W}, \ldots, x_{nW})$ is the vector $u \in \mathbb{R}^n$ defined as

$$\begin{aligned} u_B &= \frac{1}{2}(s+t) \\ u_W &= \frac{1}{2}(s-t) \\ u_i &= \frac{1}{2}(s-(x_{Bi}-x_{iW}))\,\forall i \in O, \end{aligned}$$

where $s := \frac{1}{n}(\sum_{i \in O}(x_{Bi} - x_{iW}) + 2\sigma) = u_B - u_W$ and $t := X_{BW} - \frac{1}{2} = u_B - u_W$. Additionally, the MSE weighting vector for the criteria can be computed as $W \in \mathbb{R}^n$ defined as

$$W_i = \frac{m^{2 \cdot u_i - 1}}{\sum_{k=1}^{n} m^{2 \cdot u_k - 1}}, i = 1, 2, \ldots, n.$$

4 Application in E-Democracy

Even though the most immediate application of MSE utility vector could be ranking alternatives, here we prefer highlighting how they can be applied in LSGDM to reduce computational costs. Of note, the data used in this section has been randomly generated.

4.1 Illustrative Example

In a rapidly growing city, the government aims to decide the allocation of the annual budget for an infrastructure project. The project aims to construct a new transportation system to address the increasing traffic congestion and improve the overall transportation infrastructure. To do so, a preliminary analysis has revealed that the best options for this project are

- Alternative 1: High-Speed Railway System.
- Alternative 2: Road Expansion and Improvement.
- Alternative 3: Public Transit Enhancement.
- Alternative 4: Sustainable Transportation Initiatives.

In order to make a final decision, the government decides to carry out a survey on the population, which will evaluate different alternatives using FPRs. Subsequently, an automatic CRP will be carried out to smooth out the disagreements among them.

For the sake of simplicity, we consider a Minimum Cost Consensus (MCC) model [15, 23] to find the collectively agreed solution which is closest to the original preferences. Specifically, we consider the optimization problem:

$$\begin{aligned} \min_{o \in \mathcal{M}_{m\times\hat{n}}} & \sum_{k=1}^{m}\sum_{i=1}^{\hat{n}} |o_{ki} - o_{ki}^0| \\ s.t. & \begin{cases} g_i = \frac{1}{m}\sum_{k=1}^{m} o_{ki} \\ |o_{ki} - g_i| \leq \varepsilon \end{cases} \end{aligned} \tag{MCC}$$

where ε is a consensus threshold, $g \in \mathbb{R}^{\hat{n}}$ is the collective opinion, and $o, o^0 \in \mathcal{M}_{m\times\hat{n}}$ stand for the modified and initial opinions, respectively.

Note that, if FPRs were used directly in this optimization model, for each decision-maker it would be necessary to consider at least $n \times (n-1)/2$ variables, i.e., 6 variables for the case $n = 4$.

Therefore, in order to speed up the process, the MSE utility vector will be used to transform the FPRs into vectors before running the optimization process, reducing from $n \times (n-1)/2$ to only n variables. Concretely, if a decision-maker provides the approximated FPR

$$\begin{pmatrix} 0.5 & 0.06 & 0.44 & 0.51 \\ 0.93 & 0.5 & 0.51 & 0.37 \\ 0.55 & 0.48 & 0.5 & 0.42 \\ 0.48 & 0.62 & 0.57 & 0.5 \end{pmatrix}$$

we will consider instead the corresponding MSE Utility vector for $\sigma = 1$ computed as in Definition 5:

$$(0.13, 0.33, 0.24, 0.29).$$

Therefore, in order to make a consensual decision, we can run the above-mentioned (MCC) model using the parameter $\varepsilon = 0.1$ and the corresponding MSE utility vectors as inputs.

In this case, 2354 people answered the survey. Therefore, after computing the corresponding 4-dimensional MSE utility vectors, we run the model. The computational process finishes in $34s$ and the agreed opinion is the utility vector:

$$(0.247, 0.252, 0.249, 0.247),$$

which indicates that Alternative 2: Road Expansion and Improvement is the most preferred choice for the people in the city.

4.2 *Time Cost Analysis*

In the previous example, we have shown that computing the MSE utility vector reduces the number of variables considered in the CRP. Here, we provide additional statistical insight to compare the following procedures for solving a GDM problem:

- Use FPRs as inputs in the MCC model,
- Compute the MSE utility vectors for the FPRs and then run the MCC model.

To compare these two procedures we carry out Monte Carlo simulations. Concretely, we are interested in three variables. The first one is the computational time required to solve the MCC model using FPRs, the second one is the computational time required to solve the MCC model using the corresponding MSE utilities, and the third variable is the average distance between the collective FPR and the APR obtained from the collective utility vector using the mapping μ.

Let us consider a GDM problem in which m experts must reach an agreement regarding which alternative among n possibilities is the best one. In order to simplify the analysis, let us fix $\sigma = 1$ and $\varepsilon = 0.2$.

Note that in the first case, the input of the MCC model o^0 stands for the upper triangle of the FPR, i.e., $\hat{n} = \frac{n(n-1)}{2}$, whereas when using the MSE utilities associated with the corresponding FPR, the dimension decreases to $\hat{n} = n$ because o^0 is the MSE utility vector itself. In addition, the distance between the collective FPR obtained following the first procedure and the APR obtained from the second procedure can be measured as $d = \frac{2}{mn(n-1)} \sum_{k=1}^{m} \sum_{i=1}^{n-1} \sum_{j=i+1}^{n} |G_{ij} - \mu(u)_{ij}|$.

For some configurations of m and n, we have carried out both procedures 50 times using randomly generated FPRs. The obtained confidence intervals for the above-described variables are shown in Tables 1, 2, and 3.

We have also made the simulations in LSGDM scenarios with 2560 decision-makers. For 4, 7, and 9 alternatives, the confidence intervals for the time when using the first procedure are, respectively, [419.59, 497.91], [5449.09, 6439.70], and [19906.78, 22800.53] (expressed in seconds), whereas the corresponding confidence intervals for the time when using the second procedure are [210.92, 256.28], [1110.93, 1876.86], and [1814.02, 3222.98] (expressed in seconds too). In all these cases, the confidence intervals for the distances are, approximately, [0.005, 0.006].

Table 1 Confidence interval for computational time (in milliseconds) of the first procedure

n \ m	10	40	160	640
4	[1.99, 4.21]	[9.35, 11.45]	[57.15, 61.61]	[11820.75, 13139.53]
7	[7.54, 9.38]	[38.17, 41.39]	[4404.34, 4516.94]	[276656.0, 330088.16]
9	[13.51, 16.69]	[74.4, 78.84]	[13000.76, 13243.88]	[868444.3, 1046717.26]

Table 2 Confidence interval for computational time (in milliseconds) in the second procedure

n \ m	10	40	160	640
4	[1.87, 2.01]	[6.29, 7.79]	[39.78, 42.9]	[3887.2, 4779.32]
7	[2.48, 4.36]	[12.26, 15.58]	[80.48, 86.08]	[27081.37, 38392.87]
9	[3.34, 3.66]	[15.6, 19.2]	[111.25, 118.99]	[34956.62, 93454.62]

Table 3 Confidence interval for distance

n \ m	10	40	160	640
4	[0.072, 0.086]	[0.036, 0.043]	[0.017, 0.022]	[0.009, 0.011]
7	[0.086, 0.094]	[0.045, 0.05]	[0.022, 0.025]	[0.011, 0.012]
9	[0.087, 0.094]	[0.047, 0.051]	[0.023, 0.025]	[0.011, 0.012]

In conclusion, using the MSE utilities allows for considerably reducing the computational cost (around 10 times faster when using 9 alternatives and 2560 decision-makers and even faster for lesser numbers of experts). In addition, Theorem 1 guarantees that the MSE utilities corresponding to the collective FPR can be derived as the weighted sum of the MSE utilities corresponding to the individual opinions. Furthermore, the use of the MSE utilities in the consensus process implies that it is not necessary to include additional constraints in the optimization model that guarantee a certain consistency degree.

5 Comparative Analysis

5.1 Comparison with the Eigenvalue Method for MPRs

This section compares the use of the MSE utility vector for MPRs (Definition 6) with the eigenvalue method to compute priorities from an MPR, which is widely applied in AHP [6, 16, 24].

For the sake of clarity, let us consider the MPR, on a 1–9 Saaty's scale, given as

$$Y = \begin{pmatrix} 1 & 2 & 1/3 & 4 \\ 1/2 & 1 & 5 & 6 \\ 3 & 1/5 & 1 & 1/7 \\ 1/4 & 1/6 & 7 & 1 \end{pmatrix}.$$

For this matrix, the maximum eigenvalue is 6.903, and the corresponding normalized eigenvector is $eig = (0.272, 0.365, 0.156, 0.207)$. If we compute the MSE weighting vector instead, we obtain the vector $W = (0.283, 0.435, 0.12, 0.163)$. Note that both of them provide a similar ranking for the alternatives, i.e., $Alt_2 > Alt_1 > Alt_4 > Alt_3$. In addition, it is theoretically guaranteed (see Sect. 3) that the MSE weighting vector is the closest one to the original matrix Y. This can be easily checked by noticing that $\Xi_Y(eig) = 1.014$, whereas $\Xi_Y(W) = 0.956$.

Furthermore, if it is considered a completely consistent MPR both methods are guaranteed to provide exactly the same solution. In fact, for a consistent MPR such as

$$Y = \begin{pmatrix} 1 & 2 & 3 & 4 \\ 1/2 & 1 & 3/2 & 2 \\ 1/3 & 2/3 & 1 & 4/3 \\ 1/4 & 1/2 & 3/4 & 1 \end{pmatrix},$$

both the weights obtained from the eigenvalue method and the MSE coincide, i.e., $W = eig = (0.48, 0.24, 0.16, 0.12)$.

Despite it being widely used [5], the eigenvalue method has some drawbacks in comparison to the MSE weighting vector. On the one hand, it is necessary to compute both the eigenvalues and eigenvectors to apply the method. From the computational point of view, computing eigenvectors can be done in $O(n^3)$ [25], whereas the computation of the weighting vector associated with the MSE utility vector just requires $O(n)$.

On the other hand, if the original MPR Y is not completely consistent, the eigenvalue method could not be reliable [5]. However, with the MSE utility approach, we can guarantee that the obtained weights are as consistent as possible because they minimize the function Ξ_Y.

In addition, the rationale regarding the relationship between eigenvalues and the ranking of the evaluated alternatives could not be clear at first sight for researchers without a mathematical background. In this sense, the MSE utility vector is much more interpretable because it is computed from the dominance of each alternative, which guarantees that the more preferred one alternative is over the others, the higher its weight.

5.2 Comparison with Classical BWM

In order to illustrate the performance of the weights obtained from the MSE utilities for BWM preferences (Definition 7), let us compare them with the preferences obtained by using the original BWM [7].

Let us consider the inputs $x_B = (8, 2, 1)$ and $x_W = (1, 5, 8)$, where $B = 3$ and $W = 1$. After applying the original BWM, the optimal weights are $(0.07142, 0.3387, 0.5899)$. If we apply the method here described, after remapping the multiplicative preferences into the linear scale, the MSE utilities are

$u = (0.0620, 0.4028, 0.5352)$ and the corresponding MSE weights are $W = (0.0742, 0.3319, 0.5938)$, which are similar to the ones obtained using Rezaei's method [7] and do not require solving any optimization problem.

In addition, if both methods are applied to a fully consistent evaluation, the obtained weights are exactly the same. For example, for the consistent inputs $x_B = (2, 1, 9, 5)$ and $x_W = (9/2, 9, 1, 9/5)$, where $B = 2$ and $W = 3$, the obtained weights are, in both cases, $W = (0.2761, 0.5521, 0.0613, 0.1104)$.

Note that both the classical BWM and the MSE weighting vector produce a weighting vector that aims to minimize a certain consistency function. However, the objective function considered in classical BWM is not differentiable, and it is hard to give an analytical solution for it, which forces inevitably to solve the optimization model using numerical methods. On the contrary, the MSE weighting vector is the analytical and, consequently, the exact solution to its associated optimization problem. In addition, computing the MSE weighting vector requires $O(n)$, whereas solving the nonlinear programming problem of classical BWM may take much more than $O(n^2)$, depending on the solver and the linearization process [26].

6 Conclusions

This chapter has developed a method to derive priority values from pairwise comparisons based on the MSE. Such a method is based on the analytical resolution of the optimization problem consisting of finding the utility values that generate a certain consistent pairwise comparison matrix. The discussion highlighted several significant properties and benefits of this approach, which collectively enhance its practical application in decision-making processes.

The proposed method facilitates the computation of an overall performance score for each alternative by utilizing a pairwise comparison matrix. This feature enables seamless integration into established decision-making frameworks, such as the AHP, enhancing the method's applicability in traditional settings.

The MSE utilities represent the analytical solution to an optimization problem, ensuring accuracy and reliability in their results. These utilities allow the construction of a consistent pairwise comparison matrix that minimizes the MSE with respect to the original pairwise comparisons provided by decision-makers. This fact ensures that such utilities properly reflect decision-makers' opinions.

The computation of MSE utilities involves straightforward algebraic operations, bypassing the need for complex mathematical procedures such as the resolution of an optimization model. This simplicity makes the method particularly accessible to practitioners who may not have extensive training in advanced mathematics. Moreover, these utilities are derived based on the dominance of the alternatives, which enhances their interpretability and practical relevance in decision-making contexts.

Additionally, the MSE method offers significant computational efficiency, especially pertinent in LSGDM scenarios. By reducing the dimensionality of the data, the method streamlines the computational process, making it a valuable tool for handling extensive datasets efficiently and effectively.

The flexibility and robustness of the MSE utility discussed in this chapter suggest a broad potential for future research and application. The method's adaptability to different types of preference structures opens up new avenues for extending its utility across various decision-making contexts.

References

1. Sahoo, L., Senapati, T., Yager, R.R. (eds.): Real Life Applications of Multiple Criteria Decision Making Techniques in Fuzzy Domain, Studies in Fuzziness and Soft Computing (STUDFUZZ, Vol. 420). Springer (2023)
2. Jensen, R.E.: An alternative scaling method for priorities in hierarchical structures. J. Math. Psychol. **28**(3), 317–332 (1984). https://doi.org/10.1016/0022-2496(84)90003-8
3. Miller, G.A.: The magical number seven, plus or minus two: some limits on our capacity for processing information. Psychol. Rev. **63**(2), 81–97 (1956). https://doi.org/10.1037/h0043158
4. Orlovski, S.A.: Decision-making with fuzzy preference relations. Fuzzy Sets Syst. **1**, 155–167 (1978)
5. Saaty, T.L.: How to make a decision: the analytic hierarchy process. Eur. J. Oper. Res. **48**(1), 9–26 (1990)
6. Saaty, T.L.: Relative measurement and its generalization in decision making why pairwise comparisons are central in mathematics for the measurement of intangible factors the analytic hierarchy/network process. Revista de la Real Academia de Ciencias Exactas, Fisicas y Naturales. Serie A. Matematicas **102**(2), 251–318 (2008). https://doi.org/10.1007/bf03191825
7. Rezaei, J.: Best-worst multi-criteria decision-making method. Omega **53**, 49–57 (2015)
8. Xu, Y., Dai, W., Huang, J., Li, M., Herrera-Viedma, E.: Some models to manage additive consistency and derive priority weights from hesitant fuzzy preference relations. Inform. Sci. **586**, 450–467 (2022). https://doi.org/10.1016/j.ins.2021.12.002
9. Herrera-Viedma, E., Herrera, F., Chiclana, F.: A consensus model for multiperson decision making with different preference structures. IEEE Trans. Syst. Man Cybern. Part A: Syst. Humans **32**, 394–402 (2002). https://doi.org/10.1109/TSMCA.2002.802821
10. Wan, S., Cheng, X., Dong, J.-Y.: Multiplicative consistency analysis of interval-valued fuzzy preference relations. Inform. Sci. **631**, 120–144 (2023). https://doi.org/10.1016/j.ins.2023.02.066
11. Xu, Z., Da, Q.: A least deviation method to obtain a priority vector of a fuzzy preference relation. Eur. J. Oper. Res. **164**(1), 206–216 (2005). https://doi.org/10.1016/j.ejor.2003.11.013
12. Koczkodaj, W.W., Orłowski, M.: Computing a consistent approximation to a generalized pairwise comparisons matrix. Comput. Math. Appl. **37**(3), 79–85 (1999). https://doi.org/10.1016/S0898-1221(99)00048-6
13. Li, C.-C., Dong, Y., Xu, Y., Chiclana, F., Herrera-Viedma, E., Herrera, F.: An overview on managing additive consistency of reciprocal preference relations for consistency-driven decision making and fusion: Taxonomy and future directions. Inform. Fusion **52**, 143–156 (2019). https://doi.org/10.1016/j.inffus.2018.12.004
14. Calache, L.D.D.R., Camargo, V.C.B., Osiro, L., Carpinetti, L.C.R.: A genetic algorithm based on dual hesitant fuzzy preference relations for consensus group decision making. Appl. Soft Comput. **121**, 108778 (2022). https://doi.org/10.1016/j.asoc.2022.108778
15. García-Zamora, D., Dutta, B., Massanet, S., Riera, J.V., Martínez, L.: Relationship between the distance consensus and the consensus degree in comprehensive minimum cost consensus models: A polytope-based analysis. Eur. J. Oper. Res. (2022). https://doi.org/10.1016/j.ejor.2022.08.015

16. Xu, Y., Liu, S., Wang, J., Shang, X.: Consensus checking and improving methods for ahp with q-rung dual hesitant fuzzy preference relations. Expert Syst. Appl. **208**, 117902 (2022). https://doi.org/10.1016/j.eswa.2022.117902
17. García-Zamora, D., Labella, Á., Ding, W., Rodríguez, R.M., Martínez, L.: Large-scale group decision making: a systematic review and a critical analysis. IEEE/CAA J. Autom. Sinica **9**(6), 949–966 (2022)
18. Yuan, R., Wu, Z., Tu, J.: Large-scale group decision-making with incomplete fuzzy preference relations: The perspective of ordinal consistency. Fuzzy Sets Syst. **454**, 100–124 (2023). https://doi.org/10.1016/j.fss.2022.04.021
19. Jing, D., Imeni, M., Edalatpanah, S.A., Alburaikan, A., Khalifa, H.A.E.-W.: Optimal selection of stock portfolios using multi-criteria decision-making methods. Mathematics **11**(2), 415 (2023). https://doi.org/10.3390/math11020415
20. Osborne, M.J., Rubinstein, A.: A Course in Game Theory. The MIT Press, Cambridge, USA (1994)
21. Li, P., Edalatpanah, S.A., Sorourkhah, A., Yaman, S., Kausar, N.: An integrated fuzzy structured methodology for performance evaluation of high schools in a group decision-making problem. Systems **11**(3) (2023). https://doi.org/10.3390/systems11030159
22. Xu, Y., Da, Q., Liu, L.: Normalizing rank aggregation method for priority of a fuzzy preference relation and its effectiveness. Int. J. Approximate Reasoning **50**(8), 1287–1297 (2009). https://doi.org/10.1016/j.ijar.2009.06.008. Special Section on Interval/Probabilistic Uncertainty
23. Ben-Arieh, D., Easton, T.: Multi-criteria group consensus under linear cost opinion elasticity. Decis. Support Syst. **43**(3), 713–721 (2007)
24. Zeinalnezhad, M., Alavi, S., Mousavi, E.: Prioritization of gpm activities from lean-agile-resilience perspective using fuzzy analytic hierarchy process. J. Fuzzy Extension Appl. (Online First) (2022). https://doi.org/10.22105/jfea.2022.329797.1202
25. Pan, V.Y., Chen, Z.Q.: The complexity of the matrix eigenproblem. In: Symposium on the Theory of Computing (1999)
26. Jiang, S., Song, Z., Weinstein, O., Zhang, H.: Faster dynamic matrix inverse for faster lps. arXiv preprint arXiv:2004.07470 (2020)

MCDA/F-DEMATEL/ICTs Method Under Uncertainty in Mathematics Education: How to Make a Decision with Flipped, Gamified, and Sustainable Criteria

Jin Su Jeong and David González-Gómez

Abstract Teaching and learning at universities are a significant challenge because of the worldwide COVID-19 pandemic, which demands substantial optimization of different types of education platforms. Here, as an important approach, the multi-criteria decision analysis (MCDA) method showed the possibility that could be designed and used in these uncertain environments. Even, in the mathematics education context, it is always necessary to identify the appropriate criteria and sub-criteria, which could adapt to these uncertain pedagogical changes since there are only a few cases to engage and activate its model in the academic area. Also, with amplified acceptance, a fruitful flipped and gamified methodology with sustainable development could support to advance the unreliable situations. This book chapter endeavoured to design and implement appropriate criteria for flipped, gamified, and sustainable mathematics education during uncertain environments such as COVDI-19 and possible incidents with a multi-criteria decision analysis/fuzzy-decision-making trial and evaluation laboratory (MCDA/F-DEMATEL) method. In detail, it allocates flipped classrooms, and gamified information, communication, and technology (G-ICT) with sustainable development in university mathematics education after the professionals', decision-makers, consultations, and recommendations. The definitive results after proper analyses could be appraised with the weighted linear combination (WLC) together with their weighted-coefficients' calculation, and sensitivity-analysis. Here, as a case analysis, the results of the expert panel survey demonstrated that scenario (1) which was decided by the expert panel obtained 83% positive perception. Scenario (2) that assigned equal weight was the less preferred with 47% positive perception through the same expert panel survey. Consequently, the results

---

J. S. Jeong (✉) · D. González-Gómez
Department of Science and Mathematics Education, Teacher Training College, University of Extremadura, Avd. de la Universidad s/n, 10004 Cáceres, Spain
e-mail: jin@unex.es

D. González-Gómez
e-mail: dggomez@unex.es

L. Sahoo et al. (eds.), *Decision Making Under Uncertainty Via Optimization, Modelling, and Analysis*, Studies in Systems, Decision and Control 558,
https://doi.org/10.1007/978-981-96-0085-4_5

indicated that this method proposed could be used to authenticate the most principal flipped and gamified mathematics education with sustainable development for adapting uncertain educational environments with education contexts corresponding to more long-term goals. Accordingly, decision-making glitches could be pertained by the method suggested at private industry levels, government sectors, and other academic subjects, which constantly requires continuous achievement of knowledge and dissemination for both certain and uncertain situations that might be encountered.

Keywords MCDA/F-DEMATEL/ICTs · Mathematics education · Uncertainty · Flipped classroom · Gamification · Sustainable development

1 Introduction

The worldwide COVID-19 pandemic caused the existing educational progress suspension in numerous institutes [19, 83]. Before the incidents, the conventional teaching style was altered into a model based on online, which could allow students that remain and complete the classes and activities [11, 21]. According to the United Nations Educational, Scientific, and Cultural Organization (UNESCO), various institutes have been affording over 70% through online and virtual platforms for classes and assessments [3, 69, 76]. Here, online education disclosed a growing consistent propensity (about 15.4% per year) in educational organizations [76, 78]. Yet, after the incidents and during the COVID-19 pandemic, learning circumstances were obligatory to be altered because of millions of cases according to the World Health Organization (WHO) [9, 83, 84]. At the level of universities, substantial optimization is necessary to overcome a noteworthy challenge via different types of education platforms [35]. Particularly, the multi-criteria decision analysis (MCDA) method with an analytic hierarchy process (AHP) could be considered as an alternative to being a substantial possibility, which could be devised and expended under these uncertain circumstances [15, 29, 47]. With the inconsistency of different decision-makers, it can arouse diverse decision-making through specific portrayals [38, 63]. Here, the fuzzy-decision-making trial and evaluation laboratory (F-DEMATEL) method could be leveraged [8, 26, 32]. The information, communication, and technology (ICT) combined with the MCDA/F-DEMATEL will allow us to evaluate and show whole proportions and practices under uncertainty [13, 59, 77]. This new hybrid method indicated that an MCDA/F-DEMATEL with ICTs can be exploited to improve a major/efficient connotation for a scheme proposed [36, 64, 70]. The MCDA/F-DEMATEL/ICTs method can be also examined for the flipped and gamified methods, which are associated with series to postulate whole requirements and experiences [8, 26, 37]. Hence, in the perspective of mathematics education, with this new uncertainty, the appropriate adaptations and criteria are necessary to be selected with the noticeable features of this working procedure in mathematics education [4, 11].

Deliberating the changes above-mentioned under uncertainty, in higher education, flipped-classroom instruction methodology instigates enrichment for the learning results in students' mathematics courses [25, 35, 80]. The paradigm of flipped classroom (inversed-instruction methodology (IIM)) was originally devised by Jonathan Bergmann and Aaron Sams [24, 74]. Particularly, with the constructivism and the social-learning model, this comparatively novel educational methodology has its fundamentals [28]. Thus, it shapes that direct teaching and lecture cannot be an operative instructional device in a big group learning context but can be operative if it could be dispensed to individual students [6, 65]. For example, lectures in a routine flipped classroom are dispensed to students' homes through written materials, video lectures, and online tasks/quizzes [25, 73]. In the class, students will have more student-centred learning accomplishments and, thereby, can make more interactional courses while dispensing "just-in-time" classes, which can address questions not solved in the previous sessions [55]. In the framework prearranged, as a significant idea, the ICTs could be premeditated to support the flipped classroom owing to pertinent and malleable student-centred instructional methodology [32, 41, 68]. The ICT application during the last seventy years had a continual increase, which affected tremendously into mathematics and science, technology, engineering, and mathematics (STEM) education [27, 56]. [22] and [61] indicated that the ICTs assist the flipped classroom in virtual and educational platforms along with the multi-aspect networks through MCDA-AHP, lecturer-to-lecturer, lecturer-to-student, and student-to-student. Yet, even though investigations are presently being commissioned, more investigations necessitate the capability and competence of explicit flipped classrooms with the MCDA/F-DEMATEL/ICTs [23, 43]. Despite these motivations indicated and the absence of information/knowledge, this book chapter seeks to look for the criteria of flipped mathematics education with the sustenance of the MCDA/F-DEMATEL/ICTs, which regulate current educational changes under this new uncertainty.

Considering the situations indicated previously under uncertainty, the integration of gamification methodology into the flipped mathematics classroom with the MCDA/F-DEMATEL/ICTs could be an answer for a setting corresponding. [14] mentioned that gamification is the employment and deployment of elements through a game-like strategy within a non-game setting, which can stimulate learners and resolve problems indicated [52, 89]. Particularly, there is a difference between gamification and serious games or full-fledged games, which accentuates elements and non-game settings through the previous definition proposed. Here, the elements of game strategy could build game blocks and characteristics, which could play a vital part in gameplay [14]. In detail, the elementary and particular elements of game design involve leaderboards, badges, points, etc. although element classifications for game design are differing, which can be utilized for triggering specific comportments amongst users and responding to their psychological requirements [82]. [12] mentioned that the self-determination theory (SDT) is the most referenced theory for the studies of gamification. In the context of the theory indicated, it acclaimed a user-centred framework for significant gamification strategy, which concentrates on the intrinsic inspiration for educating learners [57]. Thus, goal-setting theory (GST) in the latest years has been converted into a guiding theory [40]. In reality, throughout diverse motivation theories, there are many corresponding elements. For educators and researchers, it may be complicated to denote too many theories before scheming gamification employment [12]. For the flipped classroom for mathematics education with the MCDA/F-DEMATEL/ICTs, the most pertinent theories of rudimentary gamification can be reflected and excerpt the crucial elements to create the gamification design model corresponding together with social comparison theory, goal-setting theory, flow theory, behaviour reinforcement theory, and self-determination theory [33]. Despite these motivations indicated and the absence of information/knowledge, this book chapter pursues to check out the criteria of flipped and gamified mathematics education with the sustenance of the MCDA/F-DEMATEL/ICTs, which regulate current educational changes under this new uncertainty.

Contemplating the circumstances designated previously under uncertainty, the European Commission (EC) indicated that mathematics and science learning interest in elementary and secondary schools decreased and is decreasing (European [18]). In a higher-level learning society, it is critical that it is necessary to upsurge requests for these discoveries because of the enhancement of the information organization (European [18, 32, 72]). Here, with the increased attention to sustainable development in education, it is necessary to apply it as a long-term sequence, which has the objectives and practices of sustainability of UNESCO [66, 75]. Also, it was accomplished by the Decade of Education for Sustainable Development (DESD) in the United Nations (UN) [60, 75]. Particularly, students in these occurrences still have various barriers in the area of mathematics education [20, 28, 71]. Various authors [7, 28, 50] acclaimed the inevitability of adjustment for present educational settings under uncertainty along with new learning methodologies in mathematics education [49]. For the composition of a sustainable mathematics education regulating current educational settings under uncertainty, mathematics education theme and practice are important in these mathematics compositions along with curricular material and educational contents, which are detailed and focused on [2, 34]. Yet, since the understanding of sustainable development for mathematics education is still in its premature stage, it is essential to achieve it as a fundamental purpose [46, 54]. In several institutes, although there are few existences for the application, sustainable development for education has accomplished numerous commitments in shifting humanities by decision-makers (European [18, 46], OECD [58]). Here, [17] and [51] denoted that mathematics education for sustainable development conjoining with ICTs has been influenced by flipped and gamified methods, which can unite the education gap that is confronted. In postsecondary education, various research has stated the whole work of mathematics education for sustainable development using flipped and gamified methods [3, 43] together with various aspects through the MCDA/F-DEMATEL/ICTs. Despite these motivations indicated and the absence of information/knowledge, this book chapter pursues to check out the criteria of flipped, gamified, and sustainable mathematics education with the MCDA/F-DEMATEL/ICTs, which regulate current educational changes under this new uncertainty.

Based on the references aforementioned, it is noticeable that university teaching and learning has a noteworthy gap that demands substantial optimization of different types of education platforms during the worldwide COVID-19 pandemic. The book chapter proposed is trying to observe and organize the criteria for flipped, gamified, and sustainable mathematics education during uncertain environments and possible incidents with the MCDA/F-DEMATEL/ICTs method to solve the problem and gap indicated as a research question. Here, along with the method indicated, criteria and sub-criteria corresponding have been rendered, arbitrated, weighted, and committed such as (1) flipped and gamified technology, (2) mathematics education, and (3) sustainable development. The results' analyses for 15 sub-criteria from 3 criteria were deliberated by the weighted linear combination (WLC) together with the weight coefficients' reckoning and sensitivity analysis (scenarios (1)-(5) in this case) in the context of the MCDA/F-DEMATEL/ICTs method. Finally, an expert panel survey has been performed to check the viability of scenarios/sub-scenarios.

2 Case Analysis Through Sample and Context Provided

With an entry-level subject designated "mathematics and its didactics", this research was executed with this mandatory one for the bachelor's degree in Primary Education, Teacher Training College, the University of Extremadura (Spain). For the primary education degree, the course has a total of 210 students distributed into 3 groups (see Table 1 as a case analysis employed). As a core program, this is for second-year students and contains various themes, definitions, and theories for primary education in mathematics. Its contents are prepared in six units: presentation to mathematics education; natural numbers and enumeration methods; arithmetic operations; divisibility; fractions and decimal numbers; and integer numbers. Also, it covers the didactics of mathematics, the properties of teaching/learning, the technological practice, the history of mathematics, and plentiful thoughts on mathematics as the basic characters. In compliance with the curriculum indicated previously, the pre-service teachers require a total of 150 h (6 credits), which allows them to obtain the competencies estimated.

Table 1 Overall students' demographic data in 3 groups

Elements		Group 1	Group 2	Group 3
Total students (n)		70	70	70
Female (%)		46	47	46
Male (%)		54	53	54
Age (n)		21		
GPA (n)		6.71	6.83	6.91
Educational background (%)	Science	15	17	20
	Technology	1	2	5
	Social science	72	69	65
	Arts	1	1	0
	Others	11	11	10

3 MCDA/F-DEMATEL/ICTs Method Under Uncertainty

In the MCDA/F-DEMATEL/ICTs method, the first step consists of determining the criteria and sub-criteria that can partake a direct influence on their conceivable assessment for flipped and gamified technology, mathematics education, and sustainable development for regulating existing educational settings under uncertainty.

In Table 2, the criterion and its 5 sub-criteria for flipped and gamified technology the decision-makers (the authors) through the experts' group assistance.

In Table 3, the criterion and its 5 sub-criteria for mathematics education the decision-makers (the authors) through the experts' group assistance.

In Table 4, the criterion and its five sub-criteria for sustainable development of the decision-makers (the authors) through the experts' group assistance.

Here, to solve certain problems of decision-making, the AHP in the line of MCDA was firstly commissioned to acquire the weight as relative significance for criteria and sub-criteria and then to analyse their assessment [16, 63, 70]. Thus, Fig. 1 [88] demonstrated the pair-wise comparison method (PCM) with a rating scale of 1 to 9 and 1/9 through the decision makers' relevance and the specified criterion (see Eq. (1)) [81].

$$R = (r_{ij})_{n\times n} = \begin{bmatrix} r_{11} & \cdots & r_{1n} \\ \vdots & \ddots & \vdots \\ r_{n1} & \cdots & r_{nn} \end{bmatrix} \tag{1}$$

Throughout the PCM indicated, each criterion/sub-criterion weight can be assigned along with the consistency ratio (CR) matrix. Equation (2) revealed that the weight of a CR can be systematized merely if the weight calculated is smaller than 10% as to be suitable [5, 62]. However, if it is greater than 10%, decision-makers should regulate their appraisal obtained [36, 39]. Also, the consistency index (CI) and the random index (RI) in the PCM equation context can gauge the divergences along with the generations with two indices as shown in Eq. (3) [48, 63].

$$CR = \frac{CI}{RI} \tag{2}$$

$$CI = \frac{\lambda_{max} - n}{n - 1} \tag{3}$$

In the MCDA/F-DEMATEL/ICTs method, the second step contains the F-DEMATEL method, which is based on criteria and sub-criteria elected [79]. To determine the selected criteria and sub-criteria, this outranking technique was used to take out their coefficients, which pinpoint the right selection for flipped, gamified, and sustainable mathematics education during uncertain environments along with the environments through the MCDA-AHP system. In Fig. 2, [39] and [44] showed

Table 2 Criteria and sub-criteria were analysed and computed for flipped and gamified technology under uncertainty

Criteria	Sub-criteria	Weights	Sub-criteria implementation
Flipped and gamified technology	Contributions to technology employment	0.43	Contribution in technology employment embraces the framework/depiction in flipped, gamified, and sustainable mathematics education, which regulates the existing educational settings under uncertainty relating to the objectives implied
	Contributions to technology distribution	0.28	Contributions in technology distribution embrace the framework/depiction in flipped, gamified, and sustainable mathematics education, which regulates the existing educational settings under uncertainty relating to the objectives implied
	Contributions to user coordination	0.17	Contributions in user coordination embrace the framework/depiction in flipped, gamified, and sustainable mathematics education, which regulates the existing educational settings under uncertainty relating to the objectives implied
	Contributions to interactive support	0.07	Contributions in interactive support embrace the framework/depiction in flipped, gamified, and sustainable mathematics education, which regulates the existing educational settings under uncertainty relating to the objectives implied
	Contributions to prominent program	0.05	Contributions to prominent programs embrace the framework/depiction in flipped, gamified, and sustainable mathematics education, which regulates the existing educational settings under uncertainty relating to the objectives implied
Effectiveness	$CI = 0.048; \lambda_{max} = 5.19; CR = 0.04 < 0.1$		

a theory of fuzzy logic, which can undertake the uncertainty of decision-making by way of a sturdy method. In this same context, it can employ, control, and systematize the data under uncertainty across the criteria and sub-criteria. Thus, [85] and [87] in Eq. (4) mentioned that a scale with fuzzified Likert was pertained for emerging the criteria and sub-criteria in the average base. Here, a triangular fuzzy number (TFN) as disclosed in Fig. 2 can be untaken with the functions of fuzzy rating/membership.

Table 3 Criteria and sub-criteria analysed and computed for mathematics under uncertainty

Criteria	Sub-criteria	Weights	Sub-criteria implementation
Mathematics education	Contributions to university professor	0.45	Contribution to university professors embraces the framework/depiction in flipped, gamified, and sustainable mathematics education, which regulates the existing educational settings under uncertainty relating to the objectives implied
	Contributions to university course	0.28	Contributions to university courses embrace the framework/depiction in flipped, gamified, and sustainable mathematics education, which regulates the existing educational settings under uncertainty relating to the objectives implied
	Contributions to the university program	0.12	Contributions to university programs embrace the framework/depiction in flipped, gamified, and sustainable mathematics education, which regulates the existing educational settings under uncertainty relating to the objectives implied
	Contributions to university evaluation	0.09	Contributions to university evaluation embrace the framework/depiction in flipped, gamified, and sustainable mathematics education, which regulates the existing educational settings under uncertainty relating to the objectives implied
	Contributions to the university system	0.06	Contributions to the university system embrace the framework/depiction in flipped, gamified, and sustainable mathematics education, which regulates the existing educational settings under uncertainty relating to the objectives implied
Effectiveness	CI = 0.023; $\lambda_{max} = 5.09$; CR = 0.02 < 0.1		

Table 4 Criteria and sub-criteria analysed and computed for sustainable development under uncertainty

Criteria	Sub-criteria	Weights	Sub-criteria implementation
Sustainable development	Contributions from an environmental perspective	0.43	Contributions from an environmental perspective embrace the framework/ depiction in flipped, gamified, and sustainable mathematics education, which regulates the existing educational settings under uncertainty relating to the objectives implied
	Contributions from a physical perspective	0.28	Contributions from a physical perspective embrace the framework/ depiction in flipped, gamified, and sustainable mathematics education, which regulates the existing educational settings under uncertainty relating to the objectives implied
	Contributions to morphological perspective	0.15	Contributions to the morphological perspective embrace the framework/ depiction in flipped, gamified, and sustainable mathematics education, which regulates the existing educational settings under uncertainty relating to the objectives implied
	Contributions to social perspective	0.09	Contributions from a social perspective embrace the framework/depiction in flipped, gamified, and sustainable mathematics education, which regulates the existing educational settings under uncertainty relating to the objectives implied

(continued)

Table 4 (continued)

Criteria	Sub-criteria	Weights	Sub-criteria implementation
	Contributions from an economic perspective	0.05	Contributions from an economic perspective embrace the framework/depiction in flipped, gamified, and sustainable mathematics education, which regulates the existing educational settings under uncertainty relating to the objectives implied
Effectiveness	$CI = 0.045$; $\lambda_{max} = 5.18$; $CR = 0.04 < 0.1$		

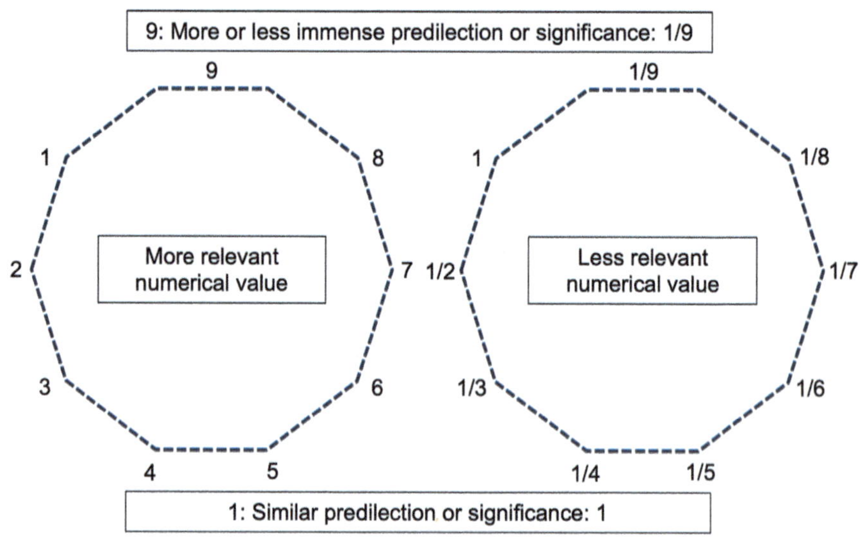

Fig. 1 [88] PCM mathematical rate scales and their proportions

$$Q = \left[q_{ij}\right]c_i \times c_i \tag{4}$$

In the same context of TFN, a triplet (*a*, *b*, *c*) describes each property: lower value (property *a*); medium value (property *b*); and upper value (property *c*) [87], which shows a $(x \leq y \leq z)$ as fuzzy set [1]. Equation (5) defines the TFN fuzzy rating and membership occasion. Here, the fuzzy-logic-DEMATEL method is merged to attain suitable criteria and sub-criteria for flipped, gamified, and sustainable mathematics education under uncertainty, which regulates present educational alterations. In Eq. (6), q_{ij}^e for the weights' assemblages: *e* entails the decision-maker's tendency, and *k* entails the overall number via decision makers.

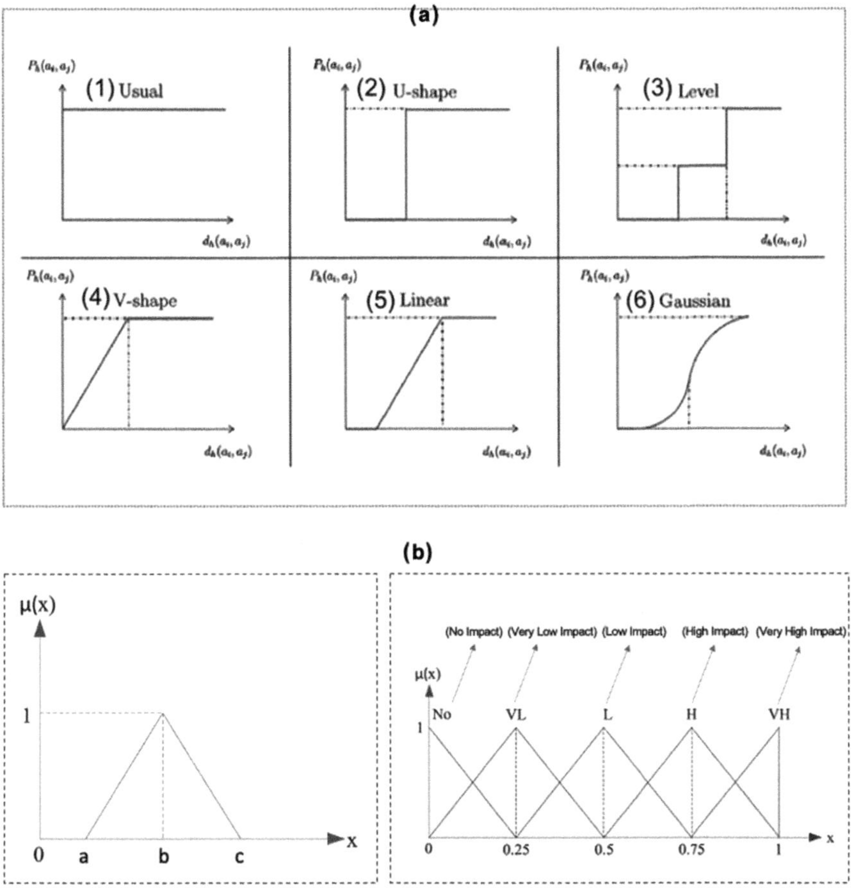

Fig. 2 **a** [39] respective functions pertained to an upsurge of the criteria and sub-criteria and **b** [87] and [1] theory of fuzzy logic through TFNs and fuzzy-rating/membership

$$\mu(x) = \begin{cases} 0, x < a \\ \frac{x-a}{b-a}, a \le x \le b \\ \frac{c-x}{c-b}, b \le x \le c \\ 0, x \ge c \end{cases} \tag{5}$$

$$q_{ij} = \left(q_{ij}^{(a)}, q_{ij}^{(b)}, q_{ij}^{(c)}\right) = \left\{ \begin{array}{c} q_{ij}^{(a)} = min(q_{ij}^{e}) \\ q_{ij}^{(b)} = \sqrt[k]{\prod_{i=1}^{k} q_{ij}^{e}} \\ q_{ij}^{(c)} = max(q_{ij}^{e}) \end{array} \right\} \tag{6}$$

[85] mentioned a cause-effect relationship diagram (CERD) as shown in Eq. (7) is the basis for rates obtained that is the weight coefficients of uncontainable criteria and sub-criteria [36]. [86] as shown in Eq. (8) spectacles firstly the criteria weight

coefficients that are constant and the reliable principal effects' matrix is appropriated. Equation (9) indicates and explains the weight coefficients through the criteria matrix for criteria and sub-criteria corresponding defuzzified by the process of quotations above-mentioned.

$$W_i = \sqrt{(D_i + R_i)^2 + (D_i - R_i)^2} \tag{7}$$

$$w_i = \frac{W_i}{\sum_{i=1}^{n} W_i} \tag{8}$$

$$A = (a^{(l)} + 4a^{(m)} + a^{(u)}) \cdot 6^{-1} \tag{9}$$

In the MCDA/F-DEMATEL/ICTs method, the third step contains the WLC performance. Here, it can blend the constant criteria proportions that can attain the highest appropriate criteria in flipped, gamified, and sustainable mathematics education under uncertainty, which regulates present educational alterations [45]. Particularly, [30] as shown in Eq. (10) specified the indicator scale (0 to 1) of appositeness on multi-criteria dilemmas. Also, it can justify where the greatest appropriate criteria can be for the complication advocated.

$$SI_i = \sum_{j=1}^{n} w_i x_{ij} \tag{10}$$

In the MCDA/F-DEMATEL/ICTs method, the fourth and final step contains a sensitivity analysis [31, 42], which can be a choice for authenticating the determination result of the response data [53, 67]. Figure 3 demonstrates the results of the sensitivity analysis with 5 dissimilar approaches.

In 5 different situations as shown in Fig. 3 and Eq. (11) demonstrated, Chen et al. [10] and [67] denoted a per cent change (PC) average, which can designate the sensitivity analysis that, here, a criterion weight (w_i) can be acquired. [30] as shown in Eq. (12) indicated the primary criterion weight w_{i0}, which can be dissimilar and the supplementary criterion weight w_j can be altered rightfully according to w_i.

$$w_i = w_{i\mp} w_{i0} \times PC \tag{11}$$

$$w_j = (1 - w_i) \times \frac{w_{j0}}{1 - w_{i0}} \tag{12}$$

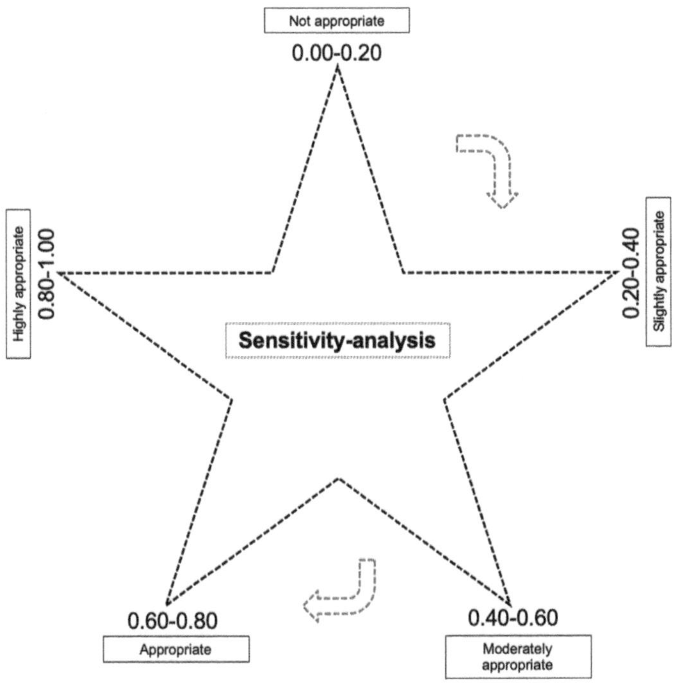

Fig. 3 Results of the sensitivity analysis of flipped, gamified, and sustainable mathematics education under uncertainty

4 MCDA/F-DEMATEL/ICTs Results to Flipped, Gamified, and Sustainable Mathematics Education Under Uncertainty

Firstly, 3 criteria and their 15 sub-criteria as shown in Tables 2, 3 and 4 were estimated and generated based on a scale of 0 to 1, which is the homogenous rates like index pertinence. Here, the criteria and sub-criteria reviewed have been described tougher with the noteworthy assets on the processes of evaluation. Expressly, the decision-makers all the time in the weighting process were intricated. Then, 3 criteria, flipped and gamified technology, mathematics education, and sustainable development, and 15 sub-criteria under uncertainty.

Besides the results via the usage of WLC performance, Table 5 shows the results of sensitivity analysis. Also, in the same table, it reveals the first (1), standing to a selection of decision-maker(s), and 4 different sensitivity-analysis (2) to (5) advances proven in Table 5.

To regulate the current educational settings under uncertainty, the primary criteria for flipped, gamified, and sustainable mathematics education were acknowledged with the 0 to 1 scale rate, which demonstrates the higher the rate the more noteworthy

Table 5 Sensitivity analysis of three criteria and its weight correspondent to criteria according to selection completed

Sensitivity-analysis	Criteria weights		
	Flipped and gamified technology	Mathematics education	Sustainable development
(1) Standing to selection of decision-maker(s) (1)	0,65	0,21	0,14
(2) Standing to match weights of criteria (2)	0.33	0.33	0.33
(3) Standing to flipped and gamified technology (3)	0.50	0.25	0.25
(4) Standing to mathematics education (4)	0.25	0.50	0.25
(5) Standing to sustainable development (5)	0.25	0.25	0.50

criteria and sub-criteria. Concerning 5 distinct scenarios (1) a (5), based on the information indicated in Table 5, scenario (1) was selected as the highest momentous criterion by the processes of weightings/rankings via decision-maker(s). Here, they were polished with the review of expert panel discussion and comprehensive bibliographical examination. The results were 0.65 for flipped and gamified technology, 0.21 for mathematics education criteria, and 0.14 for the sustainable consideration criteria (see Fig. 4). Subsequently, scenario (2) as the least favourite one had the same weight and rank for 3 criteria such as equally 0.33. The rest of scenarios ((3) to (5)) had the importance of one criterion (0.5) and two criteria had the same weight (0.25).

Regarding the sensitivity analysis, its results through the archetype and configuration shaped by the WLC method had remarkable determination/relevance. In Table 6, the expert panel survey was executed to find out the alienated and intangible advances. Here, 41 decision-makers in total participated based on a 5-point Likert scale (i.e., 'very favourable' to 'very unfavourable'). Accordingly, these labourers decided what the improved expert panel test procreated, and the 5 scenarios' potentials were scrutinized.

In detail, regarding scenario (1), it obtained 29% in the section of 'Very favourable' while partaking 83% positive consciousness along with 54% in the section of 'Favorable'. Yet, scenario (2) obtained 13% in the section of 'Very favourable' while partaking 47% positive consciousness along with a 34% in the section of 'Favorable'. Again, both manners, the weightings/rankings and survey-analysis of the criteria and sub-criteria, revealed the same results and preferences; the most preferred scenario (1), and the least preferred scenario (2). Particularly, in the section of scenario (1), flipped and gamified technology obtaining 0.65 was the most disturbed element. As shown in Fig. 5, it was monitored by scenario (3), which contemplated the most important criteria to be flipped and gamified technology.

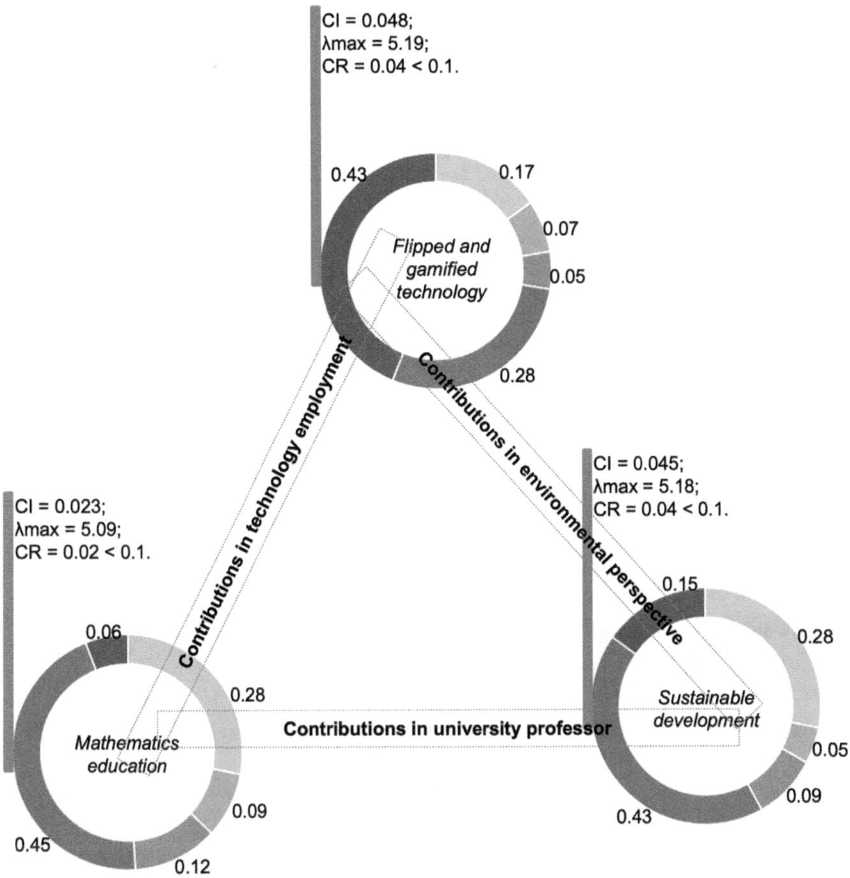

Fig. 4 Scenario (1)—0.65 for flipped and gamified technology, 0.21 for mathematics education criteria, and 0.14 for the sustainable consideration criteria under uncertainty

For the distribution of 5 sub-criteria of scenario (1), the sub-criterion 'contributions in technology employment' weighted 0.43 as the most disturbed element. Here, as shown in Fig. 6, it contains a 47% very favourable result. Furthermore, in the same figure indicated, the scenarios (3) and (5) were unlike, but not meaningfully. Therefore, the scenarios executed can readdress the opportunities of the applicable employed scenarios of flipped, gamified, and sustainable mathematics education, which familiarizes the current pedagogical settings under uncertainty.

With the WCL, sensitivity analysis, and expert panel survey analysis, it generated dissimilar patterns and preferences, which reproduced the outcomes of 15 possible impacts and 3 criteria inside of 5 scenarios. As an indicator-based model, the results denoted the criteria and sub-criteria categorization for flipped, gamified, and sustainable mathematics education, which regulates current pedagogical settings under uncertainty. In the sensitivity analysis, 5 different possible and effectual scenarios

Table 6 Survey analysis of 5 scenarios Likert collection from the WLC

Degree of answer		Scenarios				
		Scenario (1)	Scenario (2)	Scenario (3)	Scenario (4)	Scenario (5)
%	0-Empty/do not know	1	6	5	6	7
	1-Very unfavourable	1	15	8	6	11
	2-Unfarvourable	4	20	11	13	15
	3-Not unfavorable, not favourable	11	12	16	20	16
	4-Favourable	54	34	38	38	36
	5-Very favourable	29	13	22	17	15
Average mean in 5-point Likert (n)		4.01	2.93	3.40	3.29	3.08

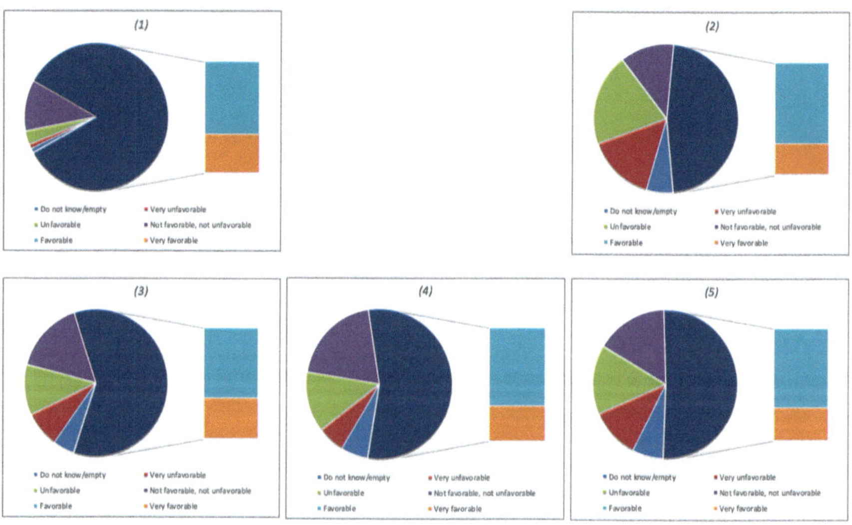

Fig. 5 Survey analysis of 5 scenarios Likert collection from the WLC

were executed. Finally, in the expert panel survey analysis, the results reiterated the same pattern of criteria and sub-criteria selection exploited for flipped, gamified, and sustainable mathematics education under uncertainty with comparable education contexts and attainable data. Thus, in the other aspects, the results illustrated conceivable weaknesses/complications through the process of their choices/actions for flipped and gamified technology contributions in sustainable mathematics education under uncertainty, which have not yet been fittingly subjugated. Therefore, for

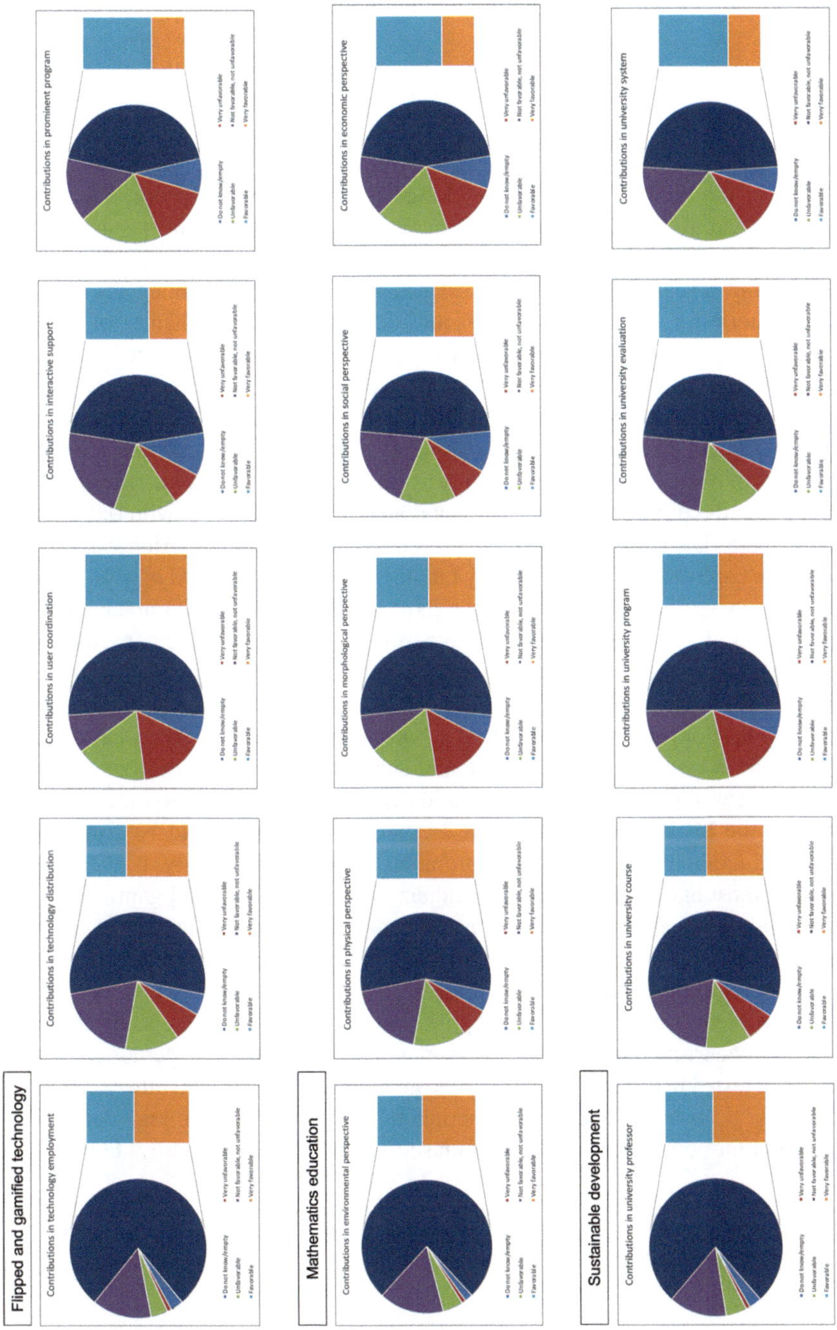

Fig. 6 Survey analysis of 15 sub-criteria Likert collection of 3 criteria through sensitivity-analysis

the objectives proposed, these results could be expended to establish the decision problems raised inside of the projected methodology.

5 Conclusions

The goal of the book chapter was to design and implement appropriate criteria for flipped, gamified, and sustainable mathematics education under uncertainty with an MCDA/F-DEMATEL/ICTs method in mathematics education. The method presented and revealed the possibility of categorizing and examining the criteria and sub-criteria for sustainable mathematics education via the schemes of flipped and gamified technology, which regulates the current pedagogical settings under uncertainty. Based on the decisions made by authors and expert panels through a wide-ranging literature assessment, the MCDA/F-DEMATEL/ICTs method identified 3 criteria and 15 sub-criteria. Subsequently, through the calculation made via the method of WLC and sensitivity analysis, firstly, the most correct criterion was designated as the flipped and gamified technology, i.e. scenario (1) with 0.65 based on a 0–1 scale. Then, the most correct sub-criteria inside of the criterion was designated as the contributions in technology employment, i.e. 0.43 based on a 0–1 scale. Finally, the expert panel survey analysis indicated that scenario (1) obtained 29% in the section of 'Very favourable' while partaking in 83% positive consciousness along with 54% in the section of 'Favorable'. Yet, scenario (2) obtained 13% in the section of 'Very favourable' while partaking 47% positive consciousness along with 34% in the section of 'Favorable'. Again, both manners, the weightings and rankings and survey-analysis of the criteria and sub-criteria, revealed the same results and preferences; the most preferred scenario (1), and the least preferred scenario (2). Therefore, the MCDA/F-DEMATEL/ICTs method proposed could be considered an essential method for the gaps and problems indicated previously. It engaged to certify the most essential sustainable mathematics with flipped and gamified technology criteria, which regulated the current pedagogical settings under uncertainty that corresponded to a contextualized educational matter and robust context of results. Likewise, for the objectives proposed, these results could be expended to establish the decision problems raised inside of the projected methodology. In the same context, the glitches occurring during the decision-making process could be pertained by the method suggested in private industry levels, government sectors, and other academic subjects that persistently necessitate continuous accomplishment of knowledge and propagation for both certain and uncertain settings that might be confronted.

Acknowledgements The authors express their appreciation to the Ministry of Science and Innovation, Research Project (PID2020-115214RB-I00/AEI/https://doi.org/10.13039/501100011033) for making this research possible.

Appendix

Nomenclature: abbreviations used throughout the book chapter.

AHP	Analytic hierarchy process
CERD	Cause-effect relationship diagram
CI	Consistency index
COVID-19	Coronavirus disease 2019
CR	Consistency ratio
DESD	Decade of education for sustainable development
EC	European commission
F-DEMATEL	Fuzzy-decision-making trial and evaluation laboratory
G-ICT	Gamified-information, communication, and technology
GST	Goal-setting theory
ICT	Information, communication, and technology
IIM	Inversed-instruction methodology
MCDA	Multi-criteria decision analysis
MCDA /F-DEMATEL	Multi-criteria decision analysis/fuzzy-decision-making trial and evaluation laboratory
OECD	Organization for economic co-operation, and development
PC	Percent change
PCM	Pair-wise comparison method
RI	Random index
SDT	Self-determination theory
STEM	Science, technology, engineering, and mathematics
TFN	Triangular fuzzy number
UN	United Nations
UNESCO	United Nations Educational, Scientific, and Cultural Organization
WHO	World Health Organization
WLC	Weighted linear combination

References

1. Akyuz, E., Celik, E.: A fuzzy DEMATEL method to evaluate critical operational hazards during gas freeing process in crude oil tankers. J. Loss Prev. Process Ind. **38**, 243–253 (2015)
2. Aunio, P., Heiskari, P., Van Luit, J.E., Vuorio, J.: The development of early numeracy skills in kindergarten in low, average and high performance groups. J. Early Childhood Res. **13**(1), 3–16 (2015)

3. Azeiteiro, U.M., Bacelar-Nicolau, P., Caetano, F.J., Caeiro, S.: Education for sustainable development through e-learning in higher education: experiences from Portugal. J. Clean. Prod. **106**, 308–319 (2015)
4. Bacelar-Nicolau, P., Caeiro, S., Martinho, A., Azeiteiro, U.M., Amador, F.: E-Learning for the environment. The Universidade Aberta (Portuguese open distance university) experience in the environmental sciences Post-Graduate courses. Int. J. Sustain. High. Educ. **10**, 354–367 (2009)
5. Balana, B.B., Mathijs, E., Muys, B.: Assessing the sustainability of forest management: an application of multi-criteria decision analysis to community forests in northern Ethiopia. J. Environ. Manage. **91**(6), 1294–1304 (2010)
6. Bergmann, J., Sams, A.: Flipped learning: maximizing face time. Training & Development: New York (2014)
7. Butterworth, B., Varma, S., Laurillard, D.: Dyscalculia: from brain to education. Science **332**, 1049–1053 (2011)
8. Büyüközkan, G., Çifçi, G.: A novel hybrid MCDM approach based on fuzzy DEMATEL, fuzzy ANP and fuzzy TOPSIS to evaluate green suppliers. Expert Syst. Appl. **39**, 3000–3011 (2012)
9. Cao, W., Fang, Z., Hou, G., Han, M., Xu, X., Dong, J., Zheng, J.: The psychological impact of the COVID-19 epidemic on college students in China. Psychiatry Res. **287**, 112934 (2020)
10. Chen-Yi, H., Ke-Ting, C., Gwo-Hshiung, T.: FMCDM with fuzzy DEMATEL approach for customers' choice behavior model. Int. J. Fuzzy Syst. **9**(4), 236–246 (2007)
11. Crawford, J., Butler-Henderson, K., Rudolph, J., Glowatz, M.: COVID-19: 20 countries' higher education intra-period digital pedagogy responses. J. Appl. Learn. Teach. **3**, 9–28 (2020)
12. Deci, E.L., Ryan, R.M.: The "what" and "why" of goal pursuits: Human needs and the self-determination of behavior. Psychol. Inq. **11**(4), 227–268 (2000)
13. Delivand, M.K., Cammerino, A.R.B., Garofalo, P., Monteleone, M.: Optimal locations of bioenergy facilities, biomass spatial availability, logistics costs and GHG (greenhouse gas) emissions: A case study on electricity productions in South Italy. J. Clean. Prod. **99**, 129–139 (2015)
14. Deterding, S., Dixon, D., Khaled, R., Nacke, L.: From game design elements to gamefulness: Defining" gamification. In Proceedings of the 15th international academic MindTrek conference: Envisioning future media environments (pp. 9–15). Tampere, Finland (2011)
15. Dias, L., Mousseau, V., Figueira, J., Clímaco, J.: An aggregation/disaggregation approach to obtain robust conclusions with ELECTRE TRI. Eur. J. Oper. Res. **138**(2), 332–348 (2002)
16. Diao, G.C.: Building a teaching quality evaluation model for universities based on hierarchical analysis. J. Kaifeng College Educ. **37**(12), 140–141 (2017)
17. Eneroth, C.: E-learning for environment. Improving e-learning as a tool for cleaner production education (Licentiate Dissertation). Lund, Sweden: Lund University (2000)
18. European Commission: Science Education for Responsible Citizenship (2015). Available online: http://goo.gl/Vm3pZ0 (accessed on 14 November 2021).
19. Fauci, A.S., Lane, H.C., Redfield, R.R.: COVID-19—Navigating the uncharted. N. Engl. J. Med. **382**, 1268–1269 (2020)
20. Fletcher, J.M., Lyon, G.R., Fuchs, L.S., Barnes, M.A.: Learning Disabilities: From Identification to Intervention. Guilford Press, New York, NY, USA (2007)
21. Garg, R., Jain, D.: Prioritizing e-learning websites evaluation and selection criteria using fuzzy set theory. Manag. Sci. Lett. **7**, 177–184 (2017)
22. Garrison, R.: Theoretical challenges for distance education in the 21st century: a shift from structural to transactional issues. Int. Rev. Res. Open Distrib. Learn. **1**(1), 1–17 (2000)
23. González-Gómez, D., Airado, D., Cañada-Cañada, F., Jeong, J.S.: A comprehensive application to assist in acid-base titration self-learning: an approach for high school and undergraduate students. J. Chem. Educ. **92**(5), 855–863 (2015)
24. González-Gómez, D., Jeong, J.S.: The flipped learning model in general science: effects on students' learning outcomes and affective dimensions. In: Mintez, J.J., Walater, E.M. (eds.), Active Learning in College Science: The Case for Evidence-Based Practice, pp. 541–549. Springer International Publishing (2020)
25. González-Gómez, D., Jeong, J.S., Airado, D., Cañada-Cañada, F.: Performance and perception in the flipped learning model: An initial approach to evaluate the effectiveness of a new teaching methodology in a general science classroom. J. Sci. Educ. Technol. **25**(3), 450–459 (2016)

26. Govindan, K., Khodaverdi, R., Vafadarnikjoo, A.: Intuitionistic fuzzy based DEMATEL method for developing green practices and performances in a green supply chain. Expert Syst. Appl. **42**, 7207–7220 (2015)
27. Hansen, D.E.: Knowledge transfer in on-line learning environments. J. Mark. Educ. **30**(2), 93–105 (2008)
28. Hill, H.C., Blunk, M., Charalambous, C., Lewis, J., Phelps, G., Sleep, L., Ball, D.L.: Mathematical knowledge for teaching and the mathematical quality of instruction: an exploratory study. Cogn. Instr.. Instr. **26**(4), 430–511 (2008)
29. Jeong, J.S.: Design of spatial PGIS-MCDA-based land assessment planning for identifying sustainable land-use adaptation priorities for climate change impacts. Agric. Syst. **167**, 61–71 (2018)
30. Jeong, J.S., García-Moruno, L., Hernández-Blanco, J.: A site planning approach for rural buildings into a landscape using a spatial multi-criteria decision analysis methodology. Land Use Policy **32**, 108–118 (2013)
31. Jeong, J.S., García-Moruno, L., Hernández-Blanco, J., Jaraíz-Cabanillas, F.J.: An operational method to supporting siting decisions for sustainable rural second home planning in ecotourism sites. Land Use Policy **41**, 550–560 (2014)
32. Jeong, J.S., González-Gómez, D.: A web-based tool framing a collective method for optimizing the location of a renewable energy facility and its possible application to sustainable STEM education. J. Clean. Prod. **251**, 119747 (2020)
33. Jeong, J.S., González-Gómez, D.: A STEM course analysis during COVID-19: a comparison study in performance and affective domain of PSTs between F2F and F2S flipped classroom. Frontiers in Psychology, 3352 (2021a)
34. Jeong, J.S., González-Gómez, D.: Flipped-OCN method in mathematics learning to analyze the attitudes of pre-service teachers. Mathematics **9**(6), 607 (2021)
35. Jeong, J.S., González-Gómez, D.: Mathematics self-belief comparison and examination of pre-service teacher (PST) through a flipped-open calculation based on numbers (ABN) learning method. Heliyon **8**(7), e09806 (2022)
36. Jeong, J.S., González-Gómez, D., Cañada-Cañada, F.: Prioritizing elements of science education for sustainable development with the MCDA-FDEMATEL method using the flipped e-learning scheme. Sustainability **11**(11), 3079 (2019)
37. Jeong, J.S., González-Gómez, D., Conde-Núñez, M.C., Gallego-Picó, A.: Examination of students' engagement with R-SPQ-2F of learning approach in flipped sustainable science course. J. Balt. Sci. Educ. **18**(6), 880–891 (2019)
38. Jeong, J.S., Ramírez-Gómez, A.: A multicriteria GIS-based assessment to optimize biomass facility sites with parallel environment—A case study in Spain. Energies **10**(12), 2095 (2017)
39. Kablan, M.M.: Decision support for energy conservation promotion: an analytic hierarchy process approach. Energy Policy **32**, 1151–1158 (2004)
40. Landers, R.N., Bauer, K.N., Callan, R.C.: Gamification of task performance with leaderboards: a goal setting experiment. Comput. Hum. Behav.. Hum. Behav. **71**, 508–515 (2017)
41. Lee, J., Lee, W.: The relationship of e-learner's self-regulatory efficacy and perception of e-learning environmental quality. Comput. Hum. Behav.. Hum. Behav. **24**(1), 32–47 (2018)
42. Leonelli, M., Görgen, C., Smith, J.Q.: Sensitivity analysis in multilinear probabilistic models. Inf. Sci. **411**, 84–97 (2017)
43. Lim, C.P., Chai, C.S.: An activity theoretical approach to research of ICT integration in Singapore schools: Orienting activities and learner autonomy. Comput. Educ.. Educ. **43**(3), 215–236 (2004)
44. Lin, C.L., Tzeng, G.H.: A value-created system of science (technology) park by using DEMETEL. Expert Syst. Appl. **36**, 9683–9697 (2009)
45. Ma, J., Cao, L.P.: Research on classroom teaching quality evaluation system based on fuzzy hierarchical analysis and fuzzy comprehensive evaluation-Northwest University for Nationalities as an example. J. Central Univ. Nat. (Natural Sci. Edn.) **25**(04), 34–40+45 (2016)
46. Maaß, K., Artigue, M.: Implementation of inquiry-based learning in day-to-day teaching: a synthesis. ZDM Mathematics Educ. **45**, 779–795 (2013)

47. Malczewski, J.: GIS and Multicriteria Decision Analysis. John Wiley and Sons, New York (1999)
48. Malczewski, J., Jackson, M.: Multicriteria spatial allocation of educational resources: an overview. Socioecon. Plann. Sci.. Plann. Sci. **34**, 219–235 (2000)
49. Martín-Sánchez, A., González-Gómez, D., Jeong, J.S.: Service learning as an education for sustainable development (ESD) teaching strategy: design, implementation, and evaluation in a STEM university course. Sustainability **14**(12), 6965 (2022)
50. Martínez-Montero, J., Sánchez, C.: Resolución de problemas y método ABN. Alphen aan den Rijn, The Netherlands: Wolters Kluwer Educación (2013)
51. McVey, M.: E-learning and education for sustainability. Int. Rev. Educ. **62**(1), 117–121 (2016)
52. Melo, L., Cañada-Cañada, F., González-Gómez, D., Jeong, J.S.: Exploring pedagogical content knowledge (PCK) of physics teachers in a Colombian secondary school. Educ. Sci. **10**(12), 362 (2020)
53. Meszaros, C., Rapcsak, T.: On sensitivity analysis for a class of decision systems. Decision Support Syst. **16**, 231–240 (1996)
54. Mora-Granados, M., González-Gómez, D., Jeong, J.S., Gallego-Picó, A.: A molecularly imprinted polymer for selective extraction of phenolic acids from human urine. Appl. Sci. **11**(4), 1577 (2021)
55. Moraros, J., Islam, A., Yu, S., Banow, R., Schindelka, B.: Flipping for success: evaluating the effectiveness of a novel teaching approach in a graduate level setting. BMC Med. Educ. **15**, 27 (2015)
56. Naccarato, E., Karakok, G.: Expectations and implementations of the flipped classroom model in undergraduate mathematics courses. Int. J. Math. Educ. Sci. Technol. **46**(7), 968–978 (2015)
57. Nicholson, S.: A user-centered theoretical framework for meaningful gamification. Paper presented at the proceedings of games + learning + society, Vol. 8, 1 (2012)
58. OECD Library: Global Teaching Insights (2021). Available online: https://www.globalteachinginsights.org. Accessed 23 October 2023
59. Parkes, M., Stein, S., Reading, C.: Student preparedness for university e-learning environments. Internet Higher Educ. **25**, 1–10 (2014)
60. Pavlova, M.: Teaching and learning for sustainable development: ESD research in technology education. Int. J. Technology Design Educ. **23**, 733–748 (2013)
61. Pereira, A., Mendes, A.Q., Morgado, L., Amante, L., Bidarra, J.: Universidade Aberta's Pedagogical Model for Distance Education. Universidade Aberta, Lisbon, Portugal (2008)
62. Perpina, C., Martínez-Llario, J.C., Pérez-Navarro, Á.: Multicriteria assessment in GIS environments for siting biomass plants. Land Use Policy **31**, 326–335 (2013)
63. Saaty, T.L.: The Analytic Hierarchy Process. McGraw-Hill, New York (1996)
64. Sahoo, L., Senapati, T., Yager, R.R.: Real life applications of multiple criteria decision making techniques in fuzzy domain, studies in fuzziness and soft computing (STUDFUZZ, volume 420), Springer Nature (2023)
65. Sams, A., Bergmann, J.: Flip your students' learning. Technol. Rich Learn. **70**(6), 16–20 (2013)
66. Segalàs, J., Ferrer-Balas, D., Svanstrom, M., Lundqvist, U., Mulder, K.F.: What has to be learnt for sustainability? A comparison of bachelor engineering education competences at three European universities. Sustain. Sci. **4**(1), 17–27 (2009)
67. Selçuk, L.: An avalanche hazard model for Bitlis Province, Turkey, using GIS based multi-criteria decision analysis. Turkish J. Earth Sci. **22**, 523–535 (2013)
68. Shee, D.Y., Wang, Y.S.: Multi-criteria evaluation of the web-based e-learning system: a methodology based on learner satisfaction and its applications. Comput. Educ.. Educ. **50**(3), 894–905 (2008)
69. Stub, S.T.: Countries face an online education learning curve (2020). Available online at: https://www.usnews.com/news/bestcountries/articles/2020-04-02/coronavirus-pandemic-tests-countries-abilities-to-createeffective-onlineeducation. Accessed October 25, 2023
70. Su, C.H., Tzeng, G.H., Hu, S.K.: Cloud e-learning service strategies for improving elearning innovation performance in a fuzzy environment by using a new hybrid fuzzy multiple attribute decision-making model. Interact. Learn. Environ. **24**(8), 1812–1835 (2016)

71. Thames, M.H., Ball, D.L.: What mathematical knowledge does teaching require? Knowing mathematics in and for teaching. Teach. Child. Math. **17**(4), 220–229 (2010)
72. Thomas, I.: Critical thinking, transformative learning, sustainable education, and problem-based learning in universities. J. Transform. Educ. **7**, 245–264 (2009)
73. Tourón, J., Santiago, R.: Flipped learning model and the development of talent at school. Revista de Educación **368**, 33–65 (2015)
74. Tucker, B.: The flipped classroom. Online instruction at home frees class for learning. Educación Next **12**(1), 82–83 (2012)
75. UNESCO: UN decade of education for sustainable development, 2005–2014: The DESD at a glance. New York, NY, USA: UNESDOC (2005)
76. UNESCO: COVID-19 Educational disruption and response (2020). Available online at: https://en.unesco.org/covid19/educationresponse. Accessed October 25, 2023
77. Wan, S., Zhong, L., Dong, J.: A new method for group decision making with hesitant fuzzy preference relations based on multiplicative consistency. IEEE Trans. Fuzzy Syst. **28**(7), 1449–1463 (2019)
78. Wang, C., Cheng, Z., Yue, X.G., McAleer, M.: Risk management of COVID-19 by 657 universities in China. J. Risk Financ. Manag. **13**, 36 (2020)
79. Wang, H.G., Wang, S.D.: Analysis of influencing factors of teachers' teaching quality in universities based on DEMATEL method. China Manag. Inform. **22**(23), 221–223 (2019)
80. Wei, X., Cheng, I., Chen, N.S., Yang, X., Liu, Y., Dong, Y., Zhai, X.: Effect of the flipped classroom on the mathematics performance of middle school students. Education Tech. Research Dev. **68**(3), 1461–1484 (2020)
81. Wei, L.R., Qin, J.B.: Application of FAHP in the evaluation of teaching quality in Guangxi Beibu Gulf universities. J. Qinzhou College **28**(02), 8–11 (2013)
82. Werbach, K., Hunter, D.: For the Win: How Game Thinking can Revolutionize Your Business. Wharton Digital Press (2012)
83. WHO: Coronavirus disease (COVID-19) advice for the public (2020a). Available online at: https://www.who.int/emergencies/diseases/novel-coronavirus-2019/advice-for public. Accessed November 29, 2023
84. WHO: Coronavirus disease (COVID-19) pandemic (2020b). Available online at: https://www.who.int/emergencies/diseases/novel-coronavirus-2019. Accessed November 22, 2023
85. Wu, W.W.: Segmenting critical factors for successful knowledge management implementation using the fuzzy DEMATEL method. Appl. Soft Comput.Comput. **12**(1), 527–535 (2012)
86. Wu, W.W., Lee, Y.T.: Developing global managers' competencies using the fuzzy DEMATEL method. Expert Syst. Appl. **32**(2), 499–507 (2007)
87. Zadeh, L.A.: Fuzzy sets. Information. Control **8**, 338–353 (1965)
88. Zeleny, M.: Multiple Criteria Decision Making. McGraw-Hill, New York, NY, USA (1982)
89. Zichermann, G., Cunningham, C.: Gamification by Design: Implementing Game Mechanics in Web and Mobile Apps. O'Reilly Media, Inc. (2011)

Uncertainty Approaches for Spatial Data Management

Frederick E. Petry

Abstract Useful spatial data representations are often critical to decision-making and planning. Many aspects of spatial information involve various factors of uncertainty, imprecise boundaries, and temporal issues, depending on the applications and decisions required. In this survey, we overview soft computing approaches for managing spatial data uncertainty including rough sets, Dempster-Shafer evidence theory and fuzzy set techniques.

Keywords Spatial data · Rough sets · Fuzzy sets · Dempster-Shafer evidence theory

1 Introduction

Useful spatial data representations are often critical to decision-making and planning. Ranging from ecological site issues, and appropriate locations for construction, to planning for operational military issues, all of these decisions involve aspects of spatial data. Many aspects of spatial information involve various factors of uncertainty are surveying, such as imprecise boundaries and temporal issues, depending on applications and decisions required.

For spatial information, the issues of dealing with its imprecision and uncertainty aspects have been considered by many researchers [11, 18, 54]. These issues arise especially in the implementation and usage of geographical information systems (GIS) [8, 26]. When employing geographical information systems in many applications the value of decisions using such tools will largely depend on decision makers' ability to assess the validity and reliability of the information as the basis for their decisions [54]. To make effective use of geographical information systems technology the user has to be able to judge the characteristics of errors and their degree in spatial databases, follow any error during GIS operations, and finally provide

F. E. Petry (✉)
Naval Research Laboratory, Stennis Space Center, Hancock, MS, USA
e-mail: fred.petry@nrlssc.navy.mil

L. Sahoo et al. (eds.), *Decision Making Under Uncertainty Via Optimization, Modelling, and Analysis*, Studies in Systems, Decision and Control 558,
https://doi.org/10.1007/978-981-96-0085-4_6

estimates of the accuracy of the results which may be both graphic and numeric. The broad term accuracy is typically used to represent the various types and characteristics of potential errors in geographical information systems. However, for our purposes in this survey paper, we are concerned with the typical aspects that can be addressed by one or more of the various soft computing techniques. Typically the operations and processes used for spatial data assume that the common attributes and features, as well as their relationships, have a priori been developed exactly and precisely. This ignores the fact that imprecision and inexactness occur for the location of features and the specification of attribute values. Such imprecision and inexactness can be produced in the processes found in several stages of the compilation of data and the development of a spatial database [42]. Spatial data uncertainty modeling has utilized a variety of techniques including concepts from natural language processing, non-monotonic logics, Dempster Shafer evidence theory, fuzzy sets such as type 2 fuzzy and intuitionistic fuzzy sets and rough set theory approaches [5, 12, 23, 44].

2 Rough Set Approaches

An important approach for uncertainty representation uses the rough set concept of indiscernibility of values. We first give a background of rough sets and spatial data [55]. Then we provide the overview needed for rough set theory and discuss the rough set model for imprecise data.

2.1 Rough Set Theory

Rough set theory [31] provides another approach for representing uncertainty. The approximation region in rough set theory provides partitions of a universe into equivalence classes. This partitioning provides a way to increase or decrease granularity, to group items considered indiscernible for a specific application.

Specifically, we have a non-empty domain Z. Then the crucial aspect of uncertainty for rough sets is the indiscernibility relation, or equivalence relation E. This defines the approximation space, $A_r = (Z, E)$ and using E the equivalence classes can be generated where $[x]E$ is an equivalence class containing $x \in Z$. These equivalence classes are termed elementary sets in A_r and the union of elementary sets is called a definable set.

A rough set $R \subseteq Z$, however, can be defined in terms of the upper $E * R$ approximation region and lower approximation region (ER), where

$$ER = \{x \in Z|[x]E \subset R\} \; and \; E * R = \{x \in Z|[x]E \cap R = \Phi\}$$

ER is the positive region, $Z - E * R$ is the negative region, and $E * R - ER$ is the boundary or borderline region of the rough set X, allowing for the distinction between

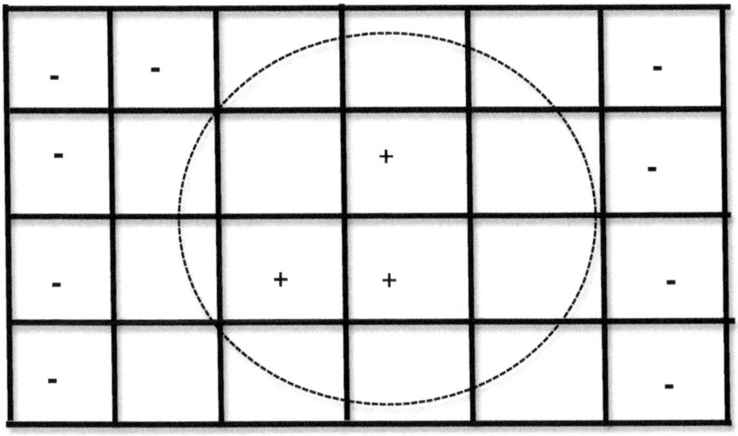

Fig. 1 Illustration of the concept of a rough set X

certain and possible inclusion in a rough set. The set approximation regions provide a mechanism for determining whether something certainly belongs to the rough set, may belong to the rough set, or certainly does not belong to the rough set. Given the upper and lower approximations $E * R$ and ER, of R a subset of Z, the R-positive region of R is $POSR(R) = ER$, the R-negative region of R is $NEGR(R) = Z - E * R$ and the boundary or R-borderline region of R is $BNR(R) = E * R - ER$. R is called R-definable if and only if $ER = E * R$. Otherwise, $ER \neq E * R$ and R is rough with respect to E.

Figure 1 provides an illustration of a rough set. The grid of rectangles represents the equivalence classes that partition the universe. The lower approximation values are indicated by '+' and those in the E-negative region by '−'. Then the boundary region of the upper approximation is all other classes.

2.2 *Rough Spatial Example*

Now we can provide an example using spatial terms such as might be found in an analyst labeling of features in the remote sensed image [49]. Let the universe Z of terms being used be Z = {bridge, hill, tor, ridge, pasture, field, farm, bayou, creek}. Now define an equivalence relation E as follows:

$$\mathrm{E} = \{[\mathrm{bridge}], [\mathrm{hill,\ tor,\ ridge}], [\mathrm{pasture,\ field,\ farm}][\mathrm{bayou,\ creek}]\}.$$

Now specifically we are given a particular set R = {bridge, hill, ridge, pasture, field, farm, bayou}, then we must determine the upper and lower approximations:

ER = {bridge, pasture, field, farm}, and.

E*R = {bridge, hill, tor, ridge, pasture, field, farm, bayou, creek}.

So the lower approximation consists of equivalence classes that are included totally in the set R. The upper approximation contains the lower approximation plus those classes that are only partially included in R. In the above data, all the values in the classes [bridge] and [pasture, field, farm] are included in R so these belong to the lower approximation region. Then [bayou, creek] is a class that isn't completely included in ER since 'creek' isn't in R. However, since 'bayou' $\in R$ then [bayou, creek] is part of the upper approximation. Finally rough set in A_r consists of a group of subsets of Z having identical upper and lower approximations. So for the example we have been considering, some representative elements of the rough set are:

{{bridge, bayou, creek, pasture, field, farm}, {bridge, bayou, creek, hill, tor, ridge}, {bridge, pasture, field, farm,}.

2.3 Rough Sets Background

An early approach for spatial data is found in the ROSE system [38]. This proposed a rough sets approach for realm-based spatial information detailing a framework for the formal modeling of such data. Imprecision in spatial data can often depend on the resolution for which the data was specified [50] and then applied to problems of data integration. Resolution consideration corresponds with the indiscernibility issue, which is a central aspect of rough set representation. In another approach, rough set theory [1] was used to develop spatial data rough classification and also to provide a representation of spatial locations. This approach also outlined a quality measure for the rough classifications in contrast to the crisp classifications. By evaluating this approach using real-world vegetation map data from map layers, they determined how to combine rough set and fuzzy set techniques. This enabled the geographical information reclassification needed for final data integration. An approach to mapping and geographical information systems [49] used the concept of a rough raster space for spatial objects' field representations. They provided an assessment of the approach using images obtained from remote sensing for a case study of their classifications. In [6] K-labeled partitions are used to represent maps. Using this they consider the partitions in their relationship as rough sets and so are capable of providing approximations to objects with vague boundaries in maps. Extensions were considered for stratified partitions. This enabled better granularity detailed levels that can provide better map scale transformations and develop the approaches for stratified rough sets.

2.4 Rough Sets for Gridded Spatial Data

To obtain possible results, in addition to the obvious, certain results encountered in querying an ordinary spatial database system, the use of the boundary region information may be used in addition to that of the lower approximation region. The results in the lower approximation region are certain and correspond to exact matches. The boundary region of the upper approximation contains those results that are possible, but not certain.

The approximation regions of rough sets are beneficial whenever information concerning spatial data regions is accessed. Consider a spatial domain with imprecise boundaries. For gridded data, this means depending on granularity, points representing different spatial classes can fall into the same grid. Consider a region such as a forest. One can reasonably conclude that any grid point identified as FOREST that is surrounded by grid points also identified as FOREST is a point represented by the feature FOREST. However, consider points identified as FOREST that are adjacent to points identified as MEADOW. It may be possible that these points represent meadow areas as well as forest areas but were identified as FOREST in the classification process. Likewise, points identified as MEADOW but adjacent to FOREST points may represent areas that contain part of the forest. This uncertainty maps naturally to the use of the approximation regions of the rough set theory, where the lower approximation region represents certain data and the boundary region of the upper approximation represents uncertain data. It applies to spatial database querying and spatial database mining operations.

If we force a finer granulation of the partitioning, a smaller boundary region results [50]. This occurs when the resolution is increased. As the partitioning becomes finer and finer, finally a point is reached where the boundary region is non-existent. Then the upper and lower approximation regions are the same and there is no uncertainty in the spatial data as can be determined by the representation of the model.

3 Dempster-Shafer Approaches

Often spatial data can be specified by locations represented as intervals. To deal with their uncertainty we will use Dempster-Shafer (DS) theory and first review the basics of the theory.

3.1 Dempster-Shafer Theory and Intervals

Dempster-Shafer's theory [41] is a general approach to representing uncertainty. Ordinarily, in probability theory, the sum of probabilities must be one. Here, D = S theory permits the representation of uncertain information and allows its incomplete specification in contrast to probability. A Dempster-Shafer (D-S) belief structure for a domain X consists of subsets called focal elements: $\{ F_1,...F_q\}$, $F_i \subset X$, $F_i \neq \Phi$, representing a variable V whose domain is X. Then the basic probability assignment, BP = m, assigns a belief mass m to each part of the power set:

$$m:2^x \rightarrow [0, 1]$$

So the specification of a value, v, is inexact where for the focal set, $F_i \subset X$, and $m(v)$ indicates the probability that $v \in F_i$. For the basic probability m, we have two properties: 1. mass of a null set is zero and 2. The sum of the masses, including any unspecified or incomplete specification, is one:

1. $m(\Phi) = 0$ m (Φ) = 0
2. $\sum m(R_i) = 1$, where $R_i \in 2^x$

Two commonly used measures for a D-S belief structure are measures of belief (best case) and plausibility (worst case) [40]. The belief for a specific set S, $Bel(S)$, is the sum of the masses of all subsets of S:

$$Bel(S) = \sum m(R_i) \; where \; \mathrm{R_i} \in \mathrm{S}$$

The plausibility, $Pl(S)$, is the sum of the masses of the sets $\mathrm{R_i}$ that intersect S:

$$Pl(S) = \sum \mathrm{m(R_i)} \; where \; R_i \cap S = \Phi$$

Here, we are considering a representation of spatial uncertainty in spatial locations by intervals instead of sets where the intervals consist of ranges of spatial values. There have been several Dempster-Shafer applications utilizing intervals [2, 24]. Now we define the basic concept of intervals [28]. Letting X be the space under consideration then an interval, L, will be represented by the lower and upper bounds, x_*, x^*:

$$L = [x_*, x^*] = \{v \in X \,|x_* \le v \le x^*\}$$

3.2 Uncertain Spatial Locations

An uncertain spatial location, $\mathrm{L_k}$, will be represented in this section as a D-S belief structure by using two nested spatial intervals, LA_k and LB_k. We let LA_k be an outer or less certain interval and LB_k a more certain innermost interval. Each spatial interval uses lower and upper spatial values, x_i, and has the associated belief masses, $m(LA_k)$, $m(LB_k)$.

$$LA_k = [x_1, x_4], \; LB_k = [x_2, x_3], \; x_1 \le x_2 \le x_3 \le x_4$$

where $LB_k \subseteq LA_k$.

Additionally for the D-S masses

$$m(LA_k) = 1 - M\,(LB_k)$$

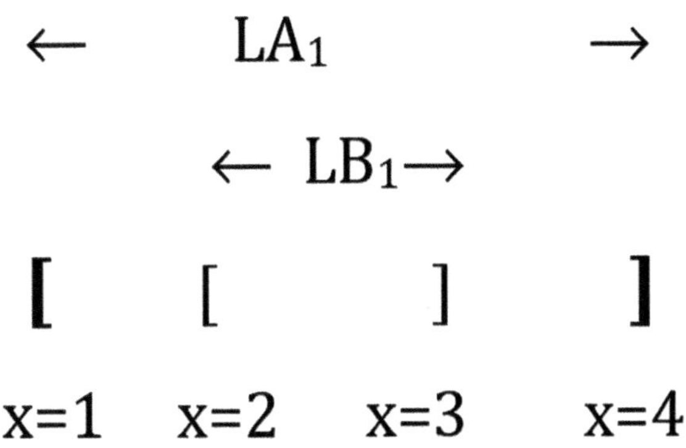

Fig. 2 Example of the location represented as nested intervals

To provide illustrative examples of uncertain locations for clarity we consider locations along the x-axis. For example, consider location one, L_1, which is illustrated in Fig. 2:

$$L_1\text{: } (\mathrm{LB}_1 = [2, 3], \mathrm{m}([2, 3]) = 0.75); (\mathrm{LA}_1 = [1, 5], \mathrm{m}([1, 5]) = 0.25)$$

The basic semantic interpretation is that an individual will specify an inner interval, LB_k, and as the best or most precise they can specify the location values with the degree of certainty, $m(LB_k)$. Then the outer interval, LA_k, represents the range of the values that are possible for the values with possibility $m(LB_k)$.

Let D be the length of an interval, e.g., LA_k, as $D(LA_k)$. D is the difference between the upper and lower values of the interval

$$D(LA_k) = |x_1 - x_4|.$$

So for $LA_1 = [1, 5][1, 5], D(LA_1) = 4$.

It can also be useful to consider the certainty of an interval length and let a weighted length, weighted by the belief mass of the interval be

$$D'(LB_k) = |x_3 - x_2| * m(LB_k).$$

The term <u>definite</u> location refers to a spatial location L_k for the special case where $LA_k = LB_k$. This simply means there is complete certainty that the values of x are totally contained in a single interval, i.e., for $LB_k, m(LB_k) = 1$. Furthermore consider the case for location L_k, where the interval LA_k is degenerate, $LA_k = (x_1, x_1)$. Here denotes the location as <u>crisp</u> and then $m(LA_k) = 1$.

3.3 Relationships Between Locations

There are a variety of approaches to evaluate the relationships between locations. Relations of spatial intervals can be considered as two aspects:

1. Topological relationships such as overlaps
2. Metric relationships which consider the distance between intervals

An approach known as the nine-intersection model provides topological relations, denoted S, between two locations to be defined [14]. For spatial location relationships, these are:

before, equal, meets, overlaps, contains, starts, finishes.

There are thirteen possible relationships including inverses. Note that the inverse of equal is also equal [46].

Two issues must be considered in describing relationships. The relationship must consider the D-S uncertainty of locations in specifying the possible relationships between locations. Furthermore, any relationship between locations must take into account relationships relative to the inner and outer intervals of locations as well [27].

In general, we will assume the inner interval does not meet the outer interval, ($NotS_m(L_2, L_1)$). So an inner interval has a strictly "inside" relationship to the outer interval and we can transitively infer certain relations. For example, let a location L_2 have an overlaps relation (but is not contained by) with the outer interval of the location L_1. Then by the transitivity, the following are relations with the inner interval of L_1:

before, meets, overlaps, finishes and contained by.

Next, we will examine the determination of relationships in a more quantitative manner. First we consider selected possible relationships of intervals, $S_{type}(L_2, L_1)$ between L_1 and L_2. Here we consider that L_2 might be moving nearer, that is becomes closer to L_1. We denote these as

$S_b(L_2, L_1)$: L_2 is in front of or before L_1.

$S_m(L_2, L_1)$: L_2 meets L_1 (touching).

$S_o(L_2, L_1)$: L_2 overlaps L_1.

$S_c(L_2, L_1)$: L_2 contains L_1 (inside of).

For simplicity to start, we let L_2 be a definite location and L_1 a fully uncertain location.

We let these be given as (illustrated in Fig. 3)

$$L_2 = \{\ \mathrm{LB}_2 = \mathrm{LA}_2 = [\mathrm{x}_1, \mathrm{x}_2];\ \mathrm{m}(\mathrm{LB}_2) = 1\}\ ;$$

$$L_1 = \{\ \mathrm{LA}_1 = [\mathrm{x}_3,\mathrm{x}_6];\ \mathrm{m}(\mathrm{LA}_1);\ \mathrm{LB}_1 = [\mathrm{x}_4\ ,\ \mathrm{x}_5\ ];\ \mathrm{m}(\mathrm{LB}_1)\}$$

The component relationship of both inner and outer intervals must be considered to determine relationships between locations. We can illustrate this in several cases.

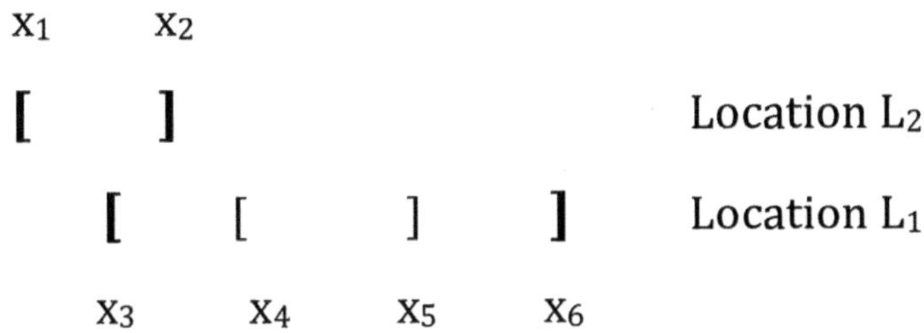

Fig. 3 Example of definite and fully uncertain location

A more complex set of potential relationships must be determined if both locations are fully uncertain.

For the following cases since we are considering relationships with the definite location L_2 we make use of the values $m(LA_2) = m(LB_2) = 1$.

Case 1: $x_2 > x_3$ and $x_2 < x_4$.

In this case, there are two relationships: overlap with the outer interval and before the inner interval.

$$\{ S_b(L_2,LB_1),m(LB_1)*1;So(L_2,LA_1),m(LA_1)*1\}$$

There are other relations possible since the condition on x_1 was unspecified.

Case 2: In particular if it is also true that $x_2 > x_3$ then

$$S_c(L_2,LA_1)\ ,m(LA_1) * 1$$

Case 3: Also if there is some imprecision in the interval bounds such as $x_2 \geq x_3$ then a meet relationship would also be possible

$$S_m(L_2,LA_1)\ ,m(LA_1) * 1$$

4 Fuzzy Set Approaches for Spatial Data

In this section, we will develop an approach for spatial data by beginning with the basic constructs of point and lines using a fuzzy set approach.

4.1 Fuzzy Set Definitions

Fuzzy set representations [22, 33, 53] provide membership degrees of data values in a set as opposed to crisp sets. For a domain D, a fuzzy set, FS, is

$$\text{FS(D)} = \{ <a_i, \mu(a_i) > |0 \leq \mu(a_i) \leq 1\} , a_i \in D, i = 1,2,\ldots,n$$

where a_i is a data value and $\mu(a_i)$ is the membership of the data value.

A number of researchers in the GIS and spatial database area have considered models of spatial data using fuzzy set approaches, as in the modeling of geographic objects with indeterminate boundaries [7, 15, 45] and on uncertainty in GIS and spatial data [9, 34, 47]. Some early work by geographers in the 1970s utilized fuzzy sets in topics such as behavioral geography and geographical decision-making [16, 25]. However, the first consistent approach to the use of fuzzy set theory as it could be applied in GIS was developed by Robinson [35–37]. More recently, there have been several efforts utilizing fuzzy sets for spatial databases including some for capturing spatial relationships [10, 19], querying spatial information [29, 49], and object-oriented model modeling [12, 17, 30].

4.2 *Fuzzy Spatial Points*

The basic structure of spatial data is considered to be the point. In this case, we are examining uncertain points using fuzzy memberships. Underlying the point definition is the structure of the composite coordinates. We next provide definitions of the types of coordinates to be considered.

An uncertain coordinate, Co_k of R^n, has its value $v(Co_k)$ represented in general as an uncertain interval $I = [x_1, x_2, ..., x_n]$. For the sake of clarity and illustration, we will use two-dimensional coordinates, $I = [x_1, x_2]$. So finally

$$v(Co_k) = \{I_k, \mu(I_k)\}$$

The basic semantic interpretation here is that an application or an analyst can specify a coordinate's values with some degree of certainty, $\mu(I_k)$.

Next, we define the concepts of certain coordinates and crisp coordinates. A coordinate Co_k is called certain for the special case where $\mu(I_k) = 1$. This is interpreted as meaning there is complete certainty that the values of the coordinate Co_k are contained in a single interval, i.e., I_k.

Now for a certain coordinate Co_k, if the interval I_k is degenerate we term this coordinate as being crisp. For example, if $I_k = (a, a)$, the coordinate value Co_k is specific, i.e. $v(Co_k) = a$, and $m(Co_k) = 1$. For clarity, Co_k will be used instead of $v(Co_k)$ subsequently.

Next, we utilize the above definitions of coordinates to develop approaches for the types of uncertain points, completely and partially uncertain. Let $S = R^n$ be the n-dimensional coordinate space forming the context of our application. Then Pt_j is an uncertain point in the space S, where

$$Pt_j = (Co_1, Co_2, ..., Co_n)$$

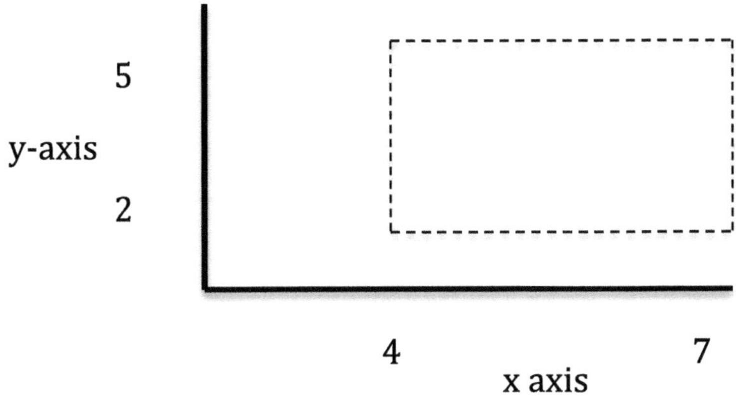

Fig. 4 Uncertain point in two-dimensional plane

where each Co_i is in general an uncertain coordinate of the space. A given point will be called completely uncertain if all coordinates Co_j are completely uncertain.

We must consider the subcases of an uncertain point termed a partially uncertain point. If some coordinates of such a point are certain coordinates or even a crisp coordinate, these sorts of points are termed partially uncertain.

Now in order to illustrate spatial concepts we will use the ordinary space R^2 in which we can give specific examples. Then for each Pt_j, we consider an ordered pair of coordinates, (Co_x, Co_y), the x and y coordinates, respectively. For example, consider a completely uncertain point, $Pt(Co_x, Co_y)$, letting the x and y coordinates both be imprecise:

$$Co_x = \{ \mathrm{I_x} = [4, 7], [4, 7] = 0.6 \}$$

$$Co_y = \{ \mathrm{I_y} = [2, 5], [2, 5] = 0.5 \}$$

When it is clear we can omit the subscripts, x or y, using a conventional presentation of using the x-coordinate first. In the ordinary two-dimensional plane, this will be presented as a rectangle as shown in Fig. 4.

4.3 *Fuzzy Spatial Lines*

So, using two uncertain points Pt_1 and Pt_2 we can provide a definition of an uncertain line Ln. Again, for simplicity, we use only two points, Pt_1, Pt_2, for specifying a line:

$$Ln(Pt_1, Pt_2)$$

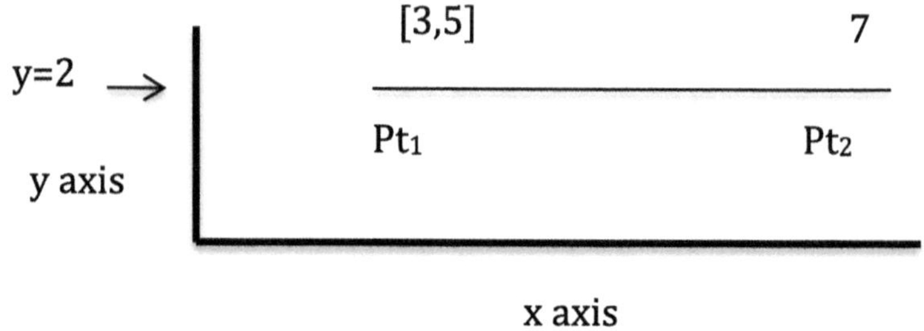

Fig. 5 Example of line: partially uncertain point, Pt_1, crisp point Pt_2

Then we must examine the several cases of uncertainty for a line based on the characteristics of the defining points of the line.

Case i: Each point Pt_1 and Pt_2 are completely uncertain.

Case ii: Either one or both of points are partially uncertain.

Case iii: A crisp line is then the case where both points are crisp points.

In order to use these ideas we will consider how to define line length for these cases. To have a well-defined or consistent length for an uncertain line overlap of the defining points Pt_1 and Pt_2 must be disallowed. This will then be valid if the intervals of their coordinates are not overlapping:

$$\forall \mathrm{j}, \mathrm{k}, \mathrm{I_j} \in \mathrm{Pt_1}, \mathrm{I_k} \in \mathrm{Pt_2}: \mathrm{I_j} \cap \mathrm{I_k} = \Phi$$

We provide one example of a line as shown in Fig. 5. This illustrates Case ii where there is a partially uncertain point, Pt_1, and a crisp point Pt_2.

$$Pt_1 = (\mathrm{Cox} = \{\ \mathrm{I_x} = [3, 5], [3, 5] = 0.7\}\ , \mathrm{Coy} = \{\ \mathrm{I_y} = [2, 2], [2, 2] = 1.0\}\ )$$

$$Pt_2 = (\mathrm{Cox} = \{\ \mathrm{I_x} = [7, 7], [7, 7] = 1.0\}\ , \mathrm{Coy} = \{\ \mathrm{I_y} = [2, 2], [2, 2] = 1.0\}\ )$$

The basic foundation measure for spatial structures is the length of a line Ln. Length is calculated using the Euclidean distance Ds between the end points of Ln, Pt_1 and Pt_2, as determined by the coordinates. For an ordinary distance Ds, three properties must be satisfied;

1. $Ds(Pt_1, Pt_2) = 0$; reflexive
2. $Ds(Pt_i, Pt_j) = Ds(Pt_j, Pt_i)$ symmetric
3. $Ds(Pt_i, Pt_j) + Ds(Pt_j, Pt_k) \geq Ds(Pt_i, Pt_k)$ triangle inequality

So then the distance between points Pt_i, Pt_j in R^n can be given by the Minkowski metric where in particular the Euclidean distance is the L_2 metric. For the line of Fig. 5 we have the interval length using $Ds(Pt_1, Pt_2)$:

$$Ln(Pt_1, Pt_2) = (([7-5], [7-2] = [2, 5]); (([2, 5]) = [2, 5] * [7, 7] = 0.7{*}1.0 = 0.7))$$

5 Summary and Directions

In this review, we have surveyed three major soft computing approaches for handling uncertain spatial data. Rough set theory uses the idea of indiscernibility of values to allow a granular approach for representing the uncertainty of spatial data. The Dempster-Shafer approach described used a nested interval representation of spatial data. It provides the mechanism to allow the uncertainty to be described as involving both certainty and possibility of the intervals. Representations of imprecise spatial concepts including points and lines were presented using fuzzy set approaches.

Several extensions of soft computing mechanisms could be applied to spatial data. One development is a combined approach of rough and fuzzy sets [13]. Another uses the specification of degrees of both membership and non-memberships known as fuzzy intuitionistic sets [3]. These have the restriction that the sum of the memberships is bounded by one. This range has been extended in the developments known as Pythagorean and Fermatean sets [39, 52].

5.1 Information Measures

Imprecise data need to have approaches to quantify the information content. Such measures are of value in evaluating combinations of spatial data where the data is reduced to allow effective decision-making [4].

An uncertainty measure for a rough set can be based on the indiscernibility relation that is defined on the domain, as this produces a partitioning into a finite set of equivalence classes. First, there is no loss of information when based on the partitioning as only one value occurs in each of the equivalence classes. However there are not as many classes in a coarser partitioning, and so every class is of a larger cardinality.

Specific assessments of the imprecision of a rough set R have been proposed [32]. The first proposed was the accuracy, that is the ratio of the count of values in the lower approximation, ER, relative to the upper approximation $E * R$ count. This provides a metric relative to the completeness of knowledge of the rough set R. So, it can be given as the set cardinalities ratio, AC_R.

$$ACR = Card(ER)/Card(E * R) \; where \; 0 \leq ACR \leq 1.$$

Correspondingly another derived measure, roughness $\mathrm{IN_R}$, provides the incompleteness of knowledge of the set's values. It is simply derived from the accuracy:

$$IN_R = 1 - AC_R$$

There have several studies interested in uncertainty measures for interval evidence theory [21, 43] including entropy metrics [51]. In particular, for the Dempster-Shafer approach, there are two functions—belief and plausibility. The belief measure, $Bel(S)$, refers to how much an evidence statement S is supported at least partially, which then provides a lower bound. Correspondingly the plausibility, $Pl(S)$, is an upper bound on the support for the truth of a statement S since there is a limited amount of evidence showing the statement is false. So plausibility is related to belief by being $1 - Bel(\sim S)$ where $\sim S$ is negation of the statement. Then it has a range in the interval [0, 1] and provides a measure of how much the evidence is in favor of $\sim S$.

Acknowledgements We would like to acknowledge the Naval Research Lab's Base Program for supporting this research.

References

1. Ahlqvist, O., Keukelaar, J., Oukbir, K.: Rough classification and accuracy assessment. Int. J. Geogr. Inf. Sci. **14**(5), 475–496 (2000)
2. Auer, E., Luther, W., Rebner, G., Limbourg, P.: A verified MATLAB toolbox for the Dempster-Shafer theory. Proceedings European Conference on Safety and Reliability, Gdansk Po., pp. 125–130 (2005)
3. Atanassov, K.: Intuitionistic fuzzy sets. Fuzzy Sets Syst. **20**, 87–96 (1986)
4. Beaubouef, T., Petry, F.: Arora G information-theoretic measures of uncertainty for rough sets and rough relational databases. Inf. Sci. **109**(1), 185–195 (1998)
5. Beaubouef, T., Petry, F., Breckenridge, J.: Rough set based uncertainty management for spatial databases and geographical information systems. In: Soft Computing in Industrial Applications (ed. Y. Suzuki), Springer-Verlag, London, Chap 6a. pp. 471–479 (2000)
6. Bittner, T., Stell, J.: Stratified Rough Sets and Vagueness. In: Kuhn, W., Worboys, M.F., Timpf, S. (eds.) COSIT 2003. LNCS, vol. 2825, pp. 286–303. Springer, Heidelberg (2003)
7. Burrough, P., Frank, A. (eds.): Geographic Objects with Indeterminate Boundaries, GISDATA Series, vol. 2. UK, Taylor and Francis, London (1996)
8. Chang, K.: Introduction to Geographic Information Systems (9th ed.). McGraw-Hill. p. 1. ISBN: 978-1-259-92964-9 (2016)
9. Cobb, M., Petry, F.: Geometric approximations of spatial boundaries and assessment of fuzzy spatial relationships. Fuzzy Sets Syst. **113**, 111–120 (2000)
10. Cobb, M., Petry, F.: Modeling spatial data within a fuzzy framework. J. Amer. Soc. Inform. Sci. **49**(3), 253–266 (1998)
11. Couclelis, H.: The certainty of uncertainty: gis and the limits of geographic knowledge. Trans. GIS **7**(2), 165–175 (2003)
12. Cross, V., Firat, A.: Fuzzy objects for geographical information systems. Fuzzy Sets Syst. **113**, 19–36 (2000)
13. Dubois, D., Prade, H.: Rough fuzzy sets and fuzzy rough sets. Int. J. General Syst., 17 (1990). https://doi.org/10.1080/03081079008935107
14. Egenhofer, M., Franzosa, R.: Point set topological spatial relations. Int. J. Geogr. Inform. Syst. **5**(2), 161–174 (1991)
15. Fisher, P.: Boolean and fuzzy regions. In: Burrough, P.A., Frank, A. (eds.) Geographic Objects with Indeterminate Boundaries, pp. 87–94. Taylor and Francis, London (1996)

16. Gale, S.: Inexactness, fuzzy sets, and the foundations of behavioral geography. Geogr. Anal. **4**, 337–349 (1972)
17. George, R., Buckles, B., Petry, F., Yazici, A.: Uncertainty modeling in object-oriented geographical information systems. In: Proceedings of 1992 Conference on Database and Expert System Applications, pp. 77–86. Springer, Berlin (1992)
18. Goodchild, M.: Twenty years of progress: GIScience in 2010. J. Spatial Inform. Sci. (1) (2010). https://doi.org/10.5311/JOSIS.2010.1.2
19. Guesgen, H., Albrecht, J.: Imprecise reasoning in geographic information systems. Inter J. Fuzzy Sets Syst. **113**, 121–131 (2000)
20. Guesgen, H.: Fuzzifying spatial relations. In: Matsakis, P., Sztandra, L. (eds.) Applying Soft Computing in Defining Spatial Relations, pp. GR 99–122. Physica Verlag, Heidelberg (2002)
21. Jiang, W.: Wang S an uncertainty measure for interval-valued evidences. Int. J. Comput. Commun. Control **12**(5), 631–644 (2017)
22. Klir, G.: Uncertainty and Information. Wiley, Hoboken NJ (2006)
23. Kruse, R., Borgelt, C., Braune, C., Mostaghim, S.: Steinbrecher M Computational Intelligence: A Methodological Introduction, 2nd edn. Springer Verlag, London (2016)
24. Lee, E., Zhu, Q.: An interval dempster-shafer approach. Computers Math. Applic. **24**(7), 89–95 (1992)
25. Leung, Y.: Locational choice: a fuzzy set approach. Geographic Bulletin **19**, 28–34 (1979)
26. Longley, P., Goodchild, M., Maguire, D., Rhind, D.: Geographic Information Systems and Science 3rd Ed., Wiley Chichester UK (2010)
27. Matsakis, P.: Understanding the Spatial Organization of Image Regions by Force Histograms. In: Matsakis, P., Sztandra, L. (eds.) Applying Soft Computing in Defining Spatial Relations, pp. 1–16. Physica Verlag, Heidelberg (2002)
28. Moore, R., Kearfott, B., Cloud, M.: Introduction to Interval Analysis. SIAM, Philadelphia, PA (2009)
29. Morris, A., Petry, F., Cobb, M.: Fuzzy object-oriented database modeling of spatial data. In: Proceedings of the IPMU Conference. Paris, EDK Press, pp. 604–11 (1998)
30. Morris, A.: A framework for modeling uncertainty in spatial databases. Trans. GIS **7**, 83–101 (2003)
31. Pawlak, Z.: Rough sets. Int. J. Man-Mach. Studies **21**, 127–134 (1984)
32. Pawlak, Z.: Rough Sets: Theoretical Aspects of Reasoning about Data. Springer, Dordrecht (1991)
33. Pedrycz, W., Gomide, F.: An Introduction to Fuzzy Sets: Analysis and Design. MIT Press, Boston MA (1996)
34. Petry, F., Cobb, M., Wen, L., Yang, H.: Design of system for managing fuzzy relationships for integration of spatial data in querying. Fuzzy Sets Syst. **140**, 51–73 (2003)
35. Robinson, V.: Interactive machine acquisition of a fuzzy spatial relation. Comput. Geosci. **6**, 857–872 (1990)
36. Robinson, V.: Implications of fuzzy set theory for geographic databases. Comput. Environ. Urban Syst. **12**, 89–98 (1988)
37. Robinson, V., Frank, A.: About different kinds of uncertainty in geographic information systems. In: Proceedings AUTOCARTO 7 Conference, pp. 440–449. American Society for Photogrammetry and Remote Sensing, Falls Church (1985)
38. Schneider, M.: Spatial Data Types for Database Systems Doctoral Thesis. Fern Universität Hagen, GR (1995)
39. Senapati, T., Yager, R.: Fermatean fuzzy sets. J. Ambient. Intell. Humaniz. Comput. **11**, 663–674 (2020)
40. Shafer, G.: Belief functions and possibility measures. In: The Analysis of Fuzzy Information vol. 1, pp. 51–84. CRC Press, Boca Raton, FL (1987)
41. Shafer, G.: A Mathematical Theory of Evidence. Princeton University Press, Princeton (1976)
42. Shekar, S., Chawla, S.: Spatial Databases: A Tour, Upper Saddle River, NJ, Prentice Hall (2003)
43. Song, Y., Wang, X., Lei, L.: Yue S Uncertainty measure for interval-valued belief structures. Measurement **80**, 241–250 (2016)

44. Stoms, D.: Reasoning with uncertainty in intelligent geographic information systems. In: Proceedings of GIS87. Falls Church, VA, American Society for Photogrammetry and Remote Sensing, pp. 69–39 (1987)
45. Varsi, A.: Vagueness in geography. Philos. Geogr. **4**, 49–65 (2001)
46. Vazirgiannis, M.: Uncertainty handling in spatial relationships. In: ACM-SAC 2000 Proceedings, Como Italy, pp. 215–221 (2000)
47. Verstraete, J.: Surface area of level-2 fuzzy regions unifying possibilistic and versitic interpretations of regions. eds. Rutkowski, L., Korytkowski, M., Scherer, R., Tadeusiewicz, R., Zadeh, L. and Zurada, J. Artifical Intelligence and Soft Computing, pp. 342–349. Springer (2012)
48. Wang, F.: A fuzzy grammar and possibility theory-based natural language user interface for spatial queries. Int. J. Fuzzy Sets Syst. **113**, 147–159 (2000)
49. Wang, S., Yuan, H., Chen, G., Li, D., Shi, W.: Rough Spatial Interpretation. In: Tsumoto, S., Słowiński, R., Komorowski, J., Grzymała-Busse, J.W. (eds.) RSCTC 2004. LNCS (LNAI), vol. 3066, pp. 435–444. Springer, Heidelberg (2004)
50. Worboys, M.: Computation with imprecise geospatial data. Comput. Environ. Urban Syst. **22**(2), 85–106 (1998)
51. Yager, R.: Interval valued entropies for Dempster Shafer structures. Knowl. Syst. **161**, 390–397 (2018)
52. Yager, R.: Pythagorean membership grades in multicriteria decision-making. IEEE Trans. Fuzzy Syst. **22**, 958–965 (2013)
53. Zadeh, L.: Fuzzy sets. Inf. Control. **8**, 338–353 (1965)
54. Zhang, J., Goodchild, M.: Uncertainty in Geographical Information, London UK, Taylor and Francis Pub (2002)
55. Zhang, Q., Xie, Q., Wang, G.: A survey on rough set theory and its applications. CAAI Trans. Intell. Tech. **1**(4), 323–333 (2016)

Evaluating Sustainable Production Barriers and Efficiency-Enhancing Techniques in Manufacturing Companies Using Bipolar Neutrosophic REF

Selçuk Korucuk and Ahmet Aytekin

Abstract The concept of sustainable production has become more important to meet customers' demands while still providing a habitable environment for future generations and not risking their ability to meet their own needs. Beyond profitability and competition, manufacturing companies, in particular, have had to develop a sustainable production approach that takes care of all users comprehensively. Making sustainable production effective has become one of the most basic indicators for manufacturing companies. However, there are numerous barriers to achieving the appropriate level of sustainable production in manufacturing companies. The immediate removal of these barriers has a favorable impact on both businesses and society. In this context, the study examined and determined the criteria that include sustainable production obstacles and production improvement techniques at manufacturing companies with corporate identity in the Ordu province. The Bipolar Neutrosophic REF methodology was used to evaluate the criteria and options for this purpose. It has been concluded that "lack of high awareness of social and environmental events in businesses" is the most significant barrier to increasing production in manufacturing companies with a corporate identity. Furthermore, it has been found that "management through value analysis" is the most effective approach for sustainable production. The desired requirements, particularly those relating to environmental protection, can be accomplished thanks to the sustainable production practices employed by manufacturing companies. These practices support the implementation of efficiency-enhancing techniques. Furthermore, the study underlines the main barriers that businesses may encounter while implementing sustainable production processes, which have acquired global attention in recent years. Users can improve efficiency by being aware of relevant problems and implementing suitable strategies.

S. Korucuk
Department of Logistics Management, Bulancak Kadir Karabaş Vocational School, Giresun University, Giresun, Türkiye
e-mail: selcuk.korucuk@giresun.edu.tr

A. Aytekin (✉)
Department of Business Administration, Hopa Faculty of Economics and Administrative Sciences, Artvin Çoruh University, 08100 Hopa, Artvin, Türkiye
e-mail: ahmetaytekin@artvin.edu.tr

L. Sahoo et al. (eds.), *Decision Making Under Uncertainty Via Optimization, Modelling, and Analysis*, Studies in Systems, Decision and Control 558,
https://doi.org/10.1007/978-981-96-0085-4_7

Keywords Production · Sustainable production barriers · Efficiency-enhancing techniques · Bipolar neutrosophic sets · Nearest solution to the references method · Bipolar neutrosophic REF

1 Introduction

Today, in harsh competitive conditions, businesses have to keep their costs at the optimum level and use their resources effectively, efficiently, and economically. Businesses that have caught up with the demands for sustainable production in the globalizing world have improved their production processes, and have started to work with elements such as zero errors, and zero risks [29]. It has been stated that unsustainable consumption and production patterns have a greater impact than ever before, not only on the environment but also on society, the economy, and businesses [16].

Sustainable production refers to the creation of products that do not pollute the environment, do not conserve energy, and do not produce natural resources through processes that are economical and safe for workers, communities, and consumers [59]. In another definition, the term refers to the production of products that are as environmentally friendly as possible, the conservation of energy and natural resources, and the creation of a reliable and economic structure for employees, society, and consumers [25]. In other words, sustainable production is explained as the development of production systems that increase resource efficiency and reduce the emission of pollutants into the environment through ecological design, the use of environmentally friendly materials, and the optimization of production processes [27].

The concept of sustainable production can be approached in two ways: the first is general, for example economic, ecological, and social, and the second is specific to manufacturing companies and concerns material selection, product design, and production process [60]. In this context, there are many reasons why companies are committed to sustainable production. The benefits of sustainable production include reduced costs through improved resource efficiency and regulatory compliance, improved brand reputation, access to new markets, reduced employee turnover through the creation of attractive jobs, and a long-term business approach that creates opportunities for access to finance and capital [38].

The main goal of sustainable production practices is the creation of products through energy-efficient and resource-saving production systems. In addition, the sustainable production approach has been determined to be an approach that is achieved not only by creating products economically but also by avoiding negative effects on the environment and being socially responsible towards society [57]. Özel [44] argues that if production is sustainable, the environment, employees, society, and organizations all benefit. He stated that under these conditions, businesses are always expected to be more cost-effective and productive in the long term and, most of the time, in the short term.

According to Rosen and Kishawy [49], the success of implementing sustainable production depends on many factors. Quantitative and qualitative information is required to evaluate production processes (e.g., amount and type of metal used in the process, type of pollutants emitted, etc.). This information is often not readily available and is difficult to obtain. A holistic sustainability approach to management can prevent inconsistent practices and the development of a sustainability-focused culture and should instead be managed by specialized departments. To achieve sustainability goals, it is necessary to effectively address sustainability issues with fewer decision-making and action procedures. However, the fundamental concept of sustainable development is to meet current needs without depleting the resources of future generations. Even as companies begin to recognize the importance of sustainable production practices, they face many challenges in adapting to these practices. Factors such as inadequate knowledge about sustainability, reluctance to adopt the concept, and financial constraints are proving to be significant problems [54].

In this context, many reasons motivate the authors to solve the studied problem. Efficient use of limited resources, creating a favorable environment for future generations, increasing competitiveness, and ensuring customer satisfaction are important concepts for all companies. Depending on the applied experience, expertise, perspective, and knowledge of the decision-makers, this study includes significant real-life benefits to beneficiaries and companies. This can highlight and enable a new connection between manufacturing companies' sensitivity to sustainable production, efforts to increase sustainable production, efforts to ensure social and environmental trust and satisfaction of beneficiaries, and results in terms of performance and improved production of new models. Furthermore, this research is regarded as an essential step in developing a real, effective, and applicable solution to a decision-making problem involving the identification of barriers to sustainable production and the selection of the most appropriate technique to increase efficiency in an important sector such as manufacturing. On the other hand, the study is significant because it provides an opportunity to develop a model that will allow manufacturing businesses to self-assess their sustainable production techniques and procedures to improve the efficiency.

Traditionally, establishing environmentally friendly manufacturing processes has been considered a barrier to company success. However, many businesses have recently discovered that these initiatives result in lower operational costs and happier personnel. On the other hand, improving efficiency and promoting sustainable production is a challenging task for society today. The situation encourages the authors to study the issues of productivity management and sustainable production. Unsustainable production patterns have a greater impact on society, the economy, businesses, and the environment than in the past [16].

Businesses are trying to develop new global policies to provide a more habitable environment for present and future generations by ensuring effective sustainable production and productivity increase management and making it a key component of their strategy. However, due to several barriers and challenges, the process of adopting and implementing sustainable production can be slow, tough, and complex. This study aims to address the challenges identified by manufacturing companies and raise awareness. Furthermore, contemporary businesses must produce not just innovative products and services but also long-term manufacturing processes. In other words, both firms and industries must recognize and incorporate social and environmental responsibilities into all manufacturing processes and decision-making procedures. This critical issue is one of the reasons for undertaking the study.

In addition, the study serves as a road map that will help draw attention to sustainable production and its barriers and production issues in the manufacturing sector, as well as help gain customer continuity and efficiency in business processes. It also examines the similarities and differences in the levels of sustainable production barriers in businesses in the relevant sector and to what extent they reflect them. Thus, the study provides a practical road map for the development process by comparing sustainable production and efficiency issues in the manufacturing sector. Another theoretical and practical significance of the study is to propose an effective, efficient, and robust decision-making model capable of dealing with the existing identified uncertainties. Moreover, the aim is to present a solid and robust methodological framework to fill the theoretical gaps present in the literature by exploiting the strengths of the methods used in the study. In this context, the study aims to summarize the factors that hinder sustainable production in manufacturing enterprises with corporate identity in Ordu Province and to select the most ideal technique to increase efficiency. For this purpose, the criteria and techniques identified during the literature review were evaluated using the Bipolar Neutrosophic (BN) Multi-Criteria decision-making (MCDM) methodology following expert judgments. In the methodology proposed in the study, the BN subjective weighting approach was preferred to weight the criteria. Besides, BN-REF was used to evaluate the options.

There are some important reasons why the BN-REF methodology is employed in this study. The reasons for this derive from the basic features of the methodology. Bipolarity is the term used to describe the human mind's tendency to dispute and make decisions based on both positive and negative impulses. Positive information conveys what is feasible, respectable, allowable, desired, or considered acceptable. Conversely, negative statements clarify what is not feasible, unacceptable, or not allowed [13, 15, 18, 31]. Deli et al. [18] proposed Bipolar Neutrosophic Sets (BNSs) as an important solution tool for decision-making problems by including the membership degrees of Neutrosophic Sets (NSs) for falsity, truthfulness, and indeterminacy, as well as the negative equivalents of these memberships. The BN subjective weighting approach enables the efficient and easy collection of expert opinions from those who are not familiar with MCDM methodologies and their application processes. It is seen as a crucial instrument for solving the problem in this context. Aytekin [7] proposed REF, which has several extensions designed for different decision-making scenarios. Among these extensions, REF-I applies to problems where criteria are measured using different scales [8, 28]. As rank options, REF-II eliminates the rank reversal problem [9, 28]. Options can be classified via REF-Sort [11]. Efficiency analysis can make use of REF-III [12, 36]. A novel extension of the REF will be proposed in this study to identify the optimal solution (reference) for each criterion and evaluate options based on these references. Therefore, the BNSs' ability to model unclear and bipolar problems will be combined with REF's reference-based solution approach. With its features, the proposed methodology will ensure that the evaluations of experts who are not familiar with the MCDM methodology are appropriately obtained, modeling conflicts and uncertainties and resulting in an efficient solution to the studied problem.

2 Literature Review

A detailed literature review on sustainable production barriers and production is given below.

- Massoud et al. [41] found in their study of Lebanese companies that barriers such as inadequate government support and incentives and insufficient information on net benefits/costs prevent sustainable production.
- Lemenih and Kassa [33] assessed the opportunities and challenges of sustainable production and marketing of rubbers and resins in Ethiopia.
- Paumgarten et al. [46] examined the benefits and barriers to promoting sustainable production and marketing of African products.
- Mathiyazhagan et al. [42] investigated sustainable manufacturing holistically throughout the supply chain and found 26 main barriers to sustainability, with suppliers' lack of environmental knowledge being the most significant.
- Alzawawi [5] suggested solutions by examining the factors and obstacles to creating sustainability in supply chain management and operations.
- Chkanikova and Mont [17] examined the factors and obstacles to sustainable food retailing within the framework of corporate supply chain responsibility.
- Söylemez et al. [56] assessed the obstacles encountered in the adoption of sustainable production and consumption in Türkiye.
- Bhanot et al. [14] used an integrated approach consisting of four methodologies to assess the drivers and barriers to sustainable production.
- Gardas et al. [20] prioritized barriers to sustainable production using Delphi and DEMATEL methods.
- Korucuk [30] evaluated factors limiting production in manufacturing companies using AHP and TOPSIS.
- Narimissa [43] examined the drivers and barriers to implementing and improving sustainable supply chain management.
- Goyal et al. [22] employed hybrid fuzzy AHP-fuzzy TOPSIS to examine barriers and solutions to ensure a green future in sustainable production and consumption.
- Liu et al. [34] examined barriers to sustainable food consumption and production in China from a circular economy perspective.
- Özgüner [45] identified obstacles to sustainable production and classified them according to their importance.

- Hariyani et al. [24] identified 31 different barriers to sustainable manufacturing based on a thorough evaluation of 118 studies on the subject. The study indicates that businesses and policymakers should take an organized approach to overcoming these barriers.
- Akar [3] identified and prioritized barriers to implementing sustainable production in an agricultural machinery company.
- Alam et al. [4] assessed barriers to integrating technological innovations in the textile sector and emphasized the importance of sustainable production.
- Sharma et al. [53] examined the barriers to implementing digital technologies to achieve sustainable production and consumption in the food sector from a circular economy perspective.

A comprehensive review of the literature indicated that no studies examined barriers to sustainable manufacturing with efficiency techniques. At this point, we believe the study will contribute to the literature and fill a significant gap in the manufacturing industry.

3 Methodology

Fuzzy MCDM methods are commonly used to solve a wide range of real-world problems [51]. Recent examples of these applications include selecting measures for the remediation of agricultural land after radioactive contamination [2], comparing mobile banking applications [50], evaluating industrial robots [21], selection of PhD supervisor [40], and solving a vending machine location selection problem [58]. These current application examples demonstrate how fuzzy MCDM methodologies are applied in a variety of real-world problems and fields, including social sciences, engineering, and educational sciences. However, the diversity of fuzzy information, different structures of decision-making problems, and ongoing advances in science provide the door to new approaches. In this context, new MCDM methods and fuzzy sets are being developed. One of the new methods is REF, and one of the new clusters is BNS.

Aytekin [7] developed the REF, which includes a structure that allows for the evaluation of both qualitative and quantitative criteria. REF has several extensions, including REF-I, REF-II, REF-III, and REF-Sort. REF-Sort is used in classification problems, while REF-III is used to evaluate the efficiency of decision-making units or options [10]. REF-I and REF-II methods can be employed to find a compromise solution in decision-making problems including criteria that have different structures. REF-I allows for evaluating options based on frequency and ranking in qualitative criteria [8, 10, 28]. REF-II eliminates the rank reversal problem and the need to recalculate for current options when new alternatives are added or eliminated from decision-making problems [9, 28].

BNSs are an extension of NSs focusing on bipolarity. In this study, a novel BN-REF method is proposed to solve decision-making problems involving uncertainty.

For this purpose, general explanations and important operations of BNSs will be given in this section of the study. Then, the steps for implementing the BN-REF methodology will be detailed.

3.1 Bipolar Neutrosophic Sets

NS, which Smarandache [55] developed, enables the modeling of uncertainty-related problems employing the concepts of falsity, truthfulness, and indeterminacy. In the following period, many NS extensions have been developed with the capacity to address a variety of problems [31]. BNS, which was proposed by Deli et al. [18], is one of them. BNS considers the human mind's propensity to discuss and base judgments on both positive and negative experiences. Positive preferences in BNS align with desires as they identify and reject objects or values that are less ideal than others. On the other hand, because they specify which things or values should be rejected, negative preferences are linked to constraints [15, 18, 31].

Bipolar Neutrosophic Sets (BNSs) contain the membership degrees for truth, indeterminacy, and falsity, as well as their opposites. Assume that X is a universe of discourse, then a BNS H is denoted as $H = \{\langle x, a^+(x), b^+(x), c^+(x), a^-(x), b^-(x), c^-(x)\rangle : x \in X\}$, where $a^+, b^+, c^+ : X \to [1,0]$ and $a^-, b^-, c^- : X \to [-1,0]$. From the positive membership degrees of an element $x \in X$ to BNS H, $a^+(x)$ is the truth membership degree, $b^+(x)$ is the indeterminacy membership degree, and $c^+(x)$ is the falsity membership degree. Also, $a^-(x)$, $b^-(x)$, $c^-(x)$ are the negative membership degrees [1, 18, 23, 31]. For the sake of simplicity, let $h_i = \langle a_i^+, b_i^+, c_i^+, a_i^-, b_i^-, c_i^-\rangle$ be one or more BN numbers (BNNs), where $i = 1,2,\ldots,m$. The basic operators, the score function $s(h_i)$, the certainty function $\xi(h_i)$, and the accuracy function $Q(h_i)$ are presented in Eqs. (1)–(8). Also, the Hamming distance ($Hd(h_1, h_2)$), the Euclidean distance ($Ed(h_1, h_2)$), and BN weighted averaging operator (BNWAO) are given in Eq. (9)–(11), respectively [6, 18, 23, 26, 31, 35, 47, 48], where $\lambda > 0$.

$$\begin{aligned} h_1 + h_2 =& \langle a_1^+ + a_2^+ - a_1^+ a_2^+, b_1^+ b_2^+, c_1^+ c_2^+, -a_1^- a_2^-, \\ & -(-b_1^- - b_2^- - b_1^- b_2^-), -(-c_1^- - c_2^- - c_1^- c_2^-)\rangle \end{aligned} \tag{1}$$

$$\begin{aligned} h_1 h_2 =& \langle a_1^+ a_2^+, b_1^+ + b_2^+ - b_1^+ b_2^+, c_1^+ + c_2^+ - c_1^+ c_2^+, \\ & -(-a_1^- - a_2^- - a_1^- a_2^-), -b_1^- b_2^-, -c_1^- c_2^-\rangle \end{aligned} \tag{2}$$

$$h_i^\lambda = \langle \left(a_i^+\right)^\lambda, 1 - \left(1 - b_i^+\right)^\lambda, 1 - \left(1 - c_i^+\right)^\lambda, -\left(1 - (1 - (-a_i^-))^\lambda\right), -\left(-b_i^-\right)^\lambda, -\left(-c_i^-\right)^\lambda\rangle \tag{3}$$

$$\lambda h_i = \langle 1 - (1 - a_i^+)^\lambda, (b_i^+)^\lambda, (c_i^+)^\lambda, -(-a_i^-)^\lambda, -(-b_i^-)^\lambda, -(1 - (1 - (-c_i^-))^\lambda)\rangle, \tag{4}$$

$$s(\mathrm{h}_i) = \frac{a_i^+ + 1 - b_i^+ + 1 - c_i^+ + 1 + a_i^- - b_i^- - c_i^-}{6} \tag{5}$$

$$\varrho(\mathrm{h}_i) = a_i^+ - c_i^+ + a_i^- - c_i^- \tag{6}$$

$$\xi(\mathrm{h}_i) = a_i^+ - c_i^- \tag{7}$$

$$\begin{aligned}\mathrm{Hd}(h_1, h_2) =& \frac{1}{6}\left(\left|a_1^+ - a_2^+\right| + \left|b_1^+ - b_2^+\right| + \left|c_1^+ - c_2^+\right|\right. \\ &\left. + \left|a_1^- - a_2^-\right| + \left|b_1^- - b_2^-\right| + \left|c_1^- - c_2^-\right|\right)\end{aligned} \tag{8}$$

$$\mathrm{Ed}(h_1, h_2) = \sqrt{\frac{1}{6}\left(\left|a_1^+ - a_2^+\right|^2 + \left|b_1^+ - b_2^+\right|^2 + \left|c_1^+ - c_2^+\right|^2 + \left|a_1^- - a_2^-\right|^2 + \left|b_1^- - b_2^-\right|^2 + \left|c_1^- - c_2^-\right|^2\right)} \tag{9}$$

$$\begin{aligned}&\mathrm{BNWAO_w}(h_1, \ldots, h_m) = \mathrm{w}_1 h_1 \oplus \ldots \oplus \mathrm{w}_1 h_m \\ &= \langle \sum_{\mathrm{i}=1}^{m} \mathrm{w}_i a_i^+, \sum_{\mathrm{i}=1}^{m} \mathrm{w}_i b_i^+, \sum_{\mathrm{i}=1}^{m} \mathrm{w}_i c_i^+, \sum_{\mathrm{i}=1}^{m} \mathrm{w}_i a_i^-, \sum_{\mathrm{i}=1}^{m} \mathrm{w}_i b_i^-, \sum_{\mathrm{i}=1}^{m} \mathrm{w}_i c_i^- \rangle\end{aligned} \tag{10}$$

where $\mathrm{w_i} = \{\mathrm{w}_1, \ldots, \mathrm{w}_m\}$, $\mathrm{w_i} \geq 0$ and $\sum_{\mathrm{i}=1}^{m} \mathrm{w}_i = 1$.

3.2 The Proposed BN-REF Methodology

The studied problem will be solved by a two-step process. The weights of the criteria will be set in the first stage, and the options will be ranked in the second stage. Expert assessments will be used to weight the criteria. In this context, the procedure for the subjective weighting approach based on BNWAO is as follows:

Step 1. To solve the decision-making problem, the criteria ($N = \{N_1, \ldots, N_n\}$), options ($O = \{O_1, \ldots, O_m\}$), and experts ($Z = \{Z_1, \ldots, Z_t\}$) are identified.

Step 2. Before assigning a preference or performance rating to each option in the criteria, experts assess the criteria's importance. Then, the preference, priority, or performance of options are evaluated by experts. For this, the linguistic terms listed in Table 1 are used [1]. As a result, the BN importance values of j-th criterion ($j = 1, \ldots, n$) determined by p-th expert ($p = 1, \ldots, t$) is shown as $\vartheta_j^{(p)}$, where $\vartheta_j^{(p)} = \langle a_{\mathrm{j}}^{+(\mathrm{p})}, b_{\mathrm{j}}^{+(\mathrm{p})}, c_{\mathrm{j}}^{+(\mathrm{p})}, a_{\mathrm{j}}^{-(\mathrm{p})}, b_{\mathrm{j}}^{-(\mathrm{p})}, c_{\mathrm{j}}^{-(\mathrm{p})} \rangle$. On the other hand, the BN rating value of i-th option ($i = 1, \ldots, \mathrm{m}$) in j-th criterion by p-th expert is denoted as $\zeta_{ij}^{(p)} = \langle a_{\mathrm{ij}}^{+(\mathrm{p})}, b_{\mathrm{ij}}^{+(\mathrm{p})}, c_{\mathrm{ij}}^{+(\mathrm{p})}, a_{\mathrm{ij}}^{-(\mathrm{p})}, b_{\mathrm{ij}}^{-(\mathrm{p})}, c_{\mathrm{ij}}^{-(\mathrm{p})} \rangle$.

Step 3. The linguistic terms in Table 1 are used to determine the important levels of expert evaluations ($\natural_p$). Then, expert weights are computed using Eq. (11).

$$\lambda_p = \frac{s(\natural_p)}{\sum_{\mathrm{p}=1}^{t} s(\natural_p)}, \tag{11}$$

Table 1 Linguistic terms for evaluating criteria and options

Notations	Linguistic terms	Bipolar Neutrosophic Numbers					
		a^+	b^+	c^+	a^-	b^-	c^-
VVL	Very Very Low	0.10	0.95	0.90	−0.85	−0.05	−0.05
VL	Very Little	0.15	0.9	0.8	−0.65	−0.1	−0.1
L	Low	0.25	0.7	0.8	−0.55	−0.15	−0.3
BM	Below Medium	0.3	0.4	0.6	−0.3	−0.2	−0.1
M	Medium	0.5	0.5	0.5	−0.5	−0.5	−0.5
AM	Above Medium	0.75	0.2	0.25	−0.25	−0.6	−0.5
H	High	0.85	0.15	0.2	−0.2	−0.7	−0.9
VH	Very High	0.95	0.10	0.15	−0.15	−0.80	−0.95
VVH	Very Very High	1	0	0.1	−0.1	−0.9	−1

where $\lambda_p \geq 0$ and $\sum_{p=1}^{t} \lambda_p = 1$.

Step 4. The integrated BN importance values of criteria are calculated via Eq. (12).

$$\vartheta_j = \langle \sum_{\text{p=1}}^{t} \lambda_p a_j^{+(\text{p})}, \sum_{\text{p=1}}^{t} \lambda_p b_j^{+(\text{p})}, \sum_{\text{p=1}}^{t} \lambda_p c_j^{+(\text{p})}, \sum_{\text{p=1}}^{t} \lambda_p a_j^{-(\text{p})}, \sum_{\text{p=1}}^{t} \lambda_p b_j^{-(\text{p})}, \sum_{\text{p=1}}^{t} \lambda_p c_j^{-(\text{p})} \rangle \tag{12}$$

Step 5. The weight coefficients of the criteria are computed by applying Eq. (13), where $w_j \geq 0$ and $\sum_{j=1}^{n} w_j = 1$.

$$w_j = \frac{\text{s}(\vartheta_j)}{\sum_{j=1}^{n} \text{s}(\vartheta_j)} \tag{13}$$

The options will be ranked using BN-REF in the second stage. The following steps will be implemented in this context.

Step 6. The integrated BN decision matrix is constructed via Eq. (14).

$$\zeta_{ij} = \langle \sum_{\text{p=1}}^{t} \lambda_p a_{ij}^{+(\text{p})}, \sum_{\text{p=1}}^{t} \lambda_p b_{ij}^{+(\text{p})}, \sum_{\text{p=1}}^{t} \lambda_p c_{ij}^{+(\text{p})}, \sum_{\text{p=1}}^{t} \lambda_p a_{ij}^{-(\text{p})}, \sum_{\text{p=1}}^{t} \lambda_p b_{ij}^{-(\text{p})}, \sum_{\text{p=1}}^{t} \lambda_p c_{ij}^{-(\text{p})} \rangle \tag{14}$$

, where $\zeta_{ij} = \langle a_{\text{ij}}^+, b_{\text{ij}}^+, c_{\text{ij}}^+, a_{\text{ij}}^-, b_{\text{ij}}^-, c_{\text{ij}}^- \rangle$.

Step 6. The reference value of each criterion is determined. Expert judgments or theoretical best and worst values can be employed to determine the reference value of each criterion. The linguistic terms in Table 2 that indicate the best value for each criterion will be employed as the reference value in this study. As a result, $\zeta_{rj} = \langle a_{\text{rj}}^+, b_{\text{rj}}^+, c_{\text{rj}}^+, a_{\text{rj}}^-, b_{\text{rj}}^-, c_{\text{rj}}^- \rangle$ shows the reference value of j-th criterion.

Table 2 List of criteria and options

	Name	Explanation	Source(s)
N1	Lack of awareness and knowledge about sustainability concepts	It represents the age of awareness and knowledge regarding sustainability	[29]
N2	Lack of resources and insufficient workforce	It refers to a lack of resources and an insufficient workforce in the business	[14, 37]
N3	Lack of a clear framework for performance measures and metrics at strategic, tactical, and operational levels for sustainable production	It is the absence of a clear framework for performance measures and metrics at strategic, tactical, and operational levels for sustainable production	[61]
N4	Lack of high awareness of social and environmental events in businesses	It refers to the lack of high awareness of social and environmental events in businesses	[39]
N5	Lack of management support	Explains the lack of management support for sustainable production	[30, 45]
N6	Low acceptance levels of innovative applications	It refers to low acceptance levels of innovative practices	[61]
N7	Lack of technological development	Explains the inadequacy of technological development in businesses	[14]
N8	Negative attitudes towards sustainability concepts	It refers to negative attitudes towards sustainability concepts. In other words, it is a belief or conviction that sustainability concepts are unimportant and worthless	[54]
N9	Ineffective policies and lack of legislation	Explains ineffective policies and lack of legislation	[30]
N10	Lack of funds and loans for green projects (environmentally friendly projects)	It refers to the lack of funds and loans for green or environmentally friendly projects	[54]
O1	Management through Value Analysis	It expresses how to make the production process more sustainable and at the lowest cost	[19]
O2	Just-in-Time Production Method	The required product explains how to produce the required quantity in the required time, using the least amount of labour and space	[52]
O3	Nominal Grouping Technique	It is a method for listing the list of 'problem causes' determined by using the decisions of all members of a group, in order of priority	[32]
O4	Cost—Efficiency Allocation	Explains the relationship between cost and efficiency	[19]

Step 7. The distances of options from the reference values $\left(\mathrm{d}(\zeta_{ij})\right)$ are computed using Eq. (15).

$$\mathrm{d}(\zeta_{ij}) = \frac{1}{6}\left(\left|a_{ij}^{+} - a_{rj}^{+}\right| + \left|b_{ij}^{+} - b_{rj}^{+}\right| + \left|c_{ij}^{+} - c_{rj}^{+}\right| + \left|a_{ij}^{-} - a_{rj}^{-}\right| + \left|b_{ij}^{-} - b_{rj}^{-}\right| + \left|c_{ij}^{-} - c_{rj}^{-}\right|\right) \quad (15)$$

Step 8. The weighted distance score of each option is calculated via Eq. (16).

$$v_i = \sum_{j=1}^{n} \mathrm{d}(\zeta_{ij}) w_j \quad (16)$$

The options are ranked based on their v_i values. The option that has the smallest v_i is ranked first place.

4 Results

Today, businesses have strategized many innovations in their production processes to survive in competitive market conditions and to make the concept of sustainability effective. The increasing environmental sensitivity of society and the state and their search for more sustainable policies have caused sustainable practices in production to gain importance. Integrated strategies, especially focusing on sustainable production and efficiency-increasing techniques, have come to the fore in manufacturing businesses [32].

Table 2 is a list of the criteria and options considered in solving the decision problem. On the other hand, three of the experts whose opinions were consulted to assess the problem are operations managers with more than ten years of experience. While one is a customs clearance manager, the other is an academician who has been conducting scientific studies in this sector for a long time and has field expertise.

Experts' ($Z = \{Z_1, \ldots, Z_t\}$) linguistic evaluations of the criteria are presented in Table 3.

The linguistic evaluations of options given by experts are displayed in Table 4.

Table 5 presents the integrated BN importance values (ϑ_j), $s(\vartheta_j)$ values, weight coefficients, and importance rankings of criteria.

Table 3 Evaluations of criteria

	N1	N2	N3	N4	N5	N6	N7	N8	N9	N10
Z1	H	H	VH	VVH	AM	AM	H	H	M	M
Z2	H	H	VH	VVH	AM	AM	H	H	M	M
Z3	M	M	AM	VH	H	H	H	AM	AM	AM
Z4	H	H	H	AM	H	VH	H	VH	M	M
Z5	H	VH	AM	AM	VVH	H	VH	H	AM	H

Table 4 Evaluations of options

		N1	N2	N3	N4	N5	N6	N7	N8	N9	N10
Z1	O1	H	H	AM	H	VH	VH	VH	H	M	AM
	O2	H	H	H	H	AM	L	AM	H	H	H
	O3	H	H	H	H	AM	AM	H	VH	VH	H
	O4	H	H	H	H	L	L	H	AM	VH	AM
Z2	O1	AM	M	AM	M	AM	M	AM	M	M	H
	O2	H	H	H	VH	AM	H	VH	H	VH	H
	O3	AM	AM	AM	H	H	H	VH	H	H	AM
	O4	VH	VH	H	H	AM	M	AM	M	H	M
Z3	O1	H	H	H	H	H	AM	AM	H	L	VH
	O2	AM	AM	AM	H	AM	AM	H	AM	H	H
	O3	H	H	H	H	AM	AM	H	H	AM	H
	O4	H	H	H	H	H	AM	M	AM	M	AM
Z4	O1	AM	AM	H	H	H	H	M	M	M	AM
	O2	H	H	AM	H	H	AM	VH	M	AM	H
	O3	M	AM	AM	M	M	AM	M	M	M	AM
	O4	H	H	VH	VH	H	M	H	H	AM	H
Z5	O1	H	VH	AM	BM	VVH	H	VH	H	AM	H
	O2	VH	VVH	AM	L	VVH	H	H	AM	L	AM
	O3	H	VH	H	H	VH	VH	H	H	AM	H
	O4	VVH	VVH	VH	H	VH	VH	VVH	VH	H	H

The most important barrier is ″N4; a lack of high awareness of social and environmental events in businesses ″ as seen in Table 5. Also, the importance ranking order of criteria is found as N4≻N7≻N8≻N3~N5≻N6≻N2≻N1≻N10≻N9.

The integrated BN decision matrix and the reference values are given in Table 6. Because all criteria in this study include a barrier characteristic, the reference values are determined by the minimal linguistic terms stated in Table 3.

Finally, the weighted distance score of each option is determined, as shown in Table 7.

The best option is "O1; management through value analysis" as seen in Table 7. The ranking order of alternatives is also found as O1≻O3≻O2≻O4.

4.1 *The Validation of Results*

The validity and reliability of the results obtained within the scope of the study were examined in two stages. The existence of the rank reversal problem will be

Table 5 The weighting results

	Criteria											
	N1						N2					
	a_j^+	b_j^+	c_j^+	a_j^-	b_j^-	c_j^-	a_j^+	b_j^+	c_j^+	a_j^-	b_j^-	c_j^-
ϑ_j	0.7800	0.2200	0.2600	−0.2600	−0.6600	−0.8200	0.8000	0.2100	0.2500	−0.2500	−0.6800	−0.8300
$s(\vartheta_j)$	0.7533						0.7667					
w_j	0.1000						0.1018					
Ranking	8						7					
	N3						N4					
	a_j^+	b_j^+	c_j^+	a_j^-	b_j^-	c_j^-	a_j^+	b_j^+	c_j^+	a_j^-	b_j^-	c_j^-
ϑ_j	0.8500	0.1500	0.2000	−0.2000	−0.7000	−0.7600	0.8900	0.1000	0.1700	−0.1700	−0.7600	−0.7900
$s(\vartheta_j)$	0.7933						0.8333					
w_j	0.1053						0.1106					
Ranking	4						1					
	N5						N6					
	a_j^+	b_j^+	c_j^+	a_j^-	b_j^-	c_j^-	a_j^+	b_j^+	c_j^+	a_j^-	b_j^-	c_j^-
ϑ_j	0.8400	0.1400	0.2000	−0.2000	−0.7000	−0.7600	0.8300	0.1600	0.2100	−0.2100	−0.6800	−0.7500
$s(\vartheta_j)$	0.7933						0.7800					
w_j	0.1053						0.1035					
Ranking	4						6					
	N7						N8					
	a_j^+	b_j^+	c_j^+	a_j^-	b_j^-	c_j^-	a_j^+	b_j^+	c_j^+	a_j^-	b_j^-	c_j^-
ϑ_j	0.8700	0.1400	0.1900	−0.1900	−0.7200	−0.9100	0.8500	0.1500	0.2000	−0.2000	−0.7000	−0.8300

(continued)

Table 5 (continued)

	Criteria											
$s(\vartheta_j)$	0.8300						0.8050					
w_j	0.1102						0.1069					
Ranking	2						3					
	N9						N10					
	a_j^+	b_j^+	c_j^+	a_j^-	b_j^-	c_j^-	a_j^+	b_j^+	c_j^+	a_j^-	b_j^-	c_j^-
ϑ_j	0.6000	0.3800	0.4000	−0.4000	−0.5400	−0.5000	0.6200	0.3700	0.3900	−0.3900	−0.5600	−0.5800
$s(\vartheta_j)$	0.5767						0.6017					
w_j	0.0765						0.0799					
Ranking	10						9					

Table 6 The integrated BN decision matrix and the reference values

	Criteria											
	N1						N2					
	a_{ij}^{+}	b_{ij}^{+}	c_{ij}^{+}	a_{ij}^{-}	b_{ij}^{-}	c_{ij}^{-}	a_{ij}^{+}	b_{ij}^{+}	c_{ij}^{+}	a_{ij}^{-}	b_{ij}^{-}	c_{ij}^{-}
O1	0.8100	0.1700	0.2200	−0.2200	−0.6600	−0.7400	0.7800	0.2200	0.2600	−0.2600	−0.6600	−0.7500
O2	0.8500	0.1500	0.2000	−0.2000	−0.7000	−0.8300	0.8600	0.1300	0.1900	−0.1900	−0.7200	−0.8400
O3	0.7600	0.2300	0.2700	−0.2700	−0.6400	−0.7400	0.8300	0.1600	0.2100	−0.2100	−0.6800	−0.7500
O4	0.9000	0.1100	0.1700	−0.1700	−0.7600	−0.9300	0.9000	0.1100	0.1700	−0.1700	−0.7600	−0.9300
ζ_{rj}	0.1000	0.9500	0.9000	−0.8500	−0.0500	−0.0500	0.1000	0.9500	0.9000	−0.8500	−0.0500	−0.0500
	N3						N4					
	a_{ij}^{+}	b_{ij}^{+}	c_{ij}^{+}	a_{ij}^{-}	b_{ij}^{-}	c_{ij}^{-}	a_{ij}^{+}	b_{ij}^{+}	c_{ij}^{+}	a_{ij}^{-}	b_{ij}^{-}	c_{ij}^{-}
O1	0.7900	0.1800	0.2300	−0.2300	−0.6400	−0.6600	0.6700	0.2700	0.3400	−0.2800	−0.5600	−0.6600
O2	0.7900	0.1800	0.2300	−0.2300	−0.6400	−0.6600	0.7500	0.2500	0.3100	−0.2600	−0.6100	−0.7900
O3	0.8100	0.1700	0.2200	−0.2200	−0.6600	−0.7400	0.7800	0.2200	0.2600	−0.2600	−0.6600	−0.8200
O4	0.8900	0.1300	0.1800	−0.1800	−0.7400	−0.9200	0.8700	0.1400	0.1900	−0.1900	−0.7200	−0.9100
ζ_{rj}	0.1000	0.9500	0.9000	−0.8500	−0.0500	−0.0500	0.1000	0.9500	0.9000	−0.8500	−0.0500	−0.0500
	N5						N6					
	a_{ij}^{+}	b_{ij}^{+}	c_{ij}^{+}	a_{ij}^{-}	b_{ij}^{-}	c_{ij}^{-}	a_{ij}^{+}	b_{ij}^{+}	c_{ij}^{+}	a_{ij}^{-}	b_{ij}^{-}	c_{ij}^{-}
O1	0.8800	0.1200	0.1800	−0.1800	−0.7400	−0.8500	0.7800	0.2200	0.2600	−0.2600	−0.6600	−0.7500
O2	0.8200	0.1500	0.2100	−0.2100	−0.6800	−0.6800	0.6900	0.2800	0.3400	−0.2900	−0.5500	−0.6200
O3	0.7600	0.2300	0.2700	−0.2700	−0.6400	−0.6700	0.8100	0.1700	0.2200	−0.2200	−0.6600	−0.6700
O4	0.7300	0.2600	0.3200	−0.2700	−0.5900	−0.7100	0.5900	0.4000	0.4400	−0.3900	−0.5100	−0.5500
ζ_{rj}	0.1000	0.9500	0.9000	−0.8500	−0.0500	−0.0500	0.1000	0.9500	0.9000	−0.8500	−0.0500	−0.0500

(continued)

Table 6 (continued)

	Criteria											
	N7						N8					
	a_{ij}^{+}	b_{ij}^{+}	c_{ij}^{+}	a_{ij}^{-}	b_{ij}^{-}	c_{ij}^{-}	a_{ij}^{+}	b_{ij}^{+}	c_{ij}^{+}	a_{ij}^{-}	b_{ij}^{-}	c_{ij}^{-}
O1	0.7800	0.2200	0.2600	−0.2600	−0.6600	−0.6800	0.7100	0.2900	0.3200	−0.3200	−0.6200	−0.7400
O2	0.8700	0.1400	0.1900	−0.1900	−0.7200	−0.8400	0.7400	0.2400	0.2800	−0.2800	−0.6200	−0.6600
O3	0.8000	0.2100	0.2500	−0.2500	−0.6800	−0.8300	0.8000	0.2100	0.2500	−0.2500	−0.6800	−0.8300
O4	0.7900	0.2000	0.2500	−0.2500	−0.6800	−0.7600	0.7600	0.2300	0.2700	−0.2700	−0.6400	−0.6700
ζ_{rj}	0.1000	0.9500	0.9000	−0.8500	−0.0500	−0.0500	0.1000	0.9500	0.9000	−0.8500	−0.0500	−0.0500
	N9						N10					
	a_{ij}^{+}	b_{ij}^{+}	c_{ij}^{+}	a_{ij}^{-}	b_{ij}^{-}	c_{ij}^{-}	a_{ij}^{+}	b_{ij}^{+}	c_{ij}^{+}	a_{ij}^{-}	b_{ij}^{-}	c_{ij}^{-}
O1	0.5000	0.4800	0.5100	−0.4600	−0.4500	−0.4600	0.8300	0.1600	0.2100	−0.2100	−0.6800	−0.7500
O2	0.7300	0.2600	0.3200	−0.2700	−0.5900	−0.7100	0.8300	0.1600	0.2100	−0.2100	−0.6800	−0.8200
O3	0.7600	0.2300	0.2700	−0.2700	−0.6400	−0.6700	0.8100	0.1700	0.2200	−0.2200	−0.6600	−0.7400
O4	0.7800	0.2200	0.2600	−0.2600	−0.6600	−0.7500	0.7400	0.2400	0.2800	−0.2800	−0.6200	−0.6600
ζ_{rj}	0.1000	0.9500	0.9000	−0.8500	−0.0500	−0.0500	0.1000	0.9500	0.9000	−0.8500	−0.0500	−0.0500

Table 7 The distance of options from the reference values and the results

	The distances of options from the local reference											
	N1	N2	N3	N4	N5	N6	N7	N8	N9	N10		
O1	0.6833	0.6583	0.6583	0.5833	0.7483	0.6583	0.6467	0.6067	0.4100	0.6967		
O2	0.7217	0.7350	0.6583	0.6383	0.6850	0.5750	0.7350	0.6200	0.6133	0.7083		
O3	0.6450	0.6967	0.6833	0.6700	0.6333	0.6717	0.6833	0.6833	0.6333	0.6833		
O4	0.7733	0.7733	0.7600	0.7467	0.6133	0.4867	0.6717	0.6333	0.6583	0.6200		
	The weighted distance score of options											
	N1	N2	N3	N4	N5	N6	N7	N8	N9	N10	v_i	Ranking
O1	0.0683	0.0670	0.0693	0.0645	0.0788	0.0682	0.0712	0.0648	0.0314	0.0556	0.6393	1
O2	0.0722	0.0748	0.0693	0.0706	0.0721	0.0595	0.0810	0.0663	0.0469	0.0566	0.6693	3
O3	0.0645	0.0709	0.0720	0.0741	0.0667	0.0695	0.0753	0.0730	0.0485	0.0546	0.6691	2
O4	0.0773	0.0787	0.0800	0.0826	0.0646	0.0504	0.0740	0.0677	0.0504	0.0495	0.6752	4

investigated first in this context. All possible scenarios were examined to remove options from the problem. Table 8 shows the results.

The results in Table X8 show that the proposed methodology does not generate the rank reversal problem. Furthermore, even when the options are removed from the problem, the weighted distance scores of the current options do not change. It can be shown that BN-REF preserves REF's feature of not requiring recalculation for options that remain in the problem after the addition or removal of options. The results of the BN-REF will be compared to the results of two methodologies used in the literature in the second stage. The rankings with BN-VIKOR, BN-SAW, BN-WPM, and BN-WASPAS will be compared in this context. The ranking results obtained via these methods are presented in Fig. 1.

When the solutions obtained by different methods in Fig. 1 are analyzed, it is found that O1 ranks first place in all rankings. In this context, it is possible to state that the results generated with BN-REF are valid and reliable.

5 Conclusions and Implications

Today, sustainable practices have become an essential issue for companies. Manufacturing companies have to create a long-term production strategy that considers all stakeholders in addition to profitability and competitiveness. Increasing the effectiveness of sustainable production has emerged as an important goal for manufacturing companies. However, achieving the necessary level of sustainable output in manufacturing companies can be difficult. Businesses and society as a whole benefit when these barriers are immediately removed. In this regard, the study examined and determined the standards for long-term production barriers and strategies to improve production in manufacturing companies that have corporate identities in the Ordu province. The criteria and alternatives for this aim were evaluated using the Bipolar Neutrosophic REF methodology.

In this respect, the study aimed to identify factors that hinder sustainable production and strategies that improve efficiency in manufacturing businesses with corporate identities in Ordu Province. A thorough review of the literature revealed that very few studies have been made on the barriers to sustainable manufacturing and the implementation of efficiency-enhancing strategies. As a result, the study proposed a methodology that will guide future research and allow for comparisons with previous findings.

According to the findings of the study, the most significant barrier to sustainable production is "a lack of high awareness of social and environmental events in businesses." This finding expresses the insufficiency of the sustainable production viewpoint in manufacturing companies. It stands out as an issue that companies must establish a social, environmental, and economic perspective on the obstacles to sustainable production and efficiency and thoroughly investigate them. It should not be forgotten that sustainable manufacturing is the responsibility of all environments, not just companies. The "lack of technological development" has been identified as

Table 8 The distance of options from the reference values

	Ranking Results										
		Excluding Options									
	Original	O1	O2	O3	O4	O1–O2	O1–O3	O1–O4	O2–O3	O2–O4	O3–O4
O1	1	–	1	1	1	1	1	–	1	1	1
O2	3	2	–	2	3	2	–	2	–	–	2
O3	2	1	2	–	2	–	2	1	–	2	–
O4	4	3	3	3	–	–	–	–	2	–	–
	Weighted Distance Scores										
		Excluding Options									
	Original	O1	O2	O3	O4	O1–O2	O1–O3	O1–O4	O2–O3	O2–O4	O3–O4
O1	0.6393	–	0.6393	0.6393	0.6393	–	–	–	0.6393	0.6693	0.6393
O2	0.6693	0.6693	–	0.6693	0.6693	–	0.6693	0.6693	–	–	0.6693
O3	0.6691	0.6691	0.6691	–	0.6691	0.6691	–	0.6691	–	0.6752	–
O4	0.6752	0.6752	0.6752	0.6752	–	0.6752	0.6752	–	0.6752	–	–

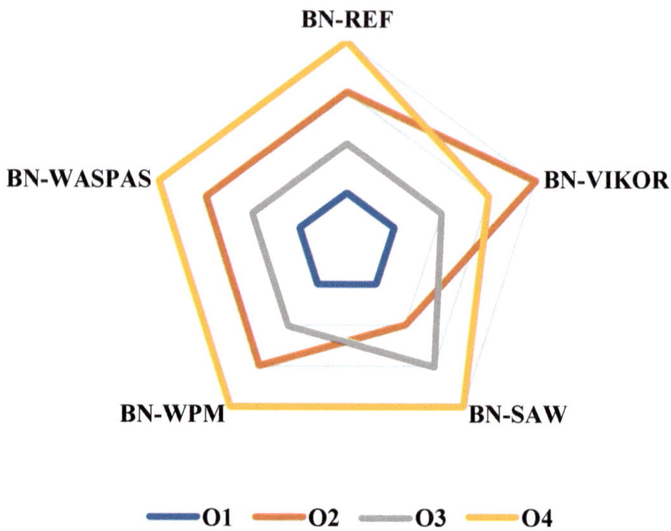

Fig. 1 The ranking results obtained via different methods

one of the most significant obstacles to sustainable production. Technological development and progress are required for effective sustainable production. The expansion of effective technology throughout the organization, in particular, enables sustainable operations to be carried out in collaboration with expert personnel. In other words, improving the effectiveness and performance of technical infrastructure is surely one strategy to improve efficiency and eliminate obstacles to sustainable production. Another obstacle to sustainable production is the "lack of a clear framework for performance measures and metrics at strategic, tactical, and operational levels for sustainable production." In this context, it should be regarded as a priority for state and local government organizations to focus more on sustainability and build effective policies and legislation. Understanding should be developed through training on the concept of sustainability both within and outside of the business, and the importance of sustainability should be highlighted not only to businesses but to the entire society.

The best strategy for enhancing efficiency has been identified as "management through value analysis." It describes which production processes are more sustainable and how to carry them out at the lowest cost using an efficiency-increasing approach that is unique to manufacturing companies. This study's findings fill a gap in the literature and differ from previous studies in both conceptual and methodological aspects.

Furthermore, due to the proposed solutions, it is anticipated that the study provide knowledge that will help decision-makers create strategies for eliminating barriers to sustainable production. Another important finding is that the study outlines measuring methods for increasing efficiency by eliminating ineffective processes in the context of sustainable manufacturing.

Businesses can use the study's findings to preserve and strengthen their power against external environmental elements such as competition by properly managing their internal structure. Productivity management may help businesses transfer funds to where they are most needed, minimize loss, increase profitability by using the correct corporate structure, and lessen the time it takes to reach objectives. Within the context of the sustainability theme, this study also helps determine the mission, vision, goals, and objectives of the future of businesses. In other words, companies have the power to influence how things are organized in the future. The study contributes to the formulation of efficiency measuring criteria and strategic plans. Businesses creating efficiency improvement strategies and measurement systems may employ a dynamic framework.

Minimizing adverse effects on the environment is the primary goal of social responsibility. Therefore, businesses and industries need to assess their production functions in conjunction with environmental protection issues and conduct a thorough analysis of their environmental sensitivity. As a result, developing industry awareness regarding the use of environmentally friendly manufacturing methods and technology in conjunction with a sustainable production strategy is an essential contribution of the study. The study will be particularly helpful because it addresses efficiency issues that fall under the umbrella of efficiency, such as the main challenges that companies may encounter in the process of sustainable production and how they should respond to these challenges, which have recently attracted the attention of governments and business organizations worldwide.

On the other hand, identifying sustainable production barriers and selecting the best efficiency-enhancing technique include several conflicts and inconsistencies within the framework of decision-makers and practitioners. Therefore, this situation allows us to evaluate the perspectives of decision-makers and practitioners on sustainable production. There are also many factors and alternatives regarding sustainable production barriers and the choice of the most ideal production improvement technique. Many conflicting qualitative and quantitative criteria need to be considered when evaluating relevant factors and alternatives. In this regard, by identifying the criteria set and efficiency-enhancing techniques, the study makes a substantial contribution to the field and literature.

BN-REF methodology was preferred in the study to obtain the evaluations of experts, and therefore to solve a group of decision-making problems that involve uncertainty. BN-REF enabled the determination of the most ideal among the options based on the references that indicate the best solution in the criteria. Another essential aspect of BN-REF is its resistance to the rank reversal problem. The proposed methodology, with its simple and flexible structure, has aspects that can be used to solve problems in similar or different fields in the future.

Finally, the evaluation of the selection processes of sustainable production barriers and techniques for increasing efficiency using the methods covered in the study also enables suppliers, customers, companies, managers, and executives to scientifically substantiate the suitable approaches presented. In other words, this study contributes to the interaction of theoretical and practical applications in terms of the results achieved.

6 Limitations and Future Research

The study examined and evaluated barriers to sustainable production and efficiency-enhancing techniques in manufacturing industries. On the other hand, the evaluation was undertaken under limitations. The study was restricted to the manufacturing industries, the data sets were not deep enough or easily accessible, and the literature review was conducted within the framework of specified ideas, publication types, and periods. Another limitation of the research is that it focuses on selecting the best efficiency-enhancing technique to overcome barriers to sustainable production, and other studies on manufacturing processes or process management applications have not been adequately investigated. All of these limitations restricted the study.

In the future, those interested in the subject and business managers can transfer their practical approaches to a scientific framework by using the study's methodology to evaluate sustainable production barriers and the scope of the best production improvement technique. Finally, future research could compare research findings in different decision-making environments. In addition, researchers can evaluate the current situation by carrying out various comparative analyses using different methods. Additionally, the BN-REF methodology and REF methods can be applied to diverse and complex decision-making problems in other industries.

Appendix

List of the Abbreviatons

Abbreviation	Word or Phrase
AHP	Analytical Hierarchy Process
AM	Above Medium
BM	Below Medium
BN	Bipolar Neutrosophic
BN-REF	Bipolar Neutrosophic Nearest Solution to References Method
BNS	Bipolar Neutrosophic Set
BNWAO	Bipolar Neutrosophic Weighted Averaging Operator
DEMATEL	The Decision Making Trial And. Evaluation Laboratory
H	High
L	Low
M	Medium
MCDM	Multi-Criteria Decision Making
N1	Lack of awareness and knowledge about sustainability concepts
N2	Lack of resources and insufficient workforce

(continued)

(continued)

Abbreviation	Word or Phrase
N3	Lack of a clear framework for performance measures and metrics at strategic, tactical, and operational levels for sustainable production
N4	Lack of high awareness of social and environmental events in businesses
N5	Lack of management support
N6	Low acceptance levels of innovative applications
N7	Lack of technological development
N8	Negative attitudes towards sustainability concepts
N9	Ineffective policies and lack of legislation
N10	Lack of funds and loans for green projects (environmentally friendly projects)
NS	Neutrosophic Set
O1	Management through Value Analysis
O2	Just-in-Time Production Method
O3	Nominal Grouping Technique
O4	Cost—Efficiency Allocation
REF	Nearest Solution to References Method
SAW	Simple Additive Weighting
TOPSIS	Technique for Order Preference by Similarity to Ideal Solution
VH	Very High
VIKOR	Multicriteria Optimization and Compromise Solution (VIseKriterijumska Optimizacija I Kompromisno Resenje in Serbian)
VL	Very Little
VVH	Very Very High
VVL	Very Very Low
WASPAS	Weighted Aggregated Sum Product Assessment
WPM	Weighted Product Method
Z1	Expert 1
Z2	Expert 2
Z3	Expert 3
Z4	Expert 4
Z5	Expert 5

References

1. Abdel-Basset, M., Gamal, A., Son, L.H., Smarandache, F.: A bipolar neutrosophic multi-criteria decision-making framework for professional selection. Appl. Sci. **10**(4), 1202 (2020). https://doi.org/10.3390/app10041202
2. Abrams, F., Hendrickx, L., Sweeck, L., Camps, J., Cattrysse, D., Van Orshoven, J.: Accounting for uncertainty and disagreement in multi-criteria decision making using triangular fuzzy numbers and Monte Carlo simulation: A case study about selecting measures for remediation of agricultural land after radioactive contamination. In: Real Life Applications of Multiple Criteria Decision Making Techniques in Fuzzy Domain, pp. 125–144, Springer Nature Singapore, Singapore (2022)
3. Akar, G.S.: Tedarik Zincirlerinde Sürdürülebilir İmalatın Önündeki Engelleyici Faktörlerin Tam Tutarlılık Yöntemiyle (FUCOM) Değerlendirilmesi. Bucak İşletme Fakültesi Dergisi, **5**(2), 298–318. https://doi.org/10.38057/bifd.1177985 (2022)
4. Alam, M.F.B., Hosen, M.I., Mridha, J.H., Chowdhury, S.E., Rahman, M.A.: Assessing the barriers of integrating technological innovations in textiles sector: Implications towards sustainable production. Green Technol. Sustain. **1**(3), 100039 (2023)
5. Alzawawi, M. (2014). Drivers and obstacles for creating sustainable supply chain management and operations. In: ASEE Zone Conference Proceedings, pp. 1–8
6. Akram, M., Shumaiza, Smarandache, F.: Decision-making with bipolar neutrosophic TOPSIS and bipolar neutrosophic ELECTRE-I. Axioms **7**(2), 33 (2018)
7. Aytekin, A.: Çok kriterli karar problemine uzaklık ve referans temelli çözüm yaklaşımı. Unpublished PhD dissertation, Anadolu University, Eskişehir, Türkiye (2020). https://openaccess.artvin.edu.tr/xmlui/handle/11494/2558
8. Aytekin, A., Durucasu, H.: Nearest solution to references method for multicriteria decision-making problems. Decision Sci. Lett. **10**(2), 111–128 (2021)
9. Aytekin, A.: Efficiency and performance analyses of food companies via IDOCRIW, REF-II, and OCRA methods. Business Studies and New Approaches, 7 (2021)
10. Aytekin, A.: Çok kriterli karar analizi. Nobel Bilimsel (2022)
11. Aytekin, A., Görçün, Ö.F., Ecer, F., Pamucar, D., Karamaşa, Ç.: Foreign market selection of suppliers through a novel REF-Sort technique. Kybernetes **52**(11), 4958–4992 (2023)
12. Aytekin, A., Korucuk, S., Karamaşa, Ç.: Ranking countries according to logistics and international trade efficiencies via REF-III. J. Intell. Manag. Decis **2**, 74–84 (2023)
13. Aytekin, A., Bozkurt, E., Orakçı, E., Uysal, M., Simic, V., Korucuk, S., Pamucar, D.: A bipolar neutrosophic combined compromise solution-based hybrid model for identifying blockchain application barriers and Benchmarking consensus algorithms. Eng. Appl. Artif. Intell. **133**, 108343 (2024)
14. Bhanot, N., Rao, P.V., Deshmukh, S.G.: An integrated approach for analysing the enablers and barriers of sustainable manufacturing. J. Clean. Prod. **142**, 4412–4439 (2017). https://doi.org/10.1016/j.jclepro.2016.11.123
15. Bosc, P., Pivert, O.: On a fuzzy bipolar relational algebra. Inf. Sci. **219**, 1–16 (2013)
16. Cao, S., Yuan, L., Zheng, H., Wang, X.: Research of the risk factors of china's unsustainable socioeconomic development: lessons for other nations. Soc. Indic. Res. **123**, 337–347 (2015)
17. Chkanikova, O., Mont, O.: Corporate supply chain responsibility: drivers and barriers for sustainable food retailing. Corp. Soc. Responsib. Environ. Manag. **22**(2), 65–82 (2015)
18. Deli, I., Ali, M., Smarandache, F.: Bipolar neutrosophic sets and their application based on multi-criteria decision making problems. Collected Papers. Volume X: On Neutrosophics, Plithogenics, Hypersoft Set, Hypergraphs, and other topics, 98 (2015)
19. Demirbaş, Z.A.: Verimlilik Arttırma Tekniği Olarak Metot Etüdünün Bir Hazır Giyim İşletmesinde Uygulanmasının İşletme Performansına Etkileri., (Yayımlanmamış Yüksek Lisans Tezi), Dokuz Eylül Üniversitesi Sosyal Bilimler Enstitüsü İzmir (2010)
20. Gardas, B.B., Raut, R.D., Narkhede, B.: Modelling the challenges to sustainability in the textile and apparel (T & A) sector: a Delphi- DEMATEL Approach. Sustain. Production Consum. **15**, 96–108 (2018)

21. Garg, C.P., Görçün, Ö.F., Kundu, P., Küçükönder, H.: An integrated fuzzy MCDM approach based on Bonferroni functions for selection and evaluation of industrial robots for the automobile manufacturing industry. Expert Syst. Appl. **213**, 118863 (2023)
22. Goyal, S., Garg, D., Luthra, S.: Sustainable production and consumption: analysing barriers and solutions for maintaining green tomorrow by using fuzzy-AHP–fuzzy-TOPSIS hybrid framework. Environ. Dev. Sustain. **23**, 16934–16980 (2021)
23. Görçün, Ö.F., Aytekin, A., Korucuk, S.: Fresh food supplier selection for global retail chains via bipolar neutrosophic methodology. J. Clean. Prod. **419**, 138156 (2023)
24. Hariyani, D., Mishra, S., Sharma, M.K., Hariyani, P.: Organizational barriers to the sustainable manufacturing system: a literature review. Environ. Challenges **9**, 100606 (2022). https://doi.org/10.1016/j.envc.2022.100606
25. Heijungs, R., Huppes, G., Guinee, J.B.: Life cycle assessment and sustainability analysis of products, materials and technologies: toward a scientific framework for sustainability life cycle analysis. Polym. Degrad. Stab. **95**(3), 422–428 (2010)
26. Irvanizam, I., Syahrini, I., Zi, N.N., Azzahra, N., Iqbal, M., Marzuki, M., Subianto, M.: An improved EDAS method based on bipolar neutrosophic set and its application in group decision-making. Appl. Comput. Intell. Soft Comput. **2021**, 1–16 (2021)
27. Kang, H.S., Lee, J.Y., Choi, S., Kim, H., Park, J.H., Son, J.Y., Kim, B.H., Do Noh, S.: Smart manufacturing: past research, present findings, and future directions. Int. J. Precision Eng. Manuf. Green Technol. **3**(1), 111–128 (2016). https://doi.org/10.1007/s40684-016-0015-5
28. Kırda, K., Aytekin, A.: Assessing industrialized countries' environmental sustainability performances using an integrated multi-criteria model and software. Environ. Develop. Sustain., 1–46 (2023)
29. Korucuk, S.: Erzurum, Erzincan ve Bayburt İllerindeki İmalat İşletmelerinde Depolama ve Satın Alma Kararlarının Üretime Etkisine Yönelik Bir Karşılaştırma, Atatürk Üniversitesi Sosyal Bilimler Enstitüsü Dergisi **22**(2), 1171–1202 (2018)
30. Korucuk, S.: Üretim İşletmelerinde Verimliliğin Önündeki Engellerin Ve Verim Artırıcı Tekniklerin Bütünleşik AHP-TOPSIS İle Sıralanması: Erzurum İli Örneği. Verimlilik Dergisi **1**(1), 219–241 (2019)
31. Korucuk, S., Tirkolaee, E.B., Aytekin, A., Karabasevic, D., Karamaşa, Ç.: Agile supply chain management based on critical success factors and most ideal risk reduction strategy in the era of industry 4.0: application to plastic industry. Oper. Manag. Res., 1–22 (2023)
32. Küçük, O., Korucuk, S.: Süreç yönetimi. Detay Yayıncılık: Ankara, Turkey (2018)
33. Lemenih, M., Kassa, H. (Eds.): Opportunities and challenges for sustainable production and marketing of gums and resins in Ethiopia (2011)
34. Liu, Y., Wood, L.C., Venkatesh, V.G., Zhang, A., Farooque, M.: Barriers to sustainable food consumption and production in China: A fuzzy DEMATEL analysis from a circular economy perspective. Sustain. Prod. Consum. **28**, 1114–1129 (2021)
35. Liu, C.: Bipolar neutrosophic distance measure in multi-attribute decision making. Infinite Study (2020)
36. Lukić, R.: Application of the REF Method in the Evaluation of Trade Efficiency in Serbia. Rev. Int. Comparat. Manag./Revista de Management Comparat International **25**(1) (2024)
37. Luthra, S., Mangla, S.K., Xu, L., Diabat, A.: Using AHP to evaluate barriers in adopting sustainable consumption and production initiatives in a supply chain. Int. J. Prod. Econ. **181**, 342–349 (2016)
38. Machado, C.G., Winroth, M.P., Ribeiro da Silva, E.H.D.: Sustainable manufacturing in Industry 4.0: an emerging research agenda. Int. J. Prod. Res. (2019), https://doi.org/10.1080/00207543.2019.1652777
39. Malek, J., Desai, T.N.: Prioritization of sustainable manufacturing barriers using best worst method. J. Clean. Prod. **226**, 589–600 (2019)
40. Mandal, S., Gazi, K.H., Salahshour, S., Mondal, S.P., Bhattacharya, P., Saha, A.K.: Application of interval valued intuitionistic fuzzy uncertain mcdm methodology for ph. d supervisor selection problem. Results Control Optim. **15**, 100411 (2024)

41. Massoud, M.A., Fayad, R., Kamleh, R., El-Fadel, M.: Environmental management system (ISO 14001) certification in developing countries: challenges and ımplementation strategies. Environ. Sci. Technol., 1884–1887 (2010)
42. Mathiyazhagan, K., Govindan, K., NoorulHaq, A., Geng, Y.: An ISM approach for the barrier analysis in implementing green supply chain management. J. Clean. Prod. **47**, 283–297 (2013). https://doi.org/10.1016/j.jclepro.2012.10.042
43. Narimissa, O., Kangarani-Farahani, A., Molla-Alizadeh-Zavardehi, S.: Drivers and barriers for implementation and improvement of Sustainable Supply Chain Management. Sustain. Dev. **28**(1), 247–258 (2020)
44. Özel, F.: Sürdürülebilir üretim / sustainable production. Erişim Tarihi: 14.12.2023 (2012), https://firatozel.wordpress.com/2012/12/11/surdurulebilir-uretimsustainable-production/
45. Özgüner, Z.: Sürdürülebilir Üretimin Önündeki Engeller ve Çözüm Önerileri. Uluslararası Sosyal Bilimler Akademi Dergisi **4**(8), 40–63 (2022). https://doi.org/10.47994/usbad.1054565
46. Paumgarten, F., Kassa, H., Zida, M., Moeliono, M.: Benefits, challenges, and enabling conditions of collective action to promote sustainable production and marketing of products from Africa's dry forests. Rev. Policy Res. **29**(2), 229–250 (2012)
47. Pramanik, S., Dalapati, S., Alam, S., Roy, T.K.: VIKOR based MAGDM strategy under bipolar neutrosophic set environment. Neutrosophic Sets Syst. **19**(1), 57–69 (2018)
48. Pramanik, S., Dalapati, S.: VIKOR-based MAGDM strategy revisited in bipolar neutrosophic set environment. J. Comput. Cognit. Eng. **2**(3), 220–225 (2023)
49. Rosen, M.A., Kishawy, H.A.: Sustainable manufacturing and design: concepts, practices and needs. Sustainability **4**(2), 154–174 (2012). https://doi.org/10.3390/su4020154
50. Roy, P., Shaw, K.: A fuzzy MCDM decision-making model for m-banking evaluations: comparing several m-banking applications. J. Ambient. Intell. Humaniz. Comput. **14**(9), 11873–11895 (2023)
51. Sahoo, L., Senapati, T., Yager, R.R. (Eds.): Real Life Applications of Multiple Criteria Decision Making Techniques in Fuzzy Domain, Studies in Fuzziness and Soft Computing (STUDFUZZ, volume 420), Springer Nature (2023)
52. Sert, M, Kesen, E.S.: Tam Zamaninda Üretim Felsefine Dayali Bir Seri Üretim Hattinin Simülasyon Tekniğiyle Performans Analizi, S.Ü. Müh. Bilim ve Tekn. Derg. **7** (1), 115–134 (2019)
53. Sharma, M., Joshi, S., Govindan, K.: Overcoming barriers to implement digital technologies to achieve sustainable production and consumption in the food sector: a circular economy perspective. Sustain. Prod. Consum. **39**, 203–215 (2023)
54. Singh, S., Srivastava, S., Jangirala, S.K.: System dynamics analysis of sugarcane supply chain in Indian sugar industry. Glob. Bus. Rev. (2021). https://doi.org/10.1177/0972150921999521
55. Smarandache, F.: Neutrosophy: Neutrosophic Probability, Set, and Logic: Analytic Synthesis & Synthetic Analysis. American Research Press, Rehoboth (1998)
56. Söylemez, Yıldırım, E., Dil, Şahin, M., Koç, Y.D.: Sürdürülebilir Üretim Ve Tüketimin Benimsenmesinde Karşılaşılan Engellerin Değerlendirilmesi, Dumlupınar Üniversitesi Sosyal Bilimler Dergisi, Avrasya Özel Sayısı, 192–204 (2016)
57. Tolga, A.,Ç., Turgut, Z.K.: Sustainable and renewable energy power plants evaluation by fuzzy TODIM technique. J. Oper. Res. Stat. Econ. Manag. Inform. Syst. **6**(1), 49–68,.50–51 (2018)
58. Yildiz, A., Ozkan, C.: A novel modified Delphi-based spherical fuzzy AHP integrated spherical fuzzy CODAS methodology for vending machine location selection problem: a real-life case study in İstanbul. Neural Comput. Appl. **36**(2), 823–842 (2024)
59. Youngjae, C., Bopaya, B.: Sustainable manufacturing and the role of the ınternational journal of production research. Int. J. Prod. Res. **51**(23–24), 7448–7455, 7448 (2013)
60. Vinodh, S., Jeya Girubha, R.: Sustainable concept selection using ELECTRE. Clean Technol. Environ. Policy **14**(651–656), 653 (2012)
61. Zarte, M., Pechmann, A., Nunes, I.L.: Decision support systems for sustainable manufacturing surrounding the product and production life cycle—a literature review. J. Clean. Prod. **219**, 336–349 (2019)

Assessment of Smart City Performance Using Multi-Criteria Decision-Making Methods

Sarbast Moslem and Gülay Demir

Abstract This study seeks to fill a significant gap in the existing literature by introducing a systematic methodology for objectively evaluating the performance of smart cities worldwide using Multi-Criteria Decision-Making (MCDM) methodologies. While various indices describing smart city rankings have been regularly published, the absence of a systematic approach has been noted. In response to this gap, our research aims to provide a structured framework for assessing smart cities' performance. Building upon the ten criteria identified in the Eden Strategy Institute's 2020/21 report, we focus on analyzing the performance of leading smart cities such as Amsterdam, Barcelona, Helsinki, London, Montreal, New York City, Seoul, Shanghai, Singapore, and Vienna. These criteria encompass a wide range of dimensions including budget and financial incentives, innovation ecosystems, leadership quality, past performance, people orientation, preparedness, support programs, smart policies, and vision. The chosen MCDM methodology offers an effective means of managing complex and multi-criteria processes. Each criterion is meticulously evaluated for every city using a specified rating scale. Subsequently, the weight of each criterion is determined utilizing the WENSLO method, and each city's performance is calculated employing the RAWEC method. By adopting this approach, our study aims to deliver a comprehensive and analytical evaluation that transcends subjective assessments. Furthermore, our research contributes a methodology for identifying essential criteria for ranking smart cities and evaluating countries based on these criteria. This methodology can serve as a valuable tool for academic researchers and policymakers alike, providing them with a systematic framework for assessing the development of smart cities and guiding decision-making processes in this domain.

Keywords Smart city · MCDM · WENSLO · RAWEC

S. Moslem
School of Architecture Planning and Environmental Policy, University College Dublin, Dublin 15, Belfield 04 V1W8, Ireland
e-mail: sarbast.moslem@ucd.ie

G. Demir (✉)
High School of Health Vocational Services, Sivas Cumhuriyet University, Sivas 58140, Turkey
e-mail: gulaydemir@cumhuriyet.edu.tr

L. Sahoo et al. (eds.), *Decision Making Under Uncertainty Via Optimization, Modelling, and Analysis*, Studies in Systems, Decision and Control 558,
https://doi.org/10.1007/978-981-96-0085-4_8

1 Introduction

Today, rapid urbanization across the world and rapid advances in technology are leading to profound changes in urban management and planning. In this context, the smart cities idea is a vital approach developed to solve the complex and growing problems faced by modern cities and to improve urban life [6, 11, 19, 23]. Smart cities aim to make progress in critical areas such as sustainability, quality of life, environmental protection, and social inclusion, centred on the use of technology and data [8], [7, 9, 10, 14, 20]. This approach integrates critical systems in a city, enabling more effective and efficient management and increasing citizens' quality of life. MCDM is a solution used to manage complex and multi-criteria processes. It is used in the assessment of smart cities because these cities are complex, consisting of multiple components, and cannot be evaluated with a single criterion. Different factors need to be brought together to assess the performance of smart cities objectively, and each factor's importance needs to be determined. The multi-criteria decision-making method provides a more robust basis for decision-making by addressing the relationships and uncertainties between these factors.

This study aims to thoroughly analyze the world's leading smart cities using the MCDM technique. Ten criteria identified in the Eden Strategy Institute's 2020/21 report will form the basis of this assessment. Each factor will be examined separately to help objectively analyze the smart city's performance. This analysis will provide an essential contribution to understanding and comparing the development of smart cities within an academic framework.

As a consequence of the outcomes of this study and the proposed technique, we make many additions to the literature. The main contributions are outlined below:

- To the best of our knowledge, this is the first time the WENSLO and RAWEC techniques have been employed to address a problem in the literature. In this regard, our work fills a gap in the MCDM literature. However, evaluating cities based on their smartness is a novel use for these systems.
- Smart cities reflect ways for making the best use of scarce resources, providing top-tier urban management and services, and improving inhabitants' quality of life. This study brings the weighting and ranking of criteria affecting the development of smart cities into a vital role focus by examining them from an MCDM perspective. This perspective offers an effective analytical tool that attracts the attention of policymakers and specialized academics.
- Sensitivity analysis of parameters assesses the ranking robustness of the decisions provided. The integrated technique produces trustworthy findings in terms of criterion weights. The comparison analysis also uses traditional MCDM approaches such as WASPAS, SAW, MARCOS, and CRADIS.
- The proposed decision support tool provides a complete and holistic approach to assisting policymakers, including local governments, in developing suitable and successful planning strategies.

The rest of the study is organized as follows. The second section provides a complete literature analysis on city rankings using MCDM methodologies. In the third section, WENSLO and RAWEC techniques are discussed. The fourth section discusses an illustrated case, while the fifth section conducts sensitivity and comparison analysis. The sixth section includes discussion and management implications.

1.1 Research Questions and Research Gaps

Research Questions:

- What are the decisive criteria and factors in the selection of smart cities?
- What methodologies and decision support models are used in the selection and evaluation of smart cities?
- What is the impact of multi-criteria decision-making (MCDM) approaches in the selection of smart cities?
- What are the differences and similarities between different methodologies and models used in the selection of smart cities?
- How are the reliability and applicability of methodologies and models used in the selection of smart cities assessed?

Research Gaps:

- There is a need for further research to standardize and determine the criteria and factors used in the selection of smart cities.
- More comprehensive studies are needed in the existing literature to effectively compare different methodologies and models for selecting smart cities.
- Further research is required on the reliability and accuracy of MCDM approaches used in the selection of smart cities.
- It is important to conduct more research on the applicability and generalizability of models and methodologies developed for the selection of smart cities.
- Future research should focus on methodologies and models used in the selection of smart cities to better consider factors such as sustainability, inclusivity, and social acceptance.

2 Literature Review

The development and implementation of smart cities have attracted wide attention in the academic literature in recent years. Studies in this field focus on defining the smart city concept, analyzing its components, examining its application areas, and evaluating its impacts. Research on the definition and various dimensions of the smart city concept emphasizes that smart cities are complex and holistic urban systems [5, 12]. These studies reveal that smart cities aim to optimize urban life by integrating information and communication technologies such as technology, data analytics, and digital innovations. There are also studies on the components and application areas of smart cities. For example, a study by [2] investigated how intelligent transport systems can improve urban mobility. Similarly, a study by [22] examined the impacts of smart environmental management systems on urban sustainability. For example, studies on the application of smart technologies in critical areas such as transport, energy, environment, security, and public services show that smart cities contribute to critical objectives such as sustainability, efficiency, quality of life, and urban well-being.

However, a multi-disciplinary approach is required to determine strategies and policies related to the development and implementation of smart cities. For this reason, studies evaluate the criteria for becoming a smart city using MCDM methods. For example, [3] used the fuzzy and fuzzy-AHP methods to assess the value of several sustainability criteria in a smart city. They created sustainability metrics for creating a smart city in a poor country. [4] examined smart urban logistics solutions in Istanbul using literature research and expert comments, and the findings were produced utilizing the Quality House matrix of the Quality Function Deployment methodology and a fuzzy MCDM method. [16] utilized ANP to assess smart and sustainable city parameters in their research. According to the findings, "smart living" was the most important criterion, with 35.459% priority, while "smart management" came last, with 5.738%. According to the TOPSIS technique, Tokyo, London, and New York are in the top three in the research's total ranking, which includes 44 cities globally. [1] assessed the smartness of cities using integrated gray-based approaches. In this fashion, they offered a gray addition to the LBWA and EDAS methods. They weighted six variables with the LBWA-G approach and ranked the performance of 17 European cities using the EDAS-G method. [13] conducted a case study in Shanghai to show the use of MCDM approaches, including GIS, for flood risk assessment of metro systems in smart city construction. [15] used PROMETHEE and TOPSIS techniques to rank Polish towns based on their sustainable development and smart city status.

The literature review on smart cities is an invaluable resource for academic researchers and policymakers. These studies help identify methods and policies for creating and implementing smart cities and motivate further study in this area.

3 Material and Method

This section describes a multi-criteria framework in which models are used to calculate the weighting factors of the criterion and assess the alternatives.

3.1 WENSLO Method for Criteria Prioritization

The weight coefficients of the criterion are determined using the WENSLO method. This method was first presented by [17]. It has a six-step process framework [17]:

Step 1. Preparing the initial decision matrix.

A1, A2,…, Am is a vector space of alternatives that represents a collection of alternatives (m alternative). C1, C2,…, Cn represents a vector space of criteria (n criteria). The combined decision matrix (Z) is obtained using Eq. (1).

$$Z = \left[z_{ij}\right]_{mxn} = \begin{bmatrix} z_{11} & \cdots & z_{1n} \\ \vdots & \ddots & \vdots \\ z_{m1} & \cdots & z_{mn} \end{bmatrix} \tag{1}$$

z_{ij} is criterion *j*. in alternative *i*.

Step 2. Obtaining normalization matrix (T).

Equation (2) is used to normalize the combined decision matrix.

$$t_{ij} = \frac{z_j}{\sum_{j=1}^{n} z_j} \tag{2}$$

Step 3. Calculate the criteria class interval (ρ_j).

The size of the jth criteria class interval is computed using Sturges' rule, Eq. (3):

$$\rho_j = \left(\frac{max(z_j) - min(z_j)}{1 + 3.322 * log(m)}\right) \tag{3}$$

Step 4. Calculate the criteria slope ($tan\varphi_j$).

The slope of the criterion is calculated by Eq. (4).

$$tan\varphi_j = \frac{\sum_{i=1}^{m} z_{ij}}{(m-1)\rho_j} \tag{4}$$

Step 5. Determination of the criterion envelope (ε_j)

The sum of the partial Euclidean distances between two successive criteria is determined using Eq. (5).

$$\varepsilon_j = \left(\sum_{i=1}^{m-1} \sqrt{\left(z_{i+1,j} - z_{ij}\right)^2 + \left(\rho_j\right)^2}\right) \tag{5}$$

Step 6. Describe the envelope–slope ratio (δ_j).

Equation (6) calculates the ratio of the entire Euclidean distance to the criteria slope.

$$\delta_j = \frac{\varepsilon_j}{tan\varphi_j} \tag{6}$$

Step 6. Obtain weights (w_j) of each of the criteria.

Equation (7) calculates the weights depending on the criteria's significance coefficient.

$$w_j = \frac{\delta_j}{\sum_{j=1}^{n}\delta_j} \tag{7}$$

3.2 RAWEC Method to Rank Alternatives

For the ranking of alternatives, the RAWEC method is used. This method was first presented by [18]. It has a four-step process framework [18]:

Step 1. Construction of the initial decision matrix.

The combined decision matrix (X) is obtained using Eq. (8).

$$X = \left[x_{ij}\right]_{mxn} = \begin{bmatrix} x_{11} & \cdots & x_{1n} \\ \vdots & \ddots & \vdots \\ x_{m1} & \cdots & x_{mn} \end{bmatrix} \tag{8}$$

x_{ij} is criterion j. in alternative i.

Step 2. Normalization of the decision matrix.

When normalizing the initial decision matrix, double normalization is performed with Eqs. (9) and (10).

$$n_{ij} = \frac{x_{ij}}{x_j^{max}}, and\ n_{ij}\prime = \frac{x_j^{min}}{x_{ij}} for\ benefit\ criteria \tag{9}$$

$$n_{ij} = \frac{x_j^{min}}{x_{ij}}, and\ n_{ij}\prime = \frac{x_{ij}}{x_j^{max}} for\ cost\ criteria \tag{10}$$

Step 3. Calculation of deviation from criterion weight.

Equations (11) and (12) provide the weighting of the normalized decision matrix as well as the computation of the deviation from the weighted criterion.

$$v_{ij} = \sum_{i=1}^{n} w_j\left(1 - n_{ij}\right) \tag{11}$$

$$v_{ij}\prime = \sum_{i=1}^{n} w_j\left(1 - n_{ij}\prime\right) \tag{12}$$

Step 4. Calculation of the value of the RAWEC method.
It is calculated according to Eq. (13).

$$Q_i = \frac{v_{ij}\prime - v_{ij}}{v_{ij}\prime + v_{ij}} \quad (13)$$

The RAWEC method returns a number ranging from −1 to 1. The superiority of an alternative is decided by how valuable the alternative's strategy is. The alternative with the highest value is the best selection.

4 WENSLO-RAWEC Framework for Judging Countries' Smart City Performance

The assessment of smart cities is a complex process and is often carried out using many different criteria and benchmarks. In this context, organizations such as the Eden Strategy Institute use specific criteria to assess and rank smart city development. This study analyses the criteria used by the Eden Strategy Institute and how these criteria contribute to the ranking of countries according to their smart city levels. The criteria used by the Eden Strategy Institute to rank countries are given in Table 1.

After assessing each of the 10 criteria on a scale of one to four, they computed each city's overall score and established its ranking position. Figure 1 depicts 50 smart cities [24].

However, each country may attribute different importance to these criteria. Therefore, this study proposes a fairer ranking of countries by determining the importance levels of the criteria. Analyses of the top ten countries on the 2020/21 Top 50 Smart City Government list indicate how the relevance of the criterion influences the rankings of the countries. These studies emphasize the relevance of criteria in the evaluation of smart cities and provide a more equitable and thorough ranking mechanism. This study highlights the relevance of the methodologies used in evaluating and rating smart cities, as well as the necessity to define the importance levels of the criteria.

4.1 Weight Coefficients of Criteria Are Determined Using the WENSLO Method

The application of the WENSLO method is presented to determine the weight coefficients of the criteria. The evaluation scores of the first 10 countries in the 2020/21 Top 50 Smart City Governments list are given in Table 2.

The normalized decision matrix obtained using Eq. (2) is given in Table 3.

Table 1 Smart city criteria

Criteria	Description
Vision (C1)	A coherent, well-defined, and updated plan for developing a "smart city"
Leadership (C2)	Dedicated city leadership guides smart city projects without big discontinuities
Budget (C3)	Sufficient and sustained funding for smart city initiatives
Financial Incentives (C4)	Financial incentives can successfully stimulate private sector engagement (e.g., grants, rebates, subsidies, competitions), particularly in smart city projects and initiatives
Support Programmes (C5)	In-kind programs attract private actors to join in smart city projects and activities, such as incubators, events, networks, and prizes
Talent Readiness (C6)	Programmes to prepare the city's talent with smart talents
People Centricity (C7)	A genuine, people-first design of the future metropolis, without any systematic disrespect for human life or basic human rights
Innovation Ecosystems (C8)	The municipal administration played a key role in maintaining innovation and collaboration by engaging a diverse spectrum of stakeholders
Smart Policies (C9)	A favourable policy and regulatory framework for smart city growth (e.g., data governance, IP protection, urban design)
Track Record (C10)	The government's experience in facilitating successful smart city programs, with no systemic or widespread failure

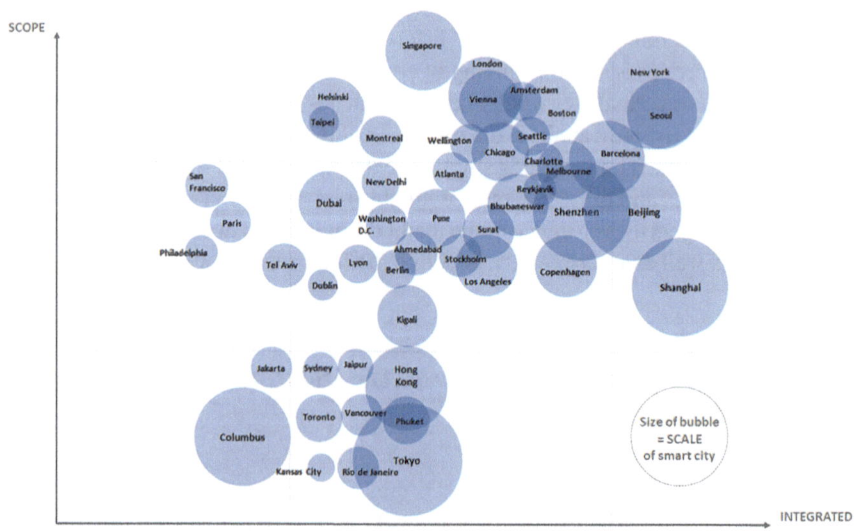

Fig. 1 Eden Strategy Institute's Top 50 Smart City Governments *Source* 2020/21 Top 50 Smart City Governments

Table 2 Initial decision-making matrix

	C1	C2	C3	C4	C5	C6	C7	C8	C9	C10
A1	3	3	3	4	4	4	4	3,9	3,9	3
A2	3	4	3	3	3	4	3	3	4	4
A3	4	3	3	3	3	3,1	4	3	3	4
A4	3	3	3	3	3	3	3,1	4	3	4
A5	3	3	4	4	4	4	3	2	3	4
A6	4	3	3	3	3	3	4	3	2,9	3
A7	3	3	3	3	3	3	2,9	3	4	3,9
A8	3	3	2,1	3,1	4	3	4	3	2,1	4
A9	4	3	3	2	3	3,1	3,1	3	4	3
A10	3	4	3	3	2	3	3,9	3,1	3	3,1

Kaynak: 2020/21 Top 50 Smart City Governments
A1: Singapore, A2: Seoul, A3: London, A4: Barcelona, A5: Helsinki, A6: New York City, A7: Montreal, A8: Shanghai, A9: Vienna, A10: Amsterdam

Table 3 Normalized decision matrix

	C1	C2	C3	C4	C5	C6	C7	C8	C9	C10
A1	0,0909	0,0938	0,0997	0,1286	0,1250	0,1205	0,1143	0,1258	0,1185	0,0833
A2	0,0909	0,1250	0,0997	0,0965	0,0938	0,1205	0,0857	0,0968	0,1216	0,1111
A3	0,1212	0,0938	0,0997	0,0965	0,0938	0,0934	0,1143	0,0968	0,0912	0,1111
A4	0,0909	0,0938	0,0997	0,0965	0,0938	0,0904	0,0886	0,1290	0,0912	0,1111
A5	0,0909	0,0938	0,1329	0,1286	0,1250	0,1205	0,0857	0,0645	0,0912	0,1111
A6	0,1212	0,0938	0,0997	0,0965	0,0938	0,0904	0,1143	0,0968	0,0881	0,0833
A7	0,0909	0,0938	0,0997	0,0965	0,0938	0,0904	0,0829	0,0968	0,1216	0,1083
A8	0,0909	0,0938	0,0698	0,0997	0,1250	0,0904	0,1143	0,0968	0,0638	0,1111
A9	0,1212	0,0938	0,0997	0,0643	0,0938	0,0934	0,0886	0,0968	0,1216	0,0833
A10	0,0909	0,1250	0,0997	0,0965	0,0625	0,0904	0,1114	0,1000	0,0912	0,0861

$$t_{11} = \frac{3}{3+3+\cdots+4+3} = 0{,}0909 \; t_{12} = \frac{3}{3+4+\cdots+3+4} = 0{,}0938$$

The calculation is exemplified as follows. Then the criterion class range is obtained according to Sturges rule with Eq. (3) and given in Table 4.

Table 4 Size of the jth criterion class interval

	C1	C2	C3	C4	C5	C6	C7	C8	C9	C10
ρ_j	0,0070	0,0072	0,0146	0,0149	0,0145	0,0070	0,0073	0,0149	0,0134	0,0064

Table 5 Criteria slope values

	C1	C2	C3	C4	C5	C6	C7	C8	C9	C10
$tan\varphi_j$	15,8473	15,3671	7,6077	7,4675	7,6836	15,9434	15,2798	7,4434	8,3154	17,2880

Table 6 Criteria envelope values

	C1	C2	C3	C4	C5	C6	C7	C8	C9	C10
ε_j	0,2077	0,1396	0,2122	0,2400	0,2500	0,1335	0,2382	0,2449	0,2751	0,1446

$$\rho_1 = \frac{0{,}1212 - 0{,}0909}{1 + 3{,}322 * log(10)} = 0{,}0070 \ldots \rho_{10} = \frac{0{,}1111 - 0{,}0833}{1 + 3{,}322 * log(10)} = 0{,}0064$$

The slope of the criterion is obtained by Eq. (4) and presented in Table 5.

$$tan\varphi_{C1} = \frac{0{,}0909 + \cdots + 0{,}0909}{(10 - 1) * 0{,}0070} = 15{,}8473 \; tan\varphi_{C10} = \frac{0{,}0833 + \cdots + 0{,}0861}{(10 - 1) * 0{,}0064} = 17{,}2880$$

The criteria envelope is obtained with Eq. (5) and given in Table 6.

$$\varepsilon_{C1} = \sqrt{(0{,}0909 - 0{,}0909)^2 + (0{,}0070)^2} + \cdots + \sqrt{(0{,}0909 - 0{,}1212)^2 + (0{,}0070)^2} = 0{,}2077$$

$$\vdots$$

$$\varepsilon_{C10} = \sqrt{(0{,}1111 - 0{,}0833)^2 + (0{,}0064)^2} + \cdots + \sqrt{(0{,}0861 - 0{,}0833)^2 + (0{,}0064)^2} = 0{,}1446$$

The envelope-slope ratios are obtained with Eq. (6) and shown in Table 7.

$$\delta_{C1} = \frac{15{,}8473}{0{,}2077}, \ldots \delta_{C10} = \frac{17{,}2880}{0{,}1446} = 119{,}5307$$

The criteria weights calculated by Eq. (7) are shown in Table 8.

Table 7 Envelope–slope ratios

	C1	C2	C3	C4	C5	C6	C7	C8	C9	C10
δ_j	76,3155	110,0722	35,8583	31,1204	30,7361	119,4206	64,1445	30,3893	30,2275	119,5307

Table 8 Final criteria weight values

	C1	C2	C3	C4	C5	C6	C7	C8	C9	C10
w_j	0,1178	0,1699	0,0554	0,0480	0,0474	0,1843	0,0990	0,0469	0,0467	0,1845

Past performance (C10) > capability readiness (C6) > leadership (C2) are the top three most important criteria. The identification of these criteria as the most important as a result of the analysis highlights the key elements needed to ensure the successful development of smart cities.

4.2 *Alternatives Are Evaluated Using the RAWEC Method*

Table 2 was used as the initial decision matrix. Two normalized decision matrices were obtained using Eqs. (9) and (10) and presented in Table 9 and Table 10.

Then deviations from the criteria weights were calculated using Eqs. (11) and (12). Finally, RAWEC values were calculated with Eq. (13) and given in Table 11.

Table 9 Benefit normalization values

	C1	C2	C3	C4	C5	C6	C7	C8	C9	C10
A1	0,75	0,75	0,75	1	1	1	1	0,975	0,975	0,75
A2	0,75	1	0,75	0,75	0,75	1	0,75	0,75	1	1
A3	1	0,75	0,75	0,75	0,75	0,775	1	0,75	0,75	1
A4	0,75	0,75	0,75	0,75	0,75	0,75	0,775	1	0,75	1
A5	0,75	0,75	1	1	1	1	0,75	0,5	0,75	1
A6	1	0,75	0,75	0,75	0,75	0,75	1	0,75	0,725	0,75
A7	0,75	0,75	0,75	0,75	0,75	0,75	0,725	0,75	1	0,975
A8	0,75	0,75	0,525	0,775	1	0,75	1	0,75	0,525	1
A9	1	0,75	0,75	0,5	0,75	0,775	0,775	0,75	1	0,75
A10	0,75	1	0,75	0,75	0,5	0,75	0,975	0,775	0,75	0,775

Table 10 Cost normalization values

	C1	C2	C3	C4	C5	C6	C7	C8	C9	C10
A1	1	1	0,7	0,5	0,5	0,75	0,725	0,5128	0,5385	1
A2	1	0,75	0,7	0,6667	0,6667	0,75	0,9667	0,6667	0,525	0,75
A3	0,75	1	0,7	0,6667	0,6667	0,9677	0,725	0,6667	0,7	0,75
A4	1	1	0,7	0,6667	0,6667	1	0,9355	0,5	0,7	0,75
A5	1	1	0,525	0,5	0,5	0,75	0,9667	1	0,7	0,75
A6	0,75	1	0,7	0,6667	0,6667	1	0,725	0,6667	0,7241	1
A7	1	1	0,7	0,6667	0,6667	1	1	0,6667	0,5250	0,7692
A8	1	1	1	0,6452	0,5	1	0,725	0,6667	1	0,75
A9	0,75	1	0,7	1	0,6667	0,9677	0,9355	0,6667	0,525	1
A10	1	0,75	0,7	0,6667	1	1	0,7436	0,6452	0,7	0,9677

Table 11 Final ranking order calculated using the RAWEC method

ID	Alternatives	v_{ij}	v'_{ij}	Q_i	Rank
A1	Singapore	0,1342	0,1821	0,1512	3
A2	Seoul	0,1036	0,2242	0,3678	1
A3	London	0,1451	0,1868	0,1259	4
A4	Barcelona	0,1897	0,1384	−0,1563	5
A5	Helsinki	0,1318	0,1835	0,1641	2
A6	New York City	0,1970	0,1336	−0,1916	8
A7	Montreal	0,1993	0,1288	−0,2148	9
A8	Shanghai	0,1890	0,1298	−0,1858	7
A9	Vienna	0,2138	0,1120	−0,3125	10
A10	Amsterdam	0,1913	0,1371	−0,1652	6

While Seoul ranked first, Vienna ranked tenth. Eden Ranking and RAWEC Rankings are given in Fig. 2.

While Eden Strategy Institute creates a ranking by summing the scores given without weighting the criteria, in this study, the criteria were weighted using the WENSLO method and these weights were used together with RAWEC to rank the smart cities. There are methodological differences between Eden Strategy Institute's direct scoring approach and ranking within the framework of WENSLO-RAWEC. The results of these different approaches also differ. In particular, rankings using the

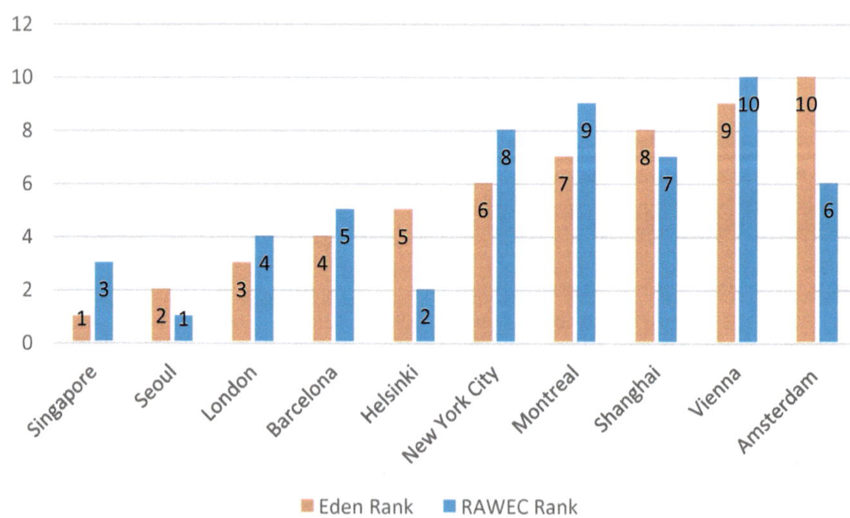

Fig. 2 Eden and RAWEC rankings

weighting method can better reflect the differences in importance between criteria and therefore provide a more comprehensive assessment.

5 Sensitivity Analysis

This section presents a two-stage sensitivity analysis to prove that the proposed method works. In this case, the first thing that was done was to see if changing the criteria weights affected the ranking results. The suggested hybrid model is then compared to the current MCDM methods.

5.1 *Ranking Sensitivity Analysis Based on Changes in Criteria Priority*

After determining the "most important criteria (Cn)" based on the weight values determined by the criterion weighting method, sensitivity analysis can be performed by changing the weight of the "most important criteria (Cn)" to see how the proposed model affects the ranking performance. In this method, a new weight factor vector is created first. In each case, the weight of the most significant criterion is lowered by 20%, and a new weight factor vector is created. Using Eq. (14), the new weight value of the criteria can be calculated as follows [21].

$$w_{n\beta} = (1 - w_{n\alpha})\left(\frac{w_\beta}{1 - w_n}\right) \tag{14}$$

where $w_{n\beta}$ indicates the new weight values calculated for the criteria. $w_{n\alpha}$ denotes the reduced value of the criteria. While the original value of the criteria is denoted by w_β, the original value of the reduced criteria is denoted by w_n.

According to the WENSLO method, the criterion with the highest weight value (0,1845) is C10. A total of 15 scenarios were obtained using Eq. (14). Among these 15 scenarios, the scenario corresponding to the original weight value of criterion C10 was accepted as Scenario 0 (S0). According to these scenarios; in Scenario 1 (S1) C10 = 0,1845, while in Scenario 15 (S15) C10 = 0,0065. After each adjustment of C10, Eq. (14) was used to derive the weights of other criteria. For example, the calculation for the first scenario (S1) was performed as follows: The actual weight value of criterion C10 (0.1845) was reduced by 20% and the weight value for scenario S1 was found to be 0.1476 (0.1845–0.1845*0.20) = 0.1476. Then, the weight values of all criteria except C10 were calculated according to 15 scenarios using Eq. (14) and given in Fig. 3.

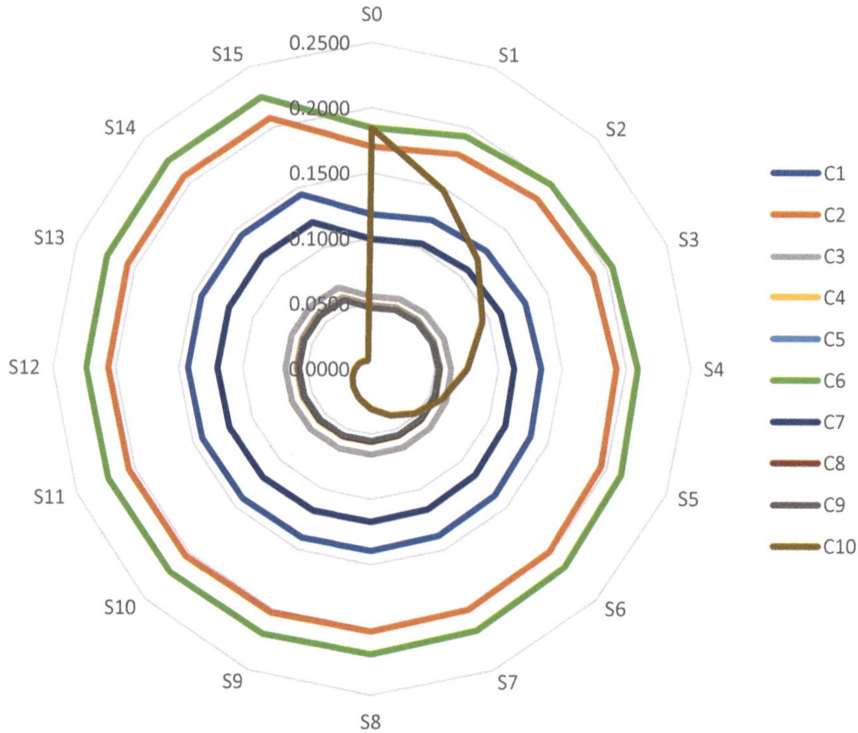

Fig. 3 Modifications in weights based on different scenarios

After obtaining the new criteria weights as shown in Fig. 3, RAWEC utility function values were obtained again for ranking the alternatives according to the 15 scenarios and the ranking values according to the scenarios are given in Fig. 4.

Figure 4 shows that the varying weight values assigned to the criterion based on the created scenarios cause no change in the ranking of all alternatives, indicating that the model is responsive to changes in the weight coefficients. As a consequence, the ranking results demonstrate that the suggested hybrid model is consistent and stable on a broad scale, confirming the efficacy, robustness, and adaptability of the proposed MCDM framework.

5.2 Compared to Another MCDM Method

The study compared the outcomes of the RAWEC method to those of other methods such as WASPAS, SAW, MARCOS, and CRADIS.

As can be seen in Fig. 5, where the comparison results are presented, the ranking positions of the alternatives other than Seoul, London, New York City, Montreal, and

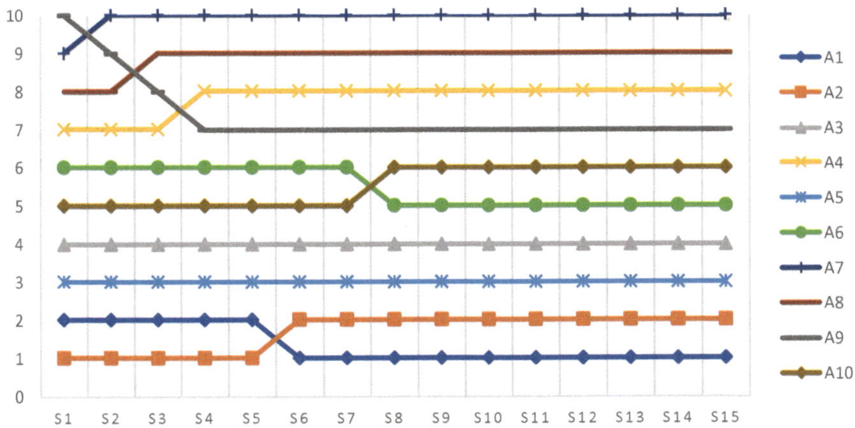

Fig. 4 Alternatives are ranked based on the new criteria weights

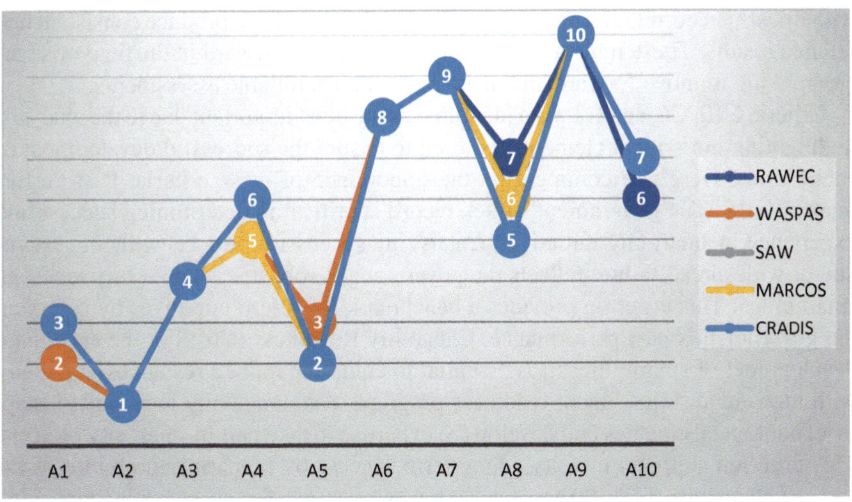

Fig. 5 Comparison between ranking performances of MCDM methods

Vienna vary very little according to all MCDM methods applied and show that the results obtained have high consistency.

In this study, the Spearman correlation coefficient was employed to statistically justify the ranking of the suggested model. The findings collected using SPSS 28 software revealed an average correlation value of 0.973 between the other four MCDM approaches and the RAWEC approach. These findings corroborate the suggested model's rating as dependable. Finally, the dependability of the suggested model is discussed using statistical analyses and findings.

6 Results and Discussion

The innovations provided by this study in the field of smart city ranking are as follows: Criteria Weighting and Objective Evaluation: This study has performed criteria weighting by taking into account the different levels of importance among the criteria. This approach makes the evaluation process more objective and provides a more comprehensive ranking. Hybrid Model Integration: By combining different MCDM methods, the hybrid model provides a more comprehensive and robust evaluation by bringing together various perspectives. Stability and Reliability: The stability and reliability of the proposed model are demonstrated by the fact that the weight values assigned to different scenarios do not change the results. This emphasizes the general validity and applicability of the model. Comparative Analysis: The study compares the proposed model with other MCDM methods. These comparative analyses more clearly reveal the effectiveness and achievements of the proposed model. Statistical Validation: Finally, statistical validation of the ranking of the proposed model solidifies the reliability and validity of the model. This validation with Spearman correlation analysis shows that the model can produce consistent and reliable results. These innovations are an essential step forward in the field of smart city ranking to move forward and make more robust, reliable assessments.

Criteria C10, C6, and C2 were identified as the most important due to the analysis, highlighting the critical elements required to ensure the successful development of smart cities. Here is a comment on the importance of these criteria: Past Performance (C10): The government's track record is critical in determining success and experience in smart city initiatives. Catalyzing successful projects without systematic or widespread failure reflects the government's robustness and effectiveness in smart cities. This criterion provides a benchmark for future initiatives by assessing the government's past performance. Capability Readiness (C6): For the successful development of smart cities, it is essential to equip the human resources in the city with innovative skills. Talent readiness programs encourage city residents to adapt to technological changes and develop the expertise to take part in smart city projects. This criterion supports the city's long-term success by preparing the future workforce. Leadership (C2): Strong leadership is essential for successfully managing and directing smart city projects. The city leadership must act decisively to ensure the continuous progress of projects and achieve targets without significant disruptions. This criterion assesses the credibility and effectiveness of city leadership, thus providing a solid foundation for smart city development. These criteria emphasize the critical elements for the successful development of smart cities, encouraging governments and city leaders to manage strategically and use resources effectively. Identifying these elements as necessary contributes to smart cities' sustainable and inclusive development.

For the rankings of the countries also Singapore, according to the Eden Strategy Institute: 1, according to WENSLO-RAWEC: 3. This difference shows how the weighting of the criteria and the assessment methodology are reflected in the results. Seoul, according to Eden Strategy Institute: 2, according to WENSLO-RAWEC:

1. Seoul ranks high both according to the Eden Strategy Institute and according to the WENSLO method and RAWEC. This shows that Seoul is a major player in terms of smart city development and has a high performance under different assessment methodologies. According to Eden Strategy Institute, London: 3, according to WENSLO -RAWEC: 4. London ranks third according to the Eden Strategy Institute and fourth according to the WENSLO method and RAWEC. This difference shows the impact of the weighting of the criteria and the evaluation methodology on the results. The results show that the proposed hybrid model is broadly consistent and stable, confirming the effectiveness, robustness, and applicability of the proposed MCDM framework.

Furthermore, the Spearman correlation coefficient analysis, which provides statistical support for the ranking of the proposed model, obtained a high correlation value of 0.973 on average between the other four MCDM methods and the RAWEC approach. These results reveal that the ranking of the proposed model is reliable. This study makes a valuable contribution to improving the effectiveness and reliability of the methods used to evaluate and rank smart cities.

This study makes an important contribution towards improving the effectiveness and reliability of the methods used to assess and rank smart cities. The proposed model can contribute to the effective utilization of resources by promoting strategic management in terms of sustainability and the development of smart cities. This study can be an important reference point for academics and experts and provide a solid foundation for future smart city development projects. This assessment shows that different ranking methods can lead to different results and that the methodology used for assessing smart cities is important in interpreting the results. It can also provide a solid basis for future smart city development projects and an important reference point for academics and experts.

Appendix

Table: Abbreviations

Abbreviation	Description	Abbreviation	Description
MCDM	Multi-Criteria Decision-Making	WENSLO	Weights by ENvelope and SLOpe
RAWEC	Ranking Alternatives with Weights of Criterion	WASPAS	Weighted Aggregated Sum Product ASsessment
SAW	Simple Additive Weighting	MARCOS	Measurement of Alternatives and Ranking according to the Compromise Solution
CRADIS	Compromise Ranking of Alternatives from Distance to Ideal Solution	ANP	Analytic Network Process

(continued)

(continued)

Abbreviation	Description	Abbreviation	Description
EDAS	Evaluation based on the Distance from Average Solution	GIS	Geographical Information Systems
TOPSIS	Technique for Order of Preference by Similarity to Ideal Solution	LBWA	Level-Based Weight Assessment
PROMETHEE	Preference Ranking Organization. Method For Enrichment Evaluations		

References

1. Adali, E.A., Öztaş, G.Z., Öztaş, T., Tuş, A.: Assessment of European cities from a smartness perspective: an integrated grey MCDM approach. Sustain. Cities Soc. **84**, 104021 (2022). https://doi.org/10.1016/j.scs.2022.104021
2. Albino, V., Berardi, U., Dangelico, R.M.: Smart cities: definitions, dimensions, performance, and initiatives. J. Urban Technol. **22**(1), 3–21 (2015). https://doi.org/10.1080/10630732.2014.942092
3. Anand, A., Winfred Rufuss, D.D., Rajkumar, V., Suganthi, L.: Evaluation of sustainability indicators in smart cities for India using MCDM approach. Energy Procedia **141**, 211–215 (2017). https://doi.org/10.1016/j.egypro.2017.11.094
4. Büyüközkan, G., Mukul, E.: Evaluation of smart city logistics solutions with fuzzy MCDM methods. Pamukkale Üniversitesi Mühendislik Bilimleri Dergisi. **25**(9), 1033–1040 (2019). https://doi.org/10.5505/pajes.2019.32956
5. Caragliu, A., Del Bo, C., Nijkamp, P.: Smart cities in Europe. In Creating Smart-er Cities, pp. 65–82. Routledge (2013)
6. Chen, Z., Gan, W., Wu, J., Lin, H., Chen, C.M.: Metaverse for smart cities: a survey. Internet Things Cyber-Phys. Syst. **4**, 203–216 (2024). https://doi.org/10.1016/j.iotcps.2023.12.002
7. Demir, G.: Evaluation of sustainable green building indicators by fuzzy multi-criteria decision making. In: Sahoo, L., Senapati, T., Yager, R.R. (eds) Real Life Applications of Multiple Criteria Decision Making Techniques in Fuzzy Domain. Studies in Fuzziness and Soft Computing, vol 420. Springer, Singapore (2023). https://doi.org/10.1007/978-981-19-4929-6_16
8. Fadhel, M. A., Duhaim, A. M., Saihood, A., Sewify, A., Al-Hamadani, M. N., Albahri, A. S., ... , Gu, Y.: Comprehensive systematic review of information fusion methods in smart cities and urban environments. Inform. Fusion, 102317 (2024).https://doi.org/10.1016/j.inffus.2024.102317
9. Ghazinoory, S., Roshandel, J., Parvin, F., Nasri, S., Fatemi, M.: Smart city maturity models: a multidimensional synthesized approach. Wiley Interdiscip. Rev. Data Mining Knowl. Discovery **14**(1), e1516 (2024). https://doi.org/10.1002/widm.1516
10. Ghosh, A., Hafnaoui, R., Mesloub, A., Elkhayat, K., Albaqawy, G., Alnaim, M.M., Mayhoub, M.S.: Active smart switchable glazing for smart city: a review. J. Build. Eng., 108644 (2024). https://doi.org/10.1016/j.jobe.2024.108644
11. Huda, N.U., Ahmed, I., Adnan, M., Ali, M., Naeem, F.: Experts and intelligent systems for smart homes' transformation to sustainable smart cities: a comprehensive review. Expert Syst. Appl. **238**, 122380 (2024). https://doi.org/10.1016/j.eswa.2023.122380
12. Hollands, R.G.: Will the real smart city please stand up? City **12**(3), 303–320 (2008). https://doi.org/10.1080/13604810802479126

13. Lyu, H.M., Yin, Z.Y., Zhou, A., Shen, S.L.: MCDM-based flood risk assessment of metro systems in smart city development: A review. Environ. Impact Assess. Rev. **101**, 107154 (2023). https://doi.org/10.1016/j.eiar.2023.107154
14. Nguyen, H., Nawara, D., Kashef, R.: Connecting the indispensable roles of IoT and artificial intelligence in smart cities: a survey. J. Inform. Intell. (2024).https://doi.org/10.1016/j.jiixd.2024.01.003
15. Ogrodnik, K.: Application of MCDM/MCDA methods in city rankings-review and comparative analysis. Econ. Environ. **86**(3), 132–151 (2023). https://doi.org/10.34659/eis.2023.86.3.689
16. Ozkaya, G., Erdin, C.: Evaluation of smart and sustainable cities through a hybrid MCDM approach based on ANP and TOPSIS technique. Heliyon **6**(10) (2020). https://doi.org/10.1016/j.heliyon.2020.e05052
17. Pamučar, D., Ecer, F., Gligorić, Z., Gligorić, M., Deveci. M.: A novel WENSLO and ALWAS multicriteria methodology and its application to green growth performance evaluation. IEEE Trans. Eng. Manag. (2023).https://doi.org/10.1109/TEM.2023.3321697
18. Puška, A., Štilić, A., Pamučar, D., Božanić, D., Nedeljković, M.: Introducing a novel multi-criteria ranking of alternatives with weights of criterion (RAWEC) model. MethodsX. **12**, 102628 (2024). https://doi.org/10.1016/j.mex.2024.102628
19. Sharifi, A., Allam, Z., Bibri, S.E., Khavarian-Garmsir, A.R.: Smart cities and sustainable development goals (SDGs): a systematic literature review of co-benefits and trade-offs. Cities **146**, 104659 (2024). https://doi.org/10.1016/j.cities.2023.104659
20. Sindiramutty, S.R., Tan, C.E., Tee, W.J., Lau, S.P., Balakrishnan, S., Kaur, S., ... , Aslam, M.F.A.: Modern smart cities and open research challenges and issues of explainable artificial intelligence. Adv. Explainable AI Appl. Smart Cities, 389–424 (2024).https://doi.org/10.4018/978-16684-6361-1.ch015
21. Vrtagić, S., Softić, E., Subotić, M., Stević, Ž, Dordevic, M., Ponjavic, M.: Ranking road sections based on MCDM model: new improved Fuzzy SWARA (IMF SWARA). Axioms **10**(2), 1–23 (2021). https://doi.org/10.3390/axioms10020092
22. Yigitcanlar, T., Kamruzzaman, M.: Does smart city policy lead to sustainability of cities? Land Use Policy **73**, 49–58 (2018). https://doi.org/10.1016/j.landusepol.2018.01.034
23. Zeng, F., Pang, C., Tang, H.: Sensors on internet of things systems for the sustainable development of smart cities: a systematic literature review. Sensors **24**(7), 2074 (2024). https://doi.org/10.3390/s24072074
24. 2020/21 Top 50 Smart City Governments, https://www.smartcitygovt.com/. Access Date: 17 April 2024

Impact of Technological Innovation on Employment Under Intuitionistic Fuzzy Einstein Aggregation Information with Z-Numbers

Shahzaib Ashraf, Maria Akram, and Chiranjibe Jana

Abstract Many are worried about the future of human jobs due to the dramatic changes brought about by the fast development of AI and automation in the industrial environment. To better comprehend the multifaceted effects of technological advancement on the labor market, this case study zeroes in on the industrial sector. To evaluate the impact of technology innovation on employment, Multi-Criteria Decision-Making (MCDM) is useful for methodically finding and choosing relevant factors. Important components outlined by MCDM, such as "Policy Efficacy and Alignment," "Technological Integration Effectiveness," "Employment Impact and Dynamics," and "Adaptation Strategies by Businesses," are among the established criteria. An invaluable idea for capturing fuzzy information in decision-making is the intuitionistic fuzzy $\hat{Z}$ number $(IF\hat{Z}N)$, which can handle higher levels of uncertainty than fuzzy and fuzzy $\hat{Z}$ numbers $(F\hat{Z}N)$. To handle higher levels of uncertainty, we offer and examine intuitionistic fuzzy $\hat{Z}$ numbers $(IF\hat{Z}N)$, which we think are a major improvement over current fuzzy set designs. The new operating guidelines and weighted aggregation operators that go along with them are the real prize here. We focus on updating the operational specifications for operators that aggregate Einstein data. We take a close look at these newly suggested rules and their features. Next, we develop new aggregation techniques for $(IF\hat{Z}N)$ data by expanding upon these operational ideas, and we investigate the characteristics and connections between these operators. In order to make possible future applications easier, we also present a method for making decisions with numerous attributes in the $IF\hat{Z}N$ setting. Lastly, a case study is provided to illustrate the approach's practical applicability.

S. Ashraf (✉) · M. Akram
Institute of Mathematics, Khwaja Fareed University of Engineering and Information Technology, Rahim Yar Khan 64200, Pakistan
e-mail: shahzaibashraf@kfueit.edu.pk

M. Akram
e-mail: mariaakram601@gmail.com

C. Jana
Saveetha School of Engineering, Saveetha Institute of Medical and Technical Sciences (SIMATS), Chennai 602105, Tamil Nadu, India
e-mail: jana.chiranjibe7@gmail.com

L. Sahoo et al. (eds.), *Decision Making Under Uncertainty Via Optimization, Modelling, and Analysis*, Studies in Systems, Decision and Control 558,
https://doi.org/10.1007/978-981-96-0085-4_9

Keywords Intuitionistic fuzzy sets · Z-numbers · Einstein operators · Multi-criteria decision-making

1 Introduction

The rapid development of automation and artificial intelligence (AI) technologies has become a transformational force in the unrelenting wave of technological advancement, altering the very fabric of industries globally. Amid all of this innovation, a serious worry—the impending threat of labor displacement of humans—becomes apparent. This case study carefully examines the complex field of manufacturing, with the main goal being to analyze and understand the multifaceted influence that technical innovation has on job dynamics. As the fourth industrial revolution approaches, automation and artificial intelligence (AI) technologies are enabling us to go to new heights. Technological improvements have driven paradigm shifts in the history of industry, but the current rate and extent of change are unparalleled. At the core of this profound shift is the manufacturing sector, which is a pillar of economic growth. Beyond normal concerns of productivity and efficiency, the transformation of robots into ever more complex creatures poses fundamental questions about the future of human labor.

A methodological strategy known as multi-criteria decision-making (MCDM) works well in complicated situations when there are several criteria or elements to consider. MCDM may be used effectively in the case study on how technological innovation affects employment in the manufacturing industry to help make well-informed and balanced choices by taking into account a variety of criteria and alternative approaches. Researchers have turned to MCGDM-based CDE approaches because of the complex information originating from multiple DMs for alternatives and assessment criteria. These approaches include a fuzzy rough number-based AHP-TOPSIS [1], modified rough VIKOR-based design concept [2], new design concept evaluation method [3], and new customer oriented design concept evaluation by using improved Z-number-based MCDM [4]. As a result, CDE becomes a typical multi-criteria group decision-making (MCGDM) issue [5]. The fuzzy set, whose membership grade whose value ranges from 0 to 1, was introduced by Zadeh [6] in 1965 as a solution to the MCDM problem. Many researchers are working on different parts of it, such as novel information measures for Fermatean fuzzy sets [7], an integrated Shannon entropy and TOPSIS [8], and group multi-criteria design concept assessment [9]. The selection of green suppliers makes use of the fuzzy AHP and TOPSIS technique [10], MCDM [11–14], and multi-criteria fuzzy decision support [15–18].

The Intuitionistic Fuzzy Sets (IFS), which deal with membership $\curlywedge_\Lambda$ and non-membership $\curlyvee_\Lambda$ grade values between 0 and 1, were created by Atanassov [19]. Although fuzzy sets are a fantastic concept for MCDM, they are limited to one membership grade. IFS is subjected to the $0 \leq \curlywedge_\Lambda + \curlyvee_\Lambda \leq$ criteria. There are many researchers working on it, including those working on a new method for generalized intuitionistic fuzzy soft sets [20], an fuzzy approach for MCGDM [21–24], a circular intuitionistic fuzzy TODIM [25], an assessment of the environmental impact of industrial pollution [26], the role of IFS in DM problems [27], similarities and distances in IFS [28], and a new similarity measure for IFS [29]. [30] and [31] examine IFS, fuzzy mathematical morphology, and extended intuitionistic fuzzy soft sets.

To improve the formalization of ambiguity, Zadeh [32] developed the concept of the $\hat{Z}$ number, which consists of two values. The first is the fuzzy value; the second is the dependability of the first. By applying the $\hat{Z}$ number to the fuzzy set, the MCDM problem may be solved with greater accuracy and less uncertainty. Techniques for converting Z-numbers to classical fuzzy numbers [33], ranking $\hat{Z}$ numbers [34], and quantifying uncertainty for $\hat{Z}$ numbers [35]. Following the fuzzy $\hat{Z}$ number, a lot of academics have concentrated on intuitionistic fuzzy $\hat{Z}$ numbers $(IF\hat{Z}N)$ [36], [37]; application of $(IF\hat{Z}N)$ [38]; and intuitive multiple centroid defuzzification of $(IF\hat{Z}N)$ [39]; development of fuzzy systems; and their applications in multi-criteria decision-making [40].

The goal of the proposed $IF\hat{Z}N$ aggregation operator is to increase the effectiveness of decision-making in fuzzy systems by drawing on Einstein's T-Conorm and T-Norm. The Einstein T-norm's fuzzy set overlap processing efficiency may be used to improve decision-making through the creation of a new aggregation operator. Gaining proficiency with this powerful and succinct operator has several benefits. It enables decision-making processes to be modified more precisely, improving their efficacy and filling in information gaps. This invention deftly handles the complex interactions between various data kinds, facilitating comparative study and advancing theoretical understanding. This empirical assessment not only validates the operator's usefulness but also propels other advancements in the fuzzy systems domain. The research proposal outlines the following primary objectives:

- The primary objective of this study is to extensively examine the suitability and efficacy of Einstein standards in the context of $IF\hat{Z}N$.
- To provide an understanding of Einstein's fundamental ideas within the context of $IF\hat{Z}N$.
- Clearly describe and examine Einstein aggregation processes in $IF\hat{Z}N$, assessing essential features to provide informative data.

The article is structured as follows: Sect. 2 introduced some basic definition, scoring, and accuracy functions. Section 3 defines Einstein operations as applied to $IF\hat{Z}N$. In Sect. 4, we introduce Einstein aggregation operators for $IF\hat{Z}N$, accompanied by their proofs and essential properties. Section 5 provides an overview of the methodology for Multiple Attributes Group Decision-Making (MAGDM). Section 6 delves into numerical aspects related to AI-enabled energy efficiency in Buildings. Finally, in Sect. 7, we offer concluding remarks and present the study's findings.

2 Preliminaries

This section introduces several fundamental definitions and operations that played a crucial role in developing the proposed tasks.

Definition 1 ([6]) The fuzzy set identified under the Entire Set $\bigvee$ is

$$\varLambda = \left\{(\pi, \curlywedge_{\varLambda}(\pi))|\pi \in \bigvee\right\}$$

where $\curlywedge_{\varLambda}(\pi)$ degrees of contentment, $\curlywedge_{\varLambda} : \bigvee \to [0, 1]$.

Definition 2 ([19]) The intuitionistic fuzzy set identified under the Entire Set $\bigvee$ is

$$\varLambda = \left\{(\pi, \curlywedge_{\varLambda}(\pi), \Upsilon_{\varLambda}(\pi))|\pi \in \bigvee\right\}$$

where $\curlywedge_{\varLambda}(\pi)$ degrees of contentment, $\curlywedge_{\varLambda} : \bigvee \to [0, 1]$ and $\Upsilon_{\varLambda}(\pi)$ degrees of discontentment, $\curlywedge_{\varLambda} : \bigvee \to [0, 1]$. Here,

$$0 \leq \curlywedge_{\varLambda} + \Upsilon_{\varLambda} \leq 1$$

Definition 3 ([32]) A $\hat{z}$-numbers is an ordered pair of fuzzy number embodied by $Z = (\curlywedge_{\varLambda}, \tau_{\curlywedge_{\varLambda}})$ the $\curlywedge_{\varLambda}$ component is the contentment. While $\tau_{\curlywedge_{\varLambda}}$ is the reliability of the $\curlywedge_{\varLambda}$.

Definition 4 ([36]) The intuitionistic fuzzy $\hat{Z}$ number $(IF\hat{Z}N)$ set is identified under the Entire Set $\bigvee$:

$$\varLambda = \left\{(\pi, (\curlywedge_{\varLambda}(\pi), \tau_{\curlywedge_{\varLambda}(\pi)}), (\Upsilon_{\varLambda}(\pi), \tau_{\Upsilon_{\varLambda}(\pi)}))|\pi \in \bigvee\right\}$$

where $\curlywedge_{\varLambda}(\pi)$ degrees of contentment and $\tau_{\curlywedge_{\varLambda}}$ is reliability of contentment, $(\curlywedge_{\varLambda}, \tau_{\curlywedge_{\varLambda}}) : \bigvee \to [0, 1]$ and $\Upsilon_{\varLambda}(\pi)$ degrees of discontentment and $\tau_{\Upsilon_{\varLambda}}$ is reliability of discontentment, $(\Upsilon_{\varLambda}, \tau_{\Upsilon_{\varLambda}}) : \bigvee \to [0, 1]$. Here,

$$0 \leq \curlywedge_{\varLambda} + \Upsilon_{\varLambda} \leq 1$$

$$0 \leq \tau_{\curlywedge_{\varLambda}} + \tau_{\Upsilon_{\varLambda}} \leq 1$$

Definition 5 ([41]) Suppose that $(a, b) \in (0, 1) \times (0, 1)$ for any real numbers. Then, Einstein norms are defined as

$$\hat{T}(a, b) = \frac{ab}{1 + (1 - a)(1 - b)},$$

$$\hat{S}(a, b) = \frac{a + b}{1 + ab}.$$

Definition 6 To compare $IF\hat{Z}N\ \Lambda_i = \{(\curlywedge_{\Lambda_i}, \tau_{\curlywedge_{\Lambda_i}}), (\Upsilon_{\Lambda_i}, \tau_{\Upsilon_{\Lambda_i}})\}$ we introduced the score function as

$$\Upsilon(\Lambda_i) = \frac{1 + ((\curlywedge_{\Lambda_i})(\tau_{\curlywedge_{\Lambda_i}}) - (\Upsilon_{\Lambda_i})(\tau_{\Upsilon_{\Lambda_i}}))}{2}$$

$\Upsilon(\Lambda_i) \in [-1, 1]$
if $\Upsilon(\Lambda_i) = \Upsilon(\Lambda_i')$ then calculate the accuracy function

$$\Gamma(\Lambda_i) = \frac{(\curlywedge_{\Lambda_i})(\tau_{\curlywedge_{\Lambda_i}}) - (\Upsilon_{\Lambda_i})(\tau_{\Upsilon_{\Lambda_i}})}{2}$$

$\Gamma(\Lambda_i) \in [0, 1]$.

3 Einstein Operation Aggregation Operators Based on Intuitionistic Fuzzy Z Number Set

Definition 7 Consider $\Lambda_1 = \{(\curlywedge_1, \tau_{\curlywedge_1}), (\Upsilon_1, \tau_{\Upsilon_1})\}$ and $\Lambda_2 = \{(\curlywedge_2, \tau_{\curlywedge_2}), (\Upsilon_2, \tau_{\Upsilon_2})\}$ be two $IF\hat{Z}N$ then

$$\Lambda_1 \oplus \Lambda_2 = \left\{\left(\frac{\curlywedge_1+\curlywedge_2}{1+\curlywedge_1\curlywedge_2}, \frac{\tau_{\curlywedge_1}+\tau_{\curlywedge_2}}{1+\tau_{\curlywedge_1}\tau_{\curlywedge_2}}\right)\left(\frac{\Upsilon_1\Upsilon_2}{1+(1-\Upsilon_1)(1-\Upsilon_2)}, \frac{\Upsilon_1\Upsilon_2}{1+(1-\Upsilon_1)(1-\Upsilon_2)}\right)\right\}.$$

$$\Lambda_1 \otimes \Lambda_2 = \left\{\left(\frac{\curlywedge_1\curlywedge_2}{1+(1-\curlywedge_1)(1-\curlywedge_2)}, \frac{\tau_{\curlywedge_1\curlywedge_2}}{1+(1-\tau_{\curlywedge_1})(1-\tau_{\curlywedge_2})}\right), \left(\frac{\Upsilon_1+\Upsilon_2}{1+\Upsilon_1\Upsilon_2}, \frac{\tau_{\Upsilon_1}+\tau_{\Upsilon_2}}{1+\tau_{\Upsilon_1}\tau_{\Upsilon_2}}\right)\right\}.$$

$$\lambda \odot \Lambda_1 = \left\{\left(\frac{(1+\curlywedge_1)^\lambda-(1-\curlywedge_1)^\lambda}{(1+\curlywedge_1)^\lambda+(1-\curlywedge_1)^\lambda}, \frac{(1+\tau_{\curlywedge_1})^\lambda-(1-\tau_{\curlywedge_1})^\lambda}{(1+\tau_{\curlywedge_1})^\lambda+(1-\tau_{\curlywedge_1})^\lambda}\right), \left(\frac{2\Upsilon_1{}^\lambda}{(2-\Upsilon_1)^\lambda+(\Upsilon_1)^\lambda}, \frac{2\tau_{\Upsilon_1}{}^\lambda}{(2-\tau_{\Upsilon_1})^\lambda+(\tau_{\Upsilon_1})^\lambda}\right)\right\}.$$

$$\Lambda_1^\lambda = \left\{\left(\frac{2\curlywedge_1{}^\lambda}{(2-\curlywedge_1)^\lambda+(\curlywedge_1)^\lambda}, \frac{2\tau_{\curlywedge_1}{}^\lambda}{(2-\tau_{\curlywedge_1})^\lambda+(\tau_{\curlywedge_1})^\lambda}\right), \left(\frac{(1+\Upsilon_1)^\lambda-(1-\Upsilon_1)^\lambda}{(1+\Upsilon_1)^\lambda+(1-\Upsilon_1)^\lambda}, \frac{(1+\tau_{\Upsilon_1})^\lambda-(1-\tau_{\Upsilon_1})^\lambda}{(1+\tau_{\Upsilon_1})^\lambda+(1-\tau_{\Upsilon_1})^\lambda}\right)\right\}.$$

4 Intuitionistic Fuzzy $\hat{Z}$ Number Operators for Aggregation

Here, we address the subject of aggregation techniques created especially for D-qROFS. We elucidate key elements that are fundamental to these operators' operations by looking at their mathematical underpinnings. This methodological introduction intends to better understanding of the effective usage of algebraic norms in the realm of aggregation operations for D-qROFS.

4.1 Hybrid Weighted Average Aggregation Operators of Intuitionistic Fuzzy $\hat{Z}$ Number Einstein ($IF\hat{Z}NEHWA$)

Within this segment, we will furnish an elucidation of the average of intuitionistic fuzzy $\hat{Z}$ number ($IF\hat{Z}N$), utilizing the Einstein norm and denoted as $IF\hat{Z}N$EWA, along with the corresponding theorem.

Definition 8 Let $\Lambda_\iota = ((\curlywedge_\iota, \tau_{\curlywedge_\iota}), (\Upsilon_\iota, \tau_{\Upsilon_\iota}))$ where $\iota = 1, 2, ..., \chi$ be a collection of $IF\hat{Z}N$. Then, $IF\hat{Z}N$EWA operator is a mathematical expression $IF\hat{Z}NEOWA$: $q^\chi \rightarrow q$ such that

$$IF\hat{Z}NEHWA(\Lambda_1, \Lambda_2, ..., \Lambda_\chi) = \sum_{\iota=1}^{\chi}(\Xi_\iota \Lambda'_{\delta(\iota)}),$$

where $\Xi = (\Xi_1, \Xi_2, ..., \Xi_\chi)^T$ is the χ vector weight of $\Lambda_{\delta(\iota)}$ where $\iota = 1, 2, ..., \chi$ with $\Xi_\iota > 0$ and $\sum_{\iota=1}^{\chi} \Xi_\iota = 1$.

Theorem 1 *$\Lambda_\iota = ((\curlywedge_\iota, \tau_{\curlywedge_\iota}), (\Upsilon_\iota, \tau_{\Upsilon_\iota}))$ where $\iota = 1, 2, ..., \chi$ be a collection of $IF\hat{Z}N$ then $IF\hat{Z}NEHWA$ is identified as*

$$IF\hat{Z}NEHWA(\Lambda_1, \Lambda_2, ..., \Lambda_\chi) = \sum_{\iota=1}^{\chi}(\Xi_\iota \Lambda'_{\delta(\iota)})$$

$$= \left\{ \begin{array}{c} \left(\frac{\prod_{\iota=1}^{\chi}(1+(\curlywedge'_{\delta(\iota)}))^{\Xi_\iota} - \prod_{\iota=1}^{\chi}(1-(\curlywedge'_{\delta(\iota)}))^{\Xi_\iota}}{\prod_{\iota=1}^{\chi}(1+(\curlywedge'_{\delta(\iota)}))^{\Xi_\iota} + \prod_{\iota=1}^{\chi}(1-(\curlywedge'_{\delta(\iota)}))^{\Xi_\iota}}, \frac{\prod_{\iota=1}^{\chi}(1+(\tau_{\curlywedge'_{\delta(\iota)}}))^{\Xi_\iota} - \prod_{\iota=1}^{\chi}(1-(\tau_{\curlywedge'_{\delta(\iota)}}))^{\Xi_\iota}}{\prod_{\iota=1}^{\chi}(1+(\tau_{\curlywedge'_{\delta(\iota)}}))^{\Xi_\iota} + \prod_{\iota=1}^{\chi}(1-(\tau_{\curlywedge'_{\delta(\iota)}}))^{\Xi_\iota}} \right), \\ \left(\frac{2\prod_{\iota=1}^{\chi}\Upsilon'_{\delta(\iota)}{}^{\Xi_\iota}}{\prod_{\iota=1}^{\chi}(2-\Upsilon'_{\delta(\iota)})^{\Xi_\iota} + \prod_{\iota=1}^{\chi}(\Upsilon'_{\delta(\iota)})^{\Xi_\iota}}, \frac{2\prod_{\iota=1}^{\chi}\tau_{\Upsilon'_{\delta(\iota)}}{}^{\Xi_\iota}}{\prod_{\iota=1}^{\chi}(2-\tau_{\Upsilon'_{\delta(\iota)}})^{\Xi_\iota} + \prod_{\iota=1}^{\chi}(\tau_{\Upsilon'_{\delta(\iota)}})^{\Xi_\iota}} \right) \end{array} \right\}.$$

Proof We will prove it by using mathematical induction method if $\chi = 2$ based upon operations (1) and (3)

$$\Xi_1 \odot \Lambda'_{\delta(1)} = \left\{ \begin{array}{c} \left(\frac{(1+(\curlywedge'_{\delta(1)}))^{\Xi_1} - (1-(\curlywedge'_{\delta(1)}))^{\Xi_1}}{(1+(\curlywedge'_{\delta(1)}))^{\Xi_1} + (1-(\curlywedge'_{\delta(1)}))^{\Xi_1}}, \frac{(1+(\tau_{\curlywedge'_{\delta(1)}}))^{\Xi_1} - (1-(\tau_{\curlywedge'_{\delta(1)}}))^{\Xi_1}}{(1+(\tau_{\curlywedge'_{\delta(1)}}))^{\Xi_1} + (1-(\tau_{\curlywedge'_{\delta(1)}}))^{\Xi_1}} \right), \\ \left(\frac{2\Upsilon'_{\delta(1)}{}^{\Xi_1}}{(2-\Upsilon'_{\delta(1)})^{\Xi_1} + (\Upsilon'_{\delta(1)})^{\Xi_1}}, \frac{2\tau_{\Upsilon'_{\delta(1)}}{}^{\Xi_1}}{(2-\tau_{\Upsilon'_{\delta(1)}})^{\Xi_1} + (\tau_{\Upsilon'_{\delta(1)}})^{\Xi_1}} \right) \end{array} \right\}.$$

$$\Xi_2 \odot \Lambda'_{\delta(2)} = \left\{ \begin{array}{c} \left(\frac{(1+(\curlywedge'_{\delta(2)}))^{\Xi_2} - (1-(\curlywedge'_{\delta(2)}))^{\Xi_2}}{(1+(\curlywedge'_{\delta(2)}))^{\Xi_2} + (1-(\curlywedge'_{\delta(2)}))^{\Xi_2}}, \frac{(1+(\tau_{\curlywedge'_{\delta(2)}}))^{\Xi_2} - (1-(\tau_{\curlywedge'_{\delta(2)}}))^{\Xi_2}}{(1+(\tau_{\curlywedge'_{\delta(2)}}))^{\Xi_2} + (1-(\tau_{\curlywedge'_{\delta(2)}}))^{\Xi_2}} \right), \\ \left(\frac{2\Upsilon'_{\delta(2)}{}^{\Xi_2}}{(2-\Upsilon'_{\delta(2)})^{\Xi_2} + (\Upsilon'_{\delta(2)})^{\Xi_2}}, \frac{2\tau_{\Upsilon'_{\delta(2)}}{}^{\Xi_2}}{(2-\tau_{\Upsilon'_{\delta(2)}})^{\Xi_2} + (\tau_{\Upsilon'_{\delta(2)}})^{\Xi_2}} \right) \end{array} \right\}.$$

$$IF\hat{Z}NEHWA(\Lambda_1, \Lambda_2, ..., \Lambda_\chi) = \sum_{\iota=1}^{\chi}(\Xi_\iota \Lambda'_{\delta(\iota)}) = \Xi_1 \Lambda'_{\delta(1)} \oplus \Xi_2 \Lambda'_{\delta(2)}$$

$$= \left\{ \begin{array}{c} \left(\frac{(1+(\curlywedge'_{\delta(1)}))^{\Xi_1} - (1-(\curlywedge'_{\delta(1)}))^{\Xi_1}}{(1+(\curlywedge'_{\delta(1)}))^{\Xi_1} + (1-(\curlywedge'_{\delta(1)}))^{\Xi_1}}, \frac{(1+(\tau_{\curlywedge'_{\delta(1)}}))^{\Xi_1} - (1-(\tau_{\curlywedge'_{\delta(1)}}))^{\Xi_1}}{(1+(\tau_{\curlywedge'_{\delta(1)}}))^{\Xi_1} + (1-(\tau_{\curlywedge'_{\delta(1)}}))^{\Xi_1}} \right), \\ \left(\frac{2\Upsilon'^{\Xi_1}_{\delta(1)}}{(2-\Upsilon'_{\delta(1)})^{\Xi_1} + (\Upsilon'_{\delta(1)})^{\Xi_1}}, \frac{2\tau_{\Upsilon'_{\delta(1)}}^{\Xi_1}}{(2-\tau_{\Upsilon'_{\delta(1)}})^{\Xi_1} + (\tau_{\Upsilon'_{\delta(1)}})^{\Xi_1}} \right) \end{array} \right\} \oplus \left\{ \begin{array}{c} \left(\frac{(1+(\curlywedge'_{\delta(2)}))^{\Xi_2} - (1-(\curlywedge'_{\delta(2)}))^{\Xi_2}}{(1+(\curlywedge'_{\delta(2)}))^{\Xi_2} + (1-(\curlywedge'_{\delta(2)}))^{\Xi_2}}, \frac{(1+(\tau_{\curlywedge'_{\delta(2)}}))^{\Xi_2} - (1-(\tau_{\curlywedge'_{\delta(2)}}))^{\Xi_2}}{(1+(\tau_{\curlywedge'_{\delta(2)}}))^{\Xi_2} + (1-(\tau_{\curlywedge'_{\delta(2)}}))^{\Xi_2}} \right), \\ \left(\frac{2\Upsilon'^{\Xi_2}_{\delta(2)}}{(2-\Upsilon'_{\delta(2)})^{\Xi_2} + (\Upsilon'_{\delta(2)})^{\Xi_2}}, \frac{2\tau_{\Upsilon'_{\delta(2)}}^{\Xi_2}}{(2-\tau_{\Upsilon'_{\delta(2)}})^{\Xi_2} + (\tau_{\Upsilon'_{\delta(2)}})^{\Xi_2}} \right) \end{array} \right\}.$$

$$\Xi_1 \Lambda'_{\delta(1)} \oplus \Xi_2 \Lambda'_{\delta(2)}$$

$$= \left\{ \begin{array}{c} \left(\frac{\prod_{\iota=1}^{2}(1+(\curlywedge'_{\delta(\iota)}))^{\Xi_\iota} - \prod_{\iota=1}^{2}(1-(\curlywedge'_{\delta(\iota)}))^{\Xi_\iota}}{\prod_{\iota=1}^{2}(1+(\curlywedge'_{\delta(\iota)}))^{\Xi_\iota} + \prod_{\iota=1}^{2}(1-(\curlywedge'_{\delta(\iota)}))^{\Xi_\iota}}, \frac{\prod_{\iota=1}^{2}(1+(\tau_{\curlywedge'_{\delta(\iota)}}))^{\Xi_\iota} - \prod_{\iota=1}^{2}(1-(\tau_{\curlywedge'_{\delta(\iota)}}))^{\Xi_\iota}}{\prod_{\iota=1}^{2}(1+(\tau_{\curlywedge'_{\delta(\iota)}}))^{\Xi_\iota} + \prod_{\iota=1}^{2}(1-(\tau_{\curlywedge'_{\delta(\iota)}}))^{\Xi_\iota}} \right), \\ \left(\frac{2\prod_{\iota=1}^{2}\Upsilon'^{\Xi_\iota}_{\delta(\iota)}}{\prod_{\iota=1}^{2}(2-\Upsilon'_{\delta(\iota)})^{\Xi_\iota} + \prod_{\iota=1}^{2}(\Upsilon'_{\delta(\iota)})^{\Xi_\iota}}, \frac{2\prod_{\iota=1}^{2}\tau_{\Upsilon'_{\delta(\iota)}}^{\Xi_\iota}}{\prod_{\iota=1}^{2}(2-\tau_{\Upsilon'_{\delta(\iota)}})^{\Xi_\iota} + \prod_{\iota=1}^{2}(\tau_{\Upsilon'_{\delta(\iota)}})^{\Xi_\iota}} \right) \end{array} \right\}.$$

for $\chi = 2$, we justified it further we consider it for $\chi = j$ is also holds correctly such as

$$IF\hat{Z}NEOWA(\Lambda_{\delta(1)}, \Lambda_{\delta(2)}, ..., \Lambda_j) = \sum_{\iota=1}^{j}(\Xi_\iota \Lambda'_{\delta(\iota)})$$

$$= \left\{ \begin{array}{c} \left(\frac{\prod_{\iota=1}^{j}(1+(\curlywedge'_{\delta(\iota)}))^{\Xi_\iota} - \prod_{\iota=1}^{j}(1-(\curlywedge'_{\delta(\iota)}))^{\Xi_\iota}}{\prod_{\iota=1}^{j}(1+(\curlywedge'_{\delta(\iota)}))^{\Xi_\iota} + \prod_{\iota=1}^{j}(1-(\curlywedge'_{\delta(\iota)}))^{\Xi_\iota}}, \frac{\prod_{\iota=1}^{j}(1+(\tau_{\curlywedge'_{\delta(\iota)}}))^{\Xi_\iota} - \prod_{\iota=1}^{j}(1-(\tau_{\curlywedge'_{\delta(\iota)}}))^{\Xi_\iota}}{\prod_{\iota=1}^{j}(1+(\tau_{\curlywedge'_{\delta(\iota)}}))^{\Xi_\iota} + \prod_{\iota=1}^{j}(1-(\tau_{\curlywedge'_{\delta(\iota)}}))^{\Xi_\iota}} \right), \\ \left(\frac{2\prod_{\iota=1}^{j}\Upsilon'^{\Xi_\iota}_{\delta(\iota)}}{\prod_{\iota=1}^{j}(2-\Upsilon'_{\delta(\iota)})^{\Xi_\iota} + \prod_{\iota=1}^{j}(\Upsilon'_{\delta(\iota)})^{\Xi_\iota}}, \frac{2\prod_{\iota=1}^{j}\tau_{\Upsilon'_{\delta(\iota)}}^{\Xi_\iota}}{\prod_{\iota=1}^{j}(2-\tau_{\Upsilon'_{\delta(\iota)}})^{\Xi_\iota} + \prod_{\iota=1}^{j}(\tau_{\Upsilon'_{\delta(\iota)}})^{\Xi_\iota}} \right) \end{array} \right\}.$$

then we prove it for $\iota = j + 1$, then

$$IF\hat{Z}NEOWA(\Lambda_{\delta(1)}, \Lambda_{\delta(2)}, ..., \Lambda_j) = \sum_{\iota=1}^{j}(\Xi_\iota \Lambda'_{\delta(\iota)}) \sum(\Xi_{j+1} \Lambda'_{j+1})$$

$$= \left\{ \begin{array}{c} \left(\frac{\prod_{\iota=1}^{j+1}(1+(\curlywedge'_{\delta(\iota)}))^{\Xi_\iota} - \prod_{\iota=1}^{j+1}(1-(\curlywedge'_{\delta(\iota)}))^{\Xi_\iota}}{\prod_{\iota=1}^{j+1}(1+(\curlywedge'_{\delta(\iota)}))^{\Xi_\iota} + \prod_{\iota=1}^{j+1}(1-(\curlywedge'_{\delta(\iota)}))^{\Xi_\iota}}, \frac{\prod_{\iota=1}^{j+1}(1+(\tau_{\curlywedge'_{\delta(\iota)}}))^{\Xi_\iota} - \prod_{\iota=1}^{j+1}(1-(\tau_{\curlywedge'_{\delta(\iota)}}))^{\Xi_\iota}}{\prod_{\iota=1}^{j+1}(1+(\tau_{\curlywedge'_{\delta(\iota)}}))^{\Xi_\iota} + \prod_{\iota=1}^{j+1}(1-(\tau_{\curlywedge'_{\delta(\iota)}}))^{\Xi_\iota}} \right), \\ \left(\frac{2\prod_{\iota=1}^{j+1} \Upsilon'^{\Xi_\iota}_{\delta(\iota)}}{\prod_{\iota=1}^{j+1}(2-\Upsilon'_{\delta(\iota)})^{\Xi_\iota} + \prod_{\iota=1}^{j+1}(\Upsilon'_{\delta(\iota)})^{\Xi_\iota}}, \frac{2\prod_{\iota=1}^{j+1} \tau_{\Upsilon'_{\delta(\iota)}}^{\Xi_\iota}}{\prod_{\iota=1}^{j+1}(2-\tau_{\Upsilon'_{\delta(\iota)}})^{\Xi_\iota} + \prod_{\iota=1}^{j+1}(\tau_{\Upsilon'_{\delta(\iota)}})^{\Xi_\iota}} \right) \end{array} \right\}.$$

This implies that $j+1$ holds. Hence it is true for all n and it completes the proof.

Property 1 (Idempotency) If $\Lambda_\iota = ((\curlywedge_\iota, \tau_{\curlywedge_\iota}), (\Upsilon_\iota, \tau_{\Upsilon_\iota}))$ where $\iota = 1, 2, ..., \chi$ as a number of $IF\hat{Z}N$s are all equal, i.e., $\Lambda_{\delta(\iota)} = \Lambda \ \forall \ \iota$, then

$$IF\hat{Z}NEHWA(\Lambda_1, \Lambda_2, ..., \Lambda_\chi) = \Lambda.$$

Proof Since $\Lambda_\iota = ((\curlywedge_\iota, \tau_{\curlywedge_\iota}), (\Upsilon_\iota, \tau_{\Upsilon_\iota}))$ where $\iota = 1, 2, ..., \chi$. Then, by equation, we have

$$IF\hat{Z}NEHWA(\Lambda_1, \Lambda_2, ..., \Lambda_\chi) = \sum_{\iota=1}^{\chi}(\Xi_\iota \Lambda'_{\delta(\iota)})$$

$$= \left\{ \begin{array}{c} \left(\frac{\prod_{\iota=1}^{\chi}(1+(\curlywedge'_{\delta(\iota)}))^{\Xi_\iota} - \prod_{\iota=1}^{\chi}(1-(\curlywedge'_{\delta(\iota)}))^{\Xi_\iota}}{\prod_{\iota=1}^{\chi}(1+(\curlywedge'_{\delta(\iota)}))^{\Xi_\iota} + \prod_{\iota=1}^{\chi}(1-(\curlywedge'_{\delta(\iota)}))^{\Xi_\iota}}, \frac{\prod_{\iota=1}^{\chi}(1+(\tau_{\curlywedge'_{\delta(\iota)}}))^{\Xi_\iota} - \prod_{\iota=1}^{\chi}(1-(\tau_{\curlywedge'_{\delta(\iota)}}))^{\Xi_\iota}}{\prod_{\iota=1}^{\chi}(1+(\tau_{\curlywedge'_{\delta(\iota)}}))^{\Xi_\iota} + \prod_{\iota=1}^{\chi}(1-(\tau_{\curlywedge'_{\delta(\iota)}}))^{\Xi_\iota}} \right), \\ \left(\frac{2\prod_{\iota=1}^{\chi} \Upsilon'^{\Xi_\iota}_{\delta(\iota)}}{\prod_{\iota=1}^{\chi}(2-\Upsilon'_{\delta(\iota)})^{\Xi_\iota} + \prod_{\iota=1}^{\chi}(\Upsilon'_{\delta(\iota)})^{\Xi_\iota}}, \frac{2\prod_{\iota=1}^{\chi} \tau_{\Upsilon'_{\delta(\iota)}}^{\Xi_\iota}}{\prod_{\iota=1}^{\chi}(2-\tau_{\Upsilon'_{\delta(\iota)}})^{\Xi_\iota} + \prod_{\iota=1}^{\chi}(\tau_{\Upsilon'_{\delta(\iota)}})^{\Xi_\iota}} \right) \end{array} \right\}.$$

$$= \left\{ \begin{array}{c} \left(\frac{(1+(\curlywedge'_{\delta(\iota)}))-(1-(\curlywedge'_{\delta(\iota)}))}{(1+(\curlywedge'_{\delta(\iota)}))+(1-(\curlywedge'_{\delta(\iota)}))}, \frac{(1+(\tau_{\curlywedge'_{\delta(\iota)}}))-(1-(\tau_{\curlywedge'_{\delta(\iota)}}))}{(1+(\tau_{\curlywedge'_{\delta(\iota)}}))+(1-(\tau_{\curlywedge'_{\delta(\iota)}}))} \right), \\ \left(\frac{2\Upsilon'_{\delta(\iota)}}{(2-\Upsilon'_{\delta(\iota)})+(\Upsilon'_{\delta(\iota)})}, \frac{2\tau_{\Upsilon'_{\delta(\iota)}}}{(2-\tau_{\Upsilon'_{\delta(\iota)}})+(\tau_{\Upsilon'_{\delta(\iota)}})} \right) \end{array} \right\} = \Lambda.$$

Thus, $IF\hat{Z}NEHWA(\Lambda_1, \Lambda_2, ..., \Lambda_\chi) = \Lambda$ holds.

Property 2 (Monotonicity) Let $\Lambda_\iota = ((\curlywedge_\iota, \tau_{\curlywedge_\iota}), (\Upsilon_\iota, \tau_{\Upsilon_\iota}))$ where $\iota = 1, 2, ..., \chi$ and $\Lambda'_\iota = ((\curlywedge_\iota, \tau_{\curlywedge_\iota}), (\Upsilon_\iota, \tau_{\Upsilon_\iota}))$ where $\iota = 1, 2, ..., \chi$ be two sets of $IF\hat{Z}N$s, if $\Lambda_{\delta(\iota)} \leq \Lambda'_{\delta(\iota)} \ \forall \ \iota$.

$$= \left\{ \begin{array}{c} \left(\frac{\prod_{\iota=1}^{\chi}(1+(\curlywedge'_{\delta(\iota)}))^{\Xi_\iota} - \prod_{\iota=1}^{\chi}(1-(\curlywedge'_{\delta(\iota)}))^{\Xi_\iota}}{\prod_{\iota=1}^{\chi}(1+(\curlywedge'_{\delta(\iota)}))^{\Xi_\iota} + \prod_{\iota=1}^{\chi}(1-(\curlywedge'_{\delta(\iota)}))^{\Xi_\iota}}, \frac{\prod_{\iota=1}^{\chi}(1+(\tau_{\curlywedge'_{\delta(\iota)}}))^{\Xi_\iota} - \prod_{\iota=1}^{\chi}(1-(\tau_{\curlywedge'_{\delta(\iota)}}))^{\Xi_\iota}}{\prod_{\iota=1}^{\chi}(1+(\tau_{\curlywedge'_{\delta(\iota)}}))^{\Xi_\iota} + \prod_{\iota=1}^{\chi}(1-(\tau_{\curlywedge'_{\delta(\iota)}}))^{\Xi_\iota}} \right), \\ \left(\frac{2\prod_{\iota=1}^{\chi} \Upsilon'^{\Xi_\iota}_{\delta(\iota)}}{\prod_{\iota=1}^{\chi}(2-\Upsilon'_{\delta(\iota)})^{\Xi_\iota} + \prod_{\iota=1}^{\chi}(\Upsilon'_{\delta(\iota)})^{\Xi_\iota}}, \frac{2\prod_{\iota=1}^{\chi} \tau_{\Upsilon'_{\delta(\iota)}}^{\Xi_\iota}}{\prod_{\iota=1}^{\chi}(2-\tau_{\Upsilon'_{\delta(\iota)}})^{\Xi_\iota} + \prod_{\iota=1}^{\chi}(\tau_{\Upsilon'_{\delta(\iota)}})^{\Xi_\iota}} \right) \end{array} \right\} \leq$$

$$= \left\{ \begin{array}{c} \left(\dfrac{\prod_{\iota'=1}^{\chi}(1+(\curlywedge'_{\delta(\iota')}))^{\Xi_{\delta(\iota')}} - \prod_{\iota'=1}^{\chi}(1-(\curlywedge'_{\delta(\iota')}))^{\Xi_{\delta(\iota')}}}{\prod_{\iota'=1}^{\chi}(1+(\curlywedge'_{\delta(\iota')}))^{\Xi_{\delta(\iota')}} + \prod_{\iota'=1}^{\chi}(1-(\curlywedge'_{\delta(\iota')}))^{\Xi_{\delta(\iota')}}}, \dfrac{\prod_{\iota'=1}^{\chi}(1+(\tau_{\curlywedge'_{\delta(\iota')}}))^{\Xi_{\delta(\iota')}} - \prod_{\iota'=1}^{\chi}(1-(\tau_{\curlywedge'_{\delta(\iota')}}))^{\Xi_{\delta(\iota')}}}{\prod_{\iota'=1}^{\chi}(1+(\tau_{\curlywedge'_{\delta(\iota')}}))^{\Xi_{\delta(\iota')}} + \prod_{\iota'=1}^{\chi}(1-(\tau_{\curlywedge'_{\delta(\iota')}}))^{\Xi_{\delta(\iota')}}} \right), \\ \left(\dfrac{2\prod_{\iota'=1}^{\chi} \Upsilon'_{\delta(\iota')}{}^{\Xi_{\delta(\iota')}}}{\prod_{\iota'=1}^{\chi}(2-\Upsilon'_{\delta(\iota')})^{\Xi_{\delta(\iota')}} + \prod_{\iota'=1}^{\chi}(\Upsilon'_{\delta(\iota')})^{\Xi_{\delta(\iota')}}}, \dfrac{2\prod_{\iota'=1}^{\chi} \tau_{\Upsilon'_{\delta(\iota')}}{}^{\Xi_{\delta(\iota')}}}{\prod_{\iota'=1}^{\chi}(2-\tau_{\Upsilon'_{\delta(\iota')}})^{\Xi_{\delta(\iota')}} + \prod_{\iota'=1}^{\chi}(\tau_{\Upsilon'_{\delta(\iota')}})^{\Xi_{\delta(\iota')}}} \right) \end{array} \right\}. \text{ Hence}$$

$$IF\hat{Z}NEHWA(\Lambda_1, \Lambda_2, ..., \Lambda_\chi) \le IF\hat{Z}NEHWA(\Lambda'_{\delta(1)}, \Lambda'_{\delta(2)}, ..., \Lambda'_\chi)$$

Property 3 (Boundedness) let $\Lambda_\iota = ((\curlywedge_\iota, \tau_{\curlywedge_\iota}), (\Upsilon_\iota, \tau_{\Upsilon_\iota}))$ where $\iota = 1, 2, ..., \chi$ is a collection of $IF\hat{Z}N$s such that such that $\Lambda^+ = \max\limits_\iota \Lambda_{\delta(\iota)}$ and $\Lambda^- = \min\limits_\iota \Lambda_{\delta(\iota)}$ then,

$$\Lambda^- \le IF\hat{Z}NEHWA(\Lambda_1, \Lambda_2, ..., \Lambda_\chi) \le \Lambda^+.$$

Proof The proof is straight forward.

4.2 Hybrid Weighted Geometric Aggregation Operators of Intuitionistic Fuzzy $\hat{Z}$ Number Einstein ($IF\hat{Z}NEHWG$)

In this section, we will furnish an elucidation of the average of intuitionistic fuzzy $\hat{Z}$ number ($IF\hat{Z}N$), utilizing the Einstein norm and denoted as $IF\hat{Z}NEHWG$, along with the corresponding theorem.

Definition 9 Let $\Lambda_\iota = ((\curlywedge_\iota, \tau_{\curlywedge_\iota}), (\Upsilon_\iota, \tau_{\Upsilon_\iota}))$ where $\iota = 1, 2, ..., \chi$ as a number of $IF\hat{Z}N$. Then, $IF\hat{Z}NEHWG$ operator is a expression $IF\hat{Z}NEHWG$: $q^\chi \to q$ such that

$$IF\hat{Z}NEHWG(\Lambda_1, \Lambda_2, ..., \Lambda_\chi) = \prod_{\iota=1}^{\chi}(\Xi_\iota \Lambda'_{\delta(\iota)}),$$

where $\Xi = (\Xi_1, \Xi_2, ..., \Xi_\chi)^T$ be the χ vector weight of $\Lambda_{\delta(\iota)}$ where $\iota = 1, 2, ..., \chi$ with $\Xi_\iota > 0$ and $\sum\limits_{\iota=1}^{\chi} = 1$.

Theorem 2 $\Lambda_\iota = ((\curlywedge_\iota, \tau_{\curlywedge_\iota}), (\Upsilon_\iota, \tau_{\Upsilon_\iota}))$ *where* $\iota = 1, 2, ..., \chi$ *as a number of* $IF\hat{Z}N$, *the set's value utilizing the* $IF\hat{Z}NEHWG$ *operator is also a* $IF\hat{Z}N$, *and*

$$IF\hat{Z}NEHWG(\Lambda_1, \Lambda_2, ..., \Lambda_\chi) = \prod_{\iota=1}^{\chi}(\Xi_\iota \Lambda'_{\delta(\iota)})$$

$$= \left\{ \begin{array}{c} \left(\frac{2\prod_{\iota=1}^{\chi} {\curlywedge'_{\delta(\iota)}}^{\Xi_\iota}}{\prod_{\iota=1}^{\chi}(2-\curlywedge'_{\delta(\iota)})^{\Xi_\iota}+\prod_{\iota=1}^{\chi}(\curlywedge'_{\delta(\iota)})^{\Xi_\iota}}, \frac{2\prod_{\iota=1}^{\chi} {\tau_{\curlywedge'_{\delta(\iota)}}}^{\Xi_\iota}}{\prod_{\iota=1}^{\chi}(2-\tau_{\curlywedge'_{\delta(\iota)}})^{\Xi_\iota}+\prod_{\iota=1}^{\chi}(\tau_{\curlywedge'_{\delta(\iota)}})^{\Xi_\iota}} \right), \\ \left(\frac{\prod_{\iota=1}^{\chi}(1+(\Upsilon'_{\delta(\iota)}))^{\Xi_\iota}-\prod_{\iota=1}^{\chi}(1-(\Upsilon'_{\delta(\iota)}))^{\Xi_\iota}}{\prod_{\iota=1}^{\chi}(1+(\Upsilon'_{\delta(\iota)}))^{\Xi_\iota}+\prod_{\iota=1}^{\chi}(1-(\Upsilon'_{\delta(\iota)}))^{\Xi_\iota}}, \frac{\prod_{\iota=1}^{\chi}(1+(\tau_{\Upsilon'_{\delta(\iota)}}))^{\Xi_\iota}-\prod_{\iota=1}^{\chi}(1-(\tau_{\Upsilon'_{\delta(\iota)}}))^{\Xi_\iota}}{\prod_{\iota=1}^{\chi}(1+(\tau_{\Upsilon'_{\delta(\iota)}}))^{\Xi_\iota}+\prod_{\iota=1}^{\chi}(1-(\tau_{\Upsilon'_{\delta(\iota)}}))^{\Xi_\iota}} \right) \end{array} \right\}.$$

Property 4 (Idempotency) Let $\Lambda_\iota = ((\curlywedge_\iota, \tau_{\curlywedge_\iota}), (\Upsilon_\iota, \tau_{\Upsilon_\iota}))$ and $\iota = 1, 2, 3..., \chi$ be a collection of $IF\hat{Z}NS$ if all the $\Lambda_{\delta(\iota)}$ are identical then

$$IF\hat{Z}NEHWG(\Lambda_1, \Lambda_2, ..., \Lambda_\chi) = \Lambda$$

Property 5 (Boundedness) Let $\Lambda_\iota = ((\curlywedge_\iota, \tau_{\curlywedge_\iota}), (\Upsilon_\iota, \tau_{\Upsilon_\iota}))$ where $\iota = 1, 2, ..., \chi$ as a number of $IF\hat{Z}N$. Then

$$\Lambda^- \leq IF\hat{Z}NEHWG(\Lambda_1, \Lambda_2, ..., \Lambda_\chi) \leq \Lambda^+$$

Property 6 (Monotonicity) Let $\Lambda_\iota = ((\curlywedge_\iota, \tau_{\curlywedge_\iota}), (\Upsilon_\iota, \tau_{\Upsilon_\iota}))$ where $\iota = 1, 2, ..., \chi$ and $\Lambda'_\iota = ((\curlywedge_\iota, \tau_{\curlywedge_\iota}), (\Upsilon_\iota, \tau_{\Upsilon_\iota}))$ where $\iota = 1, 2, ..., \chi$ be two sets of $IF\hat{Z}N$, if $\Lambda_\iota \leq \Lambda'_\iota$ $\forall\, \iota$.

$$IF\hat{Z}NEHWG(\Lambda_1, \Lambda_2, ..., \Lambda_\chi) \leq IF\hat{Z}NEHWG(\Lambda'_{\delta(1)}, \Lambda'_{\delta(2)}, ..., \Lambda'_\chi)$$

5 $IF\hat{Z}N$ Einstein Operator for Multi-attributes Group Decision-Making

In this section, we have formulated an algorithm to tackle the Multiple Attributes Decision-Making (MADM) problem using the proposed average and geometric aggregation operators. Additionally, we provided an MADM example to illustrate the application of these operators. Let's assume we have a collection of alternatives represented as $\xi = \{\xi_1, \xi_2, \xi_3, ...\xi_n\}$, and a collection of attributes represented as $\beta = \{\beta_1, \beta_2, \beta_3, ...\beta_n\}$, with their respective weight vectors $\omega = \{\omega_1, \omega_2, ...\omega_n\}$. The weight vectors must satisfy the requirement that the weights belong to a closed unit interval (i.e., ranging from 0 to 1) and that their sum must be equal to 1, ensuring a valid weighting scheme. Suppose the intuitionistic fuzzy $\hat{Z}$-number decision matrix denoted by $\Lambda^k = [\Lambda^k_{ij}]_{m\times n}$.

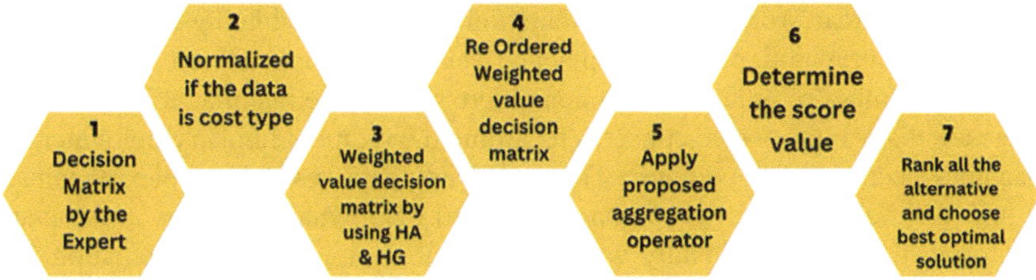

Fig. 1 Flow chart of algorithm

Algorithm

1. Decision matrices by the expert.
2. In the MADM process, two types of criteria are commonly used: positive criteria and negative criteria. If a negative criterion is measured as "cost," we convert it to "benefit" by taking the complement, making "lower cost" equivalent to "higher benefit."
3. Compute the weighted value decision matrix by averaging and geometric operator.
4. Re-ordered the weighted value decision matrix by using the score definition.
5. The MAGDM process involves integrating the attributes for each alternative using the arithmetic and geometric aggregation Einstein operators for $IF\hat{Z}NS$s.
6. To compute the score values for each alternative using score definition.
7. To rank all alternatives in descending order and choose the best one.

The flow chart of algorithm is given in Fig. 1.

6 Numerical Problem

In this segment of the article, we present a Multi-Attributes Decision-Making (MAGDM) problem to demonstrate the applicability and efficacy of this approach in tackling complex decision-making challenges. To exemplify this, we present a scenario centered on Impact of Technological Innovation on Employment. Within this context, we have deliberately selected four distinct attributes for evaluating the performance of these operations: Technological Integration Effectiveness, Employment Impact and Dynamics, Adaptation Strategies by Businesses, and Policy Efficacy and Alignment, we have identified four potential alternatives: Investment in Continuous Workforce Training, Encouraging Collaborative Innovation Hubs, Implementing Flexible Labor Policies, and Creating Economic Incentives for Sustainable Practices. The fourth industrial revolution has begun, bringing with it previously unheard-of levels of technical progress, with automation and artificial intelligence (AI) leading the way in revolutionary developments. These developments have had an influence on a number of industries, but none more so than the manufacturing industry. There are many unanswered concerns regarding the future of human employment as machines become more complex. This case study explores the complex dynamics of how technological innovation has influenced employment patterns in the manufacturing industry, specifically with regard to the integration of automation and artificial intelligence. The need for efficiency and competitiveness in industrial processes has resulted in a rethinking of the traditional workforce. This study aims to clarify the complex link between employment and technology advancement, illuminating company tactics and the function of government regulations in this changing environment.

A significant turning point in history, the beginning of the fourth industrial revolution, is marked by an unheard-of acceleration of technical progress. Artificial intelligence (AI) and automation are leading transformational advances that have an influence on many sectors. The manufacturing industry is the one where this seismic shift is most noticeable, since these breakthroughs have weaved a tapestry of deep alteration. The conventional dynamics of labor and industry are being challenged by fundamental issues regarding the future of human employment as robots reach previously unheard-of levels of complexity. Manufacturing's long history as a reliable driver of economic expansion is coming to an end. The industry has seen tremendous changes, from the automation of the first industrial revolution to the computerization of later periods. At this point, when the fourth industrial revolution is about to begin, the combination of cyber-physical systems holds great potential as well as difficulties. This development not only illustrates the cyclical nature of technology advancement but also poses important queries regarding the nature of work itself in this dynamic environment.

The pressing need to understand the complex interactions between the advancement of technology and the nature of work in the industrial sector is what spurred this case study. The attraction of enhanced productivity, accuracy, and worldwide competitiveness via automation and artificial intelligence stands in stark contrast to the concerns over job loss, shortages in skills, and social fallout. With the manufacturing sector serving as a testing ground for technological innovation, the research aims to guide stakeholders crossing unfamiliar territory by illuminating a future route as well as identifying immediate repercussions. There are several goals for the investigation. Initially, the aim is to examine the manufacturing industry's use of automation and artificial intelligence (AI) technologies, analyzing the technologies that are being used and how common they are in various sub-sectors. Two, to examine changes in employment trends by contrasting employment statistics from the past with current trends shaped by technology advancements. Thirdly, examining the adaption tactics that companies have adopted and looking into ways to lessen negative effects on employment, such as upskilling and human–machine collaboration. In the end, the assessment will focus on how government policies function and how well they shape the complex link between employment and technology advancement.

There are four Attributes

Technological Integration Effectiveness (β_1):

Evaluate how well automation and artificial intelligence have been incorporated into the industrial industry. Analyze the degree and effectiveness of technology adoption while taking increased competitiveness, cost savings, and efficiency into account.

Employment Impact and Dynamics (β_2):

Examine how technological progress affects jobs in the industrial industry. Analyze shifts in employment trends, such as the emergence and displacement of jobs, and evaluate the general characteristics of the labor force in manufacturing. Calculate the cost savings, waiting time reductions, resource optimization, and efficiency advantages brought about by the use of AI.

Adaptation Strategies by Businesses (β_3):

Examine the tactics used by manufacturing companies to capitalize on technology improvements while reducing negative job implications. Examine the efficacy of tactics including implementing collaborative human–machine processes, reskilling initiatives, and workforce upskilling.

Policy Efficacy and Alignment (β_4):

Analyze how well government policies have shaped the connection between employment in the industrial sector and technological innovation. Examine how well policies promote worker shifts, encourage technology usage, and correspond with industry demands.

There are four Alternatives

Investment in Continuous Workforce Training (ξ_1):

Make it a priority to provide the industrial staff with ongoing training so they may become more skilled and capable of adjusting to new technology. In order to lower the danger of job displacement, this option focuses on proactive steps to guarantee that the workforce stays up to date with technology improvements.

Encouraging Collaborative Innovation Hubs (ξ_2):

Encourage the creation of cooperative innovation hubs where government agencies, businesses, and academic institutions work together to integrate technology while protecting jobs. This approach seeks to encourage creativity through collaborative workplaces, which can lead to increased efficiency and new job prospects.

Implementing Flexible Labor Policies (ξ_3):

Promote labor laws that allow for flexible work arrangements, including remote work, part-time work, and flexible scheduling. In response to the changing nature of work brought about by technology improvements, flexible labor rules may help allay worries about job displacement.

Creating Economic Incentives for Sustainable Practices (ξ_4):

Provide financial incentives to manufacturing companies that integrate socially and sustainably conscious activities into their technical transformations. This substitute seeks to make sure that companies emphasize efficiency and social issues in their strategy by matching financial incentives with ethical technological adoption.

Table 1 Decision matrix by the expert

Λ_{ij}	β_1	β_2	β_3	β_4
ξ_1	((0.3, 0.4), (0.5, 0.4))	((0.5, 0.4), (0.3, 0.4))	((0.4, 0.5), (0.3, 0.3))	((0.2, 0.6), (0.3, 0.2))
ξ_2	((0.6, 0.3), (0.2, 0.1))	((0.4, 0.2), (0.2, 0.3))	((0.2, 0.4), (0.5, 0.3))	((0.3, 0.3), (0.2, 0.6))
ξ_3	((0.4, 0.2), (0.4, 0.5))	((0.3, 0.2), (0.6, 0.2))	((0.2, 0.3), (0.4, 0.2))	((0.3, 0.2), (0.6, 0.3))
ξ_4	((0.3, 0.2), (0.5, 0.3))	((0.4, 0.5), (0.1, 0.2))	((0.5, 0.3), (0.4, 0.5))	((0.7, 0.1), (0.2, 0.6))

Table 2 Weighted value decision matrix using HA

Λ_{ij}	β_1	β_2
ξ_1	((0.19659, 0.31611), (0.56960, 0.51708))	((0.59368, 0.48079), (0.18357, 0.25727))
ξ_2	((0.47588, 0.20581), (0.31688, 0.19282))	((0.47176, 0.29797), (0.11649, 0.23651))
ξ_3	((0.33976, 0.48292), (0.49367, 0.57238))	((0.37668, 0.20541), (0.48021, 0.09085))
ξ_4	((0.24502, 0.17520), (0.63955, 0.37616))	((0.49470, 0.64601), (0.07643, 0.08039))
Λ_{ij}	β_3	β_4
ξ_1	((0.45809, 0.57781), (0.22024, 0.21008))	((0.12627, 0.39346), (0.55703, 0.42264))
ξ_2	((0.28215, 0.48406), (0.38036, 0.18627))	((0.19660, 0.16039), (0.40783, 0.75115))
ξ_3	((0.28303, 0.32441), (0.32055, 0.17001))	((0.20500, 0.13929), (0.75115, 0.53172))
ξ_4	((0.57781, 0.39224), (0.31857, 0.44830))	((0.47799, 0.06913), (0.45377, 0.75115))

In this scenario, we have a collection of alternatives denoted as $\xi = \{\xi_1, \xi_2, \xi_3, \xi_4\}$. We also have a set of attributes denoted as $\beta = \{\beta_1, \beta_2, \beta_3, \beta_4\}$. we have attributes weight vector $\widetilde{\Omega} = \{0.19, 0.32, 0.30, 0.15\}$ and the associated weight vector $\widetilde{\Xi} = \{0.340.200.150.26\}$.

Step 1 Decision matrices by the expert in Table 1.

Step 2 Due to the benefit-type nature of the criteria, normalization is unnecessary.

Step 3 In Table 2, weighted value decision matrix using hybrid averaging and in Table 3 weighted value decision matrix using hybrid geometric.

Step 4 In Table 4 re-ordered weighted value decision matrix of hybrid averaging and in Table 5 re-ordered weighted value decision matrix of hybrid geometric.

Step 5 Integrate the attributes for each alternative using $IF\hat{Z}NEHWA$ and $IF\hat{Z}NEHWG$ operator in Table 6 and 7, respectively.

Step 6 Compute the score values for each alternative.

By using $IF\hat{Z}NEHWA$ operator.

$\Upsilon(\xi_1) = 0.41406$, $\Upsilon(\xi_2) = 0.39343$, $\Upsilon(\xi_3) = 0.34892$ and $\Upsilon(\xi_4) = 0.34635$

By using $IF\hat{Z}NEHWG$ operator.

$\Upsilon(\xi_1) = 0.56393$, $\Upsilon(\xi_2) = 0.54120$, $\Upsilon(\xi_3) = 0.50327$ and $\Upsilon(\xi_4) = 0.53891$

Step 7 Ranking all possibilities in descending order in Table 8.

As a result, we determine that option ξ_1 is the best optimal solution. The graphical representation of all the operators such as $IF\hat{Z}NEHWA$ and $IF\hat{Z}NEHWG$ given in Fig. 2.

Table 3 Weighted value decision matrix using HG

Λ_{ij}	β_1	β_2
ξ_1	((0.37772, 0.52243), (0.35998, 0.31129))	((0.38111, 0.27809), (0.35972, 0.45576))
ξ_2	((0.68236, 0.38990), (0.15287, 0.07611))	((0.27051, 0.14121), (0.25901, 0.42995))
ξ_3	((0.54820, 0.68871), (0.29059, 0.36263))	((0.19589, 0.08504), (0.68804, 0.21572))
ξ_4	((0.43963, 0.34863), (0.42991, 0.19542))	((0.28998, 0.43443), (0.18975, 0.19702))
Λ_{ij}	β_3	β_4
ξ_1	((0.30907, 0.42219), (0.35328, 0.34056))	((0.43150, 0.75115), (0.20918, 0.12129))
ξ_2	((0.16519, 0.33253), (0.53511, 0.31004))	((0.54015, 0.48762), (0.11322, 0.39346))
ξ_3	((0.16585, 0.19738), (0.47087, 0.28860))	((0.55150, 0.45383), (0.39346, 0.19050))
ξ_3	((0.42219, 0.25218), (0.46868, 0.60359))	((0.81639, 0.31456), (0.13926, 0.39346))

Table 4 Re-ordered weighted value decision matrix using HA

Λ_{ij}	β_1	β_2
ξ_1	((0.59368, 0.48079), (0.18357, 0.25727))	((0.45809, 0.57781), (0.22024, 0.21008))
ξ_2	((0.47176, 0.29797), (0.11649, 0.23651))	((0.28215, 0.48406), (0.38036, 0.18627))
ξ_3	((0.28303, 0.32441), (0.32055, 0.17001))	((0.37668, 0.20541), (0.48021, 0.09085))
ξ_4	((0.49470, 0.64601), (0.07643, 0.08039))	((0.57781, 0.39224), (0.31857, 0.44830))
Λ_{ij}	β_3	β_4
ξ_1	((0.12627, 0.39346), (0.55703, 0.42264))	((0.19659, 0.31611), (0.56960, 0.51708))
ξ_2	((0.47588, 0.20581), (0.31688, 0.19282))	((0.19660, 0.16039), (0.40783, 0.75115))
ξ_3	((0.33976, 0.48292), (0.49367, 0.57238))	((0.20500, 0.13929), (0.75115, 0.53172))
ξ_4	((0.24502, 0.17520), (0.63955, 0.37616))	((0.47799, 0.06913), (0.45377, 0.75115))

Table 5 Re-ordered weighted value decision matrix using HG

Λ_{ij}	β_1	β_2
ξ_1	((0.43150, 0.75115), (0.20918, 0.12129))	((0.37772, 0.52243), (0.359980.31129))
ξ_2	((0.68236, 0.38990), (0.15287, 0.07611))	((0.54015, 0.48762), (0.113220.39346))
ξ_3	((0.54820, 0.68871), (0.29059, 0.36263))	((0.55150, 0.45383), (0.393460.19050))
ξ_4	((0.81639, 0.31456), (0.13926, 0.39346))	((0.28998, 0.43443), (0.189750.19702))
Λ_{ij}	β_3	β_4
ξ_1	((0.30907, 0.42219), (0.35328, 0.34056))	((0.38111, 0.27809), (0.35972, 0.45576))
ξ_2	((0.27051, 0.14121), (0.25901, 0.42995))	((0.16519, 0.33253), (0.53511, 0.31004))
ξ_3	((0.16585, 0.19738), (0.47087, 0.28860))	((0.19589, 0.08504), (0.68804, 0.21572))
ξ_4	((0.43963, 0.34863), (0.42991, 0.19542))	((0.42219, 0.25218), (0.46868, 0.60359))

Table 6 Using $IF\hat{Z}NEHWA$ operator

Λ_i	β_j
ξ_1	((0.38439, 0.42994), (0.60763, 0.55484))
ξ_2	((0.34867, 0.27742), (0.56839, 0.54515))
ξ_3	((0.27901, 0.26649), (0.73889, 0.50957))
ξ_4	((0.45537, 0.37321), (0.64439, 0.74062))

Table 7 Using $IF\hat{Z}NEHWG$ operator

Λ_i	β_j
ξ_1	((0.40619, 0.51659), (0.29295, 0.27983))
ξ_2	((0.41839, 0.35871), (0.26443, 0.25595))
ξ_3	((0.37499, 0.33359), (0.44819, 0.26454))
ξ_4	((0.53176, 0.34407), (0.28015, 0.37529))

Table 8 Ranking of numerical problem

Sr.	Operators	Scoring
1	$IF\hat{Z}NEHWA$ Operator	$\Upsilon(\xi_1) \geq \Upsilon(\xi_2) \geq \Upsilon(\xi_3) \geq \Upsilon(\xi_4)$
2	$IF\hat{Z}NEHWG$ Operator	$\Upsilon(\xi_1) \geq \Upsilon(\xi_2) \geq \Upsilon(\xi_4) \geq \Upsilon(\xi_3)$

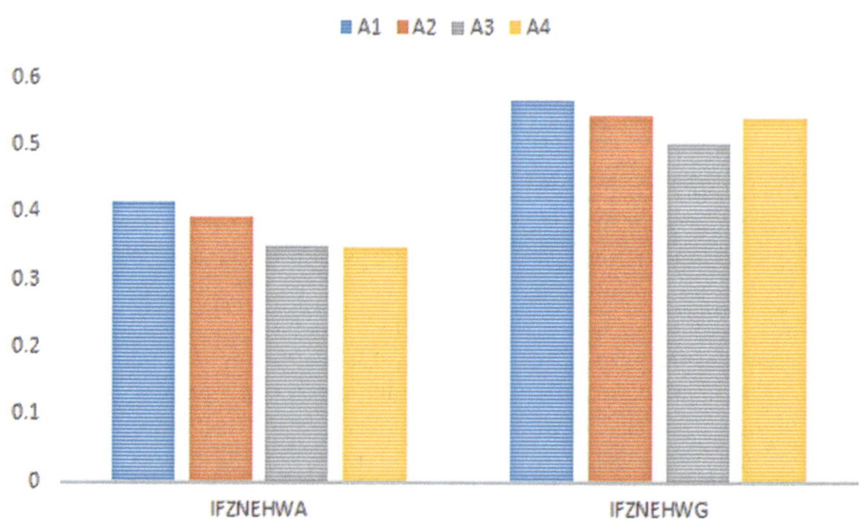

Fig. 2 Graphical representation of ranking

7 Conclusion

Finding basic operational principles for intuitionistic fuzzy $\hat{Z}$ numbers ($IF\hat{Z}N$) with the help of the Einstein operator is the main goal of this research. Through this study, we have created the $IF\hat{Z}NEHWA$ and $IF\hat{Z}NEHWG$ aggregation operators and analyzed and explained their features in depth. This research takes into consideration the fact that MAGDM situations are inherently complicated by using these operators, which permit judgments that take into consideration both positive and negative aspects by means of a parameter. Last but not least, this study's findings support the methodology's claim that it's the best technique to deal with MAGDM issues. The suggested $IF\hat{Z}N$ operators are shown to be valid and useful via thorough examination of many scenarios pertaining to the creation of a medication distribution system for cancer therapy. This paper also provides a rigorous mathematical example to back up its findings. When looking at MAGDM problems from every angle, the research concludes that the suggested strategy is the best bet.

Research in the future will concentrate on creating new decision-making methods that are specific to the $IF\hat{Z}N$ setting. In order to improve the effectiveness of decision-making, these methods will combine many approaches, including TOPSIS, the ELECTRE method, and EDAS.

References

1. Zhu, G.N., Hu, J., Ren, H.: A fuzzy rough number-based AHP-TOPSIS for design concept evaluation under uncertain environments. Appl. Soft Comput. **91**, 106228 (2020)
2. Qi, J., Hu, J., Peng, Y.: Modified rough VIKOR based design concept evaluation method compatible with objective design and subjective preference factors. Appl. Soft Comput. **107**, 107414 (2021)
3. Qi, J., Hu, J., Peng, Y.: New design concept evaluation method involving customer preferences based on rough distance to redefined ideal solution. Comput. Indus. Eng. **147**, 106677 (2020)
4. Qi, J., Hu, J., Huang, H., Peng, Y.: New customer-oriented design concept evaluation by using improved Z-number-based multi-criteria decision-making method. Adv. Eng. Inform. **53**, 101683 (2022)
5. Ernawati, D., Pujawan, I.N., Batan, I.M.L., Anityasari, M.: Evaluating alternatives of product design: a multi criteria group decision making approach. Int. J. Serv. Oper. Manag. **20**(3), 271–288 (2015)
6. Zadeh, L.A.: Fuzzy sets. Inform. Control **8**(3), 338–353 (1965)
7. Ashraf, S., Naeem, M., Khan, A., Rehman, N., Pandit, M.K.: Novel information measures for Fermatean fuzzy sets and their applications to pattern recognition and medical diagnosis. Comput. Intell. Neurosci. (2023)
8. Tiwari, V., Jain, P.K., Tandon, P.: An integrated Shannon entropy and TOPSIS for product design concept evaluation based on bijective soft set. J. Intell. Manuf. **30**, 1645–1658 (2019)
9. Shidpour, H., Da Cunha, C., Bernard, A.: Group multi-criteria design concept evaluation using combined rough set theory and fuzzy set theory. Expert Syst. Appl. **64**, 633–644 (2016)
10. Vinodh, S., Sai Balagi, T.S., Patil, A.: A hybrid MCDM approach for agile concept selection using fuzzy DEMATEL, fuzzy ANP and fuzzy TOPSIS. Int. J. Adv. Manuf. Technol. **83**, 1979–1987 (2016)

11. Gupta, S., Soni, U., Kumar, G.: Green supplier selection using multi-criterion decision making under fuzzy environment: A case study in automotive industry. Comput. Indus. Eng. **136**, 663–680 (2019)
12. Jana, C., Garg, H., Pal, M.: Multi-attribute decision making for power Dombi operators under Pythagorean fuzzy information with MABAC method. J. Ambient Intell. Humanized Comput. **14**(8), 10761–10778 (2023)
13. Jana, C., Pal, M., Liu, P.: Multiple attribute dynamic decision making method based on some complex aggregation functions in CQROF setting. Comput. Appl. Mathem. **41**(3), Art. no. 103 (2022)
14. Jana, C., Dobrodolac, M., Simic, V., Pal, M., Sarkar, B., Stević, Z.: Evaluation of sustainable strategies for urban parcel delivery: Linguistic q-rung orthopair fuzzy Choquet integral approach. Eng. Appl. Artif. Intelli., 126, Art. no. 106811 (2023)
15. Mohebbi, A., Achiche, S., Baron, L.: Multi-criteria fuzzy decision support for conceptual evaluation in design of mechatronic systems: a quadrotor design case study. Res. Eng. Design **29**, 329–349 (2018)
16. Riaz, M., Farid, H.M.A., Jana, C., Pal, M., Sarkar, B.: Efficient city supply chain management through spherical fuzzy dynamic multistage decision analysis. Eng. Appl. Artif. Intel., 126, Art. no. 106712 (2023)
17. Palanikumar, M., Arulmozhi, K., Jana, C., Pal, M.: Multiple-attribute decision-making spherical vague normal operators and their applications for the selection of farmers. Expert Syst. **40**(3), e13188 (2023)
18. Jana, C., Garg, H., Pal, M., Sarkar, B., Wei, G.: MABAC framework for logarithmic bipolar fuzzy multiple attribute group decision-making for supplier selection. Complex Intell. Syst., 1–16 (2023)
19. Atanassov, K.T., Atanassov, K.T.: Intuitionistic fuzzy sets (pp. 1–137). Physica-Verlag HD (1999)
20. Khan, M.J., Kumam, P., Liu, P., Kumam, W., Ashraf, S.: A novel approach to generalized intuitionistic fuzzy soft sets and its application in decision support system. Mathematics **7**(8), 742 (2019)
21. Hussain, A., Mahmood, T., Smarandache, F., Ashraf, S.: TOPSIS approach for MCGDM based on intuitionistic fuzzy rough Dombi aggregation operations. Comput. Appl. Mathem. **42**(4), 176 (2023)
22. Paul, T.K., Jana, C., Pal, M.: Multi-criteria group decision-making method in disposal of municipal solid waste based on cubic Pythagorean fuzzy EDAS approach with incomplete weight information. Appl. Soft Comput.,144, Art. no. 110515 (2023)
23. Paul, T.K., Jana, C., Pal, M., Simic, V.: Multi-attribute group decision making method based on Pythagorean fuzzy Einstein interactive power averaging approach for sustainable cement industry. Appl. Soft Comput., 148, Art. no. 110898 (2023)
24. Jana, C., Simic, V., Pal, M., Sarkar, B., Pamucar, D.: Hybrid multi-criteria decision-making method with a bipolar fuzzy approach and its applications to economic condition analysis. Eng. Appl. Artif. Intell., 132, Art. no. 107837 (2024)
25. Ashraf, S., Chohan, M.S., Muhammad, S., Khan, F.: Circular intuitionistic fuzzy TODIM approach for material selection for cryogenic storage tank for liquid nitrogen transportation. IEEE Access (2023)
26. Ashraf, S., Ali, M., Sohail, M., Eldin, S.M.: Assessing the environmental impact of industrial pollution using the complex intuitionistic fuzzy ELECTREE method: a case study of pollution control measures. Frontiers Environ. Sci. **11**, 610 (2023)
27. Husain, S., Ahmad, Y., Alam, M.A.: A study on the role of intuitionistic fuzzy set in decision making problems. Int. J. Comput. Appl. **48**(0975–888), 35–41 (2012)
28. Szmidt, E.: Distances and Similarities in Intuitionistic Fuzzy Sets, vol. 307. Springer International Publishing, Switzerland (2014)
29. Szmidt, E., Kacprzyk, J.: A new similarity measure for intuitionistic fuzzy sets: straightforward approaches may not work. In: 2007 IEEE International Fuzzy Systems Conference, pp. 1–6. IEEE (2007)

30. Bouchet, A., Montes, S., Ballarin, V., Díaz, I.: Intuitionistic fuzzy set and fuzzy mathematical morphology applied to color leukocytes segmentation. Signal, Image and Video Process. **14**, 557–564 (2020)
31. Garg, H., Arora, R.: Generalized and group-based generalized intuitionistic fuzzy soft sets with applications in decision-making. Appl. intell. **48**, 343–356 (2018)
32. Zadeh, L.A.: A note on Z-numbers. Inform. Sci. **181**(14), 2923–2932 (2011)
33. Kang, B., Wei, D., Li, Y., Deng, Y.: A method of converting Z-number to classical fuzzy number. J. Inform. Comput. Sci. **9**(3), 703–709 (2012)
34. Jiang, W., Xie, C., Luo, Y., Tang, Y.: Ranking Z-numbers with an improved ranking method for generalized fuzzy numbers. J. Intell. Fuzzy Syst. **32**(3), 1931–1943 (2017)
35. Kang, B., Deng, Y., Hewage, K., Sadiq, R.: A method of measuring uncertainty for Z-number. IEEE Trans. Fuzzy Syst. **27**(4), 731–738 (2018)
36. Sari, I.U., Kahraman, C.: Intuitionistic fuzzy Z-numbers. In: Intelligent and Fuzzy Techniques: Smart and Innovative Solutions: Proceedings of the INFUS 2020 Conference, Istanbul, Turkey, July 21–23, 2020 (pp. 1316–1324). Springer International Publishing (2021)
37. Qi, J., Hu, J., Peng, Y.: Conceptual design evaluation using interval intuitionistic fuzzy-z-number for multiple uncertain information from decision-maker. Int. J. Fuzzy Syst. **25**(8), 3119–3143 (2023)
38. Jaini, N.I.: Application of intuitionistic Z-numbers in supplier selection. Intell. Automat Soft Comput. **35**(1), 47–61 (2023)
39. Alam, Hakim Nik Badrul., N.M.F., Ku Khalif, K.M.N., Jaini, N.I., Abu Bakar, A.S., Abdullah, L.: Intuitive multiple centroid defuzzification of intuitionistic Z-numbers. J. Fuzzy Extension Appl. **3**(2), 126–139 (2022)
40. Sahoo, L., Senapati, T., Yager, R.R. (Eds.): Real Life Applications of Multiple Criteria Decision Making Techniques in Fuzzy Domain, Studies inFuzziness and Soft Computing (STUDFUZZ, volume 420), Springer Nature (2023)
41. Wang, W., Liu, X.: Intuitionistic fuzzy information aggregation using Einstein operations. IEEE Trans. Fuzzy Syst. **20**(5), 923–938 (2012)

Cubic Analytic Hierarchy Process with Application in Decision-Making

Ismat Beg, Muhammad Gulistan, and Muhammad Asif

Abstract In this chapter, we extend the AHP and FAHP model collectively new approach based on CAHP using TCNs values for predictor estimation. The cubic analytic hierarchy (CAHP) process can be applied to more complicated complex problems where the decision-makers have multiple uncertainties such as determining preference values regarding the considered objects. A numerical example is performed in practical life to demonstrate the application of the (CAHP) process and show the advantages of the proposed new methodology for decision-making. The effectiveness and flexibility of the present method are established through a comparative analysis with the help of several existing methods.

Keywords Analytic hierarchy process (AHP) · Cubic set · Triangular cubic number · Multi-criteria decision-making

1 Introduction

Freedom is an ever-present element in daily life, where goals are formed through decision-making. It has been meant that decision-making (DM) is quite straightforward in the case of single-criterion systems. In this case, we simply need to filter out the option with the highest star priority. Yet, in of case DM more complex approaches have to be designed which overcome these difficulties. The first is that in the problems of MCDM, it is necessary to count the number of factors or criteria the problems have, how to identify the problems and if it has to collect data it has to fit the edge with which we build the DM's will. Towards systemizing research work

I. Beg (✉)
Department of Mathematics and Statistical Sciences, Lahore School of Economics, Lahore, Pakistan
e-mail: ibeg@lahoreschool.edu.pk

M. Gulistan · M. Asif
Department of Mathematics and Statistics, Hazara University Mansehra, Mansehra, KP, Pakistan
e-mail: gulistanmath@hu.edu.pk

L. Sahoo et al. (eds.), *Decision Making Under Uncertainty Via Optimization, Modelling, and Analysis*, Studies in Systems, Decision and Control 558,
https://doi.org/10.1007/978-981-96-0085-4_10

in the MCDM field, the problems have been categorized into two types: those of the multiple-attribute decision-making (MADM) based on the distinction of purposes and types of data. First, as aforementioned, making decisions with many attributes is a very crucial concern, and perhaps, the simplest, but significant problem, or research theme in this area is how best to represent imprecise information. In practice, however, uncertainty is part of the input data set, because there are numerous mistakes during data collection for various reasons, some misconceived handling, or just misplaced self-interest. On the other hand, the concern here is that presenting information about things becomes difficult when systems are unnecessarily complex. Many theories have thus been emerging recently to address the problem of fuzziness inherent in data.

Zadeh [1], in his groundbreaking paper, introduced the concept of an FS which offers a suitable method of managing vagueness. Concerning the state of non-membership, fuzzy sets do not incorporate the state of neutrality; that is, neither support nor opposition. However, some limitations are encountered while applying these definitions, for instance, one membership function is defined for that specific object, or while working on them, some parameters are not well set. In a similar manner, the analysis involving the FS concept is insufficient to meet the decision-makers' demand to provide the right decisions under conditions of uncertainty and ambiguity. To solve such problems Jun [2], defined a cubic set abbreviated as CS. Fahmi et al. [3], have introduced triangular cubic numbers to deal with MCDM issues.

Saaty [4] comes up with a proposition in the AHP, a technique in management science and operation research for establishing priorities in several types of decision-making. It is useful for the decision-makers as it finalizes a complex problem into a hierarchal structure of the criteria, sub-criteria and alternatives and offers the simplest scale of relative size in dominance units to portray judgments on the structure of pair-wise comparison. If a relative magnitude ratio scale is thus perceived which is conceptualized in priority units by each set of comparisons, a general ratio scale of priorities is compiled to place the system's alternatives [5]. A breakdown of an AHP analysis comprises three general themes namely decomposition which focuses on breaking down into subunits, comparative analysis which involves making comparisons and finally the synthesis of priorities [6]. Since for an updated state of AHP [7], even though it is famous and easy to apply to solve multi-criteria decision-making problems, fundamentally it is being discussed for not having the capability to deal with innate uncertainty and ambiguity. When fulfilling the AHP model, the commonly intended crisp numbers from the scale 1 to 9 are employed to denote the comparison ratios to the aspects offered by the decision-makers across various factors. But in few realistic cases, it could be very hard for a decision-maker to provide sharp evaluation values for the comparative analysis results because of his or her possession of limited knowledge possibly due to qualitative assessment, possibly due to subjective deviations in the evaluating criteria and judgments. Thus, it can be noted that AHP seems to be wrong to derive the required assessments needed for arriving at the priorities in such a kind of situation. An individual decision-making process is, as a rule, affected by the possessed information to them regarding the world state. To overcome this problem, Laarhoven and Pedrycz [8], have formulated the FAHP in which the possibility of the membership was in terms of triangular fuzzy numbers, the fuzzy weights and the fuzzy scores of the alternatives were estimated with the help of logarithmic least squares method. For this reason, it can control the inaccuracy that prevails within the AHP method. Later Buckley [9], integrated the trapezoidal fuzzy number into the AHP and derived the fuzzy weight and fuzzy score in the geometric mean approach. Subsequently, Chang [10] used the row mean method in the analysis of the priority of similarity ratio to obtain the consideration of triangular fuzzy numbers. Another advantage of Chang's method, which makes it preferable is that it is less complicated than all the existing fuzzy AHP approaches. Despite the fact that the analysis of fuzzy AHP is time-consuming, it is reasonable to contain and depict human perception of vagueness while dealing with complex multiple-criteria decision-making. It has been applied as an illustration of the fuzzy AHP in so many special fields, for instance, the airlines industry [12], the public administration [11], the manufacturing industry [13], the electronic industry [15], the textile industry [14], the oil industry [16], the transportation industry [18], the entertainment industry [17], etc. Numerous researchers focused on solving decision-making problems and utilized them in practical life problems, as seen in [19–21].

In our research, we have discovered that cubic sets can be used when dealing with imprecise information from subjective measurements as well as, or instead of, the AHP used for converting words into numbers from objective measurements. Therefore it is necessary to create a combination of two hierarchies that is flexible at different levels of hierarchy within linguistic terms. Since cubic set deals with complex decision-making processes, we use a combination of cubic set and analytic hierarchy process (AHP) method for handling qualitative as well as quantitative data. Furthermore, besides a cubic set, we investigate a Fuzzy AHP model based on structural technique under the cubic set environment. We define an algorithm for the CAHP process for solving DM problems. We also provide practical examples of daily life problems and assess our results with presented methods to verify the precision of the projected algorithm. Finally, we will conclude our work with some future directions.

This paper is divided into six sections. Section 1, gives an overview of sets including their definitions and how they are used in attribute decision problems. In this part, we review some definitions. Section 3, introduces a method called the cube analytic hierarchy structure algorithm. Moving on to Sect. 4, we demonstrate the application of this methodology to solve a real-world problem using the cubic analytic hierarchy procedure. In Sect. 5, we compare our findings with those of methods and discuss both the strengths and limitations of our approach. Lastly in Sect. 6, we will wrap up this study and suggest future research directions.

2 Preliminaries

This section consists of some notations, a few pre-defined definitions and results.

First, some notations with their descriptions are shown in Table 1.

Then, the following cubic sets have been defined, where some of them are employed directly in the proofs of the given paper:

Definition 1 ([1]) A collection Υ:X → [0,1] is referred to as a fuzzy set while Υ(x) can be referred to as the membership function for an element x in X.

Definition 2 ([2]) The arrangement $C = \left\{\left(\widetilde{\Upsilon}(x), \Upsilon(x)/x \in X\right)\right\}$ is defined a cubic set in X, In which $\widetilde{\Upsilon}(x)$ is an IVFS in X and Υ(x) is a fuzzy set as well as in X.

Definition 3 ([3]) The TCN $\tilde{a} = \langle (a_1, a_2, a_3), \left[\alpha_{\tilde{a}}^-, \alpha_{\tilde{a}}^+\right], \beta_{\tilde{a}} \rangle$ is a CS on R, its (IVTFS) are defined as:

$$\tilde{I}_{\tilde{a}}(x) = \left\{ \begin{array}{c} \frac{x-a}{b-a}\left[\alpha_{\tilde{a}}^-, \alpha_{\tilde{a}}^+\right],\ a \le x < b \\ \frac{c-x}{c-b}\left[\alpha_{\tilde{a}}^-, \alpha_{\tilde{a}}^+\right],\ b < x \le c \\ 0,\ \text{otherwise} \end{array} \right\} \tag{1}$$

and (TFS) is

$$I_{\tilde{a}}(x) = \left\{ \begin{array}{c} \frac{x-a}{b-a}\beta_{\tilde{a}},\ a \le x < b \\ \frac{c-x}{c-b}\beta_{\tilde{a}},\ b < x \le c \\ 0,\ \text{otherwise} \end{array} \right\} \tag{2}$$

Definition 4 ([3]) Let $\tilde{a} = \langle (a_1, a_2, a_3), \left[\alpha_{\tilde{a}}^-, \alpha_{\tilde{a}}^+\right], \beta_{\tilde{a}} \rangle$ and $\tilde{b} = \langle (b_1, b_2, b_3), \left[\alpha_{\tilde{b}}^-, \alpha_{\tilde{b}}^+\right], \beta_{\tilde{b}} \rangle$ be two TCNs and $\gamma \neq 0$ be any real number. Then:

Table 1 Abbreviations

Definition	Abbreviation
Cubic analytic hierarchy process	AHP
Fuzzy analytic hierarchy process	FAHP
Analytic hierarchy process	CAHP
Interval-valued triangular fuzzy set	IVTFS
Triangular fuzzy set	TFS
Linguistic terms and the identical triangular cubic numbers	LT&ITCN
Decision-making	DM

1. Addition of two TCNs

$$\tilde{a}+\tilde{b}=\langle(a_1+b_1,a_2+b_2,a_3+b_3),\left[min\left(\alpha_{\tilde{a}}^{-},\alpha_{\tilde{b}}^{-}\right),min\left(\alpha_{\tilde{a}}^{+},\alpha_{\tilde{b}}^{+}\right)\right],max(\beta_{\tilde{a}},\beta_{\tilde{b}})\rangle \tag{3}$$

2. Subtraction of two TCNs

$$\tilde{a}+\tilde{b}=\langle(a_1-b_3,a_2-b_2,a_3-b_1),\left[min\left(\alpha_{\tilde{a}}^{-},\alpha_{\tilde{b}}^{-}\right),min\left(\alpha_{\tilde{a}}^{+},\alpha_{\tilde{b}}^{+}\right)\right],max(\beta_{\tilde{a}},\beta_{\tilde{b}})\rangle \tag{4}$$

3. Multiplication of a TCN by a constant value

$$\gamma\tilde{a}=\begin{Bmatrix}(\gamma a_1,\gamma a_2,\gamma a_3),\left[\alpha_{\tilde{a}}^{-},\alpha_{\tilde{a}}^{+}\right],\beta_{\tilde{a}} if\,(\gamma>0)\\(\gamma a_3,\gamma a_2,\gamma a_1),\left[\alpha_{\tilde{a}}^{-},\alpha_{\tilde{a}}^{+}\right],\beta_{\tilde{a}} if\,(\gamma<0)\end{Bmatrix} \tag{5}$$

4. The inverse of a TCN

$$\tilde{a}^{-1}=\langle\left(\frac{1}{a_3},\frac{1}{a_2},\frac{1}{a_1}\right),\left[\alpha_{\tilde{a}}^{-},\alpha_{\tilde{a}}^{+}\right],\beta_{\tilde{a}} where\ \tilde{a}\neq 0\rangle \tag{6}$$

5. Multiplication of two TCNs

$$\tilde{a}\tilde{b}=\begin{Bmatrix}\langle(a_1b_1,a_2b_2,a_3b_3),\left[min\left(\alpha_{\tilde{a}}^{-},\alpha_{\tilde{b}}^{-}\right),min\left(\alpha_{\tilde{a}}^{+},\alpha_{\tilde{b}}^{+}\right)\right],max(\beta_{\tilde{a}},\beta_{\tilde{b}})\rangle if\,(a_3>0,b_3>0)\\\langle(a_1b_3,a_2b_2,a_3b_1),\left[min\left(\alpha_{\tilde{a}}^{-},\alpha_{\tilde{b}}^{-}\right),min\left(\alpha_{\tilde{a}}^{+},\alpha_{\tilde{b}}^{+}\right)\right],max(\beta_{\tilde{a}},\beta_{\tilde{b}})\rangle if\,(a_3<0,b_3>0)\\\langle(a_1b_3,a_2b_2,a_3b_1),\left[min\left(\alpha_{\tilde{a}}^{-},\alpha_{\tilde{b}}^{-}\right),min\left(\alpha_{\tilde{a}}^{+},\alpha_{\tilde{b}}^{+}\right)\right],max(\beta_{\tilde{a}},\beta_{\tilde{b}})\rangle if\,(a_3<0,b_3<0)\end{Bmatrix} \tag{7}$$

6. Division of two TCNs

$$\frac{\tilde{a}}{\tilde{b}}=\begin{Bmatrix}\langle\left(\frac{a_1}{b_3},\frac{a_2}{b_2},\frac{a_3}{b_1}\right),\left[min\left(\alpha_{\tilde{a}}^{-},\alpha_{\tilde{b}}^{-}\right),min\left(\alpha_{\tilde{a}}^{+},\alpha_{\tilde{b}}^{+}\right)\right],max(\beta_{\tilde{a}},\beta_{\tilde{b}})\rangle if\,(a_3>0,b_3>0)\\\langle\left(\frac{a_3}{b_3},\frac{a_2}{b_2},\frac{a_1}{b_1}\right),\left[min\left(\alpha_{\tilde{a}}^{-},\alpha_{\tilde{b}}^{-}\right),min\left(\alpha_{\tilde{a}}^{+},\alpha_{\tilde{b}}^{+}\right)\right],max(\beta_{\tilde{a}},\beta_{\tilde{b}})\rangle if\,(a_3<0,b_3>0)\\\langle\left(\frac{a_3}{b_1},\frac{a_2}{b_2},\frac{a_1}{b_3}\right),\left[min\left(\alpha_{\tilde{a}}^{-},\alpha_{\tilde{b}}^{-}\right),min\left(\alpha_{\tilde{a}}^{+},\alpha_{\tilde{b}}^{+}\right)\right],max(\beta_{\tilde{a}},\beta_{\tilde{b}})\rangle if\,(a_3<0,b_3<0)\end{Bmatrix} \tag{8}$$

7. Division of a TCN by a constant value

$$\frac{\tilde{a}}{\gamma}=\begin{Bmatrix}\left(\frac{a_1}{\gamma},\frac{a_2}{\gamma},\frac{a_3}{\gamma}\right),\left[\alpha_{\tilde{a}}^{-},\alpha_{\tilde{a}}^{+}\right],\beta_{\tilde{a}} if\,(\gamma>0)\\\left(\frac{a_3}{\gamma},\frac{a_2}{\gamma},\frac{a_1}{\gamma}\right),\left[\alpha_{\tilde{a}}^{-},\alpha_{\tilde{a}}^{+}\right],\beta_{\tilde{a}} if\,(\gamma<0)\end{Bmatrix} \tag{9}$$

3 Cubic Analytic Hierarchy Process

In this section, we introduce the AHP in cubic atmosphere. The following is a step-wise procedure that is depicted for the model.

Algorithm of Proposed Method

Step (1): Identify the problem within the content, and reconstruct its structure with reference to the hierarchy, as it consists of four levels.

(i) Goal which the organization wants to attain;
(ii) Consists of the criteria;
(iii) Consists of the sub-criteria;
(iv) Finalized the different alternatives are assessed.

The general hierarchy is projected in Fig. 1.

The next step is to apply the weighting according to specialist judgment.

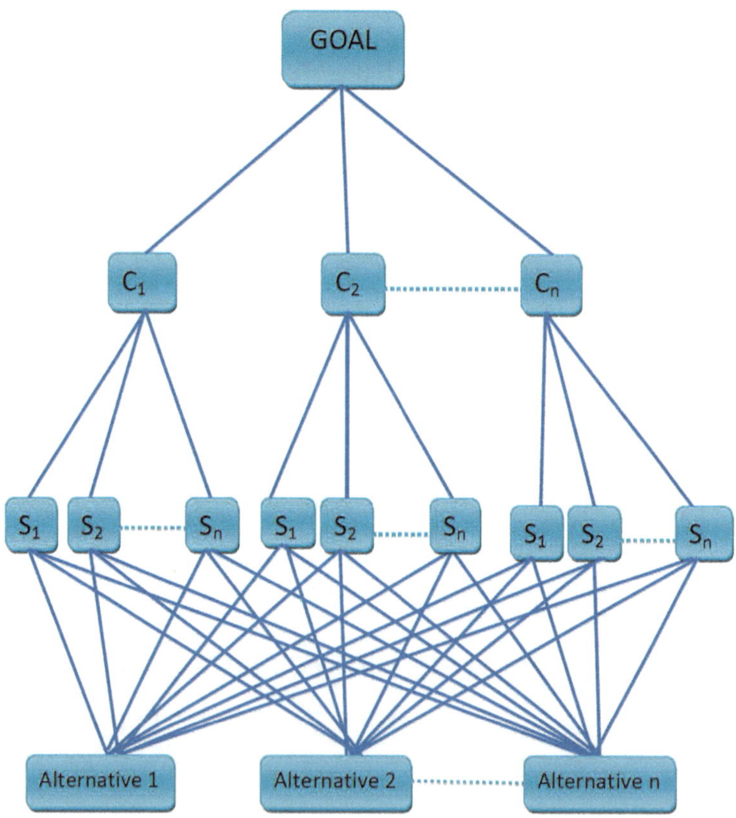

Fig. 1 Structure of hierarchal problem

Step (2): In this step, a cubic pair-wise comparison matrix is developed using the linguistic terms mentioned in the publicized Table 2.

The cubic scale is determined according to the specialist assessment. The comparison matrix is as follows:

$$\tilde{A} = \begin{bmatrix} \tilde{a}_{11} & \tilde{a}_{12} & \cdots & \tilde{a}_{1n} \\ \tilde{a}_{11} & \ddots & \cdots & \tilde{a}_{21} \\ \vdots & & & \vdots \\ \tilde{a}_{m1} & \tilde{a}_{11} & \cdots & \tilde{a}_{mn} \end{bmatrix} \tag{10}$$

where $a_{ji}=(a_{ij})^{-1}$ symbolized the TCNs used to assess the indeterminacy in the decision-making procedure.

Step (3): Check the consistency of the specialist judgments.

If the comparison matrix is the transitive relationship, specifically that, i.e., $a_{ik}=a_{ij}a_{jk}$ holds for all i, j, and k, then the comparison matrix is deemed consistent. It is important to spotlight the lower, median, and upper values of the TCNs in the comparison matrix.

Step (4): In this step, In this stage, the weights of the criteria, sub-criteria, and alternatives are computed from the comparison matrix. This is done through cubic arithmetic operations by calculating the normalized values of the row sums.

$$W_j = \frac{\sum_j^n \tilde{a}_{ij}}{\sum_i^m \tilde{a}_{ij} \sum_j^n \tilde{a}_{ij}} \tag{11}$$

Table 2 LT&ITCNs

Saaty Scale	Explanation	Cubic Triangular Scale
1	Uniformly significant	$\tilde{1}$=⟨(1,1,1);[0.50,0.50],0.50⟩
3	Faintly significant	$\tilde{3}$=⟨(2,3,4);[0.25,0.35],0.30⟩
5	Strongly significant	$\tilde{5}$=⟨(4,5,6);[0.75,0.85],0.80⟩
7	Very strongly significant	$\tilde{7}$=⟨(6,7,8);[0.85,0.95],0.90⟩
9	Absolutely significant	$\tilde{9}$=⟨(9,9,9);[1.00,1.00],1.00⟩
2	Transitional values b/w two nearest scales	$\tilde{2}$=⟨(1,2,3);[0.35,0.45],0.40⟩
4		$\tilde{4}$=⟨(3,4,5);[0.55,0.65],0.60⟩
6		$\tilde{6}$=⟨(5,6,7);[0.65,0.75],0.70⟩
8		$\tilde{8}$=⟨(7,8,9);[0.80,0.90],0.85⟩

where

$$\sum_{j}^{n} \tilde{a}_{ij} = \left(\sum_{j}^{n} \tilde{l}_{ij}, \sum_{j}^{n} \tilde{m}_{ij}, \sum_{j}^{n} \tilde{u}_{ij} \right) \tag{12}$$

$$\sum_{i}^{m} \sum_{j}^{n} \tilde{a}_{ij} = \left(\sum_{i}^{m} \sum_{j}^{n} \tilde{l}_{ij}, \sum_{i}^{m} \sum_{j}^{n} \tilde{m}_{ij}, \sum_{i}^{m} \sum_{j}^{n} \tilde{u}_{ij} \right) \tag{13}$$

and

$$(i = 1,2, \ldots, m)(j = 1,2, \ldots, n)$$

Step (5): Let $\tilde{a}_{ij} = \langle (a_1, a_2, a_3)\left[\alpha_{\tilde{a}}^{-}, \alpha_{\tilde{a}}^{+}\right], \beta_{\tilde{a}} \rangle$ be TCNs; then

$$S(\tilde{a}_{ij}) = E(\tilde{a}_{ij}) \times \left[\left(\alpha_{\tilde{a}}^{-} + \alpha_{\tilde{a}}^{+}\right) - (1 - \beta_{\tilde{a}})\right] \tag{14}$$

and

$$A(\tilde{a}_{ij}) = E(\tilde{a}_{ij}) \times \left[\left(\alpha_{\tilde{a}}^{-} + \alpha_{\tilde{a}}^{+}\right) + (1 - \beta_{\tilde{a}})\right] \tag{15}$$

which are the score and accuracy function of $\tilde{a}_{ij}$ respectively. Where $E(\tilde{a}_{ij})$ is the projected value of TCNs of $\tilde{a}_{ij}$.

$$E(\tilde{a}_{ij}) = \langle \frac{1}{9}[a_1 + a_2 + a_3] \times \left[1 + \left(\alpha_{\tilde{a}}^{-} - \beta_{\tilde{a}}\right)\right]\left[1 + \left(\alpha_{\tilde{a}}^{+} - \beta_{\tilde{a}}\right)\right] \rangle \tag{16}$$

which are the score and accuracy function of $\tilde{a}_{ij}$, respectively. To get the score function of $\tilde{a}_{ji}$, using the following equation:

$$S(\tilde{a}_{ij}) = \frac{1}{S(\tilde{a}_{ij})} \tag{17}$$

Step (6): Normalizing the score function.

Step (7): The overall weight (priority) of the alternative for the final ranking is calculated using Equation [17].

$$T_{w_{Alt}} = \sum_{j=1}^{n} \left(w_j \otimes w_{ij}\right) \tag{18}$$

where (j = 1,2,…,n)

We obtain the fragment of the CAHP from the above steps, as publicized in Fig. 2.

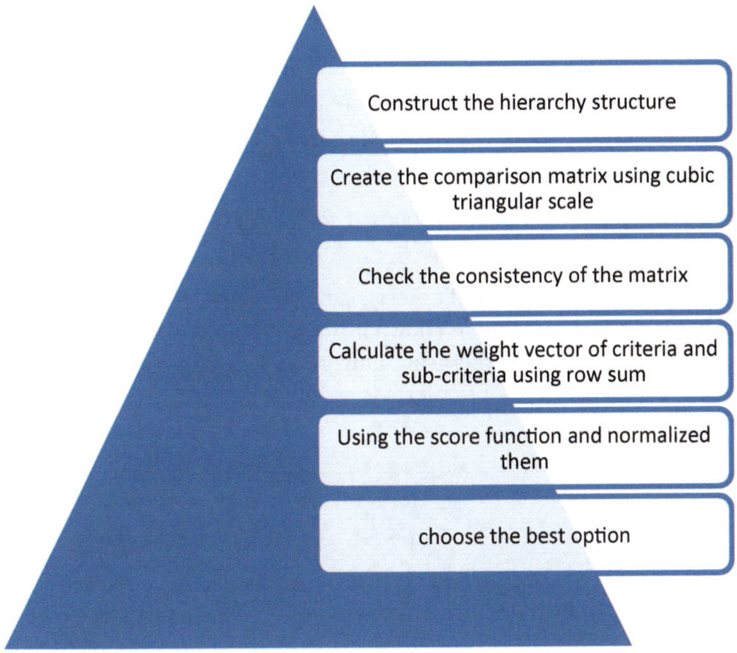

Fig. 2 Layout of cubic analytic hierarchy process (CAHP)

4 Numerical Example

This real case is solved by applying the model developed in Sect. 3. In the present age, marriage halls are preferred instead of harries for various reasons. In particular, it has been a tradition in many countries including Pakistan. In order to cut down on different expenses and to invite a limited number of guests, people often conduct marriage ceremonies in halls. For this reason, the marriage halls enhance their service quality and food quality such that many people should shift towards marriage halls and hence make them become loyal customers. The marriage halls contract the best catering firm to do that job for them and we get the higher desired output. The objective here is to choose the optimal option from the available ones. From the above decision-making problem, we use our method as follows;

Step 1. Folkdine, Chiniot and Foodnerd are the three alternatives.

The best alternative is selected on the basis of different criteria, which are defined as follows:

(1) C_1 : The first criterion is Hygiene, having the sub-criteria i.e. S_1 : Hygiene of service personnel, S_2 : Hygiene of meal, S_3 : Hygiene of service meal;

(2) C_2 : The second criterion is the Quality of the meal, having the sub-criteria i.e. S_4 : Complementary meals in a daze, S_5 : Variety of meal, S_6 : Taste of meal, S_7 : Calories of the meal;

(3) C_3 : The third criterion is Quality of service, having the sub-criteria S_8 : Service time :, S_9 : Behavior of service personnel, S_{10} : Problem solving ability, S_{11} : Communication on phone.

Step 2. Three options have been proposed for consideration. Thus, festering in a hierarchy formation can be shown as in the following Fig. 3.

Step 3. The comparison matrix of criteria, sub-criteria and alternatives is structured with the help of linguistic terms as shown in Table 1. The values presented in Table 2 are related to specialist judgments. The comparison of criteria is presented in Table 3. The cubic comparison matrix of criteria is as follows:

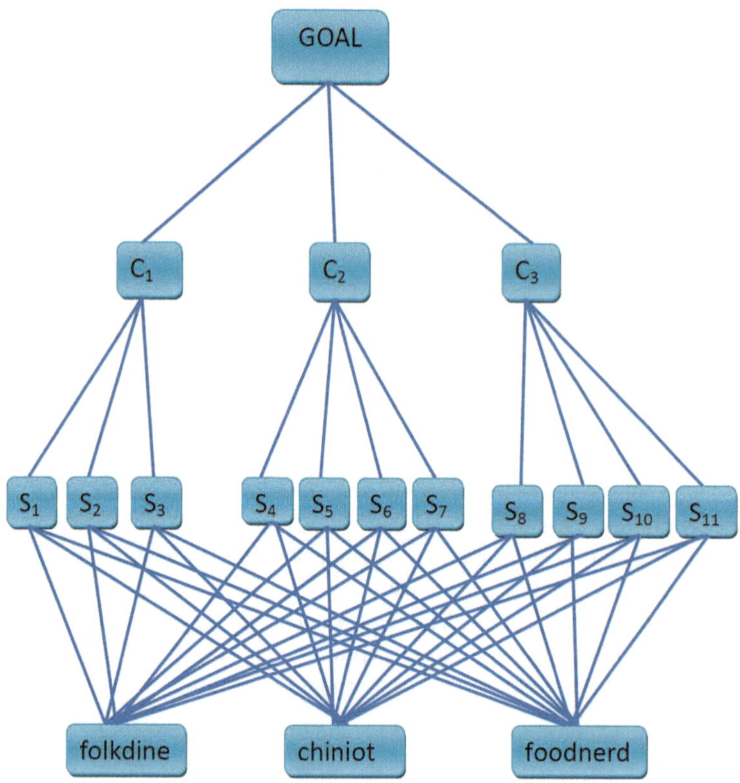

Fig. 3 Structure of hierarchy problem

Table 3 Comparison matrix for criteria

Criteria	C_1	C_2	C_3
C_1	$\langle(1,1,1);[0.50,0.50],0.50\rangle$	$\langle(2,3,4);[0.25,0.35],0.30\rangle$	$\langle\left(\frac{1}{8},\frac{1}{7},\frac{1}{6}\right);[0.85,0.95],0.30\rangle$
C_2	$\langle\left(\frac{1}{4},\frac{1}{3},\frac{1}{2}\right);[0.25,0.35],0.50\rangle$	$\langle(1,1,1);[0.50,0.50],0.50\rangle$	$\langle\left(\frac{1}{6},\frac{1}{5},\frac{1}{4}\right);[0.75,0.85],0.80\rangle$
C_3	$\langle(1,1,1);[0.50,0.50],0.50\rangle$	$\langle(4,5,6);[0.75,0.85],0.80\rangle$	$\langle(1,1,1);[0.50,0.50],0.50\rangle$

Step 4. The over-comparison matrix was consistent when applying the method proposed in [1].

Step 5. Calculating the weights.

C_1	$\langle(0.14,0.22,0.33);[0.25,0.35],0.90\rangle$
C_2	$\langle(0.07,0.09,0.11);[0.25,0.35],0.90\rangle$
C_3	$\langle(0.50,0.70,0.97);[0.25,0.35],0.90\rangle$

Step 6. Obtain the score function using Eq. (14),

C_1	$\langle 0.0061\rangle$
C_2	$\langle 0.0023\rangle$
C_3	$\langle 0.0189\rangle$

Step 7. Normalizing the weights

$$X = \begin{bmatrix} 0.20 \\ 0.10 \\ 0.70 \end{bmatrix}$$

Like a criteria and the sub-criteria calculation technique, the weight of alternatives is shown in Table 4.

Tables 4, 5 and 6 summarized the comparison matrix of decision criteria of $C_1 to C_3$ and its priority vectors.

Table 4 Comparison matrix w.r.t. C_1

Sub-criteria	S_1	S_2	S_3	Priority vector
S_1	$\left\langle \begin{matrix}(1,1,1);0.50\\ [0.50,0.50]\end{matrix} \right\rangle$	$\left\langle \begin{matrix}(4,5,6);0.80\\ [0.75,0.85]\end{matrix} \right\rangle$	$\left\langle \begin{matrix}(4,5,6);0.80\\ [0.75,0.85]\end{matrix} \right\rangle$	0.35
S_2	$\left\langle \begin{matrix}\left(\frac{1}{6},\frac{1}{5},\frac{1}{4}\right);0.80\\ [0.75,0.85]\end{matrix} \right\rangle$	$\left\langle \begin{matrix}(1,1,1);0.50\\ [0.50,0.50]\end{matrix} \right\rangle$	$\left\langle \begin{matrix}(2,3,4);0.30\\ [0.25,0.35]\end{matrix} \right\rangle$	0.25
S_3	$\left\langle \begin{matrix}\left(\frac{1}{6},\frac{1}{5},\frac{1}{4}\right);0.80\\ [0.75,0.85]\end{matrix} \right\rangle$	$\left\langle \begin{matrix}\left(\frac{1}{4},\frac{1}{3},\frac{1}{2}\right);0.50\\ [0.25,0.35]\end{matrix} \right\rangle$	$\left\langle \begin{matrix}(1,1,1);0.50\\ [0.50,0.50]\end{matrix} \right\rangle$	0.10

Table 5 Comparison matrix w.r.t. C_2

Sub-criteria	S_4	S_5	S_6	S_7	Priority vector
S_4	$\langle (1,1,1); 0.50, [0.50, 0.50] \rangle$	$\langle (4,5,6); 0.80, [0.75, 0.85] \rangle$	$\langle (\frac{1}{8}, \frac{1}{7}, \frac{1}{6}); 0.90, [0.85, 0.95] \rangle$	$\langle (6,7,8); 0.90, [0.85, 0.95] \rangle$	0.28
S_5	$\langle (\frac{1}{6}, \frac{1}{5}, \frac{1}{4}); 0.80, [0.75, 0.85] \rangle$	$\langle (1,1,1); 0.50, [0.50, 0.50] \rangle$	$\langle (\frac{1}{8}, \frac{1}{7}, \frac{1}{6}); 0.90, [0.85, 0.95] \rangle$	$\langle (9,9,9); 1.00, [1.00, 1.00] \rangle$	0.22
S_6	$\langle (6,7,8); 0.90, [0.85, 0.95] \rangle$	$\langle (6,7,8); 0.90, [0.85, 0.95] \rangle$	$\langle (1,1,1); 0.50, [0.50, 0.50] \rangle$	$\langle (6,7,8); 0.90, [0.85, 0.95] \rangle$	0.47
S_7	$\langle (\frac{1}{8}, \frac{1}{7}, \frac{1}{6}); 0.90, [0.85, 0.95] \rangle$	$\langle (\frac{1}{9}, \frac{1}{9}, \frac{1}{9}); 1.00, [1.00, 1.00] \rangle$	$\langle (\frac{1}{8}, \frac{1}{7}, \frac{1}{6}); 0.90, [0.85, 0.95] \rangle$	$\langle (1,1,1); 0.50, [0.50, 0.50] \rangle$	0.03

Table 6 Comparison matrix w.r.t. C_3

Sub-criteria	S_8	S_9	S_{10}	S_{11}	Priority vector
S_8	$\langle (1,1,1); 0.50, [0.50, 0.50] \rangle$	$\langle (\frac{1}{9}, \frac{1}{9}, \frac{1}{9}); 1.00, [1.00, 1.00] \rangle$	$\langle (9,9,9); 1.00, [1.00, 1.00] \rangle$	$\langle (\frac{1}{8}, \frac{1}{7}, \frac{1}{6}); 0.90, [0.85, 0.95] \rangle$	0.19
S_9	$\langle (9,9,9); 1.00, [1.00, 1.00] \rangle$	$\langle (1,1,1); 0.50, [0.50, 0.50] \rangle$	$\langle (9,9,9); 1.00, [1.00, 1.00] \rangle$	$\langle (6,7,8); 0.90, [0.85, 0.95] \rangle$	0.49
S_{10}	$\langle (\frac{1}{9}, \frac{1}{9}, \frac{1}{9}); 1.00, [1.00, 1.00] \rangle$	$\langle (\frac{1}{9}, \frac{1}{9}, \frac{1}{9}); 1.00, [1.00, 1.00] \rangle$	$\langle (1,1,1); 0.50, [0.50, 0.50] \rangle$	$\langle (\frac{1}{8}, \frac{1}{7}, \frac{1}{6}); 0.90, [0.85, 0.95] \rangle$	0.03
S_{11}	$\langle (6,7,8); 0.90, [0.85, 0.95] \rangle$	$\langle (\frac{1}{8}, \frac{1}{7}, \frac{1}{6}); 0.90, [0.85, 0.95] \rangle$	$\langle (6,7,8); 0.90, [0.85, 0.95] \rangle$	$\langle (1,1,1); 0.50, [0.50, 0.50] \rangle$	0.29

Tables 7, 8, 9, 10, 11, 12, 13, 14, 15, 16 and 17 summarized cubic pair-wise comparison matrix of catering firms of $S_1 to S_{11}$ and its priority vectors.

The local priority weights of the three catering firms w. r. t C_1, C_2, C_3 shown in Tables 18, 19, 20, respectively.

Global weight of the three catering firms w. r. t a goal (Table 21).

The best option is foodnerd. Hence the final ranking are shown in Fig. 4.

Table 7 Cubic pair-wise comparison matrix w.r.t S_1

Catering firms	Folkdine	Chiniot	Foodnerd	Priority vector
Folkdine	$\left\langle \begin{matrix} (1,1,1); 0.50 \\ [0.50, 0.50] \end{matrix} \right\rangle$	$\left\langle \begin{matrix} (6,7,8); 0.90 \\ [0.85, 0.95] \end{matrix} \right\rangle$	$\left\langle \begin{matrix} (4,5,6); 0.80 \\ [0.75, 0.85] \end{matrix} \right\rangle$	0.44
Chiniot	$\left\langle \begin{matrix} (\frac{1}{8}, \frac{1}{7}, \frac{1}{6}); 0.90 \\ [0.85, 0.95] \end{matrix} \right\rangle$	$\left\langle \begin{matrix} (1,1,1); 0.50 \\ [0.50, 0.50] \end{matrix} \right\rangle$	$\left\langle \begin{matrix} (\frac{1}{6}, \frac{1}{5}, \frac{1}{4}); 0.80 \\ [0.75, 0.85] \end{matrix} \right\rangle$	0.35
Foodnerd	$\left\langle \begin{matrix} (\frac{1}{6}, \frac{1}{5}, \frac{1}{4}); 0.80 \\ [0.75, 0.85] \end{matrix} \right\rangle$	$\left\langle \begin{matrix} (4,5,6); 0.80 \\ [0.75, 0.85] \end{matrix} \right\rangle$	$\left\langle \begin{matrix} (1,1,1); 0.50 \\ [0.50, 0.50] \end{matrix} \right\rangle$	0.21

Table 8 Cubic pair-wise comparison matrix w.r.t S_2

Catering firms	Folkdine	Chiniot	Foodnerd	Priority vector
Folkdine	$\left\langle \begin{matrix} (1,1,1); 0.50 \\ [0.50, 0.50] \end{matrix} \right\rangle$	$\left\langle \begin{matrix} (2,3,4); 0.30 \\ [0.25, 0.35] \end{matrix} \right\rangle$	$\left\langle \begin{matrix} (\frac{1}{9}, \frac{1}{9}, \frac{1}{9}); 1.00 \\ [1.00, 1.00] \end{matrix} \right\rangle$	0.20
Chiniot	$\left\langle \begin{matrix} (\frac{1}{4}, \frac{1}{3}, \frac{1}{2}); 0.30 \\ [0.25, 0.35] \end{matrix} \right\rangle$	$\left\langle \begin{matrix} (1,1,1); 0.50 \\ [0.50, 0.50] \end{matrix} \right\rangle$	$\left\langle \begin{matrix} (\frac{1}{6}, \frac{1}{5}, \frac{1}{4}); 0.80 \\ [0.75, 0.85] \end{matrix} \right\rangle$	0.08
Foodnerd	$\left\langle \begin{matrix} (9,9,9); 1.00 \\ [1.00, 1.00] \end{matrix} \right\rangle$	$\left\langle \begin{matrix} (4,5,6); 0.80 \\ [0.75, 0.85] \end{matrix} \right\rangle$	$\left\langle \begin{matrix} (1,1,1); 0.50 \\ [0.50, 0.50] \end{matrix} \right\rangle$	0.72

Table 9 Cubic pair-wise comparison matrix w.r.t S_3

Catering firms	Folkdine	Chiniot	Foodnerd	Priority vector
Folkdine	$\left\langle \begin{matrix} (1,1,1); 0.50 \\ [0.50, 0.50] \end{matrix} \right\rangle$	$\left\langle \begin{matrix} (2,3,4); 0.30 \\ [0.25, 0.35] \end{matrix} \right\rangle$	$\left\langle \begin{matrix} (\frac{1}{8}, \frac{1}{7}, \frac{1}{6}); 0.90 \\ [0.85, 0.95] \end{matrix} \right\rangle$	0.20
Chiniot	$\left\langle \begin{matrix} (\frac{1}{4}, \frac{1}{3}, \frac{1}{2}); 0.30 \\ [0.25, 0.35] \end{matrix} \right\rangle$	$\left\langle \begin{matrix} (1,1,1); 0.50 \\ [0.50, 0.50] \end{matrix} \right\rangle$	$\left\langle \begin{matrix} (\frac{1}{6}, \frac{1}{5}, \frac{1}{4}); 0.80 \\ [0.75, 0.85] \end{matrix} \right\rangle$	0.09
Foodnerd	$\left\langle \begin{matrix} (6,7,8); 0.90 \\ [0.85, 0.95] \end{matrix} \right\rangle$	$\left\langle \begin{matrix} (4,5,6); 0.80 \\ [0.75, 0.85] \end{matrix} \right\rangle$	$\left\langle \begin{matrix} (1,1,1); 0.50 \\ [0.50, 0.50] \end{matrix} \right\rangle$	0.69

5 Comparative Analysis

In AHP, non-fuzzy values are used for constructing the evaluation matrix which isn't precise owing to elements of insecurity, doubt, and inconsistency of decision-maker. This is then extended in a fuzzy environment to satisfy the shortcoming of the classical AHP for obtaining the desired result, where the criteria values are characterized by TCNs but without adequately addressing the comparative significance of the criteria

Table 10 Cubic pair-wise comparison matrix w.r.t S_4

Catering firms	Folkdine	Chiniot	Foodnerd	Priority vector
Folkdine	$\langle (1,1,1); 0.50, [0.50, 0.50] \rangle$	$\langle (\frac{1}{8}, \frac{1}{7}, \frac{1}{6}); 0.90, [0.85, 0.95] \rangle$	$\langle (2,3,4); 0.30, [0.25, 0.35] \rangle$	0.25
Chiniot	$\langle (6,7,8); 0.90, [0.85, 0.95] \rangle$	$\langle (1,1,1); 0.50, [0.50, 0.50] \rangle$	$\langle (2,3,4); 0.30, [0.25, 0.35] \rangle$	0.65
Foodnerd	$\langle (\frac{1}{4}, \frac{1}{3}, \frac{1}{2}); 0.30, [0.25, 0.35] \rangle$	$\langle (\frac{1}{4}, \frac{1}{3}, \frac{1}{2}); 0.30, [0.25, 0.35] \rangle$	$\langle (1,1,1); 0.50, [0.50, 0.50] \rangle$	0.10

Table 11 Cubic pair-wise comparison matrix w.r.t. S_5

Catering firms	Folkdine	Chiniot	Foodnerd	Priority vector
Folkdine	$\langle (1,1,1); 0.50, [0.50, 0.50] \rangle$	$\langle (6,7,8); 0.90, [0.85, 0.95] \rangle$	$\langle (2,3,4); 0.30, [0.25, 0.35] \rangle$	0.71
Chiniot	$\langle (\frac{1}{8}, \frac{1}{7}, \frac{1}{6}); 0.90, [0.85, 0.95] \rangle$	$\langle (1,1,1); 0.50, [0.50, 0.50] \rangle$	$\langle (1,1,1); 0.50, [0.50, 0.50] \rangle$	0.14
Foodnerd	$\langle (\frac{1}{4}, \frac{1}{3}, \frac{1}{2}); 0.30, [0.25, 0.35] \rangle$	$\langle (1,1,1); 0.50, [0.50, 0.50] \rangle$	$\langle (1,1,1); 0.50, [0.50, 0.50] \rangle$	0.15

Table 12 Cubic pair-wise comparison matrix w.r.t S_6

Catering firms	Folkdine	Chiniot	Foodnerd	Priority vector
Folkdine	$\langle (1,1,1); 0.50, [0.50, 0.50] \rangle$	$\langle (\frac{1}{9}, \frac{1}{9}, \frac{1}{9}); 1.00, [1.00, 1.00] \rangle$	$\langle (\frac{1}{8}, \frac{1}{7}, \frac{1}{6}); 0.90, [0.85, 0.95] \rangle$	0.05
Chiniot	$\langle (9,9,9); 1.00, [1.00, 1.00] \rangle$	$\langle (1,1,1); 0.50, [0.50, 0.50] \rangle$	$\langle (\frac{1}{8}, \frac{1}{7}, \frac{1}{6}); 0.90, [0.85, 0.95] \rangle$	0.38
Foodnerd	$\langle (6,7,8); 0.90, [0.85, 0.95] \rangle$	$\langle (6,7,8); 0.90, [0.85, 0.95] \rangle$	$\langle (1,1,1); 0.50, [0.50, 0.50] \rangle$	0.57

among them. As the cubic set is the union of interval-value fuzzy set and fuzzy set, they are similar in many aspects. It was found to the best of our knowledge the first time the use of analytic hierarchy process in a cubic environment. It may sound like all the problems within a given field can be solved with this method in order to meet the set goals. Table 22 presents the comparison of the result we obtained using the cubic analytic hierarchy process with other existing methods.

Table 13 Cubic pair-wise comparison matrix w.r.t S_7

Catering firms	Folkdine	Chiniot	Foodnerd	Priority vector
Folkdine	$\langle (1,1,1); 0.50, [0.50, 0.50] \rangle$	$\langle (2,3,4); 0.30, [0.25, 0.35] \rangle$	$\langle (1,1,1); 0.50, [0.50, 0.50] \rangle$	0.32
Chiniot	$\langle (\frac{1}{4}, \frac{1}{3}, \frac{1}{2}); 0.30, [0.25, 0.35] \rangle$	$\langle (1,1,1); 0.50, [0.50, 0.50] \rangle$	$\langle (\frac{1}{8}, \frac{1}{7}, \frac{1}{6}); 0.90, [0.85, 0.95] \rangle$	0.10
Foodnerd	$\langle (1,1,1); 0.50, [0.50, 0.50] \rangle$	$\langle (6,7,8); 0.90, [0.85, 0.95] \rangle$	$\langle (1,1,1); 0.50, [0.50, 0.50] \rangle$	0.58

Table 14 Cubic pair-wise comparison matrix w.r.t S_8

Catering firms	Folkdine	Chiniot	Foodnerd	Priority vector
Folkdine	$\langle (1,1,1); 0.50, [0.50, 0.50] \rangle$	$\langle (9,9,9); 1.00, [1.00, 1.00] \rangle$	$\langle (6,7,8); 0.90, [0.85, 0.95] \rangle$	0.80
Chiniot	$\langle (\frac{1}{9}, \frac{1}{9}, \frac{1}{9}); 1.00, [1.00, 1.00] \rangle$	$\langle (1,1,1); 0.50, [0.50, 0.50] \rangle$	$\langle (1,1,1); 0.50, [0.50, 0.50] \rangle$	0.10
Foodnerd	$\langle (\frac{1}{8}, \frac{1}{7}, \frac{1}{6}); 0.90, [0.85, 0.95] \rangle$	$\langle (1,1,1); 0.50, [0.50, 0.50] \rangle$	$\langle (1,1,1); 0.50, [0.50, 0.50] \rangle$	0.10

Table 15 Cubic pair-wise comparison matrix w.r.t S_9

Catering firms	Folkdine	Chiniot	Foodnerd	Priority vector
Folkdine	$\langle (1,1,1); 0.50, [0.50, 0.50] \rangle$	$\langle (4,5,6); 0.80, [0.75, 0.85] \rangle$	$\langle (\frac{1}{8}, \frac{1}{7}, \frac{1}{6}); 0.90, [0.85, 0.95] \rangle$	0.25
Chiniot	$\langle (\frac{1}{6}, \frac{1}{5}, \frac{1}{4}); 0.80, [0.75, 0.85] \rangle$	$\langle (1,1,1); 0.50, [0.50, 0.50] \rangle$	$\langle (9,9,9); 1.00, [1.00, 1.00] \rangle$	0.42
Foodnerd	$\langle (6,7,8); 0.90, [0.85, 0.95] \rangle$	$\langle (\frac{1}{9}, \frac{1}{9}, \frac{1}{9}); 1.00, [1.00, 1.00] \rangle$	$\langle (1,1,1); 0.50, [0.50, 0.50] \rangle$	0.33

6 Conclusion and Future Work

A cubic set is defined as the development of the extension of classical sets, fuzzy sets and interval value fuzzy sets. The membership function is the form of interval taken in order to improve the accuracy. In this case, real-world problems are well depicted as all the aspects of a decision are taken into account. In the proposed model

Table 16 Cubic pair-wise comparison matrix w.r.t S_{10}

Catering firms	Folkdine	Chiniot	Foodnerd	Priority vector
Folkdine	$\langle (1,1,1); 0.50, [0.50, 0.50] \rangle$	$\langle (9,9,9); 1.00, [1.00, 1.00] \rangle$	$\langle (2,3,4); 0.30, [0.25, 0.35] \rangle$	0.70
Chiniot	$\langle (\frac{1}{9}, \frac{1}{9}, \frac{1}{9}); 1.00, [1.00, 1.00] \rangle$	$\langle (1,1,1); 0.50, [0.50, 0.50] \rangle$	$\langle (2,3,4); 0.30, [0.25, 0.35] \rangle$	0.20
Foodnerd	$\langle (\frac{1}{4}, \frac{1}{3}, \frac{1}{2}); 0.30, [0.25, 0.35] \rangle$	$\langle (\frac{1}{4}, \frac{1}{3}, \frac{1}{2}); 0.30, [0.25, 0.35] \rangle$	$\langle (1,1,1); 0.50, [0.50, 0.50] \rangle$	0.10

Table 17 Cubic pair-wise comparison matrix w.r.t S_{11}

Catering firms	Folkdine	Chiniot	Foodnerd	Priority vector
Folkdine	$\langle (1,1,1); 0.50, [0.50, 0.50] \rangle$	$\langle (1,1,1); 0.50, [0.50, 0.50] \rangle$	$\langle (\frac{1}{9}, \frac{1}{9}, \frac{1}{9}); 1.00, [1.00, 1.00] \rangle$	0.10
Chiniot	$\langle (1,1,1); 0.50, [0.50, 0.50] \rangle$	$\langle (1,1,1); 0.50, [0.50, 0.50] \rangle$	$\langle (\frac{1}{8}, \frac{1}{7}, \frac{1}{6}); 0.90, [0.85, 0.95] \rangle$	0.10
Foodnerd	$\langle (9,9,9); 1.00, [1.00, 1.00] \rangle$	$\langle (6,7,8); 0.90, [0.85, 0.95] \rangle$	$\langle (1,1,1); 0.50, [0.50, 0.50] \rangle$	0.80

Table 18 Local priority vector w.r.t C_1

	S_1	S_2	S_3	Local weights
Weight	0.65	0.25	0.10	
Folkdine	0.43	0.22	0.23	0.37
Chiniot	0.33	0.09	0.08	0.26
Foodnerd	0.24	0.73	0.70	0.38

Table 19 Local priority vector w.r.t C_2

	S_4	S_5	S_6	S_7	Local weights
Weight	0.28	0.22	0.47	0.03	
Folkdine	0.26	0.72	0.06	0.33	0.27
Chiniot	0.66	0.15	0.39	0.11	0.40
Foodnerd	0.11	0.16	0.58	0.59	0.36

the cubic set played the role of the symmetric tool; the variability of the decision-maker is measured by the scale of pair-wise comparison with the triangular cubic number. Thus, considering the fact that the CAHP is rather effective in providing the

Table 20 Local priority vector w.r.t C_3

	S_8	S_9	S_{10}	S_{11}	Local weights
Weight	0.19	0.49	0.03	0.29	
Folkdine	0.81	0.26	0.71	0.11	0.33
Chiniot	0.11	0.43	0.21	0.11	0.27
Foodnerd	0.11	0.34	0.11	0.81	0.43

Table 21 Global weight

	C_1	C_2	C_3	Global weights	Local weights
Weight	0.20	0.10	0.70		
Folkdine	0.37	0.27	0.33	0.34	2nd
Chiniot	0.26	0.40	0.27	0.28	3rd
Foodnerd	0.38	0.36	0.43	0.41	1st

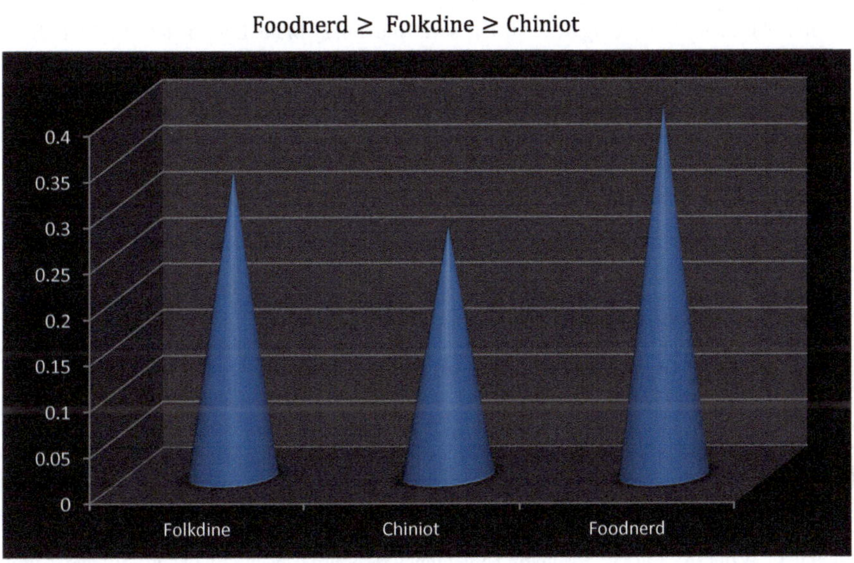

Fig. 4 The final ranking of alternatives

Table 22 Comparison of our method with other existing methods

Methods	Final ranking
AHP [4]	Folkdine $\geq$ Chiniont $\geq$ Foodnerd
FAHP [9]	Chiniont $\geq$ Folkdine $\geq$ Foodnerd
Our proposed method	Foodnerd $\geq$ Chiniont $\geq$ Folkdine

description of vagueness and uncertainty, we will arrive at a more accurate definition of the decision-making problems. A few simple operations with triangular cubic numbers are employed to sum up the priority weights from the lowest to the top level of the developed hierarchical structure, and utilizing the final general precedence weights, the discussed options are compared. An example has also been given to help explain the computational method of the CAHP proposed in this paper. In the future, we plan to apply the Cubic Analytic Hierarchy Process in the context of cubic theory which exists in real-life situations, to provide information in terms of vagueness, instability and uncertainty and employ potent tools to define and encompass complex problems.

References

1. Zadeh, L.A.: Fuzzy sets. Inf. Control. **8**, 338–353 (1965)
2. Jun, Y.B., Kim, C.S., Yang, K.O.: Cubic sets. Ann. Fuzzy Math. Inform. **4**(1), 83–98 (2012)
3. Fahmi, A., Abdullah, S., Amin, F., Ali, A.: Weighted average rating (WAR) method for solving group decision making problem using triangular cubic fuzzy hybrid aggregation (TCFHA). Punjab Univ. J. Math. **50**, 23–34 (2018)
4. Saaty, T.L.: The Analytic Hierarchy Process. McGraw-Hill International, New York, NY, U.S.A (1980)
5. Saaty, T.L.: A scaling method for priorities in hierarchical structures. J. Math. Psychology **15**, 57–68 (1977)
6. Saaty, T.L.: How to make a decision: the analytic hierarchy process, European. J Oper. Res. **48**, 9–26 (1990)
7. Saaty, T.L.: An exposition of the AHP in reply to the paper, Remarks on the analytic hierarchy process. Manage. Sci. **36**, 259–268 (1990)
8. Van Laarhoven, P.J.M., Pedrycz, W.: A fuzzy extension of Saaty.spriority theory. Fuzzy Sets Syst. **11**, 229–241 (1983)
9. Buckley, J.J.: Fuzzy hierarchical analysis. Fuzzy Sets Syst. **17**, 233–247 (1985)
10. Chang, D.Y.: Applications of the extent analysis method on fuzzy AHP. Eur. J. Oper. Res. **95**, 649–655 (1996)
11. Rezaei, J., Fahim, P.B.M., Tavasszy, L.: Supplier selection in the airline retail industry using a funnel methodology: conjunctive screeningmethod and fuzzy AHP. Expert Syst. Appl. **41**, 65–79 (2014)
12. Ju, Y., Wang, A., Liu, X.: Evaluating emergency response capacity by fuzzy AHP and 2-tuple fuzzy linguistic approach. Expert Syst. Appl. **39**, 72–81 (2012)
13. Duran, O., Aguilo, J.: Computer-aided machine-tool selection based on a Fuzzy-AHP approach. Expert Syst. Appl. **34**, 87–94 (2008)
14. Chang, C.W., et al.: Using expert technology to select unstable slicing machine to control wafer slicing quality via fuzzyAHP. Expert Syst. Appl. **34**, 10–20 (2008)
15. Cebeci, U.: Fuzzy AHP-based decision support system for selecting ERP systems in textile industry by using balanced score card, Expert Syst Appl. 36, 9 (2009)
16. Hsu, Y.L., et al.: The application of Fuzzy Delphi Method and Fuzzy AHP in lubricant regenerative technology selection. Expert Syst. Appl. **37**, 19–25 (2010)
17. Kulak, O., Kahraman, C.: Fuzzy multi-attribute selection among transportation companies using axiomatic design and analytic hierarchy process. Inf. Sci. **170**, 191–210 (2005)
18. Lo, Y.F., Wen, M.H.: A fuzzy Ahp-based technique for the decision of design feature selection in massively multiplayer online role play in game development. Expert Syst. Appl. **37**, 85–93 (2010)

19. Sahoo, L., Senapati, T., Yager, R.R. (Eds.): Real life applications of multiple criteria decision making techniques in fuzzy domain. Studies Fuzziness Soft Comput. (STUDFUZZ, volume 420). Springer Nature (2023)
20. Ying, H., Gulistan, M., Asif, M. et al.: Some aggregation operators based on dombi t-norm (TN) and t-co-norm (TCN) operations: applications in economic corridor prospective. Int. J. Fuzzy Syst. (2024). https://doi.org/10.1007/s40815-024-01702-6
21. Beg, I., Gulistan, M., Asif, M.: Neutrosophic-cubic analytic hierarchy process with applications. Infinite Study. **14**(2), 1–14 (2020)

Decision-Making Process Under Uncertain Domain of Pythagorean Fuzzy Sets Based on an Enhanced Similarity Operator

Paul Augustine Ejegwa

Abstract Similarity operator is a formidable tool for decision-making under uncertain domain like Pythagorean fuzzy set (PFS). PFS is a generalized intuitionistic fuzzy set (IFS) with better accuracy in real-world applications. Many of the discussions bordering on the applications of PFSs were carried out based on similarity operators (SOs). Several approaches of computing similarity between PFSs have been studied. Among the extant SOs, the work of Zhang et al. is deemed to contain some flaws which needed to be enhanced for reliable interpretation. To this end, this chapter explicates the SOs of Zhang et al. and develops an enhanced SO, which appropriately satisfies the similarity conditions and yields a more robust results in comparison to the SOs of Zhang et al. In order to validate the enhanced SO, we discuss its properties and find out that the SO is well applicable in decision-making problems. In addition, the enhanced SO and the SOs of Zhang et al. are compared in the context of precision, and it is substantiated that the enhanced SO can successfully measure the similarity between vastly related but inconsistent PFSs and as well yields a very precise results. Lastly, the enhanced SO is applied to decision-making problem and clustering analysis, and it is certified that the enhanced SO can deal with diverse everyday application problems more precisely than the SOs of Zhang et al.

Keywords Decision-making under uncertainty · Pythagorean fuzzy sets · Similarity operators · Intuitionistic fuzzy set

1 Introduction

Decision-makers frequently encounter uncertainty and imprecision in the process of decision-making (DM). Because of this, fuzzy set (FS) [1] was invented to reduce uncertainty. The idea of FS is described by a function called the membership degree (MD) which mapped a classical set to an interval [0, 1]. However, FS is unable to

P. A. Ejegwa (✉)
Department of Mathematics, Joseph Sarwuan Tarka University, Makurdi, Nigeria
e-mail: ejegwa.augustine@uam.edu.ng

L. Sahoo et al. (eds.), *Decision Making Under Uncertainty Via Optimization, Modelling, and Analysis*, Studies in Systems, Decision and Control 558,
https://doi.org/10.1007/978-981-96-0085-4_11

address the imprecision involved in DM. Intuitionistic fuzzy set (IFS) was created to address the issue of imprecision [2], and it has been extensively used to address DM problems in the real world. The conspicuous distinguishing feature of IFSs is that, the sum of the MD and non-membership degree (ND) cannot be greater than one. IFS has been used to tackle a number of useful issues in real-world scenarios by employing distance operators [3–5], aggregation operators [6], and partial correlation coefficient operators [7]. The process of finding the similarity index between two sets is carried out by the similarity metric, which is an essential research component in FS and its generalizations. Many methods of comparing the similarities between IFS have been created and successfully used in various fields, including pattern recognition [8], disaster management [9], medical diagnostic issues [10, 11], and various DM issues [12]. However, IFSs cannot resolve some real-world issues. For example, IFS cannot be used to model problems if the decision-maker's MD and ND are 0.7 and 0.5, because their sum exceeds 1.

The phrase "IFS of type 2," or Pythagorean fuzzy set (PFS), was created by expanding the spatial scope of IFS [13, 14]. In PFS, the square sum of MD and ND is 1, even though their sum may be greater than 1. PFS is an IFS special case that can handle certain situations where IFS is not allowed. Many applications of PFSs have been discussed. Several aggregation operators, such as Einstein operators, interactive power averaging operators, geometric aggregation operators employing Einstein t-norm, and t-conorm [15–17], have been discussed under PFSs to exemplify specific DM problems. Similarly, PFSs can be used to solve DM problems based on distance operators [18–25] and correlation coefficient operators [26–29]. Many practical applications of multiple criteria decision-making (MCDM) approaches in fuzzy domain have been explored [30]. In recent times, football analysis and disaster control have been discussed through Pythagorean fuzzy data [31, 32].

Because the scope of applications of PFSs is wider compared to IFSs, applying SOs to PFSs is crucial for practical applications of PFSs. Several SOs between PFSs were developed by adapting comparable SOs under IFSs. The axiomatic delineation of SO on PFS was introduced by Ejegwa [33]. Since both distance operator and similarity operator have dual characteristics, Zeng et al. [34] provided some techniques which utilized distance operators to create similarity operators between PFSs. Peng et al. [35] created a few SOs for PFSs and used them to discuss pattern recognition, medical diagnosis, and clustering analysis. Zhang [36] created a SO on PFSs and used it to discuss MCDM problems. With applications in pattern recognition and medical diagnosis, Wei and Wei [37] created a number of SOs on PFSs based on the cosine function. Four SOs between PFSs were recently developed by Zhang et al. [38] and used to discuss pattern recognition problems. The Pythagorean fuzzy hesitation margin is not taken into account in the first two SOs in [38], but the other two SOs take into account all of the PFSs' parameters in order to prevent any errors that might result from exclusion. Howbeit, whenever the PFSs are identical, the methods produce an identical similarity value other than one, which violates the similarity metric axioms and makes the methods unreliable.

The goal of this work is to provide a new SO between PFSs that results from the hybridization of the four SOs in [38], in order to correct the limitations of the four

SOs. For emphasis, the following are the drawbacks of the SOs in [38]: (1) if the PFSs are identical, they violated the SO's metric conditions and (2) they produce similarity values that are not defined within the similarity value range and therefore have no practical meaning. In order to achieve this, the work of Zhang et al. [38] is corrected by a hybridized SO that is proposed in this work. It is verified through comparative examples that the new SO is able to successfully solve the aforementioned setbacks.

The following is how the work is arranged: In Sect. 2, some of the features of PFSs are reviewed and the shortcomings of the SOs developed by Zhang et al. are discussed. Section 3 presents a new SO that corrects the SOs introduced by Zhang et al. and discusses its features, Sect. 4 applies the new SO to the analysis of a decision-making problem involving the selection of a house for purchase and demonstrates the new SO's superiority through comparison analysis, Sect. 5 applies the new SO to clustering analysis, and the conclusion of the chapter with suggestions for more research is presented in Sect. 6.

2 Preliminaries

Some PFS properties and the similarity functions proposed by Zhang et al. are covered in this section. Assume $S = \{s_1, s_2, \cdots, s_k\}$ to be the discourse universe. Then, we reiterate the following definitions:

Definition 1 ([2]) An IFS represented by ℓ in S is defined by

$$\ell = \{\big(s_j, M_\ell(s_j), N_\ell(s_j)\big) : s_j \in S\},$$

where $0 \leq M_\ell(s_j) + N_\ell(s_j) \leq 1$ and $M_\ell\colon S \to [0, 1]$ and $N_\ell\colon S \to [0, 1]$ are functions that describe the MD and NMD for every $s_j \in S$ for $j = 1, 2, \cdots, k$. In addition, the hesitation margin (HM) of ℓ in S is defined by $H_\ell(s_j) = 1 - M_\ell(s_j) - N_\ell(s_j)$.

By enlarging the scope of IFS, the concept of PFS is thus defined:

Definition 2 ([14]) A PFS represented by $\wp$ in S is defined by

$$\wp = \{\big(s_j, M_\wp(s_j), N_\wp(s_j)\big) : s_j \in S\},$$

where $0 \leq M^2_\wp(s_j) + N^2_\wp(s_j) \leq 1$ and $M_\wp\colon S \to [0, 1]$ and $N_\wp\colon S \to [0, 1]$ are functions that describe the MD and NMD for every $s_j \in S$ for $j = 1, 2, \cdots, k$. In addition, the HM of $\wp$ in S is defined by $H_\wp(s_j) = 1 - \big(M^2_\wp(s_j) - N^2_\wp(s_j)\big)^{0.5}$.

Now, we present some operations on PFSs as follows:

Definition 3 ([14]) If $\wp$, $\wp_1$ and $\wp_2$ are PFSs in S, then

(i) $\wp_1 \preceq \wp_2$ iff $M_{\wp_1}(s_j) \preceq M_{\wp_2}(s_j)$ and $N_{\wp_1}(s_j) \preceq N_{\wp_2}(s_j)$ $\forall s_j \in S$.
(ii) $\wp_1 = \wp_2$ iff $M_{\wp_1}(s_j) = M_{\wp_2}(s_j)$ and $N_{\wp_1}(s_j) = N_{\wp_2}(s_j)$ $\forall s_j \in S$.
(iii) $\wp_1 \subseteq \wp_2$ iff $M_{\wp_1}(s_j) \leq M_{\wp_2}(s_j)$ and $N_{\wp_1}(s_j) \geq N_{\wp_2}(s_j)$ $\forall s_j \in S$.
(iv) $\overline{\wp} = \{(s_j, N_{\wp}(s_j), M_{\wp}(s_j)) : s_j \in S\}$.
(v) $\wp_1 \cap \wp_2 = \{(s_j, \min\{M_{\wp_1}(s_j), M_{\wp_2}(s_j)\}, \max\{N_{\wp_1}(s_j), N_{\wp_2}(s_j)\}) : s_j \in S\}$.
(vi) $\wp_1 \cup \wp_2 = \{(s_j, \max\{M_{\wp_1}(s_j), N_{\wp_2}(s_j)\}, \min\{N_{\wp_1}(s_j), N_{\wp_2}(s_j)\}) : s_j \in S\}$.

Next, we present the concept of similarity operator on PFSs. The similarity measure between PFSs is one way to estimate the degree of similarity of the PFSs.

Definition 4 ([19]) If $\wp$, $\wp_1$, and $\wp_2$ are PFSs in $S = \{s_1, s_2, \cdots, s_k\}$, then the similarity metric between $\wp_1$ and $\wp_2$ represented by $\nabla(\wp_1, \wp_2)$ is a function, $\nabla\colon PFS \times PFS \to [0, 1]$ such that

(i) $0 \leq \nabla(\wp_1, \wp_2) \leq 1$,
(ii) $\nabla(\wp_1, \wp_1) = 1$, $\nabla(\wp_2, \wp_2) = 1$,
(iii) $\nabla(\wp_1, \wp_2) = 1 \Leftrightarrow \wp_1 = \wp_2$,
(iv) $\nabla(\wp_1, \wp_2) = \nabla(\wp_2, \wp_1)$,
(v) $\nabla(\wp_1, \wp) \leq \nabla(\wp_1, \wp_2) + \nabla(\wp_2, \wp)$.

In addition, $\nabla(\wp_1, \wp_2) \approx 1$ indicates that there is a high degree of similarity between $\wp_1$ and $\wp_2$, and $\nabla(\wp_1, \wp_2) \approx 0$ implies a very negligible similarity exists between $\wp_1$ and $\wp_2$.

Definition 5 Suppose

$$\wp_1 = \{(s_1, M_{\wp_1}(s_1), N_{\wp_1}(s_1)), \cdots, (s_k, M_{\wp_1}(s_k), N_{\wp_1}(s_k))\} \text{ and}$$

$$\wp_2 = \{(s_1, M_{\wp_2}(s_1), N_{\wp_2}(s_1)), \cdots, (s_k, M_{\wp_2}(s_k), N_{\wp_2}(s_k))\}$$

are PFSs in $S = \{s_1, \cdots, s_k\}$. Then, the average PFS (APFS) of $\wp_1$ and $\wp_2$ denoted by $\wp_{1,2}$ is a PFS given by

$$\wp_{1,2} = \{(s_1, M_{\wp_{1,2}}(s_1), N_{\wp_{1,2}}(s_1)), \cdots, (s_k, M_{\wp_{1,2}}(s_k), N_{\wp_{1,2}}(s_k))\},$$

where

$$\begin{aligned}
M_{\wp_{1,2}}(s_1) &= \frac{M_{\wp_1}(s_1) + M_{\wp_2}(s_1)}{2}, \\
N_{\wp_{1,2}}(s_1) &= \frac{N_{\wp_1}(s_1) + N_{\wp_2}(s_1)}{2}, \\
H_{\wp_{1,2}}(s_1) &= \frac{H_{\wp_1}(s_1) + H_{\wp_2}(s_1)}{2}, \\
\vdots \quad & \quad \vdots \qquad \vdots \qquad \vdots \\
M_{\wp_{1,2}}(s_k) &= \frac{M_{\wp_1}(s_k) + M_{\wp_2}(s_k)}{2},
\end{aligned}$$

$$N_{\wp_{1,2}}(s_k) = \frac{N_{\wp_1}(s_k) + N_{\wp_2}(s_k)}{2},$$
$$H_{\wp_{1,2}}(s_k) = \frac{H_{\wp_1}(s_k) + H_{\wp_2}(s_k)}{2}.$$

2.1 SOs of Zhang et al. Under PFSs

Zhang et al. [38] created four SOs that utilized exponential functions in their search for a trustworthy SO. Suppose

$$\wp = \{(s_1, M_\wp(s_1), N_\wp(s_1)), \cdots, (s_k, M_\wp(s_k), N_\wp(s_k))\} \text{ and}$$

$$\ell = \{(s_1, M_\ell(s_1), N_\ell(s_1)), \cdots, (s_k, M_\ell(s_k), N_\ell(s_k))\}$$

are PFSs in $S = \{s_1, \cdots, s_k\}$, then, the four exponential-based SOs developed in [38] are listed as follows:

$$\nabla_1(\wp, \ell) = \frac{1}{k}\Sigma_{j=1}^k\left[2^{1-\max\{|M_\wp^2(s_j)-M_\ell^2(s_j)|, |N_\wp^2(s_j)-N_\ell^2(s_j)|\}} - 1\right], \tag{1}$$

$$\nabla_2(\wp, \ell) = \frac{1}{k}\Sigma_{j=1}^k\left[2^{1-\frac{|M_\wp^2(s_j)-M_\ell^2(s_j)|+|N_\wp^2(s_j)-N_\ell^2(s_j)|}{2}} - 1\right], \tag{2}$$

$$\nabla_3(\wp, \ell) = \frac{1}{k}\Sigma_{j=1}^k\left[2^{1-\max\{|M_\wp^2(s_j)-M_\ell^2(s_j)|, |N_\wp^2(s_j)-N_\ell^2(s_j)|, |H_\wp^2(s_j)-H_\ell^2(s_j)|\}} - 1\right], \tag{3}$$

$$\nabla_4(\wp, \ell) = \frac{1}{k}\Sigma_{j=1}^k\left[2^{1-\frac{|M_\wp^2(s_j)-M_\ell^2(s_j)|+|N_\wp^2(s_j)-N_\ell^2(s_j)|+|H_\wp^2(s_j)-H_\ell^2(s_j)|}{2}} - 1\right]. \tag{4}$$

Because the hesitation margins are not included by the operators in (1) and (2), the SOs cannot be trusted to produce useful outcomes. However, by adding hesitation margins to (1) and (2), (3) and (4) improved (1) and (2) to produce trustworthy results. However, for smaller k and as the hesitation margins get smaller, all of these SOs will produce results that are comparable.

Assume that $\wp = \ell$ for any arbitrary k say $k = 5$, then applying (1)–(4), we obtain

$$\left.\begin{aligned}\nabla_1(\wp,\ell)&=\frac{2^{1-0}-1}{5}=0.2\\ \nabla_2(\wp,\ell)&=\frac{2^{1-0}-1}{5}=0.2\\ \nabla_3(\wp,\ell)&=\frac{2^{1-0}-1}{5}=0.2\\ \nabla_4(\wp,\ell)&=\frac{2^{1-0}-1}{5}=0.2\end{aligned}\right\},$$

which do not adhere to the similarity criterion, that says, $\nabla(\wp,\ell)=1\Leftrightarrow\wp=\ell$. This explains why the SOs must be corrected in order to meet the metric requirements. Furthermore, these SOs occasionally generate outputs that are not defined in the range $0\leq\nabla(\wp,\ell)\leq 1$. We illustrate this restriction with the help of the following example.

Example 2.1 Assume

$$\wp_1=\{(s_1,1,0),(s_2,0.8,0),(s_3,0.7,0.1)\},$$

$$\wp_2=\{(s_1,0.8,0.1),(s_2,1,0),(s_3,0.9,0.1)\},$$

$$\wp_3=\{(s_1,0.6,0.2),(s_2,0.8,0),(s_3,1,0)\}$$

are PFSs in $S=\{s_1,s_2,s_3\}$. In case, there is another PFS defined by

$$\wp=\{(s_1,0.5,0.3),(s_2,0.8,0.2),(s_3,1,0)\}.$$

The similarities between $\wp$ with each of $\wp_1$, $\wp_2$, and $\wp_3$, respectively, are then calculated using the SOs, and the following outcomes are obtained:

$$\left.\begin{aligned}\nabla_1(\wp_j,\wp)&=-0.0626,0.0142,0.2675\\ \nabla_2(\wp_j,\wp)&=0.077,0.1268,0.2887\\ \nabla_3(\wp_j,\wp)&=-0.055,0.0142,0.2844\\ \nabla_4(\wp_j,\wp)&=-0.0626,0.0142,0.2675\end{aligned}\right\},$$

for $j=1,2,3$.

3 Modification of SOs of Zhang et al. Under PFSs

Since SOs of Zhang et al. have been shown to be flawed, the SOs must be modified by creating a new SO that meets the requirements of the similarity metric and produces accurate and dependable results.

Definition 6 Assume $\wp = \{(s_j, M_\wp(s_j), N_\wp(s_j)) : s_j \in S\}$ and $\ell = \{(s_j, M_\ell(s_j), N_\ell(s_j)) : s_j \in S\}$ are PFSs for $S = \{s_1, \cdots, s_k\}$, then the new SO between $\wp$ and ℓ is defined by

$$\tilde{\nabla}(\wp, \ell) = \sum_{j=i}^{k} \left[2^{1-\frac{1}{3k}\left(\left|M_\wp^2(s_j)-M_\ell^2(s_j)\right|+\left|N_\wp^2(s_j)-N_\ell^2(s_j)\right|+\left|H_\wp^2(s_j)-H_\ell^2(s_j)\right|\right)} - 1\right]. \quad (5)$$

Suppose $\wp = \ell$, then applying (5) we obtain $\tilde{\nabla}(\wp, \ell) = \sum_{j=i}^{k} \left[2^{1-0} - 1\right] = 1$, which satisfies $\tilde{\nabla}(\wp, \ell) = 1 \Leftrightarrow \wp = \ell$.

In addition, the modified SO produces results that are within $0 \le \nabla(\wp, \ell) \le 1$. From Example 2.1, we get

$$\tilde{\nabla}(\wp_j, \wp) = 0.6371, 0.7304, 0.7956,$$

for $j = 1, 2, 3$. By comparison, these results are better than that from SOs of Zhang et al. Next, we consider some properties of the new SO.

Theorem 1 *Suppose $\wp$ and ℓ are PFSs in $S = \{s_1, s_2, \cdots, s_k\}$, then*

(i) $0 \le \tilde{\nabla}(\wp, \ell) \le 1$,
(ii) $\tilde{\nabla}(\wp, \ell) = 1 \Leftrightarrow \wp = \ell$,
(iii) $\tilde{\nabla}(\wp, \ell) = \tilde{\nabla}(\ell, \wp)$,
(iv) $\tilde{\nabla}(\wp, \ell) = \tilde{\nabla}(\overline{\wp}, \overline{\ell})$.

Proof To show $0 \le \tilde{\nabla}(\wp, \ell) \le 1$, it is sufficient to prove that $\tilde{\nabla}(\wp, \ell) \le 1$ since $\tilde{\nabla}(\wp, \ell) \ge 0$ is straightforward.

Assume that $y = \tilde{\nabla}(\wp, \ell)$ and $x = \frac{1}{3}\sum_{j=1}^{k}\left(\left|M_\wp^2(s_j) - M_\ell^2(s_j)\right| + \left|N_\wp^2(s_j) - N_\ell^2(s_j)\right| + \left|H_\wp^2(s_j) - H_\ell^2(s_j)\right|\right)$, where $x \in [0, 1]$. Then, $y = 2^{1-x} - 1$ is a curve function with values range from 0 to 1. Thus, $0 \le y \le 1$, and hence $0 \le \tilde{\nabla}(\wp, \ell) \le 1$ as expected, i.e., (i) holds.

Next, we prove (ii). Assume $\tilde{\nabla}(\wp, \ell) = 1$. Then,

$$2^{1-\frac{1}{3k}\left(\left|M_\wp^2(s_j)-M_\ell^2(s_j)\right|+\left|N_\wp^2(s_j)-N_\ell^2(s_j)\right|+\left|H_\wp^2(s_j)-H_\ell^2(s_j)\right|\right)} - 1 = 1 \Longrightarrow$$

$$2^{1-\frac{1}{3k}\left(\left|M_\wp^2(s_j)-M_\ell^2(s_j)\right|+\left|N_\wp^2(s_j)-N_\ell^2(s_j)\right|+\left|H_\wp^2(s_j)-H_\ell^2(s_j)\right|\right)} = 2 \Longrightarrow$$

$$1 - \frac{1}{3k}\left(\left|M_\wp^2(s_j) - M_\ell^2(s_j)\right| + \left|N_\wp^2(s_j) - N_\ell^2(s_j)\right| + \left|H_\wp^2(s_j) - H_\ell^2(s_j)\right|\right) = 1 \Longrightarrow$$

$$\frac{1}{3k}\left(\left|M_\wp^2(s_j) - M_\ell^2(s_j)\right| + \left|N_\wp^2(s_j) - N_\ell^2(s_j)\right| + \left|H_\wp^2(s_j) - H_\ell^2(s_j)\right|\right) = 0 \Longrightarrow$$

$$\left|M_\wp^2(s_j) - M_\ell^2(s_j)\right| + \left|N_\wp^2(s_j) - N_\ell^2(s_j)\right| + \left|H_\wp^2(s_j) - H_\ell^2(s_j)\right| = 0 \Longrightarrow$$

$$M_{\wp}(s_j) = M_{\ell}(s_j), N_{\wp}(s_j) = N_{\ell}(s_j), H_{\wp}(s_j) = H_{\ell}(s_j).$$

Hence, $\wp = \ell$.

Conversely, if $\wp = \ell$. Then, $\left|M^2_{\wp}(s_j) - M^2_{\ell}(s_j)\right| = 0$, $\left|N^2_{\wp}(s_j) - N^2_{\ell}(s_j)\right| = 0$, and $\left|H^2_{\wp}(s_j) - H^2_{\ell}(s_j)\right| = 0$. Thus,

$$\frac{1}{3k}\left(\left|M^2_{\wp}(s_j) - M^2_{\ell}(s_j)\right| + \left|N^2_{\wp}(s_j) - N^2_{\ell}(s_j)\right| + \left|H^2_{\wp}(s_j) - H^2_{\ell}(s_j)\right|\right) = 0,$$

and hence $\tilde{\nabla}(\wp, \ell) = 1$.

The proof of (iii) follows because

$$\begin{aligned}\tilde{\nabla}(\wp, \ell) &= \sum_{j=i}^{k}\left[2^{1-\frac{1}{3k}\left(\left|M^2_{\wp}(s_j)-M^2_{\ell}(s_j)\right|+\left|N^2_{\wp}(s_j)-N^2_{\ell}(s_j)\right|+\left|H^2_{\wp}(s_j)-H^2_{\ell}(s_j)\right|\right)} - 1\right] \\ &= \sum_{j=i}^{k}\left[2^{1-\frac{1}{3k}\left(\left|M^2_{\ell}(s_j)-M^2_{\wp}(s_j)\right|+\left|N^2_{\ell}(s_j)-N^2_{\wp}(s_j)\right|+\left|H^2_{\ell}(s_j)-H^2_{\wp}(s_j)\right|\right)} - 1\right] \\ &= \tilde{\nabla}(\ell, \wp).\end{aligned}$$

Similarly, (iv) holds. □

Theorem 2 *Suppose $\wp_1$, $\wp_2$, and $\wp_3$ are PFSs in $S = \{s_1, s_2, \cdots, s_k\}$ and $\wp_1 \subseteq \wp_2 \subseteq \wp_3$. Then $\tilde{\nabla}(\wp_1, \wp_3) \leq \min\left\{\tilde{\nabla}(\wp_1, \wp_2), \tilde{\nabla}(\wp_2, \wp_3)\right\}$.*

Proof To establish the proof, we need to show the following:

(i) $\tilde{\nabla}(\wp_1, \wp_3) \leq \tilde{\nabla}(\wp_1, \wp_2)$,
(ii) $\tilde{\nabla}(\wp_1, \wp_3) \leq \tilde{\nabla}(\wp_2, \wp_3)$.

Given that $\wp_1 \subseteq \wp_2 \subseteq \wp_3$, we have $M_{\wp_1}(s_j) \leq M_{\wp_2}(s_j) \leq M_{\wp_3}(s_j)$ and $N_{\wp_1}(s_j) \geq N_{\wp_2}(s_j) \geq N_{\wp_3}(s_j)\ \forall s_j \in S$. Thus,

$$\begin{aligned}\left|M^2_{\wp_1}(s_j) - M^2_{\wp_3}(s_j)\right| &\geq \left|M^2_{\wp_1}(s_j) - M^2_{\wp_2}(s_j)\right|, \\ \left|N^2_{\wp_1}(s_j) - N^2_{\wp_3}(s_j)\right| &\geq \left|N^2_{\wp_1}(s_j) - N^2_{\wp_2}(s_j)\right|, \\ \left|H^2_{\wp_1}(s_j) - H^2_{\wp_3}(s_j)\right| &\geq \left|H^2_{\wp_1}(s_j) - H^2_{\wp_2}(s_j)\right|,\end{aligned}$$

such that

$$\begin{aligned}&\left|M^2_{\wp_1}(s_j) - M^2_{\wp_3}(s_j)\right| + \left|N^2_{\wp_1}(s_j) - N^2_{\wp_3}(s_j)\right| \\ &+\left|H^2_{\wp_1}(s_j) - H^2_{\wp_3}(s_j)\right| \geq \left|M^2_{\wp_1}(s_j) - M^2_{\wp_2}(s_j)\right| \\ &+\left|N^2_{\wp_1}(s_j) - N^2_{\wp_2}(s_j)\right| + \left|H^2_{\wp_1}(s_j) - H^2_{\wp_2}(s_j)\right|.\end{aligned}$$

Clearly, $\tilde{\nabla}(\wp_1, \wp_3) \leq \tilde{\nabla}(\wp_1, \wp_2)$ which proves (i). By using the same approach, the proofs of (ii) holds. Hence,

$$\tilde{\nabla}(\wp_1, \wp_3) \leq \min\left\{\tilde{\nabla}(\wp_1, \wp_2), \tilde{\nabla}(\wp_2, \wp_3)\right\}.$$

□

4 Decision-Making via SOs

This section presents the application of the discussed SOs in DM. Here, a new approach is presented to make a suitable decision by using all the available alternatives and attributes based on SOs. Assume $A = \{a_1, \cdots, a_k\}$ is the set of attributes and

$$\wp_{jk} = \{(a_k, M_{\wp_j}(a_k), N_{\wp_j}(a_k))|a_k \in A\}, j = 1, \cdots, n, k = 1, \cdots, m$$

is the set of alternatives.
Aim: To select the best alternative (BA).
We make use of the following algorithm to determine the best alternative.
Step 1. Compute $\tilde{\nabla}(\wp_j, \wp)$ $(j = 1, \cdots, n)$, where $\wp = \{(1, 0), (1, 0), (1, 0), (1, 0)\}$ is the best alternative with m attributes.
Step 2. Find $\tilde{\nabla}^*(\wp_j, \wp) = \max_{1 \leq j \leq n}\{\tilde{\nabla}(\wp_j, \wp)\}$.
Step 3. Compute DoC (degree of confidence) using $\text{DoC} = \sum_{j=1}^{n} \left|\tilde{\nabla}^*(\wp_j, \wp) - \tilde{\nabla}(\wp_j, \wp)\right|$. The bigger the DoC, the better the SO.

Example 4.1 Assume a person wants to make a selection to buy a house from six available houses $H = \{H_1, H_2, H_3, H_4, H_5, H_6\}$. The selection criteria are property insurance (a_1), genuineness of documents (a_2), property location (a_3), and security (a_4). Table 1 shows the outcome of the selection process in Pythagorean fuzzy values.

By implementing Steps 2 and 3, we obtain the results in Tables 2 and 3, respectively.

Table 1 Six alternatives with four different attributes

H	a_1	a_2	a_3	a_4
H_1	(0.61, 0.54)	(0.59, 0.50)	(0.41, 0.66)	(0.6, 0.30)
H_2	(0.54, 0.45)	(0.50, 0.15)	(0.66, 0.12)	(0.30, 0.70)
H_3	(0.72, 0.23)	(0.53, 0.05)	(0.50, 0.03)	(0.91, 0.09)
H_4	(0.63, 0.30)	(0.59, 0.29)	(0.58, 0.07)	(0.95, 0.05)
H_5	(0.59, 0.30)	(0.41, 0.57)	(0.59, 0.19)	(0.70, 0.30)
H_6	(0.73, 0.10)	(0.54, 0.37)	(0.46, 0.24)	(0.70, 0.30)

Table 2 Results for selection

Methods	(H_1, H)	(H_2, H)	(H_3, H)	(H_4, H)	(H_5, H)	(H_6, H)
∇_1 [38]	−0.1758	−0.1845	−0.1352	−0.1264	−0.1701	−0.1600
∇_2 [38]	−0.0762	−0.1095	−0.0151	−0.0165	−0.0843	−0.0584
∇_3 [38]	−0.1758	−0.1845	−0.1352	−0.1264	−0.1701	−0.1600
∇_4 [38]	−0.1758	−0.1845	−0.1352	−0.1264	−0.1701	−0.1600
$\tilde{\nabla}$	0.4554	0.4252	0.5651	0.5845	0.4733	0.5028

Table 3 Ranking and DoC for selection

Methods	Ranking	BA	DoC
∇_1 [38]	$H_4 \succ H_3 \succ H_6 \succ H_5 \succ H_1 \succ H_2$	N/A	0.1936
∇_2 [38]	$H_3 \succ H_4 \succ H_6 \succ H_1 \succ H_5 \succ H_2$	N/A	0.2694
∇_3 [38]	$H_4 \succ H_3 \succ H_6 \succ H_5 \succ H_1 \succ H_2$	N/A	0.1936
∇_4 [38]	$H_4 \succ H_3 \succ H_6 \succ H_5 \succ H_1 \succ H_2$	N/A	0.1936
$\tilde{\nabla}$	$H_4 \succ H_3 \succ H_6 \succ H_5 \succ H_1 \succ H_2$	H_4	0.5007

Fig. 1 DoC

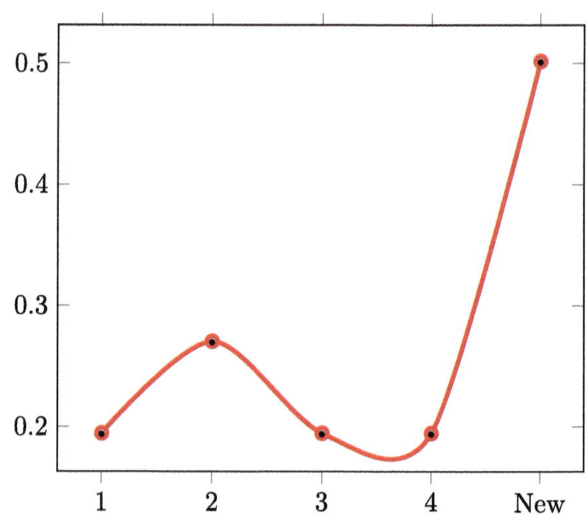

The graph of the DoCs is presented in Fig. 1.

From the results in Table 2, we see that the suitable house to purchase is house H_4 because it is the best alternative. The results from the SOs in [38] cannot be trusted for any decision because the outputs are not similarity measure values and the DoCs are very low.

5 Clustering Analysis via the New SO

Clustering analysis has to do with dividing data into various classes. We use the new SO for the clustering due to its reliability and efficiency. The following steps are adopted for the clustering analysis.

Step 1. Each of the PFSs $\wp_j$ $(j = 1, \cdots, k)$ and the whole PFSs are taken as distinct clusters.

Step 2. Find the similarities between the PFSs to obtain the Pythagorean fuzzy matrix (PFM) based on the new SO.

Step 3. The PFSs with the two largest similarities are taken as clusters.

Step 4. Find APFSs from the PFSs with the two largest similarities and compute similarities again.

Step 5. The APFSs with the largest similarity are taken as a cluster.

Example 5.1 Assume five students took a qualification examination on five subjects. We want to classify the students into classes based on their academic strength. The students $S = \{S_1, S_2, S_3, S_4, S_5\}$ are examined on English (e), Math (m), Physics (p), Biology (b), and Chemistry (c). The results of the examination are in Table 4, which are represented as PFSs in Table 5.

The first two possible clusters are $\{S_1\}$, $\{S_2\}$, $\{S_3\}$, $\{S_4\}$, $\{S_5\}$ and $\{S_1, S_2, S_3, S_4, S_5\}$. By adopting Steps 2 and 3, we get the PFM:

Table 4 Students's scores

S	e (%)	m (%)	p (%)	b (%)	c (%)
S_1	60	50	60	60	55
S_2	50	40	45	30	30
S_3	70	80	85	91	75
S_4	100	90	65	95	35
S_5	80	93	82	70	50

Table 5 Pythagorean fuzzy students's scores

S	e	m	p	b	c
S_1	(0.60, 0.40)	(0.50, 0.50)	(0.60, 0.40)	(0.60, 0.40)	(0.55, 0.45)
S_2	(0.50, 0.50)	(0.40, 0.60)	(0.45, 0.55)	(0.30, 0.70)	(0.30, 0.70)
S_3	(0.70, 0.30	(0.80, 0.20)	(0.85, 0.15)	(0.91, 0.09)	(0.75, 0.25)
S_4	(1.00, 0.00)	(0.90, 0.10)	(0.65, 0.35)	(0.95, 0.05)	(0.35, 0.65)
S_5	0.80, 0.20)	(0.93, 0.07)	(0.82, 0.18)	(0.70, 0.30)	(0.50, 0.50)

Table 6 New Pythagorean fuzzy students's scores

S	e	m	p	b	c
$S_{1,2}$	(0.550,0.450)	(0.45,0.550)	(0.525,0.475)	(0.450,0.550)	(0.425,0.575)
$S_{3,5}$	(0.750,0.250)	(0.865,0.135)	(0.835,0.165)	(0.805,0.195)	(0.625,0.375)
S_4	(1.000,0.000)	(0.900,0.100)	(0.650,0.350)	(0.950,0.050)	(0.350,0.650)

$$\begin{array}{c} \begin{array}{ccccc} S_1 & S_2 & S_3 & S_4 & S_5 \end{array} \\ \begin{bmatrix} 1 & 0.8243 & 0.7234 & 0.6586 & 0.7589 \\ 0.8243 & 1 & 0.5947 & 0.5874 & 0.6313 \\ 0.7234 & 0.5947 & 1 & 0.7420 & 0.8107 \\ 0.6586 & 0.5874 & 0.7420 & 1 & 0.7818 \\ 0.7589 & 0.6313 & 0.8107 & 0.7818 & 1 \end{bmatrix} \begin{array}{c} S_1 \\ S_2 \\ S_3 \\ S_4 \\ S_5 \end{array} \end{array}$$

From the PFM, $\tilde{\nabla}(S_1, S_2) = 0.8243$ is the largest similarity measure value, and so students S_1 and S_2 are considered as one cluster. In addition, $\tilde{\nabla}(S_3, S_5) = 0.8107$ is larger than the remaining similarity measure values, and thus students S_3 and S_5 are considered as another cluster. Hence, the results of this clustering stage are: $\{S_1, S_2\}$, $\{S_3, S_5\}$, and $\{S_4\}$.

Next, we find the APFSs of $\{S_1, S_2\}$ and $\{S_3, S_5\}$, respectively, and get the new the Pythagorean fuzzy students's scores in Table 6.

Using the new SO to find the similarities of the students in Table 6, we get the following PFM:

$$\begin{array}{c} \begin{array}{ccc} S_{1,2} & S_{3,5} & S_4 \end{array} \\ \begin{bmatrix} 1 & 0.6806 & 0.6254 \\ 0.6806 & 1 & 0.7719 \\ 0.6254 & 0.7719 & 1 \end{bmatrix} \begin{array}{c} S_{1,2} \\ S_{3,5} \\ S_4 \end{array} \end{array}$$

As $\tilde{\nabla}(S_4, S_{3,5}) = 0.7719$ is the largest similarity measure value, hence, students S_4, S_3, and S_5 are considered as an individual cluster. Thus, the clusters at this stage are: $\{S_3, S_4, S_5\}$ and $\{S_1, S_2\}$.

Therefore, the students can be clustered into the following different groups/classes as seen in Table 7.

Table 7 Students' clustering

Class	Cluster
1	$\{S_1\}, \{S_2\}, \{S_3\}, \{S_4\}, \{S_5\}$
2	$\{S_1, S_2\}, \{S_3, S_5\}, \{S_4\}$
3	$\{S_3, S_4, S_5\}, \{S_1, S_2\}$
4	$\{S_1, S_2, S_3, S_4, S_5\}$

Recall, from Table 4, we see that S_1 scored 285 out of 500 marks, S_2 scored 195 out of 500 marks, S_3 scored 401 out of 500 marks, S_4 scored 385 out of 500 marks, and S_5 scored 375 out of 500 marks. The information from Table 7 shows that the clusters are in two categories, namely, trivial classes and unique classes. The trivial classes are $\{S_1\}, \{S_2\}, \{S_3\}, \{S_4\}, \{S_5\}$ and $\{S_1, S_2, S_3, S_4, S_5\}$, respectively, while the unique classes are $\{S_1, S_2\}, \{S_3, S_5\}, \{S_4\}$ and $\{S_3, S_4, S_5\}, \{S_1, S_2\}$, respectively.

We call the first category trivial classes because grouping the students into just a class will result into ineffective teaching and learning process since their academic strengths are different. In addition, grouping the students into an individual classes may not be possible due to scarcity of trained teachers and cost implications. Nonetheless, for a better teaching and learning process, the students can be grouped into Clusters A: $\{S_3, S_4, S_5\}$ and Clusters B: $\{S_1, S_2\}$, respectively. The teachers should pay attention to the students in Cluster B due to their weak academic performance during the qualification examination.

6 Conclusion

In this chapter, a new SO is developed to correct the lapses observed in the SOs constructed by Zhang et al. [38]. The new SO is appropriately characterized to show its alliance with the similarity metric conditions. By demonstrating its real-life application potentials, we apply the new SO to selection process and clustering analysis. In the selection case, the new SO gives an interpretation with optimum degree of confidence compared to the SOs in [38]. In fact, the SOs constructed by Zhang et al. [38] fail the selection process due to their very low degrees of confidence. In the clustering analysis, the new SO shows reliability and gives an efficient clustering. We therefore assert that the new SO could be useful in decision-making under indeterminate fields since the approach is equipped to control hesitations that may hinder decision-making process. Exploring the applications of the new SO in different imprecise domains is an interesting future research direction. The construction of the new SO hinders its applications in Fermatean fuzzy environment [39, 40], picture fuzzy environment [41], q-rung orthopair fuzzy environment [42, 43], etc. However, with some adjustments, the new SO could be stretched to the aforesaid environments to enhance applications in complex fuzzy settings.

References

1. Zadeh, L.A.: Fuzzy sets. Inform. Control **8**, 338–353 (1965)
2. Atanassov, K.T.: Intuitionistic fuzzy sets. Fuzzy Sets Syst. **20**, 87–96 (1986)
3. Wang, W., Xin, X.: Distance measure between intuitionistic fuzzy sets. Pattern Recogn. Lett. **26**, 2063–2069 (2005)
4. Hatzimichailidis, A.G., Papakostas, A.G., Kaburlasos, V.G.: A novel distance measure of intuitionistic fuzzy sets and its application to pattern recognition problems. Int. J. Intell. Syst. **27**, 396–409 (2012)
5. Davvaz, B., Sadrabadi, E.H.: An application of intuitionistic fuzzy sets in medicine. Int. J. Biomathematics **9**(3), 1650037 (2016)
6. Liu, P., Chen, S.M.: Group decision making based on Heronian aggregation operators of intuitionistic fuzzy numbers. IEEE Trans. Cybern. **47**(9), 2514–2530 (2017)
7. Ejegwa, P.A., Onyeke, I.C., Kausar, N., Kattel, P.: A new partial correlation coefficient technique based on intuitionistic fuzzy information and its pattern recognition application. Int. J. Intell. Syst., Article ID 5540085, 14 p. (2023)
8. Boran, F.E., Akay, D.: A biparametric similarity measure on intuitionistic fuzzy sets with applications to pattern recognition. Inform. Sci. **255**(10), 45–57 (2014)
9. Ejegwa, P.A., Ahemen, S.: Enhanced intuitionistic fuzzy similarity operator with applications in emergency management and pattern recognition. Granular Comput. **8**, 361–372 (2023)
10. Zhou, Y., Ejegwa, P.A., Johnny, S.E.: Generalized similarity operator for intuitionistic fuzzy sets and its applications based on recognition principle and multiple criteria decision making technique. Int. J. Comput. Intell. Syst. **16**, 85 (2023)
11. Nwokoro, C.O., Inyang, U.G., Eyoh, I.J., Ejegwa, P.A.: Intuitionistic fuzzy approach for predicting maternal outcomes. In: Jana, C., Pal, M., Muhiuddin, G., Liu, P. (eds.) Fuzzy Optimization, Decision-making and Operations Research, pp. 399–421. Springer, Cham (2023)
12. Ejegwa, P.A., Agbetayo, J.M.: Similarity-distance decision-making technique and its applications via intuitionistic fuzzy pairs. J. Comput. Cognit. Eng. **2**(1), 68–74 (2023)
13. Atanassov, K.T.: Geometrical interpretation of the elements of the intuitionistic fuzzy objects. Preprint IM-MFAIS-1-89, Sofia (1989)
14. Yager, R.R., Abbasov, A.M.: Pythagorean membership grades, complex numbers and decision making. Int. J. Intelli. Syst. **28**(5), 436–452 (2013)
15. Garg, H.: A new generalized Pythagorean fuzzy information aggregation using Einstein operations and its application to decision making. Int. J. Intell. Syst. **31**(9), 886–920 (2016)
16. Memis, S., Enginoglu, S., Erkan, U.: A new classification method using soft decision-making based on an aggregation operator of fuzzy parameterized fuzzy soft matrices. Turkish J. Electr. Eng. Comput. Sci. **30**(3), 871–890 (2022)
17. Garg, H.: Generalized Pythagorean fuzzy geometric aggregation operators using Einstein t-norm and t-conorm for multicriteria decision making process. Int. J. Intell. Syst. **32**(6), 597–630 (2017)
18. Peng, X.: New similarity measure and distance measure for Pythagorean fuzzy set. Complex Intell. Syst. **5**(2), 101–111 (2019)
19. Ejegwa, P.A., Onyeke, I.C.: Some new distance and similarity algorithms for Pythagorean fuzzy sets with application in decision-making problems. In: Broumi, S. (ed.) Handbook of Research on Advances and Applications of Fuzzy Sets and Logic, pp 192–211, Hershey, Pennsylvania 17033–1240. USA (2022)
20. Li, D.Q., Zeng, W.Y.: Distance measure of Pythagorean fuzzy sets. Int. J. Intell. Syst. **33**, 348–361 (2018)
21. Wu, K., Ejegwa, P.A., Feng, Y., Onyeke, I.C., Johnny, S.E., Ahemen, S.: Some enhanced distance measuring approaches based on Pythagorean fuzzy information with applications in decision making. Symmetry **14**, 2669 (2022)
22. Zeng, W., Li, D., Yin, Q.: Distance and similarity measures of Pythagorean fuzzy sets and their applications to multiple criteria group decision making. Int. J. Intell. Syst. **33**(11), 2236–2254 (2018)

23. Ejegwa, P.A., Feng, Y., Tang, S., Agbetayo, J.M., Dai, X.: New Pythagorean fuzzy-based distance operators and their applications in pattern classification and disease diagnostic analysis. Neural Comput. Appl. **35**(14), 10083–10095 (2023)
24. Xiao, F., Ding, W.: Divergence measure of Pythagorean fuzzy sets and its application in medical diagnosis. Appl. Soft Comput. **79**, 254–267 (2019)
25. Mahanta, J., Panda, S.: Distance measure for Pythagorean fuzzy sets with varied applications. Neural Comput. Appl. **33**, 17161–17171 (2021)
26. Ejegwa, P.A., Wen, S., Feng, Y., Zhang, W.: Determination of pattern recognition problems based on a Pythagorean fuzzy correlation measure from statistical viewpoint. In: Proceeding of the 13th International Conference on Advanced Computational Intelligence, pp. 132–139. Wanzhou, China (2021)
27. Ejegwa, P.A., Wen, S., Feng, Y., Zhang, W., Liu, J.: A three-way Pythagorean fuzzy correlation coefficient approach and its applications in deciding some real-life problems. Appl. Intell. **53**(1), 226–237 (2022)
28. Yan, D., Wu, K., Ejegwa, P.A., Xie, X., Feng, Y.: Pythagorean fuzzy partial correlation measure and its application. Symmetry **15**, 216 (2023)
29. Ejegwa, P.A., Wen, S., Feng, Y., Zhang, W., Chen, J.: Some new Pythagorean fuzzy correlation techniques via statistical viewpoint with applications to decision-making problems. J. Intell. Fuzzy Syst. **40**(5), 9873–9886 (2021)
30. Sahoo, L., Senapati, T., Yager, R.R.: Real life applications of multiple criteria decision making techniques in fuzzy domain. Studies in Fuzziness and Soft Computing, Volume 420, Springer Nature (2023)
31. Li, R., Ejegwa, P.A., Li, K., Agaji 1, Feng Y, Onyeke IC,: A new similarity function for Pythagorean fuzzy sets with application in football analysis. AIMS Math. **9**(2), 4990–5014 (2024)
32. Ejegwa, P.A., Sarkar, A., Onyeke, I.C.: New methods of computing correlation coefficient based on Pythagorean fuzzy information and their applications in disaster control and diagnostic analysis. In: Jana, C., Pal, M., Muhiuddin, G., Liu, P. (eds.) Fuzzy Optimization, Decision-making and Operations Research, pp. 473–498. Springer, Cham (2023)
33. Ejegwa, P.A.: New similarity measures for Pythagorean fuzzy sets with applications. Int. J. Fuzzy Comput. Model. **3**(1), 75–94 (2020)
34. Zeng, W.Y., Li, D.Q., Yin, Q.: Distance and similarity measures of Pythagorean fuzzy sets and their applications to multiple criteria group decision making. Int. J. Intell. **11**(33), 1–19 (2018)
35. Peng, X.D., Yuan, H.Y., Yang, Y.: Pythagorean fuzzy information measures and their applications. Int. J. Intell. Syst. **32**(10), 991–1029 (2017)
36. Zhang, X.L.: A novel approach based on similarity measure for Pythagorean fuzzy multiple criteria group decision making. Int. J. Intell. Syst. **31**(6), 593–611 (2016)
37. Wei, G., Wei, Y.: Similarity measures of Pythagorean fuzzy sets based on cosine function and their applications. Int. J. Intell. Syst. **33**(3), 634–652 (2018)
38. Zhang, Q., Hu, J., Feng, J., Liu, A., Li, Y.: New similarity measures of Pythagorean fuzzy sets and their applications. IEEE Access **7**, 138192–138202 (2019)
39. Ejegwa, P.A., Sarkar, A.: Fermatean fuzzy approach of diseases diagnosis based on new correlation coefficient operators. In: Garg, H., Chatterjee, J.M. (eds.) Deep Learning in Personalized Healthcare and Decision Support, pp. 23–38. Academic Press (2023)
40. Ejegwa, P.A., Muhiuddin, G., Algehyne, E.A., Agbetayo, J.M., Al-Kadi, D.: An enhanced Fermatean fuzzy composition relation based on a maximum-average approach and its application in diagnostic analysis. J. Mathem., Volume 2022, Article ID 1786221, 12 p. (2022)
41. Cuong, B.C.: Picture fuzzy sets. J. Comput. Cybern. **30**(4), 409–420 (2014)
42. Ejegwa, P.A.: New q-rung orthopair fuzzy distance-similarity operators with applications in investment analysis, pattern recognition, clustering analysis, and selection of robot for smart manufacturing. Soft Comput. (2023). https://doi.org/10.1007/s00500-023-08799-1
43. Ejegwa, P.A., Davvaz, B.: An improved composite relation and its application in deciding patients' medical status based on a q-rung orthopair fuzzy information. Comput. Appl. Mathem. **41**, 303 (2022)

Bipolar Fuzzy Generalized Dombi Aggregation Operators for Group Decision-Making

Abhijit Saha, Abhay Kumar, Surajit Das, and Bishnupada Debnath

Abstract As an extension of fuzzy sets, bipolar fuzzy (BF) sets contain both the positive degree and negative degree of an object which makes them effectual sets for dealing with uncertainty and bipolarity. In the existing studies on bipolar fuzzy (BF) sets, (i) there is currently no aggregation operator (AO) with the required generality, and (ii) no optimization tool has been suggested as of now in order to calculate criteria weights. This motivated us to represent bipolar BF Generalized Dombi weighted operators based consensus reaching methodology to assess a collection of Enterprise Resource Planning (ERP) packages. In this approach, attribute weights are calculated using an optimization model related to information measures. To claim the superiority of our methodology, we have looked into both the comparisons to other existing methods.

Keywords Bipolar fuzzy sets · Generalized-Dombi operator · Optimization · Enterprise Resource Planning (ERP) selection

---

A. Saha (✉)
Department of Computing Technologies, SRM Institute of Science and Technology, Kattankulathur 603203, Tamil Nadu, India
e-mail: abhijit84.math@gmail.com

A. Kumar · S. Das
Department of Artificial Intelligence and Data Science, B V Raju Institute of Technology, Narsapur 502313, Telangana, India

B. Debnath
Department of Mathematics, Rajiv Gandhi University, Doimukh, Papumpare 791112, Arunachal Pradesh, India
e-mail: bishnupada.debnath@rgu.ac.in

L. Sahoo et al. (eds.), *Decision Making Under Uncertainty Via Optimization, Modelling, and Analysis*, Studies in Systems, Decision and Control 558,
https://doi.org/10.1007/978-981-96-0085-4_12

1 Introduction

In 1965, Zadeh [22] introduced the notion of fuzzy sets, which is an expansion of the traditional extract of a set. The membership of the items in a traditional set is permissible in binary form. Everything fuzzy is ambiguous, confusing, or unsure, and it is also impossible to determine if a statement or an issue will be true or false. Instead of assigning a fixed or definite value, it finds a rough value. The membership function can have values in the range [0, 1] according to fuzzy set theory. Young, tall, excellent, and high are fuzzy concepts. The term "young" is not defined by any specific or unique quantitative value. Ages 25 and 35 are considered young in certain situations. Several fields and applications regularly employ fuzzy set theory. Even while it has a wide range of uses, it also has some limitations, such as how it handles ambiguous data and takes into account human reasoning. A Bipolar fuzzy (BF) set [23], an extension of a fuzzy set, contains both "positive degree (μ^+)" and "negative degree (μ^-)", where $\mu^+ : X \to [0, 1]$ and $\mu^- : X \to [-1,0]$ are the functions. Bipolar fuzzy sets (BFSs) are effectual sets for dealing with uncertainty and bipolarity. This representation is based on the concept of "bipolar information" in the provided collection. Therefore, BFSs have a certain influence on many different fields, such as social science, medical science, quantum computing, management science, cognitive science, economics, and artificial intelligence. Recently, many aggregation operators (AOs) [5, 11, 12, 20], have been developed with BF setting. BFSs were applied in medicine [26], graph theory [18], physics and philosophy [27], and cellular combinatorics [25]. However, the deployment of ERP software packages has not yet used a BF decision support paradigm.

The combined use of BFSs, and Generalized Dombi AOs to address ambiguity issues has not yet been studied, although many works on BFSs have been published. These are the paper's contributions as a result.

(i) The research in this study adopts a methodology based on BF sets to identify and choose critical performance indicators as criteria to aid in the assessment of a wide variety of ERP software packages.
(ii) To exhibit a few AOs (BF Generalized Dombi weighted averaging (BFGDWA) and BF Generalized Dombi weighted geometric (BFGDWG) AOs) to handle concerns with BF information.
(iii) To develop an optimization model for the determination of criteria weights with partial weights information.

The rest parts are arranged as: Some preliminary ideas are furnished in the second section. The generalized Dombi operations and associated AOs namely BFGDWA and BFGDWG are introduced in Sect. 3. The created BF decision support model is shown in the following section. The examined problem is described in Sect. 5 along with a comprehensive solution. The comparison study is displayed in Sect. 6. Section 7 provides details on the conclusions.

2 Basic Concepts

Definition 2.1 [24]: A BFS ζ on U is expressed as $\zeta = \{\langle y_i, \mu^+(y_i), \mu^-(y_i)\rangle | y_i \in U\}$, $\mu^+ : U \to [0, 1]$, $\mu^- : U \to [-1, 0]$ are the "positive membership" and "negative-membership" mappings. ζ becomes a BF number (BFN) if U is singleton and we write $\zeta =< \mu^+, \mu^- >$. Let $\Re$ be the collection of all BFNs.

Definition 2.2 [12]: We have, for $\zeta \in \Re$:

$$\begin{aligned} \text{score}(\zeta) &= \frac{1}{2}(1 + \mu^+ + \mu^-), \\ \text{accuracy}(\zeta) &= \frac{1}{2}(\mu^+ - \mu^-) \end{aligned} \tag{2.1}$$

For $\zeta_1 =< \mu_1, \nu_1 >, \zeta_2 =< \mu_2, \nu_2 >\in \Re$, a ranking rule ($\mathbb{S}$: score, $\mathbb{A}$: accuracy) is:

$$\zeta_1 \succ \zeta_2 \left| \begin{array}{l} \mathbb{S}_{\zeta_1} > \mathbb{S}_{\zeta_2}, \\ \mathbb{S}_{\zeta_1} = \mathbb{S}_{\zeta_2}, \ \mathbb{A}_{\zeta_1} > \mathbb{A}_{\zeta_2} \end{array} \right.$$

$$\zeta_1 = \zeta_2 \ \left| \ \mathbb{S}_{\zeta_1} = \mathbb{S}_{\zeta_2}, \ \mathbb{A}_{\zeta_1} = \mathbb{A}_{\zeta_2}. \right.$$

Definition 2.3 [12]: The basic operations between $\zeta_1 =< \mu_1, \nu_1 >, \zeta_2 =< \mu_2, \nu_2 >\in \Re$ are:

(i) $\zeta_1^c = \langle 1 - \mu_1^+, |\mu_1^-| - 1\rangle$,
(ii) $\zeta_1 \oplus \zeta_2 = \langle \mu_1^+ + \mu_2^+ - \mu_1^+\mu_2^+, -|\mu_1^-||\mu_2^-|\rangle$,
(iii) $\zeta_1 \otimes \zeta_2 = \langle \mu_1^+\mu_2^+, \mu_1^- + \mu_2^- - \mu_1^-\mu_2^-\rangle$,
(iv) $\lambda\zeta_1 = \left\langle 1 - \left(1 - \mu_1^+\right)^\lambda, -|\mu_1^-|^\lambda \right\rangle$ $(\lambda > 0)$,
(v) $\zeta_1^\lambda = \left\langle (\mu_1^+)^\lambda, -1 + |1 + \mu_1^-|^\lambda \right\rangle$ $(\lambda > 0)$.

Definition 2.4 [4, 13–15]: The generalized Dombi operators are defined by the formula:

$$GDom_{t,k}(p_1, p_2) = \left(1 + \left(\frac{1}{t}\left(\prod_{j=1}^{2} \psi_t^k(p_j) - 1\right)\right)^{-\frac{1}{k}}\right)^{-1} \tag{2.2}$$

or

$$GDom_{t,k}(p_1, p_2) = \left(1 + \left(\frac{1}{t}\left(\prod_{j=1}^{2} \xi_t^k(p_j) - 1\right)\right)^{\frac{1}{k}}\right)^{-1} \tag{2.3}$$

where $\psi_t^k(p_j) = 1 + t\left(\frac{p_j}{1-p_j}\right)^k$, $\xi_t^k(p_j) = 1 + t\left(\frac{1-p_j}{p_j}\right)^k$ $(p_j \in (0, 1),\ j = 1, 2)$ with $t > 0$ and j.

The generalized Dombi operations are preferable to algebraic, Einstein, and Hamachar operations because of their strong precedence of change concerning parameter values "k" and "t".

3 Bipolar Fuzzy Generalized Dombi Operations

Here, we propose Generalized Dombi (GD) operations between BFNs and present their basic characteristics.

Definition 3.1: Suppose $\zeta_1 =< \mu_1^+,\ \mu_1^- >, \zeta_2 =< \mu_2^+,\ \mu_2^- >\in \Re$. Then the generalized Dombi (GD) operations between them are defined as:

(i)

$$\zeta_1 \oplus \zeta_2 = \left\langle \left(1 + \left(\frac{1}{t}\left(\prod_{j=1}^{2} \psi_t^k(\mu_j^+) - 1\right)\right)^{-\frac{1}{k}}\right)^{-1}, -\left(1 + \left(\frac{1}{t}\left(\prod_{j=1}^{2} \xi_t^k(\left|\mu_j^-\right|) - 1\right)\right)^{\frac{1}{k}}\right)^{-1} \right\rangle, \tag{3.1}$$

(ii)

$$\zeta_1 \otimes \zeta_2 = \left\langle \left(1 + \left(\frac{1}{t}\left(\prod_{j=1}^{2} \xi_t^k(\mu_j^+) - 1\right)\right)^{\frac{1}{k}}\right)^{-1}, -\left(1 + \left(\frac{1}{t}\left(\prod_{j=1}^{2} \psi_t^k(\left|\mu_j^-\right|) - 1\right)\right)^{-\frac{1}{k}}\right)^{-1} \right\rangle \tag{3.2}$$

(iii)

$$\lambda\zeta_1 = \left\langle \left(1 + \left(\frac{1}{t}((\psi_t^k(\mu_1^+))^\lambda - 1)\right)^{-\frac{1}{k}}\right)^{-1}, -\left(1 + \left(\frac{1}{t}((\xi_t^k(\left|\mu_1^-\right|))^\lambda - 1)\right)^{\frac{1}{k}}\right)^{-1} \right\rangle (\lambda > 0), \tag{3.3}$$

(iv)

$$\zeta_1^{\lambda} = \left\langle \left(1 + \left(\frac{1}{t}((\xi_t^k(\mu_1^+))^{\lambda} - 1) \right)^{\frac{1}{k}} \right)^{-1}, -\left(1 + \left(\frac{1}{t}((\psi_t^k(|\mu_1^-|))^{\lambda} - 1) \right)^{-\frac{1}{k}} \right)^{-1} \right\rangle (\lambda > 0). \tag{3.4}$$

Theorem 3.1: *For* $\zeta_j =< \mu_j^+,\ \mu_j^- >\in \Re\ (j = 1, 2)$, *we have*

(i)

$$\zeta_1 \oplus \zeta_2 = \zeta_2 \oplus \zeta_1$$

(ii)

$$\zeta_1 \otimes \zeta_2 = \zeta_2 \otimes \zeta_1$$

(iii)

$$\lambda(\zeta_1 \oplus \zeta_2) = (\lambda\zeta_1) \oplus (\lambda\zeta_2)$$

(iv)

$$(\zeta_1 \oplus \zeta_2)^{\lambda} = (\zeta_1^{\lambda}) \oplus (\zeta_2^{\lambda})$$

(v)

$$(\lambda_1 + \lambda_2)\zeta_1 = (\lambda_1\zeta_1) \oplus (\lambda_2\zeta_1)$$

(vi)

$$\zeta_1^{(\lambda_1+\lambda_2)} = (\zeta_1^{\lambda_1}) \oplus (\zeta_1^{\lambda_2})$$

***Proof*:** Follows from Definition 3.1.

4 Bipolar Fuzzy Generalized Dombi Weighted Aggregation Operators

Here, some BF Generalized Dombi weighted AOs have been developed. Throughout this section, we shall consider a collection of BFNs $\zeta_i =< \mu_i^+,\ \mu_i^- >\in \Re\ (i = 1, 2, \ldots, m)$.

4.1 Bipolar Fuzzy Generalized Dombi Weighted Averaging Operator

Definition 4.1: Suppose W_i is the weight of ζ_i and $\sum_{i=1}^{m} W_i = 1$. Then the BF Generalized-Dombi weighted averaging (BFGDWA) operator is presented as

$$BFGDWA(\zeta_1, \zeta_2, \ldots, \zeta_m) = \mathop{\oplus}_{i=1}^{m} (W_i \zeta_i) \tag{4.1}$$

Theorem 4.1: *The output value* $BFGDWA(\zeta_1, \zeta_2, \ldots, \zeta_m) \in \Re$ *and*

$$\begin{aligned} & BFGDWA(\zeta_1, \zeta_2, \ldots, \zeta_m) \\ & = \left\langle \left(\left(1 + \left(\frac{1}{t} \left(\prod_{j=1}^{n} (\psi_t^k(\mu_j^+))^{w_j} - 1 \right) \right) \right)^{-\frac{1}{k}} \right)^{-1}, \right. \\ & \left. - \left(1 + \left(\frac{1}{t} \left(\prod_{j=1}^{n} (\xi_t^k(\left|\mu_j^-\right|))^{w_j} - 1 \right) \right)^{\frac{1}{k}} \right)^{-1} \right\rangle. \end{aligned} \tag{4.2}$$

Proof: Follows from Definitions 3.1 and 4.1

The *BFGDWA* operator satisfies the following:

Theorem 4.2: *If* $\zeta_0\ (\neq \zeta_i) \in \Re$, *then* $BFGDWA(\zeta_0 \oplus \zeta_1, \zeta_0 \oplus \zeta_2, \ldots, \zeta_0 \oplus \zeta_m) = \zeta_0 \oplus BFGDWA(\zeta_1, \zeta_2, \ldots, \zeta_m)$.

Theorem 4.3: *If* $\zeta_0\ (\neq \zeta_i) \in \Re$, *then* $BFGDWA(\zeta_1, \zeta_2, \ldots, \zeta_m) = \zeta_0$.

Theorem 4.4: *For* $\zeta_i = < \mu_i^+, \mu_i^- > \in \Re$ *we have* $\zeta^- \prec BFGDWA(\zeta_1, \zeta_2, \ldots, \zeta_y) \prec \zeta^+$, *where* $\zeta^- = < \min_i \mu_i^+, \max_i \mu_i^- >$ *and* $\zeta^+ = < \max_i \mu_i^+, \min_i \mu_i^- >$.

Theorem 4.5: *If* $\xi_i = < \nu_i^+, \nu_i^- > \in \Re$ *obeying* $\mu_i^+ \leq \nu_i^+$, $\mu_i^- \geq \nu_i^-$, *then* $BFGDWA(\zeta_1, \zeta_2, \ldots, \zeta_m) \prec BFGDWA(\zeta_1, \zeta_2, \ldots, \zeta_m)$.

4.2 Bipolar Fuzzy Generalized Dombi Weighted Geometric Operator

Definition 4.2: Suppose W_i is the weight of ζ_i and $\sum_{i=1}^{m} W_i = 1$. Then the BF Generalized-Dombi weighted geometric (BFGDWG) operator is presented as

$$BFGDWA(\zeta_1, \zeta_2, \ldots, \zeta_m) = \overset{m}{\underset{i=1}{\otimes}} \left(\zeta_i^{W_i} \right). \tag{4.3}$$

Theorem 4.6: *The output value* $BFGDWG(\zeta_1, \zeta_2, \ldots, \zeta_m) \in \Re$ and

$$BFGDWG(\zeta_1, \zeta_2, \ldots, \zeta_m) = \left\langle \left(1 + \left(\frac{1}{t} \left(\prod_{j=1}^{n} (\xi_t^k(\mu_j^+))^{w_j} - 1 \right) \right)^{\frac{1}{k}} \right)^{-1}, \right.$$

$$\left. - \left(1 + \left(\frac{1}{t} \left(\prod_{j=1}^{n} (\psi_t^k(\left|\mu_j^-\right|))^{w_j} - 1 \right) \right)^{-\frac{1}{k}} \right)^{-1} \right\rangle. \tag{4.4}$$

***Proof*:** Similar to Theorem 4.1.

The *BFGDWG* operator satisfies the following:

Theorem 4.7: *If* $\zeta_0 \, (\neq \zeta_i) \in \Re$, *then* $BFGDWG(\zeta_0 \oplus \zeta_1, \zeta_0 \oplus \zeta_2, \ldots, \zeta_0 \oplus \zeta_m) = \zeta_0 \oplus BFGDWG(\zeta_1, \zeta_2, \ldots, \zeta_m)$.

Theorem 4.8: *If* $\zeta_0 \, (\neq \zeta_i) \in \Re$, *then* $BFGDWG(\zeta_1, \zeta_2, \ldots, \zeta_m) = \zeta_0$.

Theorem 4.9: *For* $\zeta_i = < \mu_i^+, \mu_i^- > \in \Re$ *we have* $\zeta^- \prec BFGDWG(\zeta_1, \zeta_2, \ldots, \zeta_y) \prec \zeta^+$, *where* $\zeta^- = < \min_i \mu_i^+, \max_i \mu_i^- >$ *and* $\zeta^+ = < \max_i \mu_i^+, \min_i \mu_i^- >$.

Theorem 4.10: *If* $\xi_i = < \nu_i^+, \nu_i^- > \in \Re$ *satisfying* $\mu_i^+ \leq \nu_i^+, \mu_i^- \geq \nu_i^-$, *then* $BFGDWG(\zeta_1, \zeta_2, \ldots, \zeta_m) \prec BFGDWG(\zeta_1, \zeta_2, \ldots, \zeta_m)$.

5 Bipolar Fuzzy Decision-Making Methodology

The proposed methodology has the steps mentioned below:

Step 1: Form the BF decision matrices (BFDMs) for all experts.

Suppose:

G_r: rth alternative ($r = 1, 2, \ldots, p$),

Q_s: sth attribute ($s = 1, 2, \ldots, q$),

E_m: mth expert (decision-maker) ($m = 1, 2, \ldots, l$),

ϖ_m: weight of E_m,

$U_m = \left[\zeta_m^{(rs)} \right]_{p \times q} = \left[< \mu_{mrs}^+, \mu_{mrs}^- > \right]_{p \times q}$: Assessment matrix of E_m in terms of BFNs.

Step 1: Construct the aggregated BFDM using the operator BFGDWA (or BFGDWG).

The A-BFDM is $\left[\zeta^{(rs)}\right]_{p\times q} = \left[< \mu_{rs}^{+}, \mu_{rs}^{-} > \right]_{p\times q}$, where:

$$\zeta^{(rs)} = BFGDWA(\zeta_1^{(rs)}, \zeta_2^{(rs)}, \ldots, \zeta_l^{(rs)}) = \bigoplus_{m=1}^{l}\left(\varpi_m \zeta_m^{(rs)}\right), \tag{5.1}$$

or

$$\zeta^{(rs)} = BFGDWG(\zeta_1^{(rs)}, \zeta_2^{(rs)}, \ldots, \zeta_l^{(rs)}) = \bigotimes_{m=1}^{l}\left(\zeta_m^{(rs)}\right)^{\varpi_m} \tag{5.2}$$

$(\varpi_m :$ weight of the mth expert$)$

Step 2: Normalize the A-BFDM.

Suppose that the normalized A-BFDM is $\left[\tilde{\xi}^{(rs)}\right]_{p\times q} = \left[< \tilde{\mu}_{rs}^{+}, \tilde{\mu}_{rs}^{-} > \right]_{p\times q}$, where:

$$\tilde{\xi}^{(rs)} = \begin{cases} \zeta^{(rs)}, & \text{if } Q_s \text{ is benefit type attribute} \\ (\zeta^{(rs)})^c, & \text{if } Q_s \text{ is cost type attribute} \end{cases}. \tag{5.3}$$

Step 3: Obtain the score matrix corresponding to the normalized A-BFDM.

The scores of the elements of the normalized A-BFDM are given by: $\left[\mathbb{S}(\tilde{\xi}^{(rs)})\right]_{p\times q} = \left[0.5\times(1+\tilde{\mu}_{rs}^{+}+\tilde{\mu}_{rs}^{-})\right]_{p\times q}$.

Step 4: Computation of the weights of attributes.

Equation (5.4) depicts the difference between the rth option and the remaining options w.r.t sth attribute.

$$DIV_{rs} = \frac{1}{p-1}\sum_{z=1}^{p} C(\mathbb{S}(\tilde{\xi}^{(rs)}), \mathbb{S}(\tilde{\xi}^{(zs)})), \tag{5.4}$$

where $C(\mathbb{S}(\tilde{\xi}^{(rs)}), \mathbb{S}(\tilde{\xi}^{(zs)}))$ signifies the cross entropy. It can be defined by

$$\begin{aligned} C(\mathbb{S}(\tilde{\xi}^{(rs)}), \mathbb{S}(\tilde{\xi}^{(zs)})) &= \mathbb{S}(\tilde{\xi}^{(rs)}) \times \ln\left(\frac{2\mathbb{S}(\tilde{\xi}^{(rs)})}{\mathbb{S}(\tilde{\xi}^{(rs)}) + \mathbb{S}(\tilde{\xi}^{(zs)})}\right) \\ &+ \mathbb{S}(\tilde{\xi}^{(zs)}) \times \ln\left(\frac{2\mathbb{S}(\tilde{\xi}^{(zs)})}{\mathbb{S}(\tilde{\xi}^{(rs)}) + \mathbb{S}(\tilde{\xi}^{(zs)})}\right) \\ &+ \left(1-\mathbb{S}(\tilde{\xi}^{(rs)})\right) \times \ln\left(\frac{1-\mathbb{S}(\tilde{\xi}^{(rs)})}{1-\frac{1}{2}\left(\mathbb{S}(\tilde{\xi}^{(rs)}) + \mathbb{S}(\tilde{\xi}^{(zs)})\right)}\right) \\ &+ \left(1-\mathbb{S}(\tilde{\xi}^{(zs)})\right) \times \ln\left(\frac{1-\mathbb{S}(\tilde{\xi}^{(zs)})}{1-\frac{1}{2}\left(\mathbb{S}(\tilde{\xi}^{(rs)}) + \mathbb{S}(\tilde{\xi}^{(zs)})\right)}\right) \end{aligned} \tag{5.5}$$

Equation (5.6) gives the total divergence w.r.t sth attribute:

$$DIV_s = \frac{1}{p-1}\sum_{r=1}^{p}\sum_{z=1}^{p} C(\mathbb{S}(\tilde{\xi}^{(rs)}), \mathbb{S}(\tilde{\xi}^{(zs)})) \tag{5.6}$$

Thus, Eq. (5.7), upon solving, gives the weights of the attributes.

$$\begin{aligned} &Max\ \chi = \sum_{s=1}^{q} \beta_s \frac{1}{p-1}\sum_{r=1}^{p}\sum_{z=1}^{p} C(\mathbb{S}(\tilde{\xi}^{(rs)}), \mathbb{S}(\tilde{\xi}^{(zs)})) \\ &\text{Subject to: } \beta_s \in \Xi,\ \sum_{s=1}^{q} \beta_s = 1,\ \beta_s \geq 0\ \forall s. \end{aligned} \tag{5.7}$$

where β_s: weight of sth attribute and Ξ: the set of all partial weights information.

Step 5: Engage the *BFGDWA* (or *BFGDWG*) operator to generate the final aggregated BFNs.

The final aggregated BFNs are $\zeta^{(r)} = <\tilde{\mu}_r^+, \tilde{\mu}_r^- >$ $(r = 1, 2, ..., p)$, where:

$$\begin{aligned} \zeta^{(r)} &= BFGDWA(\tilde{\xi}^{(r1)}, \tilde{\xi}^{(r2)}, ..., \tilde{\xi}^{(rq)}) \ (r = 1, 2, ..., p), \\ &\text{or,} \\ \zeta^{(r)} &= BFGDWG(\tilde{\xi}^{(r1)}, \tilde{\xi}^{(r2)}, ..., \tilde{\xi}^{(rq)}) \ (r = 1, 2, ..., p), \end{aligned} \tag{5.8}$$

Step 6: "Obtain the priority order of options based on the scores $\mathbb{S}(\tilde{\xi}^{(r)})$ $(r = 1, 2, ..., p)$ and pick the best one.

6 Case Study and Results

It might be difficult to choose the right ERP software for any company. ERP software works to support all processes in creating a value-added chain to meet the unique requirements of the core business operations of rapidly expanding organizations. Contrarily, conventional on-premise ERPS software is usually too rigid to keep up with evolving company practices and technological advancements. Consequently, the cloud has taken over as the preferred option for ERPS software implementation for companies of all sizes, and it is becoming more widespread in the industrial sector. An important step in the software evaluation process is the creation of carefully thought-out criteria to evaluate providers for cloud-based ERPS software. Three cloud-based ERPS vendors (alternatives) considered in this case study are: Microsoft Corporation (G_1), Systems, Applications, and Products (SAP) (G_2), Oracle Corporation (G_3), and SAGE (G_4). A variety of competing criteria must be used to evaluate ERP software. Literature has identified nine important factors that affect how an ERP package is evaluated, providing a useful foundation for the business. They are given in the following Table (Tables 1 and 2).

Suppose 0.2, 0.15, 0.35, and 0.3 are the weights of the experts. Table 3 gives the aggregated matrix (for $t = 1$, and $k = 3$).

Next, we prepare the score matrix corresponding to the aggregated BF matrix. Since the cost type attribute is absent, we have ignored normalization part. Next, assume that the criteria's partial weight information is provided by $\Xi = \{0.10 \le \beta_1 \le 0.15,\ 0.15 \le \beta_2 \le 0.25,\ 0.06 \le \beta_3 \le 0.1, 0.2 \le \beta_4 \le 0.25, 0.07 \le \beta_5 \le 0.13, 0.07 \le \beta_6 \le 0.12, 0.05 \le \beta_7 \le 0.13, 0.1 \le \beta_8 \le 0.15, 0.08 \le \beta_9 \le 0.17\}$ Then based on Eqs. (5.6)-(5.9), we have the following optimization model:

$$\text{Max}\ \chi = 0.1223\beta_1 + 0.0864\beta_2 + 0.1188\beta_3 + 0.1186\beta_4 + 0.1218\beta_5 + 0.0710\beta_6 + 0.0162\beta_7 + 0.0392\beta_8 + 0.1263\beta_9$$

Subject to:

$$\begin{cases} 0.10 \le \beta_1 \le 0.15,\ 0.15 \le \beta_2 \le 0.25,\ 0.06 \le \beta_3 \le 0.1, 0.2 \le \beta_4 \le 0.25, 0.07 \le \beta_5 \le 0.13, \\ 0.07 \le \beta_6 \le 0.12, 0.05 \le \beta_7 \le 0.13, 0.1 \le \beta_8 \le 0.15, 0.08 \le \beta_9 \le 0.17 \\ \beta_1 + \beta_2 + \beta_3 + \beta_4 + \beta_5 + \beta_6 + \beta_7 + \beta_8 + \beta_9 = 1, \\ \beta_1, \beta_2, \beta_3, \beta_4, \beta_5, \beta_6, \beta_7, \beta_8, \beta_9 \ge 0 \end{cases}$$

This gives the solution: $\beta_1 = 0.13,\ \beta_2 = 0.15,\ \beta_3 = 0.06,\ \beta_4 = 0.2,\ \beta_5 = 0.07,\ \beta_6 = 0.07,\ \beta_7 = 0.05,\ \beta_8 = 0.1,\ \beta_9 = 0.17$.

The final aggregated BFNs (choosing $t = 1$, and $k = 3$) are

$$\zeta^{(1)} =< 0.7179,\ -0.2651 >,\ \zeta^{(2)} =< 0.6413,\ -0.3105 >,$$
$$\zeta^{(3)} =< 0.7110,\ -0.2541 >,\ \zeta^{(4)} =< 0.4975,\ -0.4404 > .$$

The scores of these aggregated BFNs obtained by Eq. (2.1) are respectively 0.7264, 0.6654, 0.7285, and 0.5286. Thus the priority order is: $G_3 \succ G_1 \succ G_2 \succ G_4$ ("$\succ$" signifies "better than").

Table 1 Criteria details

Criteria	References
Information security (Q1)	She and Thuraisingham [6], Kahraman et al. [17]
User-friendliness (Q2)	Karsak and Özogul [7], Schlichter and Kraemmergaard [16]
Functionality (Q3)	Schlichter and Kraemmergaard [6], Kahraman et al. [16]
Compatibility (Q4)	Deb et al. [3]
System Reliability (Q5)	Kilic et al. [9]
Service and Support (Q6)	Alanbay [1], Kilic et al. [8]
Implementation Time (Q7)	Wei et al. [19], Ünal and Güner [21]
Cross-module integration (Q8)	Baki and Çakar [2]
Ease of customization (Q9)	Pan et al. [10]

Table 2 Experts' initial evaluation results

Criteria	Expert 1				Expert 2			
	G_1	G_2	G_3	G_4	G_1	G_2	G_3	G_4
Q_1	<0.45, –0.5>	<0.50, –0.40>	<0.10, –0.85>	<0.40, –0.55>	<0.50, –0.45>	<0.40, –0.50>	<0.85, –0.10>	<0.55, –0.40>
Q_2	<0.70, –0.25>	<0.65, –0.30>	<0.60, –0.35>	<0.35, –0.55>	<0.90, –0.10>	<0.80, –0.15>	<0.75, –0.20>	<0.30, –0.60>
Q_3	<0.70, –0.10>	<0.65, –0.15>	<0.60, –0.20>	<0.35, –0.60>	<0.70, –0.10>	<0.65, –0.15>	<0.60, –0.20>	<0.35, –0.60>
Q_4	<0.50, –0.25>	<0.40, –0.30>	<0.85, –0.35>	<0.55, –0.55>	<0.50, –0.25>	<0.40, –0.30>	<0.85, –0.35>	<0.55, –0.55>
Q_5	<0.10, –0.45>	<0.15, –0.50>	<0.20, –0.10>	<0.60, –0.40>	<0.90, –0.45>	<0.80, –0.50>	<0.75, –0.10>	<0.30, –0.40>
Q_6	<0.25, –0.90>	<0.30, –0.80>	<0.35, –0.75>	<0.55, –0.30>	<0.70, –0.10>	<0.65, –0.15>	<0.60, –0.20>	<0.35, –0.60>
Q_7	<0.90, –0.70>	<0.80, –0.65>	<0.75, –0.60>	<0.30, –0.35>	<0.10, –0.25>	<0.15, –0.30>	<0.20, –0.35>	<0.60, –0.55>
Q_8	<0.50, –0.10>	<0.40, –0.15>	<0.85, –0.20>	<0.55, –0.60>	<0.90, –0.90>	<0.80, –0.80>	<0.75, –0.75>	<0.30, –0.30>
Q_9	<0.90, –0.45>	<0.80, –0.50>	<0.75, –0.10>	<0.30, –0.40>	<0.90, –0.10>	<0.80, –0.15>	<0.75, –0.20>	<0.30, –0.60>
Criteria	Expert 3				Expert 4			
	G_1	G_2	G_3	G_4	G_1	G_2	G_3	G_4
Q_1>	<0.5, –0.45>	<0.40, –0.50>	<0.85, –0.10v	<0.55, –0.40>	<0.90, –0.10>	<0.80, –0.15>	<0.75, –0.20>	<0.30, –0.60>
Q_2	<0.10, –0.90>	<0.15, –0.80>	<0.20, –0.75>	<0.60, –0.30>	<0.90, –0.10>	<0.80, –0.15>	<0.75, –0.20>	<0.30, –0.60>
Q_3	<0.70, –0.50>	<0.65, –0.40>	<0.60, –0.85>	<0.35, –0.55>	<0.50, –0.25>	<0.40, –0.30>	<0.85, –0.35>	<0.55, –0.55>
Q_4	<0.50, –0.25>	<0.40, –0.30>	<0.85, –0.35>	<0.55, –0.55>	<0.90, –0.45>	<0.80, –0.50>	<0.75, –0.10>	<0.30, –0.40>
Q_5	<0.90, –0.45>	<0.80, –0.50>	<0.75, –0.10>	<0.30, –0.40>	<0.70, –0.10>	<0.65, –0.15v	<0.60, –0.20>	<0.35, –0.60>
Q_6	<0.70, –0.10>	<0.65, –0.15>	<0.60, –0.20>	<0.35, –0.60>	<0.50, –0.25v	<0.40, –0.30>	<0.85, –0.35>	<0.55, –0.55>
Q_7v	<0.90, –0.25>	<0.80, –0.30>	<0.75, –0.35>	<0.30, –0.55>	<0.10, –0.45>	<0.15, –0.50>	<0.20, –0.10>	<0.90, –0.40>
Q_8	<0.25, –0.10>	<0.30, –0.15>	<0.35, –0.20>	<0.55, –0.60>	<0.90, –0.90>	<0.80, –0.80v	<0.75, –0.75>	<0.30, –0.10>
Q_9	<0.45, –0.70>	<0.50, –0.65>	<0.10, –0.60>	<0.40, –0.35>	<0.70, –0.10>	<0.65, –0.15>	<0.60, –0.20>	<0.35, –0.60>

Table 3 Aggregated BF matrix

Criteria	G_1	G_2	G_3	G_4
Qv1	<0.6797, −0.3033>	<0.6065, −0.3360>	<0.7678, −0.1803>	<0.4922, −0.4634>
Q_2	<0.7609, −0.2302>	<0.6769, −0.2796>	<0.6367, −0.3229>	<0.4960, −0.4345>
Q_3	<0.6529, −0.2391v	<0.6030, −0.2655v	<0.6957, −0.3407>	<0.4573, −0.5644>
Q_4	<0.6834, −0.2992>	<0.5992, −0.3471>	<0.8241, −0.2491>	<0.5162, −0.4873>
Q_5	<0.7948, −0.2981>	<0.7064, −0.3494>	<0.6614, −0.1241>	<0.4560, −0.4376>
Q_6	<0.6085, −0.1934>	<0.5643, −0.2457>	<0.6763, −0.2935>	<0.4914, −0.4779>
Q_7>	<0.7697, −0.3497>	<0.6794, −0.3906>	<0.6418, −0.2723>	<0.6721, −0.4443>
Q_8	<0.7350, −0.2303>	<0.6493, −0.2791>	<0.6987, −0.3203>	<0.4958, −0.3095>
Q_{9v}	<0.7416, −0.2599>	<0.6736, −0.3073>	<0.6195, −0.2542>	<0.3622, −0.4359>

7 Comparative Analysis

Here, a comparative investigation is shown with the help of the existing BF methods: Jana et al.'s method [5] with BF Dombi weighted averaging (BFDWA) operator, and the method of Wei et al. [20] with BF Hamacher weighted averaging (BFHWA) operator. These tools have been applied to the case study we considered. Figure 1 gives the overall outcomes. Each method generates the same ranking order $G_1 \succ G_3 \succ G_2 \succ G_4$.

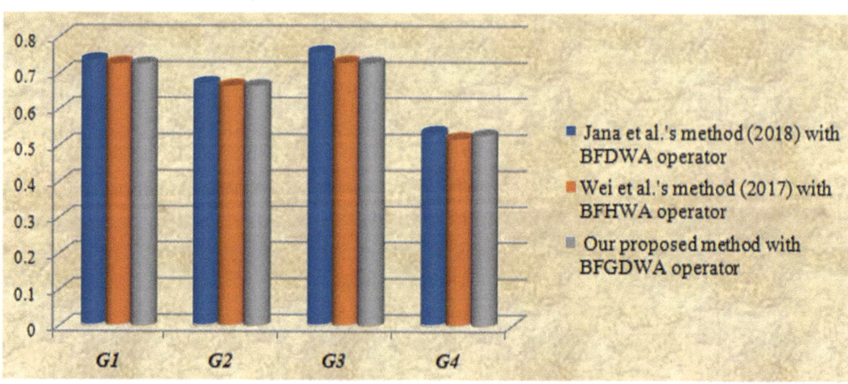

Fig. 1 Comparative study

8 Conclusions

Given that it is impacted by numerous competing criteria and extremely complex ambiguities, this ERP software assessment is a crucial MCDM problem for any organization. ERP software is used by businesses to manage and coordinate key business processes like shipping, marketing, sales, product planning, and human resources. The attribute value is appropriately represented as a bipolar fuzzy number at the time of the development of the MCDM, keeping in mind the delicate issue of developing uncertainty of real-world decision-making problems. Information that is bipolar includes both the positive and the negative aspects of the specified collection. Therefore, bipolar fuzzy sets (BFSs) have great significance. In this work, we have developed the BF Generalized-Dombi weighted averaging (BFGDWA) and BF Generalized-Dombi weighted geometric (BFGDWG) aggregation operators and presented their properties. Then, a BF group decision methodology has been shown where criteria weights are estimated using an optimization model. The comparative study confirms that our model is much more effective compared to the existing ones.

References

1. Alanbay, O.: ERP Selection Using Expert Choice Software. Honolulu, Hawaii (2005)
2. Baki, B., Çakar, K.: Determining the ERP package-selecting criteria: the case of Turkish manufacturing companies. Busi. Process Manag. **11**, 75–86 (2005)
3. Deb, P.P., Bhattacharya, D., Chatterjee, I., Saha, A., Mishra, A.R., Hassane, A.: A decision-making model with intuitionistic fuzzy information for selection of enterprise resource planning systems. IEEE Trans. Eng. Manag. (2022). https://ieeexplore.ieee.org/document/9939041/
4. Dombi, J.: The generalized Dombi operator family and the multiplicative utility function. Stud. Comput. Intell. (2009). https://doi.org/10.1007/978-3-642-00448-3-6
5. Jana, C., Pal, M., Wang, J.Q.: Bipolar fuzzy Dombi aggregation operators and its application in multiple-attribute decision-making process. J. Ambient. Intell. Humaniz. Comput. **10**, 3533–3549 (2018)
6. Kahraman, C., Beskese, A., Kaya, I.: Selection among ERP outsourcing alternatives using a fuzzy multi-criteria decision making methodology. Int. J. Prod. Res. **48**, 547–566 (2010)
7. Karsak, E.E., Özogul, C.O.: An integrated decision making approach for ERP system selection. Expert Syst. Appl. **36**, 660–667 (2009)
8. Kilic, H.S., Zaim, S., Delen, D.: Selecting “The Best” ERP system for SMEs using a combination of ANP and PROMETHEE methods. Expert Syst. Appl. **42**, 2343–2352 (2015)
9. Kilic, H.S., Zaim, S., Delen, D.: Development of a hybrid methodology for ERP system selection: the case of Turkish Airlines. Decis. Support. Syst. **66**, 82–92 (2014)
10. Pan, K., Nunes, M.B., Peng, G.C.: Risks affecting ERP post-implementation. J. Manuf. Technol. Manag. **22**, 107–130 (2011)
11. Riaz, M., Pamucar, D., Habib, A., Jamil, N.: Innovative bipolar fuzzy sine trigonometric aggregation operators and SIR method for medical tourism supply chain Math. Prob. Eng. Article ID 4182740 (2022). https://doi.org/10.1155/2022/4182740
12. Roy, A., Saha, A., Chatterjee, P., Dutta, D., Rastogi, R., Kottapalli, R.: A consensus model with bipolar fuzzy archimedean-dombi operators for group decision-making. Soft. Comput. (2023). https://doi.org/10.1007/s00500-023-08750-4

13. Saha, A., Debnath, B.K., Chatterjee, P., Panaiyappan, A.K., Das, S., Anusha, G.: Generalized Dombi based probabilistic hesitant fuzzy consensus reaching model for supplier selection under healthcare supply chain framework. Eng. Appl. Artif. Intell. **133**, 107966 (2024)
14. Saha, A., Senapati, T., Mesiar, R.: Generalized Dombi weighted aggregation operators for multi-attribute decision-making with hesitant fuzzy information. Stud. Fuzzines Soft Comput. **420**, 1–36 (2023)
15. Saha, A., Mishra, A.R., Rani, P., Senapati, T., Yager, R.R.: A dual probabilistic linguistic MARCOS approach based on generalized Dombi operator for decision-making **20**(2), 83–102 (2023)
16. Schlichter, B.R., Kraemmergaard, P.: A comprehensive literature review of the ERP research field over a decade. J. Enterp. Inf. Manag. **23**(4), 486–520 (2010)
17. She, W., Thuraisingham, B.: Security for enterprise resource planning systems. Inf. Syst. Secur. **16**, 152–163 (2007)
18. Singh, P.K., Kumar, C.A.: Bipolar fuzzy graph representation of concept lattice. Inf. Sci. **288**, 437–488 (2014)
19. Ünal, C., Güner, M.G.: Selection of ERP suppliers using AHP tools in the clothing industry. Int. J. Cloth. Sci. Technol. **21**(4), 239–251 (2009)
20. Wei, G., Alsaadi, F.E., Hayat, T., Alsaedi, A.: Bipolar fuzzy Hamacher aggregation operators in multiple attribute decision making. Int. J. Fuzzy Syst. (2017). https://doi.org/10.1007/s40815-017-0338-6
21. Wei, C.C., Chien, C.-F., Wang, M.-J.J.: An AHP-based approach to ERP system selection. Int. J. Prod. Econ. **96**, 47–62 (2005)
22. Zadeh, L.A.: Fuzzy sets. Inf. Control **8**, 338–353 (1965)
23. Zhang, W.R.: Bipolar fuzzy sets and relations, a computational frame work for cognitive modelling and multi-agent decision analysis. In: Proceedings of IEEE Conference, pp. 305–309 (1994)
24. Zhang, W.R.: Bipolar fuzzy sets In: IEEE Proceedings of FUZZY, pp. 835–840 (1998)
25. Zhang, W.R.: Bipolar quantum logic gates and quantum cellular combinatorics a logical extension to quantum entanglement. J. Quant. Inf. Sci. **3**(2), 93–105 (2013)
26. Zhang, W.R., Peace, K.E.: Causality is logically definable-toward an equilibrium based computing paradigm of quantum agent and quantum intelligence. J. Quant. Inf. Sci. **4**, 227–268 (2014)
27. Zhang, W.R.: G-CPT symmetry of quantum emergence and submergence an information conservational multi agent cellular automata unification of CPT symmetry and CP violation for equilibrium- based many world causal analysis of quantum coherence and decoherence. J. Quant. Inf. Sci. **6**(2):62–97 (2016)

Utility-Theory-Based Product Selection: Strategic Decision-Making in Automotive Industry

Chanchal, Adarsh Anand, Deepti Aggrawal, and Mohini Agarwal

Abstract The automotive industry constantly evolves, driven by technological advancements, changing consumer preferences, and regulatory requirements. New Product Development (NPD) plays a crucial role in maintaining competitiveness and meeting the demands of a dynamic market. A key aspect of NPD is product selection, which involves thorough analysis to align with market demands and organizational goals. Considering the importance of this stage, this article explores the applicability of utility theory as a filtration tool for product selection within the automotive sector, providing a detailed examination of its benefits and practical applications. When faced with multiple product options, the utility theory approach's usability can help select the product with maximum utility and align the resource management. This systematic approach allows a comprehensive analysis of attributes like performance metrics, budget and resource optimal allocation, and nuanced management preferences, thereby facilitating well-informed decision-making. Moreover, the article demonstrates the effectiveness of the proposed filtering approach through a case study. This practical illustration underscores the utility theory's efficacy and relevance in product selection in NPD.

Keywords Automotive industry · New Product Development (NPD) · Product selection · Utility theory

Chanchal · A. Anand (✉)
Department of Operational Research, University of Delhi, Delhi, India
e-mail: adarsh.anand86@gmail.com; aanand@or.du.ac.in

Chanchal
e-mail: chanchal@or.du.ac.in

D. Aggrawal
USME, Delhi Technological University, East Delhi Campus, Delhi, India
e-mail: deeptiaggrawal@dtu.ac.in

M. Agarwal
College of Humanities and Social Sciences, Grand Canyon University, Phoenix, USA
e-mail: mohini.agarwal@gcu.edu

L. Sahoo et al. (eds.), *Decision Making Under Uncertainty Via Optimization, Modelling, and Analysis*, Studies in Systems, Decision and Control 558,
https://doi.org/10.1007/978-981-96-0085-4_13

1 Introduction

The automotive industry is a crucial driver of economic growth and has a significant impact on developing the country's economic structure. The automotive sector has substantially influenced India's financial landscape with a considerable contribution of 7.1% to the Gross Domestic Product (GDP) and an impressive 49% to the manufacturing GDP [16]. This significance can be further accentuated by the industry's remarkable growth trajectory observed during the fiscal year 2023. With around 25.93 million vehicle production units, it not only heightened manufacturing activities but also signifies the present automotive industry's adaptability and resilience in response to market demands [18]. Amidst the ever-intensifying global competition, an automotive company must make necessary investments in research and development (R&D) activities. By strengthening their internal development capabilities, these enterprises can fortify their expertise in ways their competitors cannot easily emulate. A crucial aspect of this strategy lies in the realm of new product development (NPD). In the face of an increasingly intricate and competitive market, automotive companies seek innovative approaches to enhance their operations and sustain profitability [1, 13, 21]. Notably, one distinctive characteristic of the automotive sector lies in managing concurrent new product developments, often in the form of various products or projects.

NPD emerges as a strategic cornerstone for the growth and sustainability of the automotive industry. NPD process initiatives fuel the industry's innovation engine, enabling the development of cutting-edge automobiles that cater to evolving customer preferences and needs. This strategic emphasis on the NPD is rooted in the fundamental reality that the competitive advantage of most firms hinges significantly on successfully introducing innovative products to the market [7]. Furthermore, this symbiotic relationship between the automotive industry's economic or financial clout and its commitment to new product development highlights its vital role in driving sustained economic growth and development. In addition to enhancing the industry's competitiveness, NPD activities stimulate technological advancements, positioning automotive companies as global leaders. The automotive sector ensures steady growth by embracing innovation through NPD, making a substantial contribution to the advancement of the nation's economy while simultaneously shaping the future of mobility domestically and globally [14].

The NPD process is paramount in the automotive industry as companies continuously operate across diverse product lines. These companies are in a perpetual cycle of innovation, introducing new products to replace existing ones or carve out entirely new market niches. NPD is a pivotal source of competitive advantage for such firms, representing a vital process for businesses' survival, success, and rejuvenation [5]. This significance is further underscored in markets with frequent exposure to product changes and cutthroat competition. However, this pursuit of innovation comes with challenges, notably the dual hurdle of cost and risk [2]. Launching an innovation entails substantial financial, resource, workforce, machinery, and time investment, and failures in these endeavours can result in the squandering of product development funds and budget allocations [6, 23]. Therefore, the strategic imperative lies in selecting promising ideas/products based on their utmost utility measure in the early stages of the product engineering process to optimize the development and initiation of innovation.

Early product selection in the automotive industry is essential. It establishes the foundation for a successful product lifecycle, impacting profitability and market competitiveness. Early selection ensures that the product concept aligns with technology sustainability and market trends. Considering higher production costs and frequent innovation introduction in the automotive market, effective product selection minimizes resource waste and expedites development. Moreover, strategic product selection assists automotive companies in effectively investing in R&D, promoting the introduction, and providing a competitive advantage. In the automotive industry, early product selection shapes a product's design, features, and marketability trajectory. Regarding NPD, several factors must be considered while selecting a product in the initial stage. These include measures based on productive ability, marketing, and durability, such as technology feasibility, raw material and equipment, product life cycle, competitive quality-price, and supplier relationship, to name a few. In such scenarios, some decision-making technique, like utility theory analysis, proves invaluable [3]. This approach enables a systemic evaluation and comparison of the attributes mentioned above simultaneously concerning the product alternatives. The decision-makers can assign relative weights to these factors that reflect their importance. Utility theory facilitates a quantitative analysis of all product alternatives, ensuring comprehensive and unbiased assessment.

The authors have attempted to define a methodology for product selection in NPD that enhances the product's success. They have addressed a critical research gap: How can one identify or select the most promising product concept among several options generated in the first stage of product development, considering meets market demand and optimal resource allocation? To address the risks of product failure and cost-related splurge, the authors have presented the concept of utility theory analysis for product selection during the initial stage of NPD. The study aims to provide management with practical strategies to navigate the uncertainty during the innovation development process. Further, the applicability of the proposed framework has been illustrated with a case study of automotive product alternatives. It starts with identifying various product ideas in the early stage and then in the subsequent second stage. The authors have used utility theory analysis as a search and filtering tool to select the one with maximum utility for further development and planning. By doing so, this research offers management valuable tools for making informed decisions that enhance the efficacy of NPD from the outset. Moreover, it establishes a sturdy framework for automotive companies, enabling them to flourish in a swiftly changing marketplace and ensuring the successful initiation and exploration of innovative products.

1.1 New Product Development in the Automotive Industry

New product development refers to the strategic process of developing, designing, and introducing innovative automobiles, components, or technologies to the marketplace. It encompasses several interconnected stages, beginning with generating ideas and market research and culminating with introducing or launching innovative automotive products. In between, other stages include concept development, design, engineering, prototyping, testing and validation, and manufacturing [4]. NPD in the automotive industry can be driven by the constant need to meet customers' preferences, demands, technological breakthroughs, regulatory requirements, and marketing dynamics. It calls for interdisciplinary collaborative effort, technological innovation, and a comprehensive understanding of market trends and customer demands. To ensure or meet the automotive industry's continued expansion and viability in the global market, NPD's primary objective is to develop high-quality, innovative, economically sustainable, and cost-effective products that not only solely meet but exceed customer expectations [8]. The automotive industry navigates the four stages: Product idea generation, Product selection, Product evaluation, and Product development/planning [20], as shown in Fig. 1.

I. *Product Idea Generation: The Creative Genesis*

The product idea generation stage is the initial stage of NPD that serves as the creative genesis of automotive innovation; creative and innovative concepts for new products or improvements to existing products are brainstormed and developed. Expertise, visionaries, engineers, and architects converse at this stage to conceptualize and develop innovative ideas. The automotive manufacturers collaborate to identify market trends, availability or surplus of resources, emerging technologies, and demand of customers as the source of inspiration. Cross-functional teams engage in brainstorming sessions, including diverse perspectives, to produce a rich tapestry of ideas. While engineers assess the practicality of innovation, visionaries generate potential mobility alternatives, such as automated and e-automobiles. Product idea generation can be marked within the automotive sector by driving a balance between

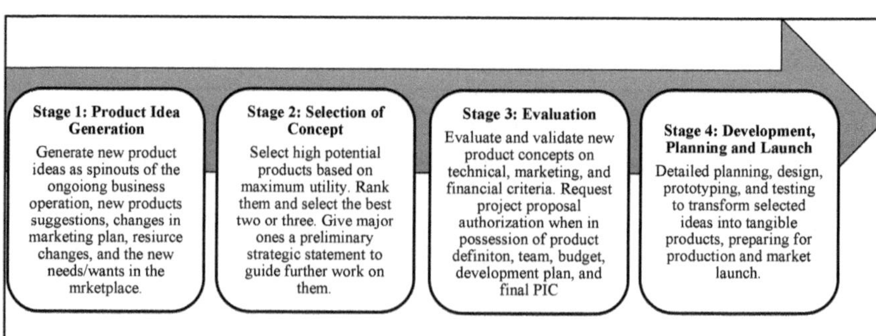

Fig. 1 Automotive new product development process

pragmatism and innovative thinking. The ultimate goal is to balance creative thinking and real-world applicability, ensuring the ideas generated have the potential to transform the industry.

II. *Product Selection*

The idea-to-reality process requires understanding the product's utility and demand. Product selection is the phase where the product ideas identified in stage 1 are subjected to thorough market analysis and research. The identified products' utility, effectiveness, and prospective demand are assessed at this stage. Analyzing utility or usefulness requires evaluating the pragmatics, efficacy, and satisfaction associated with specific characteristics, attributes, and features the product intends to provide. The utility of these factors can be evaluated through surveys among firms, focus groups, and data analysis, which aids in determining which product concepts are the most in line with the requirements or which product has the utmost utility measure per the availability of the resources and decision-makers preferences. This phase entails interpreting the intricate web of customer preferences within the automobile sector.

Utility measure analysis is the main component of this step, which measures or quantifies the appeal of various automobile features and attributes. For instance, higher priority can be allocated to innovative safety technology, fuel efficiency, or seamless connectivity. Utility Theory Analysis examines these preferences, yielding quantitative insights that can direct alternatives with higher utility for subsequent NPD stages. For the automotive industry, this stage is transformative and aligns the innovation concept and the practical utility of the vehicle.

III. *Product Evaluation: Balancing Vision and Viability*

In the product evaluation stage, the selected products or ideas from the previous stage undergo rigorous scrutiny and assessment based on their respective utility. The cross-functional teams evaluate the technical feasibility, costs related to production, regulatory compliance, and potential risks associated with products. Considering the sustainability and environmental aspects, the product evaluation stage is pivotal in the automotive industry. This step ensures that the selected product concepts are compatible with the visionary and economically and operationally viable.

IV. *Product Development, Planning and Launch*

Product development, or Product planning, is the final stage where the selected product concepts are transformed into tangible products/automotive. Based on the evaluated concepts, this stage involves extensive research and development activities, prototyping, testing, and refining product design, ensuring the automotive design aligns with the product's initial vision. Product development also encompasses creating a comprehensive manufacturing, distribution, marketing, and customer support strategy. Launching the product into the market indicates the termination of the NPD stages and marks the transition from conceptualization to a market-ready product.

Each stage of NPD marks a critical juncture in the automobile's development process, from idea generation to the final product launch [12]. Understanding the subtleties of each step is crucial for developing automobiles that satisfy customers' expectations, utilize resources effectively and efficiently, and establish new benchmarks for the corresponding sector.

1.2 Understanding Product Selection Using Utility Theory

For automotive companies, selecting practical product concepts within NPD is crucial, exerting a profound impact on all subsequent product development and market success [10]. Mitigating risk associated with the product, optimizing resources, aligning with market demands, and ensuring technological feasibility are critical considerations at this stage of NPD. Product selection encapsulates a complex process involving the evaluation of different attributes among diverse alternatives. Each attribute or factor represents a distinct dimension contributing to the integrated desired product outcome. The automotive industry is ubiquitous, with trade-offs and interdependencies among these factors. For instance, enhanced performance may entail increased production expenses, affecting affordability. Moreover, technological advancements, market dynamics, and resource availability introduce further complexities in selecting products among alternatives. As decision-makers endeavour to optimize among these diverse and often conflicting dimensions, the apparent necessity for advanced decision-making tools becomes evident. These complexities inherent in product selection make it a decision-making problem, necessitating advanced tools such as Utility Theory Analysis [9]. Utility Theory Analysis, as a filtering tool, facilitates a systematic approach while considering different attributes simultaneously and assigning appropriate weights based on their relative importance.

In the literature, utility theory has been used significantly in various decision-making contexts, including product selection, material selection, and human resource management. The foremost method proposed by Pahl and Beitz [15] is a seven-stage method that incorporates the utility theory for decision-making. Khurana et al. [11] provided a utility-based approach to evaluate the overall efficiency of HRD instruments using utility theory analysis. In the NPD domain, Ying et al. [22] proposed a strategy for product selection considering decision-maker psychology. The literature underscores the flexibility and effectiveness of utility theory analysis across different decision-making scenarios. The authors have utilized the utility theory proposed by Pahl and Beitz [15]. The authors have proposed the applicability of utility theory in the context of NPD for optimal product selection of different automotive products. The utility of the proposed approach in the early NPD stage enables companies to evaluate distinctive and complex alternatives while considering attributes.

Utility Theory Analysis is a mathematical framework that defines and assesses utility measures for different attributes. It aids in rationalizing the decision-making process by quantifying and synthesizing diverse factors. It ensures that the selected

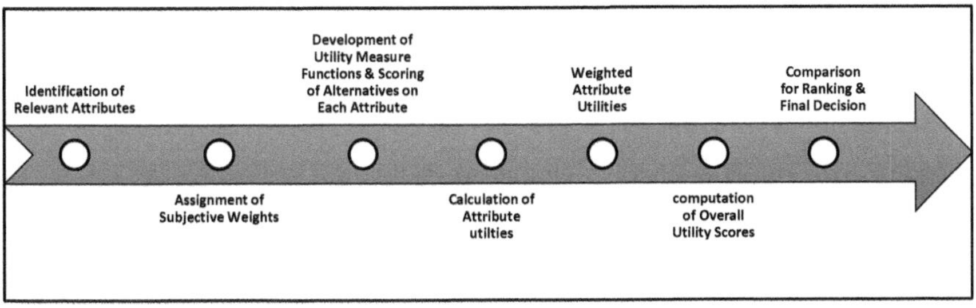

Fig. 2 Steps of utility theory-based analysis for product selection

automotive product aligns seamlessly with the strategic objectives and preferences of stakeholders, consumers, and the market. Thereby augmenting the overall success of the NPD process. The following section elaborates on the methodology of utility theory analysis that contributes significantly to product selection, as shown in Fig. 2.

Following are the notations that have been used in a later section of the analysis:

S Set of alternatives
X Number of alternatives
Y Number of attributes
f_a utility function for attribute a, $a = 1, 2, ..., Y$
w_a Weights assigned to attribute a, $a = 1, 2, ..., Y$
s_{ab} Score of alternative b on attribute a; a, $a = 1, 2, ..., Y$ and $b = 1, 2, ..., X$
u_{ab} Utility of alternative b on attribute a
U_b Overall utility of alternative b

I. *Identification of Attributes*

Initially, decision-makers identify a comprehensive set of attributes critical for evaluating alternative automotive products. The following list of attributes forms the foundation for thoroughly assessing desirable products. These attributes include measures such as performance, production, marketing, cost, design, and environmental impact.

II. *Assignment of Subjective Weights*

This step reflects each attribute's perceived importance and value in decision-making. Decision makers collaboratively assign weights, w_a to each identified attribute. Multi-criteria decision-making (MCDM) techniques, such as the Analytical Hierarchy Process (AHP), can be used for the assignment of weights concerning each attribute [17, 19]. The weight assignment process involves the engagement of experts, stakeholders, company executives, manufacturers, etc., to ensure that the weights accurately capture the shared priorities of the decision-making team.

III. *Development of Utility Functions and Scoring of Alternatives on Each Attribute*

Decision-makers' satisfaction and preferences with an alternative's performance on specific attributes are mathematically represented in terms of utility functions, f_a. These functions play a crucial role in transforming qualitative preferences into quantified values. The form of the utility function (f_a) depends on decision-makers' attributes and preferences.

Further, decision-maker assign scores, s_{ab} to asses each alternative on every identified attribute., These scales ensure consistency in the evaluation process by adhering to the units and scales defined within utility functions.

IV. *Calculation of Attribute Utilities*

Utilities corresponding to each attribute represented by u_{ab} are then calculated by applying the utility function (f_a) to the respective scores (s_{ab}). This step transforms the qualitative performance measures into quantitative utility values, quantifying the satisfaction associated with each alternative's performance on individual attributes. For each attribute $a, a = 1, 2, ..., Y$ and alternative $b, b = 1, 2, ..., X$, the utility u_{ab} can be calculated using the utility function given by: $u_{ab} = f_a(s_{ab})$.

V. *Weighted Attribute Utilities*

After calculating attribute utility measures, they are multiplied by their corresponding subjective weights (w_a). The resultant is the set of weighted attribute utility measures that represent the contribution of each attribute to the overall outcome of an alternative and is given by: $w_a * u_{ab}$.

VI. *Computation of Overall Utility Score*

The overall utility measure score, U_b for each alternative can be obtained by summing up across all computed weighted attribute utilities. The overall utility score corresponding to the alternative can be represented as

$$U_b = \sum_{a=1}^{Y} (w_a * u_{ab}) \tag{1}$$

This score denotes the comprehensive desirability of an alternative while considering all relevant attributes and their respective assigned weights.

VII. *Comparison for Ranking and Result Communication*

In this step, each alternative is ranked based on its respective utility measure scores. The most desirable product concept is the alternative with the highest or utmost utility score, aligned with the respective attribute and its assigned weights.

A well-informed final decision is communicated transparently based on the insights obtained from Utility measures.

The structured approach of Utility measure analysis, combining theoretical preferences with mathematical formulation, enhances the overall alignment of product portfolios and concepts with the market dynamics of the ever-evolving automotive industry.

2 Case Study: Automotive Product Development

Numerous factors, such as interest in expansion, assessing a product's market potential, or optimizing unused or surplus capacity within current production processes (i.e., surplus machine hours or workforce), can direct an automotive industry to innovate. The authors have presented a case study for product selection in the automotive industry to combine the proposed comprehensive approach with practical application. The case study provides a hands-on experience that makes it easier to understand the proposal's implementation in the context of practical business. This strategic approach establishes the connection between theoretical tenets and their pragmatic implementation, along with improving the selection of automotive products. The study was conducted for XYZ Engineers, a prestigious manufacturing company in India; the company's name has been masked due to the company's policy. Renowned for manufacturing premium silencer mufflers and sheet metal fabrication, XYZ Engineers is a leading business in the automotive component industry. They serve various sectors, including heavy commercial vehicles and machinery, and specialize in innovative solutions such as pipe bending and CNC services. XYZ Engineers is a well-known and reliable brand praised for its excellence in manufacturing automotive components.

In the case of XYZ Engineers, the automotive company found itself with surplus capacity in terms of machine hours and workforce, prompting an interest in utilizing this excess capacity. Additionally, the company had surplus funds, and rather than depositing them in the bank or providing loans, they decided to invest in new products to maximize profits. After conducting a thorough analysis, the firm formulated plans to introduce an innovative automotive product. Once the company decides to introduce a new product, it undergoes several stages before the final product reaches consumers. Generally, the products progress through four initial stages before moving through subsequent phases, as outlined in Sect. 1.1.

In the initial stage, management generates various ideas to determine which product to produce. The management discusses with the Research and Development (R&D) team to conceptualize a new product that aligns with available resources. Following extensive deliberations and an analysis of competitors' offerings, five essential products were identified for creation, as shown in Table 1.

Table 1 The five products finalized through Stage 1

Product code	Product name
A	Muffler of Eicher Canter
B	Turbo Connector
C	Muffler of Piaggio Ape
D	Intercooler Pipe
E	Muffler of Eicher Tractor

The muffler is an essential component of an automobile's exhaust system, reducing the noise produced by the engine's exhaust emissions. Beyond its role in noise reduction, a high-quality silencer, such as mufflers for vehicles like Canter, Piaggio Ape, and Tractor, has the potential to enhance overall engine performance and fuel efficiency. The meticulous design and construction of mufflers classify them as an essential type of sound attenuation device for automobiles. Additionally, turbo connectors belong to the components associated with the turbocharging system in certain vehicles. These connectors regulate and direct the airflow between turbocharging and the engine. High-quality connectors are essential for ensuring optimal engine performance and efficiency. Simultaneously, intercooler pipes are an integral component of the intercooler system in turbocharged engines. These pipes cool compressed air before it enters the engine, significantly improving efficiency.

The mufflers, turbo connectors, and intercooler pipes emphasize their impact on automobile noise reduction and performance optimization. The significance of each of these components is contingent on the specific goals and requirements of the automobile, with each element serving a distinct purpose that contributes to overall vehicle performance and functionality.

2.1 Product Search: Filtering or Searching Process

The search for a viable automotive product commences with a focus on the overall acceptability of the proposed ideas. The evaluation is facilitated through a filtering device utility measure. Using this measure, the company retains only a selected few ideas while discarding approximately 90% of the product concepts generated in the initial stage. Utility measure gauges how much utility a product holds for a specific firm, with a dedicated definition of the utility measure for each product concept or idea.

As the management moves from the first stage, a critical decision must be made among the five potential products. The management employs a product search and filtering procedure to streamline this decision-making process and choose from many options. This process involves discarding many product ideas and selecting only those deemed most promising. *Now the question was which product idea was better: A, B, C, D, or E?*

To answer this question, the authors have defined the Utility measure for all five product ideas and arranged them according to their utility measure.

The following attributes were identified and shortlisted. Appendix 1 provides an explanation and relative utility measure scores for each attribute.

1. *Productive ability*

    a. Equipment (A1)
    b. Personnel (A2)
    c. Raw Material (A3)

2. *Marketability*
    a. Relation to present distribution channel (A4)
    b. Competitive quality-price relation (A5)
    c. Effect on sales of existing product (A6)
3. *Durability*
    a. Size of potential market (A7)
    b. Resistance to economic fluctuation (A8)
    c. Life Cycle (A9)

To determine the relative importance of each attribute for the new product context, they were assigned weights using an MDCM approach, AHP. Analytical Hierarchy Process (AHP) is a systemic and structured decision-making methodology that is useful when faced with complex decisions involving multiple criteria and alternatives [3]. It works effectively in scenarios where qualitative factors are challenging to quantify, enabling decision-makers to incorporate subjective judgments in the decision-making process. AHP can be utilized when an extensive and comprehensive criteria evaluation is required, considering their interdependencies and relative importance. The methodology involves a hierarchical structure wherein decision-makers conduct pairwise comparisons to determine the numerical values representing the preferences for each criterion. Followed by a consistency check ensuring the judgment reliability, and using mathematical calculation, the weights for each attribute are determined and assigned for further process.

AHP has been utilized to determine attribute weights for a new product selection in NPD and is an effective methodology in addressing the complexity of evaluating several factors concerning equipment, resources, marketing, cost, and functionality, to name a few. AHP aids in making well-informed decisions in line with the objective and requirement of the product development process by systematically allocating weights to these attributes. In such instances, the utilization of AHP enhanced the transparent and objective decision-making method by effectively evaluating the most critical characteristics of the new product's success. To assign the weights, the firm manager was asked about the factors that would help launch the new automotive [19]. Table 2 shows the weights the organization has assigned to the attributes above.

A survey was conducted to derive the probabilities associated with these factors for all five product ideas, involving ten members each from the production, marketing, and durability teams. Once the various levels for each factor were defined, a numerical scaling/score was applied arbitrarily yet consistently, indicating their relative importance or value. A five-level scale was considered to ask the team members about their options. For simplicity, these weights were standardized as follows:

VG Very Good (10)
G Good (8)
A Average (6)
P Poor (4)
VP Very Poor (2)

Table 2 Attribute weights using AHP

Weights	Attributes	Assigned weights
w_1	Equipment (A1)	0.05
w_2	Personnel (A2)	0.10
w_3	Raw material (A3)	0.05
w_4	Relation to present distribution channel (A4)	0.05
w_5	Competitive quality-price relation (A5)	0.10
w_6	Effect on sales of existing product (A6)	0.15
w_7	Size of potential market (A7)	0.20
w_8	Resistance to economic fluctuation (A8)	0.10
w_9	Life cycle (A9)	0.20

For instance, they assign a score of 10 for the utility of equipment to the VG level specified when the new product utilizes the excess capacity of the existing equipment. Further, it is essential to note that these scores should remain independent of the worth of the factor itself. In assigning these weights, one should be more concerned about the relative score within the factor rather than the importance of the factor itself. The overall utility measure must also account for the relative importance of the factor.

After identifying attributes, assigning weights, and scoring, the next step is to calculate utility measures corresponding to each attribute for every product alternative.

The total expected utility measures for all five products have been calculated and shown in Tables 3, 4, 5, 6, and 7, respectively.

Table 3 provides a detailed utility evaluation of product A. In particular, concerning the muffler of Eicher Canter and its impact on the utility of equipment (A1), the analysis showed that 60% of the production team members affirm its

Table 3 Utility measure for product A

Factors	Weights	Level					Level weights (expected)	Total contribution
		10	8	6	4	2		
A1	0.15	0.6	0.2	0.2	0	0	8.8	1.32
A2	0.20	0.9	0.1	0	0	0	9.8	1.96
A3	0.10	0.4	0	0.6	0	0	7.6	0.76
A4	0.10	0.8	0.2	0	0	0	9.6	0.96
A5	0.10	0	0	0.6	0.4	0	5.2	0.52
A6	0.05	0.7	0.2	0.1	0	0	9.2	0.46
A7	0.10	0.1	0.6	0.3	0	0	7.6	0.76
A8	0.05	0.1	0.2	0.5	0.2	0	6.4	0.32
A9	0.15	0.2	0.7	0.1	0	0	8.2	1.23
Total expected utility								8.29

Table 4 Utility measure for product B

Factors	Weights	Level					Level weight (expected)	Total contribution
		10	8	6	4	2		
A1	0.15	0	0.1	0.6	0.3	0	5.6	0.84
A2	0.20	0	0.2	0.5	0.3	0	5.8	1.16
A3	0.10	0.4	0	0	0.6	0	6.4	0.64
A4	0.10	0	0	0	0.3	0.7	2.6	0.96
A5	0.10	0.1	0.5	0.4	0	0	7.4	0.74
A6	0.05	0.2	0.2	0.6	0	0	7.2	0.36
A7	0.10	0.2	0.6	0.2	0	0	6	0.60
A8	0.05	0.2	0.3	0.4	0.1	0	7.2	0.36
A9	0.15	0.3	0.2	0.3	0.2	0	7.2	1.08
Total expected utility								6.04

Table 5 Utility measure for product C

Factors	Weights	Level					Level weights (expected)	Total contribution
		10	8	6	4	2		
A1	0.15	0	0.6	0.4	0	0	7.2	1.08
A2	0.20	0.6	0.4	0	0	0	9.2	1.84
A3	0.10	0.1	0	0.2	0	0.7	3.6	0.36
A4	0.10	0	0.2	0.6	0.2	0	6	0.60
A5	0.10	0	0	0.4	0.2	0.4	4	0.40
A6	0.05	0	0.3	0.7	0	0	6.6	0.33
A7	0.10	0	0.4	0.6	0	0	6.8	0.68
A8	0.05	0	0.2	0.5	0.2	0.1	5.6	0.28
A9	0.15	0.4	0.5	0.1	0	0	8.6	1.29
Total expected utility								6.86

utilization of excess capacity. Meanwhile, 20% assert that it will leverage the existing equipment but at the expense of reducing the production of current products. Another 20% of the team members suggest additional equipment might be required. Similarly, comparable data analyses were conducted for the remaining attributes. The expected level weight for A1 can be evaluated as follows:

$$10 * 0.6 + 8 * 0.2 + 6 * 0.2 = 8.8$$

The total utility contribution corresponding to A1 for product A can be calculated as follows:

$$w_1 * U_{A1\_A} = w_1 * [f_1 * (s_{A1\_A})] = 0.15 * (10 * 0.6 + 8 * 0.2 + 6 * 0.2)$$

Table 6 Utility measure for product D

Factors	Weights	Level					Level weight (expected)	Total contribution
		10	8	6	4	2		
A1	0.15	0.4	0.2	0.4	0	0	8	1.20
A2	0.20	0	0.2	0.6	0.2	0	6	1.20
A3	0.10	0.8	0	0	0	0.2	8.4	0.84
A4	0.10	0	0	0.1	0.3	0.6	3	0.30
A5	0.10	0	0.2	0.6	0.2	0	6	0.60
A6	0.05	0.1	0.3	0.6	0	0	7	0.35
A7	0.10	0.1	0.6	0.3	0	0	7.6	0.76
A8	0.05	0.2	0.4	0.4	0	0	7.6	0.38
A9	0.15	0.3	0.7	0	0	0	8.6	1.29
Total expected utility								6.92

Table 7 Utility measure for product E

Factors	Weights	Level					Level weight (expected)	Total contribution
		10	8	6	4	2		
A1	0.15	0.2	0.6	0.2	0	0	8	1.20
A2	0.20	0.7	0.1	0.2	0	0	9	1.80
A3	0.10	0	0.1	0.2	0.2	0.5	3.8	0.38
A4	0.10	0	0	0	0.2	0.8	2.4	0.24
A5	0.10	0	0	0.9	0.1	0	5.8	0.58
A6	0.05	0.1	0.1	0.8	0	0	6.6	0.33
A7	0.10	0	0	0.8	0.2	0	5.6	0.56
A8	0.05	0.2	0.6	0.2	0	0	8	0.40
A9	0.15	0.3	0.5	0.2	0	0	8.2	1.23
Total expected utility								6.72

$$= 0.15 * 8.8 = 1.32$$

Likewise, the total contributions of respective attributes for product alternative A can be obtained. Lastly, the total expected utility measure for product A can be calculated using Eq. (1) as given:

$$U_A = \sum_{a=1}^{9} (w_A * u_{Attribute\,Aa\_A}) = 8.29$$

Therefore, the total expected utility measure for all other product alternatives can be obtained similarly.

Once the total expected utility of all the alternatives has been obtained, they can be compared concerning the ideal and worst-case scenarios.

3 Discussion

The utility measures presented in Tables 8 and 9 can be used to rank alternatives and provide essential insights into the decision-making process. Notably, the best-case scenario can be defined by the total expected utility of 10, which has been used as a benchmark for the best possible outcome. It illustrates the ideal performance when every attribute is at the highest level possible. In the worst-case scenario, having a total expected utility measure of value 2 represents the least favourable outcome when all attributes are at their lowest levels and considered the lowest possible value in the utility measurement range. For ranking alternatives, a strategic approach involves the evaluation of each alternative's performance in terms of their utility measures against these calculated benchmarks. Alternatives aligned to the best-case scenario can be ranked highest, reflecting their potential for optimal outcomes.

The alternatives have been ranked based on comparing best–worst thresholds of utility measures, outlined in Table 10.

The result table indicates that the muffler of Eicher Canter, Product A, has ranked 1. In other words, Product A has the maximum total expected utility. It is followed by the Intercooler Pipe, the Muffler of Piaggio Ape, the Muffler of Eicher Tractor, and the Turbo Connector, which ranked 2, 3, 4, and 5, respectively. Based on the table, it is evident that the company should proceed with product A, the Muffler of Eicher Cater, for the remaining two stages of product development and planning.

Table 8 Best case

Factors	Weights	Level					Level weight (expected)	Total contribution
		10	8	6	4	2		
A1	0.15	1	0	0	0	0	10	1.5
A2	0.20	1	0	0	0	0	10	2.0
A3	0.10	1	0	0	0	0	10	1.0
A4	0.10	1	0	0	0	0	10	1.0
A5	0.10	1	0	0	0	0	10	1.0
A6	0.05	1	0	0	0	0	10	0.5
A7	0.10	1	0	0	0	0	10	1.0
A8	0.05	1	0	0	0	0	10	0.5
A9	0.15	1	0	0	0	0	10	1.5
Total expected utility								10

Table 9 Worst case

Factors	Weights	Level					Level weight (expected)	Total contribution
		10	8	6	4	2		
A1	0.15	0	0	0	0	1	2	0.3
A2	0.20	0	0	0	0	1	2	0.4
A3	0.10	0	0	0	0	1	2	0.2
A4	0.10	0	0	0	0	1	2	0.2
A5	0.10	0	0	0	0	1	2	0.2
A6	0.05	0	0	0	0	1	2	0.1
A7	0.10	0	0	0	0	1	2	0.2
A8	0.05	0	0	0	0	1	2	0.1
A9	0.15	0	0	0	0	1	2	0.3
Total expected utility								2

Table 10 The final ranks

Product code	Product name	Utility	Rank
A	Muffler of Eicher Canter	0.829	1
B	Turbo Connector	0.604	5
C	Muffler of Piaggio Ape	0.686	3
D	Intercooler Pipe	0.692	2
E	Muffler of Eicher Tractor	0.672	4

4 Conclusion

The New Product Development (NPD) demands significant investments in finances, resources, workforce, machinery, and time in the automotive industry. Failure in these areas can result in the waste of product development funds and resources. Therefore, the primary objective is to identify promising ideas and products in the early stage of development, focusing on the highest utility measure. The proposed study aims to enhance product development efficiency in the early stages while mitigating risks through effective product selection and filtering. Utility theory analysis emerges as a filtering tool for selecting products based on their maximum utility. Early product selection plays a crucial role in automotive product development by laying the groundwork for a successful product lifecycle, influencing profitability and market competitiveness. The practicality of utility theory as a filtering tool has been demonstrated as a strategic tactic for effective resource allocation by management. When faced with multiple product options, this filtering tool can determine which product should progress to the subsequent development stages. Moreover, the optimal product selection in the early phases of NPD based on maximum utility improves the success rate of product development, reducing risks and cost overruns. Integrating utility theory in NPD during the early product selection ensures alignment with technology sustainability and market trends. This proactive strategic approach enhances development process efficiency and increases the likelihood of successful launches. Hence, it can be concluded that NPD can be more effective, efficient, and aligned with market demands, ultimately leading to better outcomes and reduced risks.

Acknowledgements The work in this chapter has been supported by grants sanctioned to the second author from the Institute of Eminence, DU India, as part of a Faculty Research program via Ref. No/IoE/2024-2025/12/FRP.

Appendix 1

Productive ability: For each product idea we ask different questions, i.e., regarding

(a) **Equipment**
(b) **Personnel (Technical know-how)**
(c) **Raw Materials**

What types of equipment are required?

Explanation	Rating scale
If it utilizes the excess capacity	Very good
If it utilizes the present equipment but the production of existing products reduces	Good
If some additional equipment is also required	Average
If mostly new equipment but some existing equipment can be used	Poor
If all the new equipment are required	Very poor

Personnel

Explanation	Rating scale
If it utilizes the present personnel exclusively	Very good
If few new personnel are needed at low level	Good
If many new personnel except high level	Average
If mostly new personnel except existing low level personnel	Poor
If all the new personnel are required	Very poor

Raw materials	
Explanation	Rating scale
If the same raw materials is required and can obtained from the present contractor/ supplier	Very good
If the same raw material is required but from new supplier (the existing ones refuse to supply)	Good
If some existing but some new raw material is required from the present supplier	Average
If the new raw material is obtained but some experience is available for procurement of these raw material	Poor
If the new raw material is obtained from the unknown Supplier, i.e. we don't have enough information about the supplier	Very poor

Marketability: Here one might be concerned with

(a) **Relationship to the present distribution channels**
(b) **Competitive quality-price relationship, and**
(c) **Effect on sales of existing products**

Relation to the present distribution channel	
Explanation	Rating scale
If the present channels can be used exclusively	Very good
If few new channels are required (because some dealers may refuse to distribute the new product due to space or financial or some other reason)	Good
When equal no. of new and existing channels are required	Average
When very few existing channels can be used	Poor
Entirely new channels are required	Very poor

If we produce that product whether we have to appoint new dealers or it can be sold through the same dealers

Competitive quality-price relation	
Explanation	Rating scale
Product can be priced below all competitive products of similar quality	Very good
Product can be priced below most of the competitive products of similar quality	Good
Product can be priced same as competitive products of similar quality	Average
Product can be priced above many competitive products of similar quality	Poor
Product can be priced above all the competitive products of similar quality	Very Poor

Effect on Sales of existing product	
Explanation	Rating scale
It will significantly improve the sales of the existing product	Very good
If there is possibility of small positive effect on sales of existing product	Good
No effect	Average
If there is possibility of a small negative effect on sales of the existing product	Poor
If it significantly reduces sales of the existing product	Very poor

Durability: The durability of the product may be evaluated in terms of:

(a) **Size of the potential market**
(b) **Resistance to economic fluctuations**
(c) **Life Cycle**

Size of the potential market	
Explanation	Rating scale
If there is large national and international market	Very good
Wide national market	Good
When only regional markets	Average
Very few regional markets	Poor
Very specified narrow market	Very poor

Resistance to the Economic Fluctuations	
Explanation	Rating scale
When the product can be sold in depression and inflation	Very good
If there is moderate effect of economic cycle or fluctuations	Good
If the sales will follow the economic trend	Average
Heavy dependence on economic conditions	Poor
Total dependence on economic fluctuations	Very poor

Whether the product can be sold under deflation or inflation

Life cycle	
Explanation	Rating scale
It is expected to sell the product for a very long period	Very good
It has a longer lifetime then the competitor's product	Good
Average life cycle for the industry	Average
Shorter life than competitor's products	Poor
Will rapidly become obsolete, i.e. demand for a very small period	Very poor

Whether the product is luxury or necessity

References

1. Agarwal, M., Aggrawal, D., Anand, A., Singh, O.: Modeling multi-generation innovation adoption based on conjoint effect of awareness process. Int. J. Math. Eng. Manag. Sci. **2**(2), 74 (2017)
2. Agarwal, M., Anand, A., Chanchal., Kumar, H.: Determination of launch time for a multi-generational product: a fuzzy perspective. In: Advances in Soft Computing Applications. River Publishers, pp. 105–119 (2023)
3. Anand, A., Agarwal, M., Aggrawal, D.: Multiple Criteria Decision-Making Methods: Applications for Managerial Discretion, vol. 14. Walter de Gruyter GmbH & Co KG (2022)
4. Anand, A., Aggrawal, D., Agarwal, M.: Market Assessment with OR Applications. CRC Press (2019)
5. Anand, A., Bansal, G., Agarwal, M., Aggrawal, D.: Remanufacturing: an industrial strategy for economic benefits. Diag. Tech. Indus. Eng. 137–155 (2018)
6. Anand, A., Chanchal, Kapur, P.K., Tamura, Y.: Quantification of number of adopters: a study to showcase products-sold and products-in-use. Int. J. Syst. Assur. Eng. Manag. 1–13 (2023)
7. Aronov, J., Carrión García, A., Papic, L., Galkina, N., Aggrawal, D., Anand, A.: Standards development and innovative products. when should standards be prepared? Int. J. Math. Eng. Manag. Sci. **4**(5), 1081–1093 (2019)
8. Bansal, G., Anand, A., Tiwari, S.: Role of remanufacturing in product development and related profit estimation. J. Clean. Prod. **277**, 124016 (2020)
9. Fishburn, P.C., Fishburn, P.C.: Utility Theory for Decision Making. Krieger, NY (1979)
10. Gerhard, D., Brem, A., Voigt, K.I.: Product development in the automotive industry: crucial success drivers for technological innovations. Int. J. Technol. Mark. **3**(3), 203–222 (2008)
11. Khurana, D.K., Kapur, P.K., Sachdeva, N.: Utility based tool to assess overall effectiveness of HRD instruments. Int. J. Bus. Anal. (IJBAN) **4**(2), 20–36 (2017)
12. Kotler, P., Armstrong, G.: Principles of Marketing, 15th edn. Pearson Education Limited (2014)
13. Kumar, H., Singh, O., Anand, A., Irshad, M.S.: Studying multi-stage diffusion dynamics using epidemic modeling framework. Int. J. Math. Eng. Manag. Sci. **8**(1), 105 (2023)
14. Liang, J.C.: An integrated product development process in the automotive industry. Int. J. Prod. Dev. **8**(1), 80–105 (2009)
15. Pahl, G., Beitz, W.: Engineering Design, The Design Council, vol. 12, pp. 221–226. London, Springer (1984)
16. PIB.: https://static.pib.gov.in/WriteReadData/specificdocs/documents/2023/feb/doc2023217160601.pdf (2023)
17. Saaty, T.L.: A note on the AHP and expected value theory. Socioecon. Plann. Sci. **20**(6), 397–398 (1986)
18. SIAM.: https://www.siam.in/statistics.aspx?mpgid=8&pgidtrail=15 (2023)
19. Tzeng, G.H., Huang, J.J.: Multiple Attribute Decision Making: Methods and Applications. CRC Press (2011)
20. Urban, G.L., Montgomery, D.B.: Management science in marketing (1969)
21. Vaz, C.R., Rauen, T.R.S., Lezana, Á.G.R.: Sustainability and innovation in the automotive sector: a structured content analysis. Sustainability **9**(6), 880 (2017)
22. Ying, C.S., Li, Y.L., Chin, K.S., Yang, H.T., Xu, J.: A new product development concept selection approach based on cumulative prospect theory and hybrid-information MADM. Comput. Ind. Eng. **122**, 251–261 (2018)
23. Zapata, C., Nieuwenhuis, P.: Exploring innovation in the automotive industry: new technologies for cleaner cars. J. Clean. Prod. **18**(1), 14–20 (2010)

Imperfect Production Inventory Under a Multi-Production Cycle for Non-deteriorating Items with Carbon Tax and Green Investment

Nabajyoti Bhattacharjee, Nabendu Sen, Prodosh Kiran Nath, and Laxminarayan Sahoo

Abstract In small manufacturing sectors, the storage of finished products is an essential and inevitable part of inventory management. In reality, the holding space is usually finite and due to this, a risk of early shortage is always there that could reduce the profit gain by the manufacturer. In this paper, an imperfect production inventory model is proposed to study the impact of the multi-production cycle on the profit earned for non-deteriorating items. Carbon tax is considered in the model and optimal green investment is obtained to reduce the carbon tax in both the holding and production. The formulated model is analyzed through numerical examples along with the sensitivity analysis of the parameters and managerial implications.

Keywords Production inventory model · Multi-production cycle · Carbon tax · Green investment · Non-deteriorating items

1 Introduction

The manufacturing sector producing items with negligible deterioration rate faces a real issue of storing finished products as the holding space is finite therefore it is wise to perform the production in more than one cycle. The holding capacity in a production sector is finite so during the production process the total inventory should not exceed a certain limit. So, to meet the market demand manufacturers have to produce the product more than once in a given planning horizon. Further, we consider

N. Bhattacharjee · N. Sen (✉)
Department of Mathematics, Assam University, Silchar 788011, India
e-mail: nsen08@yahoo.com

P. K. Nath
Assam University, Silchar 788011, India

L. Sahoo
Department of Computer and Information Science, Raiganj University, Raiganj 733134, India

L. Sahoo et al. (eds.), *Decision Making Under Uncertainty Via Optimization, Modelling, and Analysis*, Studies in Systems, Decision and Control 558,
https://doi.org/10.1007/978-981-96-0085-4_14

the manufacturing of ethanol-based products (e.g. biofuel, wine, beer, etc.). During the production process of ethanol base products which are usually fermentation, integrated distillation, dry mill, etc. emission of carbon-based gases is inevitable, and consequently, a carbon tax is associated with the unit of production. Moreover, to reduce the carbon tax the production house could make investments in green technologies. The manager can fix an inventory level as a threshold limit, when the inventory of finished products reaches the threshold limit the next production cycle begins. Maintaining the production cycle in this manner will overcome the shortage and further resolve the issue of finite holding space.

Much research has been done to study the production cycle under different environments. Ghosh et al. [7] provide an extensive study on different carbon policies for the production inventory model acknowledging green production. Aarya and Mukesh [1] proposed a production inventory model for deteriorating items with stock and time-dependent demand and backlogging. Singh [15] proposes a production inventory model with stock-dependent and selling price-dependent demand. Shortages are allowed with partial backlogging for deteriorating items. Tripathi and Mishra [16] developed a production inventory model with stock-dependent demand considering the effect of deterioration on demand and the production process. Halim et al. [8] developed a production inventory model for deteriorating items with both price and stock-dependent demand to study the overtime production process. The following is a research on production inventory and research gap.

References	Demand function	Threshold limit for multiple production	Carbon emission	Green investments	Multi-production cycle
Ritha and Poongodisathiya [13]	Constant	No	Yes	Yes	No
Palanivei and Gowri [12]	Quadratic and price dependent	No	No	No	No
Gautam and Kahnna [5]	Constant	No	Yes	No	No
Shen et al. [14]	Constant	No	Yes	No	No
Mashud et al. [10]	Constant	No	No	No	No
Mishra et al. [11]	Constant	No	Yes	Yes	No
Das et al. [4]	Replacement period-, stock-, price-dependent	No	No	No	No
Gautam et al. [6]	Price dependent	No	Yes	No	No
Kamna et al. [9]	Price dependent	No	No	No	No
Halim et al. [8]	Price and Stock dependent	No	No	No	Yes

(continued)

(continued)

References	Demand function	Threshold limit for multiple production	Carbon emission	Green investments	Multi-production cycle
Buttar and Sharma [3]	Time dependent	No	No	No	No
Bhattacharjee and Sen [2]	Price dependent	No	Yes	Yes	No
This paper	Price dependent	Yes	Yes	Yes	Yes

The study of the production process subject to the multiple productions under the threshold limit for non-deteriorating items with a carbon tax is not reported in the literature on the production inventory model. Further, the study on the implementation of the multi-production cycle in a given planning horizon to overcome the problem of shortage in an imperfect production model has also not been reported in the literature so far, as the author's knowledge is concerned. So, to fill the gap in the literature of production inventory an attempt has been made to develop an imperfect production inventory model to study the multi-production cycle under threshold limit for multiple productions with non-deteriorating items along with carbon tax and green investment. The description of the problem and objective of the study is illustrated in Sect. 2.

2 Description of Problem and Objective

In our present EPQ model, we consider a system producing a single non-deteriorating item. Due to the smallholding space, the manager of a production sector decides to plan the production in such a way that, whenever the inventory level is reduced to a certain limit after the production process, the next production cycle is initiated and the process continues until the planning horizon, in this way it is possible to overcome the shortage which may arise because of small storage capacity or long planning horizon. In this present model, an effort has been made to develop an imperfect production inventory model to study the impact of the multi-production cycle in the production of a single non-deteriorating item in a manufacturing sector with finite holding space. Carbon emission is considered in both holding and production also effective green investment is implemented as a measure to reduce carbon emission exponentially. The proposed model is an extension of the existing literature on production inventory models for non-deteriorating items with multiple production cycles in each planning horizon. The model formulated in this research work has not been reported in the literature of production inventory so far, the author's knowledge is a concern.

3 Notations and Assumptions

I. Notations

See Table 1.

II. Assumptions

During the formulation of the mathematical model, we consider the realistic factors but to put them in a valid mathematical expression we must set some assumptions. In our proposed model we assume the following:

(a) Lead time is zero.
(b) Demand is deterministic.
(c) The demand function is $D(p) = a - bp$

Where a is initial demand and b is price sensitivity.

(d) Deterioration is not considered. But the change in inventory concerning time is dependent on the perfect production and therefore the net production rate

Table 1 Notations

Symbols	Unit	Meaning
$I^+(t)$	Unit/unit time	Inventory level at time t during the production period
$I^-(t)$	Unit/unit time	Inventory level at time t after the production period
I^0	Unit	Threshold limit for next production
a	Unit	Initial demand
P	Unit/unit time	Net production rate
Q	Unit/unit time	Production rate
q	Constant	Percentage of imperfect production $0 < q < 1$
b	Constant	Price sensitivity $0 < b < 1$
f	Constant	Sensitivity of green investment $0 < f < 1$
m	Constant	Number of the production cycle
I	\$	Green Investment
c_e	\$/unit/unit time	Carbon emission cost in production
h	\$/unit/unit time	Holding cost
c_{eh}	\$/unit/unit time	Carbon emission cost in holding
T	Unit time	Planning horizon
T_s	Unit time	Shortage period
t_k	Unit time	Point of partition of the interval $[0, T]$
A	\$/unit/unit time	Fixed production cost
v	\$/unit/unit time	Labour cost
s	\$/unit time	Tool die cost

$P = Q(1 - q)$. Thus, the cost involved in the production of imperfect items is considered as loss cost due to imperfect production.

(e) The carbon tax is associated with both production and holding. The green investment exponentially decarbonizes the system and the carbon tax with green investment is given by; $c_e(I) = c_e e^{-fI}$, where f is the impact of green investment on carbon emission cost

(f) The planning horizon is infinite and partitioned at the points $0 = t_0 < t_1 < t_2 < t_3 < t_4 < \ldots < \ldots$, where all the points $t_{2k-1} \forall k = 1,2,3, \ldots m$ represents the termination of m number of productions.

(g) The holding capacity is finite. The threshold limit of inventory level for the initiation of another production cycle is I^0.

4 Model Formulation

Let Us Find the Inventory Level in the Initial Production Cycle $[0, t_2]$

The differential equation for the production period $[0, t_1]$ is given by

$$\frac{dI^+}{dt} = P - D(p)$$

subject to $I^+(t) = 0$ at $t = 0$

The inventory level at any time t is given by

$$I^+ = (P - D(p))t, \forall t \in [0, t_1]$$

The differential equation to represent the flow of inventory after the production period $[t_1, t_2]$ is given by;

$$\frac{dI^-}{dt} = -D(p)$$

subject to $I^-(t_2) = I^0$ at $t = t_2$

The inventory level at any time t after the production cycle is;

$$I^-(t) = I^0 + D(p)(t_2 - t), \forall t \in [t_1, t_2]$$

At time $t = t_1$, $I^+(t_1) = I^-(t_1)$ the production period is given by (Fig. 1);

$$t_{p1} = t_1 = \frac{I^0 + D(p)t_2}{P}$$

Inventory Level for First Production Cycle $[t_2, t_4]$

The differential equation to represent the inventory level for the time interval $[t_2, t_3]$ is given by

$$\frac{dI_1^+}{dt} = P - D(p)$$

subject to $I_1^+(t) = I^0$ at $t = t_2$

The inventory level at any time t during the second production cycle is given by (Fig. 2)

$$I_1^+(t) = I^0 + (P - D(p))(t - t_2), \forall t \in [t_2, t_3]$$

The differential equation to represent the inventory level for time interval $[t_3, t_4]$ is given by

$$\frac{dI_1^-}{dt} = -D(p)$$

subject to $I_1^-(t) = I^0$ at time $t = t_4$

The inventory level at any time t after the second production cycle is given by

$$I_1^-(t) = I^0 + D(p)(t_4 - t), \forall t \in [t_3, t_4]$$

At $t = t_3$, $I_1^+(t) = I_1^-(t)$ therefore the second production cycle terminates at

$$t_3 = \frac{Pt_2 + D(p)(t_4 - t_2)}{P}$$

The length of the second production period is given by

$$t_{p2} = t_3 - t_2 = \frac{D(p)(t_4 - t_2)}{P}$$

The Inventory Level of the Second Production Cycle $[t_4, t_6]$

The differential equation to represent the inventory level for the time interval $[t_4, t_5]$ is given by

$$\frac{dI_2^+}{dt} = P - D(p)$$

subject to $I_2^+(t) = I^0$ at $t = t_4$

The inventory level at any time t during the second production cycle is given by

$$I_2^+(t) = I^0 + (P - D(p))(t - t_4), \forall t \in [t_4, t_5]$$

The differential equation to represent the inventory level for time interval $[t_5, t_6]$ is given by

$$\frac{dI_2^-}{dt} = -D(p)$$

subject to $I_2^-(t) = I^0$ at time $t = t_6$

The inventory level at any time t after the second production cycle is given by

$$I_2^-(t) = I^0 + D(p)(t_6 - t), \forall t \in [t_5, t_6]$$

At $t = t_5$, $I_2^+(t) = I_2^-(t)$; therefore, the second production cycle terminates at

$$t_5 = \frac{Pt_4 + D(p)(t_6 - t_4)}{P}$$

The length of the second production period is given by

$$t_{p3} = t_5 - t_4 = \frac{D(p)(t_6 - t_4)}{P}$$

The Inventory Level of kth Production Cycle $\left[t_{2k}, t_{2k+2}\right]$ for k = 1,2, 3, …, $(m-1)$

The inventory level during the production cycle $\left[t_{2k}, t_{2k+1}\right]$ is given by

$$I_k^+(t) = I^0 + (P - D(p))(t - t_{2k}), \forall t \in \left[t_{2k}, t_{2k+1}\right]$$

And the inventory level after the production cycle $\left[t_{2k+1}, t_{2k+2}\right]$

$$I_k^-(t) = I^0 + D(p)(t_{2k+2} - t), \forall t \in \left[t_{2k+1}, t_{2k+2}\right]$$

The production period for kth production cycle is given by

$$t_{2k+1} = \frac{Pt_{2k} + D(p)(t_{2k+2} - t_{2k})}{P} \forall k = 1, 2, 3, \ldots (m-1)$$

The Inventory Level of the Final Production Cycle $[t_{2m}, T]$

The differential equation representing the inventory level during the production period $\left[t_{2m}, t_{2m+1}\right]$ is given by

$$\frac{dI_m^+}{dt} = P - D(p)$$

subject to $I_m^+(t) = I^0$ at $.t = t_{2m}$

The inventory level at any time t is given by

$$I_m^+(t) = I^0 + (P - D(p))(t - t_{2m}), \forall t \in \left[t_{2m}, t_{2m+1}\right]$$

The differential equation representing the inventory level after the production period $\left[t_{2m+1}, T\right]$

$$\frac{dI_m^-}{dt} = -D(p)$$

subject to $I_m^-(t) = 0$ at $.t = T_s$

The inventory level at any time t after the production period

$$I_m^-(t) = D(p)(T - t), \forall t \in \left[t_{2m+1}, T\right]$$

At $t = t_{2m+1}, I_m^+(t) = I_m^-(t)$, it follows that the final production period ends at

$$t_{2m+1} = \frac{Pt_{2m} + D(p)(T - t_{2m}) - I^0}{P}$$

The cost associated with production and inventory management

Setup Cost:

$$SC = O$$

Holding Cost:

$$HC = h\left[\int_0^{t_1} I^+ dt + \sum_{k=1}^{m} \int_{t_{2k}}^{t_{2k+1}} I_k^+ dt + \sum_{k=1}^{m} \int_{t_{2k-1}}^{t_{2k}} I_k^- dt + \int_{t_{2m+1}}^{T_s} I_m^- dt\right]$$

Carbon Tax in Holding:

$$CTH(I) = c_{eh} e^{-fI_1}\left[\int_0^{t_1} I^+ dt + \sum_{k=1}^{m} \int_{t_{2k}}^{t_{2k+1}} I_k^+ dt + \sum_{k=1}^{m} \int_{t_{2k-1}}^{t_{2k}} I_k^- dt + \int_{t_{2m+1}}^{T_s} I_m^- dt\right]$$

Production Cost:

$$PC = \sum_{k=1}^{m+1} \left(A + \frac{v}{Q} + sQ\right) Q(t_{2k-1} - t_{2k-2})$$

Carbon Tax during Production:

$$CT(I) = Qe_p e^{-fI_2} \sum_{k=1}^{m+1} (t_{2k-1} - t_{2k-2})$$

Green Investment:

$$GI = I_1 + I_2$$

Loss due to the Imperfect Production.

$$LIP = \sum_{k=1}^{m+1} q\left(A + \frac{v}{Q} + sQ\right) Q(t_{2k-1} - t_{2k-2})$$

Revenue:

$$TR = pD(p)T$$

Average profit with green investment:

$$AP = \frac{1}{T}[TR - (SC + HC + PTC + PC + CTP(I) + CTH(I) + LIP + I_1 + I_2)]$$

5 Objective Function and Solution Procedure

The model is formulated using a first-order differential equation with constant coefficients. There is no deterioration so the solution of the differential equation is obtained by the method of separation of variables. Since the holding space is finite and there is a second production cycle we must include a few space and time constraints together with some natural constraints associated with the EPQ model. The formulated EPQ model is an NLP and contains eight or more dependent and independent variables so to solve the problem, we use LINGO 19.0 to obtain the solution and interpret the results.

The formulated objective function and decision variables are given by

$$Z = \max AP$$

Subject to the constraints

$$I_1, I_2 > 0$$

The independent decision variables are
Production rate P
Selling price p
Production cycle end points $t_{2k-2}, k = 1, 2, 3, \ldots, (m+1)$
Green Investment $I = I_1 + I_2$
The dependent decision variable is given by
Production period $t_{2k-1}, k = 1, 2, 3, \ldots, (m+1)$
The following steps are implemented to achieve the optimal solution in LINGO 19.0Step 1Step 1

Step 1 *Set the values of the parameters* $H_0, I^0, T, O, h, A, v, s, e_p, f, a, b$ *in the DATA SET.*
Step 2 *Set the number of production cycles to be executed m.*
Step 3 *Set the search boundary of* P, p, t_{2k-2}, I_1 *and* I_2 *as @FREE subject to the constraints*
Step 4 *Obtain the value of dependent decision variables* t_{2k-1}
Step 5 *Obtain the optimal value of Z.*

6 Numerical Illustration

In the lack of real data, we assume the following value of the parameters:

Problem 6.1: Let the manufacturer produce the product in $m = 2$ three production cycles due to the lack of holding space. The manufacturer considers multiple production cycles when the inventory level reaches $I^0 = 20$ units. The period for which the demand and supply process continues is $T = 1.2$ years. The percentage of imperfect production is 5%.

Setup cost $O = \$2000$

Holding cost $h = \$0.4$

Fixed production cost $A = \$100$

Labour cost $v = \$1200$

Tool die cost $s = \$0.007$

Carbon emission cost $e_p = \$1.2$

The value of other constants associated with the model

Effect of green investment $f = 0.05$

Initial demand $a = \$200$

Price sensitivity $b = 0.25$

Solution: The optimal solution for the 3-production cycle

Profit(I)= \$25730.82

Selling price $p = \$458.6327$

Production rate $Q = 410.8828$ unit/unit time.

Perfect Production rate $P = 390.3387$ unit/unit time.

Green Investment $I_1 = \$42.2531$ and $I_2 = \$22.35132$

Timeline for multi-production inventory cycle

$t_1 = 0.1748$, $t_2 = 0.5651$ and $t_3 = 0.7258$

Problem 6.2: Let the manufacturer produce the product in $m = 3$ three production cycles due to the lack of holding space. The manufacturer considers multiple production cycles when the inventory level reaches $I^0 = 20$ units. The period for which the demand and supply process continues is $T = 1.3$ years. The percentage of imperfect production is 3%.

Setup cost $O = \$2000$

Holding cost $h = \$0.4$

Fixed production cost $A = \$100$

Labour cost $v = \$1200$

Tool die cost $\mathrm{s} = \$0.007$

Carbon emission cost $e_p = \$1.2$

The value of other constants associated with the model

Effect of green investment $f = 0.05$

Initial demand $a = 200$ units.

Price sensitivity $b = 0.25$

Solution: The optimal solution for the 3-production cycle

Profit (II)= \$25727.7

Selling price $p = \$458.65$

Production rate $Q = 411.642$ unit/unit time.

Perfect Production rate $P = 391.06$ unit/unit time.

Green Investment $I_1 = \$42.25$ and $I_2 = \$14.41$

Timeline for multi-production inventory cycle

$t_1 = 0.1701$, $t_2 = 0.5452$, $t_3 = 0.5859$, $t_4 = 0.7318$ and $t_5 = 0.8558$

7 Sensitivity Analysis

The production inventory is developed to analyze the impact of demand parameters, planning horizon, threshold limit of next production and rate of imperfect production on the decision variables to maximize the average profit. The change made on the parameters is defined as $N \rightarrow (1 + r) * N$ where $r \in \{-0.1, -0.05, 0.05, 0.1\}$. The results are interpreted and analyzed in Sect. 8.

8 Result and Discussion

The results obtained in Sect. 6, show that the implication of multiple production cycles can impact the profit earned by the seller. The profit earned for $m = 2$ is $Profit = \$25730.8$ and for $m = 3$ is $Profit = \$25727.7$, though the difference between the profit earned in both the cases is less than 2% but performing more production cycle might not help in increasing the profit rather profit reduces due to cost involved in executing multiple production cycle in a given planning horizon. From Table 2, it can be observed that average profit increases significantly with the increase in demand scaling and planning horizon. Average profit decreases with the increase in threshold limit and rate of imperfect production. From Fig. 3 a significant decrease in average profit is observed due to the increase in price sensitivity. For all the other parameters the change is either mild or insignificant therefore average profit is sensitive concerning demand scaling, price sensitivity, threshold limit, planning horizon, and rate of imperfect production.

From Table 2 it can be observed that carbon tax has almost no impact on the profit due to the implementation of green investment and the green investments increase with the increase in carbon taxes for both production and holding.

The selling price p is almost unaffected while changing the parameters except for the case of demand scaling and price sensitivity. Further, the selling price is nearly the same for both $m = 2$ and $m = 3$ thus selling price is not a sensitive parameter concerning change in the production cycle. However, the selling price is directly proportional to demand scaling and inversely to price sensitivity. Also from Fig. 3, it can be concluded that p is significantly sensitive to demand scaling and price sensitivity. Moreover, a mild change in selling price can be observed due to a change in fixed production cost.

The rate of production P is nearly constant due to changes in most of the parameters and even changes in several production cycles. But from Table 2, it can be noticed that the rate of production reduces with an increase in tool die cost which is obvious

Table 2 Sensitivity analysis for $m = 3$

Parameter	% Change	Changed value	p (\$)	P (lbs)	I_1 (\$)	I_2 (\$)	Profit (\$)
I^0	0.10	22	458.64	391.35	42.56	15.61	25,546.1
	0.05	21	458.65	391.21	42.4	15.02	25,636.9
	−0.05	19	458.65	390.91	42.1	13.8	25,818.6
	−0.10	18	458.65	390.76	41.94	13.18	25,909.4
q	0.10	0.055	459.24	289	42.32	14.35	25,609
	0.05	0.0525	458.94	390.03	42.29	14.29	25,668.5
	−0.05	0.0475	458.36	392.09	42.21	14.44	25,786.7
	−0.10	0.045	458.06	393.11	42.18	14.47	25,845.3
T	0.10	1.43	458.64	391.24	43.88	18.51	26,033.4
	0.05	1.365	458.64	391.16	43.08	16.52	25,887.8
	−0.05	1.235	458.65	390.94	41.39	12.17	25,550.8
	−0.10	1.17	458.65	390.79	40.48	9.78	25,354.3
$a = 200$	0.10	220	498.62	390.57	44.15	15.25	32,954.96
	0.05	210	478.63	390.81	43.22	14.84	29,241.3
	−0.05	190	438.65	391.29	41.23	13.95	22,414.2
	−0.10	180	418.67	391.52	40.16	13.44	19,300.9
$b = 0.25$	0.10	0.375	325.33	391.41	40.74	13.73	12,824.4
	0.05	0.3125	378.66	391.23	41.51	14.08	17,942.7
	−0.05	0.1875	591.97	390.88	42.96	14.73	38,846.1
	−0.10	0.125	858.63	390.7	43.66	15.04	65,297.9
$A = 100$	0.10	110	464.18	391.13	41.98	14.29	24,622.1
	0.05	105	461.41	391.09	42.11	14.35	25,173
	−0.05	95	455.88	391.23	42.38	14.47	26,286.25
	−0.10	90	453.12	390.99	42.52	14.54	26,848.61
$v = 1200$	0.10	1320	458.81	410.48	42.24	14.79	25,696.06
	0.05	1260	458.73	400.89	42.24	14.61	25,711.69
	−0.05	1140	458.56	380.97	42.26	14.21	25,744.16
	−0.10	1080	458.48	370.60	42.26	13.98	25,761.05
$s = 0.007$	0.10	0.0077	458.80	372.74	42.24	14.02	25,393.41
	0.05	0.00735	458.72	381.58	42.25	14.22	25,711.88
	−0.05	0.00665	458.56	401.29	42.26	14.62	25,744
	−0.10	0.0063	458.49	412.35	42.26	14.83	25,760.7
$h = 0.4$	0.10	0.44	458.65	390.85	42.25	14	25,726.9
	0.05	0.42	458.65	390.96	42.25	14.2	25,727.3
	−0.05	0.38	458.65	391.16	42.25	14.63	25,728.1
	−0.10	0.36	458.64	391.27	42.25	14.86	25,728.6

(continued)

Table 2 (continued)

Parameter	% Change	Changed value	p ($)	P (lbs)	I_1 ($)	I_2 ($)	Profit ($)
$c_e = 1.2$	0.10	1.32	458.64	391.27	44.16	14.85	25,727.1
	0.05	1.26	458.64	391.27	43.23	14.86	25,727.8
	−0.05	1.14	458.64	391.27	41.23	14.86	25,729.4
	−0.10	1.08	458.64	391.27	40.14	14.86	25,730.2
$c_{eh} = 1.5$	0.10	1.65	458.64	391.27	42.25	16.76	25,727.1
	0.05	1.575	458.64	391.26	42.25	15.84	25,727.8
	−0.05	1.425	458.64	391.27	42.25	13.83	25,729.4
	−0.10	1.35	458.64	391.27	42.25	12.75	25,730.2

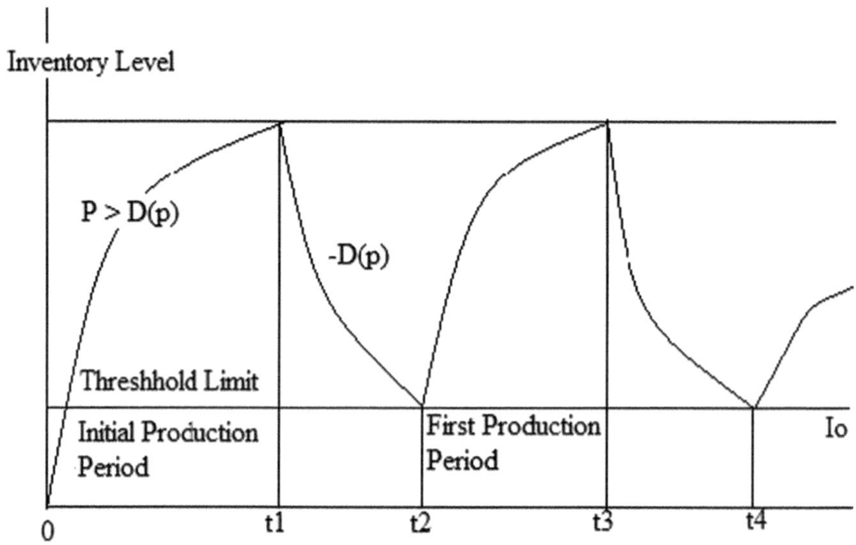

Fig. 1 Pictorial representation of multi-production with threshold limit

as to prevent machine failure the common method is to reduce the production rate. Also, the rate of production increases with the increase in labour cost and it is a common phenomenon in a manufacturing industry that, to compensate for the increase in labour cost, the manufacturer increases the production rate. Furthermore, the rate of production decreases when the percentage of defective items increases and consequently reduces the profit. Thus, the rate of production is sensitive concerning labour cost, tool die cost and percentage of defective items.

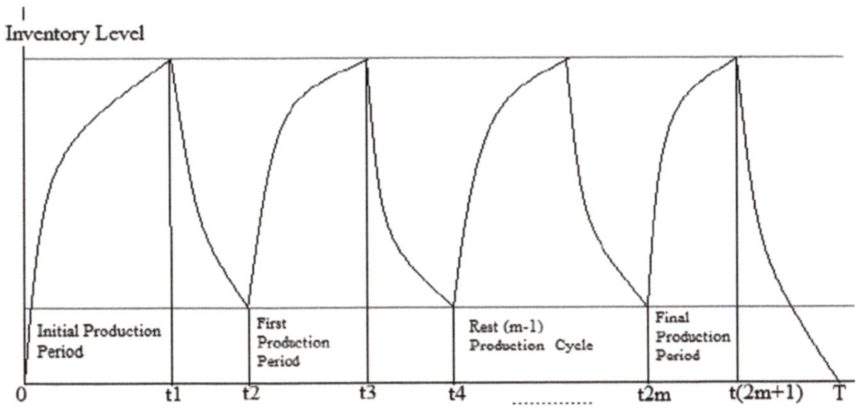

Fig. 2 Pictorial representation of multiple production cycle

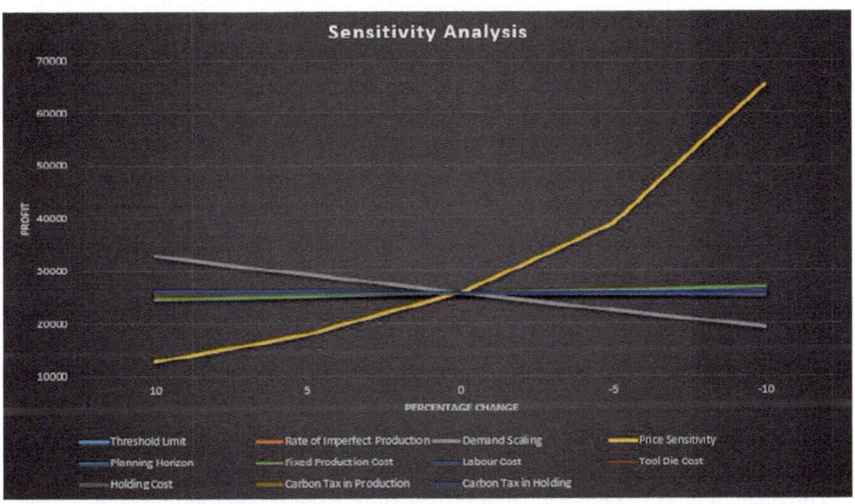

Fig. 3 Sensitivity analysis for profit

9 Managerial Implications

In the light of the solution obtained in Problems 6.1 and 6.2 and the sensitivity analysis we can describe some major highlights for the multi-production inventory model to provide precise decision-making directives;

- It is observed that increasing the number of the production cycle can reduce the profit in a given time cycle. Moreover, increasing the number of production cycles increases the production rate.

- Implementing green investment in production can increase profit. Further, the total green investment increases with the increase in the number of production cycles which is quite obvious from the fact that the number of the production cycle is directly associated with the length of production and hence needs more investment of the length of production raises.
- Increasing the fixed production cost (A) will decrease the profit and at the same time rate of production (P) and the selling price (p) increases.
- An increase in the labour cost (v) can increase the rate of production but reduce the total profit.
- An increase in tool die cost decreases the profit as well as the production rate.

10 Conclusion

The total holding space available to store finished goods in a production industry is an unavoidable problem that a manager cannot avoid. To meet the demand in given planning with the limited holding space, the production manager conducts multiple production cycles and each production in every cycle is initiated when the inventory reduces to a definite level. This is a single-item production inventory model to study the impact of multiple productions on the profit earned under a definite threshold limit of multi-production. The product considered in this model is a non-deteriorating item. The holding cost is constant and carbon tax is considered in both the production and holding separately and it is expressed as a negative exponential function of green investment. All the produced items cannot be perfect and therefore the loss due to imperfect production is associated with the total production in the total cost. The demand function considered in the model is price sensitive and the results obtained in the numerical evaluation suggest that the model is sensitive concerning demand scaling and price sensitivity. The recurrence relation for inventory level and production cycle is provided in the model so that any number of production cycles can be considered as per the requirement.

The present model is an extension of research in the production inventory model and the concept of the multi-production cycle with limited holding space is not reported in the literature earlier as far as the author's knowledge is concerned. The model can be further extended to study and obtain the recurrence relation for deteriorating items. The demand considered in this problem is price-dependent but it will be interesting to know the recurrence relation for stock-dependent demand. Further, the study can be extended to develop an inventory model with price inflation.

References

1. Aarya, D.D., Kumar, M.: A production inventory model with selling price and stock sensitive demand under partial backlogging. Int. J. Math. Oper. Res. **12**(3), 350–363 (2018)
2. Bhattacharjee, N., Sen, N.: A sustainable production inventory model for profit maximization under optimum raw material input rate during production. Opsearch (2021). https://doi.org/10.1007/s12597-021-00554-0
3. Buttar, G.S., Sharma, R.: Production inventory model under time-dependent demand, uneven manufacturing rate and manufacture time dependent delling price. J. Univ. Shanghai. Sci. Technol. **23**(4), 225–237 (2021)
4. Das, S., Manna, A.K., Mahmoud, E.E., Abualnaja, K.M., Abdel-Aty, A.-H., Shaikh, A.A.: Product replacement policy in a production inventory model with replacement period-, stock-, and price-dependent demand. J. Math. **6697279**, 1–8 (2020)
5. Gautam, P., Khanna, A.: An imperfect production inventory model with setup cost reduction and carbon emission for an integrated supply chain. Uncertain Supply Chain Manag. **6**, 271–286 (2018)
6. Gautam, P., Kamna, K.M., Jaggi, C.K.: Sustainable production policies under the effect of volume agility, preservation technology, and price reliant demand Yugoslav. J. Oper. Res. 1–18 (2020). https://doi.org/10.2298/YJOP190315018G
7. Ghosh, A., Jha, J.K., Sarmah, S.P.: Production-inventory models considering different carbon policies: a review. Int. J. Prod. Qual. Manag. **30**(1), 1–27 (2020)
8. Halim, M.A., Paul, A., Mahmoud, M., Alshahrani, B., Alazzawi, A.Y., Atheelah Y.M., Ismail, G.M.: An overtime production inventory model for deteriorating items with nonlinear price and stock dependent demand Alex. Eng. J. **60**(1), 2779–2786 (2021)
9. Kamna, K.M., Gautam, P., Jaggi, C.K.: Sustainable inventory policy for an imperfect production system with energy usage and volume agility. Int. J. Assur. Eng. Manag. **12**(1), 44–52 (2021)
10. Mashud, A.H., Md, R., Dipa, D., Ali, M.H.: A sustainable inventory model with imperfect products, deterioration, and controllable emissions. Mathematics **8**(2049), 1–21 (2020)
11. Mishra, U., Mashud, U., Tseng, M.-L., Wu, J.-Z.: Optimizing a sustainable supply chain inventory model for controllable deterioration and emission rates in a greenhouse farm. Mathematics **9**(5), 495 (2020)
12. Palanivel, M., Gowri, S.: A production-inventory model with delayed deteriorating items with quadratic and price dependent demand. Int. J. Pure Appl. Math. **119**(15), 3271–3283 (2018)
13. Ritha, W., Poongodisathiya, S.: Optimal green inventory model to regulate eco-friendly. Environment **4**(5), 13–16 (2018)
14. Shen, Y., Shen, K., Yang, C.: A production inventory model for deteriorating items with collaborative preservation technology investment under carbon tax. Sustainability **11**(5027), 1–18 (2019)
15. Singh, D.: Production inventory model of deteriorating items with holding cost, stock and selling price with backlog. Int. J. Math. Oper. Res. **14**(2), 290–305 (2019)
16. Tripathi, R.P., Mishra, S.: Innovative approach of stock-linked demand dependent production inventory model with decline deterioration. Int. J. Inventory Res. **5**(4), 251–262 (2020)

An EOQ Model of a Fresh Product Having Variable Demand Under Trade Credit Policy

Rituparna Mondal and Ranjan Kumar Jana

Abstract The product's freshness is essential in light of the current situation and the health-conscious consumer. As the product ages, its vitality diminishes, resulting in a decrease in demand at the same price. The selection of an appropriate credit strategy can generate a multitude of advantages for retailers by growing demand. Additionally, implementing a time-dependent reduction in selling price might enhance sales by serving as a type of price discount.Therefore, driven by this factor, the business provides a limited term of credit to the merchant. This study investigates the influence of customers' credit amount, credit term, and item retail price on market demand. Given that the quality and freshness of a product have a direct influence on its attractiveness and appeal. The market demand is subject to temporal fluctuations. This study introduces a supply chain model for a perishable agricultural product that encompasses suppliers, retailers, and consumers. The model considers the fluctuating demand and suggests that the merchant should implement a partial credit policy to incentivize consumers to make larger purchases. The best option is selected by maximizing the average profit derived by the system. A search operation is executed using a Particle Swarm Optimization (PSO) method that has been devised, verified, and implemented. Speculative illustrations are employed to produce a model representation.

Keywords Freshness produce · Credit policy · Time-dependent Credit amount · Time-dependent Credit period · Freshness-dependent demand

R. Mondal · R. Kumar Jana (✉)
Department of Mathematics, Sardar Vallabhbhai National Institute of Technology, Surat 395007, Gujarat, India
e-mail: rkjana2003@yahoo.com

R. Mondal
Department of Applied Science, Haldia Institute of Technology, Haldia 721657, West Bengal, India

L. Sahoo et al. (eds.), *Decision Making Under Uncertainty Via Optimization, Modelling, and Analysis*, Studies in Systems, Decision and Control 558,
https://doi.org/10.1007/978-981-96-0085-4_15

1 Introduction

1.1 Motivation

The phenomenon of products degrading over time is frequently observed, as the aging of inventory has a negative impact on demand by eroding consumer confidence in the product quality. In modern health-conscious markets, demand is significantly influenced by product freshness and decomposition. Fresh items are classified into broader categories based on their persistent use and diverse durations of viability in numerous studies. Nevertheless, this classification is inadequate in terms of practicality and logical reasoning, as the market appeal of a product is substantially determined by its shelf life. In order to prevent the rapid degradation of their freshness, color, and texture, it is imperative that retail establishments protect verdant vegetables, fruits, florals, dairy products, and other commodities in their unadulterated state. The manufacturing or harvesting of the products did not correspond with the time of purchase in the aforementioned example of the decaying commodities. This results in a significant reduction in the item's detectable freshness. The perishability of specific items is inconsistent; it decreases over time. The supplier's products are not immediately apparent in terms of their freshness. However, it becomes apparent to us after a few days. The visual and tactile qualities of these commodities, which are evaluated by human sensory perception, are the primary factors that determine their apparent freshness. The olfactory and gustatory senses may also contribute to this perception. Consequently, consumers may be discouraged by the apparent lack of novelty in these deteriorating products, which could result in a decrease in demand. Due to the increasing standard of living and health concerns, product durability has become a significant factor in consumer preference, influencing the decision to select one grocery store or retailer over another.

1.2 Literature Review

Inventory management is essential and very beneficial for firms globally. The objective of managing inventory is to handle the entire process, from acquiring materials to making sales-related decisions (such as determining the quantity and timing of purchases), to prevent excessive inventory or stock shortages. So, the store must exercise extra caution while managing the inventory of degrading items, particularly in accounting and the quantity of stock, and make the necessary tradeoffs by adjusting prices. Ghare and Schrader [1] first introduced the traditional no-shortage inventory model of deteriorating items with a constant degradation rate. Teng and Chang [2] developed an economic production quantity model for degrading items, in which the demand rate is influenced by both the price and the stock level of items on display. In addition to degradation, one of the key factors influencing consumers' decisions to buy is freshness. As the thing ages, its demand decreases since the product's quality deteriorates. Yavari et al. [3] state that a significant difficulty in inventory management is the natural tendency of many commodities to deteriorate over time, rendering them unsellable after their expiration date due to a decline in freshness and quality. As the item ages, its demand decreases due to consumers' heightened awareness of their health practices. In this particular situation, buyers desire to purchase fresh commodities far from their expiration date, allowing them to be stored for longer durations. Therefore, it is imperative for organizations to meticulously oversee and regulate their stock of perishable goods within the warehouse. Radio frequency identification smart tags are a frequently employed method for monitoring the freshness and quality of perishable commodities in warehouses. These tags are designed to reduce the likelihood of selling items that have passed their expiration date. However, there has been a dearth of research on models that incorporate the critical factor of freshness. The majority of inventory models for perishable commodities in the existing literature assume that the demand for items that are in stock and have not expired will remain constant, irrespective of their freshness state. Fujiwara and Perera [4] conducted the first investigation on the impact of utility degradation, specifically the decline in freshness. The costs of maintaining an old inventory item in stock were quantified using the penalty function. In their [5] study, Bai and Kendall examined a freshness product in which the condition of freshness declines exponentially with time and the deterministic demand is affected by both the freshness state and the inventory on display. Following that, a model for overseeing the stock of perishable goods was presented by Wu et al. in 2015 [6]. Variables such as the freshness of the product, the quantity of supply, and the expiry date have an influence on the demand, which is always changing and is considered by the model. Given the importance of freshness and stock availability, it is advantageous to maintain a high inventory level at the conclusion of the replenishment cycle. Consequently, the aforementioned study challenges the prevailing assumption that inventory always reaches zero and instead examines the possibility of a non-zero final inventory. In 2020, Shen et al. [7] developed a unified inventory model for the management of perishable agricultural goods. This model is intended for use in distribution centers, supermarkets, and farmers' professional cooperatives. The rate of demand for fresh agricultural commodities by supermarkets is determined by factors such as market demand, selling price, and freshness, according to the model.There is a clear lack of correlation between fluctuations in sales price and the maximum demand rate, as well as product freshness. Macias-Lopez et al. [8] introduced a novel inventory model in 2021 that accounts for price, stock, and time-dependent demand. This model considered six distinct functions of price-dependent demand. The ordering cost has an impact on the price, inventory cycle time, and volume ordered, as demonstrated by their model. However, the financial benefits exhibit a tendency to decrease. References [9–14] are a few of the sources that discuss product freshness.

1.3 Research Gap

The product's quality drops since it loses its freshness with time. As product freshness declines, consumer demand falls in a direct correlation. In order to keep things running smoothly and profitably, merchants adopt several policies regarding the freshness of the goods. Keeping some things from going bad can be more expensive than just selling them cheap. In order to keep those things from getting worse, the retailer lowers the price. A rise in consumer demand and a complete depletion of inventory are the goals of this strategy, which means that no goods will be discarded due to spoilage. Aside from the condition of freshness, cost is another important aspect that affects demand. When products lose some of their freshness while still being sold at the original price, sales go down since most people have preconceived notions and standards regarding the correlation between price and quality. Particularly in less developed nations, it is common practice to reduce prices in such a situation in order to increase sales and forestall decline. When carefully chosen, a lowered pricing approach may bring about a number of benefits. On the other hand, such measures may lead to wasteful spending and financial losses. In order to make the most informed decisions about purchasing and sales within the respectable retail industry, managers must have access to a decision support system. As a result, we must discuss the discounting approach for inventory models of deteriorating goods, where the demand is conditional on pricing and freshness. The present study set out to create and evaluate an inventory model in response to this occurrence. A product's freshness and its selling price are two factors that this model accounts for when predicting the product's demand. As a result, the merchant lowers the selling price to both stop the fall and increase demand. Demand from customers will rise, and inventory will run dry, according to this plan. Several researchers have devised inventory models that account for price-dependent demand for commodities that degrade [15, 16]. The concept of non-instantaneous degeneration has been contemplated by writers to a certain extent. This is illustrated in a study conducted by Wu et al. [17], which employed a two-parameter exponential degradation distribution to account for stock-dependent demand. As a result, the merchant lowers the selling price to halt the decline and increase demand. Depletion of inventories and an increase in client demand are anticipated under this strategy. Inventory models that incorporate price-dependent demand for perishable commodities have been devised by numerous authors [15, 16]. The concept of progressive deterioration has been considered by writers to a certain extent. This phenomenon is illustrated in a study conducted by Wu et al. [17], which employed a two-parameter exponential degradation distribution to account for demand that is contingent upon stock levels. Maihami and Kamalabadi [18] and Ghoreishi et al. [19] investigated the impact of price- and time-dependent demand on goods that do not degrade immediately. No one has yet conducted research on the selling price that fluctuates over time, taking into account the factor of freshness. Therefore, the work has adopted a more pragmatic perspective on time-dependent selling prices, recognizing their decline.

On the flip side, in order to stay afloat in such a cutthroat industry, businesses are turning to various profit-boosting strategies. Of them, trade credit policy is becoming increasingly popular as a means to increase demand. The supplier extends payment terms to the merchant as part of this trade credit program. You can think of this policy as an interest-free loan. Retailers are enticed to place large orders from the provider due to this facility. The policy also gives the store the opportunity to earn interest. Suppliers and retailers alike stand to gain significantly from a credit policy. This is why many stores also provide payment plans. At the outset, Goyal [20]) investigated the trade credit policy in his model, which permits the retailer to postpone payment for a predetermined period. In spite of the fact that the store has not paid interest to the supplier up to this point, the contract mandates that they do so after this period. His hypothesis is that this trade credit policy facility increases demand. Consequently, trade credit policy boosts both sales and revenue. He has assumed in his model that the vendor extends full trade credit to the merchant. So, credit policy is a key concept for increasing base demand in today's highly competitive corporate situations. Numerous publications have been discovered that examine supplier-retailer level full trade credit policies following the groundbreaking work of Goyal [20]. The supplier then gives the shop a partial credit in a more practical manner, according to some studies. The supplier's activities inspire the store to launch a credit term for customers as a demand-stimulation tactic. The subject of trade credit policy, which includes suppliers offering credit periods to retailers and retailers offering credit periods to customers, has been thoroughly researched in various studies [21–26]. The trade credit policy has been the subject of extensive consideration, whereby the supplier or retailer provides a predetermined quantity of credit and a specific repayment period to all of their customers or retailers. However, this is not always feasible due to the progressive deterioration of freshness in specific time. Therefore, it makes more sense for a store to gradually extend credit durations for certain items. In a more reasonable world, it would be ideal if stores offered progressively larger credit amounts. No one has ever thought about the time-dependent credit quantity and term before. A growing awareness of the time-dependent credit period and quantity was thus first experienced in this work. In order to overcome these limitations, our research is based on Fig. 1, which shows the shop's inventory flow.

Soft computing strategies have been effectively implemented in numerous studies to address inventory management issues over the past two decades [24, 25, 27]. The Particle Swarm Optimization (PSO) approach has garnered substantial attention among a variety of soft computing methods due to its exceptional capacity to rapidly identify optimal solutions, particularly in applications that necessitate high decimal accuracy. The authors of a recent study conducted by Guchhait and Maiti [28] employed the aforementioned approach to resolve an inventory management issue. Marketing decisions are contingent upon the PSO strategy.

Fig. 1 Flowchart of the model

1.4 Contribution and Organization

This study comes up with some new ideas to get around those restrictions.

- The inventory policy has been the subject of only a limited number of studies in relation to the product's freshness. The perishability of time-sensitive products is the primary concern in this context.
- A promotional strategy that is employed to increase demand is the gradual reduction of selling prices, as the apparent novelty of a product diminishes over time.
- Additionally, the business provides customers with an incrementally increasing credit limit and an increasingly extended credit duration as part of its marketing strategy.
- The PSO approach has been adequately adapted and implemented in the context of the analyzed issue.

The subsequent sections of this work are structured as follows: Subsequently, the pertinent mathematical symbols and assumptions are introduced. In Sect. 3, the mathematical derivation of the concept is presented in detail. Particle Swarm Optimization (PSO) is a heuristic technique that is expressly addressed in Sect. 4 of the solution process description. The model is evaluated using the dataset in Sect. 5

of the paper, which includes numerical examples. A concise summary is provided in Sect. 6. The reference list has been added at the conclusion.

2 The Necessary Notations and Assumptions of the Model

Several notations are used, which are explained here, in order to mathematically express the model. Furthermore, a set of assumptions is made.

2.1 Notations

1. The inventory system consists of a solitary decaying product, where f_r represents the freshness of the commodity.
2. $I(t)$ represents the inventory level at any instant t of cycle r.
3. H indicates the retailer's planning horizon.
4. T indicates the retailer's replenishment.
5. Q represents the retailer's order quantity.
6. A_r denotes the ordering cost in \$.
7. p_c denotes the unit purchase cost of the retailer.
8. h_r denotes the per unit holding cost of cycle r in \$.
9. m represents the mark-up of selling price.
10. s_p denotes the unit selling price of retailer in \$.
11. I_p denotes the rate of payable interest per \$ per unit time to the bank.
12. C_p denotes the credit period.
13. Z symbolize the current value of the total profit of the retailer in \$.

2.2 Assumptions

1. Due to the lack of stability, the freshness of the product diminishes with time. The freshness (f_r) of the product at any given moment t is considered to be

$$f_r = e^{-k_1 t}, \qquad 0 \le t \le T. \tag{1}$$

2. The shop offers customers a credit term $(C_p(t))$ at a specific time, t, in order to encourage demand. The credit period is directly proportional to the time, t, and can be mathematically represented as

$$C_p(t) = a + bt, \qquad 0 \le t \le T, \tag{2}$$

where a and b are a pair of positive real values that have been chosen to optimize the selling price.

3. The retailer provides the consumer with a credit amount, denoted as $\alpha(t)$, which is a function of time t and follows a specific mathematical form:

$$\alpha(t) = 1 - e^{-ct}, \qquad 0 \le t \le T, \tag{3}$$

where c is a positive real number.

4. The unit selling price is determined by adding a mark-up (represented by m) to the purchase cost. The selling price decreases exponentially over time according to the given equation:

$$s_p(t) = mp_{cr}e^{-k_2 t}, \qquad 0 \le t \le T, \tag{4}$$

where the value of k_2 is a positive real number that has been chosen to optimize the fitting of the selling price.

5. $D_m(t)$ is the rate at which the market demands a certain item at any given moment t, measured in units per unit time. This rate is influenced by factors such as the freshness of the item, the credit period offered by retailers, and the selling price. The mathematical expression for $D_m(t)$ is as follows:

$$D_m(t) = \frac{AC_p(t)}{\left[mp_c e^{-k_2 t}\right]^{\gamma}} + ke^{-k_1 t}, \qquad 0 \le t \le T. \tag{5}$$

3 Mathematical Formulation and Analysis of the Model

At the outset of each cycle, it is assumed that the retailer orders Q units of the item. The market demand slowly depletes the inventory, yet the units are restocked instantly. Product freshness declines, which in turn reduces unit demand. To keep demand steady, the store gives consumers credit amounts that vary with the passage of time. Therefore, at the beginning of each cycle, there are Q units of the item in inventory. At the end of each cycle, at $t = T$, the inventory level is zero because of the impact of market demand. Hence, the following differential equation represents the retailer's inventory level $I(t)$ at every time t in each cycle:

$$\frac{dI(t)}{dt} = -D; \qquad 0 \le t \le T, \tag{6}$$

with boundary conditions $I(T) = Q$ and $I(T) = 0$. Solving Eq. (6)

$$I(t) = \frac{A}{(mp_c)^{\gamma} k_2\gamma}\left[\left\{a - \frac{b}{k_2\gamma}\right\}\left\{e^{k_2\gamma T} - e^{k_2\gamma t}\right\} + b\left\{Te^{k_2T\gamma} - te^{k_2\gamma t}\right\}\right] + \frac{k}{k_1}\left\{e^{-k_1 t} - e^{-k_1 T}\right\}\ \forall\ 0 \le t \le T. \tag{7}$$

$$\text{Thus, } Q = I(0) = \frac{A}{(mp_c)^{\gamma}}\frac{1}{k_2\gamma}\left[\left\{a - \frac{b}{k_2\gamma}\right\}\left\{e^{k_2T\gamma} - 1\right\} + bTe^{k_2T\gamma}\right] + \frac{k}{k_1}\left\{1 - e^{-k_1T}\right\}. \tag{8}$$

The retailer's average profit is determined by the following costs and revenues:

$$\text{(i) Sales revenue, } SR = \int_0^T Ds_p(t)dt$$

$$= mp_c\left[\frac{A}{(mp_c)^{\gamma}}\left[\left\{\frac{e^{k_2(\gamma-1)T} - 1)}{k_2(\gamma-1)}\right\}\left\{a - \frac{b}{k_2(\gamma-1)}\right\} + \frac{bT}{k_2(\gamma-1)}e^{k_2(\gamma-1)T}\right]\right.$$
$$\left. + \frac{k}{(k_1+k_2)}\left\{1 - e^{-(k_1+k_2)T}\right\}\right]. \tag{9}$$

$$\text{(ii) Ordering cost, } OC = A_r. \tag{10}$$

$$\text{(iii) Purchase cost, } PC = Qp_c$$
$$= p_c\left[\frac{A}{(mp_{cr})^{\gamma}}\frac{1}{k_2\gamma}\left[\left\{a - \frac{b}{k2\gamma}\right\}\left\{e^{k_2T\gamma} - 1\right\} + bTe^{k_2T\gamma}\right] + \frac{k}{k_1}\left\{1 - e^{-k_1T}\right\}\right]. \tag{11}$$

$$\text{(iv) Holding cost, } HC = h_r\int_0^T I(t)dt$$

$$= h_r\left[\frac{A}{(mp_c)^{\gamma}(k_2\gamma)}\left[(a - \frac{b}{k_2\gamma})\left\{Te^{k_2T\gamma} - \frac{e^{k_2T\gamma}}{k_2\gamma} + \frac{1}{k_2\gamma}\right\}\right.\right.$$
$$\left.\left. + b\left\{T^2e^{k_2T\gamma} - \frac{Te^{k_2T\gamma}}{k_2\gamma} + \frac{e^{k_2T\gamma}}{(k_2\gamma)^2} - \frac{1}{(k_2\gamma)^2}\right\}\right] + \frac{k}{k_1}\left\{\frac{1}{k_1} - \frac{e^{-k_1T}}{k_1} - Te^{-k_1T}\right\}\right]$$
$$= h_c HC_1. \tag{12}$$

$$\text{where, } HC_1 = \frac{A}{(mp_c)^{\gamma}(k_2\gamma)}\left[(a - \frac{b}{k_2\gamma})\left\{Te^{k_2T\gamma} - \frac{e^{k_2T\gamma}}{k_2\gamma} + \frac{1}{k_2\gamma}\right\}\right.$$
$$\left. + b\left\{T^2e^{k_2T\gamma} - \frac{Te^{k_2T\gamma}}{k_2\gamma} + \frac{e^{k_2T\gamma}}{(k_2\gamma)^2} - \frac{1}{(k_2\gamma)^2}\right\}\right] + \frac{k}{k_1}\left\{\frac{1}{k_1} - \frac{e^{-k_1T}}{k_1} - Te^{-k_1T}\right\}. \tag{13}$$

(v) Interest amounts (payable), $IC = I_p \int_0^T \alpha(t)D(t)C_p(t)dt$

$$= \frac{A}{(mp_c)^\gamma}[IC_1 + IC_2]. \tag{14}$$

where

$$IC_1 = \int_0^T (1 - e^{-ct})(a + bt)^2 e^{k_2\gamma t} dt$$

$$= \left[(a + bT)^2 \frac{e^{k_2\gamma T}}{k_2\gamma} - \frac{a^2}{k_2\gamma} - \frac{2}{(k_2\gamma)^2}\{(a + bT)e^{k2\gamma T} - a - \frac{b}{k_2\gamma}(e^{k_2\gamma T} - 1)\}\right] -$$

$$\left[(a + bT)^2 \frac{e^{(k_2\gamma - c)T}}{k_2\gamma - c} - \frac{a^2}{(k_2\gamma - c)} - \frac{2}{(k_2\gamma)^2}\{(a + bT)e^{k2\gamma T}\right.$$

$$\left. - a - \frac{b}{k_2\gamma - c}(e^{(k_2\gamma - c)T} - 1)\}\right]. \tag{15}$$

$$IC_2 = \int_0^T [1 - e^{-ct}](a + bt)e^{-k_1 t} dt$$

$$= \left[\left(\frac{a}{k_1} + \frac{b}{k_1^2}\right)\left(1 - e^{-k_1 T}\right) - \frac{bT}{k_1}e^{-k_1 T}\right] - \left[\left(\frac{a}{k_1 + c} + \frac{b}{(k_1 + c)^2}\right)\left(1 - e^{-(k_1 + c)T}\right)\right.$$

$$\left. - \frac{bT}{k_1 + c}e^{-(k_1 + c)T}\right]. \tag{16}$$

Crisp Model: The retailer's entire profit during the planning horizon is represented by the subsequent equation:

$$Z = SR - OC - PC - HC - IC \tag{17}$$

where SR, PC, OC, HC, and IC are, respectively, given by Equations (9), (10), (11), (12), and (14). So, the mathematical problem reduces to

$$\left.\begin{array}{c} \text{determine } m,\ T \\ \text{to maximize } Z = SR - OC - PC - HC - IC \end{array}\right\} \tag{18}$$

4 Solution Methodology

Kennedy and Eberhart [29] developed the heuristic algorithm Particle Swarm Optimization (PSO) in 1995. The algorithmic approach known as Particle Swarm Optimization (PSO) is designed to replicate the collective intelligence observed in colonies of birds as they search for sustenance. Individual birds within a swarm employ the most efficient experiences of their peers and their own unique experiences to identify food sources. The PSO algorithm endeavors to tackle optimization problems in an effective manner by simulating this behavior.

The search region for the bird's sustenance is evaluated by comparing it to the search area of the optimization solution to the issue at hand. An analogy is drawn between the current location of a solution in the viable zone and the bird's positioning in the search space. The ideal solution within the search region of bird species is contrasted to the geographical location of a food supply.

The initialization of a set of potential solutions for the topic under investigation is the first step in Particle Swarm Optimization (PSO). The arrangement of each particle, which is referred to as a bird, represents a unique response, while food functions as a representation of the ideal solution. The location of each particle is updated as it traverses a multi-dimensional search space, utilizing its own historical data and the most optimal locations identified by the entire group. The position vector and velocity vector for the i-th particle in the t-th generation are denoted by $P_i(t)$ and $V_i(t)$, respectively. In the t-th generation, let $P_{besti}(t)$ denote the optimal position of the i-th particle. The optimal position vector of all particles in the t-th generation is denoted by $P_{gbest}(t)$. The formulas $P_i(t+1)$ and $V_i(t+1)$ can be used to determine the position vector and velocity vector of the particles in the $(t+1)$th generation, provided that all variables are continuous.

$$\left.\begin{aligned} V_i(t+1) &= \omega V_i(t) + \mu_1 r_1 \{P_{pbesti}(t) - P_i(t)\} + \mu_2 r_2 \{P_{gbest}(t) - P_i(t)\} \\ P_i(t+1) &= P_i(t) + V_i(t+1) \end{aligned}\right\} \tag{19}$$

The parameter ω, referred to as the "inertia weight," controls the momentum of the particle by taking into account the influence of its previous velocity. To meet the convergence condition, it is important to ascertain the value of the variable ω. The expert's investigation has found that the ideal value of ω is 0.7298. The parameters μ_1 and μ_2 are fixed values that have been assigned a value of 1.49618 in order to improve the performance of the method. The variables r_1 and r_2 are uniformly distributed between 0 and 1.

5 Numerical Illustration

Crisp Model: To evaluate the practical use of the proposed Crisp model, a specific numerical case is being examined. The precise values for various parameters are provided in the respective example.

Example: $A = 2,\ a = 0.1,\ b = 0.02,\ c = 0.5,\ p_c = 1.5,\ h_r = 0.020,\ A_r = 20,$ $k = 100, k_1 = 0.02, k_2 = 0.02, \gamma = 3.0, I_c = 0.08$

Table 1 illustrates the results of employing the proposed Particle Swarm Optimization (PSO) on the model with the specified parameter values. The results were acquired by adjusting the parameters, which revealed a positive correlation between the freshness-related metric and profit. In other words, the profit increases as the parameter k increases, and conversely, the profit decreases as the parameter k decreases. An increase in the value of k leads to a corresponding increase in market demand, which is the cause of this phenomenon.

The following is a list of notable real-life occurrences, as presented in Table 1:

(a) The profit (Z) is reduced as a result of an increase in the procurement cost (A_r). This phenomenon is graphically depicted in Fig. 2a.
(b) The profit (Z) is increased by reducing the holding cost (h_r). This phenomenon is illustrated in Fig. 2b.
(c) The market demand increases as the parameter (k) increases. Consequently, the aggregate profit (Z) experiences an increase. This phenomenon is graphically depicted in Fig. 3a.

Table 1 Outcome of Crisp model based on different parameters

Parameter	Parameter's value	T	m	Z	Parameter	Parameter's value	T	m	Z
k	60	2.1653	3.2148	733.31	k_2	0.01	2.7235	3.4665	1129.64
	70	2.5230	3.1532	858.83		0.02	2.8630	3.9670	1235.38
	80	2.3289	3.3579	984.34		0.03	2.8629	3.3669	1240.34
	90	2.2467	3.2574	1109.86		0.04	2.2622	3.3516	1245.17
	100	2.8630	3.3670	1235.37		0.05	2.8632	3.1867	1253.96
h_r	0.060	2.8863	3.2698	1229.25	A_r	15	2.6326	3.3516	1240.38
	0.050	2.6326	329516	1230.78		20	2.3262	3.3616	1235.37
	0.040	2.9632	3.3750	1232.31		25	2.9863	2.8816	1230.37
	0.030	2.2918	3.1716	1233.85		30	2.8863	3.6516	1225.37
	0.020	2.8632	3.3075	1235.38		35	2.8326	3.3516	1220.38
k_1	0.02	2.8162	3.1669	1235.38	I_p	0.12	2.8322	3.3516	1222.95
	0.03	2.8629	3.3516	1217.95		0.11	2.3262	3.1846	1225.81
	0.04	2.2162	3.2916	1200.84		0.10	2.9603	3.1186	1228.67
	0.05	2.8322	3.3516	1184.06		0.09	2.8322	3.3616	1231.39
	0.06	2.8082	3.3516	1167.58		0.08	2.6282	3.2616	1235.37

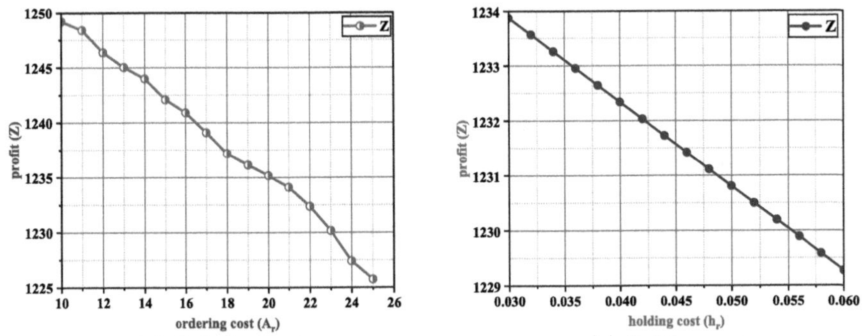

(a) Ordering cost (b) Holding cost

Fig. 2 Graphical depiction of **a** A_r versus Z and **b** h_r versus Z

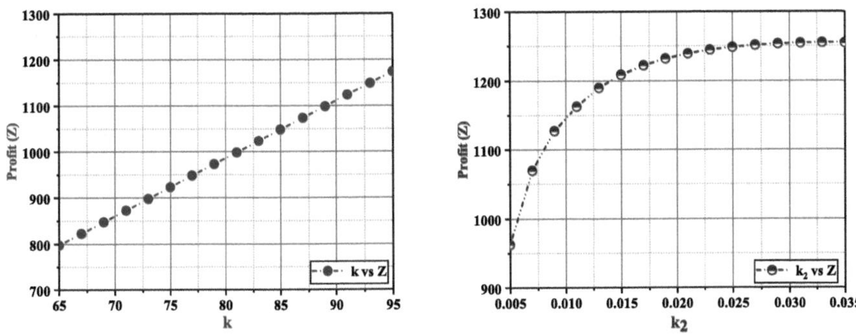

(A) Ordering cost (B) Holding cost

Fig. 3 Graphical depiction of **a** k versus Z and **b** k_2 versus Z

(d) The selling price decreases as the parameter (k_2) decreases. As a result, the retailer's profitability decreases as a result of an increase in demand for price-sensitive products. Figure 3b illustrates this phenomenon.
(e) The total profit increases as the parameter k_1 associated with the product freshness increases.
(f) The profit of the retailer (Z) increases when the rate of interest (I_p) decreases.

All of the occurrences mentioned are in accordance with reality. Thus, by taking into account these insights, a decision-maker may optimize the efficiency of business choices, as shown by this research.

6 Conclusion

Several decision-making methods are introduced in this study to address optimization challenges. In addition, the Particle Swarm Optimization (PSO) technique has been modified and verified to address optimization problems that involve both continuous and discrete variables. A mathematical framework is developed and solved using Particle Swarm Optimization (PSO) methods to effectively manage stocks of perishable commodities that degrade with time, with freshness declining at a rate that varies with time. The study examines the optimal level of demand for these products and investigates the impact of various promotional strategies, such as dynamic credit amounts and dynamic credit periods. Decision-makers can draw managerial decisions from the model with appropriate illustrations. The following is a list of several advantages of the suggested model:

- The model incorporates an exponentially decreasing selling price, allowing firms to modify prices dynamically. This can improve inventory turnover and reduce holding costs.
- Providing consumers with a credit period can improve cash flow by allowing them the option to delay payment. This can enhance sales, particularly for high-priced items or large-scale purchases.
- Companies may mitigate stockouts and excess inventory by optimizing their inventory levels to align with real-time market demand, which is influenced by product freshness.
- Businesses can effectively adapt to market situations, competitive actions, and changes in customer preferences by adjusting pricing exponentially over time.
- Providing flexible payment terms and reasonable rates for fresh products can improve consumer satisfaction and loyalty.

These benefits aid firms in overseeing their operations, diminishing their expenses, and augmenting their overall profitability. The proposed decision-making strategy can be employed to overcome any decision-making challenge with minimal modifications.

Future studies on the suggested model may incorporate several potential extensions. Here are a few potential future detections:

- The model can be expanded by incorporating situations where several perishable goods are simultaneously handled, considering the interconnections and potential cost advantages in procurement.
- The model could be extended by managing supplier relationships, manufacturing scheduling, and shipping logistics.
- Businesses may integrate sustainability factors into the EOQ model, such as reducing carbon emissions related to transportation or mitigating food waste in the supply chain.

References

1. Ghare, P.M., Schrader, G.F.: A model for exponentially decaying inventory. J. Indus. Eng. **14**(5), 238–243 (1963)
2. Teng, J.T., Chang, C.T.: Economic Production Quantity models for deteriorating items with price-and stock- dependent demand. Comput. Oper. Res. **32**(2), 297–308 (2005)
3. Yavari, M., Enjavi, H., Geraeli, M.: Demand management to cope with routes disruptions in location-inventory-routing problem for perishable products. Res. Transp. Bus. Manag. **37**, 1–14 (2020). Article ID 100552
4. Fujiwara, O., Perera, U.L.J.S.R.: EOQ models for continuously deteriorating products using linear and exponential penalty costs. Europ. J. Oper. Res. **70**(1), 104–114 (1993)
5. Bai, R., Kendall, G.: A model for fresh produce shelf-space allocation and inventory management with freshness-condition-dependent demand. INFORMS J. Comput. **20**(1), 78–85 (2008)
6. Wu, J., Chang, C.T., Cheng, M.C., Teng, J.T., Al-khateeb, F.B.: Inventory management for fresh produce when the time-varying demand depends on product freshness, stock level and expiration date. Int. J. Syst. Sci. Oper. Logist. **3**(3), 138–147 (2015)
7. Shen, L., Li, F., Li, C., Wang, Y., Qian, X., Feng, T., Wang, C.: Inventory optimization of fresh agricultural products supply chain based on agricultural Superdocking. J. Adv. Transp. (2020). https://doi.org/10.1155/2020/2724164
8. Macıas-Lopez, A., Cardenas-Barron, L.E., Peimbert-Garcia, R.E., Mandal, B.: An inventory model for perishable items with price-, stock-, and time-dependent demand rate considering shelf-life and nonlinear holding costs. Mathem. Problems Eng. (2021). https://doi.org/10.1155/2021/6630938
9. Chen, P.Y.: Economic order quality model for determining the sales prices of fresh goods at various points in time. J. Food Qual. (2017). https://doi.org/10.1155/2017/6967501
10. Yan, B., Wu, X., Ye, B., Zhang, Y.: Three-level supply chain coordination of fresh agricultural products in the Internet of Things. Indus. Manag. Data Syst. **117**(9), 1842–1865 (2017)
11. Qiu, F., Hu, Q., Xu, B.: Fresh agricultural products supply chain coordination and volume loss reduction based on strategic consumer. Int. J. Environ. Res. Public Health **17**, 7915 (2020)
12. Wang, Q., Yang, Y.: Research on secondary cold chain inventory strategy of fresh meat products. IOP Conf. Series: Mater. Sci. Eng. (2019). https://doi.org/10.1088/1757-899X/688/4/044064
13. Li, R., Teng, J.T.: Pricing and lot-sizing decisions for perishable goods when demand depends on selling price, reference price, product freshness, and displayed stocks. Europ. J. Oper. Res. (2018). https://doi.org/10.1016/j.ejor.2018.04.029
14. Banerjee, S., Agrawal, S.: Inventory model for deteriorating items with freshness and price dependent demand: Optimal discounting and ordering policies. Appl. Mathem. Model. **52**, 53–64 (2017)
15. Sana, S.: Optimal selling price and lot size with time varying deterioration and partial backlogging. Appl. Math. Comput. **217**, 185–194 (2010)
16. Sana, S.: Price-sensitive demand for perishable items-an EOQ model. Appl. Math. Comput. **217**, 6248–6259 (2011)
17. Wu, K.S., Ouyang, L.Y., Yang, C.T.: An optimal replenishment policy for non- instantaneous deteriorating items with stock-dependent demand and partial backlogging. Int. J. Prod. Econ. **101**, 369–384 (2006)
18. Maihami, R., Kamalabadi, I.N.: Joint pricing and inventory control for non- instantaneous deteriorating items with partial backlogging and time and price dependent demand. Int. J. Prod. Econ. **136**, 116–122 (2012)
19. Ghoreishi, M., Mirzazadeh, A., Weber, G.W.: Optimal pricing and ordering policy for non-instantaneous deteriorating items under inflation and customer return. Optimization **63**(12), 1785–1804 (2014)
20. Goyal, S.K.: Economic order quantity under conditions of permissible delay in payment. J. Oper. Res. Soc. **36**, 335–338 (1985)

21. Chung, K.J., Cardenas-Barron, L.E.: The simplified solution procedure for deteriorating items under stock dependent demand and two level trade credit in the supply chain management. Appl. Mathem. Model. **37**, 4653–4660 (2013)
22. Chung, K.J., Cardenas-Barron, L.E., Ting, P.S.: An inventory model with non-instantaneous receipt and exponentially deteriorating items for an integrated three layer supply chain system under two levels of trade credit. Int. J. Production Econ. **105**, 25–32 (2014)
23. Wu, J., Ouyang, L.Y., Cardenas-Barron, L.E., Goyal, S.K.: Optimal credit period and lot size for deteriorating items with expiration dates under two-level trade credit financing. Europ. J. Oper. Res. **237**(3), 898–908 (2014)
24. Guchhait, P., Maiti, M.K., Maiti, M.: Inventory model of a deteriorating item with price and credit linked fuzzy demandâŁ¯: A fuzzy differential equation approach. OPSEARCH **51**, 321–353 (2014)
25. Mondal, R., Pramanik, P., Jana, R.K., Maiti, M.K.: Credit policy for an inventory model of a deteriorating item having variable demand considering default risk. Scientia Iranica (2022). https://doi.org/10.24200/SCI.2022.56218.4607
26. Pramanik, P., Maiti, M.K.: Trade credit policy of an inventory model with imprecise variable demand: an ABC-GA approach. Soft Comput. (2019). https://doi.org/10.1007/s00500-019-04502-5
27. Pramanik, P., Maiti, M.K., Maiti, M.: A supply chain with variable demand under three level trade credit policy. Comput. Indus. Eng. **106**, 205–221 (2017)
28. Guchhait, P., Maiti, M.K., Maiti, M.: A fuzzy lifetime-based particle swarm optimisation with varying swarm size to solve a production inventory model. Int. J. Comput. Complexity Intell. Algorithms **1**(1), 68–98 (2016)
29. Kennedy, J., Eberhart, R.: Particle swarm optimization.Proceedings of the IEEE International Conference on Neural Networks, vol. 4, pp. 1942–1948 (1995). https://doi.org/10.1109/ICNN.1995.488968

Inventory Model with Negative Exponential Demand and Instantaneous Replenishment Under Preservation Technology

Sanjukta Malakar and Nabendu Sen

Abstract The present work proposes an inventory model with negative exponential demand and instantaneous replenishment. As a natural occurrence, deterioration is crucial to inventory management and can have an impact on the overall profitability of an organization. Today, researchers are working to identify ways to slow down the rate of deterioration. To slow down the rate of deterioration, we have included preservation technology in our proposed model. Moreover, carbon emissions are considered, and it is thought that the price of carbon emissions relies on green investment. This model is different from the existing inventory models with negative exponential demand, which did not incorporate carbon emission and preservation technology. Weighted particle swarm optimization (WPSO) and constriction factor particle swarm optimization (CO-PSO) are used to solve the specified model, and the optimality is demonstrated by numerical values for the system parameters and a graphical depiction of convergence characteristics. The model is then verified by sensitivity analysis of the system parameters, and managerial implications are made for efficient decision-making.

Keywords Inventory modeling · Deterioration · Instantaneous replenishment · Carbon emission · Green investment · Preservation · PSO

1 Introduction

In the twenty-first century, companies or business organizations are focusing on cost optimization through using of mathematical models and modern technologies. Considering the usefulness of inventory management in the real world, especially in the business world, inventory problems have become an exciting area of research. Inventory models are usually developed by considering various parameters like

S. Malakar · N. Sen (✉)
Assam University Silchar, Assam 788011, India
e-mail: nsen08@yahoo.com

L. Sahoo et al. (eds.), *Decision Making Under Uncertainty Via Optimization, Modelling, and Analysis*, Studies in Systems, Decision and Control 558,
https://doi.org/10.1007/978-981-96-0085-4_16

demand, deterioration, allowing shortages and without shortages, etc. Many inventory problems were developed by assuming demand as constant or stock dependent or price dependent or time-dependent. However, in reality, the demand for a product may fluctuate with time, i.e., it may increase with time or become constant, or decrease after a certain period. Thus, demand plays an important role in any inventory system.

One of the ideological tenets of the traditional inventory model is that shortages are unacceptable, but in reality, for several factors, including the need to screen out decaying items or other production problems, shortages may occur in the inventory system, which is unavoidable. Some customers in this modern era of technology might wait for the product till the inventory is replenished if the waiting period is short. Others, on the other hand, may be intolerant and choose an alternate product, which is a loss for the system. Therefore, replenishment should also be considered when determining the inventory.

Also, holding goods in inventory means the occurrence of deterioration, which is a natural phenomenon that causes damage to the products, thereby incorporating loss to the inventory system. Thus, deterioration is also an important factor in developing inventory models. To overcome the deterioration of the products, many researchers have started investing in preservation technology. While many researchers have considered deteriorating inventory systems very few have thought about slowing the rate of deterioration by using preservation technology.

The growing awareness of the environment motivates researchers to implement realistic aspects such as carbon emissions in their inventory models. Emissions can be reduced by considering green technologies. The rest parts of the present work are organized as follows. Section 2 discusses an elaborate literature review on inventory models. In Sect. 3, assumptions and notations will be used in developing the proposed model. In Sect. 4, a mathematical model is formulated and solved. Section 5 contains the description of the solution procedure. Numerical illustration is done in Sect. 6. In Sect. 7, sensitivity analysis is done. Managerial implications and limitations are discussed in Sect. 8. Section 9, contains the conclusion.

2 Literature Review

Inventory problems are almost common to all and lots of research work has already been done in the field of inventory considering its usefulness in the real world, especially in business and industries. Different inventory models were developed by different authors considering different types of parameters. Among the developed models in inventory control, models with deterioration allowing preservation technologies and green investments gain more interest from researchers. The effect of deterioration was first discussed by Ghare and Schrader [10]. Shah and Pandey [16] developed their inventory model considering time-dependent deterioration and stock-dependent demand. Another inventory model with time-dependent deterioration was developed by Sarkar and Sarkar [17]. Chen et al. [8] proposed an optimal replenishment policy for the inventory model. An EOQ model with time-varying demand and shortages was developed to describe optimal replenishment policy by Roy et al. [15]. Latha and Uthaykumar [11] developed an inventory model with different types of deterioration functions allowing shortages with partial backlogging. Also, an inventory model for deteriorating items was developed by Tripathi and Kumar [26]. Sekar and Uthayakumar [20] developed an optimal replenishment policy for minimizing the total inventory cost. Saha and Sen [18] developed an inventory model with negative exponential demand considering deterioration as a different probabilistic function. Maragatham and Palani [13] also propose an inventory model for deteriorating items. Sekar and Uthayakumar formulated [21] a production inventory model with negative exponential demand for perishable products. Sivashankari [23] proposed an inventory model with time-dependent exponential demand for deteriorative products under inflation. Singh and Rani [24] established an inventory model with exponential time-linked demand. Shah et al. [19] proposed an inventory model for deteriorating items with dynamic demand. Ahmed and Benkherouf [1] obtained a procedure for determining the optimal replenishment policy of the inventory model. Further, Taleizadeh et al. [25] studied the inventory model of imperfect items and replenishment of the imperfect items. Soni and Suthar [22] obtained the optimal replenishment policy of an inventory model with varying rates of demand. Pal et al. [14] formed an inventory model for non-instantaneous deteriorating products. In their work, they tried to determine the optimal inventory period, cycle length, and investment amount in preservation technology. Bardhan et al. [3] designed an inventory model considering preservation technology investment and stock-dependent demand. Mahato and Mahata [12] have considered preservation technology to reduce the deterioration rate in their formulated inventory model. Bhattacharjee and Sen [5] formulated a stochastic inventory model with price, stock, and emission-dependent demand. Bhattacharjee and Sen [6] proposed a non-separable inventory model to predict the shortage period with time-dependent deterioration. Barman et al. [4] established a multi-item inventory model with fixed cycle duration under preservation technology. Also, Bhattacharjee and Sen [7] studied the effect of green investment and replenishment quantity on the production cost for a production inventory model.

Motivated by the research work of Saha and Sen [18] an attempt has been made to develop an inventory model with negative exponential demand. The research gaps identified in the model are as follows: the model does not acknowledge the issue of carbon emission in the inventory. Further, Saha and Sen [18] assume shortages in their model, which is, however, not necessary if continuous replenishment is considered in the model. Furthermore, the preservation technology cost is also ignored in their model. Thus, in the era of sustainability and modern technologies, there is a clear research gap found in the work of Saha and Sen [18]. In the present paper, an attempt has been made to develop an inventory model with negative exponential demand and instantaneous replenishment. The effect of carbon emission is considered in the inventory and both green investment and preservation technology cost is taken as the decision variables in the model. To the best of the author's knowledge, the inventory model with instantaneous replenishment and negative exponential demand under preservation technology has not yet been reported in the literature of the inventory model.

3 Assumptions

i. The demand D(t) rate is exponentially declining.
ii. Shortages are not allowed.
iii. Holding cost is independent of time.
iv. Deterioration is time-proportional.
v. Replenishment is instantaneous.
vi. Carbon emission is considered and the carbon emission cost is a function of green investment.
vii. Preservation is allowed and the net deterioration in the initial deterioration $\theta(\beta) = \theta e^{-\beta\gamma}$, β is the preservation cost and γ is the sensitivity of preservation cost, $\gamma \epsilon (0,1)$.
viii. Lead time is zero.

Notations

i. h = a (a > 0) holding cost per unit per unit time
ii. θ(t) = θt, is the rate of deterioration and $0 < \theta < 1$
iii. D(t) = A $e^{-\alpha t}$
iv. C_s is the setup cost
v. C is the unit cost of the item
vi. T is the cycle length
vii. R is the replenishment rate
viii. I_g is the green investment
ix. I_0 is the maximum inventory level in (0, T)
x. β is the preservation cost.

4 Formulation and Solution of the Model

Based on the above assumptions, the differential equation governing the inventory situation in the time interval [0, T] is

$$\frac{dI}{dt} + \theta t I(t) = R - \mathrm{A}\, e^{-\alpha t}, 0 \le t \le T \tag{1}$$

$$\text{With I(0)} = I_0$$

We solve Eq. (1) by the series solution method.

$$\text{Let, I(t)} = \sum_{n=0}^{\infty} b_n t^n \tag{1a}$$

Differentiating (1a) w.r.t. t,

$$\frac{dI(t)}{dt} = \sum_{n=1}^{\infty} nb_n t^{n-1} \tag{1b}$$

Using Eq. (1a) and Eq. (1b) in Eq. (1), we have

$$\sum_{n=1}^{\infty} nb_n t^{n-1} + \theta t \sum_{n=0}^{\infty} b_n t^n = \mathrm{R} - \mathrm{A} \sum_{n=0}^{\infty} \frac{(-\alpha)^n t^n}{n!} \tag{1c}$$

Equating the coefficients of like powers of t, we get

$$b_1 = \mathrm{R} - A$$

$$b_2 = A\frac{\alpha}{2!} - \frac{\theta}{2} b_0$$

$$b_3 = -A\frac{\alpha^2}{3!} - \frac{\theta}{3}(R - A)$$

$$b_4 = A\frac{\alpha^3}{3!} - A\frac{\alpha}{2!} + \frac{\theta^2}{8} b_0$$

$$b_4 = A\frac{\alpha^4}{5!} + \frac{\theta}{5} A \frac{\alpha^2}{3!} + \frac{\theta^2}{15}(R - A)$$

Putting b_i's in Eq. 1a and neglecting the terms containing θ^2

$$\mathrm{I(t)} = b_0 + R\left(t - \frac{\theta}{3} t^3\right) - \frac{\theta}{2} b_0 t^2 + A\left[-t + \frac{\alpha t^2}{2!} - \frac{\alpha^2 t^3}{3!} + \frac{\alpha^3 t^4}{4!} - \frac{\alpha^4 t^5}{5!} + \ldots\right]$$
$$+ \mathrm{A}\left(\frac{\theta}{3} t^3 - \frac{\theta}{4}\left(\frac{\alpha}{2!}\right) t^4 + \frac{\theta}{5}\frac{\alpha^2}{3!} t^5 - \ldots\right) \tag{1d}$$

Since, $\mathrm{I(0)} = I_0$, Eq. (1d) gives

$$b_0 = I_0$$

Therefore, after simplification, Eq. (1d) can be written as

$$\mathrm{I(t)} = I_0 - \frac{\theta}{2} I_0 t^2 + \mathrm{R}\left(t - \frac{\theta}{3} t^3\right) - \frac{A}{\alpha}\left(e^{-\alpha t} - 1\right) + A\theta \sum_{n=3}^{\infty} \frac{\alpha^{n-3}}{n(n-2)!} \tag{1e}$$

Holding cost is

$$\mathrm{HC} = \int_0^T h(t) I(t) dt = \, aI_0 T + aR\left(\frac{T^2}{2} - \frac{\theta}{3}\frac{T^4}{4}\right) - \frac{\theta}{2} a I_0 \frac{T^3}{3} + \frac{aA}{\alpha^2}\left(e^{-\alpha T} - 1 + \alpha T\right)$$

$$+A a\theta \sum_{n=3}^{\infty} \frac{\alpha^{n-1}T^{n+1}}{n(n+1)(n-2)!} \tag{2}$$

Purchase cost is

$$\mathrm{PC} = \mathrm{C}\,[I_0 + RT] \tag{3}$$

The setup cost is

$$SC = C_s$$

Deterioration cost is

$$\mathrm{DC} = \mathrm{C}\left[I_0 - \int_0^T D(T)dt\right] = \mathrm{C}I_0 + \frac{A}{\alpha}(e^{-\alpha T} - 1) \tag{4}$$

Preservation cost is
PCT $= \beta$

$$\text{Carbon emission} = c_e[I_g] = \frac{c_e}{(1+I_g^2)} \tag{5}$$

$$\text{Carbon tax (CT)} = c_e[I_g]\int_0^T I(t)dt$$

$$= \frac{c_e}{1+I_g^2}\left[I_0T + R\left(\frac{T^2}{2} - \frac{\theta}{3}\frac{T^4}{4}\right) - \frac{\theta}{2}I_0\frac{T^3}{3} + \frac{A}{\alpha^2}\left(e^{-\alpha T} - 1 + \alpha T\right)\right.$$
$$\left. + A a\theta \sum_{n=3}^{\infty} \frac{\alpha^{n-1}T^{n+1}}{n(n+1)(n-2)!}\right] \tag{6}$$

The total inventory cost is

$$\mathrm{TC} = \mathrm{SC} + \mathrm{PC} + \mathrm{HC} + \mathrm{DC} + \mathrm{PCT} + \mathrm{CT}$$
$$= C_s + C[I_0 + RT] + aI_0T + aR\left(\frac{T^2}{2} - \frac{\theta}{3}\frac{T^4}{4}\right) - \frac{\theta}{2}aI_0\frac{T^3}{3} + \frac{aA}{\alpha^2}\left(e^{-\alpha T} - 1 + \alpha T\right)$$
$$+ A a\theta \sum_{n=3}^{\infty} \frac{\alpha^{n-1}T^{n+1}}{n(n+1)(n-2)!} + \mathrm{C}I_0 + \frac{A}{\alpha}\left(e^{-\alpha T} - 1\right) + \beta$$
$$+ \frac{c_e}{1+I_g^2}\left[I_0T + R\left(\frac{T^2}{2} - \frac{\theta}{3}\frac{T^4}{4}\right) - \frac{\theta}{2}I_0\frac{T^3}{3} + \frac{A}{\alpha^2}\left(e^{-\alpha T} - 1 + \alpha T\right) + A a\theta \sum_{n=3}^{\infty} \frac{\alpha^{n-1}T^{n+1}}{n(n+1)(n-2)!}\right] \tag{7}$$

5 Solution Procedure

The proposed model has been solved using two different forms of Particle Swarm Optimization (PSO) i.e., Weighted Particle Swarm Optimization (W-PSO), for this we have referred to Alam [2] and Yang [27] and Constriction Factor Particle Swarm Optimization (CO-PSO) for details, Eberhart and Shi [9]. For the adequate analysis of the solution procedure for cost minimization, all the parameters of the algorithm remain the same except for the initial weight and constriction factor. Convergence characteristic of the array of the global optimum is provided to substantiate the convergence of the solution graphically.

6 Numerical Illustration

In the lack of original data, the following hypothetical values of the parameters have been taken to solve the objective function numerically.

Problem: The following values are considered for the parameters.

$C_e = \$2.50,\ R = 250,\ I_0 = \$100000,\ \alpha = 0.2,\ A = 150,\ \theta = 0.01,\ C_s = 10000,$
$a = 1.2,\ b = 0.15,\ C = 1.50$ and $\gamma = 0.2$.

Solution: The total cost to maintain the inventory of items is \$ 307,644.25. The green investment to control carbon emissions is \$ 1409.1891 and the effective preservation to reduce the rate of deterioration is \$28.365467. For the convergence curve see Fig. 1.

7 Sensitivity Analysis

In this section, we test the flexibility of the model through sensitivity analysis of the parameters. The change made on the parameters is defined as $N \to (1 + r) * N$, where $r \in \{-0.2, -0.1, 0.1, 0.2\}$.

8 Managerial Implications and Limitations

In this section, the major high points of the results obtained in the numerical illustration are provided for the better decision-making process by the inventory manager. Further, some limitations are highlighted to provide some future scope for the researchers.

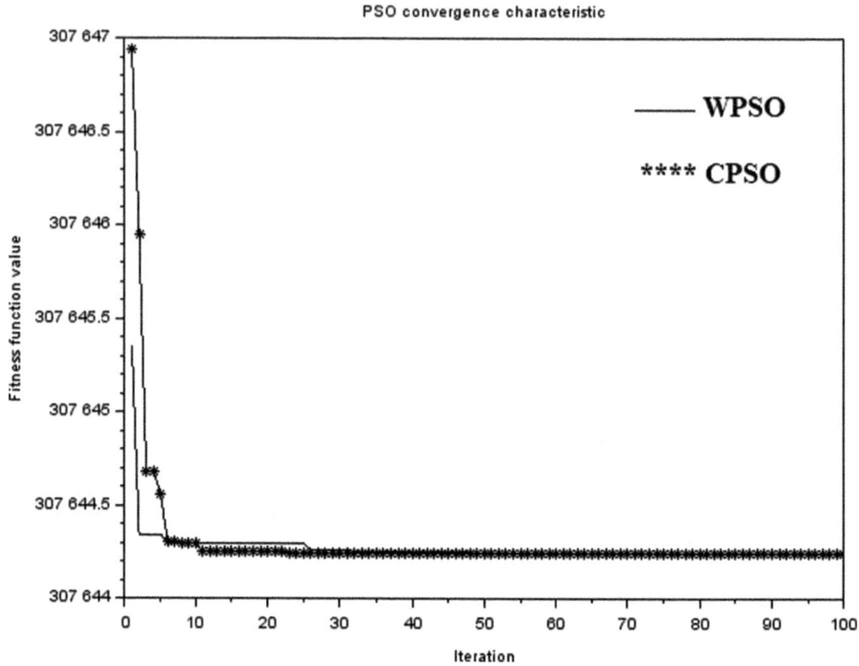

Fig. 1 Convergence curve

I. Investment in green technologies can help in reducing the carbon tax as well as the overall cost of inventory management.
II. Implementing modern preservation technologies can reduce the rate of deterioration which further reduces the deterioration cost.
III. Continuous replenishment policy can help in overcoming the shortages but the total cost of inventory management increases due to overholding of items. Further, increasing the rate of replenishment can increase the carbon footprint as a consequence of green investment increases.
IV. The results obtained in the sensitivity analysis (see Table 1), the following conclusions can be made for the parameters;

(a) *Setup cost* (C_s): Total cost increases with the increase in setup cost.
(b) *Replenishment rate* R: Total cost increases with the increase in replenishment rate. Also, a mild increase in green investment is observed with the increase in replenishment cost.
(c) *Carbon tax* (C_e): A significant increase in green investment is noticed with an increase in carbon tax. Further, a slight change in total cost is observed with the increase in carbon tax.
(d) *Initial Inventory Level* (I_0): Total cost increases with the increase in initial inventory level. Further, a significant increase in green investment is noticed with the increase in the initial inventory level.

Table1 Sensitivity analysis

Parameter	% Changed	β	I_g	TC
C_s	−20	28.365469	1409.1891	305,644.25
	−10	28.365476	1409.1893	306,644.25
	10	28.365477	1409.1892	308,644.25
	20	28.3654	1409.1891	309,644.25
R	−20	28.363977	14,008.9774	307,510.62
	−10	28.364721	1409.0832	307,577.43
	10	28.366216	1409.2952	307,711.06
	20	28.366967	1409.401	307,777.87
C_e	−20	28.366703	1259.7134	307,345.3
	−10	28.366072	1336.5321	307,498.93
	10	28.364351	1478.2995	307,782.46
	20	28.364351	1544.3258	307,914.52
I_0	−20	27.251524	1260.0803	248,544.92
	−10	27.839434	1336.7051	278,098.75
	10	28.841427	1478.1389	337,182.62
	20	29.276007	1544.0261	366,714.83
C	−20	26.940005	1409.1873	277,552.82
	−10	27.703361	1409.1884	292,598.59
	10	28.950072	1409.1898	322,689.83
	20	29.473416	1409.1903	337,735.36
a	−20	28.597919	1409.1894	278,777.31
	−10	28.483053	1409.1894	293,210.78
	10	28.245074	1409.1891	322,077.71
	20	28.121702	1409.1889	336,511.17
∝	−20	28.365538	1409.1983	307,646.13
	−10	28.365512	1409.1973	307,645.18
	10	28.365449	1409.1848	307,643.32
	20	28.365419	1409.1803	30,742.41
A	−20	28.364437	1409.0717	307,620.05
	−10	28.36494	1409.1304	307,632.15
	10	28.365991	1409.2478	307,656.35
	20	28.366523	1409.3067	307,668.45
b	−20	28.364105	1574.6417	307,976.82
	−10	28.364835	1485.0357	307,796.68
	10	28.366017	1343.9049	307,513.07
	20	28.366491	1286.9381	307,398.63
θ	−20	27.24975	14,091,892	307,643.13

(continued)

Table1 (continued)

Parameter	% Changed	β	I_g	TC
	−10	27.83868	1409.1891	307,643.72
	10	28.842034	1409.1891	307,644.72
	20	29.277076	1409.1891	307,645.16
γ	−20	34.06217	1409.1879	307,651.19
	−10	30.93186	1409.1886	307,647.37
	10	26.220047	1409.1897	307,647.37
	20	24.39759	1409.19	307,639.45

(e) *Unit cost of the item* (C): Preservation cost is increased with an increase in unit cost of the product. Also, the total cost is increased significantly increase in unit cost of the item.
(f) *Time sensitivity* (α): No change is observed in the preservation cost, green investment and total cost when the time sensitivity is increased.
(g) *Holding cost* (a): Total cost increases with the increase in holding cost.
(h) *Demand scale* (A): A mild increase in total cost is observed with an increase in demand scale.
(i) *Sensitivity of green investment* (b): A significant change in green investment is observed with an increase in the sensitivity of green investment. Further, total cost is decreased slightly with an increase in sensitivity of green investment.
(j) *Rate of deterioration* (θ): Preservation costs increased with an increase in the rate of deterioration.
(k) *Sensitivity of preservation cost* (γ): Preservation cost decreases significantly with an increase in sensitivity of preservation cost.

9 Conclusion

This is a single-item inventory model with negative exponential demand and instantaneous replenishment. The main aim of the proposed model is to study the impact of green investment and preservation technology so that total cost can be minimized. Holding cost is assumed to be independent of time and the rate of deterioration is time-proportional. Further, carbon emission cost is considered a function of green investment. As the formulated objective function is highly non-linear an algorithmic approach is considered to solve the objective function. A particle Swarm algorithm is applied to obtain the total inventory cost, an optimal value of green investment and preservation cost. The obtained result suggests that investment in green technologies can help to reduce the total inventory cost. Moreover, the implementation of preservation technologies also helps to reduce the deterioration cost. Further, an instantaneous replenishment policy can be used to overcome the shortages.

The present model is an extension of existing models in literature where the demand was considered as negative exponential but so far author's knowledge is concerned the proposed model is not reported in the literature on inventory modelling. The formulated model can be extended by considering the rate of deterioration as a linear function of time. Further, the expression of θ can be changed by considering other probability distributions and the result can be compared to analyse which distribution is fit to minimize the cost.

Acknowledgements The authors are thankful to the Department of Mathematics of Assam University, Silchar for infrastructural support. The authors are also thankful to anonymous referees for their constructive suggestions for the improvement of this paper.

Author's Contributions: SM formulated the model and performed all the calculations along with the required analysis. NS designed the manuscript and interpreted the results.

Funding: Not applicable.

Data Availability: Not applicable.

Declarations:

Conflict of Interest: The authors don't have any competing interests.

Ethics Approval and Consent to Participate: Not applicable.

Consent for publication: Not applicable.

References

1. Ahmed, B., Benkherouf, L.: On an optimal replenishment policy for inventory model for non-instantaneous deteriorating items and permissible delay in payments: revisited. Int. J. Syst. Sci. Oper. Logis. **8**(2), 132–135 (2019)
2. Alam, M.N.: Codes in MATLAB for particle Swarm optimization. Researchgate (2016). https://doi.org/10.13140/RG.2.1.4985.3206
3. Bardhan, S., Pal, H., Giri, B.C.: Optimal replenishment policy and preservation technology investment for non-instantaneous deteriorating items with stock dependent demand. Oper. Res. Int. Journal **19**, 347–368 (2019)
4. Barman, A., Das, R., De, P.K.: Optimal pricing, replenishment scheduling and preservation technology investment policy for multi-item deteriorating inventory model under shortages. Int. J. Model. Simul. Sci. Comput. **12**(05) (2021)
5. Bhattacharjee, N., Sen, N.: An inventory model to study the effect of the probabilistic rate of carbon emission on the profit earned by a supplier. Oper. Res. Decisions **4**, 5–33 (2021)
6. Bhattacharjee, N., Sen, N.: Non-separable raw material inventory model with time dependent deterioration: hybrid algorithm. Int. J. Appl. Math. Stat. **60**(1), 68–80 (2021)
7. Bhattacharjee, N., Sen, N.: A multi-item sustainable production inventory constrained model to study the effective green investment and replenishment quantity. Int. J. Math. Oper. Res. **22**(4), 528–554 (2022)
8. Cheng, M., Zhang, B., Wang, G.: Optimal policy for deteriorating items with trapezoidal type demand and partial back logging. Appl. Math. Model. **35**(7), 3552–3560 (2011)

9. Eberhart, R.C., Shi, Y.: Comparing inertial weights and constriction factor in particle swarm Optimization (2000). https://doi.org/10.1109/CEC.2000.870279
10. Ghare, P.M., Schrader, G.F.: A model for exponentially decaying inventory. J. Ind. Eng. **14**(5), 238–243 (1963)
11. Latha, K.M., Uthayakumar, R.: An inventory model for increasing demand with probabilistic deterioration, permissible delay and partial back logging. Int. J. Inf. Manag. Sci. **25**(4), 297–316 (2014)
12. Mahato, C., Mahata, G.C.: Optimal replenishment, pricing and preservation technology investment policies for non-instantaneous deteriorating items under two-level trade credit policy. J. Ind. Manag. Optim. **18**(5), 3499–34537 (2022)
13. Maragatham, M., Palani, R.: An inventory model for deteriorating items with lead time price dpendent demand and shortages. Adv. Comput. Sci. Technol. **10**(6), 1839–1847 (2017)
14. Pal, H., Bardhan, S., Giri, B.C.: Optimal replenishment policy for non-instantaneously perishable items with preservation technology and random deterioration start time. Int. J. Manag. Sci. Eng. Manag. **13**(3), 188–199 (2017)
15. Roy, T.S., Ghosh, S.K., Chaudhuri, K.: An optimal replenishment policy for EOQ models with time-varying demand and shortages. Int. J. Serv. Oper. Manag. **16**(4), 443–459 (2013)
16. Shah, N.H., Pandey, P.: Deteriorating inventory model when demand depends on advertisement and stock display. Int. J. Oper. Res. **6**(2), 33–44 (2009)
17. Sarkar, B., Sarkar, S.: An improved inventory model with partial back logging, time varying deterioration and stock-dependent demand. Econ. Model. **30**, 245–252 (2013)
18. Saha, S., Sen, N.: A study on inventory model with negative exponential demand and probabilistic deterioration under back logging. Uncertain Supply Chain Manag. **5**, 77–88 (2017)
19. Shah, N.H., Rabari, K., Patel, E.: Dynamic demand and pricing inventory model for non-instantaneous deteriorating items. Int. J. Math. Eng. Manag. Sci. **6**(2), 510–521 (2021)
20. Sekar, T., Uthayakumar, R.: A green production inventory model for multi-replenishment cycle and internally deteriorating items allowing shortages. Int. J. Adv. Res. Basic Eng. Sci. Technol. **3**(14), 394–401 (2017)
21. Sekar, T., Uthayakumar, R.: A manufacturing inventory model for exponentially increasing demand with preservation technology and shortages. Int. J. Oper. Res. **15**(2), 61–70 (2018)
22. Soni, H.N., Suthar, D.N.: Optimal pricing and replenishment policy for non-instantaneous deteriorating items with varying rate of demand and partial backlogging. Opsearch **57**, 986–1021 (2020)
23. Sivashankari, C.K.: Purchasing inventory models for exponential demand with deteriorating items and discounted cost-in third order equation. Int. J. Procure. Manag. **12**(3), 321–335 (2019)
24. Singh, S.R., Rani, M.: An inventory model exponentially time-sensitive demand and parabolic time linked holding cost under inflation and shortages with salvage value. PalArch's J. Archaeol. Egypt Egyptol. **17**(7), 10941–10949 (2020)
25. Taleizadeh, A.A., Khanbaglo, M.P.S., Cardenas-Barron, L.E.: Replenishment of imperfect items in an EOQ inventory model with Partial backordering. RAIRO Oper. Res. **54**(2), 413–434 (2018)
26. Tripathi, R., Kumar, M.: A new model for deteriorating items with inflation under permissible delay in payments. Int. J. Ind. Eng. Comput. **5**(3), 365–374 (2014)
27. Yang, X.-S.: Introduction to Computational Mathematics, 2nd edn., pp. 313–316. World Scientific Publishing, Singapore (2015)

A Real Coded Heuristic Technique in Solving a Plant Location Optimization Problem with Interval-Valued Transportation Costs and Quantities

Ranjan Kumar Gupta and Debdip Khan

Abstract In the existing literature, in all the plant location problems the per unit transportation costs and quantities have been considered to be fixed numbers at known positive levels. But, in reality, those costs and quantities can fluctuate due to several reasons and are hence not precise. Here, the impreciseness has been represented with numbers that are in intervals. Subsequently, the plant location problem has been expressed as a cost minimization problem with interval-valued coefficients. In order to solve the said interval-valued optimization problem a GA has been developed. To demonstrate the model, several examples have been considered and presented. In some examples, fixed valued costs and quantities have been considered and their results have been tested with the technique prescribed by 'least cost centre analysis' available in the literature. Finally, sensitivity analyses with respect to GA parameters have been done and exhibited graphically. The analysis displays that the GA developed by us is fairly stable. For the first time in plant location optimization, transportation costs and quantities have been considered as interval numbers. Apart from the GA developed in this paper, the formulated problem cannot be probably solved by any other existing methods. 'Least Cost Centre Analysis' provides a solution for only fixed coefficients, but our method can solve for intervals as well.

Keywords Plant location · Least cost · Interval numbers · Genetic algorithm · Order relations

R. K. Gupta (✉)
Department of Management and Marketing, West Bengal State University, Barasat, W.B, India
e-mail: rangupt@rediffmail.com

D. Khan
Department of Business Administration, Burdwan Raj College, Burdwan, W.B, India

L. Sahoo et al. (eds.), *Decision Making Under Uncertainty Via Optimization, Modelling, and Analysis*, Studies in Systems, Decision and Control 558,
https://doi.org/10.1007/978-981-96-0085-4_17

1 Introduction

One of the most critical decisions faced by modern production and operations managers during the design phase is determining the optimal location for a manufacturing plant. This decision, commonly referred to as plant location, is defined as "the process of selecting the most suitable geographic site for a firm's operations" [1]. As highlighted by Drezner and Hamacher [2], plant location is a pivotal area within operations research, encompassing various factors such as production capacity, market expansion, obsolete facilities, and potential business mergers [3].

In the face of an ever-evolving business landscape, maintaining flexibility in plant location decisions is essential. This includes considering future expansions, adding new locations, or replacing existing ones to adapt to changing conditions. Key capacity-related considerations include the volume of production, plant size, timing, and the appropriate type of capacity required [4, 5].

Given the challenges associated with relocating a plant once established, the decision regarding plant location is of paramount importance for manufacturing organizations, significantly impacting their competitive position. An optimally chosen location can provide a competitive edge and become a critical success factor [5–7].

The origins of location theory can be traced back to 1826 when German agriculturalist Johann Heinrich von Thünen first proposed a theory explaining the relationship between production location and transportation costs [8, 9]. Later, German researcher Alfred Weber formalized the study of industrial location with his seminal work, "Über den Standort der Industrien" [10]. Over nearly two centuries, research in this field has evolved considerably to address the changing business environment [11]. Numerous studies have further explored plant location, including those by Abernathy and Hersley [12], Atkine and Shriver [13], Bowersox et al. [14], Bridgeman [15], Brown and Gibson [16], Karasaka and Browball [17], Dou and Sarkis [18], Feldmann and Olhager [19], Melo et al. [20], and Theyel [21].

Several critical factors influence plant location decisions, such as the availability of raw materials, proximity to markets, access to reliable power supply, the presence of a skilled labor force, transportation infrastructure, existing facilities, and government policies. Among these, minimizing transportation costs is often a key consideration. One widely used model for optimizing plant location based on transportation costs is the 'least cost center analysis.' This model assumes that the coordinates of different markets, per-unit transportation costs, and quantities transported are known and fixed. The model identifies the optimal plant location as the median of the x and y coordinates of these markets, which is particularly useful for selecting a warehouse location relative to its principal customers or a manufacturing plant relative to its suppliers and customers [22].

Previous studies on plant location have typically treated per-unit transportation costs and quantities transported as fixed positive values. But, in practical scenarios, these costs can fluctuate due to various factors. For instance, transportation costs may rise during the rainy season compared to winter or autumn, and road conditions can further influence these costs. Additionally, the total quantity transported between the plant and each market may vary based on plant operations and market demand. Therefore, in our research, we have modeled per-unit transportation costs and quantities transported as imprecise numbers with positive values.

To address the challenges posed by such imprecise data, researchers often resort to stochastic, fuzzy, or fuzzy-stochastic approaches. However, as mentioned by Gupta et al. [23, 24], exactly specifying the membership function, or the probability distribution in these approaches is quite a difficult task for the decision-maker at times. To get rid of the difficulty, in this paper for the first time in solving a plant location problem, the imprecise numbers have been represented by interval numbers. Consequently, in this transportation cost-based location problem the objective function, which is of minimization type, is interval valued.

For solving the said problem of optimization with interval objective, a program using Genetic Algorithm (GA), which is a popular heuristic technique for optimization, has been developed. The fundamental concept of this algorithm was first developed by Prof. J. H. Holland, in the mid 60's. Since then, several researchers have worked on further development of this algorithm. Comprehensive coverage of GA can be found in works by authors such as Goldberg [25], Michalewicz [26], Sakawa [27], and Gen and Cheng [28], while its applications in various domains have been documented by Gupta and Bhunia [29], Gupta et al. [30], and Khan and Gupta [31].

Given that our plant location problem involves an interval-valued objective function, interval order relations play a crucial role in the GA's selection/reproduction process and in identifying the best chromosome in each generation. While earlier researchers like Ishibuchi and Tanaka [32] and Chanas and Kuchta [33] attempted to define order relations for interval-valued numbers, their definitions were incomplete. Sengupta and Pal [34] later proposed two approaches—deterministic and fuzzy—for comparing interval numbers. Mahato and Bhunia [35] further refined these definitions to account for the perspectives of optimistic and pessimistic decision-makers in maximization and minimization problems. Additional research on interval arithmetic has been conducted by Gupta and Khan [36, 37]. Moreover, recent studies by Roy and Khan [38], Deb and Gupta [39], and Sahoo et al. [40] have demonstrated the practical applications of heuristic algorithms and MCDM techniques in real-world scenarios.

2 Interval Arithmetic with Order Relations

If x is a set of real numbers, then an interval-valued number K may be defined as.

$\mathrm{K} = [\mathrm{k}_L, \mathrm{k}_R] = \{x : k_L \leq x \leq k_R, k_L, k_R, x \in R\}$, where k_L is the left boundary and k_R is the right boundary of interval number K. Thus, an uncertain number $\mathrm{x} \in [\mathrm{k}_L, \mathrm{k}_R]$ *is* represented by the closed interval K. It may be noted here that a point interval is a typical case when the left and right boundaries of number K coincides, i.e., $k_L = k_R = k$, and the interval number $\mathrm{K} = [\mathrm{k}_L, \mathrm{k}_R]$ reduces to the real number $\mathrm{k} = [\mathrm{k}, \mathrm{k}]$.

Some of the arithmetical operations of interval numbers are highlighted below:

Definition 1: Let us assume that $\otimes \in (+, -, ., /)$ is an arithmetic operation. Let there be two closed intervals, namely K and G, then $\mathrm{K} \otimes \mathrm{G} = \{\mathrm{k} \otimes \mathrm{g} : \mathrm{k} \in \mathrm{K} \text{ and } \mathrm{g} \in \mathrm{G}\}$.

defines a binary operation on the set of closed intervals. In case of division, it is assumed that $0 \notin G$.

Using Definition 1 the clear description/execution of the operations on two interval numbers $\mathrm{K} = [\mathrm{k}_L, \mathrm{k}_R]$ and $\mathrm{G} = [\mathrm{g}_L, \mathrm{g}_R]$ are

$$K + G = [k_L + g_L, k_R + g_R]$$

$$K - G = [k_L - g_R, k_R - g_L]$$

$$\lambda K = \begin{cases} [\lambda k_L, \lambda k_R] \text{ if } \lambda \geq 0 \\ [\lambda k_R, \lambda k_L] \text{ if } \lambda < 0 \end{cases}$$

where λ is a real number. Also,

$$K \times G = [k_L, k_R] \times [g_L, g_R]$$
$$= \begin{cases} [\mathrm{Min}(k_L g_L, k_L g_R, k_R g_L, k_R g_R), \mathrm{Max}(k_L g_L, k_L g_R, k_R g_L, k_R g_R)] \\ [k_L g_L, k_R g_R], \text{ only if } k_L \geq 0 \text{ and } g_L \geq 0 \end{cases}$$

In this chapter, the transportation costs and the transported quantities are interval valued. So, the procedure for determining the optimal solution to the plant location problem involves the concept and application of the order relations of interval numbers for minimization problems. Let two closed intervals $\mathrm{K} = [\mathrm{k}_L, \mathrm{k}_R]$ and $\mathrm{G} = [\mathrm{g}_L, \mathrm{g}_R]$ are represent the imprecise numbers belonging to the two alternatives. These two intervals K and G may be of the following three types:

Type-I: Both the intervals are disjoint.

Type-II: Intervals are partially overlapping.

Type-III: One interval is contained in the other.

Regarding optimistic and pessimistic decision-making, let us consider the definitions of order relations developed by Mahato and Bhunia [32] and used by several other researchers like Gupta et al. [23, 24]. These definitions are stated below.

In the case of optimistic decisions, the uncertainty is ignored and the decision maker selects the lowest cost/time for minimization problems. For maximization problems, the highest profit is selected by him.

Definition 2: In the case of minimization, the definition of order relation $\leq_{o\min}$ between the intervals K and G for optimistic decision-maker is given by

$$K \leq_{o\min} \mathrm{G} \text{ iff } k_L \leq g_L$$

$$K <_{o\min} G \text{ iff } K \leq_{o\min} G \text{ and } K \neq G$$

In the case of pessimistic decisions, the principle "Less uncertainty is better than more uncertainty", guides the decision-making.

Definition 3: In the case of minimization, the definition of order relation $<_{p\min}$ between the intervals $K = [k_L, k_R] = \langle k_C, k_r \rangle$ and $G = [g_L, g_R] = \langle g_C, g_r \rangle$ for a pessimistic decision maker is given by

(i) $K <_{p\min}$ G iff $k_C < g_C$ for type-I and type-II intervals
(ii) $K <_{p\min}$ G iff $k_C \leq g_C$ and $k_r > g_r$ for type-III intervals

But, If $k_C < g_C$ and $k_r < g_r$, then for type-III intervals, one cannot take a pessimistic decision. He may consider optimistic decision-making.

3 Assumptions and Notations

Assumptions

(i) The total transportation cost is to be incurred by the plant and not the markets.
(ii) Transportation between the plant and the different markets takes place along rectangular paths.
(iii) The costs of transporting a unit quantity of raw materials/finished goods through the unit distance between the plant and each market lie in the known intervals.
(iv) The quantity of raw materials/finished goods transported between the plant and each market lies in the known intervals.
(v) Once the plant is located, its position is fixed and the plant cannot be shifted to a new location because of several practical constraints.
(vi) The location decision is taken purely on the basis of the transportation cost of the plant.
(vii) While doing the sensitivity analyses after the plant is optimally located, in case there is a shift in the position of the market, then that shift takes place in the diagonal direction away from the plant.

Notations

$TC \in [TC_L, TC_R]$ Total cost of transportation of quantities (finished goods/raw materials) from/to the plant to/from the markets.
$L(X, Y)$ Location of the plant (its Cartesian coordinate).
$M_i(x_i, y_i)$ The ith market with the Cartesian coordinates of its location.
$C_i \in [C_{iL}, C_{iR}]$ Cost of transporting unit quantity through unit distance between the plant location and the ith market.

$Q_i \in [Q_{iL}, Q_{iR}]$	Quantity to be transported between the plant and the *i*th market.
D_i	Distance between the plant and the *i*th market.
N	The number of markets.
	Popsize Population size.
Pobcross	Probability of crossover.
Pobmute	Probability of mutation.
Maxgen	Maximum number of generation.

4 Mathematical Formulation

The total cost of transportation is given by

$$TC = \sum_{i=1}^{n} C_i Q_i D_i \tag{1}$$

Like the 'Least Cost Centre Analysis', discussed under the 'Introduction' section and other popular location models, here also it has been assumed that the movements between the plant and the different Markets take place along rectangular paths. Hence, rectangular distances have been considered.

$$\text{Thus,}\quad D_i = \{|X - x_i| + |Y - y_i|\} \tag{2}$$

$$\text{Hence}\ \ TC = \sum_{i=1}^{n} C_i Q_i \{|X - x_i| + |Y - y_i|\} \tag{3}$$

However, Q_i and C_i are assumed to be imprecise variables. Because of this impreciseness, Q_i and C_i are treated as interval numbers such that

$$Q_i \in [Q_{i_L}, Q_{i_R}] \text{ and } C_i \in [C_{i_L}, C_{i_R}]$$

So using Definition 1 of interval arithmetic discussed under Sect. 2 of this paper,

$$\begin{aligned} Q_i C_i &= [Q_{iL}, Q_{iR}][C_{iL}, C_{iR}] \\ &= [Q_{iL} C_{iL}, Q_{iR} C_{iR}] \end{aligned} \tag{4}$$

Hence,

$$TC = \sum_{i=1}^{n} [Q_{iL} C_{iL}, Q_{iR} C_{iR}] \{|X - x_i| + |Y - y_i|\} = [TC_L, TC_R]$$

$$where\ TC_L = \sum_{i=1}^{n} Q_{iL} C_{iL}\{|X - x_i| + |Y - y_i|\} \text{ and}$$

$$TC_R = \sum_{i=1}^{n} Q_{iR} C_{iR}\{|X - x_i| + |Y - y_i|\}$$

So, our plant location optimization problem with interval-valued C_i's and interval-valued Q_i's is

$$\text{Min } TC(X, Y) \in [TC_L, TC_R]$$

subject to,

$$0 \leq X \leq \text{Max}(x_i)$$

$$0 \leq Y \leq \text{Max}(y_i)$$

5 Solution Procedure

A real-coded GA has been developed for solving the cost minimization problem. Since the basic algorithm and the functioning principle of the Genetic Algorithm are quite popular and explained in various books, it has not been described here. However, some of its basic components like 'representation of chromosome', 'initialization of population', 'evaluation function', 'Selection process' and 'Genetic operators (crossover and mutation)' are briefly explained below.

Representation of Chromosome

One vital job to ensure the smooth and proper functioning of GA is the designing of an appropriate chromosome representation of solutions of the problem. Since our problem consists of two continuous variables, here chromosome has been represented with real numbers. Thus a chromosome has been represented by a real row matrix $W_j = [W_{j1}, W_{j2}]$, where the components W_{j1} and W_{j2} represent the decision variables, *X and Y*, respectively, of the problem.

Initialization

To ensure that the chromosomes have been properly represented, they are to be initialized in the next step. In this work, uniform distribution is used for selecting random numbers for the two components of each chromosome.

Evaluation Function

In the next step i.e. evaluation, we need to check the goodness of each of the potential solutions To do this, corresponding to each chromosome the fitness value is to be calculated. The measure for the fitness value of each chromosome is considered to be the objective function value of that chromosome.

Selection Process

Here, the tournament selection scheme of size two has been used as the selection operator. At first, two chromosomes/individuals are randomly selected. Out of these two chromosomes/individuals, the one with the better fitness value is considered to be better and is hence selected. Here, the objective function of our problem being interval valued, for the selection of better chromosome/individual Definition 3 of order relations has been used.

Crossover

After the selection process, the surviving chromosomes undergo genetic operations–crossover and mutation. The crossover operation operates on two-parent solutions at a time and generates offspring by recombining the features of both parent solutions. In this operation, the expected *pobcross*popsize* number of chromosomes will take part. So for determining the parents for crossover operation, *pobcross* popsize* number of chromosomes is selected. With the selected chromosomes, the crossover operation is done. In this work, crossover operation is done in the following manner:

Step-1: The integral value of *pobcross ∗ popsize is found* and stored in N.

Step-2: From the population, the chromosomes W_q and W_i are randomly selected for crossover.

Step-3: Generate a random real number λ *in* [0, 1].

Step-4: The rth ($r = 1, 2$) component $W_{qr}^{/}$ and $W_{ir}^{/}$ of two offspring will be created by

$$W_{qr}^{/} = \lambda W_{qr} + (1 - \lambda) W_{ir}$$

$$W_{ir}^{/} = \lambda W_{ir} + (1 - \lambda) W_{qr}$$

Step-5: Repeat Step-2 to Step-4 for $N/2$ times.

Mutation

Mutation operation is done on a sole chromosome to inject random variations inside the population. Generally, the probability of mutation is much lower as it is less dominant than selection and crossover. Here, uniform mutation has been used. If the element (gene) W_{im} of chromosome W_i is selected for mutation and the domain of W_{im} is $[l_{im}, u_{im}]$, then the reduced value of W_{im} is given by

$$W_{im}\prime = \begin{pmatrix} W_{im} + \Delta(u_{im} - W_{im}), \text{ if a random digit is 0} \\ W_{im} - \Delta(W_{im} - l_{im}), \text{ if a random digit is 1} \end{pmatrix}$$

where $m \in \{1, 2\}$ and (v) returns a value in the range $[0, v]$.

In this work, we have taken $(v) =$ a random real number between $[0, v]$.

6 Numerical Examples

Taking the help of our GA, six examples have been solved. Except for Example 4, in all other examples, the per unit transportation costs (C_i) and the quantities (Q_i) are considered interval valued. However, in Example 4, these variables are considered to be fixed. This has been done by making the left and right boundaries of the intervals take the same values. With the help of the proposed GA, 15 independent runs for each of the examples have been executed. Out of these runs, the best interval valued total transportation cost (TC) has been considered. Corresponding to each example, the best value of TC and the corresponding value of plant Location *L(X, Y)* have been obtained and displayed in Table 7. The values considered for different GA parameters are: $maxgen = 100$, $pobcross = 0.95$, $pobsize = 100$, $pobmute = 0.15$ (Tables 1, 2, 3, 4, 5 and 6).

Table 7 reveals the results of the six examples considered by us. It may be stated here that in Example 4, we deliberately made the left boundaries and the right boundaries identical to each other for each of the interval-valued quantities (i.e.,

Table 1 Example 1: Taking n = 6

Markets (x_i, y_i)	$Q_i \in [Q_{iL}, Q_{iR}]$	$C_i \in [C_{iL}, C_{iR}]$
M_1 (2, 6)	[10, 14]	[5, 7]
M_2 (10, 8)	[5, 12]	[5, 7]
M_3 (15, 7)	[7, 9]	[3, 4]
M_4 (9, 11)	[6, 7]	[4, 6]
M_5 (5, 15)	[5, 8]	[4, 5]
M_6 (10, 15)	[2, 5]	[3, 5]

Table 2 Example 2: Taking n = 6

Markets (x_i, y_i)	$Q_i \in [Q_{iL}, Q_{iR}]$	$C_i \in [C_{iL}, C_{iR}]$
M_1 (21, 62)	[101, 142]	[5, 7]
M_2 (101, 82)	[51, 122]	[5, 7]
M_3 (151, 72)	[71, 92]	[3, 4]
M_4 (91, 112)	[61, 72]	[4, 6]
M_5 (51, 152)	[51, 82]	[4, 5]
M_6 (101, 152)	[21, 52]	[3, 5]

Table 3 Example 3: Taking n = 6

Markets (x_i, y_i)	$Q_i \in [Q_{iL}, Q_{iR}]$	$C_i \in [C_{iL}, C_{iR}]$
M_1 (2, 6)	[5]	[5]
M_2 (10, 8)	[15]	[5]
M_3 (15, 7)	[12]	[3]
M_4 (9, 11)	[26]	[4]
M_5 (5, 15)	[9]	[4]
M_6 (10, 15)	[7]	[3]

Table 4 Example 4: Taking n = 6

Markets (x_i,y_i)	$Q_i \in [Q_{iL}, Q_{iR}]$	$C_i \in [C_{iL}, C_{iR}]$
M_1 (46, 22)	[35.7, 42.0]	[7, 10]
M_2 (10, 52)	[23.4, 52.2]	[6, 8]
M_3 (15, 32)	[71, 92]	[5.5, 9.7]
M_4 (1, 80)	[31.7, 72.5]	[10, 16.7]
M_5 (51, 15)	[51, 82]	[4.5, 9.0]
M_6 (70, 57)	[21, 52]	[5, 8]

Table 5 Example 5: Taking n = 4

Markets (x_i, y_i)	$Q_i \in [Q_{iL}, Q_{iR}]$	$C_i \in [C_{iL}, C_{iR}]$
M_1 (5, 52)	[11.80, 25.00]	[15.50, 16.00]
M_2 (60, 88)	[12.90, 22.50]	[13.00, 13.50]
M_3 (15, 7)	[19.90, 26.00]	[12.00, 14.50]
M_4 (57, 97)	[17.70, 18.80]	[11.00, 13.20]

Table 6 Example 6: Taking n = 8

Markets (x_i,y_i)	$Q_i \in [Q_{iL}, Q_{iR}]$	$C_i \in [C_{iL}, C_{iR}]$
M_1 (82, 56)	[15.50, 25.00]	[10.20, 15.00]
M_2 (90, 88)	[11.80, 22.50]	[15.50, 16.00]
M_3 (15, 7)	[12.90, 22.00]	[13.00, 13.50]
M_4 (49, 31)	[22.70, 26.00]	[14.00, 18.00]
M_5 (45, 15)	[19.90, 29.50]	[12.00, 14.50]
M_6 (60, 75)	[17.00, 27.00]	[13.50, 17.00]
M_7 (57, 27)	[13.40, 18.80]	[11.00, 13.20]
M_8 (97, 97)	[17.70, 20.50]	[14.50, 7.00]

Table 7 Numerical results for different examples

Examples	Optimum plant location $L\,(X, Y)$	*Minimum* $TC \in [TC_L, TC_R]$
Example 1	(9, 8)	[962, 1984]
Example 2	(91, 82)	[97710, 202620]
Example 3	(15, 32)	[53973.4, 175,120.5]
Example 4	(9, 11)	[1353, 1353]
Example 5	(15, 52)	[43097.6, 66,611.92]
Example 6	(59.66, 55.89)	[84471.17, 143,181.26]

$Q_i = Q_{iL} = Q_{iR})$ and interval-valued costs (i.e., $C_i = C_{iL} = C_{iR}$) and thus the problem has become the same as that of "Least cost Centre Analysis". It is well known that the least cost centre analysis prescribes that the optimum solution is given by the median of x_i's as the x-coordinate (X) and median of y_i's as the y-coordinate (Y) of the optimum plant location, taking C_iQ_i as the frequencies. Following this procedure, for Example 4, it can be checked that the median of x_i's is 9 and the median of y_i's is 11. This confirms that our algorithm is capable of generating perfectly accurate results for precise values of C_i *and* Q_i the least cost centre analysis. However, in addition to this, our algorithm can further provide solutions for imprecise values of C_i *and* Q_i, something which the least cost centre analysis is incapable of. Further, the interval approach presented in this paper has wider applicability.

7 Sensitivity Analysis

To study the stability of our algorithm, sensitivity analyses have been performed graphically with respect to different GA parameters by considering Example 1. From Figs. 7, 8, 9 and 10, it is revealed that over a large range of the GA parameters, the result of Example 1, generated by our GA is fairly stable.

Also, once a plant is located at its optimum location it is very difficult for the plant to relocate. However, if the location of a market gets shifted in a direction away from the plant, then it is bound to have an adverse effect on the total transportation cost to be incurred by the plant. So, it is always better for the plant if it is aware of how much its total transportation cost would be affected by the shift in each of the market's locations. To study this effect of change in the location of a market by unit distance (in the diagonal direction away from the plant) on centre of the interval valued *TC*, (i.e., $(TC_L + TC_R)/2$), sensitivity analyses by taking Example 1 have been done graphically with respect to shifting in each market. It is evident from Figs. 1, 2, 3, 4, 5 and 6 that for Example 1 the total transportation cost of the plant is most sensitive to the shift in the location of the market M_1, which is followed in order by M_2, M_4, M_5, M_3 and M_6.

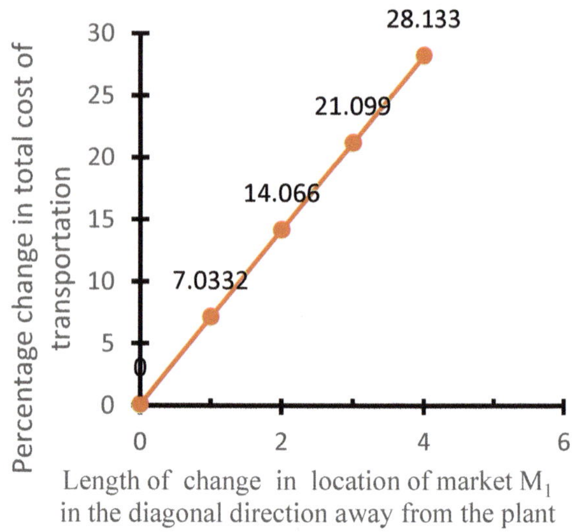

Fig. 1 '%age change in total transportation cost' versus 'change in location of market M_1'

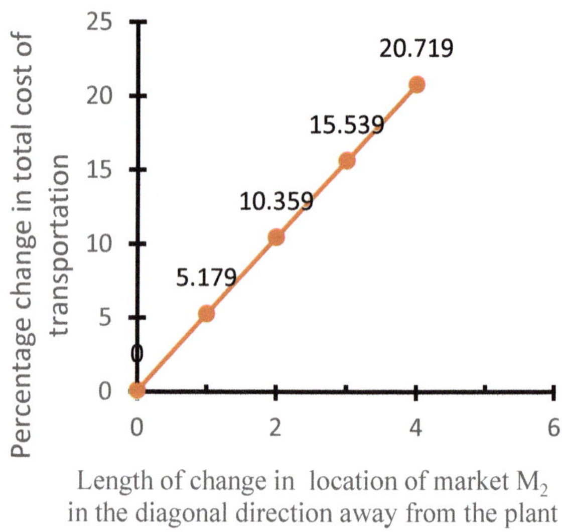

Fig. 2 '%age change in total transportation cost' versus 'change in location of market M_2'

8 Conclusions

One crucial decision in the design phase of operations management is the decision concerned with the optimum location of a plant. One of the major factors affecting this decision is the total transportation cost to be incurred by the plant. To the best of our knowledge in all the plant location problems the per unit transportation cost and quantities have been considered to be fixed numbers. However, considering the fact that in real-life situations, these costs and quantities can fluctuate due to several

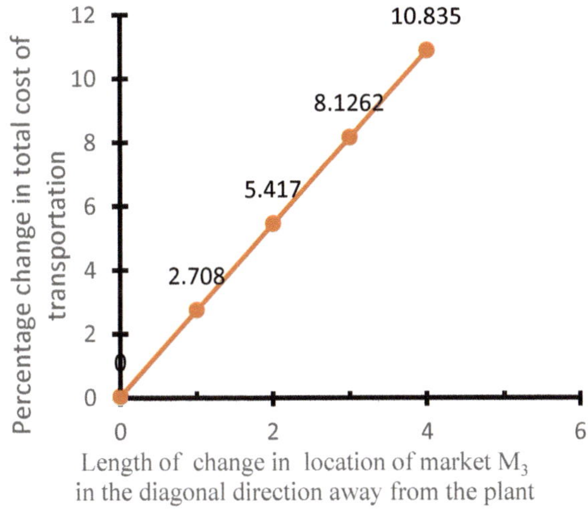

Fig. 3 '%age change in total transportation cost' versus 'change in location of market M_3'

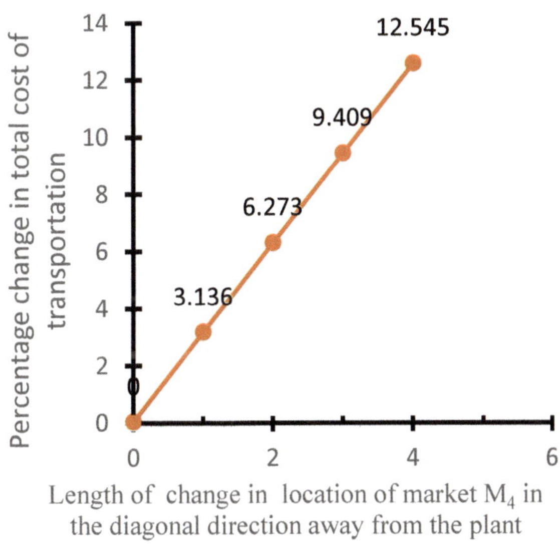

Fig. 4 '%age change in total transportation cost' versus 'change in location of market M_4'

practical reasons, they should be treated as imprecise. For the first time in a location problem we have considered per unit transportation costs and quantities as interval valued numbers. In this paper, we have formulated the problem of plant location as a constrained cost minimization problem with interval-valued coefficients and solved the same by a GA for real variables. As a special case of the problem, the upper bounds and lower bounds of the interval-valued costs and interval-valued quantities have been considered to be the same and the problem thus got transformed to the traditional fixed cost and fixed quantity based plant location problem similar to the problem of 'Least cost centre analysis'. The solution to this problem provided by our GA is

Fig. 5 '%age change in total transportation cost' versus 'change in location of market M_5'

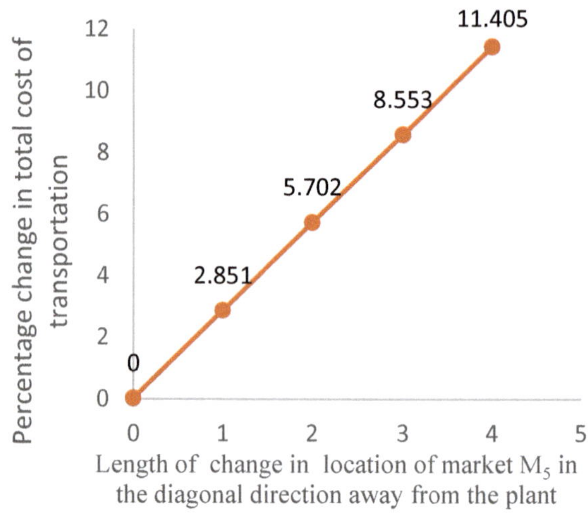

Fig. 6 '%age change in total transportation cost' versus 'change in location of market M_6'

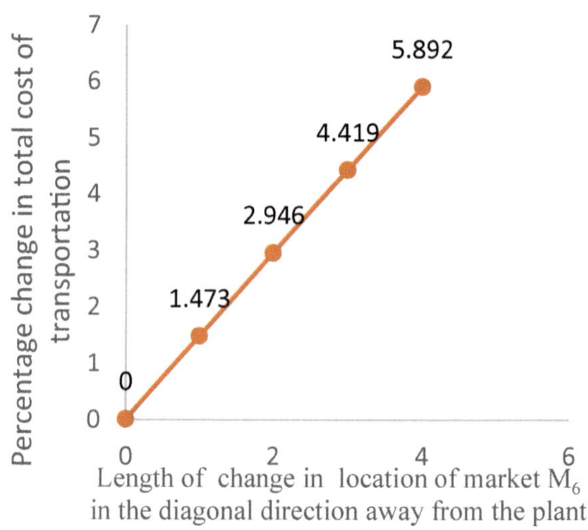

identical to the solution obtained by following the technique prescribed by 'Least cost centre analysis'. However, in addition to providing a perfectly accurate solution for precise values of per unit transportation costs and quantities, our algorithm is further capable of providing solutions for imprecise values of above-mentioned variables, something which 'Least cost centre analysis' is incapable of. Hence, the interval approach followed in this work has broader applicability. To do similar researches in future, one can formulate and solve other plant location and supply chain problems by considering variables like manufacturing costs, inventory costs, lead time, etc. as interval valued.

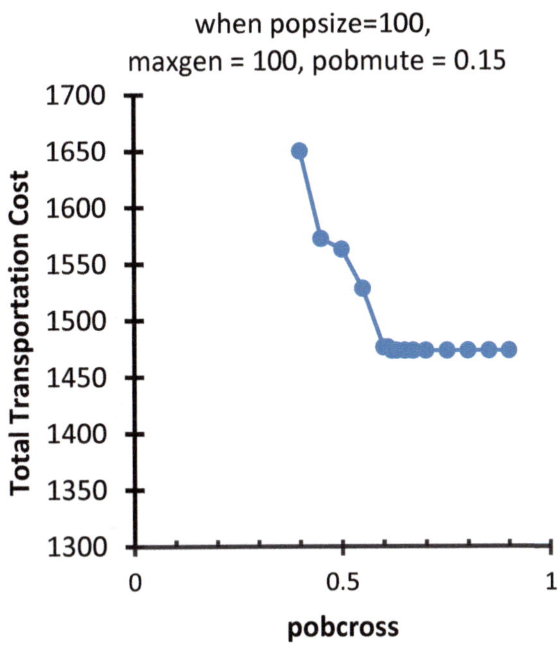

Fig. 7 'Total transportation cost' versus pobcross

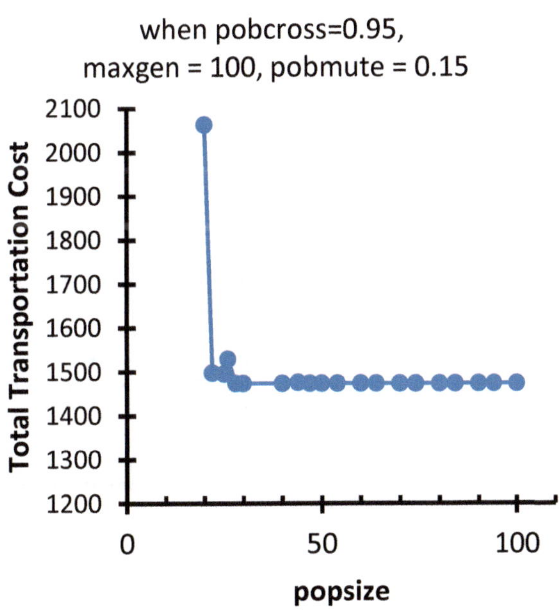

Fig. 8 'Total transportation cost' versus popsize

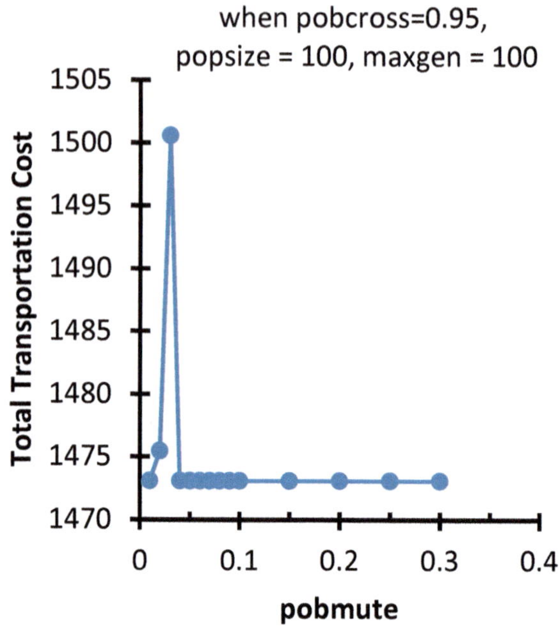

Fig. 9 'Total transportation cost' versus pobmute

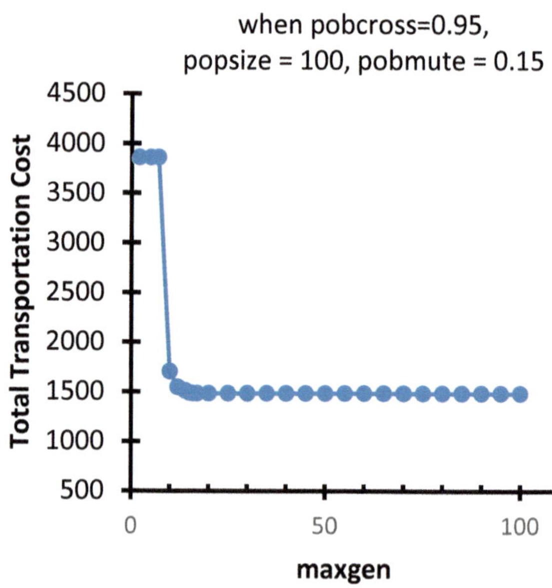

Fig. 10 'Total transportation cost' versus maxgen

References

1. Krajewski, L.J.: Operations Management, 8th edn. Prentice Hall, New Jersey (2007)
2. Drezner, Z., Hamacher, H.W.: Facility Location: Applications and Theory. Springer, New York (2004)
3. Thanh, N.P., Bostel, N., Prton, O.: A dynamic model for facility location in the design of complex supply chains. Int. J. Prod. Econ. **113**, 678–693 (2008)
4. Checherenkova, S.: Pre-Study of the Important Factors for the Factory Start-Up abroad. Mälardalen University, MSc (2008)
5. Stevenson, J.W.: Study Guide for Use with Operation Management, 9th edn. McGraw-Hill, New York (2007)
6. Yang, J., Lee, H.: Academic Papers: An AHP Decision Model for Facility Location Selection, pp. 241–254. MCB University Press (1997)
7. Kumar, A.S., Suresh, N.: Production and Operations Management, 2nd edn. New Age International, New Delhi (2008)
8. Encyclopedia Britannica.: Johann Heinrich von Thünen. Encyclopdia Britannica (2020). http://www.britannica.com/EBchecked/topic/594381/JohannHeinrich-von-Thunen. Accessed 7 May 2020
9. Wikipedia.: Location Theory. (2020). http://en.wikipedia.org/wiki/Location_theory. Accessed 7 May 2020
10. Buthe, B.V.: EinvielseitigerStandortökonom (2006). http://www.wiwi.unimuenster.de/insiwo/Download/Hausarbeiten/Hausarbeit-Buthe.pdf. Accessed 7 May 2020
11. Lalnunmawai, H.: Weber's Theory of Industrial Location (2010). http://www.trcollege.net/study-materials/102-webers-theory-of-industriallocation?catid=63%3Aeconomics. Accessed 7 May 2020
12. Abernathy, W.J., Hershey, J.C.: A spatial-allocation model for regional health service planning. Oper. Res.. Res. **20**, 629–642 (1972)
13. Atkine, R.J., Shriver, R.H.: New approach to facilities location. Harvad. Bus. Rev. **46**, 70–79 (1968)
14. Bowersox, D.J., Closs, D.J., Helferich, O.K.: Logistic Management: A system Integration pf Physical Distribution, Manufacturing Support and Materials Procurement. Macmillan, New York (1986)
15. Bridgeman, P.W.: Dimensional Analysis. Yale University Press, New Haven, Conn., U.S.A. (1963)
16. Brown, P.A., Gibson, D.F.: A quantified model for facility site selection: application to a multi-plant location problem. AIIE Trans. **4**(1), 1–10 (1972)
17. Karasaka, G.J., Browball, D.F.: Locational Analysis for Manufacturing, The MIT Press. Mass, Cambridge (1969)
18. Dou, Y., Sarkis, J.: A joint location and outsourcing sustainability analysis for a strategic offshoring decision. Int. J. Prod. Res. **48**(2), 567–592 (2010)
19. Feldmann, A., Olhager, J.: Plant roles: site competence bundles and their relationships with site location factors and performance. Int. J. Oper. Prod. Manag.Oper. Prod. Manag. **33**(6), 722–744 (2013)
20. Melo, M.T., Nickel, S., Saldanha-da-Gama, F.: Facility location and supply chain management—a review. Eur. J. Oper. Res.Oper. Res. **196**(2), 401–412 (2009)
21. Theyel, G.: Production location and sustainability. Int. J. Agile Syst. Manag. **5**(3), 267–275 (2012)
22. Wild, R.: Operations Managament, 6th ed., Thomson (2003)
23. Gupta, R.K., Bhunia, A.K., Roy, D.: A GA based penalty function technique for solving constrained redundancy allocation problem of series system with interval valued reliability of components. J. Comput. Appl. Math.Comput. Appl. Math. **232**, 275–284 (2009)
24. Gupta, R.K., Bhunia, A.K., Goyal, S.K.: An application of Genetic Algorithm in solving an inventory model with advance payment and interval valued inventory costs. Math. Comput. Model.Comput. Model. **49**, 893–905 (2009)

25. Goldberg, D.E.: Genetic Algorithms: Search, Optimization and Machine Learning, Reading. Addison Wesley, MA (1989)
26. Michalewicz, Z.: Genetic Algorithms + Data structure = Evaluation Programs. Springer, Berlin (1996)
27. Sakawa, M.: Genetic Algorithms and Fuzzy Multiobjective Optimization. Kiuwer Academic Publishers (2002)
28. Gen, M., Cheng, R.: Genetic Algorithms and Engineering Optimization. John Wiley & Sons Inc. (2000)
29. Gupta, R.K., Bhunia, A.K.: An application of real coded genetic algorithm (RCGA) for Integer Linear programming in production-transportation problems with flexible transportation cost. Adv. Modell. Optim. **8**, 73–85 (2006)
30. Gupta, R.K., Khan, D., Ghosh, P.: A study om the customers' perception of different children's health drinks. South Asian J. Market. Manag. Res. **10**, 29–39 (2020)
31. Khan, D., Gupta, R.K.: Production optimization with the maintenance of environmental sustainability based on multi-criteria decision analysis. Environ. Dev. Sustain. (2023) https://doi.org/10.1007/s10668-023-03316-8
32. Ishibuchi, H., Tanaka, H.: Multiobjective programming in optimization of the interval objective function. Eur. J. Oper. Res.Oper. Res. **48**, 219–225 (1990)
33. Chanas, S., Kuchta, D.: Multiobjective programming in the optimization of interval objective functions—a generalized approach. Eur. J. Oper. Res.Oper. Res. **94**, 594–598 (1996)
34. Sengupta, A., Pal, T.K.: Theory and methodology on comparing interval numbers. Eur. J. Oper. Res.Oper. Res. **127**, 28–43 (2000)
35. Mahato, S.K., Bhunia, A.K.: Interval-arithmetic-oriented interval computing technique for global optimization. Appl. Math. Res. eXpress 1–19 (2006)
36. Gupta, R., Khan, D.: Heuristic solutions for interval valued games. Iran. J. Numer. Anal. Optim. **12**(1), 187–200 (2022). https://doi.org/10.22067/ijnao.2021.71321.1048
37. Gupta, R.K., Khan, D.: A technique to solve mixed strategy non-cooperative zero sum games with more than two players. Int. J. Oper. Res. **49**(3), 385–402 (2024)
38. Roy, D., Khan, D.: Designing of an assembly line based on reliability approach. Int. J. Optim. Control Theor. Appl. (IJOCTA) **1**(1), 45–52 (2011). https://doi.org/10.11121/ijocta.01.2011.0045
39. Deb, I., Gupta, R.K.: A genetic algorithm based heuristic optimization technique for solving balanced allocation problem involving overall shipping cost minimization with restriction to the number of serving units as well as customer hubs. Results Control Optim. **11**, 100227 (2023)
40. Sahoo, L., Senapati, T., Yager, R.R. (eds.): Real Life Applications of Multiple Criteria Decision Making Techniques in Fuzzy Domain, Studies in Fuzziness and Soft Computing (STUDFUZZ, volume 420). Springer Nature (2023)

Interval Quadratic Programming Problem with Interval-Valued Decision Variables

Jewel Karmakar, Samiran Karmakar, and Sanat Kumar Mahato

Abstract This chapter contains an interval-oriented and multi-objective optimization-based method for interval quadratic programming problems (IQPP) with interval decision variables. Interval-valued quadratic objective functions are converted to center and width form using finite interval arithmetic. Linear interval constraints are handled using interval order relations defined in terms of decision makers' point of view. The IQPP is then converted to a non-interval (crisp) bi-objective optimization problem and solved by the Global Criterion Method (GCM). The main intention of our technique is to find an efficient solution from the decision makers' viewpoint out of uncountably many such solutions. Finally, the method is applied to several test problems and the results are compared with the existing solutions supporting the technique.

Keywords Interval quadratic programming · Interval decision variable · Decision-making · Multi-objective optimization · Efficient solution

---

Jewel Karmakar, Samiran Karmakar and Sanat Kumar Mahato authors contributed equally to this work.

J. Karmakar
Department of Mathematics, Bankura Zilla Saradamani Mahila Mahavidyapith, Nutanchati, Bankura 722101, West Bengal, India
e-mail: jewelkarmakar1991@gmail.com

S. Karmakar (✉)
Department of Mathematics, Bankura Sammilani College, Kenduadihi, Bankura 722102, West Bengal, India
e-mail: skmath.rnc@gmail.com

J. Karmakar · S. K. Mahato
Department of Mathematics, Sidho Kanho Birsha University, Ranchi-Purulia Rd, Purulia 723104, West Bengal, India
e-mail: sanatkr_mahato.math@skbu.ac.in

L. Sahoo et al. (eds.), *Decision Making Under Uncertainty Via Optimization, Modelling, and Analysis*, Studies in Systems, Decision and Control 558,
https://doi.org/10.1007/978-981-96-0085-4_18

1 Introduction

Recent trends indicate more and more involvements of future uncertainty and risk factors, impreciseness of the available data, and assumption of model parameters while formulating mathematical modeling of different real-life problems. These serious factors also affect operational Research (OR) and Optimization models. Stochastic, Fuzzy, Fuzzy-Stochastic, Rough Set Theory, and Gray optimization techniques are familiar approaches to handling these situations. However, all these well-established uncertainly handling techniques have some pros and cons. Interval-oriented optimization techniques are very fruitful in this regard. Interval-oriented methods are deterministic in nature as the uncertainty or inexactness existing in the data or model parameters are bounded by the left and right limits of the usual closed intervals, i.e., the span of uncertainty/imprecision is measured by the width of the closed intervals. Despite this big advantage, the efficiency of any interval-based optimization technique depends highly on the used interval ordering definition. Ordering of interval numbers plays an important role in making the method successful. There are many existing methods for ordering interval numbers; but, none of the definitions give a complete ordering of intervals. Moore [24] was the pioneer in this field. After Moore [24], Ishibuchi and Tanaka [7], Kundu [16], Sengupta and Pal [28], Mahato and Bhunia [23], Sahoo et al. [27], and many more have done notable works on interval ranking definitions. A survey on interval ranking definitions with detailed discussions of their acceptability has been done by Karmakar and Bhunia [9].

One of the simplest non-linear optimization problems is the Quadratic Programming Problem (QPP). It has numerous important applications in inventory management, production economical models, portfolio selection models, different engineering designs, and many more. It has already been mentioned that uncertainty and impreciseness are unavoidable in the present-day socio-economic scenario. As a result, we have to consider the Interval Quadratic Programming Problem (IQPP) instead of its crisp formulation. A lot of applications of IQPP in different fields can be cited, e.g., in planning long-term municipal solid waste (MSW) management (Li and Wang [21]), Uncertain portfolio management problems (Kumar et al. [14]; Wang et al. [33]), Power supply scheduling (Ishizaki et al. [8]), irrigation water management under uncertainty (Wang et al. [32]), etc.

In our present work, we have prescribed an interval-oriented technique for interval quadratic programming problems (IQPP) involving interval coefficients and interval decision variables. As a result, the objective function, equality/inequality constraints, and the non-negativity restrictions of the proposed QPP are all intervals. According to Ishibuchi and Tanaka [7], the center and half-width of the objective function represent the expected value and the possible extent of the uncertainty of the interval-valued function. In this respect, we have converted the objective function into its center and half-width form to transform the interval-valued problem to its equivalent crisp bi-objective optimization problem where the center value is to be optimized along with the width minimization. Interval-valued constraints are also converted to their crisp form using the interval ordering definition defined in decision makers' point of view by Sahoo et al. [27]. The classical bi-objective optimization problem is then solved by the Global Criterion Method (GCM). Several convex/non-convex test problems most of which are collected from the existing literature are solved by this technique.

2 Literature Survey

Many researchers previously formulated QPP with imprecise parameters where the impreciseness is viewed as interval numbers. Plenty of theoretical and methodological developments are seen in the relevant literature. A special class of IQPP has been considered by Liu and Wang [22] where the cost coefficients, constraint coefficients, and requirement parameters are considered as intervals. A two-level mathematical programming technique has been used by them to find the bounds of the objective function values and then transform it into a one-level traditional problem by using the duality theorem and variable transformation method. Fang et al. [2] have studied the convex IQPP with bound constraints in terms of defined membership functions. Based on the membership function of the objective function the original problem is transferred to the multi-objective problem which is later converted to semi-definite programming using the Schur Complement theorem. Li and Tian [18] have generalized the IQPP previously given by Li and Wang [22] where all the model parameters are assumed as intervals and solved by applying less computational effort. Li [17] has revised the IQPP previously studied by Li and Tian [18] by considering the equality constraints. A robust IQPP method has been developed by Li and Huang [21] which can be effective for fuzzy or fuzzy-interval-oriented QPPs. Kumar et al. [14] have investigated a more general IQPP where not only the model parameters but also the decision variables are assumed as interval-valued. A theoretical study for the existence of optimal solutions with certain level of acceptance has been incorporated. Characterization of monotonicity of the solution of IQPP has been discussed by Ishizaki et al. [8] by analyzing the derivative of the solution using Karush-Kuhn-Tucker (KKT) conditions. The bounds of the optimal solution have been obtained. A novel method for IQPP has been prescribed by Li et al. [19] for computing the upper bound of the optimal solution under weaker conditions. In another research, Li et al. [20] have presented an easy and effective method for checking the zero duality gap which is an important criterion for finding the range of optimal values for convex IQPP. In interval convex QPP theory, Hladik [5] has generalized and extended all the previously studied results and prescribed an algorithm for best and worst optimal values. Some important properties regarding the four equivalence relations of bounds of interval optimal solution of IQPP have been investigated by Xu et al. [34]. Syaripuddin et al. [29] have extended the Wolfe method for interval coefficient QPPs by transforming to its equivalent interval coefficient LPP. In another notable work, a two-level programming procedure has been discussed by Syaripuddin et al. [30] for IQPP with interval decision variables. Wang et al. [32] have developed an interval quadratic fuzzy dependent-chance programming model which has been applied extensively in irrigation water allocation planning problems with uncertainty. Nezhad et al. [31] have extended an interval linear programming handling technique to IQPP by reducing the original problem to two crisp problems which are later solved by the SQP algorithm. A new general IQPP model has been developed by Wang et al. [33] for an uncertain portfolio selection problem. An effective numerical solution technique has been developed using the Lagrange theorem and duality theorem for obtaining the effective bounds of the objective values of the model. A rigorous study about the numerical solution for IQPP has been done by Elsisy et al. [1] where the Swarm Algorithm (SA) has been incorporated with the numerical approach. The KKT conditions are obtained for a modified problem and Chaotic Particle Swarm Optimization (CPSO) and Chaotic Firefly Algorithm (CFA) are prescribed. Hormazdabadi et al. [6] have suggested a method for obtaining a complete set of optimal solutions for solving IQPP. A stochastic programming technique has been applied by Kumari and Srivastava [15] for IQPP where the objective functions are transformed to random variable functions and all constraints are into stochastic constraints using Gaussian distribution and 6σ-rule.

3 Basic Interval Mathematics

Basic arithmetical properties of interval numbers are indispensable for any interval computing technique. The present method is not a different one. To compute interval inclusion functions that identify the rigorous bounds of the objective functions of our proposed IQPP interval arithmetic is used. The pioneering work on algebraic properties of the set of interval numbers has been given by Moore [24]. After Moore [24], Ratschek and Rokne [26], Hansen and Walster [3], Kearfott [13], Mahato and Bhunia [23], and Moore et al. [25], many more have also developed modern interval arithmetic rigorously. In this section, a few of them are discussed and are used extensively in the remaining part of our discussion.

Due to Moore [24], the concept of interval numbers is the extension of real numbers in a larger sense. An interval number is a closed interval in the real line $\mathbb{R}$. An interval A is defined as

$$A = \{x \in \mathbb{R} : a_L \leq x \leq a_R\} \tag{1}$$

and is denoted by $A = [a_L, a_R]$ with width $a_R - a_L$.

Any interval $A = [a_L, a_R]$ can also be expressed in center and half-width form as

$$A = \langle a_C, a_W \rangle = \{x \in \mathbb{R} : a_C - a_W \leq x \leq a_C + a_W\} \tag{2}$$

where $a_C = \dfrac{a_L + a_R}{2}$ = center of the interval and $a_W = \dfrac{a_R - a_L}{2}$ = half-width of the interval.

Any real number x can be defined as an interval $[x, x]$ with zero width, i.e., the set of all reals can be assumed as a proper subset of the set of all interval numbers.

Definition 3.1 The addition, subtraction, scalar multiplication, multiplication, and division of two intervals $A = [a_L, a_R]$ and $B = [b_L, b_R]$ are as follows:

$$A + B = [a_L, a_R] + [b_L, b_R] = [a_L + b_L, a_R + b_R]$$

$$\lambda A = \lambda[a_L, a_R] = \begin{cases} [\lambda a_L, \lambda a_R] \text{ if } \lambda \geq 0, \\ [\lambda a_R, \lambda a_L] \text{ if } \lambda < 0 \text{ where } \lambda \in \mathbb{R} \end{cases}$$

$$A - B = [a_L, a_R] - [b_L, b_R] = [a_L, a_R] + [-b_R, -b_L] = [a_L - b_R, a_R - b_L]$$

$$A \times B = [\min(a_L b_L, a_L b_R, a_R b_L, a_R b_R), \max(a_L b_L, a_L b_R, a_R b_L, a_R b_R)]$$

$$\frac{A}{B} = A \times \left(\frac{1}{B}\right) = [a_L, a_R] \times \left[\frac{1}{b_R}, \frac{1}{b_L}\right], \quad \text{provided } 0 \notin [b_L, b_R]$$

Definition 3.2 In center and half-width form, the definition of addition, subtraction, and scalar multiplication of two intervals $A = \langle a_C, a_W \rangle$ and $B = \langle b_C, b_W \rangle$ are as follows:

$$A + B = \langle a_C, a_W \rangle + \langle b_C, b_W \rangle = \langle a_C + b_C, a_W + b_W \rangle$$

$$\lambda A = \lambda \langle a_C, a_W \rangle = \langle \lambda a_C, \mid \lambda \mid a_W \rangle$$

$$A - B = \langle a_C, a_W \rangle - \langle b_C, b_W \rangle = \langle a_C, a_W \rangle + \langle -b_C, b_W \rangle$$
$$= \langle a_C - b_C, a_W + b_W \rangle$$

Apart from these definitions, the nth power of an interval where n is a positive integer, the fractional power of an interval, modulus of an interval, and many more are given in Karmakar et al. [12] and Karmakar and Bhunia [10, 11].

4 Interval Order Relations

It is very transparent that while solving decision-making problems, especially, OR and Optimization models involving interval parameters or variables the ordering of intervals plays a vital role. Sometimes, it becomes the key factor in estimating the efficiency of the solution methodology. Many experts in this field have developed various interval ranking techniques. However, none of these definitions are complete and most of these are prescribed to solve particular types of problems which cannot be generalized. A review discussion of many existing definitions of the ordering of intervals with their acceptability in solving optimization problems is given in Karmakar and Bhunia [9]. According to them, the definitions are categorized into four groups. Here, we consider the general definitions of interval rankings.

Let $A = [a_L, a_R] = \langle a_C, a_W \rangle$ and $B = [b_L, b_R] = \langle b_C, b_W \rangle$ be a pair of arbitrary intervals. These are of the following types

Type-I: Non-overlapping intervals, i.e., either $a_R < b_L$ or $b_R < a_L$;

Type-II: Partially overlapping intervals, i.e., either $a_L < b_L < a_R < b_R$ or $b_L < a_L < b_R < a_R$;

Type-III: Fully overlapping intervals, i.e., either $b_L < a_L < a_R < b_R$ or $a_L < b_L < b_R < a_R$.

In this area, Moore [24, 25] first proposed two definitions. The first one is as follows:

$$A < B \text{ if and only if } a_R < b_L \tag{3}$$

This is a transitive order relation and applicable only for the Type-I intervals.

The second interval ranking definition which is only the interval version of the Set inclusion property is given by

$$A \subseteq B \text{ if and only if } b_L \leq a_L \wedge a_R \leq b_R \tag{4}$$

This is also a transitive relation and the limitations of these orderings are very clear from the definition.

Ishibuchi and Tanaka [7] have defined several order relations of intervals in view of decision makers' preferences. For maximization and minimization problems, they have defined separate order relations.

Mahato and Bhunia [23] have defined the modified interval ranking definitions in terms of optimistic and pessimistic decision-making. The ordering rules $\leq_{\text{omin}}$, $<_{\text{omin}}$, $\geq_{\text{omax}}$, and $>_{\text{omax}}$ are defined very straightforwardly in optimistic decision maker's standpoint where the decision maker chooses the best alternative ignoring the uncertainty.

In pessimistic decision-making the decision maker chooses the best alternative according to the principle, "Less uncertainty is better than more uncertainty", i.e., in this case, the minimum or maximum preferred interval will be the interval with less uncertainty. These ordering definitions are

Definition 4.1 For minimization problems, the order relations $<_{\text{pmin}}$ between the intervals $A = [a_L, a_R] = \langle a_C, a_W \rangle$ and $B = [b_L, b_R] = \langle b_C, b_W \rangle$ for a pessimistic decision maker are

(i) $A <_{\text{pmin}} B$ iff $a_c < b_c$, for Type-I and Type-II intervals,

(ii) $A <_{\text{pmin}} B$ iff $a_c \leq b_c$ and $a_w < b_w$, for Type-III intervals.

However, for Type-III intervals with $a_C < b_C$ and $a_W > b_W$, pessimistic decisions cannot be determined. In this case, the optimistic decision is to be considered.

Definition 4.2 For maximization problems, the order relations $>_{\text{pmax}}$ between the intervals $A = [a_L, a_R] = \langle a_C, a_W \rangle$ and $B = [b_L, b_R] = \langle b_C, b_W \rangle$ for a pessimistic decision maker are

(i) $A >_{\text{pmax}} B$ iff $a_c > b_c$, for Type-I and Type-II intervals,

(ii) $A >_{\text{pmax}} B$ iff $a_c \geq b_c$ and $a_w < b_w$, for Type-III intervals.

However, for Type-III intervals with $a_c > b_c$ and $a_w > b_w$, pessimistic decisions cannot be taken. In this case, also the optimistic decision is to be taken.

Generalized ordering definition

The interval ranking definitions in terms of pessimistic decision-making given by Mahato and Bhunia [23] are insufficient for some of the pair of Type-III intervals. According to them, those inapplicable cases may be handled by considering optimistic decision-making. To tackle those situations, some more general ordering definitions are suggested by Sahoo et al. [27] which are as follows:

Definition 4.3 For any pair of intervals A and B, the order relation for maximization problems denoted by $>_{\max}$ is defined as

$$\begin{aligned}
&\text{(i) } A >_{\max} B \Longleftrightarrow a_C > b_C \text{ for Type-I and Type-II intervals} \\
&\text{(ii) } A >_{\max} B \Longleftrightarrow \text{either } a_C \geq b_C \wedge a_W < b_W \\
&\qquad \text{or } a_C \geq b_C \wedge a_L > b_L \text{ for Type-III intervals.}
\end{aligned}$$

Definition 4.4 For the same pair of intervals A and B, the order relation for minimization problems denoted by $<_{\min}$ is defined as

$$\begin{aligned}
&\text{(i) } A <_{\min} B \Longleftrightarrow a_C < b_C \text{ for Type-I and Type-II intervals} \\
&\text{(ii) } A <_{\min} B \Longleftrightarrow \text{either } a_C \leq b_C \wedge a_W < b_W \\
&\qquad \text{or } a_C \leq b_C \wedge a_L < b_L \text{ for Type-III intervals.}
\end{aligned}$$

The order relations defined above are all partially ordered.

5 Statement of the Problems

The Interval Quadratic Programming Problem (IQPP) with interval decision variables in its most general mathematical form is stated as

$$\text{Minimize } Z = \sum_{j=1}^{n} [c_{jL}, c_{jR}][x_{jL}, x_{jR}] + \frac{1}{2}\sum_{j=1}^{n}\sum_{k=1}^{n} [d_{jkL}, d_{jkR}][x_{jL}, x_{jR}][x_{kL}, x_{kR}]$$

$$\text{Subject to} \quad \sum_{j=1}^{n} [a_{ijL}, a_{ijR}][x_{jL}, x_{jR}] \ (\leq \text{ or } = \text{ or } \geq) \ [b_{iL}, b_{iR}]; \quad i = 1, 2, \ldots, m$$

$$\text{and} \quad x_{jL} \geq 0, x_{jR} \geq x_{jL} \text{ for } j = 1, 2, \ldots, n$$

where $\left[d_{jkL}, d_{jkR}\right] = \left[d_{kjL}, d_{kjR}\right]$ $\forall$ j and k. $\left[x_{jL}, x_{jR}\right]$, $j = 1, 2, ..., n$ is an interval-valued decision variable, $\left[c_{jL}, c_{jR}\right]$, $\left[a_{ijL}, a_{ijR}\right]$, and $\left[b_{iL}, b_{iR}\right], i = 1, 2, ..., m$, $j = 1, 2, ..., n$ are, respectively, interval-valued price-parameters, activity parameters, and requirement parameters of the problem.

6 Optimal Solution

An interval-valued decision vector $\widetilde{X} = (X_1, X_2, ..., X_n)^t \in \mathbb{I}^n$ where $\mathbb{I}$ is the set of all intervals on the real line $\mathbb{R}$ and $X_j = \left[x_j^L, x_j^R\right] \in \mathbb{I}$ is the j-th interval-valued component of the vector $\widetilde{X}$. An interval vector $\widetilde{X}^*$ is said to be the minimizer vector of the objective function $Z = [Z_L, Z_R] = \langle Z_C, Z_W \rangle$ if $Z_C\left(\widetilde{X}^*\right) \leq Z_C\left(\widetilde{X}\right)$ with $Z_W\left(\widetilde{X}^*\right) \leq Z_W\left(\widetilde{X}\right)$ for any $\widetilde{X} \in \mathbb{I}^n$.

The minimum value of the objective function Z is $Z^* = \min_{\widetilde{X} \in \mathbb{I}^n} Z\left(\widetilde{X}\right)$.

7 Solution Procedure

The interval-valued objective function Z represents the function value with uncertainty. It is already pointed out that the center Z_C and the width Z_W are considered as the expected value and the possible extent of the uncertainty of the given interval-valued function Z, respectively [7]. The general structure of the proposed optimization technique is comprised of the following steps.

7.1 *Reduction of the Interval Objective Function*

To obtain the minimum value of the interval-valued quadratic objective function of the problem, initially, the said objective function is written in the center-width form of interval, and consequently, it is reduced to two deterministic (crisp) quadratic objective functions as follows:

$$Z = \sum_{j=1}^{n} \left[c_{jL}, c_{jR}\right]\left[x_{jL}, x_{jR}\right] + \frac{1}{2}\sum_{j=1}^{n}\sum_{k=1}^{n}\left[d_{jkL}, d_{jkR}\right]\left[x_{jL}, x_{jR}\right]\left[x_{kL}, x_{kR}\right] = [Z_L, Z_R],$$

where Z_L and Z_R are the lower and upper bounds of the interval objective function Z, computed by applying finite interval arithmetic. Then it is expressed in center and width form as $Z = \langle Z_C, Z_W \rangle$ where $Z_C = \dfrac{Z_L + Z_R}{2}$ and $Z_W = \dfrac{Z_R - Z_L}{2}$ are, respectively, the center and width of the interval function Z.

7.2 Interval-Valued Linear Constraint Handling Technique

The interval linear inequality constraints are tackled by the interval ordering definitions given by Sahoo et al. [27] for optimistic and pessimistic decision maker's points of view.

The ith interval-valued constraint can be rewritten in the following forms:

$$\begin{aligned} &\sum_{j=1}^{n} \left[a_{ijL}, a_{ijR}\right] \left[x_{jL}, x_{jR}\right] (\leq \text{or} = \text{or} \geq) [b_{iL}, b_{iR}] \\ \Rightarrow \quad & [g_{iL}, g_{iR}] (\leq \text{or} = \text{or} \geq) [b_{iL}, b_{iR}] \\ \Rightarrow \quad & \langle g_{iC}, g_{iW} \rangle (\leq \text{or} = \text{or} \geq) \langle b_{iC}, b_{iW} \rangle , \end{aligned}$$

where g_{iL} and g_{iR} denote the lower and upper bound of the left-hand side of i th interval constraint computed by applying finite interval arithmetic. It is then expressed in center and width form (wherever applicable) as $\langle g_{iC}, g_{iW} \rangle$ where $g_{iC} = \dfrac{g_{iL} + g_{iR}}{2}$ and $g_{iW} = \dfrac{g_{iR} - g_{iL}}{2}$ are the center and width of the interval $[g_{iL}, g_{iR}]$, and $\langle b_{iC}, b_{iW} \rangle$ denote the center and width form of the right-hand side of the interval constraint $[b_{iL}, b_{iR}]$.

Now the following cases may arise:

Case I: If the i th constraint is of the form $[g_{iL}, g_{iR}] \geq [b_{iL}, b_{iL}]$, then it is following the interval ranking definition of $[g_{iL}, g_{iR}] >_{\max} [b_{iL}, b_{iR}]$, i.e., the constraint is satisfied if

$$(g_{iC} > b_{iC}) || (g_{iC} \geq b_{iC} \wedge g_{iW} < b_{iW}) || (g_{iC} \geq b_{iC} \wedge g_{iL} > b_{iL}).$$

Case II: If the i th constraint is of the form $[g_{iL}, g_{iR}] \leq [b_{iL}, b_{iL}]$, then it is following the interval ranking definition of $[g_{iL}, g_{iR}] <_{\min} [b_{iL}, b_{iR}]$, i.e., the constraint is satisfied if

$$(g_{iC} < b_{iC}) || (g_{iC} \leq b_{iC} \wedge g_{iW} < b_{iW}) || (g_{iC} \leq b_{iC} \wedge g_{iL} < b_{iL}).$$

Case III: If the i th constraint is an equality constraint, i.e., $[g_{iL}, g_{iR}] = [b_{iL}, b_{iR}]$, then the equality constraint satisfaction rule is

$$g_{iL} = b_{iL} \wedge g_{iR} = b_{iR}.$$

7.3 Construction and Solution of Crisp Bi-Objective Optimization Problem

To minimize Z we shall consider two non-interval crisp objective functions Z_C and Z_W both are to be minimized subject to the new constraints obtained by interval-valued constraint handling techniques. The Pareto optimal solution for the constructed multi-objective optimization problem is obtained using the global criterion method. In this method, the distance from some reference point to the feasible objective region is minimized. Here the **ideal objective vector** is used as a reference point and **usual metric** is used for measuring.

8 Illustrative Numerical Examples

Example 8.1 Maximize
$$Z = [0.4, 0.6]\,[x_{3L}, x_{3R}] - [0.2, 0.3]\,[x_{1L}, x_{1R}] - [0.4, 0.6]\left\{[x_{1L}, x_{1R}]^2 + [x_{2L}, x_{2R}]^2 + [x_{3L}, x_{3R}]^2\right\}$$

subject to $[x_{1L}, x_{1R}] - [0.9, 1.12]\,[x_{2L}, x_{2R}] + [x_{3L}, x_{3R}] = [0.85, 1.05]$,

and $x_{1L}, x_{2L}, x_{3L} \geq 0, x_{1R} \geq x_{1L}, x_{2R} \geq x_{2L}, x_{3R} \geq x_{3L}$.

Solution: The interval objective function is to be rewritten in center and width form as follows:

$$\begin{aligned}
Z &= [0.4, 0.6]\,[x_{3L}, x_{3R}] - [0.2, 0.3]\,[x_{1L}, x_{1R}] \\
&\quad -[0.4, 0.6]\left\{[x_{1L}, x_{1R}]^2 + [x_{2L}, x_{2R}]^2 + [x_{3L}, x_{3R}]^2\right\} \\
&= [0.4x_{3L} - 0.3x_{1R} - 0.6x_{1R}^2 - 0.6x_{2R}^2 - 0.6x_{3R}^2, \\
&\qquad 0.6x_{3R} - 0.2x_{1L} - 0.4x_{1L}^2 - 0.4x_{2L}^2 - 0.4x_{3L}^2] \;(= [Z_L, Z_R]) \\
&= \langle 0.3x_{3R} - 0.1x_{1L} - 0.2x_{1L}^2 - 0.2x_{2L}^2 - 0.2x_{3L}^2 + 0.2x_{3L} \\
&\quad -0.15x_{1R} - 0.3x_{1R}^2 - 0.3x_{2R}^2 - 0.3x_{3R}^2, \quad 0.3x_{3R} - 0.1x_{1L} - 0.2x_{1L}^2 \\
&\quad -0.2x_{2L}^2 - 0.2x_{3L}^2 - 0.2x_{3L} + 0.15x_{1R} + 0.3x_{1R}^2 + 0.3x_{2R}^2 + 0.3x_{3R}^2\rangle \\
&= \langle Z_C, Z_W\rangle
\end{aligned}$$

where
$$Z_C = 0.3x_{3R} - 0.1x_{1L} - 0.2x_{1L}^2 - 0.2x_{2L}^2 - 0.2x_{3L}^2 + 0.2x_{3L} - 0.15x_{1R} - 0.3x_{1R}^2 - 0.3x_{2R}^2 - 0.3x_{3R}^2$$

and
$$Z_W = 0.3x_{3R} - 0.1x_{1L} - 0.2x_{1L}^2 - 0.2x_{2L}^2 - 0.2x_{3L}^2 - 0.2x_{3L} + 0.15x_{1R} + 0.3x_{1R}^2 + 0.3x_{2R}^2 + 0.3x_{3R}^2.$$

Hence, the corresponding bi-objective optimization problem is as follows:

$$\text{Maximize } Z_C = 0.3x_{3R} - 0.1x_{1L} - 0.2x_{1L}^2 - 0.2x_{2L}^2 - 0.2x_{3L}^2 + 0.2x_{3L} - 0.15x_{1R} \\ -0.3x_{1R}^2 - 0.3x_{2R}^2 - 0.3x_{3R}^2 \tag{5}$$

$$\text{Minimize } Z_W = 0.3x_{3R} - 0.1x_{1L} - 0.2x_{1L}^2 - 0.2x_{2L}^2 - 0.2x_{3L}^2 - 0.2x_{3L} + 0.15x_{1R} \\ +0.3x_{1R}^2 + 0.3x_{2R}^2 + 0.3x_{3R}^2 \tag{6}$$

subject to the constraints written in the suitable form

$$x_{1L} - 1.12x_{2R} + x_{3L} = 0.85 \tag{7}$$

$$x_{1R} - 0.9x_{2L} + x_{3R} = 1.05 \tag{8}$$

The Pareto optimal solution of the constructed bi-objective optimization problem is obtained by the Global Criterion Method. The reduced problem has been solved by **MATHEMATICA 7.0** software package. On solving the problem we get the ideal objective vector is $(0.140316, 0.0480542)^t$ and the Pareto optimal solution is $\widetilde{X}^* = ([0.160087, 0.267354], [0, 0.0827972], [0.782646, 0.782646])$ with maximum value of $Z = [-0.181669, 0.182305]$. □

Example 8.2 Maximize $Z = [2.5, 3.2]\,[x_{1L}, x_{1R}] + [5, 7]\,[x_{2L}, x_{2R}]$
$\qquad -4\,[x_{1L}, x_{1R}]\,[x_{2L}, x_{2R}] - [2.7, 3.1]\,[x_{1L}, x_{1R}]^2 - 2\,[x_{2L}, x_{2R}]^2$

subject to $3\,[x_{1L}, x_{1R}] + [1.9, 2.2]\,[x_{2L}, x_{2R}] \le [3.9, 4.25]$,

$[x_{1L}, x_{1R}] + [x_{2L}, x_{2R}] \le [0.5, 1.5]$.

and $x_{1L}, x_{2L} \ge 0, x_{1R} \ge x_{1L}, x_{2R} \ge x_{2L}$.

Solution: Here, the objective function in Center-Width form is

$$\begin{aligned} Z &= [2.5, 3.2]\,[x_{1L}, x_{1R}] + [5, 7]\,[x_{2L}, x_{2R}] \\ &\quad -4\,[x_{1L}, x_{1R}]\,[x_{2L}, x_{2R}] - [2.7, 3.1]\,[x_{1L}, x_{1R}]^2 - 2\,[x_{2L}, x_{2R}]^2 \\ &= \langle 1.6x_{1R} + 3.5x_{2R} - 2x_{1L}x_{2L} - 1.35x_{1L}^2 - x_{2L}^2 + 1.25x_{1L} + 2.5x_{2L} \\ &\quad -2x_{1R}x_{2R} - 1.55x_{1R}^2 - x_{2L}^2,\ \ 1.6x_{1R} + 3.5x_{2R} - 2x_{1L}x_{2L} \\ &\quad -1.35x_{1L}^2 - x_{2L}^2 - 1.25x_{1L} - 2.5x_{2L} + 2x_{1R}x_{2R} + 1.55x_{1R}^2 + x_{2L}^2 \rangle \\ &= \langle Z_C, Z_W \rangle \end{aligned}$$

$$\begin{aligned} \text{where } Z_C &= 1.6x_{1R} + 3.5x_{2R} - 2x_{1L}x_{2L} - 1.35x_{1L}^2 - x_{2L}^2 + 1.25x_{1L} + 2.5x_{2L} \\ &\quad -2x_{1R}x_{2R} - 1.55x_{1R}^2 - x_{2L}^2 \\ \text{and } Z_W &= 1.6x_{1R} + 3.5x_{2R} - 2x_{1L}x_{2L} - 1.35x_{1L}^2 - x_{2L}^2 - 1.25x_{1L} - 2.5x_{2L} \\ &\quad +2x_{1R}x_{2R} + 1.55x_{1R}^2 + x_{2L}^2. \end{aligned}$$

Hence, the corresponding bi-objective optimization problem is

$$\text{Maximize } Z_C = 1.6x_{1R} + 3.5x_{2R} - 2x_{1L}x_{2L} - 1.35x_{1L}^2 - x_{2L}^2 + 1.25x_{1L} + 2.5x_{2L} \\ -2x_{1R}x_{2R} - 1.55x_{1R}^2 - x_{2L}^2 \tag{9}$$

$$\text{Minimize } Z_W = 1.6x_{1R} + 3.5x_{2R} - 2x_{1L}x_{2L} - 1.35x_{1L}^2 - x_{2L}^2 - 1.25x_{1L} - 2.5x_{2L} \\ +2x_{1R}x_{2R} + 1.55x_{1R}^2 + x_{2L}^2 \tag{10}$$

subject to the given constraints.

The ideal objective vector of the produced bi-objective optimization problem is $(4.125, 0)^t$ and the corresponding Pareto optimal solution is $\widetilde{X}^* = ([0, 0], [0.891295, 0.891295])$ with $Z_{\max} = [2.86766, 4.65025]$. □

9 More Numerical Examples and Discussions

To establish the performance rate of the proposed interval computing optimization method, we have considered a few more test problems, viz., F_1–F_{10}, which are collected from the existing literature, provided in Appendix A. Apart from the problems F_1–F_8, all other test problems are slightly modified by considering interval-valued decision variables while solving those. The former pair of problems F_1–F_8 are originally with interval decision variables. The listed test problems have been chosen with different degrees of complexity level including convex QPPs. Moreover, these multi-interval variate test problems are with equality or inequality constraints and with interval or non-interval constraints. The Pareto optimal solutions of the test problems with corresponding optimizer points are given in Table 1 and the comparisons of obtained results with existing solutions are shown in Tables 2 and 3.

From the comparative results, it is very much diaphanous that most of the Pareto optimal solutions obtained by our method are far better than those obtained by other existing methods in view of reducing the uncertainty of the solution. Also, in a few cases, the solutions are almost equivalent in the sense of real-life acceptability. The deviations in the results observed in a few cases are mainly due to the variability of consideration of feasibility conditions in those model formulations. Some relaxed feasible conditions are used in some existing methods. Apart from these, our proposed solutions are well acceptable according to the interval ranking definition given in [27].

Table 1 Computational results of the test problems $F_1 - F_{10}$

Test problems	Pareto optimal solutions	Optimum value
F_1	$[x_{1L}, x_{1R}] = [0.37493, 0.375042]$, $[x_{2L}, x_{2R}] = [1.75, 1.75]$	[5.05832, 7.31666]
F_2	$[x_{1L}, x_{1R}] = [1.33064, 1.60995]$, $[x_{2L}, x_{2R}] = [0.669893, 0.669893]$	[3.21936, 5.0407]
F_3	$[x_{1L}, x_{1R}] = [0.416784, 0.416784]$, $[x_{2L}, x_{2R}] = [0, 0]$	[−1.7365, 0.729225]
F_4	$[x_{1L}, x_{1R}] = [0.367591, 0.367591]$, $[x_{2L}, x_{2R}] = [0.0482793, 0.0587955]$	[−1.55799, −0.608393]
F_5	$[x_{1L}, x_{1R}] = [0.550036, 5.66907]$, $[x_{2L}, x_{2R}] = [14.5332, 14.5332]$	[−1117.35, −248.226]
F_6	$[x_{1L}, x_{1R}] = [-0.75, -0.75]$, $[x_{2L}, x_{2R}] = [-1.37482, -1.37482]$	[−4.71875, −4.71875]
F_7	$[x_{1L}, x_{1R}] = [1.5, 1.5]$, $[x_{2L}, x_{2R}] = [0.5, 0.5]$, $[x_{3L}, x_{3R}] = [0.5, 0.5]$,	[2.75, 2.75]
F_8	$[x_{1L}, x_{1R}] = [0.268594, 0.268594]$, $[x_{2L}, x_{2R}] = [0, 0]$	[−2.39737, −0.890137]
F_9	$[x_{1L}, x_{1R}] = [0.63446, 0.63446]$, $[x_{2L}, x_{2R}] = [0, 0]$	[−2.36722, −1.0983]
F_{10}	$[x_{1L}, x_{1R}] = [1.17861, 1.18499]$, $[x_{2L}, x_{2R}] = [4.03227, 4.03227]$,	[278.83, 298.842]

10 Conclusion

This chapter presents an interval computing technique for solving interval-valued QPP. The problem's specialty is considering interval-valued decision variables, which makes the QPP models more appropriate and acceptable to the decision makers because of controlling future uncertainty affecting the model parameters and variables. The interval quadratic objective functions are initially formed in center and half-width of the intervals and then transformed into crisp bi-objective optimization

Table 2 Comparative study of interval-valued LPPs

Solutions obtained by	Problem F_1	Problem F_2	Problem F_3	Problem F_4	Problem F_5
1. Kumar et al. [14]	$Z_{min} = [4.49, 16.75] = \langle 10.62, 6.13 \rangle$	...	...	...	...
2. Elsisy et al. [1]	...	$Z_{min} = [1.025, 74] = \langle 37.5125, 36.4875 \rangle$	$Z_{min} = [-3.5, -0.75] = \langle -2.125, 1.375 \rangle$	$Z_{min} = [-3.4922, 0.5217] = \langle -1.48525, 2.00695 \rangle$	...
3. Li et al. [19]	...	...	...	$Z_{min} = [-2.90625, -0.5217] = \langle -1.71395, 1.1923 \rangle$	...
4. Hladik [4]	...	...	...	$Z_{min} = [-2.90625, -0.52165] = \langle -1.71395, 1.1923 \rangle$	...
5. Nehi et al. [6]	...	...	...	...	$Z_{min} = [-7497, -191.75] = \langle -3844.375, 3652.625 \rangle$.
6. Proposed method	$Z_{min} = [5.05832, 7.31666] = \langle 6.18749, 1.12917 \rangle$	$Z_{min} = [3.21936, 5.0407] = \langle 4.13003, 0.91067 \rangle$	$Z_{min} = [-1.7365, 0.729225] = \langle -0.503638, 1.23286 \rangle$	$Z_{min} = [-1.55799, -0.608393] = \langle -1.08319, 0.474799 \rangle$	$Z_{min} = [-1117.35, -248.226] = \langle -682.788, 434.562 \rangle$

Table 3 Comparative study of Interval-valued LPPs

Solutions obtained by	Problem F_6	Problem F_7	Problem F_8	Problem F_9	Problem F_{10}
1. Fang et al. [2]	$Z_{\min} = -4.7267$	...	...	...	...
2. Nezhad et al. [31]	...	$Z_{\max} = [2.42, 2.75] = \langle 2.585, 0.165 \rangle$	.....	....	...
3. Suprajitno et al. [29]	...	...	$Z_{\min} = [-6.25, -0.9] = \langle -3.575, 2.675 \rangle$.	...	....
4. Suprajitno et al. [30]	...	...	$Z_{\min} = [-4.64, 0.7] = \langle -1.97, 2.67 \rangle$.	...	....
5. Kumari and Srivastava [15]	...	...	$Z_{\min} = [-4.38408, -0.18706] = \langle -2.28557, 2.09851 \rangle$.	...	....
6. Li and Tian[18]	...	...	$Z_{\min} = [-6.25, -0.9] = \langle -3.575, 2.675 \rangle$.	$Z_{min} = [-3, -1.125] = \langle -2.0625, 0.9375 \rangle$.	....
7. Liu and Wang [22]	...	...	...	$Z_{\min} = [-3, -1.125] = \langle -2.0625, 0.9375 \rangle$.	....
8. Nehi et al. [6]	...	...	...	...	$Z_{\min} = [226.4321, 377.63] = \langle 302.03105, 75.59895 \rangle$.
9. Proposed method	$Z_{\min} = [-4.71875, -4.71875] = \langle -4.71875, 0 \rangle$	$Z_{\max} = [2.75, 2.75] = \langle 2.75, 0 \rangle$	$Z_{\min} = [-2.39737, -0.890137] = \langle -1.6437535, 0.7536165 \rangle$	$Z_{\min} = [-2.36722, -1.0983] = \langle -1.73276, 0.63446 \rangle$.	$Z_{\min} = [278.83, 298.842] = \langle 288.836, 10.006 \rangle$

problems to obtain Pareto optimal solutions. Interval-valued linear constraints are handled by using a specific interval ranking definition. The main goal of this technique is to find the most acceptable and efficient solution for the proposed IQPP by deploying less interval computational effort. The comparative discussions of the experimental results also reflect the same. For future research, the method can be extended for general non-linear optimization problems and used for different application problems arising in our daily lives.

Appendix A: List of Test Problems

$$\begin{aligned}
F_1 : \text{Minimize } Z = & \frac{1}{2}\Big\{[3.5, 4.5]\,[x_{1L}, x_{1R}]^2 + 2[1.5, 2.5]\,[x_{1L}, x_{1R}]\,[x_{2L}, x_{2R}] \\
& + [2.5, 3.5]\,[x_{2L}, x_{2R}]^2\Big\} \\
\text{subject to } & [1.5, 2.5]\,[x_{1L}, x_{1R}] + [2.5, 3.5]\,[x_{2L}, x_{2R}] \geq [5.5, 6.5], \\
& [1.5, 2.5]\,[x_{1L}, x_{1R}] + [0.5, 1.5]\,[x_{2L}, x_{2R}] \geq [1.5, 3.5], \\
\text{and } & x_{1L}, x_{2L} \geq 0, x_{1R} \geq x_{1L}, x_{2R} \geq x_{2L}. \\
F_2 : \text{Minimize } Z = & [x_{1L}, x_{1R}]^2 + [x_{2L}, x_{2R}]^2 + [1, 2] \\
\text{subject to } & [1, 6]\,[x_{1L}, x_{1R}] + [1, 2]\,[x_{2L}, x_{2R}] \geq [1, 12], \\
\text{and } & x_{1L}, x_{2L} \geq 0, x_{1R} \geq x_{1L}, x_{2R} \geq x_{2L}. \\
F_3 : \text{Minimize } Z = & [-5, -3]\,[x_{1L}, x_{1R}] + [1, 2]\,[x_{2L}, x_{2R}] \\
& - 2\,[x_{1L}, x_{1R}]\,[x_{2L}, x_{2R}] + [2, 3]\,[x_{1L}, x_{1R}]^2 + 2\,[x_{2L}, x_{2R}]^2 \\
\text{subject to } & [1, 2]\,[x_{1L}, x_{1R}] + [x_{2L}, x_{2R}] \leq [2, 4], \\
& [2, 3]\,[x_{1L}, x_{1R}] + [-1, -0.5]\,[x_{2L}, x_{2R}] \leq [3, 4], \\
\text{and } & x_{1L}, x_{2L} \geq 0, x_{1R} \geq x_{1L}, x_{2R} \geq x_{2L}. \\
F_4 : \text{Minimize } Z = & [-5, -3]\,[x_{1L}, x_{1R}] + [1, 2]\,[x_{2L}, x_{2R}] \\
& - 2\,[x_{1L}, x_{1R}]\,[x_{2L}, x_{2R}] + [2, 3]\,[x_{1L}, x_{1R}]^2 + 2\,[x_{2L}, x_{2R}]^2 \\
\text{subject to } & [1, 2]\,[x_{1L}, x_{1R}] + [x_{2L}, x_{2R}] \leq [2, 4], \\
& [2, 3]\,[x_{1L}, x_{1R}] + [-1, -0.5]\,[x_{2L}, x_{2R}] \leq [3, 4], \\
& [4, 5]\,[x_{1L}, x_{1R}] + [-8, -7]\,[x_{2L}, x_{2R}] = [1, 1.5], \\
\text{and } & x_{1L}, x_{2L} \geq 0, x_{1R} \geq x_{1L}, x_{2R} \geq x_{2L}. \\
F_5 : \text{Minimize } Z = & [-23, -15]\,[x_{1L}, x_{1R}] + [7, 9]\,[x_{2L}, x_{2R}] \\
& + [-3, -1]\,[x_{1L}, x_{1R}]\,[x_{2L}, x_{2R}] + [11, 15]\,[x_{1L}, x_{1R}]^2 - 4\,[x_{2L}, x_{2R}]^2 \\
\text{subject to } & [-4, -2]\,[x_{1L}, x_{1R}] + [1.8, 2]\,[x_{2L}, x_{2R}] \leq [15, 17], \\
& [5.5, 6]\,[x_{1L}, x_{1R}] + [-2.2, -2]\,[x_{2L}, x_{2R}] \leq [-13, -11], \\
\text{and } & x_{1L}, x_{2L} \geq 0, x_{1R} \geq x_{1L}, x_{2R} \geq x_{2L}. \\
F_6 : \text{Minimize } Z = & 2\,[x_{1L}, x_{1R}] + 4\,[x_{2L}, x_{2R}] \\
& - 2\,[x_{1L}, x_{1R}]\,[x_{2L}, x_{2R}] + [x_{1L}, x_{1R}]^2 + 2\,[x_{2L}, x_{2R}]^2
\end{aligned}$$

$$\text{subject to } [-1, -0.5] \le [x_{1L}, x_{1R}] \le [3, 5],$$
$$[-8, -6] \le [x_{2L}, x_{2R}] \le [4, 8],$$
$$\text{and } \quad x_{1R} \ge x_{1L}, x_{2R} \ge x_{2L}.$$

$$F_7 : \text{Maximize } Z = 3\,[x_{1L}, x_{1R}] - [x_{1L}, x_{1R}]^2 + [x_{2L}, x_{2R}] - [x_{2L}, x_{2R}]^2 + [x_{3L}, x_{3R}] - [x_{3L}, x_{3R}]^2$$
$$\text{subject to } [x_{1L}, x_{1R}] + [x_{2L}, x_{2R}] + [x_{3L}, x_{3R}] \le [1.5, 4.5],$$
$$\text{and } \quad x_{1L}, x_{2L}, x_{3L} \ge 0, x_{1R} \ge x_{1L}, x_{2R} \ge x_{2L}, x_{3R} \ge x_{3L}.$$

$$F_8 : \text{Minimize } Z = [-10, -6]\,[x_{1L}, x_{1R}] + [2, 3]\,[x_{2L}, x_{2R}] + [-1, 1]\,[x_{1L}, x_{1R}]\,[x_{2L}, x_{2R}] + [4, 10]\,[x_{1L}, x_{1R}]^2 + [10, 20]\,[x_{2L}, x_{2R}]^2$$
$$\text{subject to } [1, 2]\,[x_{1L}, x_{1R}] + 3\,[x_{2L}, x_{2R}] \le [1, 10],$$
$$[-2, 8]\,[x_{1L}, x_{1R}] + [4, 6]\,[x_{2L}, x_{2R}] \le [4, 6].$$
$$\text{and } \quad x_{1L}, x_{2L} \ge 0, x_{1R} \ge x_{1L}, x_{2R} \ge x_{2L}.$$

$$F_9 : \text{Minimize } Z = [-5, -3]\,[x_{1L}, x_{1R}] + [1, 2]\,[x_{2L}, x_{2R}] - 2\,[x_{1L}, x_{1R}]\,[x_{2L}, x_{2R}] + 2\,[x_{1L}, x_{1R}]^2 + 2\,[x_{2L}, x_{2R}]^2$$
$$\text{subject to } [1, 2]\,[x_{1L}, x_{1R}] + [x_{2L}, x_{2R}] \le [2, 4],$$
$$[2, 3]\,[x_{1L}, x_{1R}] + [-1, -0.5]\,[x_{2L}, x_{2R}] \le [3, 4].$$
$$\text{and } \quad x_{1L}, x_{2L} \ge 0, x_{1R} \ge x_{1L}, x_{2R} \ge x_{2L}.$$

$$F_{10} : \text{Minimize } Z = [-7, -4]\,[x_{1L}, x_{1R}] + [10, 14]\,[x_{2L}, x_{2R}] + 10\,[x_{1L}, x_{1R}]\,[x_{2L}, x_{2R}] + 3\,[x_{1L}, x_{1R}]^2 + 12\,[x_{2L}, x_{2R}]^2$$
$$\text{subject to } [-15, -13]\,[x_{1L}, x_{1R}] + [1, 2]\,[x_{2L}, x_{2R}] \le [-11, -10],$$
$$[1, 2]\,[x_{1L}, x_{1R}] + [-9, -8]\,[x_{2L}, x_{2R}] \le [-33, -32].$$
$$\text{and } \quad x_{1L}, x_{2L} \ge 0, x_{1R} \ge x_{1L}, x_{2R} \ge x_{2L}.$$

References

1. Elsisy, M., Hammad, D., El-Shorbagy, M.A.: Solving interval quadratic programming problems by using the numerical method and swarm algorithms. Complexity **2020**, 1–11 (2020)
2. Fang, L., Sun, L., He, G., et al.: A new algorithm for quadratic programming with interval-valued bound constraints. In: 2008 Fourth International Conference on Natural Computation, pp. 433–437. IEEE (2008)
3. Hansen, E., Walster, G.W.: Global Optimization Using Interval Analysis: Revised and Expanded, vol. 264. CRC Press, New York (2003)
4. Hladik, M.: Optimal value bounds in nonlinear programming with interval data. Top **19**(1), 93–106 (2011)
5. Hladík, M.: Interval convex quadratic programming problems in a general form. Central Eur. J. Oper. Res. **25**(3), 725–737 (2017)
6. Hormazdabadi, M.G., Nehi, H.M., Allahdadi, M.: Optimal solution set in interval quadratic programming problem. J. Math. Exten. **14**, 189–205 (2019)
7. Ishibuchi, H., Tanaka, H.: Multiobjective programming in optimization of the interval objective function. Eur. J. Oper. Res. **48**(2), 219–225 (1990)

8. Ishizaki, T., Koike, M., Ramdani, N., et al.: Monotonicity characterization of interval quadratic programming. In: Proceedings of the 21st International Symposium on Mathematical Theory of Networks and Systems, pp. 792–299 (2014)
9. Karmakar, S., Bhunia, A.: On constrained optimization by interval arithmetic and interval order relations. Opsearch **49**, 22–38 (2012)
10. Karmakar, S., Bhunia, A.K.: A new multi-section based technique for constrained optimization problems with interval-valued objective function. Appl. Math. Comput. **225**, 487–502 (2013)
11. Karmakar, S., Bhunia, A.K.: An efficient interval computing technique for bound-constrained uncertain optimization problems. Optimization **63**(11), 1615–1636 (2014)
12. Karmakar, S., Mahato, S., Bhunia, A.: Interval oriented multi-section techniques for global optimization. J. Comput. Appl. Math. **224**(2), 476–491 (2009)
13. Kearfott, R.B.: Rigorous global search: continuous problems Kluwer academic publishers. Dordrecht, Netherlands (1996)
14. Kumar, P., Panda, G., Gupta, U.C.: Generalized quadratic programming problem with interval uncertainty. In: 2013 IEEE International Conference on Fuzzy Systems (FUZZ-IEEE), pp. 1–7. IEEE (2013)
15. Kumari, S., Srivastava, S.: Application of stochastic programming technique to solve interval quadratic programming problem. Int. J. Appl. Comput. Math. **7**, 1–16 (2021)
16. Kundu, S.: Min-transitivity of fuzzy leftness relationship and its application to decision making. Fuzzy Sets Syst. **86**(3), 357–367 (1997)
17. Li, W.: Algorithm for interval quadratic programming with equality constraints. In: 2009 Sixth International Conference on Fuzzy Systems and Knowledge Discovery, pp. 78–81. IEEE (2009)
18. Li, W., Tian, X.: Numerical solution method for general interval quadratic programming. Appl. Math. Comput. **202**(2), 589–595 (2008)
19. Li, W., Xia, M., Li, H.: New method for computing the upper bound of optimal value in interval quadratic program. J. Comput. Appl. Math. **288**, 70–80 (2015)
20. Li, W., Xia, M., Li, H.: Some results on the upper bound of optimal values in interval convex quadratic programming. J. Comput. Appl. Math. **302**, 38–49 (2016)
21. Li, Y., Huang, G.: Robust interval quadratic programming and its application to waste management under uncertainty. Environ. Syst. Res. **1**(1), 1–16 (2012)
22. Liu, S.T., Wang, R.T.: A numerical solution method to interval quadratic programming. Appl. Math. Comput. **189**(2), 1274–1281 (2007)
23. Mahato, S., Bhunia, A.: Interval-arithmetic-oriented interval computing technique for global optimization. Appl. Math. Res. Express **2006**, 69642 (2006)
24. Moore, R.E.: Methods and Applications of Interval Analysis. SIAM, Philadelphia (1979)
25. Moore, R.E., Kearfott, R.B., Cloud, M.J.: Introduction to Interval Analysis/Ramon E. Moore, R Baker Kearfott, Michael J Cloud Philadelphia (2009)
26. Ratschek, H., Rokne, J.: New Computer Methods for Global Optimization. Halsted Press, Chichester (1988)
27. Sahoo, L., Bhunia, A.K., Kapur, P.K.: Genetic algorithm based multi-objective reliability optimization in interval environment. Comput. Ind. Eng. **62**(1), 152–160 (2012)
28. Sengupta, A., Pal, T.K.: On comparing interval numbers. Eur. J. Oper. Res. **127**(1), 28–43 (2000)
29. Suprajitno, H., et al.: Extension of Wolfe method for solving quadratic programming with interval coefficients. J. Appl. Math. (2017)
30. Suprajitno, H., et al.: Solution of quadratic programming with interval variables using a two-level programming approach. J. Appl. Math. (2018)
31. Taghi-Nezhad, N., Taleshian, F., Shahini, M.: A new approach for solving interval quadratic programming problem. Iran. J. Optim. **11**(1), 67–75 (2019)
32. Wang, H., Zhang, C., Guo, P.: An interval quadratic fuzzy dependent-chance programming model for optimal irrigation water allocation under uncertainty. Water **10**(6), 684 (2018)
33. Wang, J., He, F., Shi, X.: Numerical solution of a general interval quadratic programming model for portfolio selection. PloS One **14**(3), e0212913 (2019)
34. Xu, Q., Jia, S., Li, H., et al.: Some properties of interval quadratic programming problem. Int. J. Syst. Sci. Appl. Math. **2**, 105–109 (2017)

A Case Study on Medical Company Selection in the Health Sector by Using an Integrated Fuzzy AHP-MOORA Method

Brajamohan Sahoo and Bijoy Krishna Debnath

Abstract Choosing the appropriate medical company within the healthcare industry is crucial for maintaining the quality of patient care, operational effectiveness, and the reputation of the organization. It directly influences the availability of top-notch products, efficient procedures, and inventive approaches, thereby improving the delivery of healthcare services and overall outcomes. Collaborating with reputable companies cultivates confidence among stakeholders and stimulates innovation in healthcare, ultimately propelling advancements in patient care. Incorporating fuzziness in the decision-making process allows for the consideration of ambiguity and uncertainty inherent in real-world healthcare circumstances. To enhance the process of choosing medical companies within the healthcare industry, this study recommends determining the best option using the fuzzy Analytic Hierarchy Process (fuzzy AHP) and fuzzy Multi-Objective Optimization on the Basis of Ratio Analysis (fuzzy MOORA) method. In this chapter, we employ the fuzzy AHP to determine the normalized weight attribute and subsequently utilize the fuzzy MOORA approach, for the selection process in the health sector. The best medical company selection procedure involves the assessment of seven criteria, each with four alternatives. The normalization process is applied to obtain the weighted normalized fuzzy decision matrix. Beneficial and non-beneficial criteria are computed for each choice, and an overall performance index is derived. Based on the descending values of the overall performance index, the alternatives are ranked from best to worst in the final table. The study concludes that the alternative with the highest overall performance index is the most favorable choice. This chapter also discusses the implications of the case study and sensitivity analysis.

Keywords Fuzzy AHP · Fuzzy MOORA · MCDM · Medical company selection

B. Sahoo · B. K. Debnath (✉)
Department of Applied Sciences, School of Engineering, Tezpur University, Napaam, Sonitpur 784028, Assam, India
e-mail: bijoy91@tezu.ernet.in

L. Sahoo et al. (eds.), *Decision Making Under Uncertainty Via Optimization, Modelling, and Analysis*, Studies in Systems, Decision and Control 558,
https://doi.org/10.1007/978-981-96-0085-4_19

1 Introduction

Medical companies play a crucial role in the healthcare sector, leading the way in advancements for treatment, diagnosis, and patient well-being. They engage in extensive research and development to pioneer pharmaceuticals, medical equipment, and biotechnological solutions. These innovations not only cater to current medical requirements but also pioneer answers to emerging health issues. It's imperative for these companies to manufacture top-notch products to ensure safety and effectiveness in healthcare provision. Efficient management of complex supply chains ensures the global distribution of medical resources, bolstering healthcare systems worldwide. Adherence to clinical trial protocols and regulatory standards guarantees that products meet stringent criteria before reaching patients. Additionally, some companies offer healthcare services like diagnostics and telemedicine, enhancing conventional healthcare approaches. Overall, these collective endeavors fuel progress, enhance healthcare accessibility, and ultimately save lives, cementing medical companies as indispensable pillars of the healthcare landscape.

Selecting medical companies in the health sector is crucial because of their significant influence on patient care, treatment results, and the overall effectiveness of healthcare systems. Opting for the appropriate medical company guarantees access to top-notch products, services, and innovations necessary for diagnosing, treating, and preventing illnesses. Driven by the desire for improved health outcomes, healthcare providers, policymakers, and patients prioritize choosing medical companies with proven excellence, innovative practices, and adherence to regulations. These selection standards not only ensure patient safety and effectiveness but also stimulate advancements in medical science and healthcare provision. Thus, the meticulous selection of medical companies plays a fundamental role in shaping the quality and efficiency of healthcare systems, ultimately benefiting individuals and communities.

Along with the pharmaceutical industry, the medical equipment sector is an important component of the health industry. Because of advances in medical technology the medical device sector, like the pharmaceutical industry, has seen an increase in utilization, which has resulted in rising healthcare costs around the world [1].

Choosing the right medical company for the health sector can be challenging due to various factors. Firstly, there may be misunderstandings about the specific needs of healthcare organizations or communities, leading to mismatches with the services provided by the company. Moreover, insufficient information or transparency regarding the company's track record, quality standards, or financial stability can hinder decision-making. Additionally, regulatory complexities and compliance issues in healthcare can create obstacles in selecting the most suitable company. Lastly, the rapidly evolving nature of healthcare technologies makes it hard to determine which companies offer the most innovative solutions. Bridging these gaps necessitates better communication, collaboration, and transparency between healthcare stakeholders and medical companies, along with a deep understanding of the healthcare environment's needs and challenges. The primary research questions (RQs) that served as the basis for this study are as follows:

- RQ1: What are the primary difficulties facing in choosing the medical industry?
- RQ2: What connections exist between these difficulties?
- RQ3: Why do we choose to rank the medical company using this specific integrated fuzzy AHP-MOORA approach?
- RQ4: What positive effects and applications might the research's findings have on the medical sector?

Overcoming the challenges in selecting the right medical company for the health sector requires a systematic approach that addresses various aspects of decision-making. Firstly, improved communication and collaboration between healthcare stakeholders and medical companies are crucial to gain a clear understanding of the healthcare organization's needs and criteria. Transparency regarding the performance history, quality standards, financial strength, and regulatory compliance of medical companies is vital for making well-informed decisions. Conducting thorough due diligence, which includes comprehensive evaluations and assessments, can help mitigate risks and identify the most suitable medical companies.

The chapter is motivated by the urgent need for more sophisticated decision-making procedures within the healthcare realm, where the selection of medical companies holds significant sway over both patient well-being and the effectiveness of organizations. Acknowledging the complex nature of healthcare decisions, which are often marked by a multitude of factors and uncertainties, the chapter aims to confront this challenge by merging the fuzzy AHP and fuzzy MOORA techniques. This fusion intends to furnish a structured yet adaptable framework capable of handling imprecise data and varied decision objectives. Through a practical application in a real-world case study, the chapter not only seeks to validate its proposed methodology but also illuminate its practical value in enhancing decision outcomes across healthcare establishments. Hence, the chapter's impetus lies in bolstering decision-making efficiency, promoting well-informed choices, and ultimately raising the standards of patient care within the healthcare sector.

Multi-Criteria Decision-Making (MCDM) techniques provide valuable resources for addressing the intricacies of choosing the best medical company. They enable decision-makers to assess numerous factors concurrently and assign significance based on the particular circumstances. Through structured assessment and comparison, considering elements like quality, expense, innovation, regulatory adherence, and compatibility with organizational objectives, MCDM approaches aid in pinpointing the most advantageous medical company.

In addressing this challenge, we utilized an Integrated MCDM approach, blending two robust methodologies: fuzzy AHP and fuzzy MOORA. Fuzzy AHP facilitated the calculation of normalized weight attributes, allowing for a thorough evaluation of various criteria. Conversely, fuzzy MOORA aided in pinpointing the most appropriate medical company by simultaneously considering multiple objectives, such as quality, cost, technical service, diversity, customer representative, and brand. Both fuzzy AHP and fuzzy MOORA are highly regarded in the field of MCDM for their capacity to handle uncertainties and complexities inherent in decision-making processes, making them optimal selections for assessing and choosing medical companies within the healthcare sector.

Our contribution to this manuscript comprises several key features. Firstly, we identify the essential criteria for selecting the optimal medical company for the health sector. To address this case study, we utilize an integrated Multi-Criteria Decision-Making (MCDM) method within a fuzzy environment. In this chapter, we employ the fuzzy AHP method to determine the weight of the criteria. Additionally, we utilize the fuzzy MOORA method to determine the overall performance index (P_i) for each alternative and rank them according to their performance index.

To accomplish these goals, the remainder of this chapter follows a structured approach. Section 2 conducts a comprehensive review of pertinent literature. In Sect. 3, the intricacies of the fuzzy AHP and fuzzy MOORA methods are explored. The case study and results discussion are outlined in Sect. 4. Section 5 presents a sensitivity analysis. The implications of the study are discussed in Sect. 6. Finally, Sect. 7 offers concluding remarks and highlights the significance of this chapter.

2 Literature Review

Medical firms play a crucial role in the healthcare sector by providing indispensable products, services, and technologies that directly influence patient care and healthcare results. Opting for the appropriate medical company is essential as it guarantees access to top-tier medical equipment, drugs, and innovations that can greatly improve patient treatment and safety. Moreover, collaborating with reputable medical firms cultivates trust among healthcare practitioners, patients, and stakeholders, thus bolstering the credibility and standing of healthcare institutions. Furthermore, choosing the right medical company facilitates the adoption of streamlined processes, cost-effective solutions, and state-of-the-art advancements, thereby promoting operational excellence and fostering innovation in healthcare.

Nowadays, market demands have driven industries to give quick and dependable service in order to meet sensitive customer expectations [2, 3]. When it comes to a company's strategies for staying competitive in the global marketplace, the management of supply chains has garnered a lot of attention in this respect. Given how important it is to select the best supplier or company, there are several multi-criteria solutions available to aid in the challenging work of choosing one. Such approaches are employed by buying managers in businesses, including those in the top medical and industrial sectors. The following are some of the most frequently cited approaches in the literature. The selection of suppliers is based on fuzzy AHP and intuitive fuzzy MOORA introduced by Domínguez et al. [4].

To select the best firm, Seyma employs the integrated MCDM method [5]. Sahoo et al. utilize multiple criteria decision-making techniques in the fuzzy domain for real-life applications [6]. Liu uses an integrated MCDM model for sustainable supplier selection in an interval-valued intuitionistic uncertain linguistic environment [7]. Maitra et al. utilized Multi-criteria Decision Making (MCDM) and its application in selecting online learning platforms during the COVID-19 pandemic, employing the TOPSIS method [8]. Rashidi compares fuzzy DEA and fuzzy TOPSIS in the selection of sustainable suppliers [9]. Rasmussen chooses suppliers for the aerospace and defense sectors using the MCDM method [10]. Sahoo et al. utilized a Score Function-Based Effective Ranking method for Interval-Valued Fermatean Fuzzy Sets and applied it to solve multi-criteria decision-making problems [11]. In this Chapter, we use the fuzzy AHP and fuzzy MOORA method, a very good and simple MCDM method, to choose the correct medical company selection based on the assumed criteria. Liu used fuzzy AHP methods for decision-making with subjective judgments [12]. Haji uses the fuzzy AHP method in educational and vocational guidance [13]. Ayhan used a fuzzy AHP approach for supplier selection problems in a case study of a Gear motor company [14]. Previously various researchers have used this method for different selection processes. In 2015 [15], Zeeshan used the fuzzy MOORA approach to rank the components in a software system reliability estimation. Ikuobase used the fuzzy MOORA technique to develop and build an automated hammering machine [16].

Masoud also used the application of fuzzy MOORA in solar power plant location selection in 2022 [17]. In 2018, Feyza and Gizem used the fuzzy MOORA method with a new combined weight approach for selecting the best hotel [18]. Although earlier studies focused on choosing the most suitable medical company using sophisticated mathematical methods, it is still necessary to adopt a straightforward and understandable MCDM technique to deal with system selection difficulties. Fuzzy AHP and fuzzy MOORA are the most popular combined strategies for selecting the best option. It is a decision-making strategy for evaluating and selecting alternatives based on multiple factors. It helps in making the optimal choice out of a range of options that consider a variety of factors that could serve different objectives. Normalizing the criteria, allocating weights, and calculating the performance ratings of each alternative are all part of the technique. The choice with the greatest score is then considered the best option.

3 Methodology

In this study, both the fuzzy AHP and fuzzy MOORA approaches are utilized to develop an MCDM standard for selecting pharmaceutical businesses in the healthcare sector. Here, first, we use the fuzzy AHP method to find the normalized weight attribute. Next, we employ fuzzy MOORA to determine which medical company is best for the health industry.

The decision-maker contrasts the options or requirements using the linguistic phrases in Table 1.

Table 1 Linguistic terms and the corresponding triangular fuzzy numbers

Linguistic variable	Triangular fuzzy number
Equal importance	(1,1,1)
Very low importance	(1,1,2)
Low importance	(1,2,3)
Moderate low importance	(2,3,4)
Moderate importance	(3,4,5)
Moderately high importance	(4,5,6)
High importance	(6,7,8)
Very high importance	(8,9,9)

3.1 Fuzzy AHP Method

The Analytical Hierarchy Process (AHP) is a technique within the field of multi-criteria decision-making, assisting decision-makers in selecting appropriate solutions for complex, multi-objective problems. Developed by Professor Thomas L. Saaty during the 1970s and 1980s, AHP is rooted in the concept of measurement through pairwise comparisons. The fuzzy fuzzy AHP is an advanced version of the AHP that introduces the idea of fuzziness and uncertainty into decision-making. While AHP offers a systematic structure for decision-making by assigning clear-cut values to criteria and alternatives, fuzzy AHP recognizes the inherent vagueness and imprecision in decision-making. It allows for the expression of subjective judgments using fuzzy numbers. Through the utilization of fuzzy numbers and fuzzy arithmetic operations, fuzzy AHP enables a more realistic portrayal of decision-maker's judgments and establishes a more resilient framework for decision-making [19]. The process consists of the following steps:

Step 1: First, we write the decision matrix which is given in below [20]

$$D = \begin{bmatrix} 1 & a_{12} & \ldots\ldots & a_{1n} \\ a_{21} & 1 & \ldots\ldots & a_{2n} \\ \ldots\ldots & \ldots & \ldots\ldots & \ldots\ldots \\ a_{n1} & a_{n2} & \ldots\ldots & 1 \end{bmatrix}$$

In this context, each element a_{ij} is expressed as a triangular fuzzy number such as $a_{ij} = [a_{ij}^l, a_{ij}^m, a_{ij}^u]$

Step 2:
The fuzzy comparison value for each criterion's geometric mean is derived as follows [14, 20]:

$r_i = (\prod_{j=1}^{n} a_{ij})^{1/n}$
where $i = 1, 2, \ldots n$

Step 3:
Next, we determine each r_i's vector sum.

Step 4:
Find the summation vector's (-1) power. To create a rising order, replace the triangular fuzzy number.

Step 5:
Each r_i should be multiplied by the reverse vector to find the fuzzy weight of the criterion [14, 20].

$$W_i = r_i * (r_1 + r_2 + \cdots + r_n)^{-1}$$

$$= (lw_i, mw_i, uw_i)$$

Step 6:
W_i need to be defuzzified using the center area approach because they are still triangular fuzzy numbers [14].

$$M_i = \frac{lw_i + mw_i + uw_i}{3}$$

Step 7:
The non-fuzzy number, M_i, can be normalized using the given formula [14].

$$N_i = \frac{M_i}{\sum_{i=1}^{n} M_i}$$

Then we get the required normalized weight attribute.

3.2 *Fuzzy MOORA Method*

In 2006, Brauers and Zavadskas introduced a method called Multi-Objective Optimization based on Ratio Analysis (MOORA) [21]. This method serves as the foundation for choice comparison, which in turn makes it easier to identify the best option. The fuzzy MOORA is an advanced version of the MOORA. The fuzzy MOORA appears to be a suitable method for prioritizing or choosing one or more options from a range of feasible alternatives, taking into account various and often conflicting attributes. The fuzzy MOORA technique ostensibly broadens this approach by introducing fuzzy logic to handle ambiguity and incertitude in decision-making.

The steps to be followed are as follows [22].

Step 1:

The fuzzy decision matrix is first created, with the triangle membership function used to calculate each criterion value [22].

$$\begin{bmatrix} [a^l_{11}, a^m_{11}, a^u_{11}] & [a^l_{12}, a^m_{12}, a^u_{12}] & \dots & [a^l_{1n}, a^m_{1n}, a^u_{1n}] \\ [a^l_{21}, a^m_{21}, a^u_{21}] & [a^l_{22}, a^m_{22}, a^u_{22}] & \dots & [a^l_{2n}, a^m_{2n}, a^u_{2n}] \\ \dots & \dots & \dots & \dots \\ [a^l_{m1}, a^m_{m1}, a^u_{m1}] & [a^l_{m2}, a^m_{m2}, a^u_{m2}] & \dots & [a^l_{mn}, a^m_{mn}, a^u_{mn}] \end{bmatrix} \quad (1)$$

where a^l_{ij} , a^m_{ij} and a^u_{ij} stands for the lower, middle, and top values of a triangular membership function for (i, j), respectively.

Step 2:
The fuzzy decision matrix is then normalized using the vector normalization method [23].

$$r^l_{ij} = \frac{a^l_{ij}}{\sqrt{\sum_{i=1}^{m}[(a^l_{ij})^2 + (a^m_{ij})^2 + (a^u_{ij})^2]}} \quad (2)$$

$$r^m_{ij} = \frac{a^m_{ij}}{\sqrt{\sum_{i=1}^{m}[(a^l_{ij})^2 + (a^m_{ij})^2 + (a^u_{ij})^2]}} \quad (3)$$

$$r^u_{ij} = \frac{a^u_{ij}}{\sqrt{\sum_{i=1}^{m}[(a^l_{ij})^2 + (a^m_{ij})^2 + (a^u_{ij})^2]}} \quad (4)$$

Step 3:
Next, the weighted normalized fuzzy decision matrix is computed using the following formulas [22]:

$$v^l_{ij} = N_j * r^l_{ij} \quad (5)$$

$$v^m_{ij} = N_j * r^m_{ij} \quad (6)$$

$$v^u_{ij} = N_j * r^u_{ij} \quad (7)$$

The weighted normalized fuzzy decision matrix can also be produced using fuzzy criteria weights.

Step 4:
Then, we determine the total ratings for both beneficial and non-beneficial categories for each option. The lowest, middle, and highest values of the triangular membership function are utilized to determine an option's overall rating for the favorable criterion [22].

$$p^{+l}_i = \sum_{j=1}^{n} v^l_{ij} | j \in J^{\max} \quad (8)$$

$$p_i^{+m} = \sum_{j=1}^{n} v_{ij}^m \,|j \in J^{\max} \quad (9)$$

$$p_i^{+u} = \sum_{j=1}^{n} v_{ij}^u |j \in J^{\max} \quad (10)$$

The overall scores of an option are calculated as follows for non-beneficial criteria:

$$p_i^{-l} = \sum_{j=1}^{n} v_{ij}^l |j \in J^{\min} \quad (11)$$

$$p_i^{-m} = \sum_{j=1}^{n} v_{ij}^m |j \in J^{\min} \quad (12)$$

$$p_i^{-u} = \sum_{j=1}^{n} v_{ij}^u |j \in J^{\min} \quad (13)$$

Step 5:
Next, the overall performance indicator (P_i) for each alternative is calculated [24].

$$P_i(p_i^+, p_i^-) = \sqrt{\frac{1}{3}[(p_i^{+l} - p_i^{-l})^2 + (p_i^{+m} - p_i^{-m})^2 + (p_i^{+u} - p_i^{-u})^2]} \quad (14)$$

Step 6:
Following that, the choices are rated from best to worst based on the decreasing values of the total performance indicator. The most advantageous choice is one with the highest index of overall performance.

4 Case Study and Result Discussions

In this book chapter, we have used four alternatives A_1, A_2, A_3, A_4 and seven criteria utilized for evaluation are as follows:

- **Quality**: It is essential to a medical company's operations because it guarantees that its goods and services satisfy strict requirements for dependability, safety, and efficacy [25].
- **Cost**: It is essential to medical companies' operations and decision-making processes. The company's financial survival and the provision of high-quality healthcare services depend on efficient cost management [25].

- **Time**: It is crucial to a medical business's operations, supporting efficacy, patient care quality, compliance with legal requirements, and general success [26].
- **Brand**: It is required for several reasons. In the medical industry, building a reputable and recognizable brand can be advantageous in many ways.
- **Technical service**: Choosing a medical company for technical services is an important choice that can have a big impact on the general efficacy, efficiency, and standard of treatment. It is vital to establish unambiguous technical service criteria to guarantee that the selected enterprise fulfills the distinct requirements and benchmarks of the healthcare establishment.
- **Customer representative**: A medical company's choice of customer representatives is an important one since they are the ones that make sure that customers are satisfied, that communications are clear, and that their experiences are positive all around [27].
- **Diversity**: There are many advantages to using diversity criteria in a medical company's selection process, and they can help create a more welcoming and productive healthcare workplace.

Figure 1 illustrates that the decision-maker is tasked with selecting the optimal medical company. This decision-making process involves considering seven criteria, with each criterion having four alternative options.

First, we find the normalized weight attribute by using the fuzzy AHP method. The Table 2 illustrates the averaged pairwise comparison of the criteria.

The geometric mean of each criterion's fuzzy comparison values is then determined by using step 2 of Sect. 3.1

$r_i = (\prod_{j=1}^{n} a_{ij})^{1/n}$=$[(\prod_{j=1}^{n} a_{ij}^{l})^{1/n}, (\prod_{j=1}^{n} a_{ij}^{m})^{1/n}, (\prod_{j=1}^{n} a_{ij}^{u})^{1/n}]$

For $i = 1$, we get

$r_1 = [(1*4*1*6*4*6*3)^{1/7}, (1*5*1*7*5*7*4)^{1/7}, (1*6*1*8*6*8*5)^{1/7}] = [2.9, 3.4, 3.8]$.

Similar calculations are made for other values of i, and the results are then recorded in Table 3.

Next, we calculate the weight of each criterion by using step 5 of the Fuzzy AHP method 3.1 and recorded in Table 4 (Table 6).

For that, we multiply each r_i with the inverse of increasing order.

The average of the fuzzy numbers for each criterion is then used to determine the relative non-fuzzy weight of each M_i using step 6 of Sect. 3.1 After that, the normalized weights for each criterion are determined using non-fuzzy M_i and recorded in Table 5

From Table 5, we get the required normalized weight attribute. The fuzzy decision matrix containing the criteria values supplied as triangular fuzzy numbers is first normalized using Eqs. (2–4) before being used to solve the same issue using the fuzzy MOORA method. Table 7 displays this normalized fuzzy decision matrix. The decision matrix's components must be normalized for an MCDM technique to render the decision matrix's elements without dimensions and comparative.

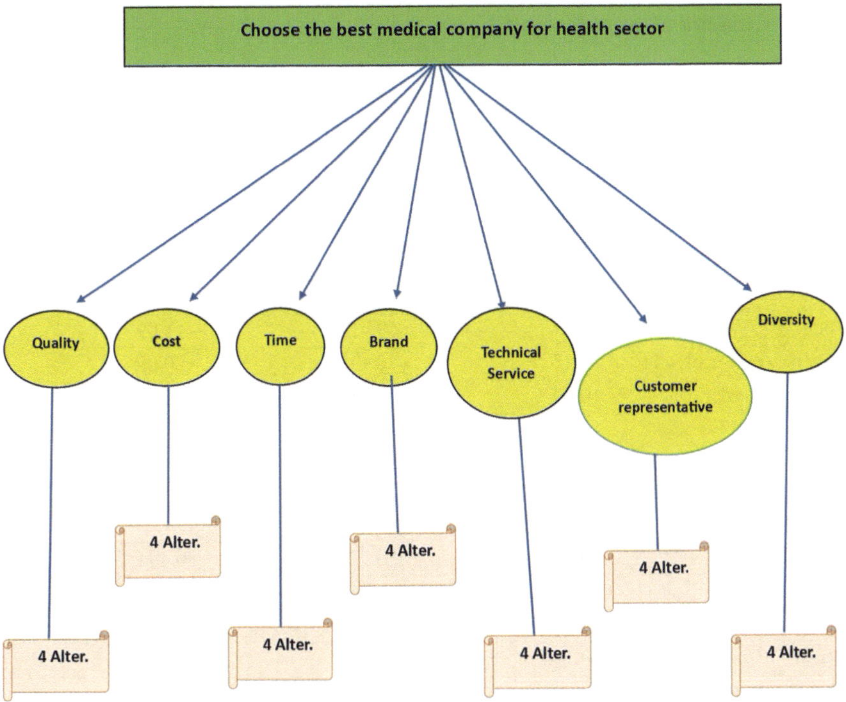

Fig. 1 Model flowchart

Table 2 Comparison matrices for criteria

Criteria	Quality	Cost	Time	Brand	Technical service	Customer representative	Diversity
Quality	(1,1,1)	(4,5,6)	(1,1,1)	(6,7,8)	(4,5,6)	(6,7,8)	(3,4,5)
Cost	(1/6,1/5,1/4)	(1,1,1)	(1/5,1/4,1/3)	(1/6,1/5,1/4)	(1/5,1/4,1/3)	(1/5,1/4,1/3)	(2,3,4)
Time	(1,1,1)	(3,4,5)	(1,1,1)	(1/5,1/4,1/3)	(1,1,1)	(2,3,4)	(4,5,6)
Brand	(1/8,1/7,1/6)	(4,5,6)	(3,4,5)	(1,1,1)	(3,4,5)	(2,3,4)	(3,4,5)
Technical service	(1,1,1)	(3,4,5)	(1,1,1)	(1/5,1/4,1/3)	(1,1,1)	(3,4,5)	(4,5,6)
Customer representative	(1/8,1/7,1/6)	(3,4,5)	(1/4,1/3,1/2)	(1/4,1/3,1/2)	(1/5,1/4,1/3)	(1,1,1)	(2,3,4)
Diversity	(1/5,1/4,1/3)	(1/4,1/3,1/2)	(1/6,1/5,1/4)	(1/5,1/4,1/3)	(1/6,1/5,1/4)	(1/4,1/3,1/2)	(1,1,1)

The normalized fuzzy criteria values are multiplied by the appropriate crisp criteria weights to produce the weighted normalized fuzzy decision matrix, which is shown in Table 8.

Table 3 Geometric means of fuzzy comparison values

Criteria	r_i		
Quality	2.9	3.4	3.8
Cost	0.33	0.41	0.51
Time	1.25	1.47	1.69
Brand	1.6	2.01	2.43
Technical service	1.03	1.22	1.43
Customer representative	0.51	0.65	0.83
Diversity	0.25	0.31	0.4
Total	7.87	9.47	11.09
Reverse (Power of −1)	0.13	0.11	0.09
Increasing order	0.09	0.11	0.13

Table 4 Relative fuzzy weights of each criterion

Criteria	W_i		
Quality	0.261	0.374	0.494
Cost	0.0297	0.0451	0.0663
Time	0.1125	0.1617	0.2197
Brand	0.144	0.2211	0.3159
Techinical service	0.0927	0.1342	0.1859
Customer representative	0.0459	0.0715	0.1079
Diversity	0.0225	0.0341	0.052

Table 5 Averaged and normalized relative weights of criteria

Criteria	M_i	N_i
Quality	0.376	0.3537
Cost	0.047	0.0442
Time	0.165	0.1547
Brand	0.227	0.2134
Technical service	0.138	0.1293
Customer representative	0.075	0.0706
Diversity	0.036	0.034

Using Eqs. (8–10), the overall ratings of the beneficial criteria for the choices under investigation are now calculated. The overall ratings of the non-beneficial criteria for the alternatives are calculated using Eqs. (11–13). Table 9 lists the overall rankings for the four different alternatives' favorable and unfavorable features. To determine the values of the general performance indicator for all choices, the vertex technique is now employed to denazify the overall ratings for beneficial and non-beneficial

Table 6 Fuzzy decision matrix

Alternative	Quality	Cost	Time	Brand	Technical service	Customer representative	Diversity
A_1	(6,7,8)	(3,4,5)	(4,5,6)	(6,7,8)	(6,7,8)	(4,5,6)	(3,4,5)
A_2	(4,5,6)	(3,4,5)	(3,4,5)	(4,5,6)	(8,9,9)	(4,5,6)	(6,7,8)
A_3	(4,5,6)	(4,5,6)	(4,5,6)	(4,5,6)	(6,7,8)	(6,7,8)	(6,7,8)
A_4	(2,3,4)	(4,5,6)	(2,3,4)	(3,4,5)	(3,4,5)	(3,4,5)	(4,5,6)

criteria. We needed less cost and time to solve this challenge. Higher values are always preferred for customer representation, quality, branding, technical service, and diversity. Because of this, we consider time and cost as non-beneficial criteria. Rest is considered as beneficial.

The highest value of the general performance indicator is recorded for the first alternative medical company.

To solve this problem, we use the following fuzzy normalized weight of the decision criterion from Table 4.

$N_Q = 0.3537, N_C = 0.0442, N_T = 0.1547, N_B = 0.2134, N_{TS} = 0.1293, N_{CR} = 0.0706, N_D = 0.034$

To write the fuzzy decision matrix in Table 6, we use the linguistic terms with triangular fuzzy numbers that are given in Table 1.

According to the information in Table 9, the first medical company is the best option for the health sector, based on the diminishing values of the overall performance indicator used to rank the options from best to worst, and the fourth medical company is the worst option for the health sector under the given assumptions.

5 Sensitivity Analysis

Sensitivity analysis is crucial for ranking alternatives due to various factors. Initially, it identifies essential elements, elucidating which variables significantly impact rankings. Subsequently, it evaluates model stability by examining the repercussions of altering parameters, ensuring consistency under different circumstances. This process aids in managing risks by scrutinizing uncertainties linked to alternatives. Furthermore, sensitivity analysis enhances outcomes by refining models according to variable influences. Its transparency improves stakeholders' comprehension of decision-making procedures and instills trust in chosen alternatives. In dynamic settings, sensitivity analysis enables adaptation to evolving conditions. Finally, it validates findings by scrutinizing their resilience against parameter adjustments, thus bolstering confidence in decision models. In this chapter, by Following normalization,

Table 7 Normalized fuzzy decision matrix

Alternative	Quality	Cost	Time	Brand	Technical service	Customer representative	Diversity
A_1	(0.329,0.384,0.439)	(0.159,0.212,0.265)	(0.262,0.327,0.393)	(0.319,0.372,0.426)	(0.250,0.292,0.333)	(0.213,0.266,0.319)	(0.145,0.194,0.242)
A_2	(0.220,0.274,0.329)	(0.159,0.212,0.265)	(0.197,0.262,0.327)	(0.213,0.266,0.319)	(0.333,0.376,0.376)	(0.213,0.266,0.319)	(0.290,0.339,0.388)
A_3	(0.220,0.274,0.329)	(0.212,0.265,0.319)	(0.262,0.327,0.393)	(0.213,0.266,0.319)	(0.250,0.292,0.333)	(0.319,0.372,0.426)	(0.290,0.339,0.388)
A_4	(0.110,0.164,0.220)	(0.212,0.265,0.319)	(0.131,0.197,0.262)	(0.160,0.213,0.266)	(0.125,0.167,0.208)	(0.160,0.213,0.266)	(0.194,0.242,0.290)

Table 8 Weighted normalized fuzzy decision matrix

Alternative	Quality	Cost	Time	Brand	Technical service	Customer representative	Diversity
A_1	(0.116,0.136,0.155)	(0.007,0.009,0.012)	(0.041,0.051,0.061)	(0.068,0.079,0.091)	(0.032,0.037,0.043)	(0.015,0.019,0.023)	(0.005,0.007,0.008)
A_2	(0.078,0.097,0.116)	(0.007,0.009,0.012)	(0.031,0.041,0.051)	(0.045,0.057,0.068)	(0.043,0.049,0.049)	(0.015,0.019,0.023)	(0.01,0.012,0.013)
A_3	(0.078,0.097,0.116)	(0.009,0.012,0.014)	(0.041,0.051,0.061)	(0.045,0.067,0.068)	(0.032,0.037,0.043)	(0.023,0.026,0.03)	(0.01,0.012,0.013)
A_4	(0.039,0.058,0.078)	(0.009,0.012,0.014)	(0.02,0.031,0.041)	(0.034,0.045,0.057)	(0.016,0.022,0.027)	(0.011,0.015,0.019)	(0.007,0.008,0.01)

Table 9 Ranking for the best medical company

Alternative	P^+			P^-			p	Rank
	l	m	u	l	m	u		
A_1	0.277	0.329	0.381	0.007	0.009	0.012	0.320597	(1)
A_2	0.222	0.275	0.32	0.007	0.009	0.012	0.264403	(3)
A_3	0.229	0.28	0.331	0.009	0.012	0.014	0.26988	(2)
A_4	0.127	0.179	0.232	0.009	0.012	0.014	0.171701	(4)

Table 10 Sensitivity analysis

Alt.	Original	0.05	0.1	0.15	0.2	0.25	0.3	0.35	0.4
A1	0.312296	0.05	0.1	0.15	0.2	0.25	0.3	0.35	0.4
A2	0.257557	0.355791	0.337065	0.318339	0.299614	0.280888	0.262162	0.243436	0.22471
A3	0.262892	0.363161	0.344047	0.324934	0.30582	0.286706	0.267592	0.248479	0.22937
A4	0.167255	0.231048	0.218887	0.206727	0.194566	0.182406	0.170246	0.158085	0.14593
Total	1	1	1	1	1	1	1	1	1

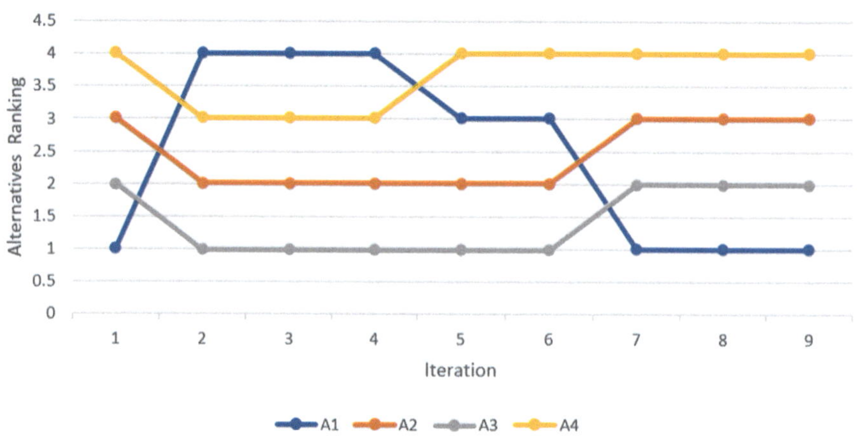

Fig. 2 Graphical representation of sensitivity analysis

the permanent values stand as: $A_1 = 0.312296$, $A_2 = 0.257557$, $A_3 = 0.262892$, and $A_4 = 0.167255$. These normalized values can be regarded as the weights of the alternatives, and sensitivity analysis is conducted. The values for each step of sensitivity analysis are provided in Table 10.

Based on Table 10, it is evident that the ranking remains stable from the 7th iteration with a value of 0.3 onwards. Again it is observed that the ranking of alternatives remains consistent, with A_1 consistently maintaining the top position from the 7th iteration onward. Consequently, this validates the model's resilience and effectiveness. Additionally, Graphical and Radar diagrams are provided below to illustrate the sensitivity analysis (Figs. 2 and 3).

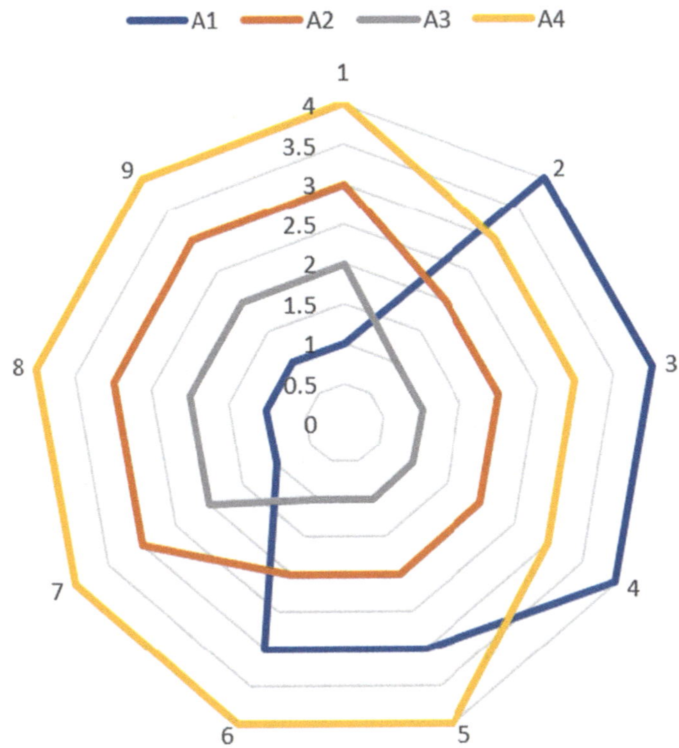

Fig. 3 Radar diagram of sensitivity analysis

6 Implications of This Study

6.1 Industrial Implication

The choice of a medical company has important commercial ramifications in the health sector. The health industry includes a wide range of businesses, including healthcare providers, biotechnology companies, and pharmaceutical and medical device manufacturers. Take into account the following important considerations and their industrial ramifications before making this decision.

1. Quality and safety compliance:
   Make sure that the company conforms to strict quality and safety standards, such as FDA regulations (where applicable). The consequences of non-compliance could include legal issues, product recalls, and reputational damage.
2. Innovation and research capabilities:
   It is important to assess the organization's commitment to research and development. Innovative products could provide your company with a market advantage.

3. Cost and pricing structure:
   Investigate the company's pricing strategy to determine whether it adheres to your budgetary restrictions. Always keep in mind that choosing cheaper products may not always be the best choice if quality is sacrificed.
4. Supply chain reliability:
   Consider how well the company can maintain a stable supply chain. Disruptions may cause delays in the industrial process or patient care.
5. Scalability and growth potential:
   Think about the company's ability to expand alongside your company. It's essential to partner with a company that can help you achieve your expansion goals.

Careful consideration of these traits and their business implications is necessary when selecting the greatest medical company in the health sector. A successful partnership in the heavily regulated and significant healthcare sector requires due diligence, open communication, and a strong working relationship.

6.2 *Academic Implications*

For educational institutions, students, and researchers in the healthcare and medical areas, there are several academic implications when selecting a medical corporation in the healthcare market. The selection of a medical corporation in the healthcare sector has the following academic repercussions.

1. Clinical training opportunities:
   Academic institutions that provide their students with clinical training opportunities, including medical schools and nursing programs, usually depend on partnerships with pharmaceutical companies. The medical corporations selected may have an impact on the standard and range of these training opportunities.
2. Access to medical technologies:
   Students and academic researchers may need access to state-of-the-art medical technology and equipment for their studies and projects. Collaboration with medical enterprises that offer cutting-edge solutions can enhance academic research and education.
3. Research collaboration:
   Researchers from academic institutions frequently collaborate with pharmaceutical companies on research projects, clinical trials, and product development. The partners that medical corporations select to collaborate with them may have an impact on the breadth and impact of university research initiatives.
4. Educational resources and materials:
   Healthcare training programs are frequently used by academic institutions with the aid of textbooks, educational materials, and other resources provided by medical firms. By selecting reputable companies, the availability of top-notch educational resources can be secured.

5. Industry-ready graduates:
   Medical schools and healthcare training programs aim to produce graduates who are capable of working in the healthcare sector. By working with medical companies that have the same objectives as academic institutions, professionals who are ready for the workplace can be produced.
6. Curriculum development:
   The selection of healthcare organizations may impact the creation of healthcare-related curricula. Academic programs can be improved by using case studies from the real world, guest speakers, or training relevant to a given sector.
7. Student internships and externships:
   Many academic programs require students to complete internships or externships in healthcare settings. Working with medical companies could give students the ability to gain practical experience.
8. Funding and scholarships:
   To promote academic research, student scholarships, or the building of academic institution infrastructure, medical corporations may offer funding, scholarships, or grants.
9. Continuing medical education (CME):
   Healthcare professionals like doctors and nurses are required to maintain their licenses by staying in school. Medical companies may offer continuing medical education (CME) programs, and the quality of these programs may affect career advancement.
10. Impact on medical research and innovation:
    Academic research is essential to the development of medicine. Working with forward-thinking medical companies can help to promote innovation and the translation of research.

The selection of a medical company in the healthcare sector can have a big impact on academics, influencing clinical training, research opportunities, curriculum development, and the overall quality of healthcare education. Academic institutions must conduct a thorough analysis of potential partners to ensure that they satisfy their educational goals and standards. Through this collaboration, it will be possible to enhance the educational process, the findings of research, and the preparedness of students and healthcare professionals for the hurried healthcare industry.

7 Conclusion

This chapter focuses on selecting the optimal medical company for the healthcare sector. Decision-makers have identified seven criteria, each with four alternatives. Additionally, the criteria have been categorized by decision-makers into two types: beneficial and non-beneficial. To address this uncertainty, an integrated MCDM method has been employed, which combines fuzzy AHP and fuzzy MOORA. Following the calculations, an overall performance index (S_i) has been established for each alternative, allowing for their subsequent ranking based on these index values. The results, presented in Table 9, reveal that A_1 emerges as the optimal medical company for the healthcare sector. It's like a useful tool that works well not just in healthcare but in different areas too. It shows how we can use these methods to figure out the right pharmaceutical companies to choose. By looking at all aspects, decision-makers in the medical field can pick companies that match the special needs and priorities of healthcare. The proper pharmaceutical business for the health industry can be found using various MCDM methods, similarly.

This chapter holds significant relevance in healthcare decision-making processes. It presents a structured method for navigating the complexities associated with choosing the most suitable medical company, considering various factors such as quality, cost, time, technical service, customer representative, brand, and diversity. By merging fuzzy AHP with fuzzy MOORA, the chapter ensures a thorough assessment, manages uncertainties, and efficiently ranks alternatives. This approach improves decision-making, resulting in enhanced patient care, treatment effectiveness, and overall healthcare system performance. It furnishes stakeholders with a valuable framework to make well-informed decisions and optimize resource allocation within the healthcare sector.

Appendix

- MCDM = Multi-Criteria Decision-Making.
- AHP = Analytic Hierarchy Process.
- MOORA = Multi-Objective Optimization on the Basis of Ratio Analysis.
- RQ = Research Questions

Acknowledgements The authors would like to express their sincere appreciation to the editors and reviewers for their invaluable comments and suggestions, which greatly enhance this chapter.

References

1. Göncü, K.K., Çetin, O.: A decision model for supplier selection criteria in healthcare enterprises with DEMATEL ANP method. Sustainability **14**(21), 13912 (2022)
2. Chou, S.-Y., Chang, Y.-H.: A decision support system for supplier selection based on a strategy-aligned fuzzy smart approach. Expert Syst. Appl. **34**(4), 2241–2253 (2008)
3. Arango-Serna, M.D., Adarme-Jaimes, W., Zapata-cortes, J.A.: Collaborative inventory in supply chain optimization. Dyna **80**(181), 71–80 (2013)
4. Pérez-Domínguez, L., Alvarado-Iniesta, A., Rodríguez-Borbón, I., Vergara-Villegas, O.: Intuitionistic fuzzy MOORA for supplier selection. Dyna **82**(191), 34–41 (2015)
5. Emec, S., Turanoglu, B., Oztas, S., Akkaya, G.: An integrated MCDM for a medical company selection in health sector. Int. J. Sci. Technol. Res. **5**(4), 77–89 (2019)
6. Sahoo, L., Senapati, T., Yager, R.R.: Real Life Applications of Multiple Criteria Decision Making Techniques in Fuzzy Domain, vol. 420. Springer Nature (2022)
7. Liu, H.-C., Quan, M.-Y., Li, Z., Wang, Z.-L.: A new integrated MCDM model for sustainable supplier selection under interval-valued intuitionistic uncertain linguistic environment. Inf. Sci. **486**, 254–270 (2019)

8. Maitra, S., Sahoo, L., Lahiri Dey, J., Tiwary, K.: Multi-criteria decision making and its application to online learning platform selection during the COVID-19 pandemic based on TOPSIS method. In: Real Life Applications of Multiple Criteria Decision Making Techniques in Fuzzy Domain, pp. 507–519. Springer (2022)
9. Rashidi, K., Cullinane, K.: A comparison of fuzzy DEA and fuzzy TOPSIS in sustainable supplier selection: implications for sourcing strategy. Expert Syst. Appl. **121**, 266–281 (2019)
10. Rasmussen, A., Sabic, H., Saha, S., Nielsen, I.E.: Supplier selection for aerospace & defense industry through MCDM methods. Clean. Eng. Technol. **12**, 100590 (2023)
11. Sahoo, L., Rana, A., Senapati, T., Yager, R.R.: Score function-based effective ranking of interval-valued fermatean fuzzy sets and its applications to multi-criteria decision making problem. In: Real Life Applications of Multiple Criteria Decision Making Techniques in Fuzzy Domain, pp. 435–451. Springer (2022)
12. Liu, Y., Eckert, C.M., Earl, C.: A review of fuzzy AHP methods for decision-making with subjective judgements. Expert Syst. Appl. **161**, 113738 (2020)
13. El Haji, E., Azmani, A., El Harzli, M.: Using the FAHP method in the educational and vocational guidance. Int. J. Mod. Educ. Comput. Sci. **10**(12) (2018)
14. Ayhan, M.B.: A fuzzy AHP approach for supplier selection problem: a case study in a gear motor company, CoRR abs/1311.2886 (2013). http://arxiv.org/abs/1311.2886 arXiv:1311.2886
15. Siddiqui, Z., Tyagi, K.: Application of fuzzy-MOORA method: ranking of components for reliability estimation of component-based software systems. Decis. Sci. Lett. **5**(1), 169–188 (2016)
16. Emovon, I., Okpako, O.S., Edjokpa, E.: Application of fuzzy MOORA method in the design and fabrication of an automated hammering machine. World J. Eng. **18**(1), 37–49 (2021)
17. Khorshidi, M., Erkayman, B., Albayrak, Ö., Kılıç, R., Demir, H.İ: Solar power plant location selection using integrated fuzzy DEMATEL and fuzzy MOORA method. Int. J. Ambient Energy **43**(1), 7400–7409 (2022)
18. Gürbüz, F., Erdinç, G.: Selecting the best hotel using the fuzzy-MOORA method with a new combined weight approach. In: 2nd International Symposium on Multidisciplinary Studies and Innovative Technologies (ISMSIT), vol. 2018, pp. 1–8. IEEE (2018)
19. Hosseinzadeh Lotfi, F., Allahviranloo, T., Pedrycz, W., Shahriari, M., Sharafi, H., Razipour GhalehJough, S.: Analytical hierarchy process (AHP) in fuzzy environment. In: Fuzzy Decision Analysis: Multi Attribute Decision Making Approach, pp. 215–237. Springer (2023)
20. Cebeci, U.: Fuzzy AHP-based decision support system for selecting ERP systems in textile industry by using balanced scorecard. Expert Syst. Appl. **36**(5), 8900–8909 (2009)
21. Brauers, W.K., Zavadskas, E.K.: The MOORA method and its application to privatization in a transition economy. Control Cybern. **35**(2), 445–469 (2006)
22. Karande, P., Chakraborty, S.: A fuzzy-MOORA approach for ERP system selection. Decis. Sci. Lett. **1**(1), 11–21 (2012)
23. Stanujkic, D., Magdalinovic, N., Stojanovic, S., Jovanovic, R.: Extension of ratio system part of MOORA method for solving decision-making problems with interval data. Informatica **23**(1), 141–154 (2012)
24. Huiqun, H., Guang, S.: ERP software selection using the rough set and TPOSIS methods under fuzzy environment. Adv. Inf. Sci. Serv. Sci. **4**(3), 111–118 (2012)
25. Sahoo, S.K., Goswami, S.S., Halder, R.: Supplier selection in the age of industry 4.0: a review on MCDM applications and trends. Decis. Mak. Adv. (2024)
26. Hosseinzadeh, M., Hama, H.K., Ghafour, M.Y., Masdari, M., Ahmed, O.H., Khezri, H.: Service selection using multi-criteria decision making: a comprehensive overview. J. Netw. Syst. Manag. **28**, 1639–1693 (2020)
27. Lotfi, F.H.: Fuzzy Decision Analysis: Multi Attribute Decision Making Approach. Springer Nature

Novel Pythagorean Fuzzy Hamacher Aggregation Operator and Its Application to Green Supplier Selection in Pharmaceutical Industry

Tapas Kumar Paul and Madhumangal Pal

Abstract This study introduces an innovative Advanced Pythagorean Fuzzy Hamacher Weighted Geometric (APyFHWG) operator. The APyFHWG operator addresses inherent limitations in several existing geometric operators (GOs), including the Pythagorean Fuzzy Dombi weighted Geometric (PyFDWG) operator, Pythagorean Fuzzy Hamacher Power Weighted Geometric (PyFHPWG) operator, Symmetric Pythagorean Fuzzy Weighted Geometric (SPyFWG) operator, Pythagorean Fuzzy Hamacher Prioritized weighted Geometric (PyFHPrWG) operator, and Pythagorean Fuzzy Interactive Hamacher Power Weighted Geometric (PyFIHPWG) operator. These shortcomings primarily involve the indeterminacy of aggregated results in specific scenarios when a Pythagorean fuzzy number contains zero membership grade and its corresponding weight component equals zero. Furthermore, the aggregated result remains indeterminate when the indeterminacy degree of a component Pythagorean fuzzy number reaches zero alongside its corresponding weight component. Additionally, MCDM methods utilizing these GOs struggle to discern the ranking order of alternatives. The proposed APyFHWG operator effectively mitigates these drawbacks. Finally, we present a case study focused on the selection of green suppliers within the pharmaceutical industry in India as a practical application of our methodology.

Keywords Advanced Pythagorean fuzzy Hamacher weighted geometric (APyFHWG) operator · Multi-criteria decision making (MCDM) · Green supplier selection · Pharmaceutical industry

T. K. Paul · M. Pal (✉)
Department of Applied Mathematics, Vidyasagar University, Midnapore 721102, India
e-mail: mmpalvu@gmail.com

M. Pal
Department of Mathematics and Innovation, Saveetha School of Engineering, Saveetha Institute of Medical and Technical Sciences, Chennai 602105, Tamilnadu, India

L. Sahoo et al. (eds.), *Decision Making Under Uncertainty Via Optimization, Modelling, and Analysis*, Studies in Systems, Decision and Control 558,
https://doi.org/10.1007/978-981-96-0085-4_20

1 Introduction

In our daily lives, numerous decisions must be made instantaneously or within specific time constraints. These decisions are either made independently, with limited knowledge, or in collaboration with individuals possessing expertise in a particular field. Decision-making relies on available information, often presented in a fuzzy, non-crisp format to align with human behavior realistically. To address the concept of fuzziness in human opinions, Zadeh introduced the fuzzy set (FS) [1]. Unlike Boolean logic, this set gauges the degree of belongingness through a membership function (MF), ranging from 0 to 1. The intuitionistic fuzzy set (IFS) was later introduced by Atanassov [2], incorporating a non-membership function (NMF) alongside the FS's MF. This extension considers an element's non-membership degree (NMD) in a way that the sum of membership degree (MD) and NMD is less than or equal to 1. Further development led to the Pythagorean fuzzy set (PyFS) by Yager [3], where the sum of squares of MD and NMD for an element is less than or equal to 1. Many works have been made related to fuzzy and other sets as they have many applications [4–17].

Aggregation operators (AOs) play a crucial role in aggregating sets of Pythagorean fuzzy numbers (PyFNs) and are integral to multi-criteria decision-making processes [18]. Jana et al. introduced the PyFDWG operator, applied in the selection of enterprise resource planning (ERP) systems for multi-attribute decision-making (MADM) [19]. Other AOs, such as the operator [20] and the SPyFWG operator [21], have found applications in diverse MADM scenarios. Researchers have explored various applications of these operators, including bipolar fuzzy environments [22, 23, 29]. Additionally, Wu et al. developed the Pythagorean fuzzy Hamacher hybrid geometric (PyFHHG) operator [24] for MADM in selecting emerging technology enterprises. Gao proposed the PyFHPrWG operator [25], applied in selecting outstanding teachers for Chinese universities. Further contributions include the PyFIHPWG operator [26], employed in MADM for selecting express companies. Several research articles [27, 30–32] have utilized Dombi operational rules, proposing hybrid MADM methods and the EDAS method on bipolar fuzzy environments. Paul et al. addressed the drawbacks of existing MADM methods by introducing an advanced Pythagorean fuzzy weighted geometric operator [18]. The ensemble approach of MADM by Paul et al. [33] combined fuzzy AHP and gray relational analysis, applied in finding the best portfolio in Indian mutual funds based on customer requirements. In another study [34], VIKOR and DEMATEL methods were hybridized to identify the optimal geological sequestration option for CO_2 storage. The application of Pythagorean fuzzy Hamacher AOs to MADM methods was proposed by Paul et al. [35], while municipal solid management in a cubic Pythagorean fuzzy environment using the EDAS method was addressed in [36]. Again Paul et al. [37] introduced the Pythagorean fuzzy fuzzy Einstein interactive power AO for selecting the best cement industry in India, aiming to eliminate decision makers' personal biases. Jana et al. [38] used MABAC method as MCDM in Pythagorean fuzzy environment. The aforementioned aggregation operators (AOs) and their associated multi-attribute decision-making (MADM) methods exhibit certain limitations in specific scenarios. Notably, the PyFDWG operator [19] produces indeterminate outcomes when both the membership component of a Pythagorean fuzzy number (PyFN) and its corresponding weight component are zero. Similarly, the PyFHPWG operator [20] yields indeterminate results when a PyFN contains 0 as the membership degree (MD) and 1 as the non-membership degree (NMD), along with its corresponding weight component being 0. The SPyFWG operator [21] exhibits indeterminacy when a weight component becomes zero, and the corresponding MD is also zero. Furthermore, the PyFHPrWG operator [25] delivers an indeterminate result under specific conditions, namely when the MD of a PyFN is zero, and both the weight component and its associated specific weight component are zero. Likewise, the PyFIHPWG operator [26] is susceptible to indeterminate outcomes. This occurs when the indeterminacy degree of a PyFN is 1, and its weight component is 0. Addressing these limitations is crucial for enhancing the reliability and applicability of these AOs in various decision-making scenarios.

This chapter introduces a novel aggregation operator, namely the APyFHWG, and develops a multi-attribute decision-making (MADM) method utilizing this operator. The proposed AO aims to overcome the limitations identified in the existing aggregation operators discussed within this chapter. Moreover, the MADM method incorporating the newly developed AO is designed to address the drawbacks associated with MADM methods utilizing the existing AOs discussed in this chapter.

The organization of the chapter is structured as follows: Sect. 2 provides an in-depth exploration of preliminary notions. Section 3 meticulously discusses the drawbacks inherent in the existing AOs considered throughout this chapter. In Sect. 4, the novel aggregation operator is introduced, highlighting its fundamental properties and demonstrating its superiority over existing AOs. The section also delves into the enhanced performance of the MADM method when employing the proposed AO. Section 5 presents a case study focusing on the green supplier selection methodology within the pharmaceutical industry. The concluding remarks are presented in Sect. 6, summarizing the key findings and contributions of the proposed AO and MADM method. The comprehensive examination and validation of the APyFHWG operator in the context of green supplier selection contribute to the advancement of decision-making methodologies in complex industrial scenarios.

2 Preliminary

This section elucidates foundational concepts essential for a comprehensive grasp of the ensuing chapter.

Definition 1 (PyFS) [39]
Let X be the universe of discourse. A PyFS $\tilde{P}$ on X is characterized as $\tilde{P} = \{\langle x, \varrho_{\tilde{P}}(x), \vartheta_{\tilde{P}}(x)\rangle : 0 \le \varrho^2_{\tilde{P}}(x) + \vartheta^2_{\tilde{P}}(x) \le 1 \;\forall\; x \in X\}$ where, $\varrho_{\tilde{P}} : X \to [0, 1]$ and $\vartheta_{\tilde{P}} : X \to [0, 1]$ are membership function and non-membership function

of $\tilde{P}$. Another function, $\pi_{\tilde{P}}: X \to [0, 1]$ arisen is called the function of hesitation margin or indeterminacy function such that $\pi_{\tilde{P}}(x) = \sqrt{1 - \varrho^2_{\tilde{P}}(x) - \vartheta^2_{\tilde{P}}(x)}$, for all x ∈ X. $\varrho_{\tilde{P}}(x)$ represents the degree of belongingness or membership degree (MD) of x in $\tilde{P}$ for all x in X and $\vartheta_{\tilde{P}}(x)$ represents the degree of non-belongingness or non-membership degree (NMD) of x in $\tilde{P}$ for all x in X. The Pythagorean fuzzy number, $\tilde{P}$ is denoted as $\langle \varrho_{\tilde{P}}, \vartheta_{\tilde{P}} \rangle$ in short.

Peng and Ma [40] proposed a score function for the comparison of two PFNs.

Definition 2 (Score function)
Let $\mathcal{F} = \langle \varrho_{\mathcal{F}}, \vartheta_{\mathcal{F}} \rangle$ is a PFN in Φ. The novel score function is defined as follows:

$$\mathfrak{s}(\mathcal{F}) = (\varrho^2_{\mathcal{F}} - \vartheta^2_{\mathcal{F}})(1 + \pi^2_{\mathcal{F}})$$

Definition 3 (Comparison rule)
Let $\mathcal{F}_1 = \langle \varrho_{\mathcal{F}_1}, \vartheta_{\mathcal{F}_1}, \pi_{\mathcal{F}_1} \rangle$, $\mathcal{F}_2 = \langle \varrho_{\mathcal{F}_2}, \vartheta_{\mathcal{F}_2}, \pi_{\mathcal{F}_2} \rangle$ are two PyFNs.
If $\mathfrak{s}(\mathcal{F}_1) > \mathfrak{s}(\mathcal{F}_2)$ then $\mathcal{F}_1 > \mathcal{F}_2$.
If $\mathfrak{s}(\mathcal{F}_1) < \mathfrak{s}(\mathcal{F}_2)$ then $\mathcal{F}_1 < \mathcal{F}_2$.
If $\mathfrak{s}(\mathcal{F}_1) = \mathfrak{s}(\mathcal{F}_2)$ then
if $\pi_{\mathcal{F}_1} < \pi_{\mathcal{F}_2}$ then $\mathcal{F}_1 > \mathcal{F}_2$,
if $\pi_{\mathcal{F}_1} > \pi_{\mathcal{F}_2}$ then $\mathcal{F}_1 < \mathcal{F}_2$,
if $\pi_{\mathcal{F}_1} = \pi_{\mathcal{F}_2}$ then $\mathcal{F}_1 = \mathcal{F}_2$.

3 Drawbacks of the Existing AOs

This section critically examines the limitations of five existing aggregation operators (AOs) within the context of this research. The operators under scrutiny include the PyFDWG [19] operator, PyFHPWG [20] operator, SPyFWG [21] operator, PyFH-PrWG [25] operator, and PyFIHPWG [26] operator.

3.1 Drawbacks of PyFDWG Operator

Definition 4 (PyFDWG) [19]
Let $P = \{\tilde{\alpha_\hbar} = \langle \varrho_{\tilde{\alpha_\hbar}}, \vartheta_{\tilde{\alpha_\hbar}} \rangle : \hbar = 1, 2, ..., n\}$ be the set of Pythagorean fuzzy numbers in PyFN(X). PyFDWG operator is a mapping $PyFDWG_\epsilon : P^n \to P$ which is defined below.

$$PyFDWG_{\epsilon}(\tilde{\alpha_1}, \tilde{\alpha_2}, ..., \tilde{\alpha_n}) = \bigotimes_{\hbar=1}^{n}(\tilde{\alpha_\hbar})^{\epsilon_\hbar},$$

$\left\langle \sqrt{\frac{1}{1+\left\{\sum_{\hbar=1}^{n}\epsilon_\hbar\left(\frac{1-\varrho_\hbar^2}{\varrho_\hbar^2}\right)^k\right\}^{\frac{1}{k}}}}, \sqrt{1-\frac{1}{1+\left\{\sum_{\hbar=1}^{n}\epsilon_\hbar\left(\frac{\vartheta_\hbar^2}{1-\vartheta_\hbar^2}\right)^k\right\}^{\frac{1}{k}}}}\right\rangle$, where $\epsilon = (\epsilon_1, \epsilon_2, ..., \epsilon_n)^T$ is a weight vector such that $\epsilon_\hbar \in [0, 1], \hbar = 1, 2, ..., n$ and $\sum_{\hbar=1}^{n}\epsilon_\hbar = 1, k \geq 1$.

The drawbacks of this operator is that it is too sensitive to the MD '0'. The instance is shown in Example 1.

Example 1 Let $\tilde{\alpha_1} = \langle 0.0, 0.7\rangle$, $\tilde{\alpha_2} = \langle 0.5, 0.6\rangle$ and $\tilde{\alpha_3} = \langle 0.7, 0.3\rangle$ with the weight vector $\epsilon = (0.0, 0.4, 0.6)^T$. Therefore $PyFDWG_{\epsilon}(\tilde{\alpha_1}, \tilde{\alpha_2}) =$ $\left\langle \sqrt{\frac{1}{1+\left\{\sum_{\hbar=1}^{3}\epsilon_\hbar\left(\frac{1-\varrho_\hbar^2}{\varrho_\hbar^2}\right)^k\right\}^{\frac{1}{k}}}}, \sqrt{1-\frac{1}{1+\left\{\sum_{\hbar=1}^{3}\epsilon_\hbar\left(\frac{\vartheta_\hbar^2}{1-\vartheta_\hbar^2}\right)^k\right\}^{\frac{1}{k}}}}\right\rangle$. In this case we will have an indeterminate form gets appear '$\frac{0}{0}$' during the aggregation process.

3.2 Drawbacks of PyFHPWG Operator

Definition 5 (Normalized Hamming distance) ch20wei2019pythagorean
Let $p_{\iota\hbar} = \langle\varrho_{\iota\hbar}, \vartheta_{\iota\hbar}\rangle$ and $p_{\iota k} = \langle\varrho_{\iota k}, \vartheta_{\iota k}\rangle$ are two PyFNs. The normalized Hamming distance (d) is defined as

$$d(p_{\iota\hbar}, p_{\iota k}) = \frac{1}{2}(|\varrho_{\iota\hbar}^2 - \varrho_{\iota k}^2| + |\vartheta_{\iota\hbar}^2 - \vartheta_{\iota k}^2|).$$

Definition 6 (Total weighted support) [20]
The total weighted support $T(p_{\iota\hbar}) = \sum_{k=1,k\neq\hbar}^{n}\epsilon_\hbar supp(p_{\iota\hbar}, p_{\iota k})$, $k = 1, ..., n$ and $\hbar = 1, ..., n$, $k \neq \hbar$, where $supp(p_{\iota\hbar}, p_{\iota k}) = 1 - d(p_{\iota\hbar}, p_{\iota k})$ satisfying the conditions:
(i) $supp(p_{\iota\hbar}) \in [0, 1]$, (ii) $supp(p_{\iota\hbar}, p_{\iota k}) = supp(p_{\iota k}, p_{\iota\hbar})$ and (iii) $supp(p_{\iota\hbar}, p_{kl})$ $\geq supp(p_{qr}, p_{st})$ if $d(p_{\iota\hbar}, p_{kl}) \leq d(p_{qr}, p_{st})$.

Definition 7 (PyFHPWG operator) [20]
Let $P = \{\tilde{\alpha_\hbar} = \langle\varrho_{\tilde{\alpha_\hbar}}, \vartheta_{\tilde{\alpha_\hbar}}\rangle : \hbar = 1, 2, ..., n\}$ be the set of PyFNs in PyFN(X). Then PyFHPWG operator is a mapping $PyFHPWG_{\epsilon}: P^n \rightarrow P$ which is defined below:

$$PyFHPWG_{\epsilon}(\tilde{\alpha_1}, \tilde{\alpha_2}, ..., \tilde{\alpha_n}) = \bigotimes_{\hbar=1}^{n}(\tilde{\alpha_\hbar})^{\frac{\epsilon_\hbar(1+T(\alpha_\hbar))}{\sum_{\hbar=1}^{n}\epsilon_\hbar(1+T(\alpha_\hbar))}} =$$

$$\left\langle \frac{\sqrt{\lambda}\prod_{\hbar=1}^{n}\varrho_{\tilde{\alpha_\hbar}}^{\frac{\epsilon_\hbar(1+T(\alpha_\hbar))}{\sum_{\hbar=1}^{n}\epsilon_\hbar(1+T(\alpha_\hbar))}}}{\sqrt{\prod_{\hbar=1}^{n}\{1+(\lambda-1)(1-\varrho_{\tilde{\alpha_\hbar}}^2)\}^{\frac{\epsilon_\hbar(1+T(\alpha_\hbar))}{\sum_{\hbar=1}^{n}\epsilon_\hbar(1+T(\alpha_\hbar))}}+(\lambda-1)\prod_{\hbar=1}^{n}\varrho_{\tilde{\alpha_\hbar}}^{\frac{2\epsilon_\hbar(1+T(\alpha_\hbar))}{\sum_{\hbar=1}^{n}\epsilon_\hbar(1+T(\alpha_\hbar))}}}}, \sqrt{\frac{\prod_{\hbar=1}^{n}\{1+(\lambda-1)\vartheta_{\tilde{\alpha_\hbar}}^2\}^{\frac{\epsilon_\hbar(1+T(\alpha_\hbar))}{\sum_{\hbar=1}^{n}\epsilon_\hbar(1+T(\alpha_\hbar))}}-\prod_{\hbar=1}^{n}(1-\vartheta_{\tilde{\alpha_\hbar}}^2)^{\frac{\epsilon_\hbar(1+T(\alpha_\hbar))}{\sum_{\hbar=1}^{n}\epsilon_\hbar(1+T(\alpha_\hbar))}}}{\prod_{\hbar=1}^{n}\{1+(\lambda-1)\vartheta_{\tilde{\alpha_\hbar}}^2\}^{\frac{\epsilon_\hbar(1+T(\alpha_\hbar))}{\sum_{\hbar=1}^{n}\epsilon_\hbar(1+T(\alpha_\hbar))}}+(\lambda-1)\prod_{\hbar=1}^{n}(1-\vartheta_{\tilde{\alpha_\hbar}}^2)^{\frac{\epsilon_\hbar(1+T(\alpha_\hbar))}{\sum_{\hbar=1}^{n}\epsilon_\hbar(1+T(\alpha_\hbar))}}}} \right\rangle,$$

where $\epsilon = (\epsilon_1, \epsilon_2, ..., \epsilon_n)^T$ is a weight vector such that $\epsilon_\hbar \in [0, 1], \hbar = 1, 2, ..., n$ and $\sum_{\hbar=1}^{n} \epsilon_\hbar = 1$ and $T(\alpha_\hbar) = \sum_{k=1,k\neq\hbar}^{n} \epsilon_\hbar supp(\alpha_\hbar, \alpha_k), k = 1, ..., n$.

The drawbacks of this operator are it is too sensitive to the MD '0'. The instance is shown in Example 2.

Example 2 Let $\tilde{\alpha_1} = \langle 0.0, 0.5\rangle$, $\tilde{\alpha_2} = \langle 0.6, 0.7\rangle$ and $\tilde{\alpha_3} = \langle 0.5, 0.4\rangle$ with the weight vector $\epsilon = (0.0, 0.5, 0.5)^T$ then $PyFHPWG(\tilde{\alpha_1}, \tilde{\alpha_2}) =$

$$\left\langle \frac{\sqrt{\lambda}\prod_{\hbar=1}^{3}\varrho_{\tilde{\alpha_\hbar}}^{\frac{\epsilon_\hbar(1+T(\alpha_\hbar))}{\sum_{\hbar=1}^{n}\epsilon_\hbar(1+T(\alpha_\hbar))}}}{\sqrt{\prod_{\hbar=1}^{3}\{1+(\lambda-1)(1-\varrho_{\tilde{\alpha_\hbar}}^2)\}^{\frac{\epsilon_\hbar(1+T(\alpha_\hbar))}{\sum_{\hbar=1}^{3}\epsilon_\hbar(1+T(\alpha_\hbar))}}+(\lambda-1)\prod_{\hbar=1}^{3}\varrho_{\tilde{\alpha_\hbar}}^{\frac{2\epsilon_\hbar(1+T(\alpha_\hbar))}{\sum_{\hbar=1}^{3}\epsilon_\hbar(1+T(\alpha_\hbar))}}}}, \sqrt{\frac{\prod_{\hbar=1}^{3}\{1+(\lambda-1)\vartheta_{\tilde{\alpha_\hbar}}^2\}^{\frac{\epsilon_\hbar(1+T(\alpha_\hbar))}{\sum_{\hbar=1}^{3}\epsilon_\hbar(1+T(\alpha_\hbar))}}-\prod_{\hbar=1}^{3}(1-\vartheta_{\tilde{\alpha_\hbar}}^2)^{\frac{\epsilon_\hbar(1+T(\alpha_\hbar))}{\sum_{\hbar=1}^{3}\epsilon_\hbar(1+T(\alpha_\hbar))}}}{\prod_{\hbar=1}^{3}\{1+(\lambda-1)\vartheta_{\tilde{\alpha_\hbar}}^2\}^{\frac{\epsilon_\hbar(1+T(\alpha_\hbar))}{\sum_{\hbar=1}^{3}\epsilon_\hbar(1+T(\alpha_\hbar))}}+(\lambda-1)\prod_{\hbar=1}^{3}(1-\vartheta_{\tilde{\alpha_\hbar}}^2)^{\frac{\epsilon_\hbar(1+T(\alpha_\hbar))}{\sum_{\hbar=1}^{3}\epsilon_\hbar(1+T(\alpha_\hbar))}}}} \right\rangle.$$

During the aggregation process we will have an indeterminate form '0^0'.

3.3 Drawbacks of SPyFWG Operator

Definition 8 (SPyFWG)[21]
Let $P = \{\tilde{\alpha_\hbar} = \langle \varrho_{\tilde{\alpha_\hbar}}, \vartheta_{\tilde{\alpha_\hbar}}\rangle : \hbar = 1, 2, ..., n\}$ be the set of PyFNs in PyFN(X). Then SPyFWG operator is a mapping $SPyFWG_\epsilon : P^n \to P$ which is defined below $SPyFWG_\epsilon(\tilde{\alpha_1}, \tilde{\alpha_2}, ..., \tilde{\alpha_n}) =$

$$\left\langle \frac{\prod_{\hbar=1}^{n}\varrho_{\tilde{\alpha_\hbar}}^{\epsilon_\hbar}}{\sqrt{\prod_{\hbar=1}^{n}(1-\varrho_{\tilde{\alpha_\hbar}}^2)^{\epsilon_\hbar}+\prod_{\hbar=1}^{n}\varrho_{\tilde{\alpha_\hbar}}^{2\epsilon_\hbar}}}, \frac{\prod_{\hbar=1}^{n}\vartheta_{\tilde{\alpha_\hbar}}^{\epsilon_\hbar}}{\sqrt{\prod_{\hbar=1}^{n}(1-\vartheta_{\tilde{\alpha_\hbar}}^2)^{\epsilon_\hbar}+\prod_{\hbar=1}^{n}\vartheta_{\tilde{\alpha_\hbar}}^{2\epsilon_\hbar}}} \right\rangle,$$

where $\epsilon = (\epsilon_1, \epsilon_2, ..., \epsilon_n)^T$ is a weight vector such that $\epsilon_\hbar \in [0, 1], \hbar = 1, 2, ..., n$ and $\sum_{\hbar=1}^{n} \epsilon_\hbar = 1$.

The drawback of this operator is it is too sensitive to the MD ‘0’ and NMD ‘0’. The instances are shown in Examples 3 and 4 respectively.

Example 3 Let $\tilde{\alpha_1} = \langle 0.0, 0.5 \rangle$, $\tilde{\alpha_2} = \langle 0.6, 0.7 \rangle$ and $\tilde{\alpha_3} = \langle 0.4, 0.5 \rangle$ with the weight vector $\epsilon = (0.0, 0.5, 0.5)^T$ then $SPyFWG(\alpha_1, \alpha_2) = \left\langle \frac{\prod_{\hbar=1}^{3} \varrho_{\tilde{\alpha_\hbar}}^{\epsilon_\hbar}}{\sqrt{\prod_{\hbar=1}^{3} (1 - \varrho_{\tilde{\alpha_\hbar}}^2)^{\epsilon_\hbar} + \prod_{\hbar=1}^{3} \varrho_{\tilde{\alpha_\hbar}}^{2\epsilon_\hbar}}}, \frac{\prod_{\hbar=1}^{3} \vartheta_{\tilde{\alpha_\hbar}}^{\epsilon_\hbar}}{\sqrt{\prod_{\hbar=1}^{3} (1 - \vartheta_{\tilde{\alpha_\hbar}}^2)^{\epsilon_\hbar} + \prod_{\hbar=1}^{3} \vartheta_{\tilde{\alpha_\hbar}}^{2\epsilon_\hbar}}} \right\rangle$. In this case an indeterminate form 0^0 appears during the aggregation process.

Example 4 Let $\tilde{\alpha_1} = \langle 0.5, 0.0 \rangle$, $\tilde{\alpha_2} = \langle 0.6, 0.7 \rangle$ and $\tilde{\alpha_3} = \langle 0.5, 0.4 \rangle$ with the weight vector $\epsilon = (0.0, 0.5, 0.5)^T$ then $SPyFWG(\alpha_1, \alpha_2) = \left\langle \frac{\prod_{\hbar=1}^{2} \varrho_{\tilde{\alpha_\hbar}}^{\epsilon_\hbar}}{\sqrt{\prod_{\hbar=1}^{2} (1 - \varrho_{\tilde{\alpha_\hbar}}^2)^{\epsilon_\hbar} + \prod_{\hbar=1}^{2} \varrho_{\tilde{\alpha_\hbar}}^{2\epsilon_\hbar}}}, \frac{\prod_{\hbar=1}^{2} \vartheta_{\tilde{\alpha_\hbar}}^{\epsilon_\hbar}}{\sqrt{\prod_{\hbar=1}^{2} (1 - \vartheta_{\tilde{\alpha_\hbar}}^2)^{\epsilon_\hbar} + \prod_{\hbar=1}^{2} \vartheta_{\tilde{\alpha_\hbar}}^{2\epsilon_\hbar}}} \right\rangle$. In this case an indeterminate form 0^0 appears during the aggregation process.

3.4 PyFHPrWG Operator

Definition 9 (Score number) [25]
Let $p_{\iota\hbar} = \langle \varrho_{\iota\hbar}, \vartheta_{\iota\hbar} \rangle$ is a PyFN. The score number of $p_{\iota\hbar}$ is defined as

$$s(p_{\iota\hbar}) = \frac{1 + \varrho_{\iota\hbar}^2 - \vartheta_{\iota\hbar}^2}{2}$$

Definition 10 (PyFHPrWG) [25]
Let $P = \{\tilde{\alpha_\hbar} = \langle \varrho_{\tilde{\alpha_\hbar}}, \vartheta_{\tilde{\alpha_\hbar}} \rangle : \hbar = 1, 2, ..., n\}$ be the set of PyFNs in PyFN(X). Then PyFHPrWG operator is a mapping $PyFHPrWG_\epsilon : P^n \to P$ which is defined below

$$PyFHPrWG_{\epsilon}(\tilde{\alpha_1}, \tilde{\alpha_2}, ..., \tilde{\alpha_n}) = \bigotimes_{\hbar=1}^{n} (\tilde{\alpha_\hbar})^{\frac{\epsilon_\hbar T_\hbar}{\sum_{\hbar=1}^{n} \epsilon_\hbar T_\hbar}} =$$

$$\left\langle \frac{\sqrt{\lambda}\prod_{\hbar=1}^{n} \varrho_{\tilde{\alpha_\hbar}}^{\frac{\epsilon_\hbar T_\hbar}{\sum_{\hbar=1}^{n}\epsilon_\hbar T_\hbar}}}{\sqrt{\prod_{\hbar=1}^{n}\{1+(\lambda-1)(1-\varrho_{\tilde{\alpha_\hbar}}^2)\}^{\frac{\epsilon_\hbar T_\hbar}{\sum_{\hbar=1}^{n}\epsilon_\hbar T_\hbar}} + (\lambda-1)\prod_{\hbar=1}^{n}\varrho_{\tilde{\alpha_\hbar}}^{\frac{2\epsilon_\hbar T_\hbar}{\sum_{\hbar=1}^{n}\epsilon_\hbar T_\hbar}}}}, \sqrt{\frac{\prod_{\hbar=1}^{n}\{1+(\lambda-1)\vartheta_{\tilde{\alpha_\hbar}}^2\}^{\frac{\epsilon_\hbar T_\hbar}{\sum_{\hbar=1}^{n}\epsilon_\hbar T_\hbar}} - \prod_{\hbar=1}^{n}(1-\vartheta_{\tilde{\alpha_\hbar}}^2)^{\frac{\epsilon_\hbar T_\hbar}{\sum_{\hbar=1}^{n}\epsilon_\hbar T_\hbar}}}{\prod_{\hbar=1}^{n}\{1+(\lambda-1)\vartheta_{\tilde{\alpha_\hbar}}^2\}^{\frac{\epsilon_\hbar T_\hbar}{\sum_{\hbar=1}^{n}\epsilon_\hbar T_\hbar}} + (\lambda-1)\prod_{\hbar=1}^{n}(1-\vartheta_{\tilde{\alpha_\hbar}}^2)^{\frac{\epsilon_\hbar T_\hbar}{\sum_{\hbar=1}^{n}\epsilon_\hbar T_\hbar}}}} \right\rangle,$$

where $\epsilon = (\epsilon_1, \epsilon_2, ..., \epsilon_n)^T$ is a weight vector such that $\epsilon_\hbar \in [0, 1]$, $\hbar = 1, 2, ..., n$ and $\sum_{\hbar=1}^{n} \epsilon_\hbar = 1$. $T_\hbar = \prod_{k=1}^{\hbar-1} s(\alpha_k)$, $\hbar = 2, ..., n$ and $T_1 = 1$.

This operator exhibits a notable drawback in its sensitivity to the membership degree (MD) being '0'. This issue is demonstrated in Example 5.

Example 5 Let $\alpha_1 = \langle 0.0, 1.0\rangle, \alpha_2 = \langle 0.5, 0.7\rangle$ and $\alpha_3 = \langle 0.6, 0.2\rangle$ with the weight vector $\epsilon = (0.0, 0.4, 0.6)^T$. In this case $T_1 = 1$, $T_2 = \prod_{\hbar=1}^{1} s(\alpha_\hbar) = 0.0$ and $T_3 = \prod_{\hbar=1}^{2} s(\alpha_\hbar) = 0.0$. Therefore $\frac{\epsilon_1 T_1}{\epsilon_1 T_1 + \epsilon_2 T_2 + \epsilon_3 T_3} = \frac{0}{0}$, an indeterminate form.

3.5 PyFIHPWG Operator

Definition 11 [26]
The distance d^w between PyFNs $p_{\iota\hbar} = \langle \varrho_{\iota\hbar}, \vartheta_{\iota\hbar}\rangle$ and $p_{ik} = \langle \varrho_{ik}, \vartheta_{ik}\rangle$ is defined as:

$$d^w(p_{\iota\hbar}, p_{ik}) = \frac{1}{2}(|\varrho_{\iota\hbar}^2 - \varrho_{ik}^2| + |\vartheta_{\iota\hbar}^2 - \vartheta_{ik}^2| + |\pi_{\iota\hbar}^2 - \pi_{ik}^2|).$$

Definition 12 (Total support) [26]
The total support $T^w(p_{\iota\hbar}) = \sum_{k=1,k\neq\hbar}^{n} supp^w(p_{\iota\hbar}, p_{ik})$, $k = 1, ..., n$ and $\hbar = 1, ..., n, k \neq \hbar$, where $supp^w(p_{\iota\hbar}, p_{ik}) = 1 - d^w(p_{\iota\hbar}, p_{ik})$ satisfying the conditions:
(i) $supp^w(p_{\iota\hbar}) \in [0, 1]$, (ii) $supp^w(p_{\iota\hbar}, p_{ik}) = supp^w(p_{ik}, p_{\iota\hbar})$ and (iii) $supp^w(p_{\iota\hbar}, p_{kl}) \geq supp^w(p_{qr}, p_{st})$ if $d^w(p_{\iota\hbar}, p_{kl}) \leq d^w(p_{qr}, p_{st})$.

Definition 13 (PyFIHPWG) [26] Let $P = \{\tilde{\alpha_\hbar} = \langle \varrho_{\tilde{\alpha_\hbar}}, \vartheta_{\tilde{\alpha_\hbar}}\rangle : \hbar = 1, 2, ..., n\}$ be the set of PyFNs in PyFN(X). Then PyFHPWG operator is a mapping $PyFHPWG_\epsilon : P^n \to P$ which is defined below.

$$PyFHPWG_{\epsilon}(\tilde{\alpha_1}, \tilde{\alpha_2}, ..., \tilde{\alpha_n}) = \bigotimes_{\hbar=1}^{n} (\tilde{\alpha_\hbar})^{\frac{\epsilon_\hbar(1+T^w(\alpha_\hbar))}{\sum_{\hbar=1}^{n}\epsilon_\hbar(1+T^w(\alpha_\hbar))}} =$$

$$\left\langle \sqrt{\frac{\lambda\prod_{\hbar=1}^{n}(1-\vartheta_{\tilde{\alpha_\hbar}}^2)^{\frac{\epsilon_\hbar(1+T^w(\alpha_\hbar))}{\sum_{\hbar=1}^{n}\epsilon_\hbar(1+T^w(\alpha_\hbar))}} - \lambda\prod_{\hbar=1}^{n}(1-\varrho_{\tilde{\alpha_\hbar}}^2-\vartheta_{\tilde{\alpha_\hbar}}^2)^{\frac{\epsilon_\hbar(1+T^w(\alpha_\hbar))}{\sum_{\hbar=1}^{n}\epsilon_\hbar(1+T^w(\alpha_\hbar))}}}{\prod_{\hbar=1}^{n}\{1+(\lambda-1)\vartheta_{\tilde{\alpha_\hbar}}^2\}^{\frac{\epsilon_\hbar(1+T^w(\alpha_\hbar))}{\sum_{\hbar=1}^{n}\epsilon_\hbar(1+T^w(\alpha_\hbar))}} + (\lambda-1)\prod_{\hbar=1}^{n}(1-\vartheta_{\tilde{\alpha_\hbar}}^2)^{\frac{\epsilon_\hbar(1+T^w(\alpha_\hbar))}{\sum_{\hbar=1}^{n}\epsilon_\hbar(1+T^w(\alpha_\hbar))}}}}, \right.$$

$$\left. \sqrt{\frac{\prod_{\hbar=1}^{n}\{1+(\lambda-1)\vartheta_{\tilde{\alpha_\hbar}}^2\}^{\frac{\epsilon_\hbar(1+T^w(\alpha_\hbar))}{\sum_{\hbar=1}^{n}\epsilon_\hbar(1+T^w(\alpha_\hbar))}} - \prod_{\hbar=1}^{n}(1-\vartheta_{\tilde{\alpha_\hbar}}^2)^{\frac{\epsilon_\hbar(1+T^w(\alpha_\hbar))}{\sum_{\hbar=1}^{n}\epsilon_\hbar(1+T^w(\alpha_\hbar))}}}{\prod_{\hbar=1}^{n}\{1+(\lambda-1)\vartheta_{\tilde{\alpha_\hbar}}^2\}^{\frac{\epsilon_\hbar(1+T^w(\alpha_\hbar))}{\sum_{\hbar=1}^{n}\epsilon_\hbar(1+T^w(\alpha_\hbar))}} + (\lambda-1)\prod_{\hbar=1}^{n}(1-\vartheta_{\tilde{\alpha_\hbar}}^2)^{\frac{\epsilon_\hbar(1+T^w(\alpha_\hbar))}{\sum_{\hbar=1}^{n}\epsilon_\hbar(1+T^w(\alpha_\hbar))}}}} \right\rangle,$$

where $\epsilon = (\epsilon_1, \epsilon_2, ..., \epsilon_n)^T$ is a weight vector such that $\epsilon_\hbar \in [0, 1], \hbar = 1, 2, ..., n$ and $\sum_{\hbar=1}^{n} \epsilon_\hbar = 1$. $T^w(\alpha_\hbar) = \sum_{k=1,k\neq\hbar}^{n} supp^w(\alpha_\hbar, \alpha_k), \hbar = 1, ..., n.$

This operator exhibits drawbacks, particularly its heightened sensitivity to a non-membership degree (NMD) of '1' and a hesitation margin of '0'. These issues are demonstrated in Examples 6 and 7, respectively.

Example 6 Let $\alpha_1 = \langle 0, 1\rangle$ and $\alpha_2 = \langle 0.3, 0.7\rangle$ with the weight vector $(0, 1)^T$. In this case we get 0^0, an indeterminate form using Definition 13.

Example 7 Let $\alpha_1 = \langle 0.6, 0.8\rangle$ and $\alpha_2 = \langle 0.3, 0.7\rangle$ with the weight vector $(0, 1)^T$. In this case we get 0^0, an indeterminate form using Definition 13.

4 APyFHWG Operator

Definition 14 Let $P = \{\tilde{\alpha_\hbar} = \langle \varrho_{\tilde{\alpha_\hbar}}, \vartheta_{\tilde{\alpha_\hbar}}\rangle : \hbar = 1, 2, ..., n\}$ be the set of PyFNs in PyFN(X). Then AdPyFHWG operator is a mapping $AdPyFHWG_\epsilon : P^n \to P$ which is defined below. $AdPyFHWG(\alpha_1, \alpha_2, ..., \alpha_n) =$

$$\left\langle \sqrt{\frac{k\left[1-\frac{1}{\lambda}\left\{1-\prod_{\hbar=1}^{n}\left(1-\lambda(1-\varrho_\hbar^2)\right)^{\epsilon_\hbar}\right\}\right]}{\prod_{\hbar=1}^{n}\left\{1+(k-1)(1-\varrho_\hbar^2)\right\}^{\epsilon_\hbar} + (k-1)\left[1-\frac{1}{\lambda}\left\{1-\prod_{\hbar=1}^{n}\left(1-\lambda(1-\varrho_\hbar^2)\right)^{\epsilon_\hbar}\right\}\right]}}, \right.$$

$$\left. \sqrt{\frac{\prod_{\hbar=1}^{n}\left\{1+(k-1)\vartheta_\hbar^2\right\}^{\epsilon_\hbar} - \left[1-\frac{1}{\lambda}\left\{1-\prod_{\hbar=1}^{n}(1-\lambda\vartheta_\hbar^2)^{\epsilon_\hbar}\right\}\right]}{\prod_{\hbar=1}^{n}\left\{1+(k-1)\vartheta_\hbar^2\right\}^{\epsilon_\hbar} + (k-1)\left[1-\frac{1}{\lambda}\left\{1-\prod_{\hbar=1}^{n}(1-\lambda\vartheta_\hbar^2)^{\epsilon_\hbar}\right\}\right]}} \right\rangle, \quad \lambda \in (0, 1)$$

and $k > 0$.

4.1 Fundamental Properties of the Proposed AO

Some fundamental properties like idempotency, boundedness, monotonicity are discussed in this section.

Definition 15 (Idempotency)
Let $\tilde{\alpha_1}, \tilde{\alpha_2}, ..., \tilde{\alpha_n}$ are PyFNs such that $\tilde{\alpha_\hbar} = \langle \varrho_\hbar, \vartheta_\hbar \rangle$, $\hbar = 1, 2, ..., n$. If $\tilde{\alpha_\hbar} = \tilde{\alpha}$ for all $\hbar = 1, 2, ..., n$ then
$AdPyFHWG(\tilde{\alpha_1}, \tilde{\alpha_2}, ..., \tilde{\alpha_n}) = \tilde{\alpha}$, where $\tilde{\alpha} = \langle \varrho_{\tilde{\alpha}}, \vartheta_{\tilde{\alpha}} \rangle$.

Definition 16 (Boundedness)
Let $\tilde{\alpha_1}, \tilde{\alpha_2}, ..., \tilde{\alpha_n}$ are PyFNs such that $\tilde{\alpha_\hbar} = \langle \varrho_\hbar, \vartheta_\hbar \rangle$, $\hbar = 1, 2, ..., n$. If $\tilde{\alpha}^- = min\{\tilde{\alpha_1}, \tilde{\alpha_2}, ..., \tilde{\alpha_n}\}$ and $\tilde{\alpha}^+ = max\{\tilde{\alpha_1}, \tilde{\alpha_2}, ..., \tilde{\alpha_n}\}$ then $\tilde{\alpha}^- \leq AdPyFHWG(\tilde{\alpha_1}, \tilde{\alpha_2}, ..., \tilde{\alpha_n}) \leq \tilde{\alpha}^+$.

Definition 17 (Monotonicity)
Let $\tilde{\alpha_1}, \tilde{\alpha_2}, ..., \tilde{\alpha_n}$ are PyFNs such that $\tilde{\alpha_\hbar} = \langle \varrho_\hbar, \vartheta_\hbar \rangle$, $\hbar = 1, 2, ..., n$. Let $\{\tilde{\alpha_1}, \tilde{\alpha_2}, ..., \tilde{\alpha_n}\}$ and $\{\tilde{\alpha_1}^*, \tilde{\alpha_2}^*, ..., \tilde{\alpha_n}^*\}$ are two sets of PyFNs such that $\alpha_\hbar \leq \alpha^*_\hbar$ for $\hbar = 1, 2, ..., n$ then $AdPyFHWG(\tilde{\alpha_1}, \tilde{\alpha_2}, ..., \tilde{\alpha_n}) \leq AdPyFHWG(\tilde{\alpha_1}^*, \tilde{\alpha_2}^*, ..., \tilde{\alpha_n}^*)$.

The newly proposed aggregation operator (AO) presented in this research has the capability to address and overcome the limitations identified in the existing AOs discussed in Examples 1, 2, 3, 4, 5, 6, and 7.

4.2 Superiority of the Proposed AO

The application of the proposed aggregation operator yields an aggregated result of $\langle 0.61231, 0.45756 \rangle$ in Example 1 when using $k = 1$ and $\lambda = 0.99$. This demonstrates the superiority of the proposed AO over the PyFDWG, as it transforms an indeterminate case into a determinate outcome under these specific conditions.

In Example 2, utilizing the proposed AO with $k = 1$ and $\lambda = 0.99$ results in an aggregated outcome of $\langle 0.54787, 0.58751 \rangle$. This further emphasizes the superiority of the proposed AO over the PyFHPWG, as it produces a determinate result.

Example 3 yields an aggregated result of $\langle 0.49072, 0.61753 \rangle$ when employing the proposed AO with $k = 1$ and $\lambda = 0.99$.

Similarly, in Example 4, the proposed AO generates an aggregated outcome of $\langle 0.54787, 0.58751 \rangle$ with $k = 1$ and $\lambda = 0.99$. The comparison with Examples 3 and 4 establishes the superior performance of the proposed AO over the SPyFWG.

Example 5 showcases the proposed AO's superiority, producing an aggregated result of $\langle 0.55794, 0.50408 \rangle$ with $k = 1$ and $\lambda = 0.99$, while the PyFHPrWG yields an indeterminate result in this specific case.

In Example 6, the proposed AO results in an aggregated outcome of $\langle 0.34697, 0.56421 \rangle$ with $k = 1$ and $\lambda = 0.99$.

Finally, in Example 7, the proposed AO produces an aggregated result of $\langle 0.34697, 0.61753 \rangle$ with $k = 1$ and $\lambda = 0.99$. The issues highlighted in Examples 6 and 7 underscore the superiority of the proposed AO over the PyFIHPWG operator.

4.3 MADM Method Using AdPyFHWG Operator

In this section, we present the development of an algorithm for a Multi-Attribute Decision-Making (MADM) method utilizing the novel APyFHWG operator. Additionally, we conduct a comparative analysis to demonstrate the superiority of the proposed MADM method using the AdPyFHWG operator in contrast to existing MADM methods considered, namely the MADM method using the PyFDWG operator [19], MADM method using the PyFHPWG operator [20], MADM method using the SPyFWG operator [21], MADM method using the PyFHPrWG operator [25], and MADM method using the PyFIHPWG operator [26].

The algorithm for the MADM method using the proposed AdPyFHWG operator is outlined as follows:

Let $\mathfrak{A} = \mathfrak{A}_1, \mathfrak{A}_2, ..., \mathfrak{A}_m$ represent the set of alternatives, and $\mathcal{C} = \mathcal{C}_1, \mathcal{C}_2, ..., \mathcal{C}_n$ denote the set of attributes. Decision makers (DMs) are required to provide their opinions or information in a Pythagorean fuzzy environment concerning alternative $\mathfrak{A}_i$ $(\iota = 1, 2, ..., m)$ based on predefined attribute $\mathcal{C}_\hbar$ $(\hbar = 1, 2, ..., n)$.

We assume that $\tilde{d_{\iota\hbar}} = \langle \varrho_{\tilde{d_{\iota\hbar}}}, \vartheta_{\tilde{d_{\iota\hbar}}} \rangle$ represents the opinion provided by the decision makers (DMs) regarding alternative $\mathfrak{A}_i$ $(\iota = 1, 2, ..., m)$ based on attribute $\mathcal{C}_\hbar$ $(\hbar = 1, 2, ..., n)$.

Let $D = [\tilde{d_{\iota\hbar}}]_{m\times n}$ be the decision matrix, where each element represents the Pythagorean fuzzy number (PyFN) in aggregated form provided by the DMs. Therefore,

Let $\epsilon = (\epsilon_1, \epsilon_2, ..., \epsilon_n)^T$ represent the weight vector assigned to the attribute vector $\mathcal{C} = (\mathcal{C}_1, \mathcal{C}_2, ..., \mathcal{C}_n)$, where $\epsilon_\hbar$ $(\hbar = 1, 2, ..., n)$ signifies the weight assigned to the attribute $\mathcal{C}_\hbar$ $(\hbar = 1, 2, ..., n)$. The MADM method algorithm utilizing the proposed Advanced Pythagorean Fuzzy Weighted Geometric (APyFWG) operator is outlined as follows.

Step 1: Consider the decision matrix D in Table 1, where $d_{\iota\hbar} = \langle \varrho_{\tilde{d}_{\iota\hbar}}, \vartheta \tilde{d_{\iota\hbar}} \rangle$. Normalize matrix D using the formula in Eq. 1.

Table 1 Decision matrix

$\langle \varrho_{\tilde{d}_{11}}, \vartheta_{\tilde{d}_{11}} \rangle$	$\langle \varrho_{\tilde{d}_{12}}, \vartheta_{\tilde{d}_{12}} \rangle$	...	$\langle \varrho_{\tilde{d}_{1n}}, \vartheta_{\tilde{d}_{1n}} \rangle$
$\langle \varrho_{\tilde{d}_{21}}, \vartheta_{\tilde{d}_{21}} \rangle$	$\langle \varrho_{\tilde{d}_{22}}, \vartheta_{\tilde{d}_{22}} \rangle$	...	$\langle \varrho_{\tilde{d}_{2n}}, \vartheta_{\tilde{d}_{2n}} \rangle$
...	...	...	...
$\langle \varrho_{\tilde{d}_{m1}}, \vartheta_{\tilde{d}_{m1}} \rangle$	$\langle \varrho_{\tilde{d}_{m2}}, \vartheta_{\tilde{d}_{m2}} \rangle$	...	$\langle \varrho_{\tilde{d}_{mn}}, \vartheta_{\tilde{d}_{mn}} \rangle$

$$N\tilde{d_{\iota\hbar}} = \begin{cases} \langle \varrho_{\tilde{d_{\iota\hbar}}}, \vartheta_{\tilde{d_{\iota\hbar}}} \rangle & C_\hbar \; is \; benefit \; type \\ \langle \vartheta_{\tilde{d_{\iota\hbar}}}, \varrho_{\tilde{d_{\iota\hbar}}} \rangle & C_\hbar \; is \; cost \; type \end{cases} \quad (1)$$

Step 2: The PyFNs, $N\tilde{d_{\iota\hbar}} = \langle N\varrho_{\tilde{d_{\iota\hbar}}}, N\vartheta_{\tilde{d_{\iota\hbar}}} \rangle$ $(\iota = 1, 2, ..., m; \hbar = 1, 2, ..., n)$ proposed by the DMs are to be aggregated by the proposed AdPyFHWG operator corresponding to ι-th row of the decision matrix D, where $\iota = 1, 2, ..., m$ based on the weight vector $\epsilon = (\epsilon_1, \epsilon_2, ..., \epsilon_n)^T$ such that $\sum_{\hbar=1}^{n} \epsilon_\hbar = 1$ and $0 \le \epsilon_\hbar \le 1$. Let $\tilde{a_i}$ be the aggregated PyFN obtained due to aggregation of PyFNs of ι-th row of matrix D, where $\iota = 1, 2, ..., m$. Therefore

$N\tilde{d_i} = AdPyFHWG(N\tilde{d_{i1}}, N\tilde{d_{i2}}, ..., N\tilde{d_{in}})$, $\iota = 1, 2, ..., m$, where $AdPyFHWG(N\tilde{d_{i1}}, N\tilde{d_{i2}}, ..., N\tilde{d_{in}}) =$

$$\left\langle \sqrt{\frac{k\left[1 - \frac{1}{\lambda}\left\{1 - \prod_{\hbar=1}^{n} (1 - \lambda(1 - \varrho^2_{N\tilde{d_{\iota\hbar}}}))^{\epsilon_\hbar}\right\}\right]}{\prod_{\hbar=1}^{n} \left\{1 + (k-1)(1 - \varrho^2_{N\tilde{d_{\iota\hbar}}})\right\}^{\epsilon_\hbar} + (k-1)\left[1 - \frac{1}{\lambda}\left\{1 - \prod_{\hbar=1}^{n} (1 - \lambda(1 - \varrho^2_{N\tilde{d_{\iota\hbar}}}))^{\epsilon_\hbar}\right\}\right]}}, \sqrt{\frac{\prod_{\hbar=1}^{n} \left\{1 + (k-1)\vartheta^2_{N\tilde{d_{\iota\hbar}}}\right\}^{\epsilon_\hbar} - \left[1 - \frac{1}{\lambda}\left\{1 - \prod_{\hbar=1}^{n} (1 - \lambda\vartheta^2_{N\tilde{d_{\iota\hbar}}})^{\epsilon_\hbar}\right\}\right]}{\prod_{\hbar=1}^{n} \left\{1 + (k-1)\vartheta^2_{N\tilde{d_{\iota\hbar}}}\right\}^{\epsilon_\hbar} + (k-1)\left[1 - \frac{1}{\lambda}\left\{1 - \prod_{\hbar=1}^{n} (1 - \lambda\vartheta^2_{N\tilde{d_{\iota\hbar}}})^{\epsilon_\hbar}\right\}\right]}} \right\rangle,$$

$\lambda \in (0, 1)$, $k > 0$ and $\iota = 1, 2, ..., m$. In this chapter we have assumed the value of the parameter λ as 0.99.

Step 3: Score values are computed for each aggregated Pythagorean fuzzy number $\tilde{a_i}$, where $\iota = 1, 2, ..., m$, utilizing the formula specified in Definition 2.

Step 4: Applying the comparison technique outlined in Definition 3, we aim to determine the ranking of the Pythagorean fuzzy numbers (PyFNs).

Step 5: End.

4.4 Comparison of the Proposed MADM Method with the Other Existing MADM Methods Considered

The proposed Multi-Attribute Decision-Making (MADM) method is compared with the other existing MADM methods discussed in this chapter. The superiority of the proposed MADM method is demonstrated in Examples 8 and 9.

Example 8 Let the decision matrix D is given in Table 2

The ranks of the alternatives $\mathfrak{A}_1$ and $\mathfrak{A}_2$ in the MADM method using D as decision matrix and $\epsilon = (0.5, 0.5)^T$ as weight vector are shown in the Table 3.

Observing Table 3 in Example 8, it becomes evident that the Multi-Attribute Decision-Making (MADM) methods utilizing the PyFIHPWG operator and the proposed operator share an identical ranking of alternatives. In contrast, other MADM

Table 2 Decision matrix

	$\mathcal{C}_1$	$\mathcal{C}_2$
$\mathfrak{A}_1$:	$\langle 0.0, 0.5\rangle$	$\langle 0.6, 0.5\rangle$
$\mathfrak{A}_2$:	$\langle 0.0, 0.5\rangle$	$\langle 0.7, 0.5\rangle$

Table 3 Comparison table

MADM methods	Aggregated PyFNs	Score value	Rank
MADM method using PyFDWG [19]	$d_1 = \langle 0.0, 0.5\rangle$, $d_2 = \langle 0.0, 0.5\rangle$	$sc(d_1) = -0.25$, $sc(d_2) = -0.25$	$\mathfrak{A}_2 = \mathfrak{A}_1$
MADM method using PyFHPWG operator [20]	$d_1 = \langle 0.0, 0.5\rangle$, $d_2 = \langle 0.0, 0.5\rangle$	$sc(d_1) = 0.375$, $sc(d_2) = 0.375$	$\mathfrak{A}_2 = \mathfrak{A}_1$
MADM method using SPyFWG operator [21]	$d_1 = \langle 0.0, 0.5\rangle$, $d_2 = \langle 0.0, 0.5\rangle$	$sc(d_1) = -0.25$, $sc(d_2) = -0.25$	$\mathfrak{A}_2 = \mathfrak{A}_1$
MADM method using PyFHPrWG [25]	$d_1 = \langle 0.0, 0.5\rangle$, $d_2 = \langle 0.0, 0.5\rangle$	$sc(d_1) = 0.375$, $sc(d_2) = 0.375$	$\mathfrak{A}_2 = \mathfrak{A}_1$
MADM method using PyFIHPWG [26]	$d_1 = \langle 0.45735, 0.5\rangle$, $d_2 = \langle 0.55535, 0.5\rangle$	$sc(d_1) = -0.04083$, $sc(d_2) = 0.05841$	$\mathfrak{A}_2 > \mathfrak{A}_1$
MADM method using proposed AO, i.e., AdPyFHWG operator ($k = 2, \lambda = 0.99$)	$d_1 = \langle 0.23414, 0.5\rangle$, $d_2 = \langle 0.26037, 0.5\rangle$	$sc(d_1) = -0.33086$, $sc(d_2) = -0.30651$	$\mathfrak{A}_2 > \mathfrak{A}_1$

Table 4 Decision matrix

	$\mathcal{C}_1$	$\mathcal{C}_2$
$\mathfrak{A}_1$:	$\langle 0.8, 0.6\rangle$	$\langle 0.5, 0.3\rangle$
$\mathfrak{A}_2$:	$\langle 0.8, 0.6\rangle$	$\langle 0.7, 0.3\rangle$

methods fail to differentiate the ranking order of alternatives. Consequently, the proposed MADM method exhibits superiority over MADM methods utilizing PyFDWG, PyFHPWG, SPyFWG, and PyFHPrWG.

Example 9 Let the decision matrix D is given Table 4.

The ranks of the alternatives $\mathfrak{A}_1$ and $\mathfrak{A}_2$ in the MADM method using D as decision matrix and $\epsilon = (0.5, 0.5)^T$ as weight vector are shown in the Table 5.

Examining Table 5 in Example 9, it is evident that the Multi-Attribute Decision-Making (MADM) methods employing PyFDWG operator, PyFHPWG operator, SPyFWG operator, PyFHPrWG, and the proposed operator share an identical ranking of alternatives. However, the MADM method using PyFIHPWG fails to differentiate the ranking order of alternatives. Consequently, the proposed MADM method demonstrates superiority over MADM methods utilizing PyFIHPWG.

Table 5 Comparison table

MADM methods	Aggregated PyFNs	Score value	Rank
MADM method using PyFDWG [19]	$d_1 = \langle 0.56270, 0.53635\rangle$, $d_2 = \langle 0.73790, 0.53635\rangle$	$sc(d_1) = 0.02896$, $sc(d_2) = 0.25682$	$\mathfrak{A}_2 > \mathfrak{A}_1$
MADM method using PyFHPWG operator [20]	$d_1 = \langle 0.64171, 0.47897\rangle$, $d_2 = \langle 0.74964, 0.47897\rangle$	$sc(d_1) = 0.59119$, $sc(d_2) = 0.66627$	$\mathfrak{A}_2 > \mathfrak{A}_1$
MADM method using SPyFWG operator [21]	$d_1 = \langle 0.65952, 0.43686\rangle$, $d_2 = \langle 0.75268, 0.43686\rangle$	$sc(d_1) = 0.24412$, $sc(d_2) = 0.37567$	$\mathfrak{A}_2 > \mathfrak{A}_1$
MADM method using PyFHPrWG [25]	$d_1 = \langle 0.67545, 0.50891\rangle$, $d_2 = \langle 0.76065, 0.50891\rangle$	$sc(d_1) = 0.59862$, $sc(d_2) = 0.65980$	$\mathfrak{A}_2 > \mathfrak{A}_1$
MADM method using PyFIHPWG [26]	$d_1 = \langle 0.87783, 0.47897\rangle$, $d_2 = \langle 0.87783, 0.47897\rangle$	$sc(d_1) = 0.54118$, $sc(d_2) = 0.54118$	$\mathfrak{A}_2 = \mathfrak{A}_1$
MADM method using proposed AO, i.e., AdPyFHWG operator ($k = 2, \lambda = 0.99$)	$d_1 = \langle 0.64242, 0.47887\rangle$, $d_2 = \langle 0.74968, 0.47887\rangle$	$sc(d_1) = 0.24904$, $sc(d_2) = 0.40213$	$\mathfrak{A}_2 > \mathfrak{A}_1$

5 Case Study on Green Supplier Selection in Pharmaceutical Industry in India

Green supplier selection in the pharmaceutical industry in India involves a strategic approach to identifying and partnering with suppliers who adhere to environmentally sustainable practices. The pharmaceutical sector in India, like globally, recognizes the importance of integrating green criteria into supplier selection processes. Here is an overview of the key aspects:

Regulatory Compliance:
Suppliers in the pharmaceutical industry in India are expected to comply with environmental regulations set by governmental bodies. Green supplier selection involves evaluating whether potential suppliers meet or exceed these regulatory requirements. Green Certification and Standards:

Pharmaceutical companies in India may prioritize suppliers with recognized green certifications such as ISO 14001 (Environmental Management System) or other relevant certifications. Compliance with international environmental standards signals a commitment to sustainability. Environmental Impact Assessment:

Evaluating the environmental impact of suppliers' operations is crucial. This involves assessing factors such as energy consumption, waste generation, and emissions. Suppliers implementing sustainable practices, such as waste reduction and energy efficiency, are preferred. Material Sourcing and Packaging:

Green supplier selection includes assessing the sustainability of raw material sourcing and packaging. Suppliers using eco-friendly materials and packaging that minimizes waste are likely to be preferred. Energy Efficiency:

Pharmaceutical companies in India may prioritize suppliers that demonstrate a commitment to energy efficiency. This can include measures such as the use of renewable energy sources, energy-efficient manufacturing processes, and the implementation of green technologies. Waste Management Practices:

Proper disposal and management of waste, including hazardous waste, are critical considerations. Suppliers implementing effective waste management practices, recycling initiatives, and pollution prevention measures are viewed favorably. Collaboration and Communication:

Green supplier selection involves open communication and collaboration with suppliers. Companies may engage with suppliers to understand and support their efforts toward environmental sustainability. This includes sharing best practices and encouraging continuous improvement. Corporate Social Responsibility (CSR):

Companies in the pharmaceutical industry in India often integrate green supplier selection into their broader Corporate Social Responsibility (CSR) initiatives. Aligning with suppliers who share a commitment to social and environmental responsibility contributes to a positive industry image. Green supplier selection in the pharmaceutical industry in India reflects a broader global trend towards sustainability and responsible business practices. The integration of environmental criteria into supplier selection processes helps drive positive environmental impacts while supporting the industry's growth and development.

The attributes (criteria) based on which the green supplier selection has been made in this chapter are described in Table 6. The vector $\epsilon = (0.15, 0.12, 0.20, 0.15, 0.18, 0.10, 0.10)^T$ is considered as weight vector of the criteria for determining the rank of the alternatives $\mathfrak{A}_\iota$, $\iota = 1, 2, 3, 4, 5$.

The decision matrix D is provided in Table 7 after proper discussion among the decision makers considered in this chapter. The alternatives considered in this chapter are Dr. Reddy's Laboratories Ltd. ($\mathfrak{A}_1$), Biocon Ltd. ($\mathfrak{A}_2$), Sun Pharmaceutical Industries Ltd. ($\mathfrak{A}_3$), Lupin Ltd. ($\mathfrak{A}_4$), Cipla Ltd. ($\mathfrak{A}_5$).

Therefore normalized decision matrix is obtained in Table 8 using Eq. 1.

From Table 9, it is clear that $\mathfrak{A}_5$ possesses the first rank in all the MADM methods including the present MADM method and rest alternatives have the identical ranking orders like as proposed method. Therefore it can be said that the proposed method is feasible and Sun Pharmaceutical Industries Ltd. ($\mathfrak{A}_3$) stands in the first position with respect to the green supplier in pharmaceutical industry in India followed by Dr. Reddy Laboratories Ltd. ($\mathfrak{A}_1$), Cipla Ltd. ($\mathfrak{A}_5$), Biocon Ltd. ($\mathfrak{A}_2$), Lupin Ltd. ($\mathfrak{A}_4$).

Table 6 Criteria description

Criteria	Elaboration of the criteria	Criteria type with weights
Environmental compliance and certification (C_1) [41, 42]	Suppliers should comply with environmental regulations set by relevant authorities. Additionally, having certifications such as ISO 14001 for Environmental Management Systems indicates a commitment to meeting recognized international standards for environmental performance	Benefit (0.15)
Waste management practices (C_2) [43]	Evaluating a supplier's waste management practices is crucial. Suppliers should have effective systems in place to minimize, segregate, and responsibly dispose of waste, including hazardous materials. A commitment to recycling and reducing overall waste generation is favorable	Benefit (0.12)
Supply chain sustainability (C_3) [44]	Examine the overall sustainability of the supplier's supply chain, from raw material sourcing to product delivery. Consider suppliers who prioritize environmentally friendly practices in transportation, packaging, and logistics	Benefit (0.20)
Raw material sourcing and sustainable packaging (C_4) [45]	Suppliers should prioritize the use of environmentally friendly raw materials and packaging. Sustainable sourcing practices and the use of recyclable or biodegradable packaging materials contribute to reducing the environmental impact of the pharmaceutical supply chain	Benefit (0.15)
Product life cycle assessment (C_5) [46]	Analyze the environmental impact of the supplier's products throughout their life cycle, from manufacturing to disposal. Prefer suppliers who invest in sustainable product design, using eco-friendly materials and reducing the overall environmental footprint	Benefit (0.18)
Chemical management and green chemistry (C_6) [47]	Ensure suppliers follow green chemistry principles, minimizing the use of harmful chemicals and prioritizing safer alternatives. Evaluate the transparency of the supplier regarding chemical usage and their commitment to adopting environmentally friendly practices	Benefit (0.10)
Total Cost of Ownership (C_7) [48]	Include the total cost of ownership in the evaluation, considering not only the purchase price but also the long-term costs associated with the product or logistic service. Evaluate factors such as maintenance costs, energy efficiency, waste management costs, and potential regulatory compliance costs to get a comprehensive view of the total cost	Cost (0.10)

Table 7 Decision matrix

	C_1	C_2	C_3	C_4	C_5	C_6	C_7
$\mathfrak{A}_1$	$\langle 0.5, 0.3\rangle$	$\langle 0.6, 0.7\rangle$	$\langle 0.4, 0.1\rangle$	$\langle 0.8, 0.2\rangle$	$\langle 0.7, 0.3\rangle$	$\langle 0.5, 0.6\rangle$	$\langle 0.6, 0.5\rangle$
$\mathfrak{A}_2$	$\langle 0.5, 0.1\rangle$	$\langle 0.4, 0.6\rangle$	$\langle 0.7, 0.2\rangle$	$\langle 0.4, 0.5\rangle$	$\langle 0.5, 0.4\rangle$	$\langle 0.7, 0.3\rangle$	$\langle 0.4, 0.2\rangle$
$\mathfrak{A}_3$	$\langle 0.6, 0.4\rangle$	$\langle 0.5, 0.6\rangle$	$\langle 0.7, 0.5\rangle$	$\langle 0.7, 0.2\rangle$	$\langle 0.6, 0.1\rangle$	$\langle 0.5, 0.2\rangle$	$\langle 0.6, 0.3\rangle$
$\mathfrak{A}_4$	$\langle 0.2, 0.3\rangle$	$\langle 0.5, 0.8\rangle$	$\langle 0.4, 0.6\rangle$	$\langle 0.5, 0.6\rangle$	$\langle 0.7, 0.2\rangle$	$\langle 0.5, 0.1\rangle$	$\langle 0.7, 0.2\rangle$
$\mathfrak{A}_5$	$\langle 0.5, 0.1\rangle$	$\langle 0.6, 0.4\rangle$	$\langle 0.4, 0.3\rangle$	$\langle 0.7, 0.3\rangle$	$\langle 0.4, 0.3\rangle$	$\langle 0.3, 0.4\rangle$	$\langle 0.7, 0.5\rangle$

Table 8 Normalized decision matrix

	C_1	C_2	C_3	C_4	C_5	C_6	C_7
$\mathfrak{A}_1$	$\langle 0.5, 0.3\rangle$	$\langle 0.6, 0.7\rangle$	$\langle 0.4, 0.1\rangle$	$\langle 0.8, 0.2\rangle$	$\langle 0.7, 0.3\rangle$	$\langle 0.5, 0.6\rangle$	$\langle 0.5, 0.6\rangle$
$\mathfrak{A}_2$	$\langle 0.5, 0.1\rangle$	$\langle 0.4, 0.6\rangle$	$\langle 0.7, 0.2\rangle$	$\langle 0.4, 0.5\rangle$	$\langle 0.5, 0.4\rangle$	$\langle 0.7, 0.3\rangle$	$\langle 0.2, 0.4\rangle$
$\mathfrak{A}_3$	$\langle 0.6, 0.4\rangle$	$\langle 0.9, 0.1\rangle$	$\langle 0.7, 0.5\rangle$	$\langle 0.7, 0.2\rangle$	$\langle 0.6, 0.1\rangle$	$\langle 0.5, 0.2\rangle$	$\langle 0.3, 0.6\rangle$
$\mathfrak{A}_4$	$\langle 0.2, 0.3\rangle$	$\langle 0.5, 0.8\rangle$	$\langle 0.4, 0.6\rangle$	$\langle 0.5, 0.6\rangle$	$\langle 0.7, 0.2\rangle$	$\langle 0.5, 0.1\rangle$	$\langle 0.2, 0.7\rangle$
$\mathfrak{A}_5$	$\langle 0.5, 0.1\rangle$	$\langle 0.6, 0.4\rangle$	$\langle 0.4, 0.3\rangle$	$\langle 0.7, 0.3\rangle$	$\langle 0.4, 0.3\rangle$	$\langle 0.3, 0.4\rangle$	$\langle 0.5, 0.7\rangle$

5.1 *Farther Comparative Study*

The superiority of the proposed MADM method is also exhibited in Example 10.

Example 10 Let the decision matrix D' be given in Table 10 and the aggregated results using the existing AOs and the proposed AO are shown in Table 11.

From Example 10, it is clear that the five considered MADM methods can't discriminate the ranking order of alternatives for the specific decision matrix D' in Table 10. But the proposed MADM method can differentiate the raking order of alternatives and $\mathfrak{A}_5$ gets the rank one in this problem. Therefore the proposed method is superior to the other MADM methods considered. The existing MADM methods can't discriminate the ranking order of alternatives due to the presence of PyFN $\langle 0.0, 1.0\rangle$ in every row of decision matrix in Table 10.

6 Conclusion

The evaluation of green supplier selection in India based on the specified criteria has yielded valuable insights into the environmental practices and sustainability efforts of prominent pharmaceutical companies. This chapter introduces a novel AO called APyFHWG operator which can eliminate the drawbacks of the existing AOs namely, PyFDWG operator, PyFHPWG operator, SPyFWG operator, PyFHPrWG operator, and PyFIHPWG operator. The MADM method using the proposed operator can overcome the drawbacks of the MADM methods using the mentioned existing AOs.

Table 9 Aggregated results and rank of alternatives in different methods and proposed method

MADM methods	Aggregated PyFNs	Score value	Rank
MADM method using PyFDWG [19]	$d_1 = \langle 0.49701, 0.54448 \rangle$, $d_2 = \langle 0.32850, 0.45228 \rangle$, $d_3 = \langle 0.46917, 0.48802 \rangle$, $d_4 = \langle 0.27348, 0.65817 \rangle$, $d_5 = \langle 0.41357, 0.49514 \rangle$	$sc(d_1) = -0.04963$, $sc(d_2) = -0.09665$, $sc(d_3) = 0.02367$, $sc(d_4) = -0.35840$, $sc(d_5) = -0.07413$	$\mathfrak{A}_3 > \mathfrak{A}_1 > \mathfrak{A}_5 > \mathfrak{A}_2 > \mathfrak{A}_4$
MADM method using PyFHPWG operator [20]	$d_1 = \langle 0.56093, 0.41978 \rangle$, $d_2 = \langle 0.48131, 0.38240 \rangle$, $d_3 = \langle 0.62012, 0.34987 \rangle$, $d_4 = \langle 0.40816, 0.54050 \rangle$, $d_5 = \langle 0.47094, 0.36496 \rangle$	$sc(d_1) = 0.56921$, $sc(d_2) = 0.54271$, $sc(d_3) = 0.63138$, $sc(d_4) = 0.43723$, $sc(d_5) = 0.54430$	$\mathfrak{A}_3 > \mathfrak{A}_1 > \mathfrak{A}_5 > \mathfrak{A}_2 > \mathfrak{A}_4$
MADM method using SPyFWG operator [21]	$d_1 = \langle 0.57486, 0.30474 \rangle$, $d_2 = \langle 0.49225, 0.30818 \rangle$, $d_3 = \langle 0.63897, 0.25106 \rangle$, $d_4 = \langle 0.42019, 0.42451 \rangle$, $d_5 = \langle 0.47905, 0.30403 \rangle$	$sc(d_1) = 0.23759$, $sc(d_2) = 0.14734$, $sc(d_3) = 0.34525$, $sc(d_4) = -0.00365$, $sc(d_5) = 0.13706$	$\mathfrak{A}_3 > \mathfrak{A}_1 > \mathfrak{A}_2 > \mathfrak{A}_5 > \mathfrak{A}_4$
MADM method using PyFHPrWG [25]	$d_1 = \langle 0.53305, 0.42246 \rangle$, $d_2 = \langle 0.49190, 0.35815 \rangle$, $d_3 = \langle 0.66803, 0.36204 \rangle$, $d_4 = \langle 0.28620, 0.52632 \rangle$, $d_5 = \langle 0.49928, 0.27801 \rangle$	$sc(d_1) = 0.55284$, $sc(d_2) = 0.55685$, $sc(d_3) = 0.65760$, $sc(d_4) = 0.40245$, $sc(d_5) = 0.58600$	$\mathfrak{A}_3 > \mathfrak{A}_5 > \mathfrak{A}_2 > \mathfrak{A}_1 > \mathfrak{A}_4$
MADM method using PyFIHPWG [26]	$d_1 = \langle 0.62022, 0.42254 \rangle$, $d_2 = \langle 0.53061, 0.38362 \rangle$, $d_3 = \langle 0.66877, 0.35053 \rangle$, $d_4 = \langle 0.49899, 0.53860 \rangle$, $d_5 = \langle 0.51585, 0.35838 \rangle$	$sc(d_1) = 0.20614$, $sc(d_2) = 0.13438$, $sc(d_3) = 0.32438$, $sc(d_4) = -0.04110$, $sc(d_5) = 0.13767$	$\mathfrak{A}_3 > \mathfrak{A}_1 > \mathfrak{A}_5 > \mathfrak{A}_2 > \mathfrak{A}_4$
MADM method using proposed AO, i.e., AdPyFHWG operator ($k = 2, \lambda = 0.99$)	$d_1 = \langle 0.56441, 0.41890 \rangle$, $d_2 = \langle 0.48542, 0.38229 \rangle$, $d_3 = \langle 0.62022, 0.35392 \rangle$, $d_4 = \langle 0.41624, 0.53788 \rangle$, $d_5 = \langle 0.47438, 0.37075 \rangle$	$sc(d_1) = 0.21548$, $sc(d_2) = 0.14481$, $sc(d_3) = 0.38654$, $sc(d_4) = -0.17844$, $sc(d_2) = 0.14341$	$\mathfrak{A}_3 > \mathfrak{A}_1 > \mathfrak{A}_2 > \mathfrak{A}_5 > \mathfrak{A}_4$

Table 10 Normalized decision matrix

	$\mathcal{C}_1$	$\mathcal{C}_2$	$\mathcal{C}_3$	$\mathcal{C}_4$	$\mathcal{C}_5$	$\mathcal{C}_6$	$\mathcal{C}_7$
$\mathfrak{A}_1$	$\langle 0.0, 1.0 \rangle$	$\langle 0.6, 0.7 \rangle$	$\langle 0.4, 0.1 \rangle$	$\langle 0.8, 0.2 \rangle$	$\langle 0.7, 0.3 \rangle$	$\langle 0.5, 0.6 \rangle$	$\langle 0.5, 0.6 \rangle$
$\mathfrak{A}_2$	$\langle 0.5, 0.1 \rangle$	$\langle 0.4, 0.6 \rangle$	$\langle 0.0, 1.0 \rangle$	$\langle 0.4, 0.5 \rangle$	$\langle 0.5, 0.4 \rangle$	$\langle 0.7, 0.3 \rangle$	$\langle 0.2, 0.4 \rangle$
$\mathfrak{A}_3$	$\langle 0.6, 0.4 \rangle$	$\langle 0.9, 0.1 \rangle$	$\langle 0.7, 0.5 \rangle$	$\langle 0.7, 0.2 \rangle$	$\langle 0.0, 1.0 \rangle$	$\langle 0.5, 0.2 \rangle$	$\langle 0.3, 0.6 \rangle$
$\mathfrak{A}_4$	$\langle 0.2, 0.3 \rangle$	$\langle 0.5, 0.8 \rangle$	$\langle 0.0, 1.0 \rangle$	$\langle 0.5, 0.6 \rangle$	$\langle 0.7, 0.2 \rangle$	$\langle 0.5, 0.1 \rangle$	$\langle 0.2, 0.7 \rangle$
$\mathfrak{A}_5$	$\langle 0.5, 0.1 \rangle$	$\langle 0.6, 0.4 \rangle$	$\langle 0.4, 0.3 \rangle$	$\langle 0.7, 0.3 \rangle$	$\langle 0.4, 0.3 \rangle$	$\langle 0.0, 1.0 \rangle$	$\langle 0.5, 0.7 \rangle$

Table 11 Aggregated results and rank of alternatives in different methods and proposed method

MADM methods	Aggregated PyFNs	Score value	Rank
MADM method using PyFDWG [19]	$d_1 = \langle 0, 1\rangle$, $d_2 = \langle 0, 1\rangle$, $d_3 = \langle 0, 1\rangle$, $d_4 = \langle 0, 1\rangle$, $d_5 = \langle 0, 1\rangle$	$sc(d_1) = -1$, $sc(d_2) = -1$, $sc(d_3) = -1$, $sc(d_4) = -1$, $sc(d_5) = -1$	$\mathfrak{A}_1 = \mathfrak{A}_2 = \mathfrak{A}_3 = \mathfrak{A}_4 = \mathfrak{A}_5$
MADM method using PyFHPWG operator [20]	$d_1 = \langle 0, 1\rangle$, $d_2 = \langle 0, 1\rangle$, $d_3 = \langle 0, 1\rangle$, $d_4 = \langle 0, 1\rangle$, $d_5 = \langle 0, 1\rangle$	$sc(d_1) = 0$, $sc(d_2) = 0$, $sc(d_3) = 0$, $sc(d_4) = 0$, $sc(d_5) = 0$	$\mathfrak{A}_1 = \mathfrak{A}_2 = \mathfrak{A}_3 = \mathfrak{A}_4 = \mathfrak{A}_5$
MADM method using SPyFWG operator [21]	$d_1 = \langle 0, 1\rangle$, $d_2 = \langle 0, 1\rangle$, $d_3 = \langle 0, 1\rangle$, $d_4 = \langle 0, 1\rangle$, $d_5 = \langle 0, 1\rangle$	$sc(d_1) = -1$, $sc(d_2) = -1$, $sc(d_3) = -1$, $sc(d_4) = -1$, $sc(d_5) = -1$	$\mathfrak{A}_1 = \mathfrak{A}_2 = \mathfrak{A}_3 = \mathfrak{A}_4 = \mathfrak{A}_5$
MADM method using PyFHPrWG [25]	$d_1 = \langle 0, 1\rangle$, $d_2 = \langle 0, 1\rangle$, $d_3 = \langle 0, 1\rangle$, $d_4 = \langle 0, 1\rangle$, $d_5 = \langle 0, 1\rangle$	$sc(d_1) = 0$, $sc(d_2) = 0$, $sc(d_3) = 0$, $sc(d_4) = 0$, $sc(d_5) = 0$	$\mathfrak{A}_1 = \mathfrak{A}_2 = \mathfrak{A}_3 = \mathfrak{A}_4 = \mathfrak{A}_5$
MADM method using PyFIHPWG [26]	$d_1 = \langle 0, 1\rangle$, $d_2 = \langle 0, 1\rangle$, $d_3 = \langle 0, 1\rangle$, $d_4 = \langle 0, 1\rangle$, $d_5 = \langle 0, 1\rangle$	$sc(d_1) = -1$, $sc(d_2) = -1$, $sc(d_3) = -1$, $sc(d_4) = -1$, $sc(d_5) = -1$	$\mathfrak{A}_1 = \mathfrak{A}_2 = \mathfrak{A}_3 = \mathfrak{A}_4 = \mathfrak{A}_5$
MADM method using proposed AO, i.e., AdPyFHWG operator ($k = 2, \lambda = 0.99$)	$d_1 = \langle 0.44837, 0.71649\rangle$, $d_2 = \langle 0.32244, 0.76827\rangle$, $d_3 = \langle 0.45806, 0.73694\rangle$, $d_4 = \langle 0.31038, 0.80156\rangle$, $d_5 = \langle 0.42476, 0.60976\rangle$	$sc(d_1) = -0.40152$, $sc(d_2) = -0.63497$, $sc(d_3) = -0.41562$, $sc(d_4) = -0.68880$, $sc(d_5) = -0.27708$	$\mathfrak{A}_5 > \mathfrak{A}_1 > \mathfrak{A}_3 > \mathfrak{A}_2 > \mathfrak{A}_4$

The effectiveness the proposed MADM method is elaborated by assessment of green supplier selection methodology in pharmaceutical industry in India. The assessment encompassed key factors such as Environmental Compliance and Certification, Waste Management Practices, Supply Chain Sustainability, Raw Material Sourcing and Sustainable Packaging, Product Life Cycle Assessment, Chemical Management, and Green Chemistry, as well as the Total Cost of Ownership as the primary criteria in the MADM method.

The key findings of this chapter are as follows:

1. In scenarios where the MD or NMD of a component PyFN becomes zero alongside its corresponding weight, the resulting aggregation becomes indeterminate within existing AOs. However, the proposed operator effectively addresses this challenge.

2. Existing MADM methods struggle to differentiate the ranking order of alternatives in certain special cases outlined in this chapter. Conversely, the proposed MADM method adeptly handles these situations by providing distinct rankings.

3. PyFN serves as a valuable tool for expressing the fuzziness in the opinions of the DEs. Utilizing the AdPyFHWG operator in the MADM method proves to be advantageous compared to other AOs considered in this chapter. The presence of parameters k and λ enhances the flexibility of the AO in decision-making. In this chapter, λ is set to 0.99, but it can range between (0, 1). Additionally, k is set to 2 in this chapter, although it can vary between $(0, \infty)$. Different combinations of k and λ offer avenues for future research.

4. The implementation of the APyFHWG operator introduces a systematic and robust approach to multi-criteria decision-making, enabling a nuanced ranking of alternatives. Sun Pharmaceutical Industries Ltd. emerges as the top contender, securing the highest position in the evaluation, underscoring its commendable dedication to environmental responsibility and sustainable business practices.

The ranking order of the alternatives, with Dr. Reddy Laboratories Ltd. following closely, succeeded by Cipla Ltd., Biocon Ltd., and Lupin Ltd., reflects the varying degrees of dedication to green initiatives within the pharmaceutical sector. This analysis not only aids in identifying industry leaders in sustainable practices but also serves as a guide for stakeholders seeking environmentally conscious and responsible partners in the pharmaceutical supply chain.

It is evident from the outcomes that Sun Pharmaceutical Industries Ltd. has demonstrated a comprehensive and superior approach to environmental sustainability, waste management, supply chain sustainability, raw material sourcing, sustainable packaging, product life cycle assessment, chemical management, and green chemistry. This company's noteworthy performance underscores the importance of holistic environmental considerations in supplier selection.

As industries worldwide increasingly embrace the imperative of sustainable practices, the pharmaceutical sector in India can draw inspiration from the evaluated companies' achievements. Stakeholders, including policymakers, customers, and investors, can use these findings to promote and reward sustainability efforts, fostering a culture of responsible business practices in the pharmaceutical supply chain. The journey towards a greener and more sustainable future for the industry requires continuous commitment and collaboration, and this research provides a valuable foundation for making informed decisions in aligning with eco-friendly suppliers in the pharmaceutical domain. Therefore the proposed chapter encourages the green supplier selection methodology in the pharmaceutical industry in Indian context.

Appendix

See Table 12.

Table 12 Abbreviations and its explanation

Abbreviation	Description
FS	Fuzzy set
IFS	Intuitionistic fuzzy set
PyFS	Pythagorean fuzzy set
PyFN	Pythagorean fuzzy number
AO	Aggregation operator
GO	Geometric operator
MCDM	Multi-criteria decision-making
MADM	Multi-attribute decision-making
MF	Membership function
NMF	Non-membership function
MD	Membership degree
NMD	Non-membership degree
APyFHWG	Advanced Pythagorean fuzzy Hamacher weighted geometric
PyFDWG	Pythagorean fuzzy Dombi weighted geometric
PyFHPWG	Pythagorean fuzzy Hamacher power weighted geometric
SPyFWG	Symmetric Pythagorean fuzzy weighted geometric
PyFHPrWG	Pythagorean fuzzy Hamacher prioritized weighted geometric
PyFIHPWG	Pythagorean fuzzy interactive Hamacher power weighted geometric

References

1. Zadeh, L.A.: Fuzzy sets. Inf. Control **8**(3), 338–353 (1965)
2. Atanassov, K.T.: New operations defined over the intuitionistic fuzzy sets. Fuzzy Sets Syst. **61**(2), 137–142 (1994)
3. Yager, R.R.: Pythagorean membership grades in multicriteria decision making. IEEE Trans. Fuzzy Syst. **22**(4), 958–965 (2013)
4. Samanta, S., Sarkar, B., Shin, D., Pal, M.: Completeness and regularity of generalized fuzzy graphs. SpringerPlus **5**, 1–14 (2016). https://doi.org/10.1186/s40064-016-3558-6
5. Senapati, T., Kim, C.S., Bhowmik, M., Pal, M.: Cubic subalgebras and cubic closed ideals of B-algebras. Fuzzy Inf. Eng. **7**(2), 129–149 (2015)
6. Sahoo, S., Pal, M., Rashmanlou, H., Borzooei, R.A.: Covering and paired domination in intuitionistic fuzzy graphs. J. Intell. Fuzzy Syst. **33**(6), 4007–4015 (2017)
7. Ghorai, G., Pal, M.: Certain types of product bipolar fuzzy graphs. Int. J. Appl. Comput. Math. **3**, 605–619 (2017). https://doi.org/10.1007/s40819-015-0112-0
8. Shyamal, A.K., Pal, M.: Triangular fuzzy matrices. Iran. J. Fuzzy Syst.
9. Mahapatra, R., Samanta, S., Pal, M., Xin, Q.: RSM index: a new way of link prediction in social networks. J. Intell. Fuzzy Syst. **37**(2), 2137–2151 (2019)
10. Rashmanlou, H., Muhiuddin, G., Amanathulla, S.K., Mofidnakhaei, F., Pal, M.: A study on cubic graphs with novel application. J. Intell. Fuzzy Syst. **40**(1), 89–101 (2021)
11. Nayak, P.K., Pal, M.: Linear programming technique to solve two person matrix games with interval pay-offs. Asia-Pac. J. Oper. Res. **26**(02), 285–305 (2009)
12. Das, A., Pal, M.: A mathematical study of an imprecise SIR epidemic model with treatment control. J. Appl. Math. Comput. **56**, 477–500 (2018). https://doi.org/10.1007/s12190-017-1083-6

13. Shyamal, A.K., Pal, M.: Two new operators on fuzzy matrices. J. Appl. Math. Comput. **15**, 91–107 (2004). https://doi.org/10.1007/BF02935748
14. Mahapatra, R., Samanta, S., Pal, M.: Applications of edge colouring of fuzzy graphs. Informatica **31**(2), 313–330 (2020)
15. Mahapatra, R., Samanta, S., Pal, M.: Generalized neutrosophic planar graphs and its application. J. Appl. Math. Comput. **65**(1), 693–712 (2021)
16. Mandal, S., Sahoo, S., Ghorai, G., Pal, M.: Genus value of m-polar fuzzy graphs. J. Intell. Fuzzy Syst. **34**(3), 1947–1957 (2018)
17. Saha, A., Pal, M., Pal, T.K.: Selection of programme slots of television channels for giving advertisement: a graph theoretic approach. Inf. Sci. **177**(12), 2480–2492 (2007)
18. Paul, T.K., Pal, M., Jana, C.: Multi-Attribute decision making method using advanced Pythagorean fuzzy weighted geometric operator and their applications for real estate company selection. Heliyon **7**(6), e07340 (2021)
19. Jana, C., Senapati, T., Pal, M.: Pythagorean fuzzy Dombi aggregation operators and its applications in multiple attribute decision-making. Int. J. Intell. Syst. **34**(9), 2019–2038 (2019)
20. Wei, G.W.: Pythagorean fuzzy Hamacher power aggregation operators in multiple attribute decision making. Fundam. Inform. **166**(1), 57–85 (2019)
21. Ma, Z., Xu, Z.: Symmetric Pythagorean fuzzy weighted geometric/averaging operators and their application in multicriteria decision-making problems. Int. J. Intell. Syst. **31**(12), 1198–1219 (2016)
22. Jana, C., Pal, M., Wang, J.: A robust aggregation operator for multi-criteria decision-making method with bipolar fuzzy soft environment 2019. Iran. J. Fuzzy Syst. **16**(6), 1–16. https://doi.org/10.1504/IJCSM.2016.077854
23. Jana, C., Pal, M.: Assessment of enterprise performance based on picture fuzzy Hamacher aggregation operators 2019. Symmetry **11**(1), 75. https://doi.org/10.1007/s00500-015-2026-5
24. Wu, S.J., Wei, G.W.: Pythagorean fuzzy Hamacher aggregation operators and their application to multiple attribute decision making. Int. J. Knowl. Based Intell. Eng. Syst. **21**(3), 189–201 (2017)
25. Gao, H.: Pythagorean fuzzy Hamacher prioritized aggregation operators in multiple attribute decision making. J. Intell. Fuzzy Syst. **35**(2), 2229–2245 (2018)
26. Wang, L., Garg, H., Li, N.: Pythagorean fuzzy interactive Hamacher power aggregation operators for assessment of express service quality with entropy weight. Soft Comput. **25** 973–993 (2021). https://doi.org/10.1007/s00500-020-05193-z
27. Jana, C., Pal, M.: Multi-criteria decision making process based on some single-valued neutrosophic Dombi power aggregation operators. Soft Comput. **25**(7), 5055–5072 (2021). https://doi.org/10.1016/j.engappai.2021.104203
28. Jana, C., Muhiuddin, G., Pal, M.: Multi-criteria decision making approach based on SVTrN Dombi aggregation functions 2021. Artif. Intell. Rev. **54**(5), 3685–3723. https://doi.org/10.3233/JIFS-16433
29. Jana, C., Garg, H., Pal, M.: Multi-attribute decision making for power Dombi operators under Pythagorean fuzzy information with MABAC method 2023. J. Ambient Intell. Human. Comput. **14**(8), 10761–10778. https://doi.org/10.1007/s12652-022-04348-0
30. Jana, C., Pal, M.: A dynamical hybrid method to design decision making process based on GRA approach for multiple attributes problem. Eng. Appl. Artif. Intell. **100**(104203) (2021). https://doi.org/10.3390/sym11010075
31. Jana, C., Muhiuddin, G., Pal, M., Al-Kadi, D.: Intuitionistic fuzzy Dombi hybrid decision-making method and their applications to enterprise financial performance evaluation. Math. Prob. Eng. **3218133** (2021). https://doi.org/10.1155/2021/3218133
32. Jana, C., Pal, M.: Extended bipolar fuzzy EDAS approach for multi-criteria group decision-making process 2021. Comput. Appl. Math. **40**(1), 9. https://doi.org/10.1007/s12652-019-01568-9
33. Paul, T.K., Pal, M., Jana, C.: Portfolio selection as a multicriteria group decision making in Pythagorean fuzzy environment with GRA and FAHP framework. Int. J. Intell. Syst. **37**(1), 478–515 (2021)

34. Paul, T.K., Jana, C., Pal, M., Simic, V.: Sustainable carbon-dioxide storage assessment in geological media using modified Pythagorean fuzzy VIKOR and DEMATEL approach. Int. J. Hydrogen Energy **48**(25), 9474–9497 (2023)
35. Paul, T.K., Jana, C., Pal, M.: Enhancing multi-attribute decision making with Pythagorean fuzzy Hamacher aggregation operators. J. Ind. Intell. **1**(1), 30–54 (2023). https://doi.org/10.56578/jii010103
36. Paul, T.K., Jana, C., Pal, M.: Multi-criteria group decision-making method in disposal of municipal solid waste based on cubic Pythagorean fuzzy EDAS approach with incomplete weight information. Appl. Soft Comput. **144**, 110515 (2023). https://doi.org/10.1016/j.asoc.2023.110515
37. Paul, T.K., Jana, C., Pal, M., Simic, V.: Multi-attribute group decision making method based on Pythagorean fuzzy Einstein interactive power averaging approach for sustainable cement industry. Appl. Soft Comput. **148**, 110898 (2023). https://doi.org/10.1016/j.asoc.2023.110898
38. Jana, C., Garg, H., Pal, M.: Multi-attribute decision making for power Dombi operators under Pythagorean fuzzy information with MABAC method. J. Ambient Intell. Human. Comput. **14**(8), 10761–10778 (2023)
39. Yager, R.R.: Pythagorean Fuzzy Subsets, 57–61 (2013)
40. Peng, X., Ma, X.: Pythagorean fuzzy multi-criteria decision making method based on CODAS with new score function. J. Intell. Fuzzy Syst. **38**(3), 3307–3318 (2020)
41. Younis, H., Sundarakani, B.: The impact of firm size, firm age and environmental management certification on the relationship between green supply chain practices and corporate performance. Benchmark. Int. J. **27**(1), 319–346 (2020)
42. Gao, H., Ju, Y., Gonzalez, E.D.S., Zhang, W.: Green supplier selection in electronics manufacturing: an approach based on consensus decision making. J. Clean. Prod. **245**, 118781 (2020). https://doi.org/10.1016/j.jclepro.2019.118781
43. Khokhar, M., Zia, S., Islam, T., Sharma, A., Iqbal, W., Irshad, M.: Going green supply chain management during COVID-19, assessing the best supplier selection criteria: a triple bottom line (TBL) approach. Probl. Ekorozwoju **17**(1) (2022)
44. Mardani, A., Kannan, D., Hooker, R.E., Ozkul, S., Alrasheedi, M., Tirkolaee, E.B.: Evaluation of green and sustainable supply chain management using structural equation modelling: a systematic review of the state of the art literature and recommendations for future research. J. Clean. Prod. **249**, 119383 (2020). https://doi.org/10.1016/j.jclepro.2019.119383
45. Roy, S.A., Ali, S.M., Kabir, G., Enayet, R., Suhi, S.A., Haque, T., Hasan, R.: A framework for sustainable supplier selection with transportation criteria. Int. J. Sustain. Eng. **13**(2), 77–92 (2020)
46. Kilic, H.S., Yalcin, A.S.: Modified two-phase fuzzy goal programming integrated with IF-TOPSIS for green supplier selection. Appl. Soft Comput. **93**, 106371 (2020). https://doi.org/10.1016/j.asoc.2020.106371
47. Wu, C., Lin, Y., Barnes, D.: An integrated decision-making approach for sustainable supplier selection in the chemical industry. Expert Syst. Appl. **184**, 115553 (2021). https://doi.org/10.1016/j.eswa.2021.115553
48. Li, Y., Diabat, A., Lu, C.C.: Leagile supplier selection in Chinese textile industries: a DEMATEL approach. Ann. Oper. Res. **287** 303–322 (2020). https://doi.org/10.1007/s10479-019-03453-2

On the Maintenance Oversight of the Healthcare Sector Based on Artificial Intelligence

Sovan Bhattacharya, Dola Sinha, Chandan Bandyopadhyay, Saibal Majumder, and Arindam Biswas

Abstract A new way of thinking about health care has emerged thanks to the use of artificial intelligence (AI). In this review paper, we identify five primary applications of artificial intelligence (AI) in health care by reviewing the relevant literature. The applications are as follows: (1) patient digital care, (2) pharmaceutical and clinical research, (3) patient involvement and compliance, (4) rehabilitation, and (5) additional administrative tasks. Artificial intelligence (AI) may have several advantages for the medical sector. Among its many capabilities are EHR management, new vaccination and therapy discovery, medical prescription error detection, data storage and analysis, and technology-assisted rehabilitation. It might also motivate patients to engage fully in their treatment and adhere to the recommended schedules. It may help to manage the 2019 coronavirus disease (COVID-19) epidemic by early detection, reduce the administrative load placed on those working in the healthcare sector, and identify clinical challenges in imaging and diagnostic services. All the same,

---

S. Bhattacharya · C. Bandyopadhyay · S. Majumder
Department of Computer Science and Engineering (Data Science), Dr. B. C. Roy Engineering College, Durgapur, India
e-mail: sovan.cse@gmail.com

C. Bandyopadhyay
e-mail: chandanb.iiest@gmail.com

S. Majumder
e-mail: saibal.majumder.1729@gmail.com

D. Sinha
Department of Electrical Engineering, Dr. B.C. Roy Engineering College, Durgapur, India
e-mail: dola.sinha@bcrec.ac.in

C. Bandyopadhyay
Department of Computer Science and Engineering, University of Bremen, Bremen, Germany

A. Biswas (✉)
Center for IOT, AI Integration with Education-Industry-Agriculture, Kazi Nazrul University, Asansol, India
e-mail: mailarindambiswas@yahoo.co.in

School of Mines and Metallurgy, Kazi Nazrul University, Asansol, India

L. Sahoo et al. (eds.), *Decision Making Under Uncertainty Via Optimization, Modelling, and Analysis*, Studies in Systems, Decision and Control 558,
https://doi.org/10.1007/978-981-96-0085-4_21

the scientific presentation covers a lot of social, ethical, and technological concerns related to AI in health care. Aspects to take into account are privacy, safety, autonomy in decision-making, cost, information and consent, accessibility, and effectiveness. The need to establish adequate rules for AI applications is justified by several reasons. Among these are improving adoption, holding healthcare professionals responsible, ensuring patient safety, and achieving notable health benefits. The adoption and use of AI must be promoted by prompt and effective leadership that also attends to issues of ethics, regulatory compliance, and trust. Since the COVID-19 pandemic shocked the global health system, the idea of artificial intelligence has revolutionized health care, and this kind of uprising could be a further step toward meeting future healthcare needs.

Keywords Artificial intelligence · Health care · Healthcare personnel · GAN · e-Health service

1 Introduction

Healthcare expenditures have grown at an exponential rate, outpacing GDP growth rates, and putting health systems throughout the world in a precarious financial position [1]. Long-term illnesses are on the rise, the population is becoming older, and healthcare institutions are already struggling to meet the demand for their services due to limited funding and overburdened infrastructure. Countries like Indonesia, Brazil, and India are seeing their health systems collapse as a result of the COVID-19 epidemic [2]. The progress of biomedical research—encompassing genomics, digital medicine, AI, and its subset, machine learning (ML)—lays the groundwork for healthcare reform, which in turn necessitates a different kind of workforce and different standards of practice. Innovative new technologies are developing with this shift. Diagnostics, therapies, care delivery, regenerative therapy, and precision medicine models might all be improved and transformed by genomics and related technologies like biometrics, tissue engineering, and the vaccine industry [3]. Table 1 displays the definitions of ideas related to AI.

1.1 Background

The introduction of AI into health care has the potential to greatly improve patient outcomes. The use of AI in health care has several potential benefits, including better patient diagnoses, more effective treatment and prevention, more cost efficiency, and the possibility of universal access to quality health care [4, 5]. Advancements in AI—which encompasses techniques such as deep learning, reinforcement learning, neural networks, and machine learning—have allowed for the development of new information, resources, and ideas [6] by simulating human intelligence in machines.

Table 1 Characteristics of the relevant terms in the context of AI

Term	Explanation of
Machine learning (ML)	• An area of AI research involves developing tools to help doctors work more quickly and accurately. Moreover, it is a catchall for a bunch of statistical techniques that let computers learn by doing, rather than just being programmed to do so • In most cases, the learning takes the form of changes to how each algorithm functions [137] • Additionally, it is a tool that doctors and other medical staff use to better manage patient information and care plans. When computers are taught to learn and reason like humans, we say that they have artificial intelligence [138]
Distributed ledger technology (DLT)	• To record and distribute data across a variety of data stores (ledgers), distributed ledger technology (DLT) is a novel and rapidly expanding solution [139]. In addition to being safe and unchangeable, it is easily accessible [140] • It has the potential to enable patients to take ownership of their data, which will ultimately lead to the development of trust in an industry that is important to all of us [141] • An innovative and cutting-edge approach to the intelligent, robust, and secure management of electronic health record data is described by the integration of DLT with artificial intelligence
Natural language processing (NLP)	• The various methods in which computers and people interact about language are the topic of natural language processing, commonly called NLP [142] • With the use of natural language processing (NLP) methods, it is feasible to collect unstructured medical records, analyze their grammar, determine their meaning, and even translate them. Because of this, electronic healthcare systems have no trouble understanding the data. By using these measures, healthcare costs can be reduced without sacrificing quality [115]
Metaverse	• The term "metaverse" describes a three-dimensional environment that is based on AR and VR. People may work, play, and communicate with one another in real time using their avatars in this virtual world [143]. Personalized to each patient's tastes and requirements, it offers health care in a way that is interesting, engaging, and fun • Some of the most recent innovations in this field include digital twinning, blockchain, telepresence, VR, AR, and AI. These technologies have far-reaching effects on the healthcare sector [144]. Academics have a "niche theme" in the healthcare-related usage of the metaverse, which encompasses all aspects of the field, including instruction, study, training, and illness management and prevention. It has evolved into a cutting-edge tool that improves the skills of future doctors in recent years • Furthermore, patients' health status may be closely monitored from the comfort of their own homes. The use of digital twins, a versatile technology, allows for the integration of the physical and virtual worlds, thus expanding the possibilities [145, 146]

(continued)

Table 1 (continued)

Term	Explanation of
Chat generative pretrained transformer (ChatGPT)	• ChatGPT is a conversational agent that is powered by artificial intelligence that uses natural language processing and machine learning methods to imitate human-like discussions [147] • The key uses of this technology in the healthcare industry, which include practice, teaching, and research, have the potential to be fruitful provided the validity problems that accompany it are proactively examined and addressed • It performs the job of a chatbot, which is computer software that can interpret and generate replies through the use of a text-based interface [148]. The application of chatbots in the healthcare industry was detailed by Xu et al. [149]. This application included patient assistance, monitoring, and administration, as well as tumour diagnosis, screening, and treatment for patients
Transformer	• Computer vision (CV), natural language processing (NLP), and speech processing are just a few of the areas that make extensive use of the Transformer model, which is a significant deep learning model that is widely applied in a range of disciplines [150] • Applications that make use of transformers include the detection of COVID-19, electronic health records, and medical imaging [40, 151, 152]. Other examples of applications include medical imaging

When it comes to health care, AI innovation is vital for finding new ways to solve problems, facilitate patient care, and uncover possibilities.

With few exceptions, AI has had mixed results in the healthcare industry despite the widespread use of the technology over the last six decades [7–9]. The lack of access to electronic data (e-data) was the main cause of this restriction, as AI relies on accurate and trustworthy e-data to inform the creation of algorithms [10]. There are a variety of medical databases, such as electronic health records (EHR), that hold data that is essential to the treatment of the patient [11]. These databases include the patient's demographics, radiology reports, medical history, vital signs, medications, laboratory results, vaccines, progress notes, and insurance information. The administrative and care-related efficiencies that may be brought about by AI may be beneficial to a variety of parties, including patients, payers, government agencies, pharmaceutical companies, and pharmaceutical providers. This possible future growth is being driven by the increasing complexity of the field's data as well as the expansion of electronically recorded patient medical records [12]. Artificial intelligence is already on par with or even exceeding human capabilities in areas such as autonomous mobility [13] and speech recognition [14]. This is not limited to the healthcare industry. Among the many applications of artificial intelligence that are now being used in the medical field, some examples include automated skin cancer imaging analyses [15], surgery [16], oncology [17], psychiatry [18], and clinical decision support [19]. Given the current challenges that the healthcare industry is facing, such as inadequate access to health care for all [20], shortages of human

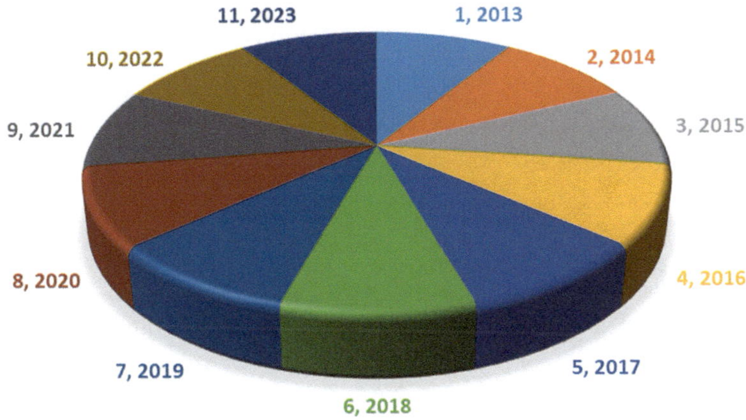

Fig. 1 Annual volume of AI-related healthcare journal articles published over the past decade

workforces [21], high service prices [22], and poor acceptance of information technology [23], the adoption of artificial intelligence will face future challenges that will make it difficult to improve the quality of health care.

When it comes to the future of artificial intelligence in health care, experts have come up with three different possibilities [19]. The first possible outcome is that artificial intelligence will one day be able to diagnose all patients, doing away with the necessity for medical professionals. One of the primary motivations for this stance is to reduce costs without compromising the level of care that is provided to patients. The ability of artificial intelligence to evaluate a greater number of patients, so enabling professionals to analyze fewer cases, is yet another possible effect. This would result in a reduction in the total amount of money spent on health care as well as the number of people who are treated by medical professionals [24]. Last but not least, artificial intelligence may be of assistance to medical professionals by improving clinical decision-making, which in turn may enhance patient outcomes and cost-effectiveness. Considering the number of related literary works the pie chart in Fig. 1 shows the number of publications done in a particular year. Here, each sector of the pie chart is represented by a pair (x,y), where x is the total publications and y is the year of publication. Subsequently, in Fig. 2, the country-wise publications are depicted during the year 2013–2023. Here, it is observed that the USA dominates other countries which reveals the importance of AI in the field of health care.

1.2 *Investigational Concerns*

As AI continues to advance at a dizzying rate, we set out to investigate current AI healthcare practices by looking at the most recent algorithms and trends in the

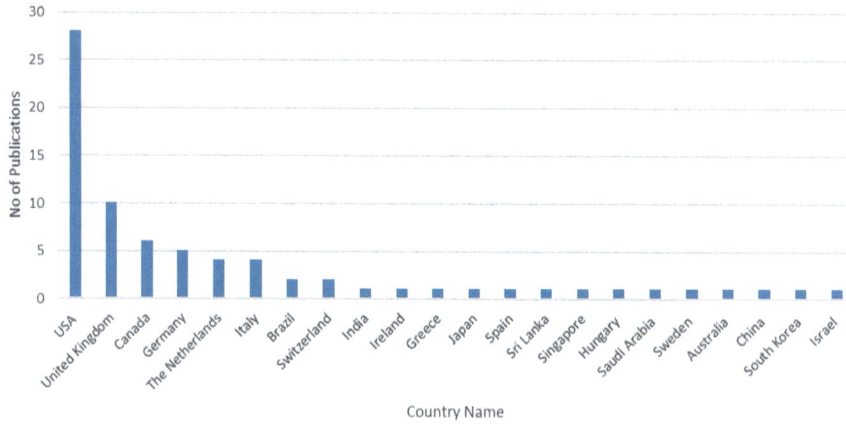

Fig. 2 Number of publications per country

industry, and we hope to offer a thorough and up-to-date overview of the cutting-edge approaches used by AI-driven healthcare solutions. We offer the first research question because this work is important for many people—researchers, healthcare providers, and lawmakers—because it sheds light on the best AI tools and methodologies for improving patient care, automating processes, and making the most of available resources.

In health care, what kinds of AI methods and algorithms are now in use, and what kinds of trends may emerge from this field?

Considering the difficulties and opposition to AI adoption in many industries, including health care [25], it is essential to study these issues and find ways to overcome them. To better assist medical professionals thinking about or already using AI technology in their organizations, we need to understand the challenges that have been encountered throughout AI adoption. Because of this, we are making an effort to address the following inquiry.

Our second research question is: what are the most significant obstacles to using AI in health care, and how can we overcome them?

Innovation is greatly influenced by startups [26, 27]. Our objective is to investigate the present status of research on AI healthcare companies and how these startups contribute to AI healthcare innovation. From the perspective of startup activity, this understanding helps founders, healthcare experts, and policymakers appreciate the AI healthcare environment. In light of this, we pose the following inquiry to go more into this subject.

Thirdly, what are the most salient features of the present climate for artificial intelligence healthcare companies concerning finance, concentration, and location?

The goal of our literature review is to identify papers that deal with these issues so that we may learn more about them.

1.3 Contribution

The increasing focus on AI advancements in health care is bolstered in three ways by this SLR. It offers.

(i) A more in-depth understanding of the influence that innovations in artificial intelligence have had on healthcare research: We discuss the most often discussed aspects of the topic, which include the use of AI in healthcare research, the adoption of AI in health care by a variety of stakeholders, and the challenges associated with the incorporation of AI in health care, such as anxiety around privacy, ethics, and trust. Subsequently, these results provide our first and second study questions with answers to their respective queries. The following are examples of ecosystem factors for AI healthcare startups: To accomplish our goal of answering the third research question, we want to enhance the findings of our SLR with a study of more than 2,000 AI healthcare enterprises located all over the globe. This will demonstrate that there is a mismatch between the geographical distribution of AI healthcare businesses and the lack of external investment that is available. A possible path for additional research into the factors that contribute to the success or failure of AI-based healthcare innovation companies is provided by the findings of our study.

(ii) Where healthcare industry executives and academics should go from here: We provide an overview of the field's potential and three distinct points of view on research topics and their prospects. We begin by trying to lay out some potential paths that academics may take in the future to add to the theoretical understanding of AI health care. Our second point is that there are chances for healthcare executives, financiers, and other stakeholders to take action by using AI to reduce expenses and increase efficiency.

(iii) There is a dearth of research on artificial intelligence healthcare firms, and no previous studies have linked this to their rapid expansion. The findings of this study shed light on the problem and point the way for more studies in this area.

As a whole, we're hoping that medical institutions, clinics, engineers, and governments may use this assessment as a starting point for developing strategies to increase AI innovation in health care for developing economies. The suggested fields of research would benefit from the increased availability of data brought about by the proliferation of AI-powered healthcare companies.

2 What Artificial Intelligence Can Do for Health Care?

By providing practitioners with access to patient information, digital health technologies make it possible for them to get a more holistic understanding of the health of their patients. A further advantage is that they make it possible for medical professionals to provide patients with more information on their health. The treatment outcomes and efficacy may be significantly improved with the use of these modalities. However, there are worries that these approaches may have a more substantial psychological impact. This is particularly true when taking into consideration the widespread usage of internet-mediated interventions (IMAs) and social media (SM) not only by professionals but also by patients and the general public [28]. Health information systems (HISs), wearable devices, telemedicine, mobile health, telehealth, MIDs, and other medical technology powered by AI combine to produce massive amounts of data.

This data makes it easier to implement machine learning (ML) and AI in healthcare systems. Research information, user experience, and the analysis of enormous datasets are all examples of the types of data that these systems learn from as they are acquired via various sources. In addition, electronic health records, often known as EHRs, include a wide variety of pertinent healthcare information relevant to individuals. With the help of cutting-edge artificial intelligence technology, health datasets may be connected to acquire precise insights into patient treatment. When it comes to the implementation of big data applications in the healthcare industry, artificial intelligence has emerged as a preferred alternative [29]. To add insult to injury, the usage of big data analytics gives medical professionals the ability to increase the quality of their clinical services by using analytical algorithms to enhance electronic health records (EHRs) [30]. To further improve the precision and efficiency of data analysis, this analytics technique makes use of recent developments in AI to rapidly sort through vast amounts of data depending on a variety of criteria [31]. One of the most common applications of artificial intelligence in the healthcare industry is to improve the health outcomes of patients and to provide healthcare services at lower costs. By focusing specifically on the following major characteristics (cf. Fig. 3), the purpose of this study is to shed light on its function in the healthcare industry. Medical imaging and diagnostics, e-health services for patients, medical research and drug invention, patient engagement and compliance, restorative measures, and other administrative applications are some of the areas that are covered by this category. The writers also cover a variety of challenges that are linked to the use of artificial intelligence in the medical field. These results contribute to the existing corpus of study by furthering the many benefits that artificial intelligence technologies may provide in the medical field.

To improve early disease diagnosis and decrease diagnostic errors for prevention, radiologists are utilizing AI for image analysis. By evaluating electrocardiogram (ECG) and echocardiography (ECHO) records, AI can aid cardiologists in making educated decisions. According to reports from a hospital in Oxford, the Ultromics

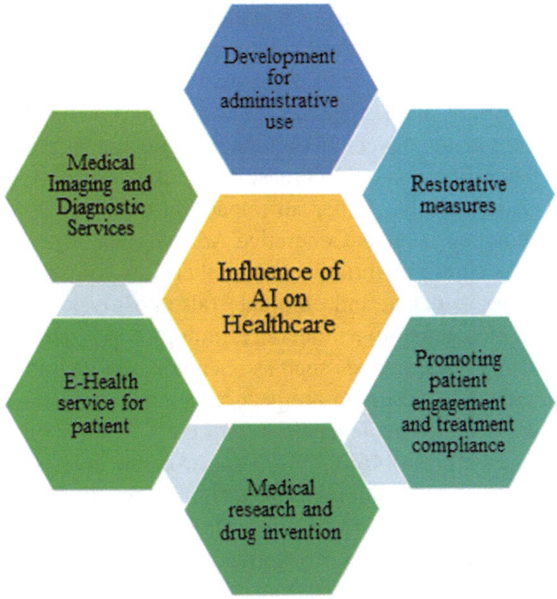

Fig. 3 Block diagram of the constituents of the AI which are applicable in health care

platform analyzes echocardiography scans for signs of ischemic heart disease and irregular heartbeats using artificial intelligence [32].

According to studies, AI can use body imaging to detect diseases such as skin and breast cancer, eye diseases, and pneumonia at an early stage [33–35]. By analyzing speech patterns, AI tools can screen for neurological diseases like Parkinson's and predict psychotic episodes [36, 37]. To foretell when diabetes may begin, a recent study employed ML models. The best model for predicting diabetes variables was a two-class augmented decision tree [38]. Medical imaging tools enhanced with artificial intelligence, such as X-ray, CT, and US, have aided in the fight against COVID-19 by allowing for earlier diagnosis, according to Gudigar et al. [39]. Predictions of COVID-19 cases were found to be accurate using all HCFL, DNN, and hybrid methods in their study. To diagnose COVID-19, a recent review outlined the use of X-rays, CT scans, MRI, and ultrasound. Using AI has been critical in the battle against the virus, according to the article [40]. The transformer deep learning model is also used for video applications, image-to-image translation, segmentation, detection, classification, and registration in medical imaging analysis [41]. To efficiently manage cases, previous research has proposed using transformers to differentiate between COVID-19 and pneumonia using X-ray and CT images [42, 43]. Another study also used chest X-ray patches and the ImageNet-trained ViT-B/32 network to detect COVID-19 [44]. One study found that COVID-19 could be automatically detected using a hybrid chest CT-based approach (Wang et al., 45). Wavelet Renyi entropy (WRE) and a three-segment biogeography-grounded optimization (3SBBO) algorithm are utilized in this computer vision-based diagnostic approach. The 3SBBO algorithm, a FNN, and WRE are all part of the system. For picture classification, we have FNN, 3SBBO for optimizing network biases and weights, and WRE for extracting image features. Its performance in identifying COVID-19 was superior to that of radial basis function neural networks, extreme learning machines with bat algorithms, and kernel-based approaches. Gheflati et al. [45] state that the ViT can differentiate between benign, malignant, and normal breast tissues using ultrasound images. In terms of US breast image classification, it performed better than CNNs.

Additionally, AI has an impact on radiology through the use of deep learning techniques, such as Generative Adversarial Networks (GANs). In GANs, two artificial neural networks work together: one to create images that look realistic and another to detect and eliminate false positives. Using the features of the training dataset, a generative model in radiology can generate new images or make copies of existing ones. Images such as pneumonia radiographs are classified by the discriminant model. Training the generator model with the discriminator model enhances radiological activities such as abnormal detection, image synthesis, and cross-domain image synthesis [46]. Using GANs to create images of lung cancer nodules was difficult for expert radiologists [47]. GANs enhance medical research and education. Get simulations and learning materials made for students in no time. Pupils can learn to differentiate between "consolidation" and "lower lobe collapse" by seeing examples of both. Learners benefit from edge-case resources derived from synthetic data. Clinical trial costs can be reduced and the number of treatment arms increased by using synthetic control arms based on historical data instead of real-life placebo groups [48]. Concerns have been raised about the public's use of ChatGPT, a language model based on deep learning, for medical advice. Instead of seeking medical advice, members of the public can use models to get treatment suggestions or diagnose themselves [49]. Research shows that one-third of adults in the US use the Internet to make their diagnoses. A significant portion of the participants sought medical advice after seeing results online [50]. A metaverse devoted to "medical technology and AI" enables AI-powered health care, such as imaging-guided diagnosis and treatment. Raw data sharing, metaverse medicine, augmented regulatory science, and virtual comparative scanning all rely on MeTAI. Model of the MeTAI ecosystem: Patients are put through virtual machine simulations before CT scans to improve imaging quality. This information is used to create a genuine scan. Upon patient approval, the physician's team is provided with metaverse images. Following security checks, medical researchers are granted access to tomographic raw data and images. Clinical trials that incorporate metaverse technology can merge virtual and real-world medical evidence. The last step is for the patient to have metaverse therapy and remote robotic surgery. Nevertheless, concerns regarding security, inequality, investment, and privacy are encountered by MeTAI [51]. Gathered, archived, and made publicly accessible for use in training AI systems are medical scans [52]. Medical scans could be accelerated and made more affordable with the help of AI, enabling more scans for more precise treatment [53]. AI affects illness diagnosis and clinical decision-making. It handles a wide variety of data, analyzes it, and then reports its findings for use in clinical decision-making and illness diagnosis. In therapy, it can replace them or help doctors make better clinical decisions [54]. The public's health might be enhanced by computer-aided diagnostics due to their sensitivity, accuracy, and specificity in identifying small radiographic abnormalities. Research using AI for imaging frequently uses lesion detection as an outcome measure, which might skew AI results because it ignores biological factors like severity and type of lesion. While non-patient-related endpoints have the potential to boost sensitivity, they also carry the risk of overdiagnosis and false positives caused by small abnormalities that mimic subclinical disease [55].

2.1 e-Health Services for Patients

According to Baig et al. [56], research has been conducted on the medical applications of machine learning and artificial intelligence, as well as wearable technology. The use of wearable technology enables virtual patient monitoring and management, which is necessary for the fulfilment of care requirements. Through the use of non-invasive sensors that patients need to wear, AI is taking control of conditions such as diabetes, hypertension, sleep apnea, and chronic bronchial asthma. If one is to believe the recent advancements and reported successes, non-invasive epidermal electrochemical glucose sensing systems could revolutionize diabetes management by providing a convenient, pain-free, and effective method for continuous glucose monitoring, thereby enhancing glycemic control and patient quality of life [57]. The results of earlier research suggested the implementation of a smart sensor system that would make use of a combined network to monitor a person's health and activities while they are at home. A platform for wearing sensors that is both biomedical and inconspicuous concerning its design is being proposed. Some of the vital indicators that are recorded by sensors include the heart rate, the breathing rate, the blood pressure, and the electrocardiogram (ECG). One way to interact with sensors is via the use of a smart device that resembles a tablet. Storage and processing of information about care for the elderly takes place in the cloud. You have it correctly. Even though the results of the examination were negative, Patel and Tarakji [58] concluded that atrial fibrillation was the most probable cause of the stroke. Patients were advised to wear a digital electrocardiogram (ECG) bracelet since this was our recommendation. Professionals in the field of electrophysiology verified the signals. Wearable digital gadgets for consumers are used to guide diagnostic procedures. Sukei et al. [59] showed that machine learning models are capable of handling various data with missing information to predict emotional states by using data from mobile sensors. This demonstration was made about mental health conditions. Using these models, medical professionals can evaluate the emotional states of their patients. It is necessary to do more research to address sparse and lost tagged data to better promote innovation in model building.

As a consequence of the COVID-19 pandemic that was brought on by SARS-CoV-2 [60], advancements have been made in biometric and online patient monitoring wearable techniques. Based on the findings of Bogu and Snyder [61], wearable sensors might be used for early COVID-19 prediction. The clinical diagnosis of COVID-19 and the detection of outbreaks will both benefit from research conducted in real-time on the virus. Both the diagnosis of SARS-CoV-2 infection and the prediction of the evolution of solid organ transplantation illness are enabled by artificial intelligence, machine learning, and large amounts of data [62]. Yu et al. [63] conducted research on artificial intelligence-enabled point-of-care apps and the implications of implementing AI into bedside care during and after the COVID-19 epidemic. The COVID-19 epidemic necessitates the use of telehealth services. There is a significant difference between telemedicine which is based on videoconferencing and apps that use the metaverse [64]. During COVID-19, there was a 38-fold growth in the use of telemedicine, which coincided with the emergence of the metaverse [51]. There is a possibility that this expansion was brought about by the decrease in the number of in-person consultations and viral control measures that were implemented during the COVID-19 pandemic [65, 66]. Two examples of new metaverse tools that have the potential to be trustworthy, user-friendly, and economical are raw data sharing and virtual comparison scanning [51]. Both of these technologies are examples of modern metaverse tools. The usage of augmented reality eyewear might potentially be used by metaverse systems to provide real-time audio and video interactions with medical professionals. AR allows for the coordination of the treatment of critical circumstances in real time by medical specialists who are located at a distance [67]. Artificial intelligence, telepresence, blockchain, virtual reality, augmented reality, and digital twinning are all means by which patient outcomes may be improved via innovative and cost-effective management. The metaverse is present on the internet and imitates human behaviour and feelings. [68] The concept encompasses facets of society and money, both offline and online, and also includes aspects of both. AI has the potential to increase the quality of important services and enrich the 3D immersive experiences offered by virtual worlds by strengthening the metaverse [69]. By using telehealth remote patient monitoring (RPM), medical professionals can remotely monitor, evaluate, and record the health state of their patients. RPM makes medical treatments more efficient by using sensors and technology that allows for communication. Examining patient difficulties and health data from a distant location is straightforward. An understanding of the participant's current state of health may be gained [70]. There is a correlation between the time management and workload of healthcare professionals and the reliability of conventional patient monitoring systems.

Skin-to-skin contact is the method that is used in this invasive monitoring method to assess health. Wearables, telehealth, and contact-based sensors are a few examples of Internet of Things (IoT) approaches that are used in remote patient monitoring (RPM) in the healthcare industry. It is often used to assess physiological variables and vital signs, such as motion detection, to assist with medical judgment and the treatment of mental and motor disorders [71]. This also helps with the treatment of mental and motor disorders. Because of the availability of RPM platforms, medical professionals were able to ensure that their patients remained healthy during the COVID-19 epidemic. Recently, a comparison was made between Care Simple COVID and Telecare COVID to monitor COVID-19. Take into consideration the possibility of employing these platforms after the epidemic and hospitalization have ended [72]. By using both classic machine learning and deep learning, applications for RPM can identify and anticipate vital signs, as well as classify patient movements. Robotic process management (RPM) systems that are driven by artificial intelligence can customize health factors, recognize patterns of behaviour, and detect early indicators of patient deterioration, therefore changing healthcare monitoring. Artificial intelligence has the potential to be beneficial to RPM facilities; nevertheless, several hurdles must be overcome. These problems include concerns about privacy, signal processing, data volume, uncertainty, unbalanced datasets, feature extraction, and explainability [71]. OpenAI.com is responsible for the development of the ChatGPT virtual language model. Because it is driven by artificial intelligence, this chatbot can interpret natural language and deliver more specific solutions to questions posed by users. ChatGPT chatbots provide information on medical issues and the treatment of such illnesses. There are numerous languages available for patients to get replies that are accurate and up-to-date about their queries about clinical characteristics, prescription drugs, and therapies. A patient's health and RPM records may be found in this document, which is accessible to those who provide medical treatment to patients. It is also important for patients to monitor their vital signs and communicate with their healthcare professionals if they see anything that looks out of the ordinary. Patients can plan their medical visits. It is [73]. ChatGPT can reply to a computer program to assist patients in remembering to take their prescriptions and to offer information about their health. By using ChatGPT, it is possible to construct virtual aides for patients. In addition to treating chronic ailments such as diabetes, virtual assistants are also able to administer over-the-counter or home medicines for those who are suffering from acute illnesses such as the common cold and influenza. Voice assistants, smartphone applications, and websites are all mediums via which you may make use of virtual assistants. As a result of the use of ChatGPT in the healthcare industry, concerns about privacy, security, liability, consent, and ethics emerge [73].

When it comes to safeguarding patients in fixed areas utilizing low-Bluetooth-range devices, one of the limitations of wearable patient monitoring systems is the absence of data connection. This presents an issue since it makes it very difficult to monitor patients. A similar level of importance is placed on the acceptability of WPM systems by end users. User knowledge, as well as consent from both the physician and the patient, are requirements. Mobile data transmission can become expensive owing to the acquisition of a variety of data over time [56].

2.2 Research in the Medical Field and the Creation of Novel Medications

Artificial intelligence is well suited to analyze the massive and intricate datasets that are employed in medical research [74]. Additionally, it assists in the development of new drugs, combines various types of data, and may be used to locate work in the field of scientific study [75]. The concentration of pharmaceutical agencies on artificial intelligence is resulting in the streamlining of drug development procedures. Scientists can apply predictive analytics [74] to select appropriate candidates for clinical trials and to construct accurate models of biological processes. In the beginning stages of clinical trials, machine learning may help assist with activities such as choosing a cohort, recruiting participants, and analyzing data via its assistance. With its assistance, clinical trials have the potential to enhance their patient-centeredness, generalizability, efficacy, and results. Despite this, there is a need for more attention to be paid to the conceptual and functional issues that machine learning encountered in clinical trials. It is not known how these methods will affect the overall quality of clinical trials or the experience of the participants; nonetheless, both machine learning and natural language processing have shown that they have the potential to enhance participant management in clinical trials. It is possible to do more research that compares different methods to improve participant management [76]. By synthesizing fresh data, generative artificial intelligence has the potential to improve the quality of databases and increase the variety of clinical research [64]. Additionally, as a result of the use of metaverse apps, researchers can carry out tests in an environment that is both regulated and immersive. The use of the metaverse has several benefits, one of which is that it makes it easier for researchers who are located in different locations to collaborate on their study. Through the use of the metaverse, they can do research together as if they were in the same physical room [77]. During clinical trials, an additional artificial intelligence-based technology known as ChatGPT might be implemented to provide information on clinical studies and to assist with the gathering of data [73]. Summary of relevant articles and identification of significant results are two of how it might assist medical researchers in effectively managing massive volumes of material that can be obtained online [78]. Chatbots that are constructed using ChatGPT can translate medical jargon, which is beneficial to researchers working in the medical field. However, there may be additional ethical considerations associated with the use of chatbots in the field of healthcare research [73].

2.3 Promoting Active Involvement and Treatment Compliance Among Patients

When it comes to health care, the ultimate obstacle is patient participation and compliance, which may determine the success or failure of health results. The term "non-compliance" refers to the situation in which a patient does not adhere to the specified therapy or drug regimen. An increased level of patient participation in health care may lead to improvements in health outcomes, such as usage, cost, and the overall patient experience [79]. A study that was conducted on healthcare executives and leaders discovered that less than fifty percent of patients were actively engaged in the formulation of treatment programs [80].

To develop treatment strategies for patients' acute or chronic health conditions, providers use the services of clinical specialists. It seldom makes a difference when patients fail to make essential behavioral adjustments, such as maintaining a healthy weight, arranging follow-up appointments, and adhering to treatment regimens [79]. Under these circumstances, the adoption of AI proved effective in improving patient engagement. To drive composite interventions and care spectrum, they are increasingly being driven by machine learning and workflow engines [81]. Several studies have shown that it is advantageous to use alerts and individualized materials to motivate behavior at certain periods [79]. Using applications and online portals for communication between healthcare providers and patients may raise engagement rates by at least sixty percent, according to research. Cloud storage allows healthcare applications to store and distribute patient information. Users can access data whenever and wherever they choose, which ultimately leads to better health outcomes for patients. AI applications designed for medical consultations provide patients with information that is not considered an emergency. Patients may be monitored by some applications, and they can also offer warnings for medicine ingestion [82]. In addition, ChatGPT is used in healthcare applications to automate time-consuming processes such as summarizing, creating notes, and producing reports, which ultimately results in greater productivity with the application [73]. It assists patients in managing their symptoms, appointments, and prescriptions, hence improving compliance, education, and self-management of chronic conditions among the patient population.

2.4 Restorative Measures

AI offers new possibilities for rehabilitation. This concept includes both the real world (robotics) and the abstract (informatics). In addition, ML is a branch of AI that makes use of exact techniques to construct algorithms that get better with time. Many fields make use of ML, including symbiotic neuro-prosthetics, myoelectric control, brain-computer interface technology, perioperative medicine rehabilitation, and more. The musculoskeletal system makes use of ML methods for diagnostic imaging, clinical decision support, and patient data evaluation. Therapy evaluations of rehabilitation exercises utilizing machine signals were conducted with the help of an artificial cognitive application [83]. AI and robotics are revolutionizing rehabilitation research and practice due to technological advancements. For example, smart homes can alert caregivers when a resident needs help with everyday tasks. Data collected and made available by smart mobile and wearable devices can help evaluate health improvement and the attainment of individual rehabilitation objectives [84]. Individuals' adherence to exercise routines can be tracked using wearable technology equipped with inertial sensors [85]. Participants in the study wore Apple iWatches while they performed rotator cuff exercises, and their exercise adherence was evaluated. The exercise accuracy across all algorithms was classified using multiple supervised learning methods [86]. The neural network's 99.4 percent detection accuracy demonstrates the promise of ML and wearables for fitness monitoring. Because of adherence issues that vary according to the effectiveness of exercise, it is possible that using wearable devices to evaluate performance will not increase adherence [87, 88]. As a social and physical support system, robots help people recover from injuries and illnesses. Robots can assist with motor, sensory, and cognitive impairments. These technological advancements improve autonomy, functionality, and health [84]. Patients with musculoskeletal dysfunction were mobilized employing robotic hands that were either dexterous or supple. Researchers are currently looking at how well this treatment works in the long run [89]. Robots powered by artificial intelligence may soon be able to track patients' precise movements and boost productivity, according to recent studies [90]. According to Frackiewicz [91], an AI-driven tool called ChatGPT can assist healthcare personnel (HCPs) in matching rehabilitation needs with the availability of therapists. Patients may benefit from a combination of conventional treatment and AI. Patients remain actively involved and enthusiastic throughout their recovery with ChatGPT's personalized and collaborative support. Fitness guidance, progress monitoring, and constructive criticism are all ways ChatGPT aids injured patients' recoveries. The language and speech skills of people who have suffered a stroke or head injury are also enhanced through conversation. This tool can be accessed through digital devices whether you're at home or on the move. Researchers taught the language model ChatGPT to rewrite texts with empathy in mind. Effective peer-to-peer mental health support is enhanced through the development of non-expert conversational skills. Community-based cognitive behavioral therapy may benefit from human-AI partnerships [92].

AI is utilized in metaverse neurorehabilitation through the use of reward-based materials, virtual character movement based on weight shift, and movement assessment based on deep learning. We aim to reduce the risk of COVID-19 transmission while providing entertainment and remedial exercises based on artificial intelligence [93, 94]. Compared to traditional physiotherapy, metaverse physiotherapy (MPT) improved cardiopulmonary and gross motor function in children with CP and decreased perceived COVID-19 infection [95]. The gait analysis of patients with osteoarthritis and Parkinson's disease has made use of ML-driven video analysis to identify anomalies and diseases [96]. With the help of digital therapists, home physiotherapy patients can receive counseling, live monitoring, and remote tracking [97]. To treat stroke patients both at home and in the hospital, Lambercy et al. [98] propose using digital interventions such as minimally supervised assisted therapy. When designing remote neurorehabilitation technologies with AI, keep patient training and home needs in mind while prioritizing utility, safety, and robustness. In neurorehabilitation, patients, families, and healthcare providers must all work together to use scalable technologies in a way that is both clinically motivated and transparent. Patients with this condition may have more faith in home-based rehabilitation programs that use technology. These characteristics also increase the likelihood that patients with neurological disorders will engage in therapy and accept technology-enhanced rehabilitation [99]. By delivering high-quality, sustained, and high-dose therapy, technology-assisted home rehabilitation helped stroke patients regain their independence and improve their quality of life both during and after the COVID-19 pandemic [98].

Wearables for sports medicine might benefit from artificial intelligence, according to a recent review. Athletes can benefit from AI's ability to analyze sensor data for patterns in their physiological responses, spatial relationships, and kinematics. AI enhances models for injury prediction, risk stratification, monitoring of patient health data, and the patient experience. Wearable device challenges may restrict the use of artificial intelligence, despite its potential in sports medicine [100]. Some problems with wearable tech include insufficient data, social bias, insecure data, anomalies, background noise, and poor data quality [100, 101]. Movements of the arms while exercising may be picked up by heart rate monitors. Robust sensors capable of gathering and transmitting accurate data can manage this [100]. Receiving approval from patients is another challenge with adoption. Half of those who used wearable devices stopped doing so within six months, and a third did so altogether [102, 103]. Previous research found that although half of patients found wearable AI insightful, 11% found it harmful. People who needed medical treatment were worried that the technology might compromise their privacy. Outlining the benefits to clinicians and the limitations of AI can help patients accept and use the technology more readily [104].

2.5 Development for Administrative Use

To streamline administrative tasks, AI can extract data from medical records, document patient encounters, and populate structured data fields in therapeutic notes [105]. In the United States, administrative and regulatory duties take up about a quarter of a nurse's time [106]. Medical professionals can save time by using voice texting [105]. Although ML-based algorithmic systems are more accurate, rule-based EHR systems are still in widespread use [80]. According to Wang et al. [51], Amazon is working on a cutting-edge ML solution to mine unstructured electronic health record data and research notes for useful insights. Additionally, BEHRT, a model for EHRs that utilizes bidirectional encoder deep neural sequence transduction, was presented by Li et al. [106]. With the use of embeddings for age, position, visits, and events, the BEHRT can describe a patient's medical history. In the future, 301 conditions are expected to accompany a person's visits. It considerably improves task precision scores in comparison to deep EHR models. Its adaptable architecture allows for the integration of evaluation, diagnosis, and medication to enhance the precision of predictions. The Hi-BEHRT risk prediction model is a relatively new development that relies on transformers. Patients with diabetes, heart failure, stroke, or chronic renal disease were better served by Hi-BEHRT than by deep-learning models when it came to risk prediction [107]. Numerous healthcare operations make use of RPA. The following are included: clinical records, medical record management, claim processing, and revenue cycle management [108]. Chatbots are also utilized by healthcare organizations (HCOs) for patient interfaces, telehealth, and mental health services. Natural language processing can improve plain-old appointment scheduling and medication refills [109]. ML can compare data from websites to process claims and payments [105]. It is necessary to verify the agency for many insurance claims. The costs associated with incorrect claims can be mitigated through data reconciliation and claim scrutiny [109]. In addition, a hybrid ML-based decision support system (ML rule-based expert system) greatly enhanced clinical prescribing error detection, according to Corny et al. [110]. Researchers have found that ML techniques, such as deep learning and natural language processing, are widely used in AI tools for clinical pharmaceutical services. According to the research, AI-based clinical pharmacy applications and resources are in their infancy. To evaluate AI tools in clinical pharmacy services in actual environments, data professionals will need to work together. Recognizing these benefits is essential for pharmacists to use them effectively and to maintain good relations with patients and the healthcare team [111].

3 Research on the Application of Artificial Intelligence in New Healthcare Businesses

To solve the research gap in artificial intelligence healthcare startups, a worldwide firm dataset was evaluated. Drug development, mental health, imaging, genetics, nutrition, hospital decision support, fertility, diagnostics, and pathology were some of the areas in which these businesses successfully used artificial intelligence. The objective is to get an understanding of the many uses of artificial intelligence and its consequences for health care. Both Crunchbase [85] and CB Insights [86] were used in the study. A variety of factors, including patents, companies, and venture money, assist CB Insights in forecasting market developments. The private startups are scouted by Crunchbase. The acquisition of data was then followed by a comprehensive cleansing. It was necessary to thoroughly identify and eliminate duplicate items to guarantee the uniqueness of the dataset. To make future analyses more accurate, we eliminated items that lacked descriptions and nations. Because of a comprehensive screening, non-AI firms were removed from the dataset, which made it eligible for AI research in the healthcare industry. When we had finished cleansing the data thoroughly, we had 2747 AI-based healthcare firms all around the globe. Through this in-depth research, we will demonstrate how AI-driven healthcare firms have the potential to revolutionize the market.

Artificial intelligence is being used by the most successful pathology and diagnostics firms for diagnosis and early detection. These new businesses are highlighted by the examples that follow. Galen is an image identification platform that is offered by Ibex, a major artificial intelligence business in the field of pathology. This platform uses AI algorithms to classify cancer, which improves diagnosis accuracy and saves time for pathologists. The camera on the phone is used by Skin-Vision to identify cancer and lesions. Utilizing visual data, medical imaging companies can identify problems in images. Through the use of machine learning, artificial intelligence assists medical professionals in identifying anomalies in brain scans. Located in Singapore Through the use of medical imaging, artificial intelligence, and computational fluid dynamics, See-Mode can prevent strokes and save lives. Genetic research is funded by several large corporations. Lifebit is the first life-genomics platform to derive personal inferences from DNA data. The FDNA developed by Lifebit, which was sponsored by USD35 million, uses artificial intelligence to identify disease-causing physiological changes. From the year 2000 to the year 2022, the graphic illustrates some fascinating patterns in the number of AI healthcare startups. In the years 2000–2002, there were very few AI healthcare businesses, which is indicative of minimal investment.

Rapid expansion from 12 in 2007 to 116 in 2013 was driven by the promise of artificial intelligence technologies to revolutionize health care. Throughout 2018, interest in AI health care increased from 139 in 2014 to 395. The number decreased from 289 to 62 between the years 2020 and 2022, which may have been caused by COVID-19, the current economic climate, the priorities of investors, or the consolidation of the sector. US, UK, and China-based artificial intelligence healthcare firms dominate the global market. There are the most startups in the United States (1197), the United Kingdom (177), and China (145). This concentration might be explained by the robust ecosystems of technology and innovation in these nations, as well as the trained personnel and assistance from the government. The United States of America is the leader in artificial intelligence healthcare startup investment, with $32.5 billion, showing a robust ecosystem. China is the second largest investor with $7.1 billion, followed by the United Kingdom with $2.9 billion. However, compared to the United States, these numbers indicate that investments in AI healthcare startups are smaller in these nations.

4 Obstacles that Have Been Faced with the Introduction of AI in Health Care

The difficulties associated with the introduction of AI in healthcare sectors are depicted in the following subsections.

4.1 Dilemmas of a Social and Ethical Nature

Concerns about the social and ethical implications of AI-like levels of dependence on technology, automation, data use, and assistive technologies are not new. As AI advances, ethical concerns about accountability, incorrect judgments, authentication, and data protection arise. It's possible that using AI for healthcare decisions, treatments, and equipment control poses risks and lacks reliability. Unintentional mistakes or negative effects caused by AI could have major repercussions [68]. Incorrectly advising doctors to discharge asthmatic patients, the AI app that predicted pneumonia complications lacked context [112]. Issues of transparency and accountability include the responsibility of AI-made decisions and the compensation of victims. ML technologies, in particular, self-review their learning boundaries and guidelines, making data bias recognition and AI output authentication particularly challenging [68]. The practical applications of AI in various fields are hindered by its explainability. Improving comprehension and acceptability of AI-based decision-critical applications is the goal of a subfield of artificial intelligence known as "explainable artificial intelligence (XAI)" [113]. Many of the workings of AI are cryptic and difficult to understand. Decisions made by AI may lose some of the trust and understanding of humans. XAI aims to increase human trust and understanding of AI by making its internal mechanisms more transparent and explainable [114]. Its methods include analysis and trusting the results of ML algorithms [115]. Together, doctors and patients can make sense of XAI-based diagnostic decisions [116]. By giving visual feedback on model prediction metrics, XAI increases user trust in AI, according to a recent study [117]. Radiologists may have more faith in the classification predictions made by automated workflows using XAI methods if they incorporate visual insights [118].

AI might potentially encounter problems with digital data collection or data scarcity. Rare medical conditions or individuals participating in understated clinical trials could be impacted [52]. Applications of data-driven AI mitigate bias and human error, but they also strengthen data biases [119]. It is possible that AI training data does not fairly reflect society at large, leading to biased decisions [53]. When legal panels have access to private medical records, security and privacy concerns surface. While AI has the potential to safeguard healthcare desktops from cyber attacks, it also has the risk of being hacked to access sensitive data or inundated with biased or false information, which makes it difficult to trace [52]. Sunarti et al. [120] state that when it comes to clinical AI applications, there are ethical concerns regarding privacy, safety, security, choice, cost, information, consent, access, and efficacy. The principles of beneficence, autonomy, equity, and non-maleficence should take precedence when integrating AI into healthcare systems [97].

4.2 Governance-Related Concerns

National and international rules are required to guarantee that applications controlled by artificial intelligence in the healthcare industry adhere to the norms of medical ethics. The European Union (EU) enacted the General Data Protection Regulation (GDPR) in 2018 to exercise control over artificial intelligence. When subject data is handled or kept by organizations situated in the EU, the General Data Protection Regulation (GDPR) protects it. Additionally, it sets the framework for significant changes in both the United States of America and Canada [121]. The Artificial Intelligence Act, also known as the AIA, was recently drafted by the European

Commission to address the concerns that have been raised by society over the use of AI. A reduction in the dangers associated with certain technology applications and an increase in the usage of artificial intelligence are the aims of these guidelines. The proposed act mandates conformity evaluations before deployment and post-market observational study [122], to ensure that high-risk artificial intelligence systems are by the criteria of the AIA standards.

4.3 Barriers Associated with Technology

For the operation of HCPs to be technically possible, AI models must have attributes and functions that are straightforward [123]. Some of the challenges that are preventing the widespread use of artificial intelligence in the healthcare industry include the high cost of improving data validity, the growing cost of storing and backing up data for research reasons, and the insufficiency of IT infrastructure to support AI [124]. Other possible downsides of artificial intelligence systems include bias, brittleness (the tendency to be readily deceived), and inapplicability outside of the training area [125]. Brittleness was defined as the tendency to be quickly misled. Things like shifting datasets, using random matching of confounders instead of real signals, the frequency of unintended biases in clinical practice, making algorithms understandable, creating trustworthy confidence measures for the model, and testing the generalizability to other populations are all important considerations that should be taken into account [126]. In light of the fact that cost, technological infrastructure, and the use of AI systems by healthcare professionals are all significant factors, healthcare providers must develop and implement a viable plan for the deployment of AI in health care. The widespread skepticism or lack of knowledge that healthcare professionals have about AI-based clinical decision support systems is another significant obstacle that stands in the way of their general implementation. This mistrust or lack of understanding is caused by the unknown hazards that these systems may offer. In these kinds of circumstances, XAI solutions are given priority to boost end-user confidence and overcome poor adoption of artificial intelligence [117]. In addition, Choudhury and Asan [127] found that physicians' perspectives on the possible advantages and disadvantages of artificial intelligence in the healthcare industry were influenced by several factors, such as the amount of work they had to do, their level of trustworthiness and the dangers connected with AI, and their willingness to get training in AI. The lack of accountability for artificial intelligence is another element that is considered to be a barrier to the deployment of AI. The introduction of artificial intelligence into the curriculum of medical and nursing schools is key to ensuring the safe usage of AI in the future. In a method that is comparable to that of the human brain, artificial intelligence systems take in data and create outcomes.

5 Shortcomings of AI in the Healthcare Industry

Machine learning and deep learning models need enormous datasets to accurately identify or anticipate the many available occupations. However, owing to the sensitive nature of patient records and the typical unwillingness of healthcare organizations (HCOs) to disclose patient information, data accessibility is a challenging problem in the healthcare business. In addition, after an algorithm has utilized the data for the first time, it is becoming more difficult to access the data. It is difficult to create this condition because of opposition from inside the company [128], but machine learning-based systems can make continual improvements as additional data is introduced to their training set.

The use of AI-powered applications raises concerns over the protection of personal information and privacy. When a data breach occurs, hackers often target health records because of the value of this information and the vulnerability they provide. Because of this, maintaining the confidentiality of medical records is of the highest significance [129].

The process of overfitting occurs when the algorithm internalizes the links that exist between the characteristics of the patient and the results. Because of all the many factors that might affect the outcomes, the algorithm generates predictions that are not accurate. The phenomenon known as data leaking is another cause for concern. This happens when an artificial intelligence system grows increasingly accurate at predicting outcomes outside of the limited training dataset [130–132].

In addition, artificial intelligence systems that use deep learning are unable to give as much context for their projections as they earlier were able to. It becomes more difficult for algorithms to defend themselves legally when they produce recommendations that are not very good. Because of this, the experts are unable to comprehend the relationship that exists between the data and their projections, which makes the situation more complicated. The public may lose faith in healthcare systems as a result of the black-box issue that exists in artificial intelligence systems [133].

Those who work in the healthcare industry have a genuine fear over the possibility of job loss and replacement as a result of artificial intelligence. In addition to that, they need to be re-engineered. Another issue that arises with artificial intelligence is the expense of educating medical professionals to make efficient use of it [134]. A fundamental barrier that stands in the way of the efficient use of artificial intelligence in the medical field is the absence of enough experimental data to confirm the efficacy of AI-based drugs in clinical trials that are now in the planning stages. The majority of artificial intelligence research has been conducted in environments that do not include real patients. Consequently, it could be difficult to generalize the findings and conclusions. In addition, academic institutions are cautious about applying AI-based solutions since there is a dearth of data and the quality of research is subpar [135]. Several downsides are associated with AI as a whole. These include the high cost of building applications based on AI, the fact that it causes people to become lazy, the fact that it causes people to lose their employment since AI can perform monotonous activities, and the fact that robots are unable to sense emotions or become creative [136].

6 Epilogue

There are numerous uses for artificial intelligence in the healthcare industry are discussed. In this paper we conduct a comprehensive review of the existing literature on the intersection of artificial intelligence and health care, identifying some primary applications of this technology. The following applications are focused on providing digital healthcare services to patients, conducting research in the pharmaceutical and clinical fields, promoting patient participation and adherence to treatment plans, assisting with patient rehabilitation, and performing additional administrative tasks. The technologies are useful in the following areas of technology-assisted rehabilitation: medical imaging, diagnosis, virtual patient care, patient engagement, treatment adherence, administrative load reduction, innovation in medication and vaccines, exercise compliance monitoring, gait analysis, and pandemic prevention. The administration of treatment also benefits from these technological advancements. On top of that, these technologies are useful when administering treatments. The healthcare industry is quickly embracing artificial intelligence (AI), but there are some obstacles to overcome in terms of ethics, technology management, and governance. Data security and privacy issues may arise from the use of personally identifiable information that is subject to oversight by legal panels. The usage of such content can lead to these issues. There may be limitations to how much problem-solving power artificial intelligence has. This is because it cannot replicate human traits like empathy and the health data it contains is of poor quality. While AI has many positive uses and has been quite successful, it cannot, in the context of mutual aid, replace human relationships. Robots fail miserably at managing and collaborating with humans because they can't form meaningful relationships with them. This completely ruins their chances of accomplishing their goals. Ensuring that AI development and deployment are in line with people's best interests while also taking technical, ethical, and social concerns into account is crucial for the effective administration of AI-based technologies. Medical imaging, diagnostics, virtual patient care, drug development, patient engagement, rehabilitation, and executive functions are some of the fields that might benefit from AI, and this study adds to what is already known about these areas. The most current literature review addresses the ethical, social, governance-related, and technical issues that arise for healthcare practitioners when implementing AI into patient care. These challenges are examined in the context of medical application.

References

1. Snowdon, A.: Digital Health: A Framework for Healthcare Transformation (2020) Available from: https://www.gs1ca.org/documents/digital_health-affht.pdf (accessed on 23 January 2023)
2. Williams, O.D.: COVID-19 and private health: Market and governance failure. Development 63, 181–190 (2020)
3. Ng, R., Tan, K.B.: Implementing an individual-centric discharge process across Singapore public hospitals. Int. J. Environ. Res. Public Health **18**, 8700 (2021)
4. Pee, L.G., Pan, S.L., Cui, L.: Artificial intelligence in healthcare robots: a social informatics study of knowledge embodiment, J. Assoc. Inf. Sci. Technol. **70**(4):351–369 (2019) https://doi.org/10.1002/asi.24145
5. Sunarti, S., Fadzlul Rahman, F., Naufal, M., Risky, M., Febriyanto, K., Masnina, R.: Artificial intelligence in healthcare: opportunities and risk for future. Gac. Sanit. **35**, S67–S70 (2021) https://doi.org/10.1016/j.gaceta.2020.12.019
6. du Plessis, M.: The role of knowledge management in innovation. J. Knowl. Manag. **11**(4), 0–29 (2007). https://doi.org/10.1108/13673270710762684
7. Wahl, B., Cossy-Gantner, A., Germann, S., Schwalbe, N.R.: Artificial intelligence (AI) and global health: how can AI contribute to health in resource-poor settings? BMJ Glob. Health **3**(4), e000798 (2018). https://doi.org/10.1136/bmjgh-2018-000798
8. Hajkowicz, S., Sanderson, C., Karimi, S., Bratanova, A., Naughtin, C.: Artificial intelligence adoption in the physical sciences, natural sciences, life sciences,social sciences and the arts and humanities: A bibliometric analysis of research publications from 1960–2021. Technol. Soc. 102260 (2023) https://doi.org/10.1016/j.techsoc.2023.102260
9. Hoffmann, C.H.: Is AI intelligent? An assessment of artificial intelligence, 70 years after Turing. Technol. Soc. **68**, 101893 (2022). https://doi.org/10.1016/j.techsoc.2022.101893
10. Yang, X., Wang, Y., Byrne, R., Schneider, G., Yang, S.: Concepts of artificial intelligence for computer-assisted drug discovery. Chem. Rev. **119**(18), 10520–10594 (2019)
11. Sánchez Sánchez, P.M., et al.: Intelligent user profiling based on sensors and location data to detect intrusions on mobile devices (2020) https://www.igi-global.com/gateway/chapter/251794
12. Davenport, T., Kalakota, R.: The potential for artificial intelligence in healthcare. Future Healthc. J. **6**(2), 94–98 (2019). https://doi.org/10.7861/futurehosp.6-2-94
13. Yimer, T.H., Wen, C., Yu, X., Jiang, C.: A study of the minimum safe distance between human driven and driverless cars using safe distance model. ArXiv200607022 Phys (2020) [Online]. Available from: http://arxiv.org/abs/2006.07022. (Accessed 26 May 2021)
14. de Barcelos Silva, A., et al.: Intelligent personal assistants: a systematic literature review. Expert Syst. Appl. **147**, 113193 (2020). https://doi.org/10.1016/j.eswa.2020.113193
15. Hosny, A., Parmar, C., Quackenbush, J., Schwartz, L.H., Aerts, H.J.W.L.: Artificial intelligence in radiology. Nat. Rev. Cancer **18**(8) (2018) https://doi.org/10.1038/s41568-018-0016-5
16. Zhou, X.-Y., Guo, Y., Shen, M., Yang, G.-Z.: Application of artificial intelligence in surgery. Front. Med. **14**(4), 417–430 (2020). https://doi.org/10.1007/s11684-020-0770-0
17. French, J., et al.: Identification of patient prescribing predicting cancer diagnosis using boosted decision trees. In: Riaño, D., Wilk, S., Ten Teije, A. (Eds.) Artificial intelligence in medicine. Lecture Notes in Computer Science, Springer International Publishing, Cham, pp. 328–333, 11526 (2019) https://doi.org/10.1007/978-3-030-21642-9_42
18. Fakhoury, M.: Artificial intelligence in psychiatry. In: Kim, Y.-K. (Ed.) Frontiers in Psychiatry: Artificial Intelligence, Precision Medicine, and Other Paradigm Shifts, Advances in Experimental Medicine and Biology, Springer, Singapore, pp. 119–125 (2019) https://doi.org/10.1007/978-981-32-9721-0_6
19. Neves, M.R., Marsh, D.W.R.: Modelling the impact of AI for clinical decision support. In: Ria˜no, D., Wilk, S., Ten Teije, A. (Eds.) Artificial Intelligence in Medicine, Lecture Notes in

Computer Science, vol. 11526, Springer International Publishing, Cham, pp. 292–297 (2019) https://doi.org/10.1007/978-3-030-21642-9_37
20. Bhatia, R.: Telehealth and COVID-19: using technology to accelerate the curve on access and quality healthcare for citizens in India. Technol. Soc. **64**, 101465 (2021) https://doi.org/10.1016/j.techsoc.2020.101465
21. Kasthuri, A.: Challenges to healthcare in India - the five A's. Indian J. Community Med. Off. Publ. Indian Assoc. Prev. Soc. Med. **43**(3), 141–143 (2018) https://doi.org/10.4103/ijcm.IJCM_194_18
22. I. of M. (US) C. on the Summit, H.P.E., Greiner, A.C., Knebel, E.: Challenges facing the health system and implications for educational reform, National Academies Press (US) (2003) https://www.ncbi.nlm.nih.gov/books/NBK221522/
23. Zhidkoff, R.: the future of healthcare with AI, medium 23 (2020) https://becominghuman.ai/the-future-of-healthcare-with-ai-f8593f877b08
24. Sahni, N.R., Stein, G., Zemmel, R., Cutler, D.M.: The potential impact of artificial intelligence on healthcare spending. In: The Economics of Artificial Intelligence: Health Care Challenges, University of Chicago Press (2023) https://www.nber.org/books-and-chapters/economics-artificial-intelligence-health-care-challenges/potential-impact-artificial-intelligence-healthcare-spending
25. Radhakrishnan, J., Chattopadhyay, M.: Determinants and barriers of artificial intelligence adoption—a literature review. In: Sharma S.K., Dwivedi, Y.K., Metri, B., Rana, N.P. (Eds.) Re-imagining Diffusion and Adoption of Information Technology and Systems: A Continuing Conversation, IFIP Advances in Information and Communication Technology, Springer International Publishing, Cham, pp. 89–99 (2020) https://doi.org/10.1007/978-3-030-64849-7_9
26. Colombo, M.G., Piva, E.: Strengths and weaknesses of academic startups: a conceptual model. IEEE Trans. Eng. Manag. **55**(1), 37–49 (2008). https://doi.org/10.1109/TEM.2007.912807
27. Davila, A., Foster, G., Gupta, M.: Venture capital financing and the growth of startup firms. J. Bus. Ventur. **18**(6), 689–708 (2003). https://doi.org/10.1016/S0883-9026(02)00127-1
28. World Health Organization. Global strategy on digital health 2020–2025. pp. 7–13 (2021). Available online: https://www.who.int/docs/defaultsource/documents/gs4dhdaa2a9f352b0445bafbc79ca799dce4d.pdf (accessed on 10 December 2022)
29. Dash, S., Shakyawar, S.K., Sharma, M., Kaushik, S. Big data in healthcare: Management, analysis and future prospects. J. Big Data **6**, 54 (2019)
30. Zhang, C., Ma, R., Sun, S., Li, Y., Wang, Y., Yan, Z.: Optimizing the electronic health records through big data analytics: a knowledge-based view. IEEE Access **7**, 136223–136231 (2019)
31. Rawat, S.: How is big data analytics using AI? 2021. Available from: https://www.analyticssteps.com/blogs/how-big-dataanalytics-using-ai (accessed on 11 January 2023)
32. Ghosh, P.: AI early diagnosis could save heart and cancer patients. science correspondent. BBC News (2018). Available from: https://www.bbc.com/news/health-42357257 (accessed on 11 July 2021)
33. Wang, D., Khosla, A., Gargeya, R., Irshad, H., Beck, A.H.: Deep learning for identifying metastatic breast cancer. arXiv (2016) arXiv:1606.05718
34. Esteva, A., Kuprel, B., Novoa, R.A., Ko, J., Swetter, S.M., Blau, H.M., Thrun, S.: Dermatologist-level classification of skin cancer with deep neural networks. Nature **542**, 115–118 (2017)
35. Rajpurkar, P., Irvin, J., Zhu, K., Yang, B., Mehta, H., Duan, T., Ding, D., Bagul, A., Langlotz, C., Shpanskaya, K.: Chexnet: Radiologist-level pneumonia detection on chest x-rays with deep learning. arXiv (2017) arXiv:1711.05225
36. Bedi, G., Carrillo, F., Cecchi, G.A., Slezak, D.F., Sigman, M., Mota, N.B., Ribeiro, S., Javitt, D.C., Copelli, M., Corcoran, C.M.: Automated analysis of free speech predicts psychosis onset in high-risk youths. NPJ Schitzophrenia **1**, 15030 (2015)
37. IBM Research. IBM 5 in 5: With AI, our words will be a window into our mental health. 2017. Available from: https://www.ibm.com/blogs/research/2017/1/ibm-5-in-5-our-words-will-be-the-windows-to-our-mental-health/ (accessed on 25 December 2022)

38. Chou, C.-Y., Hsu, D.-Y., Chou, C.-H.: Predicting the onset of diabetes with machine learning methods. J. Pers. Med. **13**, 406 (2023)
39. Gudigar, A., Raghavendra, U., Nayak, S., Ooi, C.P., Chan, W.Y., Gangavarapu, M.R., Dharmik, C., Samanth, J., Kadri, N.A., Hasikin, K., et al: Role of Artificial Intelligence in COVID-19 Detection. Sensors **21**, 8045 (2021)
40. He, K., Gan, C., Li, Z., Rekik, I., Yin, Z., Ji, W., Gao, Y., Wang, Q., Zhang, J., Shen, D: Transformers in medical image analysis. Intell. Med. **3**, 59–78 (2023)
41. Khanna, V.V., Chadaga, K., Sampathila, N., Prabhu, S., Chadaga, R., Umakanth, S.: Diagnosing COVID-19 using artificial intelligence: A comprehensive review. Netw. Model Anal Health Inf. Bioinforma 11, 25 (2022)
42. Costa, G.S.S., Paiva, A.C., Junior, G.B., Ferreira, M.M.: COVID-19 automatic diagnosis with ct images using the novel transformer architecture. In: Proceedings of the 21st Brazilian Symposium on Computing Applied to Health, Rio de Janeiro, Brazil, 15–18 June 2021; pp. 293–301 (2021)
43. van Tulder, G., Tong, Y., EMarchiori, E.: Multi-view analysis of unregistered medical images using cross-view transformers. In: Proceedings of the Medical Image Computing and Computer Assisted Intervention–MICCAI 2021: 24th International Conference, Part III 24, Strasbourg, France, 27 September–1 October 2021; Springer Nature: Basel, Switzerland, 2021; pp. 104–113
44. Krishnan, K.S., Krishnan, K.S.: Vision transformer based COVID-19 detection using chest x-rays. In: Proceedings of the 2021 6th International Conference on Signal Processing, Computing and Control (ISPCC), Solan, India, 7–9 October 2021; pp. 644–648
45. Gheflati, B., Rivaz, H.: Vision transformer for classification of breast ultrasound images. arXiv, arXiv:211014731 (2021)
46. Wolterink, J.M., Mukhopadhyay, A., Leiner, T., Vogl, T.J., Bucher, A.M., Išgum, I.: Generative adversarial networks: A primer for radiologists. Radio Graphics **41**, 840–857 (2021)
47. Chuquicusma, M.J.M., Hussein, S., Burt, J., Bagci, U.: How to fool radiologists with generative adversarial networks? A visual turing test for lung cancer diagnosis. In: Proceedings of the IEEE 15th International Symposium on Biomedical maging, Washington, DC, USA, 4–7 April 2018; pp. 240–244 (2018)
48. Arora, A., Arora, A.: Generative adversarial networks and synthetic patient data: Current challenges and future perspectives. Futur. Healthcare J. **9**, 190–193 (2022)
49. Will ChatGPT transform healthcare? Nat. Med. **29**, 505–506 (2023)
50. Kuehn, B.M.: More than one-third of US individuals use the internet to self-diagnose. JAMA **309**, 756–757 (2013)
51. Wang, G., Badal, A., Jia, X., Maltz, J.S., Mueller, K., Myers, K.J., Niu, C., Vannier, M., Yan, P., Yu, Z., et al.: Development of metaverse for intelligent healthcare. Nat. Mach. Intell. **4**, 922–929 (2022)
52. Nuffield Council on Bioethics. Artificial Intelligence (AI) in Healthcare and Research. Nuffield Council on Bioethics (2018). Available from: https://www.nuffieldbioethics.org/assets/pdfs/Artificial-Intelligence-AI-in-healthcare-and-research.pdf (accessed on 11 December 2022)
53. House of Lords. AI in the UK: Ready, willing and able? house of lords select committee on artificial intelligence: report of session 2017–2019. Authority of the House of Lords (2018). Available online: https://publications.parliament.uk/pa/ld201719/ldselect/ldai/100/100.pdf (accessed on 16 December 2022)
54. Secinaro, S., Calandra, D., Secinaro, A., Muthurangu, V., Biancone, P.: The role of artificial intelligence in healthcare: A structured literature review. BMC Med. Inform. Decis. Mak. 21, 1–23 (2021)
55. Oren, O., Gersh, B.J., Bhatt, D.L.: Artificial intelligence in medical imaging: Switching from radiographic pathological data to clinically meaningful endpoints. Lancet Digit. Health **2**, e486–e488 (2020)
56. Baig, M.M.; Gholam Hosseini, H.; Moqeem, A.A.; Mirza, F.; Lindén, M.: A systematic review of wearable patient monitoring systems–current challenges and opportunities for clinical adoption. J. Med. Syst. **41**, 115 (2017)

57. Kim, J., Campbell, A.S., Wang, J.: Wearable non-invasive epidermal glucose sensors: A review. Talanta 177, 163–170 (2018)
58. Patel, D., Tarakji, K.G.: Smartwatch diagnosis of atrial fibrillation in patient with embolic stroke of unknown source: A case report. Cardiovasc. Digit. Health J. 2, 84–87 (2021)
59. Sükei, E., Norbury, A., Perez-Rodriguez, M.M., Olmos, P.M., Artés, A.: Predicting emotional states using behavioral markers derived from passively sensed data: data-driven machine learning approach. JMIR mHealth uHealth **9**, e24465 (2021)
60. Natarajan, A., Su, H.-W., Heneghan, C.: Assessment of physiological signs associated with COVID-19 measured using wearable devices. NPJ Digit. Med. **3**, 1–8 (2020)
61. Bogu, G., Snyder, M.: Deep learning-based detection of COVID-19 using wearables data. Deep Learning-Based Detection of COVID-19 Using Wearables Data. MedRxiv (2021)
62. Tschopp, J., L'Huillier, A.G., Mombelli, M., Mueller, N.J., Khanna, N., Garzoni, C., Meloni, D., Papadimitriou-Olivgeris, M., Neofytos, D., Hirsch, H.H., et al.: First experience of SARS-CoV-2 infections in solid organ transplant recipients in the Swiss transplant cohort study. Am. J. Transplant. **20**, 2876–2886 (2020)
63. Yu, M., Tang, A., Brown, K., Bouchakri, R., St-Onge, P., Wu, S., Reeder, J., Mullie, L., Chassé, M.: Integrating artificial intelligence in bedside care for COVID-19 and future pandemics. BMJ 375, e068197 (2021)
64. Garavand, A., Aslani, N.: Metaverse phenomenon and its impact on health: A scoping review. Inform. Med. Unlocked **32**, 101029 (2022)
65. Ganapathy, K.: Telemedicine and neurological practice in the COVID-19 Era. Neurol. India **68**, 555–559 (2020)
66. Lukas, H., Xu, C., Yu, Y., Gao, W.: Emerging telemedicine tools for remote COVID-19 diagnosis, monitoring, and management. ACS Nano **14**, 16180–16193 (2020)
67. Chen, D., Zhang, R.: Exploring research trends of emerging technologies in health metaverse: A bibliometric analysis; Elsevier: Amsterdam, The Netherlands (2022)
68. Gadekallu, T.R., Huynh-The, T., Wang,W., Yenduri, G., Ranaweera, P., Pham, Q.-V., da Costa, D.B., Liyange, M.: Blockchain for the metaverse: A review. arXiv, arXiv:2203.09738 (2022)
69. Chengoden, R., Victor, N., Huynh-The, T., Yenduri, G., Jhaveri, R.H., Alazab, M., Bhattacharya, S., Hegde, P., Maddikunta, P.K.R., Gadekallu, T.R.: Metaverse for healthcare: A survey on potential applications, challenges and future directions. IEEE Access **11**, 12765–12795 (2023)
70. Futurside. Remote Patient Monitoring Devices, Technology, and Its Future (2022). Available from: https://futurside.com/howwearable-medical-device-is-reshaping-remote-patient-monitoring-rpm/ (accessed on 11 January 2023)
71. Shaik, T., Tao, X., Higgins, N., Li, L., Gururajan, R., Zhou, X., Acharya, U.R.: Remote patient monitoring using artificial intelligence: Current state, applications, and challenges. WIREs Data Min. Knowl. Discov. **13**, 1485 (2023)
72. Bouabida, K., Malas, K., Talbot, A., Desrosiers, M., Lavoie, F., Lebouché, B., Taguemout, M., Rafie, E., Lessard, D., Pomey, M.-P.: Remote patient monitoring program for COVID-19 patients following hospital discharge: A cross-sectional study. Front. Digit. Health **3**, 153 (2021)
73. Javaid, M., Haleem, A., Singh, R.P.: ChatGPT for healthcare services: An emerging stage for an innovative perspective. BenchCouncil Trans. Benchmarks stand. Eval. **3**, 100105 (2023)
74. Academy of Royal Medical Colleges. Artificial intelligence in healthcare (2019). Available from: https://www.aomrc.org.uk/wpcontent/uploads/2019/01/Artificial_intelligence_in_healthcare_0119.pdf (accessed on 11 January 2023)
75. O'mara-Eves, A., Thomas, J., McNaught, J., Miwa, M., Ananiadou, S.: Using text mining for study identification in systematic reviews: A systematic review of current approaches. Syst. Rev. **4**, 5 (2015)
76. Weissler, E.H., Naumann, T., Andersson, T., Ranganath, R., Elemento, O., Luo, Y., Freitag, D.F., Benoit, J., Hughes, M.C., Khan, F., et al.: The role of machine learning in clinical research: Transforming the future of evidence generation. Trials **22**, 1–15 (2021)

77. Suh, I., McKinney, T., Siu, K.-C.: Current perspective of Metaverse application in medical education, research and patient care. Virtual Worlds **2**, 115–128 (2023)
78. Khan, R.A., Jawaid, M., Khan, A.R., Sajjad, M.: ChatGPT-Reshaping medical education and clinical management. Pak. J. Med. Sci. **39**, 7653 (2023)
79. Davenport, T., Kalakota, R.: The potential for artificial intelligence in healthcare. Future Healthcare J. **6**, 94–98 (2019)
80. Davenport, T.H., Hongsermeier, T., Mc Cord, K.A.: Using AI to improve electronic health records. Harvard Business Review (2018). Available online: https://hbr.org/2018/12/using-ai-to-improve-electronic-health-records (accessed on 28 December 2022)
81. Volpp, K., Mohta, S.: Improved engagement leads to better outcomes, but better tools are needed. Insights Report. NEJM Catalyst. 2016. Available from: https://catalyst.nejm.org/patient-engagementreport-improved-engagement-leads-betteroutcomes-better-toolsneeded (accessed on 15 January 2023)
82. Hukunda, B., Rau, A., Upadhyay, P.: Reimaging healthcare opportunities with artificial intelligence. Infosys Navigate Your Next (2018). Available from: https://www.infosys.com/industries/healthcare/featuresopinions/Documents/reimaginінghealthcare-opportunities.pdf (accessed on 10 January 2023)
83. Anderson, D.: Artificial intelligence and applications in PM&R. Am. J. Phys. Med. Rehabil. **98**, e128–e129 (2019)
84. Luxton, D.D., Riek, L.D.: Artificial intelligence and robotics in rehabilitation. In: Brenner, L.A., Reid-Arndt, S.A., Elliott, T.R., Frank, R.G., Caplan, B. (Eds.) Handbook of Rehabilitation Psychology. American Psychological Association: Washington, DC, USA, pp. 507–520 (2019)
85. Goldzweig, C.L., Orshansky, G., Paige, N.M., Towfigh, A.A., Haggstrom, D.A., Miake-Lye, I., Beroes, J.M., Shekelle, P.G.: Electronic patient portals: evidence on health outcomes, satisfaction, efficiency, and attitudes. Ann. Intern. Med. **159**, 677–687 (2013)
86. Sinsky, C.A., Willard-Grace, R., Schutzbank, A.M., Margolius, D., Bodenheimer, T.: In search of joy in practice: A report of 23 high-functioning primary care practices. Ann. Fam. Med. **11**, 272–278 (2013)
87. Sinsky, C., Colligan, L., Li, L., Prgomet, M., Reynolds, M.S., Goeders, M.L., Westbrook, J., Tutty, M., Blike, G.: Allocation of physician time in ambulatory practice: A time and motion study in 4 specialties. Ann. Intern. Med. **165**, 753–760 (2016)
88. Lin, S., Khoo, J., Schillinger, E.: Next big thing: Integrating medical scribes into academic medical centres. BMJ Simul. Technol. Enhanc. Learn. **2**, 27–29 (2016)
89. Choi, C., Schwarting, W., DelPreto, J., Rus, D.: Learning object grasping for soft robot hands. IEEE Robot. Autom. Lett. **3**, 2370–2377 (2018)
90. Aggarwal, R., Ganvir, S.S.: Artificial intelligence in physiotherapy. Physiother. J. Indian Assoc. Physiother. **15**, 55 (2021)
91. Frackiewicz, M.: ChatGPT and the future of rehabilitation therapy: An AI-driven approach. TS2 Space (2023). Available from: https://ts2.space/en/chatgpt-and-the-future-of-rehabilitation-therapy-an-ai-driven-approach/ (accessed on 10 May 2023)
92. Sharma, A., Lin, I.W., Miner, A.S., Atkins, D.C., Althoff, T.: Human–AI collaboration enables more empathic conversations in text-based peer-to-peer mental health support. Nat. Mach. Intell. **5**, 46–57 (2023)
93. Chung, E.J., Lee, B.-H.: The effects of flipped learning on learning motivation and attitudes in a class of college physical therapy students. J. Probl. Learn. **5**, 29–36 (2018)
94. Nadarzynski, T., Miles, O., Cowie, A., Ridge, D.: Acceptability of artificial intelligence (AI)-led chatbot services in healthcare: A mixed-methods study. Digit. Health **5**, 2055207619871808 (2019)
95. Moon, I., An, Y., Min, S., Park, C.: Therapeutic effects of metaverse rehabilitation for cerebral palsy: A randomized controlled trial. Int. J. Environ. Res. Public Health **20**, 1578 (2023)
96. Kidziński, Ł., Delp, S., Schwartz, M.: Automatic real-time gait event detection in children using deep neural networks. PLoS ONE **14**, e0211466 (2019)

97. Farhud, D.D., Zokaei, S.: Ethical issues of artificial intelligence in medicine and healthcare. Iran. J. Public Health **50**, i–v (2021)
98. Lambercy, O., Lehner, R., Chua, K., Wee, S.K., Rajeswaran, D.K., Kuah, C.W.K., Ang, W.T., Liang, P., Campolo, D., Hussain, A., et al.: Neurorehabilitation from a distance: Can intelligent technology support decentralized access to quality therapy? Front. Robot. **8**, 612415 (2021)
99. Neibling, B.A., Jackson, S.M., Hayward, K.S., Barker, R.N.: Perseverance with technology-facilitated home-based upper limb practice after stroke: A systematic mixed studies review. J. Neuroeng. Rehabil. **18**, 1–26 (2021)
100. Chidambaram, S., Maheswaran, Y., Patel, K., Sounderajah, V., Hashimoto, D.A., Seastedt, K.P., McGregor, A.H., Markar, S.R., Darzi, A.: Using artificial intelligence-enhanced sensing and wearable technology in sports medicine and performance optimisation. Sensors **22**, 6920 (2022)
101. Gichoya, J.W., McCoy, L.G., Celi, L.A., Ghassemi, M.: Equity in essence: A call for operationalising fairness in machine learning for healthcare. BMJ Health Care Inform. **28**, e100289 (2021)
102. Soliño-Fernandez, D., Ding, A., Bayro-Kaiser, E., Ding, E.L.: Willingness to adopt wearable devices with behavioral and economic incentives by health insurance wellness programs: Results of a US cross-sectional survey with multiple consumer health vignettes. BMC Public Health **19**, 1649 (2019)
103. Gao, Y., Li, H., Luo, Y.: An empirical study of wearable technology acceptance in healthcare. Ind. Manag. Data Syst. **115**, 1704–1723 (2015)
104. Tran, V.-T., Riveros, C., Ravaud, P.: Patients' views of wearable devices and AI in healthcare: Findings from the ComPaRe e-cohort. NPJ Digit. Med. **2**, 53 (2019)
105. Wani, S.U.D., Khan, N.A., Thakur, G., Gautam, S.P., Ali, M., Alam, P., Alshehri, S., Ghoneim, M.M., Shakeel, F.: Utilization of artificial intelligence in disease prevention: Diagnosis, treatment, and implications for the healthcare workforce. Healthcare **10**, 608 (2022)
106. Berg, S.: Nudge theory explored to boost medication adherence. Chicago: American Medical Association (2018). Available from: www.ama-assn.org/delivering-care/patient-support-adv ocacy/nudge-theory-exploredboost-medication-adherence (accessed on 1 January 2023)
107. Li, Y., Rao, S., Solares, J.R.A., Hassaine, A., Ramakrishnan, R., Canoy, D., Zhu, Y., Rahimi, K., Salimi-Khorshidi, G.: BEHRT: Transformer for electronic health records. Sci. Rep. **10**, 7155 (2020)
108. Li, Y., Mamouei, M., Salimi-Khorshidi, G., Rao, S., Hassaine, A., Canoy, D., Lukasiewicz, T., Rahimi, K.: Hi-BEHRT: Hierarchical transformer-based model for accurate prediction of clinical events using multimodal longitudinal electronic health records. IEEE J. Biomed. Health Inform. **27**, 1106–1117 (2023). Available from: https://ieeexplore.ieee.org/stamp/stamp.jsp?arnumber=99 64038 (accessed on 10 May 2023)
109. Utermohlen, K.: Four robotic process automation (RPA) applications in the healthcare industry. Medium (2018). Available from: https://medium.com/@karl.utermohlen/4-robotic-process-automation-rpa-applications-inthe-healthcare-industry-4d449b24b613 (accessed on 29 December 2022)
110. Corny, J., Rajkumar, A., Martin, O., Dode, X., Lajonchère, J.-P., Billuart, O., Bézie, Y., Buronfosse, A.: A machine learning–based clinical decision support system to identify prescriptions with a high risk of medication error. J. Am. Med. Inform. Assoc. **27**, 1688–1694 (2020)
111. Ranchon, F., Chanoine, S., Lambert-Lacroix, S., Bosson, J.-L., Moreau-Gaudry, A., Bedouch, P.: Development of artificial intelligence powered apps and tools for clinical pharmacy services: A systematic review. Int. J. Med. Inform. **172**, 104983 (2023)
112. Caruana, R., Lou, Y., Gehrke, J., Koch, P., Sturm, M., Elhadad, N.: Intelligible models for healthcare: Predicting pneumonia risk and hospital 30-day readmission. In: Proceedings of the 21th ACM SIGKDD International Conference on Knowledge Discovery and Data Mining, Sydney, Australia, 10 August 2015; ACM Press: Sydney, NSW, Australia, pp. 1721–1730 (2015)
113. Clement, T., Kemmerzell, N., Abdelaal, M., Amberg, M.: XAIR: A systematic Metareview of explainable AI (XAI) aligned to the software development process. Mach. Learn. Knowl. Extr. **5**, 78–108 (2023)

114. Tiwari, R.: Explainable AI (XAI) and its applications in building trust and understanding in AI decision making. International J. Sci. Res. Eng. Manag. **7**, 1–13 (2023)
115. Iroju, O.G., Olaleke, J.O.: A systematic review of natural language processing in healthcare. Int. J. Inf. Technol. Comput. Sci. **7**, 44–50 (2015)
116. Alvarez-Melis, D., Jaakkola, T.S.: Towards robust interpretability with self-explaining neural networks (2018). arXiv, arXiv:1806.07538
117. Giuste, F., Shi, W., Zhu, Y., Naren, T., Isgut, M., Sha, Y., Tong, L., Gupte, M., Wang, M.D.: Explainable artificial intelligence methods in combating pandemics: A systematic review. IEEE Rev. Biomed. Eng. **16**, 5–21 (2022)
118. Jadhav, S., Deng, G., Zawin, M., Kaufman, A.E.: COVID-view: Diagnosis of COVID-19 using chest CT. IEEE Trans. Vis. Comput. Graph. **28**, 227–237 (2021)
119. Dilsizian, S.E., Siegel, E.L.: Artificial intelligence in medicine and cardiac imaging: harnessing big data and advanced computing to provide personalized medical diagnosis and treatment. Curr. Cardiol. Rep. **16**, 1–8 (2013)
120. Sunarti, S., Rahman, F.F., Naufal, M., Risky, M., Febriyanto, K., Masnina, R.: Artificial intelligence in healthcare: Opportunities and risk for future. Gac. Sanit. **35**, S67–S70 (2021)
121. Forcier, M.B., Gallois, H., Mullan, S., Joly, Y.: Integrating artificial intelligence into health care through data access: Can the GDPRact as a beacon for policymakers? J. Law Biosci. **6**, 317–335 (2019)
122. Schaake, M.: The European commission's artificial intelligence act. Standford univeristy human-centered artificial intelligence (BHAI), Standford, Canada (2021). Available from: https://hai.stanford.edu/sites/default/files/2021-06/HAI_Issue-Brief_The-European-Commissions-Artificial-Intelligence-Act.pdf (accessed on 5 May 2023)
123. Brown, R.: Challenges to successful AI implementation in healthcare. Data Science Central (2022). Available from: https://www.datasciencecentral.com/challenges-to-successful-ai-implementation-in-healthcare/ (accessed on 30 March 2023)
124. Tachkov, K., Zemplenyi, A., Kamusheva, M., Dimitrova, M., Siirtola, P., Pontén, J., Nemeth, B., Kalo, Z., Petrova, G.: Barriers to use artificial intelligence methodologies in health technology assessment in central and east European Countries. Front. Public Health **10**, 921226 (2022)
125. Marcus, G.: Deep learning: A Critical Appraisal (2018). arXiv. Available from: https://arxiv.org/abs/1801.00631 (accessed on 1May 2019). 158. Kelly, C.J., Karthikesalingam, A., Suleyman, M., Corrado, G., King, D.: Key challenges for delivering clinical impact with artificial intelligence. BMC Med. **17**, 195 (2019)
126. Kelly, C.J., Karthikesalingam, A., Suleyman, M., Corrado, G., King, D.: Key challenges for delivering clinical impact with artificial intelligence. BMC Med. **17**, 195 (2019)
127. Choudhury, A., Asan, O.: Impact of accountability, training, and human factors on the use of artificial intelligence in healthcare: Exploring the perceptions of healthcare practitioners in the US. Hum. Factors Healcare **2**, 100021 (2022)
128. Joshi, N.: 5 AI implementation challenges in healthcare (2020). Available from: https://www.allerin.com/blog/5-aiimplementation-challenges-in-healthcare (accessed on 31 March 2023)
129. Khan, B., Fatima, H., Qureshi, A., Kumar, S., Hanan, A., Hussain, J., Abdullah, S.: Drawbacks of artificial intelligence and their potential solutions in the healthcare sector. Biomed. Mater. Devices 1–8 (2023)
130. Baowaly, M.K., Lin, C.-C., Liu, C.-L., Chen, K.-T.: Synthesizing electronic health records using improved generative adversarial networks. J. Am. Med. Inform. Assoc. **26**, 228–241 (2018)
131. Neill, D.B.: Using artifcial intelligence to improve hospital inpatient care. IEEE Intell. Syst. **28**, 92–95 (2013)
132. Fernandes, M., Vieira, S.M., Leite, F., Palos, C., Finkelstein, S., Sousa, J.M.: Clinical decision support systems for triage in the emergency department using intelligent systems: A review. Artif. Intell. Med. **102**, 101762 (2020)
133. Gama, F., Tyskbo, D., Nygren, J., Barlow, J., Reed, J., Svedberg, P.: Implementation frameworks for artificial intelligence translation into health care practice: Scoping review. J. Med. Internet Res. **24**, e32215 (2022)

134. Wolff, J., Pauling, J., Keck, A., Baumbach, J.: Systematic review of economic impact studies of artificial intelligence in health care. J. Med. Internet Res. **22**, e16866 (2020)
135. Alami, H., Lehoux, P., Denis, J.L., Motulsky, A., Petitgand, C., Savoldelli, M., Rouquet, R., Gagnon, M.P., Roy, D., Fortin, J.P.: Organizational readiness for artifcial intelligence in health care: Insights for decision-making and practice. J. Health Organ. Manag. **35**, 106–114 (2021)
136. Duggal, N.: Advantages and disadvantages of artificial intelligence. Simplilearn (2023). Available from: https://www.simplilearn.com/advantages-and-disadvantages-of-artificial-intelligence-article (accessed on 11 May 2023)
137. Javaid, M., Haleem, A., Singh, R.P., Suman, R., Rab, S.: Significance of machine learning in healthcare: Features, pillars and applications. Int. J. Intell. Networks **3**, 58–73 (2022)
138. Coursera. What is machine learning in health care? Applications and opportunities (2022). Available from: https://www.coursera.org/articles/machine-learning-in-health-care (accessed on 30 March 2023)
139. Hashimy, L., Treiblmaier, H., Jain, G.: Distributed ledger technology as a catalyst for open innovation adoption among small and medium-sized enterprises. J. High Technol. Manag. Res. **32**, 100405 (2021)
140. Stampa, K.: How distributed ledger technology will transform health data. Healthcare (2020). Available from: https://healthcaredigital.com/technology-and-ai/how-distributed-ledger-technology-will-transform-health-data (accessed on 30March 2023)
141. Alruwaili, F.F.: Artificial intelligence and multi agent based distributed ledger system for better privacy and security of electronic healthcare records. PeerJ. Comput. Sci. **6**, e323 (2020)
142. Sadiku, M.N.O., Zhou, Y., Musa, S.M.: Natural language processing. Int. J. Adv. Sci. Res. Eng. **4**, 68–70 (2018)
143. Trunfio, M., Rossi, S.: Advances in Metaverse investigation: streams of research and future agenda. Virtual Worlds **1**, 103–129 (2022)
144. Park, S.-M., Kim, Y.-G.: A metaverse: Taxonomy, components, applications, and open challenges. IEEE Access **10**, 4209–4251 (2022)
145. Petrigna, L., Musumeci, G.: The Metaverse: A new challenge for the healthcare system: A scoping review. J. Funct. Morphol. Kinesiol. **7**, 63 (2022)
146. Thomason, J.: MetaHealth-how will the Metaverse change health care? J. Metaverse **1**, 13–16 (2021). Available from: https://dergipark.org.tr/en/download/article-file/2167692 (accessed on 31 December 2022)
147. Hassani, H., Silva, E.S: The Role of ChatGPT in data science: How AI-assisted conversational interfaces are revolutionizing the field. Big Data and Cogn. Comput. **7**, 62 (2023)
148. Sallam, M.: ChatGPT utility in healthcare education, research, and practice: systematic review on the promising perspectives and valid concerns. Healthcare **11**, 887 (2023)
149. Xu, L., Sanders, L., Li, K., Chow, J.C.L.: Chatbot for health care and oncology applications using artificial intelligence and machine learning: systematic review. JMIR Cancer **7**, e27850 (2021)
150. Lin, T., Wang, Y., Liu, X., Qiu, X.: A survey of transformers. AI Open **3**, 111–132 (2022)
151. Shome, D., Kar, T., Mohanty, S.N., Tiwari, P., Muhammad, K., AlTameem, A., Zhang, Y., Saudagar, A.K.J.: COVID-transformer: Interpretable COVID-19 detection using vision transformer for healthcare. Int. J. Environ. Res. Public Health **18**, 11086 (2021)
152. Shamshad, F., Khan, S., Zamir, S.W., Khan, M.H., Hayat, M., Khan, F.S., Fu, H.: Transformers in medical imaging: A survey. Med. Image Anal. 102802 (2023)

Early Diagnosis of Medical Images in Healthcare Management by Artificial Intelligence

Rakib Hasan, Moddassir Khan Nayeem, P. Santhiya, and Amrit Das

Abstract Medical imaging is crucial for the early diagnosis and management of diseases, offering detailed insights into patient conditions. With the advent of artificial intelligence (AI) and machine learning (ML), the potential to automate and refine these diagnostic processes has expanded significantly. This book chapter outlines advanced machine learning techniques applied to detect abnormalities in medical images, enhancing healthcare management systems. Key algorithms such as Logistic Regression (LR), Random Forest (RF), and Decision Trees (DT) are explored for their efficacy in differentiating COVID-19 from healthy chest X-ray images. The integration of these AI techniques not only improves clinical workflows by reducing costs but also enhances patient outcomes by providing quick and accurate diagnoses. Evidence from this study indicates that deep learning models can identify COVID-19 cases from images with over 99% accuracy, matching the expertise of radiologists. The application of AI in this context demonstrates its capacity to relieve the heavy burdens on healthcare systems by automating complex diagnostic tasks and optimizing the allocation of critical resources. Furthermore, this chapter discusses the broader implications of AI in healthcare administration, highlighting how intelligent algorithms contribute to strategic decision-making and the resilience of healthcare infrastructures during crises like the COVID-19 pandemic. The insights gained underscore AI's transformative potential in augmenting human capabilities and shaping future healthcare solutions.

R. Hasan · M. K. Nayeem
Department of Mechanical, Aerospace and Industrial Engineering, University of Texas, San Antonio, TX, USA
e-mail: rakib.hasan@my.utsa.edu

M. K. Nayeem
e-mail: moddassirkhan.nayeem@utsa.edu

P. Santhiya · A. Das (✉)
Department of Mathematics, School of Advanced Sciences, Vellore Institute of Technology, Vellore, Tamil Nadu, India
e-mail: das.amrit12@gmail.com; amrit.das@vit.ac.in

P. Santhiya
e-mail: santhiyanathan8@gmail.com

L. Sahoo et al. (eds.), *Decision Making Under Uncertainty Via Optimization, Modelling, and Analysis*, Studies in Systems, Decision and Control 558,
https://doi.org/10.1007/978-981-96-0085-4_22

Keywords Healthcare management · Artificial intelligence · Machine learning · COVID-19 · X-ray · Logistic regression · Random forest · Decision tree · CNN · TensorFlow

1 Introduction

The COVID-19 pandemic has presented unprecedented challenges for healthcare systems and providers worldwide. However, amidst these challenges, there is a beacon of hope in the form of artificial intelligence (AI) and machine learning. These technologies are proving to be invaluable in augmenting human capabilities and efficiency, particularly in the fields of medical image analysis and predictive diagnostics [1]. Their potential to revolutionize healthcare management is immense, offering a promising future for healthcare professionals, researchers, and policymakers.

Medical imaging is an indispensable tool healthcare professionals use for screening, diagnosis, treatment planning, and assessing response to therapy in most diseases. Different imaging modalities like X-ray, CT (computed tomography), MRI (magnetic resonance imaging), ultrasound, and nuclear imaging provide detailed anatomical and functional information about tissues in the human body [2]. However, manual interpretation of medical images is a challenging task requiring extensive training of radiologists. At the same time, due to the increasing burden of both communicable and non-communicable diseases, the workload on radiology departments has increased manifold. As per studies, the global volume of medical imaging studies reached 4.6 billion in 2014, growing rapidly at 7% per year [3]. To tackle the problem of overwhelming imaging volumes combined with a shortage of qualified radiologists, computer-aided diagnosis (CAD) systems are being developed actively in healthcare. CAD systems aim to support radiologists by automatically detecting clinically significant features in medical images that may indicate the presence of tumours, lesions, or other abnormalities [4]. Machine learning and artificial intelligence algorithms play a pivotal role in building robust CAD solutions. They have the potential to significantly optimize radiology workflows by fast screening of regular studies, bringing suspicious cases to attention, and reducing errors or fatigue-related interpretation inaccuracies. With rising evidence around safety, time, and cost benefits, the integration of AI-based decision support is also gaining acceptance in clinical settings [5].

Besides improving efficiency, machine learning systems impact patient outcomes when incorporated into critical diagnosis tasks. Early detection of disorders leads to timely therapeutic interventions, which will enhance prognosis and survival rates. For instance, vast research in building CAD systems is driven by the fact that the 5-year survival rate for lung cancer patients improves from 16 to 70% when diagnosed at an early stage [6]. A similar rationale exists behind machine learning research for brain, cardiac, breast, and several other disorders diagnosed using medical imaging.

This chapter delves into the research conducted at the intersection of AI, machine learning, and COVID-19 diagnostics, serving as a case example of how intelligent algorithms can bolster healthcare management under pressure. We have developed a prediction model based on a Convolutional Neural Network (CNN) implemented with TensorFlow, using a dataset containing 3661 COVID-19 and 10,192 regular (healthy) chest X-ray images. The study's performance was measured by classification accuracy. This research not only demonstrates the potential of AI and machine learning in COVID-19 diagnostics but also highlights the broader need within healthcare administration for reliable, automatable diagnostic aids and quantitative data to guide resource allocation amidst the pandemic, providing reassurance about the ongoing efforts in this field.

Building upon the foundation laid by our CNN-based predictive model, this chapter progresses to scrutinize the broader spectrum of artificial intelligence applications in healthcare crisis management. The initial success in utilizing CNN for diagnosing COVID-19 from chest X-ray images prompts a deeper exploration into how these technologies can be scaled and optimized across different scenarios within healthcare systems. Thus, we are compelled to ask critical research questions that not only assess the efficacy of such AI models in clinical accuracy but also examine their impact on operational efficiency during healthcare emergencies.

As we navigate the complexities of integrating artificial intelligence with clinical practices, this study is driven by pivotal questions that seek to harness the potential of machine learning in enhancing the diagnostic processes of healthcare systems. Firstly, how do advanced machine learning algorithms enhance the accuracy of COVID-19 diagnosis from chest X-ray images compared to traditional diagnostic methods? Moreover, can these AI-driven methods expedite the diagnosis, thereby improving the efficiency of healthcare responses during emergencies such as pandemics? This investigation also compares the effectiveness of several algorithms—Decision Trees, Random Forests, Logistic Regression, and Convolutional Neural Networks—in accurately classifying COVID-19 cases. Additionally, we evaluate the reliability and practical applicability of these models by analyzing some metrics like sensitivity, specificity, and accuracy. Ultimately, this study explores the broader implications of deploying AI in healthcare, particularly its role in strategic decision-making and managing healthcare resources during crises. By addressing these questions, our research aims to provide substantive insights into the deployment of AI technologies for more resilient and responsive healthcare infrastructures.

2 Literature Review

Artificial intelligence (AI) has revolutionized medical image analysis, offering significant potential for early diagnosis and healthcare management. The application of AI in medical imaging has garnered substantial attention from researchers and health professionals [7]. AI algorithms, particularly deep learning, have demonstrated remarkable progress in image-recognition tasks, leading to a reduction in the burden on physicians treating various medical conditions, including cancer, and gliomas [8–10] and during pandemic situations like COVID-19 [11]. Leveraging deep learning methods in medical image analysis could accelerate the acceptance of AI applications among clinicians and patients, emphasizing the need for medical professionals to understand the underlying principles and techniques of AI [12, 13]. Furthermore, machine learning (ML) algorithms have made a tremendous impact in the field of medical imaging, with weakly supervised deep learning models gaining popularity in medical image segmentation [14, 15].

The use of CNN and deep architectures has attracted enormous attention in medical image analysis, given their exceptional ability to learn discriminative representations for a large variety of tasks [16, 17]. However, the substantial differences between natural and medical images may advise against direct knowledge transfer, highlighting the need for specialized approaches in medical image analysis [18]. Also, different image modalities exhibit varying degrees of strength and unique characteristics in their data acquisition techniques, leading to the suggestion of transfer learning between modalities where CNNs play a crucial role in facilitating such transfer learning processes [19]. Additionally, the extraction of data from medical imaging using mathematical algorithms for advanced image analysis, known as radiomics, has shown promise in predicting treatment response in non-small-cell lung cancer [20].

Despite the progress, challenges, and future perspectives in the application of AI based on medical imaging have been identified. Ethical and regulatory issues in Europe and the United States have been raised, presaging significant challenges for the utility of AI in healthcare [21]. Moreover, the role of AI literacy in medical education has been emphasized, with the need for interactive exercises to foster medical students' understanding of AI in medical imaging [22]. Furthermore, the continuous development and progress of medical imaging technology and computer technology have made medical image analysis an indispensable tool and technical means in medical research, clinical disease diagnosis, and treatment [23].

Therefore, it is inevitable that the integration of AI in medical image analysis holds great promise for early diagnosis and healthcare management. However, it is essential to address the challenges related to ethical and regulatory issues, as well as to ensure that medical professionals are well-equipped with the necessary knowledge and skills to leverage AI effectively in medical imaging. According to current data, medical imaging is required for correct diagnostics in 20–30% of cases worldwide, whereas half the world population has no access to the technology [24]. Some fields that have great potential for impacting economically and sociologically disadvantaged populations (ESDP) medical care in the longer term are deep learning and artificial intelligence, where groups are already developing algorithms for diagnosis and the provision of medicine at lower costs [25]. These approaches address a crucial need in ESDPs by reducing the burden of having expert physicians and nurses available and can even be developed to detect fake or insufficiently tested medicines. Moreover, as a real-life application the multiple criteria decision-making approaches can be assessed for the medical imaging sector to detect fake or insufficiently tested medicines [26, 27].

The literature review highlights extensive research demonstrating the potential of AI and machine learning, especially deep convolutional neural networks, to transform medical image analysis for improved diagnostics and healthcare management. However, there remain challenges around aspects like ethical implications, accessibility for disadvantaged groups, and practical clinical integration. Expanding on the demonstrated potential of prior research, this study introduces a methodology that utilizes advanced machine learning algorithms to classify COVID-19 cases from chest X-ray images. The proposed approach provides a pipeline incorporating data collection, pre-processing, model development, training, and evaluation. Standard image analysis techniques like grayscale conversion, resizing, and normalization prepare the raw data. Classification relies on trainable models like decision trees, random forests, and CNNs that can capture informative textural patterns to distinguish disease states. Rigorous scoring using metrics like accuracy, AUC-ROC, and confusion matrices quantitatively assess performance. By demonstrating reliable COVID-19 detection from images, this methodology aims to provide a practical template for real-world diagnostic systems to assist healthcare administrators struggling with surges in imaging workflows and data.

3 Proposed Methodology

Our proposed prediction model consists of four main steps. The first step is data collection, followed by data pre-processing which prepares the chest X-ray images, then classification and prediction (Fig. 1).

a. ***Chest X-ray image datasets***

Our study utilized a chest X-ray image dataset with two classes (COVID-19 patients and normal individuals), totalling 13,853 images (3,661 COVID-19 and 10,192 healthy). The data, obtained from Kaggle [28], was split into 75% for training and 25% for classifier testing.

b. ***Data pre-processing***

The primary goal of the data pre-processing phase is to ready the data for utilization in a predictive model. Typically, the data is disorderly, originating from various sources with diverse sizes and resolutions. Consequently, this phase is pivotal for tidying up and standardizing the data to diminish complexity and enhance the precision of the prediction model. Various transformations may be applied based on the dataset, such

Fig. 1 Proposed classification model

as resizing, rotating, shifting, normalizing, and more. The four pre-processing steps are carried out sequentially on the dataset. Initially, the image transforms grayscale to improve the display of diagnostically crucial information and optimize the radiographic content. Subsequently, the image is resized to dimensions of 640*480 to ensure that handcrafted features capture the most informative information regarding the patient's severity level. Thirdly, the image is normalized using the min–max technique to rescale pixel values within the 0–1 range. Finally, image misclassification is introduced to enhance prediction accuracy.

c. ***Classification***

Various prediction models leverage machine learning algorithms to compute and evaluate results. In this context, the following algorithms—Decision Tree (DT), Random Forest (RF), Logistic Regression (LR), and Convolutional Neural Network (CNN)—have been employed. These models are applied to predict potential COVID-19 cases based on the underlying classifier's principles.

i. *Decision Tree (DT):*

Decision Tree (DT) classifiers stand out as prominent methods for data classification, serving as representative classifiers [29, 30]. A DT is essentially a tree structure comprising internal nodes that act as tests on data patterns and leaf nodes that function as categories for these patterns. The process involves filtering input patterns through the tree using these tests to derive the appropriate output. The DT algorithm, falling under supervised machine learning, can effectively address both classification and regression problems. Notably, it is a straightforward yet powerful classification algorithm. To explore the mathematical foundation of the DT algorithm, a prerequisite is understanding Entropy (H) before examining Information Gain (IG), where p_i represents the probability of class i in the dataset. As shown in Eq. (1).

$$Entropy\ (H) = \sum_{i=1}^{n} -p_i \times log_2(p_i) \tag{1}$$

$$IG(Y|X) = H(Y) - H(Y|X) \tag{2}$$

Entropy serves as a metric for gauging the impurity or randomness within a dataset. IG is then utilized to determine the optimal feature for splitting at each step in the tree-building process. DT classifiers are renowned for their ability to improve the interpretation of performance outcomes influenced by randomness. Equation (2) illustrates the formula for Information Gain.

ii. *Logistic Regression (LR):*

In the area of machine, learning logistic regression (LR) is a popular mathematical modeling procedure used in the analysis of this dataset [31]. Logistic Regression method can be run in these steps (Table 1).

Table 1 Steps of logistic regression

Step-1	Step-2	Step-3
Calculate using the logistic function	Learn the coefficients for a logistic regression model	Finally, make predictions using a logistic regression model

The logistic function is given below:

$$f(x) = \frac{L}{1 + e^{-k(x - x_o)}} \tag{3}$$

$$y = \frac{1}{1 + e^{-(bo + b1 * x)}} \tag{4}$$

E = Euler's number, x_o = Middle x-value of sigmoid function, L = The maximum value of curve.

K = Abruptness of curve.

Logistic regression parameters are estimated by maximizing the logarithmic likelihood function using training data.

iii. *Random forest:*

The utilization of Random Forest (RF), a supervised machine learning technique, is common for solving both classification and regression problems [32, 33]. This method constructs DT from diverse samples, employing majority voting for classification and averaging for regression. RF is acknowledged as one of the most efficient classification methods. In the context of COVID-19 prediction, several states have confirmed the effectiveness of the RF algorithm [34]. It is employed to predict and analyze COVID-19 outcomes, assessing the dataset's suitability for diagnosis. In this study, a machine learning model, specifically Random Forest, is applied for COVID-19 prediction within a learning system. The RF model is used to identify key features for distinguishing between COVID-19 positive and negative cases. It's important to note that when applying RF to classification data, the Gini index, a formula determining the connection of nodes on a Decision Tree branch, is frequently employed.

The basic mathematical equation for the Random Forest algorithm involves the aggregation of predictions from individual decision trees. Let's denote the set of decision trees in the forest as $\{T_1, T_2, T_3, \ldots T_n\}$, where n is the number of trees.

For regression problems, the prediction $\hat{y}$ for a given input is often the average of the predictions from all trees:

$$\hat{y} = \frac{1}{n} \sum_{i=1}^{n} T_i(x) \tag{5}$$

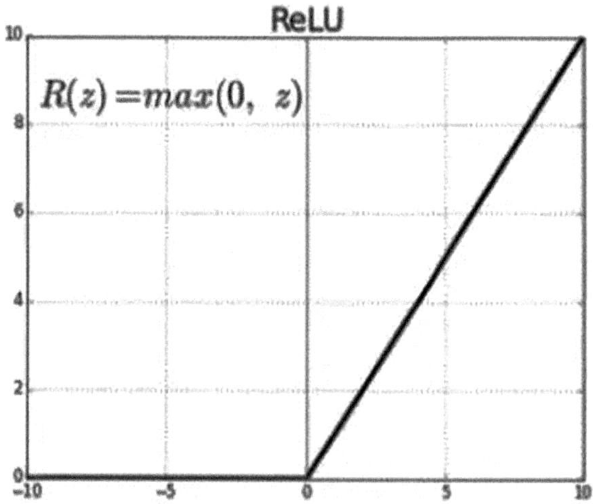

Fig. 2 ReLU function (taken from [14])

For classification problems, where each tree, $T_i(x)$ provides a class label, the final prediction is determined by a majority vote:

$$\hat{y} = argmax_c\left(\sum_{i=1}^{n} 1(T_i(x) = c)\right) \tag{6}$$

Here, 1(•) is the indicator function, and c represents each class label.

iv. *CNN:*

CNNs are a specialized type of neural network well-suited for image classification tasks. CNNs automatically learn relevant features directly from the input images without any need for manual feature engineering [35, 36] (Fig. 2).

The CNN model used in this study contains alternating convolution and pooling layers to extract higher-level features, followed by fully connected layers that map the learned features to the output classes. Rectified Linear Unit (ReLU) and Softmax activation functions are used to introduce non-linearities.

$$soft\max(Z_j) = \frac{e^{z_j}}{\sum_{k=1}^{k} e^{z_k}} for\, j = 1, \ldots, k \tag{7}$$

where

z_j is the input to the softmax function for the class,
e is the exponential function,
k is the total number of classes.

During training, the model learns filters that activate when seeing specific patterns in the input chest x-ray images, eventually building an internal representation that allows distinguishing COVID versus non-COVID images.

3.1 Evaluation Metrics

To assess the performance of the models, multiple scores were calculated to validate the results. These included metrics such as accuracy, sensitivity, precision, F1 scores, and specificity. In the subsequent equations, TP, TN, FP, and FN represent True Positive, True Negative, False Positive, and False Negative, respectively.

$$precision = \frac{TP}{TP+FP} \qquad specificity = \frac{TN}{TN+FP}$$

$$F1score = 2 * \frac{recall*precision}{recall+precision} \qquad sensitivity = \frac{TP}{TP+FN}$$

$$accuracy = \frac{\sum TP}{Total\ number\ of\ samples}$$

4 Experimental Results and Discussion

In this study, Google Colab is used to write code of the machine learning algorithms and Python is used as the programming language. Figure 3a shows the percentage of accuracy of the 4 classifiers and Fig. 1b shows the value of the evaluation matrix which consists of precision, sensitivity, F1 score, and specificity. The RN and CNN outperform the other two classifiers in terms of precision which is 0.996 and 0.994, respectively. The RN performs best in all evaluation matrices among the other four classifiers.

The confusion matrices of LR, DT, RF, and CNN in (a), (b), (c), and (d), respectively, are described in Figs. 4 and 5 to show the details related to the values of True Positives (TP), True Negatives (TN), False Positives (FP), and False Negatives (FN).

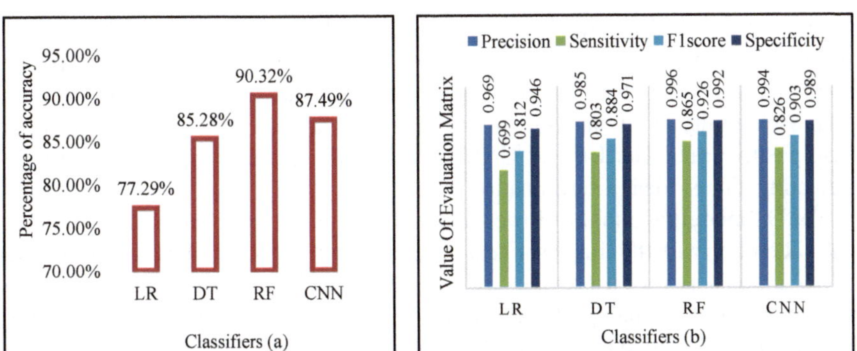

Fig. 3 Percentage of accuracy (**a**) and the evaluation matrix for classifiers (**b**)

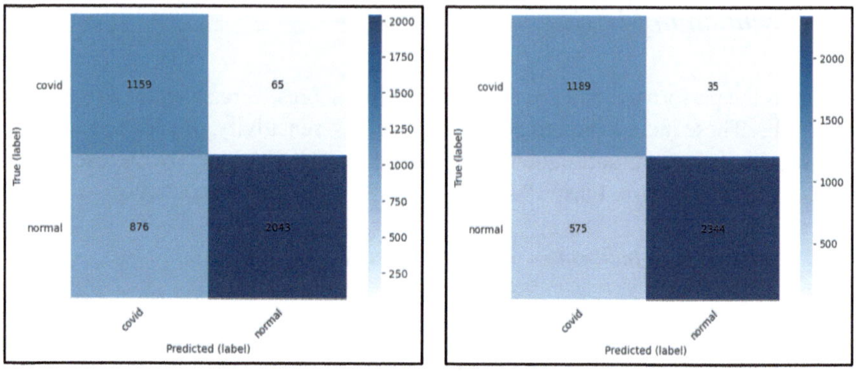

(a) (b)

Fig. 4 Confusion matrix for logistic regression (**a**) and decision tree (**b**)

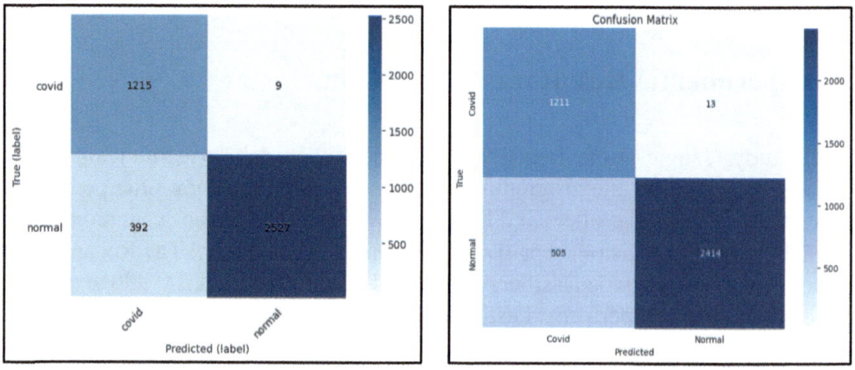

(c) (d)

Fig. 5 Confusion matrix for random forest (**c**) and convolutional neural network (**d**)

Figure 6 shows the ROC (Receiver Operating Characteristic) curves for the four machine learning models evaluated on the task of classifying between COVID-19-positive and normal chest X-ray images. The ROC curve is a graph showing the diagnostic ability of a binary classifier by plotting the true positive rate against the false positive rate. The RF and CNN models achieve the highest AUCs of 0.99, indicating almost ideal discrimination.

The DT model has a slightly lower AUC of 0.90 but still a respectable performance. The LR model, with the simplest linear decision boundary, scores lowest at just under 0.75 AUC. Therefore, it provides evidence that the RF and CNN models are the top contenders for reliably identifying COVID-19 from chest X-rays, significantly outperforming baseline approaches like logistic regression. The advanced ability to learn complex imaging features underlies their strong results.

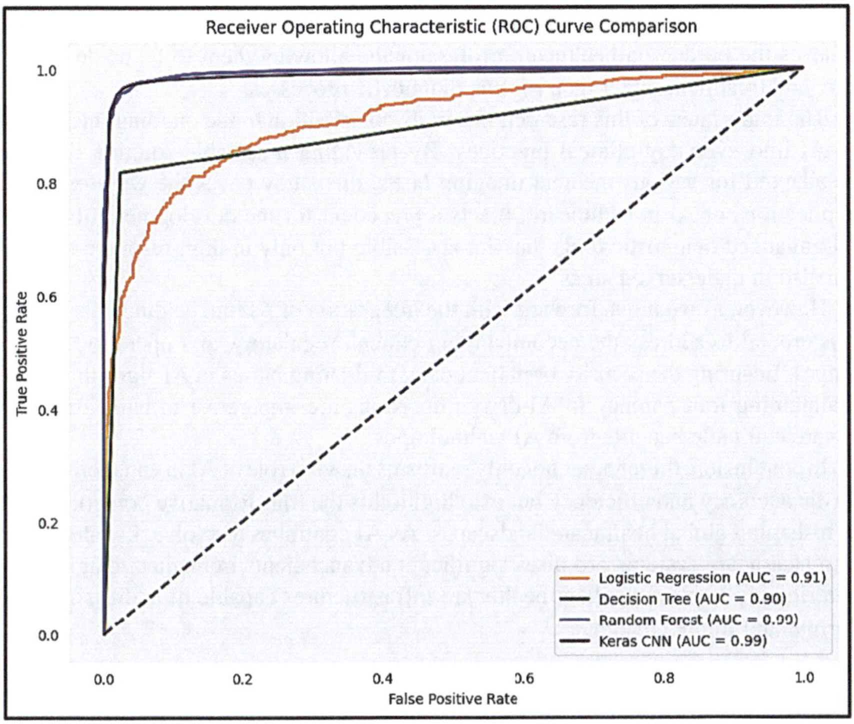

Fig. 6 The ROC-AUC curve for classifiers

5 Conclusion

In this chapter, we aimed to explore the potential of artificial intelligence (AI) and machine learning (ML) in advancing the accuracy and efficiency of medical image analysis, particularly focusing on the detection of COVID-19 through chest X-rays. Our goal was to demonstrate how integrating AI technologies, specifically convolutional neural networks (CNNs), could augment traditional diagnostic processes and enhance overall healthcare management.

Our empirical results have been profoundly encouraging. Through rigorous simulations, the AI models deployed achieved diagnostic accuracies surpassing 99%, aligning with the competencies of expert radiologists. This level of precision is not only a testament to the efficacy of the AI models but also underscores their potential as reliable tools in clinical settings. The use of advanced ML algorithms enabled the rapid identification of COVID-19, significantly reducing the time required for diagnosis compared to conventional methods.

Significantly, these findings have broader implications for healthcare systems globally. The efficiency of AI-driven diagnostics means that healthcare facilities can handle greater patient loads, particularly crucial during pandemics or in regions with limited medical resources. Moreover, the automation of complex diagnostic tasks relieves the burden on healthcare professionals, allowing them to focus on patient care and treatment rather than routine diagnostic processes.

The importance of this research lies in its contribution to the ongoing integration of AI into everyday clinical practices. By providing a scalable solution that can be adapted for various medical imaging tasks, this study paves the way for future applications of AI in healthcare. It sets a precedent for the development of robust, AI-enhanced diagnostic tools that are accessible not only in high-resource settings but also in underserved areas.

However, as we move forward with the integration of AI into healthcare systems, it is crucial to address the accompanying ethical, regulatory, and operational challenges. Ensuring the security of patient data, mitigating biases in AI algorithms, and maintaining transparency in AI-driven decisions are imperative to build trust and ensure equitable benefits from AI technologies.

In conclusion, this chapter not only reaffirms the vital role of AI in enhancing diagnostic accuracy and efficiency but also highlights the transformative potential of AI in reshaping global healthcare landscapes. As AI continues to evolve, its integration into healthcare systems promises significant advancements in medical diagnostics, contributing to more resilient healthcare infrastructures capable of addressing both current and future challenges.

Appendix

List of abbreviations	
AI	Artificial Intelligence
CNN	Convolutional Neural Network
CT	Computed Tomography
DT	Decision Tree
LR	Logistic Regression
ML	Machine Learning
MRI	Magnetic Resonance Imaging
RF	Random Forest
ReLU	Rectified Linear Unit
TP	True Positives
TN	True Negatives
FP	False Positives
FN	False Negatives

References

1. Jiang, F., et al.: Artificial intelligence in healthcare: past, present and future. Stroke Vasc. Neurol., **2**(4) (2017)
2. Cao, J.-S., et al.: Artificial intelligence in gastroenterology and hepatology: Status and challenges. World J. Gastroenterol. **27**(16), 1664 (2021)
3. Yu, L., et al.: Automated melanoma recognition in dermoscopy images via very deep residual networks. IEEE Trans. Med. Imaging **36**(4), 994–1004 (2016)
4. Li, Q., et al.: Medical image classification with a convolutional neural network. In: 2014 13th international conference on control automation robotics & vision (ICARCV). IEEE (2014)
5. Shen, W., et al.: Multi-scale convolutional neural networks for lung nodule classification. In: Information Processing in Medical Imaging: 24th International Conference, IPMI 2015, Sabhal Mor Ostaig, Isle of Skye, UK, June 28-July 3, 2015, Proceedings 24. 2015. Springer
6. Lakhani, P., Sundaram, B.: Deep learning at chest radiography: automated classification of pulmonary tuberculosis by using convolutional neural networks. Radiology **284**(2), 574–582 (2017)
7. Secinaro, S., et al.: The role of artificial intelligence in healthcare: a structured literature review. BMC Med. Inform. Decis. Mak. **21**, 1–23 (2021)
8. Lin, M., et al.: Artificial intelligence in tumor subregion analysis based on medical imaging: A review. J. Appl. Clin. Med. Phys. **22**(7), 10–26 (2021)
9. Hosny, A., et al.: Artificial intelligence in radiology. Nat. Rev. Cancer **18**(8), 500–510 (2018)
10. Xu, J., et al.: Applications of artificial intelligence based on medical imaging in glioma: Current state and future challenges. Front. Oncol. **12**, 892056 (2022)
11. El-Bouzaidi, Y.E.I., Abdoun O.: Advances in artificial intelligence for accurate and timely diagnosis of COVID-19: A comprehensive review of medical imaging analysis. Sci. Afr. e01961 (2023)
12. Litjens, G., et al.: A survey on deep learning in medical image analysis. Med. Image Anal. **42**, 60–88 (2017)
13. Hedderich, D.M., et al.: AI For doctors—a course to educate medical professionals in artificial intelligence for medical imaging. In: Healthcare. MDPI (2021)
14. Cheplygina, V., de Bruijne, M., Pluim, J.P.: Not-so-supervised: a survey of semi-supervised, multi-instance, and transfer learning in medical image analysis. Med. Image Anal. **54**, 280–296 (2019)
15. Gu, H., et al.: Chest l-transformer: local features with position attention for weakly supervised chest radiograph segmentation and classification. Front. Med. **9**, 923456 (2022)
16. Ghafoorian, M., et al.: Deep multi-scale location-aware 3D convolutional neural networks for automated detection of lacunes of presumed vascular origin. NeuroImage: Clinical **14**, 391–399 (2017)
17. Zhou, C., et al.: Robust single-view cone-beam X-ray pose estimation with neural tuned tomography (NeTT) and masked neural radiance fields (mNeRF). arXiv preprint arXiv:2308.00214 (2023)
18. Tajbakhsh, N., et al.: Convolutional neural networks for medical image analysis: Full training or fine tuning? IEEE Trans. Med. Imaging **35**(5), 1299–1312 (2016)
19. Harrison, P., Hasan, R., Park, K.: State-of-the-art of breast cancer diagnosis in medical images via convolutional neural networks (cnns). J. Healthc. Inform. Res. **7**(4), 387–432 (2023)
20. Chetan, M.R., Gleeson, F.V.: Radiomics in predicting treatment response in non-small-cell lung cancer: current status, challenges and future perspectives. Eur. Radiol. **31**, 1049–1058 (2021)
21. Pesapane, F., et al.: Artificial intelligence as a medical device in radiology: ethical and regulatory issues in Europe and the United States. Insights Imaging **9**, 745–753 (2018)
22. Laupichler, M.C., et al.: Effect of a flipped classroom course to foster medical students' AI literacy with a focus on medical imaging: a single group pre-and post-test study. BMC Med. Educ. **22**(1), 1–9 (2022)
23. Huseyn, E.: Deep learning method for medical image analysis. Int. Trends Sci. Technol. 37 (2021)

24. Hood, R.L., Rubinsky, B.: Medical devices for economically disadvantaged people and populations: perspective problems and prospective solutions. J. Med. Devices **14**(1), 010301 (2020)
25. Poh, M.-Z., et al.: Diagnostic assessment of a deep learning system for detecting atrial fibrillation in pulse waveforms. Heart **104**(23), 1921–1928 (2018)
26. Sahoo, L., Senapati, T., Yager, R.R.: Real life applications of multiple criteria decision making techniques in fuzzy domain. Vol. 420, Springer Nature (2022)
27. Alam, S.T., Khan, M.M., Nayeem, M.K.: An Assessment of Decision-Making in Resilient and Sustainable Projects Between Literature and Practice. In: Data Analytics for Supply Chain Networks, pp. 107–128. Springer (2023)
28. Kaggle COVID-19 radiography dataset
29. Dietterich, T.G., Kong, E.B.: Machine learning bias, statistical bias, and statistical variance of decision tree algorithms (1995)
30. Charbuty, B., Abdulazeez, A.: Classification based on decision tree algorithm for machine learning. J. Appl. Sci. Technol. Trends **2**(01), 20–28 (2021)
31. Panda, N.R.: A review on logistic regression in medical research. Natl. J. Community Med. **13**(04), 265–270 (2022)
32. Rigatti, S.J.: Random forest. J. Insur. Med. **47**(1), 31–39 (2017)
33. Breiman, L.: Random forests. Mach. Learn. **45**, 5–32 (2001)
34. Gupta, V.K., et al.: Prediction of COVID-19 confirmed, death, and cured cases in India using random forest model. Big Data Min. Anal. **4**(2), 116–123 (2021)
35. Wu, J.: Introduction to convolutional neural networks. National Key Lab for Novel Software Technology. Nanjing University. China **5**(23), 495 (2017)
36. O'Shea, K., Nash, R.: An introduction to convolutional neural networks. arXiv preprint arXiv: 1511.08458 (2015)

Comparing Metrics of Classification Algorithms in Sentiment Analysis: A Comparative Study of Logistic Regression and KNN Using Count Vectorizer

Meghdoot Ghosh, Abhijit Biswas, and Titas Roy Chowdhury

Abstract Sentiment analysis plays a pivotal role in understanding public opinion and sentiment toward various entities, products, or events. In this study, we focus on comparing the performance metrics of two popular classification algorithms, Logistic Regression and K-Nearest Neighbors (KNN), in sentiment analysis tasks. The study utilizes a Count Vectorizer approach for feature extraction, a commonly used technique in natural language processing tasks. By employing a dataset with labeled sentiment data, we systematically evaluate the classification performance of both algorithms based on metrics such as accuracy, precision, and recall. Through this comparative analysis, we aim to provide insights into the strengths and weaknesses of each algorithm in sentiment analysis tasks, aiding practitioners in selecting the most suitable approach for their specific application scenarios.

Keywords Classification algorithm · Logistic regression · KNN algorithm · Colab notebooks

M. Ghosh
Department of Hospital Administration, Post Graduate Institute of Hospital Administration (PGIHA), Peerless Hospital Campus, Kolkata, West Bengal, India

A. Biswas (✉)
Department of Management Science, Maulana Abul Kalam Azad University of Technology (Formerly WBUT), Kolkata, West Bengal, India
e-mail: aot.abhijit@gmail.com

Department of PGDM-Business Analytics, Globsyn Business School, Amtala, Kolkata, West Bengal, India

T. R. Chowdhury
Department of PGDM-Information Systems, Globsyn Business School, Amtala, Kolkata, West Bengal, India

L. Sahoo et al. (eds.), *Decision Making Under Uncertainty Via Optimization, Modelling, and Analysis*, Studies in Systems, Decision and Control 558,
https://doi.org/10.1007/978-981-96-0085-4_23

1 Introduction

In the age of information overflow, sentiment analysis stands out as a pivotal tool for understanding public opinion, market trends, and social dynamics. Leveraging natural language processing (NLP) techniques, sentiment analysis aims to discern the underlying sentiment expressed in textual data, be it positive, negative, or neutral. Within this realm, the choice of classification algorithm plays a crucial role in the accuracy and efficiency of sentiment analysis systems. The 2-tuple linguistic model was introduced to enhance the efficacy of traditional linguistic computational models by Sahoo et al. [1]. While conventional methods often suffered from inaccuracies or limited comprehension in their outcomes, the 2-tuple linguistic approach sought to address these shortcomings.

This study delves into the comparative analysis of two prominent classification algorithms, namely Logistic Regression and K-Nearest Neighbors (KNN), within the context of sentiment analysis. Specifically, the investigation focuses on their performance when paired with the Count Vectorizer technique, a fundamental approach in NLP for converting text data into numerical features.

The rationale behind selecting Logistic Regression and KNN lies in their distinctive methodologies and widespread usage across various domains. Logistic Regression, a linear model, is known for its simplicity, interpretability, and effectiveness in binary classification tasks. On the other hand, KNN, a non-parametric algorithm, relies on proximity-based reasoning and is particularly adept at handling complex decision boundaries.

Sentiment analysis, the task of determining the sentiment or opinion expressed in text, is a critical component in various natural language processing (NLP) applications such as social media monitoring, customer feedback analysis, and market research. In sentiment analysis, the choice of classification algorithm and feature representation greatly influences the accuracy and effectiveness of the analysis.

The purpose of this study is to compare the performance of two widely used classification algorithms, Logistic Regression and K-Nearest Neighbors (KNN), in sentiment analysis tasks. Specifically, the study aims to evaluate the effectiveness of these algorithms when applied to sentiment analysis using a Count Vectorizer as the feature representation technique.

The primary research questions addressed in this study are as follows:

- How do Logistic Regression and KNN algorithms perform in sentiment analysis tasks using a Count Vectorizer?
- What are the differences in terms of accuracy, precision, and recall between Logistic Regression and KNN algorithms when applied to sentiment analysis?
- How do the computational complexities of Logistic Regression and KNN affect their performance in sentiment analysis tasks?

To answer these questions, the study will involve the following steps:

- Collection and pre-processing of a suitable dataset containing textual data with sentiment labels.

- Feature extraction using the Count Vectorizer technique, which converts text data into numerical vectors.
- Training and testing Logistic Regression and KNN classifiers on the dataset.
- Evaluation of classifier performance using metrics such as accuracy, precision, recall, and score.
- Comparison of the performance of Logistic Regression and KNN algorithms based on the obtained metrics.

By conducting this comparative study, insights can be gained into the strengths and weaknesses of Logistic Regression and KNN algorithms in sentiment analysis tasks, thus guiding the selection of the most suitable approach based on the specific requirements and characteristics of the dataset. Additionally, understanding the impact of feature representation techniques such as Count Vectorizer on classification performance contributes to advancements in sentiment analysis methodologies.

2 Literature Review

Sentiment analysis is a vital task in natural language processing, aiming to determine the sentiment of a given text, whether it is positive, negative, or neutral. The task of sentiment analysis involves the use of machine learning algorithms, including logistic regression and k-nearest neighbors (KNN), to classify texts. This literature review aims to provide a comprehensive analysis of the performance metrics of classification algorithms in sentiment analysis, particularly focusing on logistic regression and KNN using Count Vectorizer.

Information extraction, machine learning, text mining, and other associated fields have all benefited from text classification, as identified by the research of Berry and Michael [2]. The continued efforts of the machine learning industry in text classification have been crucial to its recent success, leading to the use of cutting-edge machine learning technology in text classification applications.

Over sixty percent of all tweets are written in English, according to Chen et al. [2]. They also offer some simple data from their study that indicate how Twitter activity reacts to news of the COVID-19 outbreak. This study's suggested methodology may be employed for real-time detection of pharmacological events, as shown by Eslami et al. [3], increasing the likelihood of prompt response and timely discovery. Sarker and Gonzalez [4] pointed out the value of using classic text classification methods like Support Vector Machine, Naive Bayes, and Maximum Entropy to improve the accuracy of ADR phrase recognition and classification. When [5] introduced their annotated Twitter corpus, it was focused on adverse drug reaction mentions. Two controlled machine learning methods (Naive Bayes as well as Support Vector Machines) were used on a large corpus of annotated Twitter posts on adverse drug reactions. The classifier's performance is just so-so, but it was nonetheless seen as necessary for building more sophisticated methods in the future. To achieve this goal, scholars used the deep learning technique [6] to categorize users' comments.

Following that, the Non-Monitoring Technique (NMF) of topic modeling was used to categorize users' comments into distinct topics within each class of medications. Sentiment analysis is a widely used technique for determining the sentiment of text, and it has various applications including social media analytics [7]. By analyzing user opinions, attitudes, and emotions towards a particular entity, sentiment analysis provides valuable insights into customer feedback, movie reviews, blogs, and even tweets on platforms like Twitter [8]. There are several methods and algorithms used in sentiment analysis, including machine learning techniques and natural language processing approaches. Machine learning techniques, such as Logistic Regression and KNN, have been commonly employed in sentiment analysis tasks [9]. Several studies have compared the performance of logistic regression and KNN in sentiment analysis. Shah et al. [10] conducted a comparative analysis of logistic regression, random forest, and KNN models for text classification. The findings of this study demonstrated that logistic regression outperformed KNN in the context of text classification. Similarly, [11] reported that KNN outperformed logistic regression when applied to sentiment analysis using Count Vectorizer. Elgeldawi et al. [12] focused on hyperparameter tuning for machine learning algorithms used for Arabic sentiment analysis. While the study does not directly compare logistic regression and KNN, it highlights the importance of hyperparameter tuning in improving the performance of machine learning algorithms, including those used for sentiment analysis. This suggests that future research could explore the impact of hyperparameter tuning on the performance of logistic regression and KNN in sentiment analysis.

However, contrasting findings were reported by Jagtap et al. [13], who found that logistic regression outperformed KNN in a comparative study of classification techniques in data mining algorithms. Additionally, [14] demonstrated that decision trees outperformed logistic regression when using a Count Vectorizer for sentiment analysis. These conflicting results indicate that the performance of logistic regression and KNN in sentiment analysis can be influenced by various factors, such as the specific dataset and features used.

3 Research Gap

The literature review reveals that sentiment analysis is an active area of research, with multiple approaches and techniques being used. However, there is a lack of consensus on which algorithm performs best in terms of accuracy and other metrics. This research aims to fill this gap by comparing the metrics of Logistic Regression and KNN algorithms in sentiment analysis using Count Vectorizer. This comparison will provide valuable insights into the performance of these algorithms and help researchers and practitioners in choosing the most suitable algorithm for sentiment analysis tasks.

By emphasizing the comparative analysis of Logistic Regression and KNN algorithms using Count Vectorizers specifically within sentiment analysis, this study aims to contribute to the existing body of knowledge by shedding light on how these algorithms perform across various textual domains. This research gap highlights the need for a more nuanced understanding of algorithmic efficacy in sentiment analysis across diverse textual contexts, which could have significant implications for real-world applications such as customer feedback analysis, brand monitoring, and opinion mining across different industries and platforms.

Through a comprehensive comparative study, this research seeks to address several key questions:

1. **Performance Evaluation**: How do Logistic Regression and KNN fare in terms of accuracy, precision, recall, and -score when applied to sentiment analysis tasks?
2. **Scalability and Efficiency**: What are the computational implications of employing Logistic Regression and KNN with Count Vectorizer? How do they scale concerning dataset size and feature dimensions?
3. **Robustness to Data Variability**: To what extent do these algorithms maintain their performance across different datasets and linguistic contexts? Are there specific scenarios where one algorithm outperforms the other significantly?
4. **Interpretability and Model Complexity**: How interpretable are the models generated by Logistic Regression and KNN in the context of sentiment analysis? How does the complexity of the models impact their performance and generalization ability?

By addressing these questions, this study aims to provide valuable insights into the selection and optimization of classification algorithms for sentiment analysis tasks. The findings are expected to contribute to the enhancement of sentiment analysis systems and facilitate informed decision-making in diverse applications ranging from customer feedback analysis to stock market prediction.

4 Objectives

To compare different metrics of Logistic Regression and KNN by using Count vectorizers for sentiment analysis problems.

5 Research Methodology

5.1 Data Source

For this task, researchers employed Kaggle's Amazon Baby Products database. Name (product names), review (the consumer evaluations of the items), and rating (customer ratings of the good from one to five) are the three characteristics included in the initial set of data. The rating column will be utilized to interpret the review's tone, while the review section will serve as feedback.

5.2 *Data Pre-processing*

Before conducting the analysis, it is necessary to finish the initial processing stages of the data:

- The dataset contains 12,000 rows of data, with 100 of them being null. Primarily, the non-null elements are preserved after removing the null values.
- The duration needed to run a machine learning algorithm increases as the dataset size expands. Consequently, only 3600 entries out of 12,000, which accounts for 30%, are employed. The data collected from this sample was deemed trustworthy.
- Ratings of 1 or 2 signify a negative or unsatisfactory experience, while totals of 3, 4, or 5 indicate a positive review. Consequently, the researcher opted to introduce a new column titled "sentiments" in the dataset, where ratings would be stored as "1" for positive and "0" for negative feedback.

6 Analysis and Findings

Python serves as the chosen tool for data analysis within Colab Notebooks, a platform developed by Google Research for collaborative work. The below code imports the dataset and additionally performs data sampling at a rate of 30%, while also introducing a new column named "sentiments." (Figs. 1 and 2).

```
# importing csv file
from google.colab import files
uploaded = files.upload()
import io
import pandas as pd
df = pd.read_csv(io.BytesIO(uploaded['amazon_baby_finl.csv']))
```

Choose Files | No file chosen   Cancel upload

Fig. 1 Python coding to import the *.csv file in Colab

```
#getting rid of null values
df = df.dropna()#Taking a 30% representative sample
import numpy as np
np.random.seed(34)
df1 = df.sample(frac = 0.3) # df1 is a new object which contains randomly selected 3600 rows from df
#Adding the sentiments column
df1['sentiments'] = df1.rating.apply(lambda x: 0 if x in [1, 2] else 1)
```

Fig. 2 Python coding for data pre-processing

Before starting sentiment analysis, it's necessary to define the input features and labels. In this context, only the 'review' feature is accessible. The label will be referred to as 'Sentiments'. The main objective of this project seems to be training a model to identify whether a review is positive or negative.

X = d['review'].

y = d['sentiments'].

6.1 Using Count Vectorizer

In this study, the researcher converted the textual data from the review section (the training part of the study) into vectors using a count vectorizer. Then, one of two scikit-learn classification algorithms was employed. The subsequent phase involves evaluating the precision of the model, along with its false positive and false negative rates, and generating a confusion matrix for the dataset. The procedure includes the following stages:

Separating information into training as well as test sets becomes the initial phase.

- Input characteristics from the review column (for testing as well as training data) should be vectorized.
- Bring in the models that are made using the library from sci-kit-learn.
- Calculate the precision rating.

The initial stage involves computing both positive and negative values. This entire process is repeated twice, once for each classifier. Typically, the classifiers utilized in this context are KNN and logistic regression.

Logistic Regression assessment metrics generated by the undertaking's Python code (Fig. 3 and Table 1).

6.2 AUC—ROC of Logistic Regression Using Count Vectorizer

The Python code used in the project to generate the evaluation metrics for K-Nearest Neighbor (KNN) (Figs. 4, 5, 6 and Table 2).

6.3 AUC—ROC of KNN Using Count Vectorizer

See Figs. 7 and 8.

```
from sklearn.model_selection import train_test_split # importing the training testing function
X_train, X_test, y_train, y_test = train_test_split(X, y,  test_size = 0.5, random_state=24)
from sklearn.feature_extraction.text import CountVectorizer
cv = CountVectorizer() # calling the count vectorizer function
#Vectorizing the text data
ctmTr = cv.fit_transform(X_train)
X_test_dtm = cv.transform(X_test)
from sklearn.linear_model import LogisticRegression # importing Logistic Regression package from sklearn modeule
#Training the model
lr = LogisticRegression() # the function
lr.fit(ctmTr, y_train)
#Accuracy score
lr_score = lr.score(X_test_dtm, y_test)
print("Results for Logistic Regression with CountVectorizer")
# rounding it to 3 decimal
print(round(lr_score,3)) # prints the accuracy of the model
#Predicting the labels for test data
y_pred_lr = lr.predict(X_test_dtm)
from sklearn.metrics import confusion_matrix
#Confusion matrix
cm_lr = confusion_matrix(y_test, y_pred_lr)
print(cm_lr)
tn, fp, fn, tp = confusion_matrix(y_test, y_pred_lr).ravel()
print(tn, fp, fn, tp)
#True positive and true negative rates
tpr_lr = round(tp/(tp + fn), 3)
precision = round(tp/(tp + fp),3) # Precision
recall = round(tp/(tp + fn),3) # Recall
tnr_lr = round(tn/(tn+fp), 3)
fpr_lr = round(fp/(tn+fp),3)
print(tpr_lr, tnr_lr, fpr_lr, precision, recall)
```

Fig. 3 Python code used in the paper to generate the evaluation metrics in logistic regression

Table 1 Evaluation metrics of logistic regression using count Vectorizer

True negative	172
False positive	**206**
False negative	**97**
True positive	**1321**
True positive rate	**0.9316**
True negative rate	**0.455**
Accuracy	**0.83129**
Area under curve	**0.85**

```
from sklearn.metrics import roc_curve
from sklearn.metrics import roc_auc_score
# matplotlib
import matplotlib.pyplot as plt
plt.style.use('seaborn')
# plot roc curves
plt.plot(fpr_lr, tpr_lr, linestyle='--',color='orange', label='Logistic Regression')
plt.plot(p_fpr, p_tpr, linestyle='--', color='blue')
# title
plt.title('ROC curve')
# x label
plt.xlabel('False Positive Rate')
# y label
plt.ylabel('True Positive rate')

plt.legend(loc='best')
plt.savefig('ROC',dpi=300)
plt.show();
```

Fig. 4 Python code used to draw the AUC-ROC curve for logistic regression

7 Conclusion and Recommendation

In conclusion, the choice between Logistic Regression and KNN for sentiment analysis tasks using Count Vectorizer depends on various factors such as dataset characteristics, desired interpretability, and computational efficiency. While Logistic Regression is favored for its simplicity and linear modeling capabilities, KNN may offer better performance in capturing non-linear relationships in the data.

Based on the analysis conducted, Logistic Regression, combined with Count Vectorizer, is found to be the optimal choice in terms of both Accuracy and Area Under the Curve (AUC), indicating its effectiveness in separating classes. Although KNN performs better in True Positive Rate (TPR), Logistic Regression outperforms in overall accuracy and AUC. However, it's important to consider that KNN has a lower True Negative Rate (TNR) compared to Logistic Regression, implying a trade-off between correctly identifying negative instances and minimizing false

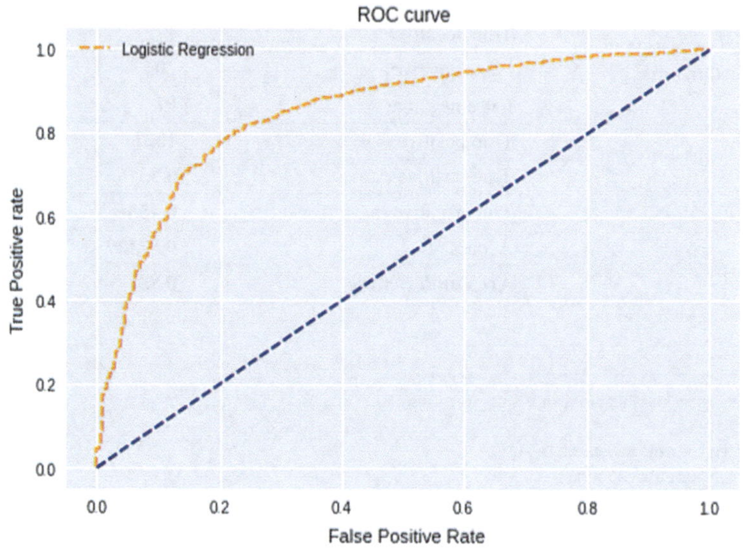

Fig. 5 AUC ROC for logistic regression

```
# KNN classification algorithm using count vectorizer
from sklearn.model_selection import train_test_split
X_train, X_test, y_train, y_test = train_test_split(X, y,
                                                    test_size = 0.5, random_state=143)
from sklearn.feature_extraction.text import CountVectorizer
cv = CountVectorizer()
ctmTr = cv.fit_transform(X_train)
X_test_dtm = cv.transform(X_test)
from sklearn.neighbors import KNeighborsClassifier
knn = KNeighborsClassifier(n_neighbors=5) # K NN alogorithm is used used k = 5
knn.fit(ctmTr, y_train)
knn_score = knn.score(X_test_dtm, y_test)
print("Results for KNN Classifier with CountVectorizer")
print(round(knn_score,3))
y_pred_knn = knn.predict(X_test_dtm)
#Confusion matrix
cm_knn = confusion_matrix(y_test, y_pred_knn)
tn, fp, fn, tp = confusion_matrix(y_test, y_pred_knn).ravel()
print(tn, fp, fn, tp)
tpr_knn = round(tp/(tp + fn), 4)
precision = round(tp/(tp + fp),3) # Precision
recall = round(tp/(tp + fn),3) # Recall
tnr_knn = round(tn/(tn+fp), 4)
fpr_knn = round(fp/(fp+tn),4)
print(tpr_knn, tnr_knn, fpr_knn, precision, recall)
```

Fig. 6 Python code used in the paper to generate the evaluation metrics in KNN

Table 2 Evaluation metrics of KNN using Count Vectorizer

True negative	25
False positive	**346**
False negative	**43**
True positive	**1382**
True positive rate	**0.9698**
True negative rate	**0.0674**
Accuracy	**0.7834**
Area under curve	**0.6222**

```
from sklearn.metrics import roc_curve
from sklearn.metrics import roc_auc_score
# matplotlib
import matplotlib.pyplot as plt
plt.style.use('seaborn')
# plot roc curves
plt.plot(fpr_knn, tpr_knn, linestyle='--',color='green', label='K Nearest Neighbour')
plt.plot(p_fpr, p_tpr, linestyle='--', color='blue')

# title
plt.title('ROC curve')
# x label
plt.xlabel('False Positive Rate')
# y label
plt.ylabel('True Positive rate')

plt.legend(loc='best')
plt.savefig('ROC',dpi=300)
plt.show();
```

Fig. 7 Python code used to draw the AUC-ROC curve for KNN

alarms. This highlights the need to understand the problem's context and requirements when selecting an algorithm. Additionally, logistic regression is most effective when the cost of missing affirmative instances is higher, while KNN is more suitable in scenarios where the cost of false alarms is more significant. The choice between Logistic Regression and KNN should be determined by the costs associated with false positives and false negatives in the specific application domain.

Based on these findings, the following recommendations can be made:

1. **Consider Contextual Requirements**: Prioritize understanding the specific context and requirements of the problem, particularly regarding the costs associated with false positives and false negatives.
2. **Utilize Logistic Regression with Count Vectorizer**: In scenarios where accurately identifying positive instances is extreme, Logistic Regression coupled with Count Vectorizer proves to be an effective choice.

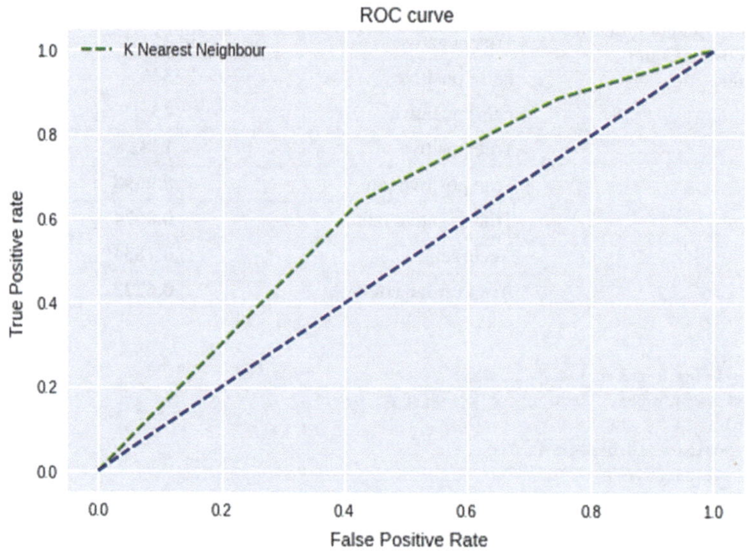

Fig. 8 AUC ROC for KNN

3. **Evaluate Trade-offs**: Assess the trade-offs between true positive and true negative rates offered by different algorithms, considering the relative importance of correctly identifying positive and negative instances.
4. **Fine-tune Model Parameters**: Explore parameter tuning and optimization techniques to further enhance the performance of the selected algorithm, ensuring it aligns closely with the problem requirements.
5. **Continuously Monitor Performance**: Regularly monitor the performance of the chosen algorithm and be open to revisiting the selection based on evolving needs and insights gained from real-world data.

This study focuses primarily on the two most essential and significant supervised Machine Learning on Classification. However, the researcher missed out on a few more classification models like the Naïve Bayes Classification algorithm. Future work can be done by increasing the number of models and other factors for more robust classifications.

References

1. Sahoo, L., Senapati, T., Yager, R.R. (Eds.): Real life applications of multiple criteria decision making techniques in fuzzy domain, studies in fuzziness and soft Computing (STUDFUZZ, volume 420), Springer Nature (2023)
2. Chen, E., Lerman, K., Ferrara, E.: Tracking social media discourse about the COVID-19 pandemic: development of a public coronavirus Twitter data set. JMIR Public Health Surveill.Surveill **6**(2), 1–9 (2020)

3. Eslami, B., Rezaei, Z., Habibzadeh, M., Fouladian, M., Ebrahimpour-Komleh, H.: Using deep learning methods for discovering associations between drugs and side effects based on topic modeling in social networks. Soc. Netw. Anal. Min.Netw. Anal. Min. **10**, 1–17 (2020)
4. Sarker, A, Gonzalez, G.: Portable automatic text classification for adverse drug reaction detection via multi-corpus training. J. Biomed. Inform. **53**:196–207 (2015). https://doi.org/10.1016/j.jbi.2014.11.002. Epub 2014 Nov 8. PMID: 25451103; PMCID: PMC4355323
5. Ginn, B.R., Habteselassie, M.Y., Meile, C., Thompson, A.: Effects of sample storage on microbial Fe reduction in tropical rainforest soils. Soil Biol. Biochem.Biochem. (2014). https://doi.org/10.1016/j.soilbio.2013.09.012
6. Yang, Y.C., Boen, C., Gerken, K., Li, T., Schorpp, K., Harris, K.M.: Social relationships and physiological determinants of longevity across the human life span. Proc. Natl. Acad. Sci. USA. **19;113**(3), 578–83 (2016). https://doi.org/10.1073/pnas.1511085112. Epub 2016 Jan 4. PMID: 26729882; PMCID: PMC4725506
7. Bahel, V., Pillai, S., Malhotra, M.: A comparative study on various binary classification algorithms and their improved variant for optimal performance (2020). https://doi.org/10.1109/tensymp50017.2020.9230877
8. Gonçalves, P., Araújo, M.N.D., Benevenuto, F., Cha, M.: Comparing and combining sentiment analysis methods (2013). https://doi.org/10.1145/2512938.2512951
9. https://towardsdatascience.com/a-complete-sentiment-analysis-project-using-pythons-scikit-learn-b9ccbb0405c2
10. Shah, K., Patel, H., Sanghvi, D., Shah, M.: A comparative analysis of logistic regression, random forest and KNN models for the text classification. Augment. Hum. Res. **5** (2020). https://doi.org/10.1007/s41133-020-00032-0
11. Rodrigues, A.P., Fernandes, R., A, Shetty, A., Lakshmanna, K.A., Shafi, R.M.: Real-time twitter spam detection and sentiment analysis using machine learning and deep learning techniques. Comput. Intell. Neurosci. (2022). https://doi.org/10.1155/2022/5211949
12. Elgeldawi, E., Sayed, A., Galal, A.R., Zaki, A.M.: Hyperparameter tuning for machine learning algorithms used for Arabic sentiment analysis. Informatics **8**, 79 (2021). https://doi.org/10.3390/informatics8040079
13. Jagtap, N., Shevatekar, P.P., Mustary, N.: A comparative study of classification techniques in data mining algorithms (2017). https://doi.org/10.21884/ijmter.2017.4211.vxayk
14. Itoo, F., Meenakshi, Singh, S.: Comparison and analysis of logistic regression, Naïve Bayes and KNN machine learning algorithms for credit card fraud detection. Int. J. Inf. Technol.**13**, 1503–1511 (2020). https://doi.org/10.1007/s41870-020-00430-y

Healthcare Waste Disposal Location Selection Using q-Rung Orthopair Fuzzy MEREC—CoCoSo Technique—A Case Study

Saima Debbarma, Sayanta Chakraborty, and Apu Kumar Saha

Abstract Wastes generated from the health care centres are considered healthcare wastes (HCWs), which must be properly disposed of, as they may cause harmful effects to the environment and human health. The process of choosing a location for HCW disposal (HCWD) is rather complicated since there are a lot of factors to consider, including social, environmental, economic, etc., and strict government laws regarding HCWD too. Proper HCWD location (HCWDL) selection is required for dumping the HCWs, which can reduce the pollution as well as protect the environment. The present study aims to conduct a case study on identifying the best location for HCWD in Tripura, India under q-rung orthopair fuzzy environment (q-ROFE), using the concept of multi-criteria group decision making (MCGDM) considering the socio-economic condition of Tripura. To aggregate various decision-makers' (DMs) assessments, q-ROF weighted geometric operator (q-ROFWGO) has been used. The notions of Methods based on the Removal Effects of Criteria (MEREC) and Combined Compromise Solution (CoCoSo) have been redefined under q-ROFE with the aid of q-ROFWGO and integrated to identify the best HCWDL. To overcome the limitations of existing q-ROF score function (q-ROFSF), an improved q-ROFSF has been proposed, which can successfully rank q-ROF numbers (q-ROFNs). To validate the uniformity and dependability of the proposed integrated technique (PIT), comparisons are executed. The PIT has identified the "Biomedical waste treatment plant, Mohanpur" as the optimal location.

Keywords q-rung orthopair fuzzy sets · Score function · HCWDL selection · MEREC · CoCoSo

S. Debbarma (✉) · S. Chakraborty · A. K. Saha
Department of Mathematics, National Institute of Technology Agartala, Tripura 799046, India
e-mail: debbarmasaima123@gmail.com

L. Sahoo et al. (eds.), *Decision Making Under Uncertainty Via Optimization, Modelling, and Analysis*, Studies in Systems, Decision and Control 558,
https://doi.org/10.1007/978-981-96-0085-4_24

1 Introduction

The rapid growth in population has a significant impact on the rise of hospital waste which can be harmful to both human health and the environment. One of the most important concerns is the disposal of HCWs, which are mainly generated from hospitals, dispensaries, nursing home, dental clinic and several other healthcare centres. WHO estimates that 85% of HCWs are not hazardous and the remaining 15% may be radioactive, toxic, or contagious [1, 2]. These 15% of HCWs present various ecological and health risks if these wastes are not appropriately managed or disposed of [3, 4]. The critical strategic issue faced by the healthcare specialist and municipality is the selection of a disposal location for HCW. Therefore, proper HCWDL selection is a complex decision-making process [5, 6]. In the present study, a case study has been executed on Tripura, the north-eeastern state of India, where four (04) locations of Tripura have been selected as HCWDL alternatives based on nine (09) social, economical and environmental parameters. The present treatise models the HCWDL selection problem under q-ROFE to address uncertainty present in the HCWDL selection process.

Uncertainty is a significant issue in real-life decision-making problems due to a lack of knowledge and human limitations. Fuzzy set (FS) [7] is a useful tool for handling ambiguity. q-ROF sets [8] is more superior to Intuitionistic Fuzzy Set (IFS) [9] and Pythagorean Fuzzy Set (PFS) [10] at handling uncertainties in real-world decision-making. To aggregate DMs' assessments, q-ROFWGO [11] has been used. An essential component of ranking fuzzy numbers (FNs) is their score function (SF) [12]. The q-ROFSF that is now in use [8] can rank q-ROFNs, but sometimes it becomes incapable of ranking them, necessitating the use of an accuracy function (AF) in order to rank q-ROFNs. In particular, the current q-ROFSF scores 0 when q-ROFNs have the same membership degree (MD) and non-MD (NMD). Therefore, an improved SF under q-ROFE that can rank any kind of q-ROFNs is required to devise. In order to avoid the need for AFs to further rank the numbers, the current work attempts to create an enhanced q-ROFSF that can rank any type of q-ROFNs.

In the present study, traditional MEREC [13] and CoCoSo [14] methods are extended under q-ROFE with the aid of q-ROFWGO. MEREC is an objective weighing technique that avoids inconsistencies and computational complexity, making it suitable for solving complex MCDM problems. On the other hand, CoCoSo [14], is a method that selects the best alternative based on the aggregated simple additive weighting and exponentially weighted product model. It is a simple method with less computational complexity and is user-friendly. Shanmugasundar et al. [15] used the MEREC method in selecting the most suitable robot for spray-painting applications. Qiyas et al. [16] proposed an improved CoCoSo method to identify the ranking of alternatives with the introduction of a fresh logarithmic picture fuzzy score function, for dealing with the problem of comparison. Naz et al. [17] modified the EDAS method under q-ROFE to solve the MAGDM problem. Wang et al. [18] redefined the CoCoSo method based on Frank's operational laws and softmax function. However, traditional MEREC and CoCoSo methods are not capable of dealing with the uncertainty associated with GDM situations. Four new MCDM methods have been developed for the optimization of two PAC processes by Das and Chakraborty [19] and weights of criteria have been calculated by using MEREC. Tripura et al. [20] redefined MEREC under picture fuzzy environment to identify the optimal third-party logistics. So in the present study, MEREC and CoCoSo have been redefined under q-ROFE with the aid of q-ROFWGO and integrated to solve the HCWDL selection problem.

1.1 Literature Review

The following is a discussion of a thorough literature analysis on HCWDL selection and management:

Yazdani et al. [5] redefined the best–worst method with interval rough numbers for HCWDL decisions. TAŞ [21] used the fuzzy PIPRECIA method to select pre-disposal temporary landfill sites for medicinal products in Turkey. [22] evaluated landfill sites for medical waste created during the COVID-19 pandemic using circular IFS. [23] established a case study of the HCWDL selection problem on FFSs. Chakraborty and Saha [1] used LR-fuzzy AHP and WASPAS to identify HCW recycling technology. Rao and Sujatha [24] carried out a study on selecting the best HCW treatment method based on Fermatean Fuzzy WASPAS. The application of Fermatean fuzzy IWP-TOPSIS-GRA in HCWT technology selection has been proposed by Chen et al. [25]. Beheshtinia et al. [26] introduced an enhanced MCDM method to deal with the healthcare waste disposal method selection (HCWDMS) problem. Chakraborty et al. [27] proposed grey OPA to solve an HCWDL selection problem. Chakraboty and Saha [2] solved the HCW management problem under the Fermatean fuzzy environment. Debbarma et al. [28] proposed an extended FF-SWARA-MABAC method to identify the optimal HCW recycling technology selection.

1.2 Research Question

During the study of the present work, mainly two research questions have been raised. Firstly, which location in Tripura is the best for waste disposal concerning environmental safety and public health? Secondly, as existing q-ROFSF fails to rank all types of q-ROFNs, how to rank those q-ROFNs? For this purpose, a new q-ROFWGO-based MCGDM method has been mathematically modeled to identify the best HCWDL Selection with the aid of novel q-ROFSF.

1.3 Research Gap and Motivation

From the literature, it could be observed that there is very little research that focuses on HCWDL Selection, through a mathematical approach. On the other hand, q-ROFS is an excellent tool to overcome the drawbacks of IFN and PFN theories and is very useful for making decisions in complicated and ambiguous circumstances. The existing q-ROFSF is not competent enough to rank any type of q-ROFNs, especially when both MDs and NMDs are the same in q-ROFNs. In order to address these gaps, the current work developed an integrated q-ROFWGO-based MCGDM technique to handle the problem of choosing the best HCWDL selection using the proposed q-ROFSF.

1.4 Novelty

The current study is innovative in basically three ways:

- The shortcoming of existing SF has been overcome by introducing novel q-ROFSF.
- With the help of q-FOFWGO, the concepts of MEREC and CoCoSo have been expanded under q-ROFE to address group decision-making concerns.
- For the first time, the problem of identifying the most appropriate HCWDL of Tripura, a North-Eastern state of India, has been structured as an MCGDM model considering the socio-economic condition of the state and the location has been identified by using the proposed framework.

1.5 Contribution of the Study

The key contributions are:

- To overcome the shortcomings of the current q-ROFSF, a new SF has been designed under q-ROFE by considering all of its advantageous aspects. The important characteristics of the suggested SF have been explored theoretically.
- MEREC and CoCoSo methods have been extended under q-ROFE blending with q-ROFWGO.
- Using the q-ROFWGO based proposed integrated extended MEREC—CoCoSo technique, optimal HCWD selection problem has been solved.
- The integrity of the results obtained has been validated through comparison.

The other sections are arranged as follows: Sect. 2 contains preliminary information on q-ROFSs, their operational laws and q-ROFWGO. The proposed extended MEREC and CoCoSo approach has been presented in Sect. 3. The case study has been discussed in Sect. 4 to simulate and resolve the issue of choosing the most appropriate location of Tripura for HCWD. The consistency of the ranking results has been discussed using a variety of result analyses in Sect. 5. Section 3 provides a final summary of the work and a recommendation for future research.

2 Basic Concepts

In this section, the basic definitions of q-ROFS, operational laws and q-ROFWGO are discussed.

2.1 q-Rung Orthopair Fuzzy Set (q-ROFS)

Let X be a fixed set. A q-ROFS Q on X is defined by $Q = \{\langle x, \alpha_Q(x), \beta_Q(x)\rangle\}, q \geq 1$, where $\alpha_Q : X \to [0, 1]$ and $\beta_Q : X \to [0, 1]$ denote MD and NMD, respectively, which satisfies $0 \leq (\alpha_Q(x))^q + (\beta_Q(x))^q \leq 1$. The indeterminacy degree is given as $(\pi_Q(x))^q = 1 - ((\alpha_Q(x))^q + (\beta_Q(x))^q)$. For convenience, $Q = (\alpha, \beta)$ is called a q-ROFN.

2.2 Score Function and Accuracy Function

Let $Q = (\alpha, \beta)$ be a q-ROFN, then the q-ROF $S(Q)$ and $A(Q)$ proposed by [11] is defined as

$$S(Q) = (\alpha_Q(x))^q - (\beta_Q(x))^q, S(Q) \in [-1, 1] \tag{1}$$

$$A(Q) = (\alpha_Q(x))^q + (\beta_Q(x))^q, A(Q) \in [0,1] \tag{2}$$

2.3 Operational Rules

Let $Q_1 = (\alpha_1, \beta_1)$, $Q_2 = (\alpha_2, \beta_2)$ and $Q = (\alpha, \beta)$ be three q-ROFNs, then

i. $\oplus Q_2 = \left(\sqrt[q]{(\alpha_1)^q + (\alpha_2)^q - (\alpha_1)^q(\alpha_2)^q}, \beta_1\beta_2\right)$;

ii. $Q_1 \otimes Q_2 = \left(\alpha_1\alpha_2, \sqrt[q]{(\beta_1)^q + (\beta_2)^q - (\beta_1)^q(\beta_2)^q}\right)$;

iii. $\lambda Q = \left(\sqrt[q]{1 - (1 - \alpha^q)^\lambda}, \beta^\lambda)\right), \lambda > 0$;

iv. $Q^{\lambda} = \left(\alpha^{\lambda}, \sqrt[q]{1 - (1 - \beta^{q})^{\lambda}}\right), \lambda > 0;$
v. The $Q^{C} = (\beta, \alpha)$.

2.4 q-ROFWGO

Let $Q_j = (\alpha_j, \beta_j)(j = 1, 2, \ldots, n)$ be q-ROFNs and $w = (w_1, w_2, \ldots, w_n)^T$ be the weight vector for Q_j with $\sum_{i=1}^{n} w_i = 1$, then the q-ROFWGO [11] can be expressed as

$$q - ROFWGO(Q_1, Q_2, \ldots, Q_n) = \prod_{j=1}^{n} (Q_j)^{w_j} = \left(\prod_{j=1}^{n} (\alpha_{Q_j})^{w_j}, \left(1 - \prod_{j=1}^{n} (1 - (\beta_{Q_j})^{q})^{w_j}\right)^{\frac{1}{q}} \right) \quad (3)$$

3 Proposed q-ROFWGO Based MCGDM

In this section, the proposed q-ROFSF and q-ROFWGO-based extended MEREC and CoCoSo techniques have been discussed.

3.1 Proposed q-ROFSF

SF is an important measure to rank fuzzy numbers. But sometimes, Eq. (1) fails to rank q-ROFNs precisely, especially when the MD and NMD of q-ROFNs are equal. A novel q-ROFSF has been developed to address the error present in Eq. (1). The following example illustrates the limitation of existing q-ROFSF.

Let $Q_1 = (0.5, 0.5)$ and $Q_2 = (0.2, 0.2)$ be two q-ROFNs. Existing q-ROFSF defined in Eq. (1) is unable to distinguish between considered q-ROFNs, since $S(Q_1) = S(Q_2) = 0$. Therefore, to deal with the drawback of the existing SF, an improved SF has been proposed below.

Suppose $Q_j = (\alpha_j, \beta_j)$ be a q-ROFN. Now, SF for a q-ROFN Q_j is defined as follows:

$$S(Q_j) = \frac{1}{5}\left(e^{\alpha_j^q + \beta_j^q - 1} + 4\alpha_j^q - \beta_j^q\right), \quad \text{where } S(Q_j) \in [0, 1]. \quad (4)$$

For $Q_1 = (0.5,0.5)$ and $Q_2 = (0.2,0.2)$, corresponding to Eq. (4), $S(Q_1) = 0.169473$ and $S(Q_2) = 0.079563$ consider $q = 3$ and consequently, $Q_1 > Q_2$. Therefore, regardless of the MD and NMD of q-ROFNs, the suggested SF can efficiently rank q-ROFNs. The following theoretical validations have been made on the efficacy of the enhanced SF.

Theorem 3.1 Equation (4) satisfies

(P1) $S(0,1) = 0$ and $S(1,0) = 1$.

(P2) $0 \le S(Q_j) \le 1$.

Proof: (P1). From Eq. (4), it is clear that $S(Q_j)_{min} = 0$ and $S(Q_j)_{max} = 1$. Thus P1 holds.

(P2). $0 \le \alpha_j^q \le 1$ and $0 \le \beta_j^q \le 1$ imply $0 \le \frac{1}{5}\left(e^{\alpha_j^q+\beta_j^q-1} + 4\alpha_j^q - \beta_j^q\right) \le 1$

Hence $0 \le S(Q_j) \le 1$

Theorem 3.2 $S(Q_j)$ increases and decreases monotonically with respect to α_j and β_j, respectively.

Proof Differentiation of Eq. (4) partially with respect to α_j and β_j give

$$\frac{\partial S(Q_j)}{\partial \alpha_j} = \frac{1}{5} q\alpha_j^{q-1}\left(e^{\alpha_j^q+\beta_j^q-1} + 4\right) \ge 0, \forall \alpha_j \in [0, 1]$$

$$\frac{\partial S(Q_j)}{\partial \beta_j} = -\frac{1}{5} q\beta_j^{q-1}\left(1 - e^{\alpha_j^q+\beta_j^q-1}\right) \le 0, \forall \beta_j \in [0, 1]$$

Hence, the theorem is proved.

Theorem 3.3 For any two q-ROFNs $Q_1 = (\alpha_1, \beta_1)$ and $Q_2 = (\alpha_2, \beta_2)$, if $\alpha_1 > \alpha_2$ and $\beta_1 < \beta_2$, then $S(Q_1) > S(Q_2)$.

Proof: For $\alpha_1 > \alpha_2$ and $\beta_1 < \beta_2$, $e^{\alpha_1^q+\beta_1^q-1} + 4\alpha_1^q - \beta_1^q > e^{\alpha_2^q+\beta_2^q-1} + 4\alpha_2^q - \beta_2^q$

That is, $\frac{1}{5}\left(e^{\alpha_1^q+\beta_1^q-1} + 4\alpha_1^q - \beta_1^q\right) > \frac{1}{5}\left(e^{\alpha_2^q+\beta_2^q-1} + 4\alpha_2^q - \beta_2^q\right)$

That is, $S(Q_1) > S(Q_2)$.

Hence the theorem.

3.1.1 Q-ROFWGO-Based Proposed Extended MEREC Technique

MEREC finds criteria weights (CWs) by utilizing variations in the performance of each option for individual criteria. With the use of q-ROFWGO, MEREC has been expanded under q-ROFE in the current study to handle MCGDM difficulties. The proposed q-ROFWGO-based MEREC is discussed below.

Step 1 Construct fuzzy decision matrix $\tilde{Q} = [A_{ij}^l]_{m\times n}$ corresponding to each DM as follows

$$\tilde{Q} = [A^l_{ij}]_{m\times n} = \begin{array}{c} \\ A_1 \\ A_2 \\ \vdots \\ A_m \end{array} \begin{array}{c} \begin{array}{cccc} C_1 & C_2 & \cdots & C_n \end{array} \\ \begin{bmatrix} (\alpha^l_{11}, \beta^l_{11}) & (\alpha^l_{12}, \beta^l_{12}) & \cdots & (\alpha^l_{1n}, \beta^l_{1n}) \\ (\alpha^l_{21}, \beta^l_{21}) & (\alpha^l_{22}, \beta^l_{22}) & \cdots & (\alpha^l_{2n}, \beta^l_{2n}) \\ \vdots & \vdots & \vdots & \vdots \\ (\alpha^l_{m1}, \beta^l_{m1}) & (\alpha^l_{m2}, \beta^l_{m2}) & \cdots & (\alpha^l_{mn}, \beta^l_{mn}) \end{bmatrix} \end{array} \tag{5}$$

where $A^l_{ij} = \left(\alpha^l_{ij}, \beta^l_{ij}\right)$ denotes q-ROFN with alternatives A_i (i = 1,2, ... m) and criteria C_j (j = 1,2, ... n) by l th DM.

Step 2 Aggregate DMs' evaluations using Eq. (3) as

$$[\tilde{A}_{ij}]_{m\times n} = \begin{array}{c} A_1 \\ A_2 \\ \cdots \\ A_m \end{array} \begin{bmatrix} C_1 & \ldots & C_n \\ (\alpha_{11}, \beta_{11}) & \ldots & (\alpha_{1n}, \beta_{1n}) \\ (\alpha_{21}, \beta_{21}) & \ldots & (\alpha_{2n}, \beta_{2n}) \\ \ldots & \ldots & \ldots \\ (\alpha_{m1}, \beta_{m1}) & \ldots & (\alpha_{mn}, \beta_{mn}) \end{bmatrix}$$

Step 3 Normalize the aggregated matrix $\tilde{A}_{ij}$ as follows:

$$\tilde{N} = \left. \begin{array}{c} \tilde{\tilde{A}}_{ij} = (\alpha_{ij}, \beta_{ij}) forBC \\ \tilde{A}^C_{ij} = (\beta_{ij}, \alpha_{ij}) forNBC \end{array} \right\} \tag{6}$$

where $\tilde{N}$ denotes the normalized matrix, BC and NBC stands for beneficiary and non-beneficiary criteria.

Step 4 Using Eq. (4), derive A_{ij}.

Step 5 Calculate the complete performance of alternatives $Ş_i$ as

$$Ş_i = \ln\left(1 + \left(\frac{1}{n}\sum_j |\ln(A_{ij})|\right)\right) \tag{7}$$

Step 6 Calculate performance of alternatives $Ş'_{ij}$ as.

$$Ş'_{ij} = \ln\left(1 + \left(\frac{1}{n}\sum^{n}_{k,k\neq j} |\ln(A_{ik})|\right)\right) \tag{8}$$

Step 7 Calculate aggregated absolute deviations. The removal effect of criterion j is determined as.

$$d_j = \sum^{m}_{i=1} |Ş'_{ij} - Ş_i| \tag{9}$$

Step 8 Identify the final CW

$$w_j = \frac{\daleth_j}{\sum_{k=1}^{m} \daleth_k} \tag{10}$$

3.1.2 Q-ROFWGO Based Proposed Extended CoCoSo Method

CoCoSo is an efficient model to deal with MCDM problems. In the current study, q-ROFWGO is used in conjunction with the extension of the CoCoSo idea under q-ROFE to solve MCGDM difficulties. The suggested q-ROFWGO-based CoCoSo approach is discussed herewith:

Step 1 Construct a fuzzy decision matrix corresponding to each DM as follows:

$$\tilde{Q} = [A_{ij}^l]_{m\times n} = \begin{array}{c} \\ A_1 \\ A_2 \\ \vdots \\ A_m \end{array} \begin{array}{c} \begin{array}{cccc} C_1 & C_2 & \ldots & C_n \end{array} \\ \begin{bmatrix} (\alpha_{11}^l, \beta_{11}^l) & (\alpha_{12}^l, \beta_{12}^l) & \cdots & (\alpha_{1n}^l, \beta_{1n}^l) \\ (\alpha_{21}^l, \beta_{21}^l) & (\alpha_{22}^l, \beta_{22}^l) & \cdots & (\alpha_{2n}^l, \beta_{2n}^l) \\ \vdots & \vdots & \vdots & \vdots \\ (\alpha_{m1}^l, \beta_{m1}^l) & (\alpha_{m2}^l, \beta_{m2}^l) & \cdots & (\alpha_{mn}^l, \beta_{mn}^l) \end{bmatrix} \end{array} \tag{11}$$

where $A_{ij}^l = \left(\alpha_{ij}^l, \beta_{ij}^l\right)$ denotes q-ROFN with alternatives A_i and criteria C_j by lth DM.

Step 2 Aggregate decision matrices using Eq. (3) and find $\tilde{A}_{ij}$.

Step 3 Normalize $\tilde{A}_{ij}$ using Eq. (6) and obtain $\tilde{N}$.

Step 4 Estimate S_i and P_i matrices of each alternative as given below:

$$S_i = \oplus_{j=1}^{n} w_j \tilde{N} \tag{12}$$

$$P_i = \oplus_{j=1}^{n} \tilde{N}^{w_j} \tag{13}$$

Step 5 Find the crisp scores of S_i and P_i using Eq. (4) and hence compute the relative weights of the alternative using three appraisal score strategies which are derived using the formula given as

$$K_{ia} = \frac{P_i + S_i}{\sum_{i=1}^{m}(P_i + S_i)} \tag{14}$$

$$K_{ib} = \frac{S_i}{\min\limits_i S_i} + \frac{P_i}{\min\limits_i P_i} \tag{15}$$

$$K_{ic} = \frac{\lambda(S_i) + (1-\lambda)(P_i)}{\left(\lambda \max\limits_i S_i + (1-\lambda)\max\limits_i P_i\right)} \tag{16}$$

Step 6 Rank alternatives by calculating K_i as

$$K_i = (K_{ia}K_{ib}K_{ic})^{\frac{1}{3}} + \frac{1}{3}(K_{ia} + K_{ib} + K_{ic}) \tag{17}$$

4 Application of Proposed Technique in Selection of HCWDL in Tripura—A Case Study

This section provides a summary of the different HCWDLs of Tripura, India, highlights key elements for choosing the appropriate HCWDL and evaluates those elements. It also shows how the suggested q-ROFWGO-based extended MEREC—CoCoSo technique can be applied to selecting the best HCWDL.

Direct dumping the waste into the landfills and open areas may cause harmful effects to both the environment as well as living beings. This may also cause a rise in environmental pollution. Therefore, a proper disposal location should be considered to dump the garbage generated mainly from the health care centers. Here four (04) locations of Tripura have been identified as HCWD stations, which are presented in Table 1. To identify the best location of disposal, nine (09) factors are identified based on the socio-economic condition of Tripura and are furnished in Table 2. Three experts, namely, DM1, DM2, and DM3 have been assembled for the panel based on their qualifications and experience in the relevant field. A questionnaire is used to gather the linguistic evaluations of the selected alternatives in relation to several contributing criteria for HCWDL selection. Their professional judgments are included in the questionnaire and the 9-level q-ROFN scale, shown in Table 3, is used to translate linguistic assessments into q-ROFNs.

To identify the optimal location of Tripura for HCWD, the proposed technique is executed as follows.

Step 1 For HCWDL selection, four (04) locations of Tripura have been chosen based on nine (09) factors, presented in Tables 1 and 2, respectively.

Step 2 Linguistic assessments of experts are converted to q-ROFNs using Table 3 and presented in Table 4.

Step 3 Assigning weights (0.25, 0.4, 0.35) to the DMs, respectively, and using Eq. (3), decision matrices corresponding to criteria preferences assigned by DMs are aggregated. Here, $q = 3$ is considered during the calculation.

Table 1 HCWD locations

Sl. No	HCWD locations
*L*1	Bio-medical waste treatment plant, Mohanpur
*L*2	AMC Garbage Dump, Abhaynagar
*L*3	Temporary RRR Collection Centre, Kailashahar
*L*4	Sewage Treatment Plant, Chandinamura

Table 2 Criteria associated with HCWDL

Sl. No	Criteria
C1	Possible risk of intrusion and emission
C2	Distance to the urban and city infrastructure
C3	Geographic and geologic circumstances
C4	Eco-friendly services
C5	Land, Transportation and Maintenance expenditure
C6	Possibility of future development
C7	Availability of employees
C8	Sensitivity towards environment, local and territorial rules
C9	Satisfaction levels of residents towards location selection

Step 4 Normalized matrix is shown in Table 5.

Step 5 Using Eqs. (7–10), CWs are determined (Table 6).

Step 6 Using Eqs. (12–17), K_i values are calculated and alternatives are ranked, which are presented in Table 7 and shown in Fig. 1.

Table 7 indicates that L1 that is, "Biomedical waste treatment plant, Mohanpur" is the best location for HCWD in Tripura.

5 Results and Discussions

The suggested extended MEREC-CoCoSo approach has illustrated several noteworthy outcomes in selecting the best HCWDL according to the model formulation. L1 that is, "Biomedical waste treatment plant, Mohanpur", has been determined to be the best location alternative for disposing of the HCW in this application. To examine the trustworthiness of ranking outcomes obtained through the proposed methodology, comparative and sensitivity studies are shown.

5.1 Comparative Study

The proposed approach has been compared with ARAS, EDAS, WSM, WPM and WASPAS under q-ROFE. L1 has been identified as the best disposal location by EDAS and WPM method, whereas ARAS, WSM and WASPAS methods have chosen L2 as the best alternative. All the methods have identified L4 as the least desirable disposal location, which is shown in Table 8 (Fig. 2).

A small variation in ranking results (RRs) could be found in Table 8. So to check the correlation between the RRs obtained by various MCDM methods under q-ROFE and the proposed method, Spearman's Rank Correlation Coefficients between the

Table 3 The 9-level scale of q-ROFNs

Extreme Low	Very Low	Low	Moderately Low	Moderate	Moderately High	High	Very High	Extreme High
(0.1,0.99)	(0.15,0.8)	(0.25,0.77)	(0.37,0.69)	(0.5,0.65)	(0.65,0.6)	(0.77,0.58)	(0.86,0.45)	(0.9,0.25)

Table 4 Assessments of DMs

		C1	C2	C3	C4	C5	C6	C7	C8	C9
DM1	*L*1	(0.77,0.58)	(0.5,0.65)	(0.5,0.65)	(0.77,0.58)	(0.5,0.65)	(0.77,0.58)	(0.5,0.65)	(0.77,0.58)	(0.5,0.65)
	*L*2	(0.86,0.45)	(0.77,0.58)	(0.77,0.58)	(0.86,0.45)	(0.86,0.45)	(0.86,0.45)	(0.86,0.45)	(0.77,0.58)	(0.5,0.65)
	*L*3	(0.77,0.58)	(0.5,0.65)	(0.5,0.65)	(0.77,0.58)	(0.77,0.58)	(0.86,0.45)	(0.77,0.58)	(0.77,0.58)	(0.5,0.65)
	*L*4	(0.77,0.58)	(0.5,0.65)	(0.5,0.65)	(0.5,0.65)	(0.5,0.65)	(0.5,0.65)	(0.5,0.65)	(0.77,0.58)	(0.5,0.65)
DM2	*L*1	(0.65,0.6)	(0.5,0.65)	(0.65,0.6)	(0.77,0.58)	(0.77,0.58)	(0.65,0.6)	(0.5,0.65)	(0.86,0.45)	(0.65,0.6)
	*L*2	(0.77,0.58)	(0.5,0.65)	(0.5,0.65)	(0.77,0.58)	(0.86,0.45)	(0.5,0.65)	(0.77,0.58)	(0.86,0.45)	(0.5,0.65)
	*L*3	(0.86,0.45)	(0.77,0.58)	(0.77,0.58)	(0.5,0.65)	(0.77,0.58)	(0.65,0.6)	(0.77,0.58)	(0.77,0.58)	(0.5,0.65)
	*L*4	(0.77,0.58)	(0.5,0.65)	(0.5,0.65)	(0.77,0.58)	(0.5,0.65)	(0.5,0.65)	(0.77,0.58)	(0.86,0.45)	(0.65,0.6)
DM3	*L*1	(0.77,0.58)	(0.65,0.6)	(0.77,0.58)	(0.5,0.65)	(0.86,0.45)	(0.77,0.58)	(0.5,0.65)	(0.65,0.6)	(0.65,0.6)
	*L*2	(0.77,0.58)	(0.77,0.58)	(0.5,0.65)	(0.77,0.58)	(0.86,0.45)	(0.86,0.45)	(0.65,0.6)	(0.5,0.65)	(0.65,0.6)
	*L*3	(0.86,0.45)	(0.5,0.65)	(0.65,0.6)	(0.86,0.45)	(0.77,0.58)	(0.65,0.6)	(0.5,0.65)	(0.77,0.58)	(0.5,0.65)
	*L*4	(0.86,0.45)	(0.77,0.58)	(0.5,0.65)	(0.65,0.6)	(0.86,0.45)	(0.5,0.65)	(0.65,0.6)	(0.86,0.45)	(0.5,0.65)

Table 5 Normalized matrix

	C1	C2	C3	C4	C5	C6	C7	C8	C9
L1	(0.588,0.72)	(0.548,0.634)	(0.646,0.607)	(0.662,0.607)	(0.566,0.718)	(0.72,0.588)	(0.5,0.65)	(0.758,0.546)	(0.609,0.614)
L2	(0.554, 792)	(0.648,0.612)	(0.557,0.635)	(0.792,0.554)	(0.45,0.86)	(0.692,0.553)	(0.746,0.562)	(0.692,0.57)	(0.548,0.634)
L3	(0.491,0.837)	(0.594,0.625)	(0.651,0.606)	(0.673,0.579)	(0.58,0.77)	(0.697,0.571)	(0.662,0.571)	(0.77,0.58)	(0.5,0.65)
L4	(0.543,0.8)	(0.582,0.628)	(0.5,0.65)	(0.651,0.606)	(0.599,0.604)	(0.5,0.65)	(0.651,0.606)	(0.837,0.491)	(0.555,0.631)

Table 6 Weights of criteria

	C1	C2	C3	C4	C5	C6	C7	C8	C9
Weights	0.152707	0.124717	0.126681	0.088569	0.88569	0.103633	0.10889	0.066143	0.14009

Table 7 Ranking of alternatives

Alternatives	K_{ia}	K_{ib}	K_{ic}	K_i	Ranking
L1	0.266641	2.206759	0.998266	1.994705	1
L2	0.246593	2.182947	0.923211	1.909675	2
L3	0.244793	2.095268	0.91647	1.863043	3
L4	0.241974	2	0.905916	1.808973	4

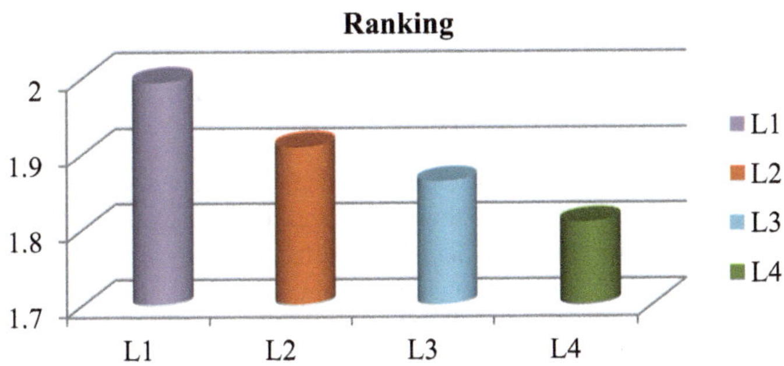

Fig. 1 Ranking of HCWDLs

Table 8 Comparison of MCDMs

MCDM	Ranking
Proposed method	$L1 > L2 > L3 > L4$
q-ROF-ARAS	$L2 > L1 > L3 > L4$
q-ROF—EDAS	$L1 > L2 > L3 > L4$
q-ROF—WSM	$L2 > L1 > L3 > L4$
q-ROF—WPM	$L1 > L2 > L3 > L4$
q-ROF—WASPAS	$L2 > L1 > L3 > L4$

results obtained by the proposed method and other MCDMs have been calculated and obtained as (0.8,1.00,0.8,1.00,0.8). As all these values are greater than 0.7, the ranking findings are therefore consistent and the proposed method is able to tackle every important MCDM concern.

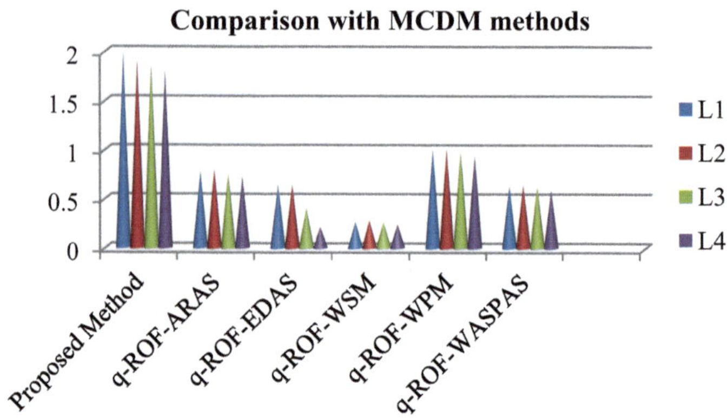

Fig. 2 Comparison with various MCDM methods

5.2 *Sensitivity Analysis with Variation in Parameter Q*

As q has a substantial impact on the findings, it is very pertinent to check the sensitivity [29, 30] of the results with respect to this parameter. It is thus examined how different values of the parameter q affect the decision results. To do so, the K_i values for each alternative have been recalculated by changing the values of q from 1 to 10. Table 9 shows the outcomes of the rankings. On changing the value q from 1 to 10, it could be observed that the results obtained by different values of q differ from each other and a significant change is found in RRs. So, the result obtained is sensitive with respect to variation in q (Fig. 3).

Table 9 Variation q

q	Ranking
1	$L2 > L1 > L3 > L4$
2	$L1 > L2 > L3 > L4$
3	$L1 > L2 > L3 > L4$
4	$L1 > L2 > L3 > L4$
5	$L1 > L2 > L3 > L4$
6	$L1 > L2 > L3 > L4$
7	$L1 > L2 > L3 > L4$
8	$L1 > L2 > L3 > L4$
9	$L1 > L2 > L3 > L4$
10	$L1 > L2 > L3 > L4$

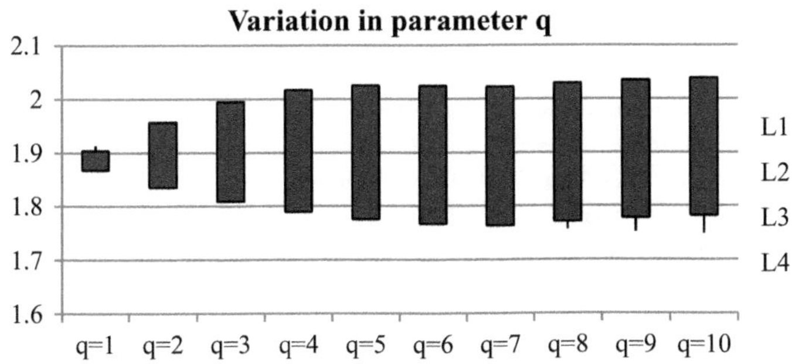

Fig. 3 Variation in q

5.3 *Variation in CWs*

To test the resilience of the suggested technique, SA has been conducted with respect to variation in CWs. Table 10 shows how changes in CWs impact the ranking results. It has been observed that L1 is selected as the best HCWDL alternative and L4 as the least desirable one for Case I, Case II, Case VIII to Case XI and L2 has been identified as the worst alternative for Case III to Case VII. A significant change could be observed in RRs during variation of CWs based on BC and NBC. It is possible to conclude that the suggested method is sensitive to variations in CWs because it yields findings that are considerably impacted by changes in CWs (Fig. 4).

Table 10 CW variation

Cases	Variation in weights		Ranking
Case I	Weights obtained through proposed approach		*L*1 > *L*2 > *L*3 > *L*4
Case II	Equal weights		*L*1 > *L*2 > *L*3 > *L*4
Case III	50% to NBC	50% to BC	*L*1 > *L*4 > *L*3 > *L*2
Case IV	60% to NBC	40% to BC	*L*1 > *L*4 > *L*3 > *L*2
Case V	70% to NBC	30% to BC	*L*1 > *L*4 > *L*3 > *L*2
Case VI	80% to NBC	20% to BC	*L*1 > *L*4 > *L*3 > *L*2
Case VII	90% to NBC	10% to BC	*L*1 > *L*4 > *L*3 > *L*2
Case VIII	10% to NBC	90% to BC	*L*1 > *L*2 > *L*3 > *L*4
Case IX	20% to NBC	80% to BC	*L*1 > *L*2 > *L*3 > *L*4
Case X	30% to NBC	70% to BC	*L*1 > *L*2 > *L*3 > *L*4
Case XI	40% to NBC	60% to BC	*L*1 > *L*2 > *L*3 > *L*4

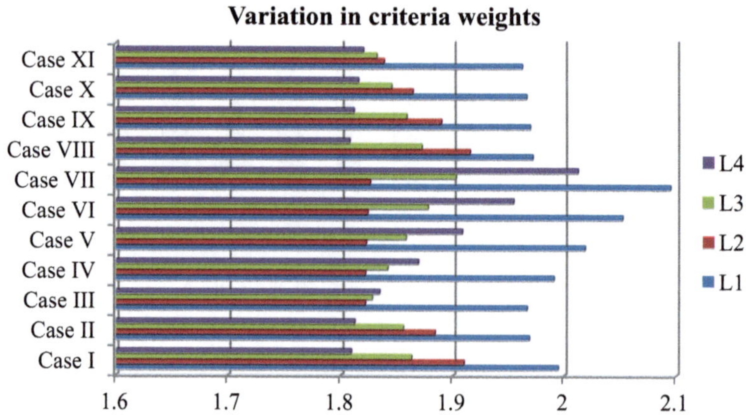

Fig. 4 Variation in CWs

5.4 *Variation in Parameter Λ*

As the parameter λ has a significant influence on the findings, it is very pertinent to check the flexibility and sensitivity of the results with respect to this parameter. It is examined how different values of λ are affecting the decision results. To do so, K_i has been recalculated for each alternative by changing λ from 0 to 1. Table 11 depicts that the ranking remains the same for all the cases. For all values of λ, L1 is the best option for HCWD and followed by L2, L3 and L4. Thus the results are consistent irrespective of various values of the parameter λ. Figure 5 portrays how variations in λ affect the ranking order of alternatives.

Table 11 Variation in parameter λ

λ	Ranking
0	*L1 > L2 > L3 > L4*
0.1	*L1 > L2 > L3 > L4*
0.2	*L1 > L2 > L3 > L4*
0.3	*L1 > L2 > L3 > L4*
0.4	*L1 > L2 > L3 > L4*
0.5	*L1 > L2 > L3 > L4*
0.6	*L1 > L2 > L3 > L4*
0.7	*L1 > L2 > L3 > L4*
0.8	*L1 > L2 > L3 > L4*
0.9	*L1 > L2 > L3 > L4*
1	*L1 > L2 > L3 > L4*

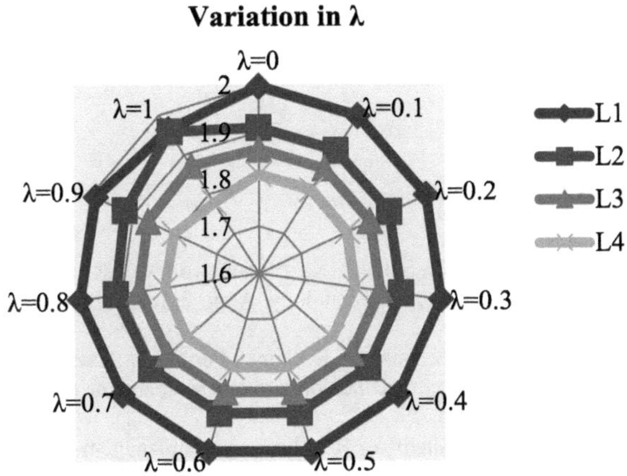

Fig. 5 Variation in λ

6 Conclusion

The proper management of HCW is very crucial for the protection of the environment and public health from pollution. An essential component of managing municipal solid waste is the HCWDL selection. Disposing of HCW in a proper disposal location can reduce environmental pollution, soil pollution, etc. For the sake of the safety of public health and environmental sustainability, it is essential to identify a proper HCWDL. In the present study, an MCGDM model is developed to find the best HCWDL by creating a new q-ROFSF. The improved q-ROFSF is able to encounter the deficits of the present SF as well as to rank the alternatives irrespective of their MDs and NMDs. According to the recommended approach, L1 is the most effective choice for reaching the given objective. Comparing the recommended integrated technique to several MCDMs, the results demonstrate that it generates consistent results. To evaluate the robustness of the proposed methodology, sensitivity analyses have been carried out and the findings reveal that the ranking outcomes are consistent with modifications in the criteria's weights. The suggested method is essentially beneficial in three ways. Firstly, the new q-ROFSF can rank q-ROFNs regardless of their MD and NMD while also getting around the restrictions of the current SF. Secondly, using q-ROFWGO in conjunction with the suggested methodology, two well-known MCDM techniques—MEREC and CoCoSo—have been expanded under q-ROFE to enable them to handle scenarios including collective decision-making in the face of uncertainty. Thirdly, taking into account Tripura's socioeconomic circumstances, a case study has been conducted to determine the best HCWDL selection. The results of this study will assist local authorities in the proper disposal of wastes.

As there are so many alternatives, criteria, and DMs in the problem, it takes a lot of hands-on work, especially while gathering information. In our upcoming project, we will develop a computer-assisted solution method for high-dimensional hierarchical problems while making sure that the number of variables, options and experts do not influence the process of solving the problem. Moreover, assessments from professionals with industrial experience can also be used to incorporate commercial knowledge.

References

1. Chakraborty, S., Saha, A.K.: A framework of LR fuzzy AHP and fuzzy WASPAS for health care waste recyclingtechnology. Appl. Soft Comput. (2022a). https://doi.org/10.1016/j.asoc.2022.109388
2. Chakraborty, S., Saha, A.K.: Novel Fermatean fuzzy Bonferroni Mean aggregation operators for selecting optimal health care waste treatment technology. Eng. Appl. Artif. Intell. **119**, 105752 (2023)
3. Chauhan A., Singh A.: Healthcare waste management: a state-of-the-art literature review. Int. J. Environ. Waste. Manag. **18**(2), 120–144 (2016)
4. IndiaStat: http://www.indiastat.com/table/environmentandpollution/11/solidwaste/261/910950/data.aspx. Accessed 06 Feb 2016 (2013)
5. Yazdani, M., Tavana, M., Pamučar, D., Chatterjee, P.: A rough based multi-criteria evaluation method for healthcare waste disposal location decisions. Comput. & Ind. Eng. **143**, 106394 (2020)
6. Sahoo, L., Senapati, T., Yager, R.R. (Eds.): Real life applications of multiple criteria decision making techniques in fuzzy domain (Vol. 420). Springer Nature (2022)
7. Zadeh, L.A.: Fuzzy sets. Inf. Control. **8**(3), 338–353 (1965)
8. Yager, R.R.: Generalized orthopair fuzzy Sets. IEEE Trans Fuzzy Syst. (2016)
9. Atanassov, K.T.: Intuitionistic Fuzzy Sets. Fuzzy Sets Syst. **20**, 87–96 (1986). https://doi.org/10.1016/S0165-0114(86)80034-3
10. Yager, R.R.: Pythagorean membership grades in multi criteria decision making. Technical report MII-3301. Machine Intelligence Institute. Iona College. New Rochelle (2013)
11. Liu, P., Wang, P.: Some q-rung orthopair fuzzy aggregation operators and their applications to multiple-attribute decision making. Int. J. Intell. Syst. **33**(2), 259–280 (2018)
12. Debbarma, S., Chakraborty, S., Saha, A.K.: Information aggregation based group decision making under Fermatean fuzzy environment for spent lithium-ion battery recycling techniques evaluation. Int. J. Hydrog. Energy **65**, 891–904 (2024a)
13. Keshavarz-Ghorabaee, M., Amiri, M., Zavadskas, E.K., Turskis, Z., Antucheviciene, J.: Determination of objective weights using a new method based on the removal effects of criteria (MEREC). Symmetry, **13**(4), 525 (2021)
14. Yazdani, M., Zarate, P., Kazimieras Zavadskas, E., Turskis, Z.: A combined compromise solution (CoCoSo) method for multi-criteria decision-making problems. Manag. Decis. **57**(9), 2501–2519 (2019)
15. Shanmugasundar, G., Sapkota, G., Čep, R., Kalita, K.. Application of MEREC in multi-criteria selection of optimal spray-painting robot. Processes **10**(6), 1172 (2022)
16. Qiyas, M., Naeem, M., Khan, S., Abdullah, S., Botmart, T., Shah, T.: Decision support system based on CoCoSo method with the picture fuzzy information. J. Math. **2022**, 1–11 (2022)
17. Naz, S., Akram, M., Fatima, A., Nadeem, A.: q-Rung orthopair fuzzy 2-tuple linguistic hamy mean operators for MAGDM with modified EDAS method. In: Real Life Applications of Multiple Criteria Decision Making Techniques in Fuzzy Domain (pp. 369–415). Singapore: Springer Nature Singapore (2022)
18. Wang, H., Mahmood, T., Ullah, K.: Improved CoCoSo method based on frank softmax aggregation operators for T-spherical fuzzy multiple attribute group decision-making. Int. J. Fuzzy Syst. **25**(3), 1275–1310 (2023)
19. Das, P.P., Chakraborty, S.: A comparative assessment of multicriteria parametric optimization methods for plasma arc cutting processes. Decis. Anal. J. **6**, 100190 (2023)

20. Tripura, C., Chakraborty, S., Bhattacharya, B.: Picture fuzzy aggregation operator based integrated MEREC-WASPAS technique for video conferencing tool selection. J. Uncertain Syst. (2024)
21. TAŞ, M.A.: Assessment of site selection criteria for medical waste during COVID-19 pandemic. Avrupabilimveteknolojidergisi (28), 63–69 (2021)
22. Çakir, E., Taş, M.A., Ulukan, Z.: A new circular intuitionistic fuzzy MCDM: A case of Covid-19 medical waste landfill site evaluation. In: 2021 IEEE 21st international symposium on computational intelligence and informatics (CINTI) (pp. 000143–000148). IEEE (2021)
23. Mishra, A.R., Rani, P.: Multi-criteria healthcare waste disposal location selection based on Fermatean fuzzy WASPAS method. Complex & Intell. Syst. **7**(5), 2469–2484 (2021)
24. Rao, C.N., Sujatha, M.: A consensus-based Fermatean fuzzy WASPAS methodology for selection of healthcare waste treatment technology selection. Decis. Mak.: Appl. Manag. Eng. **6**(2), 600–619 (2023)
25. Chen, W., Zeng, S., Zhang, E.: Fermatean fuzzy IWP-TOPSIS-GRA multi-criteria group analysis and its application to healthcare waste treatment technology evaluation. Sustainability **15**(7), 6056 (2023)
26. Beheshtinia, M.A., JafariKahoo, S., Fathi, M.: Prioritizing healthcare waste disposal methods considering environmental health using an enhanced multi-criteria decision-making method. Environ. Pollut. Bioavailab. **35**(1), 2218568 (2023)
27. Chakraborty, S., Raut, R.D., Rofin, T.M., Chakraborty, S.: A grey ordinal priority approach for healthcare waste disposal location selection. Grey Syst.: Theory Appl. **13**(4), 767–784 (2023)
28. Debbarma, S., Chakraborty, S., Saha, A.K.: Health care waste recycling concerning circular economy: a Fermatean fuzzy aggregation operator-based SWARA–MABAC approach. Environ., Dev. Sustain. 1–40 (2024b)
29. Chakraborty, S., Saha, A.K.: Selection of optimal lithium ion battery recycling process: A multi-criteria group decision making approach. J. Energy Storage **55**, 105557 (2022b)
30. Chakraborty, S., Saha, A.K.: Selection of Forklift unit for transport handling using integrated MCDM under neutrosophic environment. FactaUniversitatis, Ser.: Mech. Eng. (2022c)

Optimal Plastic Waste Recycling Technology Selection Using Picture Fuzzy SWARA–MARCOS Technique

Chayel Tripura, Sayanta Chakraborty, and Baby Bhattacharya

Abstract Recycling is gaining popularity as a practical solution to the plastic waste (PW) problem, as landfill disposal and burning have detrimental impacts on human and environmental health. The research focuses on developing a mathematical model to determine the best plastic waste recycling technique (PWRT) considering social, environmental, and financial factors, as effective management of PW is crucial for environmental and public health. In this sense, the challenge of optimal PWRT selection has been modelled as a multi-criteria group decision-making (MCGDM) problem under a picture fuzzy environment (PFE). The concepts of conventional Stepwise Weight Assessment Ratio Analysis (SWARA) and Measurement of Alternatives and Ranking according to COmpromise Solution (MARCOS) methodologies have been expanded under PFE with the help of Picture fuzzy weighted geometric operator (PFWGO) to determine the ideal PWRT. An enhanced picture fuzzy scoring function (PFSF) has been created to rank picture fuzzy numbers (PFNs) regardless of their membership and non-membership grades. The key characteristics of the suggested PFSF have been verified theoretically. The proposed technique has identified CLR as the optimal PWRT. To assess the integrity and resilience of the suggested approach, sensitivity and comparison studies have been conducted.

Keywords PWRT · PFWGO · SWARA · MARCOS · PFSF

1 Introduction

PWs are plastic objects that pile up in the environment and disturb the ecological environment causing threats to life. Plastics are cheap and durable which make up for their demand in the market which in turn contributes to the generation of PWs. PWs are dumped in landfills or the oceans which severely affect aquatic life. Plastic is a polymeric material made up of long interconnected chains of molecules that cannot

C. Tripura (✉) · S. Chakraborty · B. Bhattacharya
Department of Mathematics, National Institute of Technology Agartala, Tripura 799046, India
e-mail: chyeltripura@gmail.com

L. Sahoo et al. (eds.), *Decision Making Under Uncertainty Via Optimization, Modelling, and Analysis*, Studies in Systems, Decision and Control 558,
https://doi.org/10.1007/978-981-96-0085-4_25

be broken down. Moreover, the disposal of PWs happens so casually like thrown out of a moving car, piled up on the roadsides, carelessly scattered in large gatherings, and many more incidents. So disposal of plastic waste is inevitable because of improper collection of waste followed by a failed segregation system. The world produces 450 million tons of plastic waste (Sustainable Development blog) every year in the form of chocolate wrappers, plastic carry bags, plastic utensils, etc. These PWs mostly go untreated and unmanaged and are dumped in various landfills and water bodies that pollute the environment. Out of the total proportion of plastic waste produced, only 15% of the waste undergo recycling with less than a 9% recovery rate. Various government bodies and non-government bodies have come up with measures to curb and manage plastic waste (Plastic Waste Management, 2019). Plastics are extremely durable accounting for their popularity among consumers due to it being cheaply accessible. Plastic pollution is a major concern globally and one of the best ways to manage the existing plastic waste is to recycle them and increase their life span to avoid landfiling. Firstly, recycling of PWs reduces pollution and improves the ecosystem [1, 2]. Secondly, recycling PWs requires fewer raw materials required to produce new plastics, thus conserving natural resources. Thirdly, landfilling is controlled by recycling the PWs, which otherwise erode the soil and the chemicals seep into the groundwater. Fourthly, this promotes a sustainable lifestyle.

Pyrolysis, gasification, chemical recycling, mechanical recycling, and others are some of the PWRTs. Recycling PW is now essential, and researching the most effective technique is advised to raise public awareness. Since experts disagree on this, it is a challenge requiring GDM from a variety of DMs. Numerous factors, including cost, health risk, working environment, and so forth, are taken into account while evaluating this issue. The purpose of this study is to evaluate the most sustainable PWRT to mitigate the impact of PWs on life. However, a lot of uncertainty is present in selecting the optimal PWRT considering various social, economic, and environmental factors. Thus the problem of optimal PWTR selection should be considered under uncertainty.

Zadeh [3] established the fuzzy sets (FSs), which describe the membership function (MF)—a measure of an element's degree of belongingness in the set, where non-MF and MF add up to exactly one. Later, as an extension of FSs, [4] proposed Intuitionistic fuzzy sets (IFSs), which define membership degree (MD) and non-membership degree (NMD), the total of which is between 0 and 1. IFS is limited to situations where the choice is "this" or "that" and is unable to handle additional uncertainty. Cuong and Kreinovich's [5] introduction of the picture fuzzy set (PFS), a new fuzzy extension, was made possible by this, which takes into consideration three membership degrees: positive, negative, and neutral. In situations where the information presented is ambiguous, it offers greater flexibility and accuracy when evaluating a decision-making model. Consider a democratic election in which voters of various opinions cast ballots. Consequently, the options are "vote for," "vote against," "abstain," or "refusal to vote." It appears that the opinion "abstain," which denotes a voter's neutrality or refusal to vote for either party, is taken into consideration but not for any of the parties. Picture fuzzy sets can be useful in these situations. This facilitates a more comprehensive evaluation while addressing uncertainty, which is advantageous for making more accurate decisions. Cuong and Kreinovich [5] defined the SF, which fails to rank PFNs when the MD and NMD of the PFNs are the same. In this case, the scores of PFNs become zero and hence one has to check the accuracy value. An essential component of ranking fuzzy numbers (FNs) is their SF. Therefore, it is necessary to construct an improved SF under PFE that is capable of ranking any kind of PFN, regardless of their MD and NMD. Hence to avoid the need for an accuracy function (AF) to further rank the numbers, the current work attempts to create an improved PFSF that can rate any type of PFN and it lies in [0, 1] satisfying all the prominent properties of an ideal SF. In the existing literature, there are many aggregation operators (AOs) defined under PFE, like picture fuzzy arithmetic operators, Einstein operators and many more but PFWGO [6] has been chosen in the present study for its simplicity yet wider application in the field of uncertainty. Previously, Tripura et al. [7] used PFWAO-based MCGDM techniques for the selection of an ideal video conferencing tool. PFWGO is used here for aggregating various decision makers' (DMs') assessments under uncertainty. On the other hand, there are several MCDM methods for figuring out criteria weights (CWs) and ranking alternatives. In the present study, SWARA (Kersuline et al., 2010) and MARCOS (Stevic et al., 2020) have been chosen as weighing and ranking methods, respectively. Majeed et al. [8] estimated the criteria weights for the selections of landfill sites in the Baghdad governorate by applying the SWARA technique. Vrtagić et al. [9] implemented fuzzy SWARA for road sections based on the MCDM model. Ecer et al. [10] assessed the private health insurance companies that provided health insurance during the COVID-19 outbreak using the MARCOS technique under the Intuitionistic Fuzzy environment to rank the health insurance companies. Sahoo et al. [11] compiled a book including different MCDM techniques in a fuzzy environment. Kavyasree et al. [12] worked in cubic hesitant fuzzy to understand different MCDM techniques and their working procedure. Zarhaoui et al. [13] applied the MARCOS method in an Intuitionistic Fuzzy environment for the Green Supplier Selection problem. Duc [14] evaluated machining processes involving milling, grinding, and turning using the MARCOS method to examine different parameters in a multi-criteria decision-making model. Ghoushchi et al. [15] worked on the evaluation of clean energy barriers with the help of the SWARA-CODAS algorithm in a spherical fuzzy environment. Debbarma et al. [16] studied Healthcare Waste Recycling with the help of SWARA and MABAC under a Fermetean Fuzzy environment. The traditional SWARA and MARCOS are incapable of dealing with group decision-making (GDM) situations under uncertainty. So the notions of these two methods have been extended under PFE with the aid of PFWGO to deal with group decision-making problems under uncertainty.

2 Literature Review

To carry out the research, brief literature on PWR techniques and their implementation for sustainable environmental development has been reviewed and discussed below.

Vanapalli et al. [17] carried out a study on plastic waste management during and after post-COVID-19 when the use of PPE and such plastics increased. Geetha et al. [18] introduced a novel HPF-ELECTRE by extending ELECTRE-III to apply to the plastic recycling problem. Dai et al. [19] reviewed the pyrolysis technology for plastic waste recycling to fuel and energy. Santos et al. [20] analyzed the plastic wastes in marine regions using a multi-criteria decision approach. Shahsavar et al. [21] modelled a bio-recovery of municipal different plastic waste management based on a new integrated Multi-Criterion.

Decision-Making (MCDM) approach through a case study in Mashahd, Iran. Jeon et al. [22] implemented PHFE-based WASPAS-AHP perspective to select flexible packaging bags with the ban on single-use plastics. Jaisankar et al. [23] explored the therapeutic plastic waste technology using hesitant fuzzy incorporating EWM and ARAS. Kang et al. [24] identified appropriate plastic waste disposal techniques during this COVID-19 crisis in an interval-valued q-rung orthopair fuzzy environment devising FUCOM for calculating criteria weights and MABAC to rank the waste disposal techniques. Sasirekha et al. [25] determined the best recycling technology using the MABAC method in a single-valued neutrosophic fuzzy approach. Soni et al. [26] developed various composite materials using PW and agro waste in different compositions for building construction and identified the best material using the fuzzy MCDM technique.

2.1 Research Question

The present study mainly focuses on answering the following research questions:

- The main research question is to perceive which PWRT is efficient in terms of environmental impact as well as costs.
- The existing picture fuzzy score function fails to rank all the types of PFNs. For this purpose, PFWGO-based SWARA and MARCOS have been modelled to solve this problem with the help of improved PFSF.

2.2 Research Gap and Motivation

Based on the extant research, it is evident that proper PW management is essential for both community safety and sustainable environmental development. One of the best uses of PWM in terms of financial and environmental benefits is waste recycling.

However, as far as research is concerned, very little of it has concentrated on determining the best suitable PWRT by analytical methods. However, PFS is a useful tool for managing uncertainty in real life. However, not all types of PFN can be ranked using the existing PFSF [5]. Only a small number of studies have focused on creating an improved PFSF. Furthermore, it is important to expand the notions of conventional SWARA and MARCOS within PFE through the assistance of AO, to enable these techniques to be applied to scenarios involving GDM under uncertainty. The need to secure both public health safety and sustainable environmental development has led us to model the PWRT selection problem as an MCGDM problem under PFE and solve it utilizing extended MCGDM methodologies with the help of improved PFSF.

2.3 *Novelty of the Present Study*

The novelty of the present research is threefold.

- To overcome the limitations of the existing PFSF, an improved PFSF has been introduced to rank PFNs appropriately irrespective of MD and NMD of PFNs. The salient properties of the improved SF have been authenticated through theoretical and numerical validations.
- The notions of traditional SWARA and MARCOS methods have been extended under PFE with the aid of PFWGO to make them capable enough to deal with GDM problems.
- For the first time, the problem of PWRT selection has been modelled as an MCGDM problem under PFE and solved using the proposed integrated PFWGO-based SWARA-MARCOS technique aided with the proposed improved PFSF.

2.4 *Contribution of the Present Study*

PFN tackles the ambiguity in decision-making by defining positive, neutral, and negative degrees whose sums fall within the interval [0, 1]. The present treatise has employed an MCGDM approach to determine the optimal PWRT, which is urgently needed to stop the Earth's plastic pollution. The following are the primary contributions of the current study:

- To rank alternatives, an updated PFSF has been devised, which overcomes the limitations of the previous one. Theoretical validations have been utilized to analyze the noteworthy characteristics of the enhanced PFSF.
- Using PFWGO, the traditional SWARA approach has been extended under PFE to determine the preference vector for criteria.

- To rank the recycling alternatives, the traditional MARCOS method has been extended under PFE with the aid of PFWGO.
- PFWGO-based integrated MCGDM has been used to model and solve the PWRT selection problem.
- CA and SA have been used to verify the consistency, integrity, and robustness of the proposed integrated approach.

The rest of the manuscript has been organizes as follows:

Section 2 contains the preliminaries and operations of PFNs and PFWGO. The proposed PFSF, related theorems, PFWGO-based extended SWARA, and MARCOS methods followed by their integration are discussed in Sect. 3. Section 4 contains the application of the proposed technique in PWRT selection. Results obtained through the proposed method and their validations are presented in Sect. 5. Section 6 has been accomplished with concluding remarks and future research directions.

3 Mathematical Background

In this section, the basic concepts of PFSs, its operational laws and PFWGO have been discussed.

3.1 Picture Fuzzy Set

A PFS A on the Universal Set X can be defined as follows:

$$A = \{(x, \mu_A(x), \eta_A(x), \nu_A(x))|x \in X\}$$

where $\mu_A(x) \in [0,1],$, $\eta_A(x) \in [0,1],$, $\nu_A(x) \in [0,1]$ are called positive MD, neutral MD, and negative MD of x in A, respectively, satisfying: $0 \leq \mu_A(x)+\eta_A(x)+\nu_A(x) \leq 1$. Also, $\forall x \in X$, $1 - (\mu_A(x) + \eta_A(x) + \nu_A(x))$ is called refusal MD of x in A.

3.2 Score and Accuracy Function

Let $A(\mu_A, \eta_A, \nu_A)$ be a PFN, then SF and AF of A can be respectively defined as

$$S(A) = \mu_A - \nu_A, S(A)\epsilon[-1,1]$$

$$H(A) = \mu_A + \eta_A + \nu_A, H(A) \in [0,1]$$

3.3 Operations of PFNs

Let A and B be two PFNs, then the operations on them are defined as

(1) **Complement:** $\overline{A} = \{(x, \nu_A(x), \eta_A(x), \mu_A(x))|x \in X\}$
(2) $A \oplus B = (\mu_A + \mu_B - \mu_A\mu_B, \eta_A\eta_B, \nu_A\nu_B)$
(3) $A \otimes B = (\mu_A\mu_B, \eta_A + \eta_B - \eta_A\eta_B, \nu_A + \nu_B - \nu_A\nu_B)$
(4) $tA = \left(1 - (1 - \mu_A)^t, {\eta_A}^t, {\nu_A}^t\right), t > 0$
(5) $A^t = \left({\mu_A}^t, 1 - (1 - \eta_A)^t, 1 - (1 - \nu_A)^t\right), t > 0$

3.4 Picture Fuzzy Weighted Geometric Operator

Let $A_k (k = 1,2, \ldots, n)$ be a collection of PFNs. The PFWGO is a mapping $P^n \rightarrow P$ such that

$$PFWG_A(A_1, A_2, \ldots, A_3) = \underset{j=1}{\overset{n}{\oplus}} (A_k)^{\omega_k}$$
$$= \left(\prod_{k=1}^{n} (\mu_{A_k})^{\omega_k}, 1 - \prod_{k=1}^{n} (1 - \eta_{A_k})^{\omega_k}, 1 - \prod_{k=1}^{n} (1 - \nu_{A_k})^{\omega_k}\right) \quad (1)$$

where $\omega = (\omega_1, \omega_2, \ldots, \omega_n)^T$ is the weight vector of $A_k (k = 1, 2, \ldots, n)$ and $\omega_k > 0, \sum_{k=1}^{n} \omega_k = 1$.

4 PFWGO-Based GDM Using Improved PFSF

In this section, the traditional MCDM techniques SWARA and MARCOS have been extended under PFE to enable precise GDM with the help of PFWGO and also an improved PFSF has been developed and its salient features are theoretically explained.

4.1 Improved Picture Fuzzy Score Function

SF is the measure through which a fuzzy number can be ranked and lies in [−1,1]. The existing PFSF [5] can rank the PFNs only when the MDs and NMDs of PFNs are not equal. But if equal, then the scores become zero and then the numbers are ranked using AF which lies in [0,1]. For example, if $A = (0.4, 0.1, 0.4)$ and $B = (0.3,0.2,0.3)$

are two PFNs, then existing SF results $S(A) = 0$ and $S(B) = 0$, that is both of the PFNs are having the same score, so the numbers cannot be ranked based on existing PFSF. To overcome this limitation of existing PFSF, a novel improved PFSF has been developed in the present study which lies between [0,1] and could easily rank any type of PFNs irrespective of the values of MD and NMD. Moreover, all the properties of an ideal SF also prevail in the proposed SF. The proposed PFSF and its properties are discussed below:

4.2 Improved PFSF

Let $A(\mu_A, \eta_A, \nu_A)$ be a PFN, then SF of A is denoted by $S(A)$ and is defined as

$$S(A) = \frac{\exp\{(2\mu_A - \nu_A + 1)/3\} - 1}{\exp(1) - 1}, SF(A)\epsilon[0,1] \tag{2}$$

Let us consider the above example, then use Eq. (2) it is seen that $S(A) = 0.59$ and $S(B) = 0.556$. and hence $A > B$. So the proposed PFSF can rank any type of PFNs and the scores lie in [0,1]. The properties of the proposed PFSF are discussed below:

Theorem 3.1.1. S(A) increases and decreases monotonically concerning μ and ν, respectively.

Proof. The partial derivative of Eq. (2) with respect to μ and ν are, respectively, as follows:

$$\frac{\partial S}{\partial \mu} = \frac{2[\exp\{(2\mu_A - \nu_A + 1)/3\}]}{3[\exp(1) - 1]} > 0$$

$$\frac{\partial S}{\partial \nu} = -\frac{1}{3}\left[\frac{\exp\{(2\mu_A - \nu_A + 1)/3\}}{\exp(1) - 1}\right] < 0$$

Hence the theorem is proved.

Theorem 3.1.2 The SF given by Eq. (2) satisfies the following:

(i) $S(1,0,0) = 1$
(ii) $S(0,0,1) = 0$
(iii) $0 \le S(A) \le 1$

Proof: Using theorem 3.1.1, $S(A)$ acquires the least value '0' or utmost value '1', for $S(A) = (0,0,1)$ or $S(A) = (1,0,0)$ respectively. Thus, $(S(A))_{min} = 0$ and $(S(A))_{max} = 1$. Hence $0 \le S(A) \le 1$.

Theorem 3.1.3 For any two given PFNs, $A(\mu_A, \eta_A, \nu_A)$ and $B(\mu_B, \eta_B, \nu_B)$, if $\mu_A > \mu_B$ and $\nu_A < \nu_B$, then $S(A) > S(B)$.

Proof: $\mu_A > \mu_B$and $\nu_A < \nu_B$ imply $(2\mu_A - \nu_A + 1) > (2\mu_B - \nu_B + 1)$

That is, $\exp\left\{\frac{(2\mu_A - \nu_A + 1)}{3}\right\} - 1 > \exp\{\frac{(2\mu_B - \nu_B + 1)}{3}\} - 1$

That is, $\left[\frac{\exp\left\{\frac{(2\mu_A - \nu_A + 1)}{3}\right\} - 1}{\exp(1) - 1}\right] > \frac{\exp\left\{\frac{(2\mu_B - \nu_B + 1)}{3}\right\} - 1}{\exp(1) - 1}$

Hence $S(A) > S(B)$.

4.3 PFWGO-Based Proposed Extended SWARA Method

SWARA (Kersuline, 2010) is a simple weighing method that considers the criteria in ascending order as per the experts' opinion. However, as the traditional SWARA method is not capable of dealing with GDM situations under uncertainty, so in the present study, SWARA has been extended under PFE with the aid of PFWGO. The proposed algorithm is discussed below.

Step 1: Rank criteria according to their preference. The more the importance of the criteria more the value assigned to it and so is the value assigned to the less important criteria. For l DMs and n number of criteria, assign priority p_j^k to criterion j for kth DM.

Step 2: Aggregate DMs' assessments using Eq. (1) assigning weights $\omega = (\omega_1, \omega_2, \ldots, \omega_k)^T$ to DMs.

Step 3: Calculate the score $p_{j'}$ of each criterion using Eq. (2), where $p_{j'}$ denotes the merged relative importance score for each criterion.

Step 4: Rank the criterion with the highest score value first and accordingly criteria with their relative importance in descending order. Calculate the relative significance of criterion j concerning the previous criterion $(j - 1)$ and denote it by s_j and find

$$s_j = p_{(j-1)'} - p_{j'}, j = 2, \ldots, n. \tag{3}$$

Step 5: Obtain the coefficient of each criterion by binary comparison and denote it as C_j. The coefficients show the importance of criterion $(j + 1)$ concerning criterion j. Calculate C_j for each criterion as follows:

$$C_j = \begin{cases} 1, \ j = 1; \\ s_j + 1, \ j = 2, \ldots, n \end{cases} \tag{4}$$

Step 6: Calculate corrected weights $(s_{j'})$ for all criteria as follows:

$$s_{j'} = \begin{cases} 1, j = 1 \\ \frac{s_{j-1'}}{C_j}, j = 2, \ldots, n \end{cases} \tag{5}$$

Step 7: Calculate the final CW w_j as follows:

$$w_j = \frac{s_j\prime}{\sum_{j=1}^{n} s_j\prime}, j = 1,2, \ldots, n \quad (6)$$

4.4 PFWGO-Based Proposed Extended MARCOS Method

MARCOS was introduced by Stevic et al. (2020), which is a ranking method of alternatives. Utility degree is calculated based on the ideal and anti-ideal solutions. Further, the utility function is evaluated according to which the alternatives are ranked. However, MARCOS is not capable of dealing with GDM situations under uncertainty in its traditional form. Thus in the present study, traditional MARCOS had been extended under PFE using PFWGO, and the proposed algorithm is discussed below:

Step 1: Construct linguistic decision matrices (DMTs) involving DM's linguistic assessments. In the given MCGDM problem, let $A = \{A_1, A_2, \ldots, A_n\}$ be the set of alternatives against $C = \{C_1, C_2, \ldots, C_m\}$ set of criteria. Let $D = \{D_1, D_2, \ldots, D_l\}$ be the DMs who give their opinion on the given set of problems. Let $X = [x_{ij}^k]_{n \times m}$ be the DMT provided by the DMs, where x_{ij}^k denotes the performance of the alternative A_i with respect to each criterion C_j, $i = 1,2, \ldots, n, j = 1,2, \ldots, m$. Meanwhile, AAI and AI are evaluated, where AAI and AI are ideal and anti-ideal solutions respectively. AI is maximum in the case of beneficial criterion (BC) and minimum in the case of cost-criteria (CC). The reverse exists in the case of AAI. Now the matrix has two additional rows of $x_{aai} = \max(x_{ij})$ and $x_{ai} = \min(x_{ij})$ for $i = 1,2, \ldots, mandj = 1,2, \ldots, n$

Step 2: Assign PFNs to linguistic terms given by DMs. Let $\omega = (\omega_1, \omega_2, \ldots, \omega_k)^T$ be the weights of the DMs.

Step 3: Utilizing Eq. (1), aggregate PFDMTs.

Step 4: Normalize aggregated DMT considering the complement of the corresponding PFN for CC and let $\hat{x}_{ij}$ be the normalized elements.

Step 5: Compute the weighted normalized matrix.

$$\overline{x}_{ij} = \omega_k \hat{x}_{ij}, i = 1,2, \ldots, n, j = 1,2, \ldots, m \quad (7)$$

Using Eq. (2), find crisp scores.

Step 6: Determine the utility degree.

$$K_i^+ = \frac{S_i}{S_{ai}}, K_i^- = \frac{S_i}{S_{aai}} \quad (8)$$

$$where,\ S_i = \sum_{j=1}^{n} \overline{x}_{ij}, S_{aai} = \sum_{j=1}^{n} x_{aai},\ S_{ai} = \sum_{j=1}^{n} x_{ai}$$

Step 7: Calculate the utility function as

$$f(K_i^+) = \frac{K_i^+}{K_i^+ + K_i^-}, f(K_i^-) = \frac{K_i^-}{K_i^+ + K_i^-}$$

$$f(K_i) = \frac{K_i^+ + K_i^-}{1 + \frac{1-f(K_i^+)}{f(K_i^+)} + + \frac{1-f(K_i^-)}{f(K_i^-)}} \tag{9}$$

Step 8: Rank the alternatives according to the utility function.

5 Application of Proposed Integrated MCGDM in PWRT Selection

PWR (Ogundairo, 2021) is urgently needed since it will lower pollution and the amount of waste dumped in the ocean and on land. Recycling plastic trash entails the time-consuming processes of collecting, sorting, and recycling plastic garbage. These procedures also need human participation, which creates jobs for lots of people. Polyethylene terephthylate (PET), high-density polyethylene (HDPE), low-density polyethylene (LDPE), polypropylene (PP), polystyrene (PS), and other plastic polymers are recyclable. A thorough explanation of the different PWRTs is discussed below.

5.1 Closed-Loop Recycling (CLR)

A product or material can be used, recycled into a new product, and then continuously recycled back into raw materials without losing any of its original attributes. This recycling process is known as closed-loop recycling. The goal of closed-loop recycling systems is to prolong a resource's availability to an economy. A closed-loop recycling process uses the recycled material produced to build a new product within the same category. The most well-known instances of this are milk bottles made of HDPE and PET. In a closed-loop recycling system, materials or products can be recycled again and again into new products or raw materials without degrading.

5.2 Open-Loop Recycling (OLR)

Reprocessing products and using the resulting recyclate to create new products is known as "open loop" recycling. Often, this is made into a product with a longer lifespan. Making plastic pipes out of recycled HDPE bottles is an example of open-loop recycling. With a lifespan of several decades, these pipes prolong the useful life of this recovered material.

5.3 Chemical Recycling (CR)

The procedure modifies a polymer's chemical structure to create monomers and other chemical building blocks that can be utilized as raw materials in other processes. Chemical recycling processes can break down plastics into their component parts, which can subsequently be transformed into high-value secondary raw materials. These components can then be used to generate new compounds and polymers.

5.4 Thermal Depolymerization (TD)

TDP is the process of converting a polymer primarily using heat energy into a monomer or a mixture of monomers. It is not the same as other forms of depolymerization, which might be catalyzed or uncatalyzed and depend on the use of chemicals or biological action.

5.5 Polymer Degradation (PD)

The loss of a polymer's physical properties, such as strength, due to modifications in its chemical composition is known as polymer degradation. Polymers, especially plastics, are prone to degradation at every stage of a product's life cycle, including initial production, usage, environmental disposal, and recycling. The rate at which this deterioration occurs varies substantially; some industrial procedures can completely break down a polymer in a matter of hours, whereas biodegradation may take decades.

5.6 *Pyrolysis (PY)*

PY is the process of breaking down waste plastic by heat at various temperatures (between 300 and 900 °C) without the presence of oxygen, to produce liquid oil. Pyrolysis is a common technique used to transform PW into solid, liquid, and gaseous fuels.

5.7 *Gasification (GS)*

The chemical reaction in which PW is heated to the point of breaking down into its component molecules in a low-oxygen environment. Syngas, or synthesis gas, is produced by the high-temperature reaction of PW with a gasifying agent.

In this study, three DMs have been considered to assess seven PWRTs under twelve different criteria (Table 1) assigning weights (0.34, 0.33, 0.33) to the DMs. Using Table 2, the linguistic assessments are converted into PFNs.

PFWGO-based integrated MCGDM technique for PWRT selection is described as follows:

Step 1: DMTs are formulated and presented in terms of PFNs in Table 3.

Step 2: DMTs are aggregated using Eq. (1) and presented in Table 4.

Step 3: CWs are calculated by applying Eqs. (3)–(6) and presented in Table 5 (Fig. 1).

Step 4: The utility degree of alternatives is calculated using Eq. (9) and ranked accordingly.

The results are presented in Table 6 (Fig. 2).

From Table 6, it can be observed that the PIT has identified CLR as the best PWRT. The main advantages of using CLR are as follows:

i. It decreases pollution by producing meagre amounts of greenhouse gas emissions.
ii. It uses less virgin plastics thus preserving natural resources.
iii. It makes use of non-recyclable plastics and hence does not contribute to landfills.

Table 1 Description of criteria

Criteria	CWT1	CWT2	CWT3	CWT4	CWT5	CWT6	CWT7	CWT8	CWT9	CWT10	CWT11	CWT12
Type	BC	CC	CC	CC	CC	CC	BC	CC	BC	BC	BC	BC

Table 2 Scale of PFNs

Linguistic terms	Excessively high (ExHi)	Really high (RHi)	High (Hi)	Mediocrity high (McH)	Mediocre (Mc)	Mediocrity low (McLo)	Low (Lo)	Really low (RLo)	Excessively low (ExLo)
PFN	(0.9, 0, 0.05)	(0.85, 0.04, 0.09)	(0.75, 0.05, 0.1)	(0.6, 0.0, 0.3)	(0.5, 0.1, 0.4)	(0.3, 0, 0.6)	(0.25, 0.05, 0.6)	(0.2, 0.04, 0.5)	(0.1, 0, 0.85)

Table 3 Decision matrices

Criteria →	Alternatives ↓	CWT1	CWT2	CWT3	CWT4	CWT5	CWT6	CWT7	CWT8	CWT9	CWT10	CWT11	CWT12
DM1	CLR	(0.5, 0.1, 0.4)	(0.3, 0, 0.6)	(0.6, 0, 0.3)	(0.5, 0.1, 0.4)	(0.5, 0.1, 0.4)	(0.75, 0.05, 0.1)	(0.6, 0, 0.3)	(0.75, 0.05, 0.1)	(0.5, 0.1, 0.4)	(0.75, 0.05, 0.1)	(0.5, 0.1, 0.4)	(0.75, 0.05, 0.1)
	OLR	(0.85, 0.04, 0.09)	(0.3, 0, 0.6)	(0.3, 0, 0.6)	(0.75, 0.05, 0.1)	(0.5, 0.1, 0.4)	(0.85, 0.04, 0.09)	(0.5, 0.1, 0.4)	(0.75, 0.05, 0.1)	(0.75, 0.05, 0.1)	(0.75, 0.05, 0.1)	(0.5, 0.1, 0.4)	(0.75, 0.05, 0.1)
	CR	(0.25, 0.05, 0.6)	(0.85, 0.04, 0.09)	(0.85, 0.04, 0.09)	(0.75, 0.05, 0.1)	(0.85, 0.04, 0.09)	(0.6, 0, 0.3)	(0.75, 0.05, 0.1)	(0.9, 0, 0.05)	(0.5, 0.1, 0.4)	(0.5, 0.1, 0.4)	(0.25, 0.05, 0.6)	(0.85, 0.04, 0.09)
	TD	(0.85, 0.04, 0.09)	(0.75, 0.05, 0.1)	(0.75, 0.05, 0.1)	(0.75, 0.05, 0.1)	(0.5, 0.1, 0.4)	(0.25, 0.05, 0.6)	(0.85, 0.04, 0.09)	(0.85, 0.04, 0.09)	(0.5, 0.1, 0.4)	(0.75, 0.05, 0.1)	(0.5, 0.1, 0.4)	(0.85, 0.04, 0.09)
	PD	(0.85, 0.04, 0.09)	(0.85, 0.04, 0.09)	(0.85, 0.04, 0.09)	(0.85, 0.04, 0.09)	(0.6, 0, 0.3)	(0.5, 0.1, 0.4)	(0.85, 0.04, 0.09)	(0.9, 0, 0.05)	(0.5, 0.1, 0.4)	(0.5, 0.1, 0.4)	(0.75, 0.05, 0.1)	(0.85, 0.04, 0.09)
	PY	(0.85, 0.04, 0.09)	(0.5, 0.1, 0.4)	(0.3, 0, 0.6)	(0.75, 0.05, 0.1)	(0.3, 0, 0.6)	(0.5, 0.1, 0.4)	(0.85, 0.04, 0.09)	(0.85, 0.04, 0.09)	(0.25, 0.05, 0.6)	(0.85, 0.04, 0.09)	(0.85, 0.04, 0.09)	(0.85, 0.04, 0.09)
	GS	(0.1, 0, 0.85)	(0.9, 0, 0.05)	(0.9, 0, 0.05)	(0.85, 0.04, 0.09)	(0.85, 0.04, 0.09)	(0.5, 0.1, 0.4)	(0.9, 0, 0.05)	(0.9, 0, 0.05)	(0.1, 0, 0.85)	(0.1, 0, 0.85)	(0.9, 0, 0.05)	(0.9, 0, 0.05)
DM2	CLR	(0.9, 0, 0.05)	(0.5, 0.1, 0.4)	(0.6, 0, 0.3)	(0.6, 0, 0.3)	(0.85, 0.04, 0.09)	(0.3, 0, 0.6)	(0.1, 0, 0.85)	(0.5, 0.1, 0.4)	(0.5, 0.1, 0.4)	(0.75, 0.05, 0.1)	(0.1, 0, 0.85)	(0.25, 0.05, 0.6)

(continued)

Table 3 (continued)

Criteria → Alternatives ↓		CWT1	CWT2	CWT3	CWT4	CWT5	CWT6	CWT7	CWT8	CWT9	CWT10	CWT11	CWT12
	OLR	(0.85, 0.04, 0.09)	(0.5, 0.1, 0.4)	(0.5, 0.1, 0.4)	(0.75, 0.05, 0.1)	(0.5, 0.1, 0.4)	(0.75, 0.05, 0.1)	(0.5, 0.1, 0.4)	(0.5, 0.1, 0.4)	(0.75, 0.05, 0.1)	(0.85, 0.04, 0.09)	(0.85, 0.04, 0.09)	(0.75, 0.05, 0.1)
	CR	(0.75, 0.05, 0.1)	(0.5, 0.1, 0.4)	(0.5, 0.1, 0.4)	(0.25, 0.05, 0.6)	(0.5, 0.1, 0.4)	(0.1, 0, 0.85)	(0.5, 0.1, 0.4)	(0.5, 0.1, 0.4)	(0.25, 0.05, 0.6)	(0.5, 0.1, 0.4)	(0.85, 0.04, 0.09)	(0.25, 0.05, 0.6)
	TD	(0.85, 0.04, 0.09)	(0.75, 0.05, 0.1)	(0.75, 0.05, 0.1)	(0.75, 0.05, 0.1)	(0.5, 0.1, 0.4)	(0.25, 0.05, 0.6)	(0.85, 0.04, 0.09)	(0.85, 0.04, 0.09)	(0.5, 0.1, 0.4)	(0.75, 0.05, 0.1)	(0.5, 0.1, 0.4)	(0.85, 0.04, 0.09)
	PD	(0.75, 0.05, 0.1)	(0.5, 0.1, 0.4)	(0.5, 0.1, 0.4)	(0.85, 0.04, 0.09)	(0.5, 0.1, 0.4)	(0.75, 0.05, 0.1)	(0.75, 0.05, 0.1)	(0.9, 0, 0.05)	(0.3, 0, 0.6)	(0.75, 0.05, 0.1)	(0.75, 0.05, 0.1)	(0.75, 0.05, 0.1)
	PY	(0.75, 0.05, 0.1)	(0.9, 0, 0.05)	(0.75, 0.05, 0.1)	(0.9, 0, 0.05)	(0.75, 0.05, 0.1)	(0.75, 0.05, 0.1)	(0.75, 0.05, 0.1)	(0.75, 0.05, 0.1)	(0.75, 0.05, 0.1)	(0.6, 0, 0.3)	(0.75, 0.05, 0.1)	(0.75, 0.05, 0.1)
	GS	(0.75, 0.05, 0.1)	(0.9, 0, 0.05)	(0.75, 0.05, 0.1)	(0.9, 0, 0.05)	(0.75, 0.05, 0.1)	(0.75, 0.05, 0.1)	(0.9, 0, 0.05)	(0.9, 0, 0.05)	(0.9, 0, 0.05)	(0.75, 0.05, 0.1)	(0.75, 0.05, 0.1)	(0.9, 0, 0.05)
DM3	CLR	(0.75, 0.05, 0.1)	(0.25, 0.05, 0.6)	(0.25, 0.05, 0.6)	(0.25, 0.05, 0.6)	(0.25, 0.05, 0.6)	(0.25, 0.05, 0.6)	(0.5, 0.1, 0.4)	(0.25, 0.05, 0.6)	(0.75, 0.05, 0.1)	(0.75, 0.05, 0.1)	(0.75, 0.05, 0.1)	(0.25, 0.05, 0.6)
	OLR	(0.75, 0.05, 0.1)	(0.5, 0.1, 0.4)	(0.5, 0.1, 0.4)	(0.75, 0.05, 0.1)	(0.75, 0.05, 0.1)	(0.75, 0.05, 0.1)	(0.5, 0.1, 0.4)	(0.5, 0.1, 0.4)	(0.5, 0.1, 0.4)	(0.5, 0.1, 0.4)	(0.25, 0.05, 0.6)	(0.75, 0.05, 0.1)

(continued)

Table 3 (continued)

Criteria → Alternatives ↓		CWT1	CWT2	CWT3	CWT4	CWT5	CWT6	CWT7	CWT8	CWT9	CWT10	CWT11	CWT12
	CR	(0.75, 0.05, 0.1)	(0.75, 0.05, 0.1)	(0.75, 0.05, 0.1)	(0.75, 0.05, 0.1)	(0.75, 0.05, 0.1)	(0.5, 0.1, 0.4)	(0.75, 0.05, 0.1)	(0.75, 0.05, 0.1)	(0.5, 0.1, 0.4)	(0.5, 0.1, 0.4)	(0.5, 0.1, 0.4)	(0.25, 0.05, 0.6)
	TD	(0.75, 0.05, 0.1)	(0.25, 0.05, 0.6)	(0.25, 0.05, 0.6)	(0.75, 0.05, 0.1)	(0.25, 0.05, 0.6)	(0.5, 0.1, 0.4)	(0.5, 0.1, 0.4)	(0.6, 0, 0.3)	(0.5, 0.1, 0.4)	(0.5, 0.1, 0.4)	(0.5, 0.1, 0.4)	(0.5, 0.1, 0.4)
	PD	(0.5, 0.1, 0.4)	(0.75, 0.05, 0.1)	(0.75, 0.05, 0.1)	(0.75, 0.05, 0.1)	(0.75, 0.05, 0.1)	(0.5, 0.1, 0.4)	(0.75, 0.05, 0.1)	(0.75, 0.05, 0.1)	(0.75, 0.05, 0.1)	(0.75, 0.05, 0.1)	(0.5, 0.1, 0.4)	(0.5, 0.1, 0.4)
	PY	(0.75, 0.05, 0.1)	(0.75, 0.05, 0.1)	(0.75, 0.05, 0.1)	(0.25, 0.05, 0.6)	(0.5, 0.1, 0.4)	(0.5, 0.1, 0.4)	(0.5, 0.1, 0.4)	(0.75, 0.05, 0.1)	(0.75, 0.05, 0.1)	(0.75, 0.05, 0.1)	(0.75, 0.05, 0.1)	(0.75, 0.05, 0.1)
	GS	(0.75, 0.05, 0.1)	(0.5, 0.1, 0.4)	(0.5, 0.1, 0.4)	(0.75, 0.05, 0.1)	(0.5, 0.1, 0.4)	(0.5, 0.1, 0.4)	(0.5, 0.1, 0.4)	(0.75, 0.05, 0.1)	(0.6, 0, 0.3)	(0.6, 0, 0.3)	(0.5, 0.1, 0.4)	(0.75, 0.05, 0.1)

Table 4 Aggregated DMT

Criteria → Alternatives↓	CWT1	CWT2	CWT3	CWT4	CWT5	CWT6	CWT7	CWT8	CWT9	CWT10	CWT11	CWT12
CLR	(0.696756, 0.050869, 0.198541)	(0.333741, 0.050869, 0.542733)	(0.445533, 0.017289, 0.421285)	(0.419517, 0.050869, 0.449987)	(0.470617, 0.063569, 0.400245)	(0.381522, 0.033783, 0.477267)	(0.312205, 0.035189, 0.600457)	(0.451575, 0.0668, 0.402428)	(0.573906, 0.083302, 0.311312)	(0.75, 0.05, 0.1)	(0.33748, 0.050869, 0.564145)	(0.359244, 0.05, 0.477267)
OLR	(0.814587, 0.043412, 0.093412)	(0.433435, 0.068158, 0.475143)	(0.433435, 0.068158, 0.475143)	(0.75, 0.05, 0.1)	(0.573906, 0.083302, 0.311312)	(0.781627, 0.046712, 0.096712)	(0.5, 0.1, 0.4)	(0.571584, 0.083798, 0.314099)	(0.653417, 0.06304, 0.2159)	(0.68097, 0.064076, 0.213036)	(0.470617, 0.063569, 0.400245)	(0.75, 0.05, 0.1)
CR	(0.521929, 0.05, 0.311312)	(0.683737, 0.063569, 0.209838)	(0.683737, 0.063569, 0.209838)	(0.521929, 0.05, 0.311312)	(0.68373, 0.063569, 0.209838)	(0.312205, 0.035189, 0.600457)	(0.656071, 0.0668, 0.212714)	((0.696756, 0.050869, 0.198541)	(0.397768, 0.083798, 0.475143)	(0.5, 0.1, 0.4)	(0.473891, 0.064076, 0.397808)	(0.374393, 0.046712, 0.475357)
TD	(0.814587, 0.043412, 0.093412)	(0.516226, 0.05, 0.316875)	(0.516226, 0.05, 0.316875)	(0.75, 0.05, 0.1)	(0.395021, 0.083302, 0.477267)	(0.316439, 0.067304, 0.540875)	(0.709686, 0.060836, 0.210161)	(0.755071, 0.026583, 0.16766)	(0.5, 0.1, 0.4)	(0.653417, 0.06304, 0.2159)	(0.5, 0.1, 0.4)	(0.709686, 0.060836, 0.210161)
PD	(0.68097, 0.0640076, 0.213036)	(0.683737, 0.063569, 0.209838)	(0.683737, 0.063569, 0.209838)	(0.814587, 0.043412, 0.093412)	(0.609496, 0.050869, 0.375373)	(0.571584, 0.083798, 0.314099)	(0.781627, 0.046712, 0.096712)	(0.845904, 0.017289, 0.067304)	(0.484876, 0.050869, 0.397562)	(0.656071, 0.0668, 0.212714)	(0.653417, 0.067304, 0.2159)	(0.68097, 0.064076, 0.213036)
PY	(0.781627, 0.046712, 0.096712)	(0.696756, 0.050869, 0.198541)	(0.554295, 0.033783, 0.311312)	(0.537995, 0.046712, 0.314379)	(0.482914, 0.051382, 0.4)	(0.571584, 0.083798, 0.314099)	(0.68097, 0.064076, 0.213036)	(0.781627, 0.046712, 0.096712)	(0.521929, 0.05, 0.311312)	(0.726138, 0.030438, 0.168603)	(0.781627, 0.046712, 0.096712)	(0.781627, 0.046712, 0.096712)
GS	(0.385737, 0.033783, 0.501744)	(0.73697, 0.035189, 0.187415)	(0.693937, 0.051382, 0.201784)	(0.830097, 0.030438, 0.080451)	(0.68097, 0.064076, 0.213036)	(0.571584, 0.083798, 0.314099)	(0.73697, 0.035189, 0.187415)	(0.845904, 0.017289, 0.067304)	(0.379727, 0, 0.534315)	(0.357554, 0.016784, 0.54255)	(0.693937, 0.051382, 0.201784)	(0.845904, 0.017289, 0.067304)

Table 5 CWs obtained by the proposed extended SWARA method

Sl. No	CWT1	CWT2	CWT3	CWT4	CWT5	CWT6	CWT7	CWT8	CWT9	CWT10	CWT11	CWT12
Criteria Weights	0.115041	0.111263	0.107927	0.095607	0.093396	0.087926	0.080953	0.080149	0.079759	0.074894	0.073084	0.070257

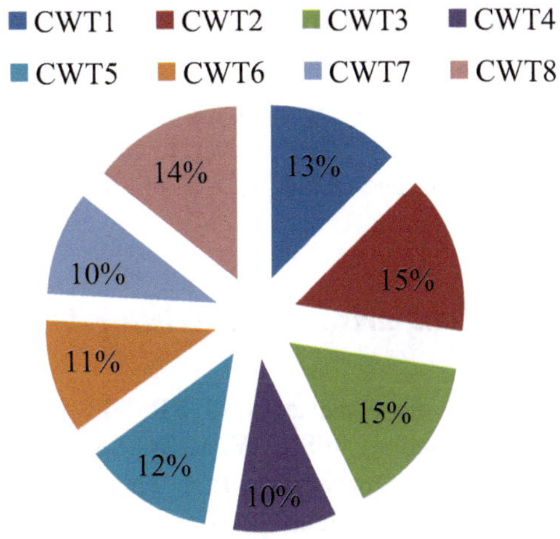

Fig. 1 Distribution of criteria weights via extended SWARA method

Table 6 Ranking of PWRTs

Alternatives	PFWGO-based extended MARCOS	Rank
CLR	0.183107	1
OLR	0.17504	2
CR	0.138147	6
TD	0.173616	3
PD	0.147338	5
PY	0.171139	4
GS	0.128912	7

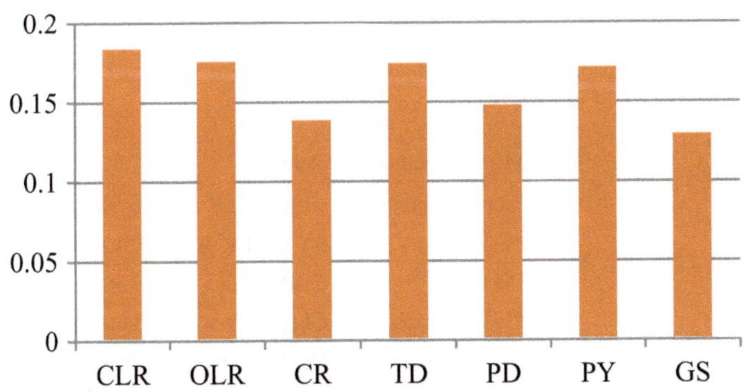

Fig. 2 Ranking of PWRTs

6 Results Analysis

From the ranking result, CLR is the best PWRT evaluated using PFWGO-based SWARA-MARCOS. However, to check the consistency, robustness, and reliability of the integrated technique, CA and SA have been carried out as follows:

6.1 Comparison with Existing MCDM Techniques

To check the consistency of RR, PIT is compared with other existing MCDM methods, namely, WSM, WPM, WASPAS, MOORA, ARAS, and EDAS under PFE. The results are furnished in Table 7 and shown in Fig. 3. Table 7 shows that CLR is the best PWRT in the majority of the aforementioned scenarios, while there are differences in the ranking that follows. The variations in ranking to distinct procedures used by the various MCDM approaches account for the small differences in ranking. In order to examine the relationships between ranking results (RRs), Spearman's rank correlation coefficients (SRCCs) have been computed. The correlation between RRs as measured by SRCCs, is produced using different techniques. In general, a value of CC larger than 0.7 indicates a higher degree of correlation between the RRs. About PF-MARCOS, the SRCCs of PF-WSM, PF-WPM, PF-WASPAS, PF-MOORA, PF-ARAS, and PF-EDAS are $(0.76, 0.95, 0.88, 0.93, 0.79, 0.95)$. Because every value is higher than 0.75, the outcomes are reliable. Thus, the suggested model is dependable, strong, and efficient.

Table 7 CA with other MCDM methods

MCDM methods	Ranking
Proposed method	CLR > OLR > TD > PY > PD > CR > GS
PF-WSM	PY > OLR > TD > CLR > PD > GS > CR
PF-WPM	CLR > TD > OLR > PY > CR > PD > GS
PF-WASPAS	CLR > PY > TD > OLR > PD > GS > CR
PF-MOORA	CLR > TD > PY > OLR > PD > CR > GS
PF-ARAS	PY > OLR > TD > CLR > PD > CR > GS
PF-EDAS	CLR > TD > OLR > PY > CR > PD > GS

Fig. 3 Comparison with other MCDM methods

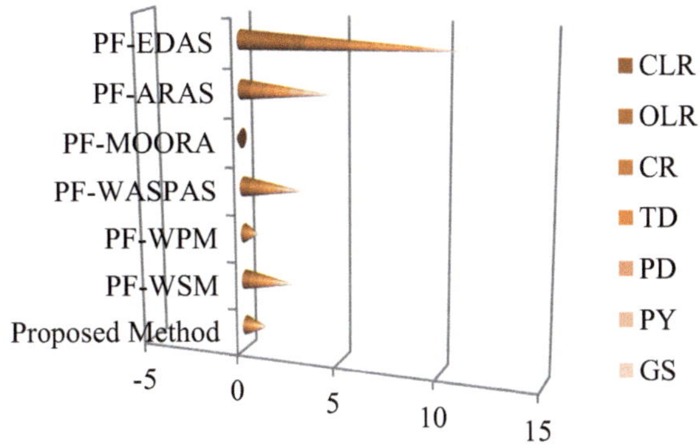

Fig. 4 Graphical representation of ranking concerning various PFSF

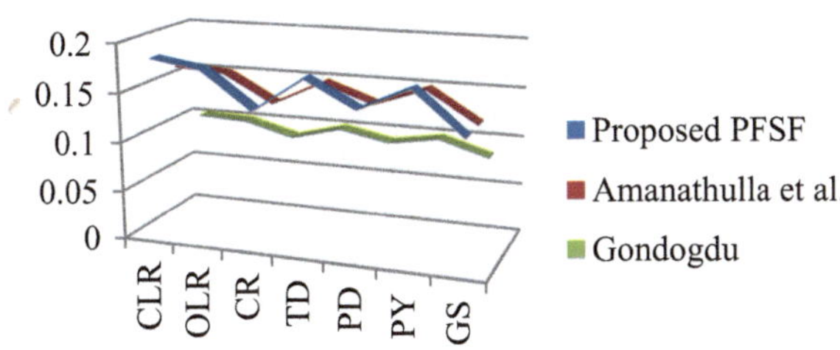

6.2 Comparison with Other Existing Score Functions

For further validation of the obtained results, the suggested approach has also been evaluated against many score functions in a fuzzy setting. Comparing the ranking result to Gondogdu's SF (2021), there is complete consistency. PY is favoured over TD in the third and fourth rankings of Amanathulla (2021), where there is a small difference in ranking. Table 8 indicates that, overall, the suggested strategy is fairly consistent because the ranking is nearly constant across all situations.

Table 8 Comparison with various PFSF

Various PFSF	Ranking
Proposed PFSF	CLR > OLR > TD > PY > PD > CR > GS
[27]	CLR > OLR > PY > TD > PD > CR > GS
Gondogdu (2021)	CLR > OLR > TD > PY > PD > CR > GS

6.3 Sensitivity Analysis

SA [28], [29] is the process of comparing results when variations in the problem parameters result in appreciable changes to the final results. SA has been done on the changes in the criteria weight (CW) values in this work.

6.4 Variation in CWs

CWs have a big influence on MCDM. In this case, SA has been carried out by assigning different weights to different kinds of criteria. Table 9 displays the RR (Fig. 5). Based on the suggested model, it is evident that CLR is the best PWRT. Furthermore, when giving CCs higher weights than BCs, significant variation can be seen in the ranking order of the lower-ranked alternatives. PY is the best PWRT when the CCs and BCs are given equal weight, but PY stands out from the other PWRTs when the BCs are given greater weight in the criteria. When CCs are given greater weight, CLR maintains its position as the best PWRT. As a result, the variation in CWs may be seen to have a noticeable impact on the ranking pattern. Thus, it is possible to conclude that PIT is extremely sensitive to changes in CWs.

Table 9 Variation in CW

Cases	Variation in weights	Ranking
C I	Weights obtained from SWARA	CLR > OLR > TD > PY > PD > CR > GS
C II	Equal weights	CLR > PY > TD > OLR > PD > CR > GS
	BC-CC	
C III	50–50	CLR > PY > TD > OLR > PD > CR > GS
C IV	60–40	PY > TD > OLR > CLR > PD > CR > GS
C V	70–30	PY > TD > OLR > PD > CLR > GS > CR
C VI	80–20	PY > OLR > TD > PD > CLR > GS > CR
C VII	90–10	PY > OLR > TD > PD > GS > CLR > CR
C VIII	10–90	CLR > TD > OLR > PY > CR > PD > GS
C IX	20–80	CLR > TD > OLR > PY > CR > PD > GS
C X	30–70	CLR > TD > OLR > PY > CR > PD > GS
C XI	40–60	CLR > TD > PY > OLR > PD > TD > GS

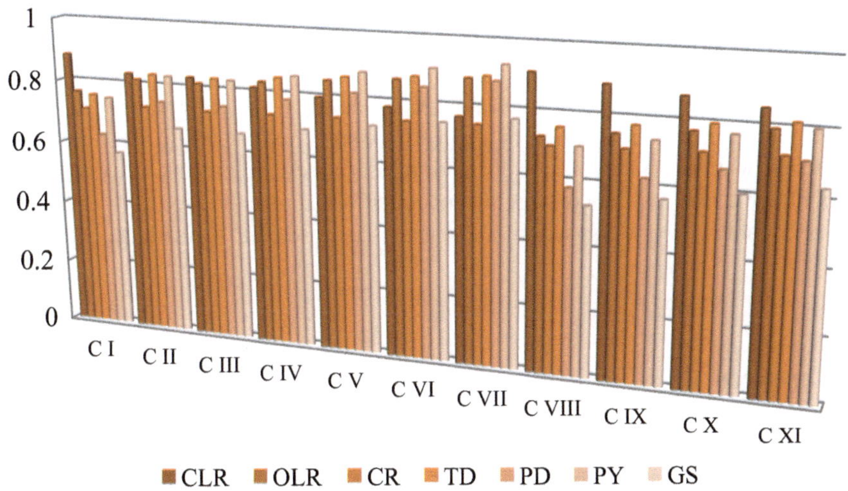

Fig. 5 Variation in criteria weights

7 Conclusion

PWM and PWR are interconnected, as recycling and reuse of PWs reduce pollution and conserve resources. PWR has grown with technology advancements, and many PWRTs are used to recycle PW and produce new value-added products. The present study has focused on constructing a hierarchical model using multiple PWR methods considering all the influential factors and the aim of the study is to identify the best PWRT. PFWGO's assistance extended the traditional SWARA approach under PFE to handle MCGDM issues. A PFWGO-based MARCOS technique was developed using improved innovative PFSF. All the essential properties of the proposed PFSF have been confirmed through numerical testing and theoretical study.

The recommended approach comes with several advantages. Firstly, this work uses AO to extend the idea of conventional SWARA under PFE, which helps manage scenarios involving GDM under uncertainty. Secondly, the extended MARCOS method that has been proposed is computationally efficient and effective, and it can also handle challenges involving GDM under uncertainty. Thirdly, the proposed augmented PFSF retains all the salient features of an efficient SF and can easily rank any type of PFNs, independent of their MD and NMD. In the end, for the first time, the PWRT selection conundrum has never before been articulated as an MCGDM problem under PFE and analytically answered. With the aid of the recommended technique, waste management authorities may find it simpler to make prudent administrative judgments during PWM.

However, there are some limitations to the proposed approach, mainly during the aggregation process and assigning weights to DMs. To overcome these shortfalls, a mathematical computing method can be devised to improve the randomness of the

DM's weights. Moreover, a more extensive computer-aided programming method can be created for the aggregation phase, guaranteeing that the solution procedure remains unaltered by the inclusion or omission of criterion, option, or DM.

Other practical problems, such as transportation, wastewater treatment, waste management in the medical field, etc., can also be solved with the help of the proposed approach. Moreover, alternative information measures based on ranking systems and weighing methods can be constructed under additional extended forms of PFS, such as spherical fuzzy sets, T-spherical fuzzy sets, PF soft sets, etc., with the help of other AOs.

References

1. Chakraborty, S., Saha, A.K.: A framework of LR fuzzy AHP and fuzzy WASPAS for health care waste recycling technology. Appl. Soft Comput. (2022a). https://doi.org/10.1016/j.asoc.2022.109388
2. Chakraborty, S., Saha, A.K.: Selection of optimal lithium ion battery recycling process: A multi-criteria group decision making approach. J. Energy Storage **55**, 105557 (2022)
3. Zadeh, L.A.: Fuzzy Set. Inf control. **8**(3), 338–353 (1965)
4. Atanassov, K.T.: Intuitionistic fuzzy sets. Fuzzy Sets Syst. **20**, 87–96 (1986)
5. Cuong, B.C., Kreinovich, V.: Picture fuzzy sets-a new concept for computational intelligence problems. In: 2013 third world congress on information and communication technologies (WICT 2013) (pp. 1–6). IEEE (2013)
6. Si, A., Das, S., Kar, S.: An approach to rank picture fuzzy numbers for decision making problems. Decis. Mak.: Appl. Manag. Eng. **2**(2), 54–64 (2019)
7. Tripura, C., Chakraborty, S., Bhattacharya, B.: Picture fuzzy aggregation operator based integrated MEREC-WASPAS technique for video conferencing tool selection. J. Uncertain Syst. (2024)
8. Majeed, R. A., Breesam, H. K.: Application of SWARA technique to find criteria weights for selecting landfill site in Baghdad governorate. In: IOP conference series: Materials science and engineering (vol. 1090, No. 1, p. 012045). IOP Publishing (2021)
9. Vrtagić, S., Softić, E., Subotić, M., Stević, Ž., Dordevic, M., Ponjavic, M.: Ranking road sections based on MCDM model: New improved fuzzy SWARA (IMF SWARA). Axioms 10(2), 92 (2021)
10. Ecer, F., Pamucar, D.: MARCOS technique under intuitionistic fuzzy environment for determining the COVID-19 pandemic performance of insurance companies in terms of healthcare services. Appl. Soft Comput. **104**, 107199 (2021)
11. Sahoo, L., Senapati, T., Yager, R. R. (Eds.).: Real Life applications of multiple criteria decision making techniques in fuzzy domain, studiesin fuzziness and soft computing (STUDFUZZ, volume 420), Springer Nature (2023)
12. Kavyasree, P.R., Surender Reddy, B.: Cubic picture hesitant fuzzy linear spaces and their applications in multi criteria decision making. In: Real Life Applications of Multiple Criteria Decision Making Techniques in Fuzzy Domain (pp. 533–557). Singapore: Springer Nature Singapore (2022)
13. Zarhaoui, O., Oufaska, K., Oudani, M., El Yassini, K.: Intuitionistic fuzzy marcos method for the green supplier selection problem. In: Proceedings of the First Australian International Conference on Industrial Engineering and Operations Management, Sydney, Australia (2022)
14. Duc Trung, D.: Multi-criteria decision making under the MARCOS method and the weighting methods: applied to milling, grinding and turning processes. Manuf. Rev. **9**, 3 (2022)

15. Ghoushchi, S.J., Garg, H., Bonab, S.R., Rahimi, A.: An integrated SWARA-CODAS decision-making algorithm with spherical fuzzy information for clean energy barriers evaluation. Expert Syst. Appl. **223**, 119884 (2023)
16. Debbarma, S., Chakraborty, S., Saha, A. K.: Health care waste recycling concerning circular economy: a Fermatean fuzzy aggregation operator-based SWARA–MABAC approach. Environ., Dev. Sustain., 1–40 (2024)
17. Vanapalli, K.R., Sharma, H.B., Ranjan, V.P., Samal, B., Bhattacharya, J., Dubey, B.K., Goel, S.: Challenges and strategies for effective plastic waste management during and post COVID-19 pandemic. Sci. Total. Environ. **750**, 141514 (2021)
18. Geetha, S., Narayanamoorthy, S., Kureethara, J.V., Baleanu, D., Kang, D.: The hesitant Pythagorean fuzzy ELECTRE III: an adaptable recycling method for plastic materials. J. Clean. Prod. **291**, 125281 (2021)
19. Dai, L., Zhou, N., Lv, Y., Cheng, Y., Wang, Y., Liu, Y., Ruan, R.: Pyrolysis technology for plastic waste recycling: A state-of-the-art review. Prog. Energy Combust. Sci. **93**, 101021 (2022)
20. Santos, M.R., Dias, L.C., Cunha, M.C., Marques, J.R.: Multicriteria decision analysis addressing marine and terrestrial plastic waste management: a review. Front. Mar. Sci. **8**, 747712 (2022)
21. Shahsavar, M.M., Akrami, M., Kian, Z., Gheibi, M., Fathollahi-Fard, A.M., Hajiaghaei-Keshteli, M., Behzadian, K.: Bio-recovery of municipal plastic waste management based on an integrated decision-making framework. J. Ind. Eng. Chem. **108**, 215–234 (2022)
22. Jeon, J., Krishnan, S., Manirathinam, T., Narayanamoorthy, S., Nazir Ahmad, M., Ferrara, M., Ahmadian, A.: An innovative probabilistic hesitant fuzzy set MCDM perspective for selecting flexible packaging bags after the prohibition on single-use plastics. Sci. Rep. **13**(1), 10206 (2023)
23. Jaisankar, R., Murugesan, V., Narayanamoorthy, S., Ahmadian, A., Suvitha, K., Ferrara, M., Kang, D.: Integrated MCDM approaches for exploring the ideal therapeutic plastic disposal technology: probabilistic hesitant fuzzy domain. Water Air Soil Pollut. **234**(2), 71 (2023)
24. Kang, D., Anuja, A., Ahmadian, A., Manirathinam, T., Shanmugam, P., Narayanamoorthy, S.: Sustainable assessment of plastic and mixed waste disposal problem during COVID-19 pandemic: an integrated multi-criteria decision-making approach. Environ., Dev. Sustain. 1–26 (2023)
25. Sasirekha, D., Senthilkuma, P.: Determining the Best Plastic Recycling Technology Using the MABAC Method in a Single-Valued Neutrosophic Fuzzy Approach. Neutrosophic Sets and Systems **58**(1), 12 (2023)
26. Soni, A., Chakraborty, S., Das, P.K., Saha, A.K.: Material selection of sustainable composites by recycling of waste plastics and agro-industrial waste for structural applications: A fuzzy group decision-making approach. J. Build. Eng. **73**, 106787 (2023)
27. Amanathulla, S., Muhiuddin, G., Al-Kadi, D., Pal, M.: Multiple attribute decision-making problem using picture fuzzy graph. Math. Probl. Eng. **2021**, 1–16 (2021)
28. Chakraborty, S., Saha, A.K.: Selection of Forklift unit for transport handling using integrated MCDM under neutrosophic environment. Facta Universitatis, Series, Mechanical Engineering (2022)
29. Chakraborty, S., Saha, A.K.: Novel Fermatean Fuzzy Bonferroni Mean aggregation operators for selecting optimal health care waste treatment technology. Eng. Appl. Artif. Intell. **119**, 105752 (2023)
30. Cuong, B.: Picture fuzzy sets. J. Comput. Sci. Cybern. 30 (2015)
31. Ghenai, C., Albawab, M., Bettayeb, M.: Sustainability indicators for renewable energy systems using multi-criteria decision-making model and extended SWARA/ARAS hybrid method. Renewable Energy **146**, 580–597 (2020)
32. Gundogdu, F.K.: Picture fuzzy linear assignment method and its application to selection of pest house location. In: Intelligent and Fuzzy Techniques: Smart and Innovative Solutions: Proceedings of the INFUS 2020 Conference, Istanbul, Turkey, July 21–23, 2020 (pp. 101–109). Springer International Publishing (2021)

33. Deva, K., Mohanaselvi, S.: Picture fuzzy Einstein geometric aggregate Operators and their Application to Multiple Attribute Decision Making 1–7 (2022) https://doi.org/10.1109/ICEEICT53079.2022.9768550
34. Keršuliene, V., Zavadskas, E.K., Turskis, Z.: Selection of rational dispute resolution method by applying new step-wise weight assessment ratio analysis (SWARA). J. Bus. Econ. Manag. **11**(2), 243–258 (2010)
35. Kibria, M.G., Masuk, N.I., Safayet, R., Nguyen, H.Q., Mourshed, M.: Plastic waste: challenges and opportunities to mitigate pollution and effective management. Int. J. Environ. Res. **17**(1), 20 (2023)
36. Riaz, M., Pamucar, D., Habib, A., Riaz, M.: A new TOPSIS approach using cosine similarity measures and cubic bipolar fuzzy information for sustainable plastic recycling process. Mathematical Problems in Engineering, vol. 2021, Article ID 4309544, 18 (2021)
37. Stević, Ž, Pamučar, D., Puška, A., Chatterjee, P.: Sustainable supplier selection in healthcare industries using a new MCDM method: Measurement of alternatives and ranking according to COmpromise solution (MARCOS). Comput. Ind. Eng. **140**, 106231 (2020)
38. Wei, G.W.: Picture fuzzy aggregation operators and their application to multiple attribute decision making. J. Intell. & Fuzzy Syst.: Appl. Eng. Technol. 713–724 (2017)
39. Wei, G.: Picture fuzzy Hamacher aggregation operators and their application to multiple attribute decision making 271–320 (2018)
40. https://www.un.org/sustainabledevelopment/blog/2023/08/explainer-what-is-plastic-pollution/
41. http://swachhbharaturban.gov.in/writereaddata/SBM%20Plastic%20Waste%20Book

MCDM PROMETHEE Method in Identifying Best Teacher Awardee

Kala Raja Mohan, R. Narmada Devi, Regan Murugesan, Sathish Kumar Kumaravel, and S. Kalaiselvi

Abstract In this competitive world, every individual runs for their livelihood. In that busy pathway, each field requires recognition to prove its uniqueness in its relevant field. Awards are the specified recognition presented as a token, to recognize the excellence of the receiver in any field. It also serves as an encouraging element to make this work involved fruitfully in the future. Now, it is the responsibility of the organization which honors the awardee, to see that it reaches the suitable person. Various factors are involved in the recognition process of the awardee. The multi-criteria decision-making algorithm finds its application in many real-life situations in finding a suitable solution. This provides a solution to situations with uncertainty. This chapter aims to bring about a method to list out various factors in which the award selection is made and the criteria related to it. Then in the process of identifying the best teacher awardee the procedure of PROMETHEE method is applied. The selection of awardees is done based on the ranks obtained at the end of the process. PROMETHEE method of the MCDM algorithm is thus applied in finding the awardee of the teacher.

Keywords MCDM · Award · Best teacher · PROMETHEE · Decision-making

---

K. R. Mohan (✉)
Department of Mathematics, Panimalar Engineering College, Chennai, India
e-mail: kalamohan24@yahoo.co.in

R. N. Devi · R. Murugesan · S. K. Kumaravel
Department of Mathematics, Vel Tech Rangarajan Dr, Sagunthala R&D Institute of Science and Technology, Avadi, Chennai, Tamil Nadu 600 062, India

S. Kalaiselvi
Department of Computer Science and Engineering, National Engineering College, KR Nagar, Kovilpatti, Tamil Nadu 628503, India
e-mail: sks@nec.edu.in

L. Sahoo et al. (eds.), *Decision Making Under Uncertainty Via Optimization, Modelling, and Analysis*, Studies in Systems, Decision and Control 558,
https://doi.org/10.1007/978-981-96-0085-4_26

1 Introduction

Out of various professions, teaching is a profession, which is considered a most respectful and noble profession. In every field, people think of their individual development. But only in the teaching field, do teachers show their involvement and dedication in the development of the students rather than themself. The real happiness of the teachers lies in seeing their students in a higher position than them. They encourage their students in every aspect and they help them mold themselves to prove themselves the best in society.

These types of encouragement are needed for teachers to do their responsibilities without any hurdles. The encouragement appears in the form of awards. Awards are recognition given in almost all professions which boost up energy in further development of the individual. In the teaching field, it is provided considering various aspects of teaching such as experience, results, publications, projects, funding, and many other criteria.

The organization that felicitates with the award comes across various profiles and applicants. Selecting an efficient profile suitable for the award is a tedious process. They fix grading for each criterion and according to the scoring, they filter the profile in each level and finalize the awardee. It involves multiple levels of selection. In this technological world, there are various methods of criteria selection. The method of Multi-Criteria Decision-Making method has attracted most researchers, since it serves as an efficient method for identifying the selection of finalists. This decision-identifying method comes with different forms of analysis. A few of them are AHP, ANP, TOPSIS, CBA, ranking and correlation. The major focus of this chapter is to identify the best teacher awardee making use of the MCDM PROMETHEE method.

Many researchers have made use of MCDM methods in finding decisions in various real-life situations. A few of them are narrated here for reference.

Halouani et al. [1], made use of the PROMETHEE method for identifying a suitable project. Athawale et al. [2] made their selection of location as per required facilities applying the PROMETHEE method in the year 2010. Turcksina [3] used the AHP PROMETHEE method in identifying a suitable policy for energizing vehicle fleets. Velasquez et al. [4] have made a study on various multi-criteria decision-making models. Danielle Costa Morais et al. [5] made their work in energy generation with the help of the PROMETHEE method in the assessment of technology. Reverse logistics has been made as a study in the work done by Jafar [6] with the help of multi-criteria decision-making. Ozsahi et al. [7] analyzed cancer treatment by applying the fuzzy PROMETHEE method. Hanifah et al. [8] presented their decision-making work on the banking sector.

Ade Parlaungan Nasution et al. [9] various applications of the PROMETHEE method in decision-making. Lazim Abdullah et al. [10] in the same year made a comparative analysis on green supplier selection making use of the preference function. Akmaludin et al. [11] made a selection of credit loans during the Covid pandemic period using multi-criteria decision-making methods. Bettina Zargini [12] in the same year applied alternative preference functions to decision-making problems. In the same year, Chao Tian et al. [13] applied AHP and the fuzzy picture method to identify a suitable tourism environment. Ikuobase et al. [14] applied a multi-criteria decision-making model in selecting material that gives optimal output. Zhang et al. [15] applied a multi-criteria decision-making method in the selection of material for the GGPFWA operator.

Sánchez et al. [16] performed an analysis of statistics in the development of education. Oubahman et al. [17] made a review on transportation using the PROMETHEE method. The selection of suppliers using the AHP PROMETHEE method was done by Isa [18]. The selection of elective courses making use of the PROMETHEE method has been done by Setiadi [19]. Martina [20] made a comparative study on energy requalification by applying the PROMETHEE method The fuzzy PROMETHEE method in the selection of media platforms was done by Bozdemir et al. [21]. Sarwar et al. [22] made use of the fuzzy PROMETHEE method in selecting an optimal supplier for industrial management. Ramadan et al. [23] performed the fuzzy PROMETHEE method in the selection of rail projects that involve priority. Hamed Taherdoost and Mitra Madanchian [24] applied the multi-criteria decision-making method in applications together with its procedures in the year 2023. Kala Raja Mohan et al. [25] analyzed the gold rate using statistics. Mayang [26] made use of the PROMETHEE method in identifying the priority of the road improvement system. The collection of real-time applications of MCDM techniques in the fuzzy domain has been published under the editorship of Sahoo et al. [27]. Amal Kumar Adak et al. [28] have authored a chapter in the book that speaks about the intuitionistic fuzzy approach in bringing about solutions to MCDM problems. Aniruddha Samant et al. [29] have authored a chapter on uncertain environments to find the solution to the airport gate reassignment problem

Most of the researchers have made use of the decision-making algorithm in finding solutions to industrial decisions like supplier selection, stock maintenance, loan, and many. In the field of education, very less decision-making processes are carried out. The usage of the PROMETHEE method has not been applied in the selection of awardees.

The literature survey shows the various applications and methods of MCDM. These methods help to find solutions to various real-time situations. In this chapter, the method of PROMETHEE is proposed to find the best teacher awardee based on the various criteria related to teaching. This method can be applied to various organizations in tracing out the best employee, best student, best woman, and many more. Hence, this analysis would benefit the readers to apply the methodology in various other sectors.

2 Standard Definitions

Various definitions involved in the decision-making algorithm are listed below for reference.

2.1 Attributes

Attributes refer to the objective responsible for the selection of the decision. The criteria are fixed based on the requirement of the solution required. The important objectives which are required for the decision are termed beneficial attributes. The attribute that requires very little weightage are termed as non-beneficial attributes.

2.2 Criteria

In providing a fair selection of decisions, factors related to the decision problem are termed as criteria. Various rules, values, and conditions are represented as criteria. This helps to improve the fairness of the selection.

2.3 Decision Matrix

A matrix of dimension $m \times n$ in which the values are fixed based on the relationship between the attributes and criteria is a decision matrix. Here m represents the number of attributes and n is the number of criteria.

2.4 Normalized Decision Matrix

The matrix obtained from the decision matrix by dividing the elements by its Euclidian norm is termed a normalized Norm. Euclidian norm is found by taking the square root of the sum of squares of the elements.

2.5 Weighted Normalized Matrix

The measure of preference given for each attribute is termed weight. The matrix obtained from the product of the normalized decision matrix and the weighted matrix is termed as weighted normalized matrix.

2.6 Preference Function

A preference function takes values from 0 to 1, which helps to know the variations between two criteria. It is assessed according to the criteria chosen related to the decision which is to be made.

2.7 Aggregate Preference Function

The aggregate preference function depends on the preference function taking into consideration the weights allotted for each criterion.

2.8 Leaving and Entering Flow

The term used to determine how an attribute shows positive outranking related to all other attributes is termed leaving flow.

2.9 Net Outranking Flow

The term used to determine how an attribute shows positive outranking related to all other attributes is termed entering flow.

2.10 Rank

The order of the values which are arranged either in ascending or descending order of preference is termed as rank.

3 PROMETHEE Algorithm

The various steps involved in the process of outranking algorithm are listed below.

- Decision Matrix
- Normalized Decision Matrix
- Weights
- Preference Function
- Aggregate Preference Function

- Leaving and Entering Flow
- Net Outranking Flow
- Rank.

4 Selection of Awardee

Recognition of teachers in the name of awards brings happiness and confidence for their grooming work. The selection of awardees from teachers for an award ceremony is considered for decision-making in this chapter. For this process nomination of candidates from the Department of Mathematics has been taken into consideration. For this decision-making process, after consulting with decision-makers, four attributes and four criteria are considered. Four applicants are termed as P1, P2, P3, P4.

Teachers are the building pillars of students. Hence, they keep on developing themselves to keep them updated. For a recent update in their field, they involve themselves in various factors. They keep on studying throughout their life. They publish journals and patents. They give guidance to other scholars in research work. They involve themselves in training programs. Based on the factors involved in teachers, four different criteria are fixed for the decision-making process.

Ten different attributes considered for the selection are education, experience, publication, research guidance, feedback, additional courses completed, patent, funding projects, international collaboration, and clearance of eligibility examinations. They are symbolized as A1, A2, …, A10, respectively. Each of these attributes is assigned criteria with its score as given in Table 1.

The data related to each of the four alternatives are given in Table 2.

Corresponding to the data obtained from four alternatives, the decision matrix is obtained relating to the score function given in Table 1. The decision matrix obtained is shown in Table 3.

The normalized decision matrix is obtained using the relation with respect to beneficial criteria.

$$R_{ij} = \left[X_{ij} - \min\left(X_{ij}\right)\right]/[\max\left(X_{ij}\right) - \min(X_{ij})]$$

For non-beneficial criteria, it is found using the relation.

$$R_{ij} = \left[\max(X_{ij}) - X_{ij}\right]/[\max\left(X_{ij}\right) - \min(X_{ij})]$$

Since all the attributes defined are equally important in the decision-making of awardee, the criteria are considered to be beneficial and a normalized decision matrix is calculated the results are shown in Table 4.

All couple combinations of alternatives are formed and to each of the combination preference functions is calculated. The values are obtained using the relation given below.

Table 1 Attribute with its determination value

Sl. No	Attribute	Criteria	Score
1	Education	M.Phil.	0.3
		Ph.D.	0.4
		PDF	0.6
2	Experience	0–5	0.2
		6–10	0.4
		11–20	0.6
		> 20	0.9
3	Publication	1–5	0.2
		6–10	0.4
		11–20	0.6
		21–25	0.8
		> 25	0.9
4	Research guidance	0	0.1
		1–5	0.4
		6–10	0.6
		> 10	0.9
5	Feedback	Good	0.7
		Average	0.5
		Poor	0.2
		Excellent	0.9
6	Additional courses completed	0–5	0.2
		6–10	0.4
		11–15	0.6
		> 15	0.8
7	Patent	0	0.1
		1	0.3
		2–5	0.5
		6–10	0.7
		> 10	0.9
8	Funding projects	0	0.1
		1	0.3
		2	0.5
		3	0.6
		4	0.7
		> 4	0.9
9	International collaboration	0	0.1
		1	0.2

(continued)

Table 1 (continued)

Sl. No	Attribute	Criteria	Score
		2	0.3
		3	0.5
		4	0.7
		>4	0.9
10	Clearance of eligibility examinations	SET	0.4
		NET	0.6
		SET & NET	0.9
		NIL	0.1

Table 2 Data on alternatives

	A1	A2	A3	A4	A5	A6	A7	A8	A9	A10
P1	M.Phil.	4	3	0	Average	10	1	0	0	SET
P2	PDF	21	25	6	Excellent	6	6	2	3	NET
P3	Ph.D.	6	7	4	Poor	5	5	1	1	SET & NET
P4	Ph.D.	12	16	2	Good	12	3	0	2	SET

Table 3 Decision matrix

	A1	A2	A3	A4	A5	A6	A7	A8	A9	A10
P1	0.3	0.2	0.2	0.1	0.5	0.4	0.3	0.1	0.1	0.4
P2	0.6	0.9	0.8	0.6	0.9	0.4	0.7	0.5	0.5	0.6
P3	0.4	0.4	0.4	0.4	0.2	0.2	0.5	0.3	0.2	0.9
P4	0.4	0.6	0.8	0.4	0.7	0.6	0.5	0.1	0.3	0.4

Table 4 Normalized decision matrix

	A1	A2	A3	A4	A5	A6	A7	A8	A9	A10
P1	0	0	0	1	0.429	0.5	0	0	0	0
P2	1	1	1	0	1	0.5	1	1	1	0.4
P3	0.333	0.286	0.333	0.4	0	1	0.5	0.5	0.25	1
P4	0.333	0.571	1	0.4	0.714	0	0.5	0	0.5	0

$$P_n(i, k) = 0 if R_{ij} \leq R_{kj}$$

and

$$P(i, k) = 0 if R_{ij} > R_{kj}$$

The values are calculated corresponding to each row and the values are listed in Table 5.

The weights are calculated based on the number of criteria assigned for each attribute and are listed in Table 6.

The aggregate preference function given in Table 7 is calculated using the relation

$$\pi(i,k) = \sum\left[w_j P(j,k)\right] / \sum w_j$$

Leaving flow also known as positive flow is calculated using the relation

$$\varnothing_i^+ = \frac{1}{n-1}\sum \pi(i,k) when i \neq k$$

Table 5 Preference function

	A1	A2	A3	A4	A5	A6	A7	A8	A9	A10
P1, P2	0	0	0	1	0	0	0	0	0	0
P1, P3	0	0	0	0.6	0.429	0	0	0	0	0
P1, P4	0	0	0	0.6	0	0.5	0	0	0	0
P2, P1	1	1	1	0	0.571	0	1	1	1	0.4
P2, P3	0.667	0.714	0.667	0	1	0	0.5	0.5	0.75	0
P2, P4	0.667	0.429	0	0	0.286	0.5	0.5	1	0.5	0.4
P3, P1	0.333	0.286	0.333	0	0	0.5	0.5	0.5	0.25	1
P3, P2	0	0	0	0.4	0	0.5	0	0	0	0.6
P3, P4	0	0	0	0	0	1	0	0.5	0	1
P4, P1	0.333	0.571	1	0	0.294	0	0.5	0	0.5	0
P4, P2	0	0	0	0.4	0	0	0	0	0	0
P4, P3	0	0.285	0.667	0	0.714	0	0	0	0.25	0

Table 6 Weights for attributes

	A1	A2	A3	A4	A5	A6	A7	A8	A9	A10
W	3	4	5	4	4	4	5	6	6	4
	0.067	0.089	0.111	0.089	0.089	0.089	0.111	0.133	0.133	0.089

Table 7 Aggregate preference function

	P1	P2	P3	P4
P1	–	0.089	0.092	0.098
P2	0.731	–	0.495	0.422
P3	0.373	0.134	–	0245
P4	0.333	0.036	0.014	–

Table 8 Leaving, entering, net outranking flow, rank

	$\varnothing_i^+$	$\varnothing_i^+$	$\varnothing$	Rank
P1	0.093	0.479	0.386	4
P2	0.549	0.086	0.463	1
P3	0.251	0.2	0.051	3
P4	0.383	0.255	0.128	2

Entering flow also known as negative flow is calculated using the relation

$$\varnothing_i^+ = \frac{1}{n-1}\sum \pi(k, i) wheni \neq k$$

Net outranking flow $\varnothing$ is calculated by subtracting negative flow from positive flow related to each of the alternatives. From the net outranking values, the ranks are assigned which is used to determine the decision on the awardee. All these values are listed in Table 8.

The outcome received from the rank is written as $P2 > P4 > P3 > p1$. It is seen that the person with second data ranks first. The fourth person ranks second, the third person ranks third and the first person ranks fourth. Thus PROMETHEE method is applied in identifying the best teacher awardee.

5 Conclusion

The various uses of Multi-Criteria Decision-Making methods help in bringing about conclusions in various situations which involve uncertainty. The selection process provides an eminent solution. Among various MCDM methods, the PROMETHEE method to applied to identify the awardee for an award ceremony. In this chapter, four decision-makers are compared in identifying the best faculty. The various criteria and attributes related to the study have been identified. The scoring system has been performed. The ranking obtained by the method follows the order of 4,1,2,3. This brings about the result that the second persons rank top and becomes eligible for the award. The identification of the awardee has been done by applying the PROMETHEE method and the decision based on the algorithm is identified. This method can be extended to many decision-making problems such as the selection of players for team matches, the selection of the best course of study, the selection of a shop for purchase, and many more, thus providing the best solution in an uncertain environment.

References

1. Halouani, N., Chabchoub, H., Martel, J.-M.: PROMETHEE-MD-2T method for project selection. Eur. J. Oper. Res. **195**, 841–849 (2009)
2. Athawale, V.M., Chakraborty, S.: Facility location selection using PROMETHEE II method. In: Proceedings of the 2010 International Conference on Industrial Engineering and Operations Management Dhaka, Bangladesh, January 9–10, 59–64 (2010)
3. Turcksina, L., Bernardini, A., Macharis, C.: A combined AHP-PROMETHEE approach for selecting the most appropriate policy scenario to stimulate a clean vehicle fleet. Procedia Soc. Behav. Sci.Behav. Sci. **20**, 954–965 (2011)
4. Velasquez, M., Hester, P.T.: An analysis of multi-criteria decision making methods. Int. J. Oper. Res. **10**(2), 56–66 (2013)
5. Morais, D.C., de Almeida, A.T., Alencar, L.H., Clemente, T.R.N., Cavalcanti, C.Z.B.: PROMETHEE-ROC model for assessing the readiness of technology for generating energy. Math. Probl. Eng., Article ID 530615, 1–11 (2015)
6. Rezaei, J.: A systematic review of multi-criteria decision-making applications in reverse logistics. Transp. Res. Procedia **10**, 766–776 (2015)
7. Ozsahin, D.U., Uzun, B., Musa, M.S., Helwan, A., Wilson, C.N., Nurcin, F.V., Şenturk, N., Ozsahin, I.: Evaluating cancer treatment alternatives using fuzzy PROMETHEE method. Int. J. Adv. Comput. Sci. Appl. **8**(10), 177–182 (2017)
8. Hanifah, R.: Multi criteria decision making for bank rating based on health level of bank. Int. J. Prog.Ive Sci. Technol. **5**(2), 53–62 (2017)
9. Nasution, A.P., Harahap, D.A., Watrianthos, R.: Application decision support system using PROMETHEE method. J. Adv. Res. Dyn. & Control. Syst. **11**, 506–511 (2019)
10. Abdullah, L., Chan, W., Afshari, A.: Application of PROMETHEE method for green supplier selection: a comparative result based on preference functions. J. Ind. Eng. Int. **15**, 271–285 (2019)
11. Akmaludin, Sihombing, E.G., Dewi, L.S., Rinawati, Ester Arisawati STMIK Nusa Mandiri: Credit loan selection during the pandemic recommendation MCDM- PROMETHEE method. Sinkron **5**(1), 122–128 (2020)
12. Zargini, B.: Multicriteria decision making problems using variable weights of criteria based on alternative preferences. Am. Sci. Res. J. Eng., Technol. **74**(1), 1–14 (2020)
13. Tian, C., Peng, J.-J., Zhang, W.-y, Zhang, S., Wang, J.-Q.: Tourism environmental impact assessment based on improved AHP AND picture fuzzy PROMETHEE II methods. Technol. Econ. Dev. Econ. **26**(2), 355–378 (2020)
14. Emovon, I., Oghenenyerovwho, O.S.: Application of MCDM method in material selection for optimal design: A review. Results Mater. **7**, 1–21 (2020)
15. Zhang, Q., Junhua, Hu., Feng, J., Liu, An.: A novel multiple criteria decision making method for material selection based on GGPFWA operator. Mater. Des. **195**, 1–14 (2020)
16. Sánchez, F.C., Blacio, J.H.A., Bracho, M.G.F.: Danilo rafael andrade santamaria and ricardo sánchez casanova, neutrosophic and plithogenic statistical analysis in educational development. Neutrosophic Sets Syst. **44**, 223–234 (202)
17. Oubahman, L., Duleba, S.: Review of PROMETHEE method in transportation. Prod. Eng. Arch. **27**(1), 69–74 (2021)
18. Md Isa, M.K., Saharudin, N.S., Anuar, N.B. Mahad, N.F.: The application of AHP-PROMETHEE II for supplier selection. J. Phys.: Conf. Ser. 1988, 1–16 (2021)
19. Setiadi, H., Suni, N.A.M., Wardani, D.W.: Decision support system to select elective courses using hybrid AHP-PROMETHEE method. Performa: Media Ilm. Tek. Ind. **21**(1), 105–112 (2022)
20. Bertoncini, M., Boggio, A., Dell'Anna, F., Becchio, C., Bottero, M.: An application of the PROMETHEE II method for the comparison of energy requalification strategies to design Post-Carbon Cities. AIMS Energy **10**(4), 553–581 (2022)
21. Bozdemir, M.K.E., Alkan, A.: Selection of social media platforms using fuzzy PROMETHEE method with different scenario types. J. Eng. Stud. Res. **28**(4), 41–50 (2022)

22. Sarwar, M., Zafar, F., Majeed, I.A., Javed, S.: Selection of suppliers in industrial manufacturing: A fuzzy rough PROMETHEE approach. Math. Probl. Eng., Article ID 6141225, 1–19 (2022)
23. Ramadan, Ö., Ozdemir, Y.S.: Prioritization of rail system projects by using FUZZY AHP and PROMETHEE. J. Optim. & Decis. Mak. **1**(2), 114–122 (2022)
24. Taherdoost, H., Madanchian, M.: Using PROMETHEE method for multi-criteria decision making: applications and procedures. Iris J. Econ. & Bus. Manag. 1–7 (2023)
25. Mohan, K.R., Narmada Devi, R., Nagaram, N.B., Bharathi, T., Rasappan, S.: Neutrosophic Statistical Analysis on Gold Rate. Neutrosophic Sets Syst. **60**, 113–123 (2023)
26. Sari, M., Kosim, K., Saputra, A.: Application of the PROMETHEE II method for determining road improvement priorities. J. Math. Instr., Soc. Res. Opin. **2**(1), 11–22 (2023)
27. Sahoo, L., Senapati, T., Yager, R.R. (Eds.): Real life applications of multiple criteria decision making techniques in fuzzy domain, studies in fuzziness and soft computing (STUDFUZZ, Volume 420), Springer Nature
28. Adak, A.K., Nilkamal, Srivastava, K.N.: New ranking approach to solve MCDM problems with Generalized Intuitionistic Fuzzy Information. R. Life Appl. Mult. Criteria Decis. Mak. Tech. Fuzzy Domain, STUDFUZZ, **420**, 307–318 (2023)
29. Samanta, A., Kar, S., Kaur, D., Basu, K.: Solution of airport gate reassignment problem under uncertain environment. R. Life Appl. Mult. Criteria Decis. Mak. Tech. Fuzzy Domain, STUDFUZZ **420**, 559–566 (2023)

Evaluating Challenges in Smart Transportation with Grey Theory-Based Multi-criteria Decision-Making

Gülay Demir and Sarbast Moslem

Abstract Smart transportation is a dynamic and comprehensive approach aimed at addressing the transportation challenges faced by modern societies. This approach integrates cutting-edge ICTs to enhance transportation efficiency, promote sustainability, and improve urban living. However, for this innovative approach to be successfully implemented, the challenges encountered need to be identified, and prioritized, and effective solutions devised. This study provides a framework for understanding and addressing the primary challenges in smart transportation. These challenges include data security and privacy, reduction of environmental impacts, infrastructure deficiencies, integration challenges, societal acceptance and security concerns, costs, technological transformation, and legal regulations. To effectively address these challenges, this study employs the G-DEMATEL approach. This approach combines grey theory and DEMATEL methods to determine the importance levels of the challenges. The integration of these methods helps identify priority areas to be considered in the planning and implementation of smart transportation projects. The findings obtained serve as a strategic guide to enhance the success of smart transportation projects. Factors such as increasing societal acceptance, improving infrastructure, reducing costs, managing technological transformation, and minimizing environmental impacts are critical for the sustainability and effectiveness of these projects. Therefore, a multi-stakeholder approach should be adopted, and these factors should be addressed in a balanced manner. In conclusion, this study provides a comprehensive resource that appeals to a broad audience, ranging from academic researchers to policymakers and industry leaders. By offering a robust framework for understanding and addressing the challenges faced by smart transportation, this study contributes to the development of more effective and sustainable transportation strategies in the industry.

G. Demir (✉)
High School of Health Vocational Services, Sivas Cumhuriyet University, Sivas 58140, Turkey
e-mail: gulaydemir@cumhuriyet.edu.tr

S. Moslem
School of Architecture Planning and Environmental Policy, University College Dublin, Dublin 15, Belfield 04 V1W8, Ireland
e-mail: sarbast.moslem@ucd.ie

L. Sahoo et al. (eds.), *Decision Making Under Uncertainty Via Optimization, Modelling, and Analysis*, Studies in Systems, Decision and Control 558,
https://doi.org/10.1007/978-981-96-0085-4_27

Keywords Smart transportation · Grey theory · DEMATEL

Abbreviations

AHP	Analytical Hierarchy Process
ANP	Analytic Network Process
ARAS	Additive Ratio ASsessment
BWM	Best Worst Method
DANP	DEMATEL-based ANP
G-DEMATEL	Grey-The Decision-Making Trial and Evaluation Laboratory
GIS	Geographical Information Systems
Grey-ISM	Grey Interpretive Structural Model
GST	Grey system theory
ICTs	Information and Communication Technologies
ISM	Interpretive Structural Model
MCDM	Multi-criteria decision-making
MICMAC	Matrix-based Multiplication Applied to a Classification
SWARA	Step-wise Weight Assessment Ratio Analysis
TOPSIS	Technique for Order of Preference by Similarity to Ideal Solution
VIKOR	In Serbian "Viekriterijumsko KOmpromisno Rangiranje"

1 Introduction

Smart transport is essential to meet modern cities' complex transport needs and make city life more sustainable. Smart transport systems shorten travel times by reducing traffic congestion and create an environmentally friendly transport environment by reducing emissions [1]. By integrating ICTs, these systems provide various benefits in areas such as traffic management, public transport optimization, ride-sharing, and autonomous vehicles [2]. At the same time, solutions such as ride-sharing and integrated public transport systems can make urban transport more accessible and affordable while reducing traffic accidents [3].

Grey theory may be used to cope with uncertainties and inadequate knowledge, emphasizing complex system analysis [4]. When assessing the challenges of smart transport, grey theory is expected to be helpful in:

- Analysing Uncertainties: The complexity of smart transport systems introduces uncertainties and variability. Grey theory can be used to identify and manage these uncertainties.
- Working with Uncertain Information: Many variables in the field of smart transport are based on imprecise information. Grey theory can address how to work with such information and make decisions.

- Perspectives of Different Stakeholders: Grey theory can provide a more comprehensive analysis by considering the perspectives and interests of different stakeholders. This can increase the success of smart transport projects.

Smart transport is an essential set of challenges facing cities and societies. However, it is possible to overcome these challenges using analytical tools such as grey theory. A grey theory approach can more effectively address challenges such as data integrity, infrastructure integration, social acceptance, and sustainability. The future of smart transport will depend on the use of such analytical approaches.

This study provides both a framework for examining the issues of smart transportation and suggestions on how analytical methods such as grey theory might be applied.

The main objective of this study is to analyze the challenges encountered in smart transportation using grey theory and to propose solutions to these challenges. This study aims to clearly define how grey theory can be applied in the field of smart transportation and how uncertainties in this area can be addressed. The methodology and approach of the study will be elaborated in detail. The findings of the study are intended to assist industry experts in identifying priority areas and policy recommendations for the advancement of smart transportation. Prioritizing challenges in smart transportation poses a significant decision-making challenge for industry experts. This paper aims to address this challenge by providing a case study based on appropriate criteria for identifying the difficulties in smart transportation. Additionally, it aims to provide managerial and practical information to industry decision-makers to promote the smart transportation sector and establish a sustainable transportation system more effectively.

1.1 Research Questions and Research Gaps

The examination of challenges in smart transportation as a decision problem has not been sufficiently explored in the literature. Existing research generally addresses the complexity of smart transportation and its challenges in a holistic manner, but efforts to prioritize and manage these challenges by forming a decision-making model are limited. Research in this area emphasizes the necessity of systematically evaluating the challenges faced by smart transportation using MCDM models or similar analytical approaches. However, there is a lack of detailed guidance on how such an approach can be implemented and which criteria should be considered. Therefore, treating the challenges in smart transportation as a decision problem represents a new and significant research area in the literature. In this context, the development of MCDM models and similar analytical approaches, as well as the creation of new strategies for effectively addressing the challenges in smart transportation, is crucial. The primary aim of this research is to answer the identified questions:

RQ1. Are any mathematical tools or decision-support models used to examine the problems faced by the transportation sector?

RQ2. What are the key challenges in smart transportation?

RQ3. How are these challenges prioritized?

RQ4. What are the relationships between the lack of regulations, rapid technological development, investment costs, and other factors?

RQ4. What are the compatibility and integration challenges of smart transportation components (e.g., transportation vehicles, infrastructure, ICT)?

RQ5. What are the priority areas for developing smart transportation, and what policy recommendations can be made?

RQ6. What strategies can be proposed to improve smart transportation in terms of sustainability and efficiency?

These questions could serve as a starting point for examining and understanding the gaps in smart transportation through MCDM methods. The research gap in this area could focus on how the decision-making processes regarding the challenges in smart transportation can be enhanced using MCDM models or how the effectiveness of these models can be improved. In this context, a gap could arise around the following questions:

G1. Has the multidimensional and complex nature of challenges in smart transportation been sufficiently understood in the existing literature?

G2. Are MCDM methods comprehensive enough to accurately assess the challenges faced by smart transportation?

G3. To what extent need new MCDM models for prioritizing and managing challenges in smart transportation been identified?

G4. Can MCDM models adequately address the complexity and interactions of different components (such as technology, infrastructure, and social acceptance) in smart transportation?

G5. Is there sufficient research on how policy and strategies used in addressing challenges in smart transportation are evaluated and optimized by MCDM models?

G6. How can MCDM models be updated or adapted to consider future trends and changing conditions in smart transportation?

G7. What are the most effective MCDM approaches to achieving significant goals in smart transportation, such as sustainability, security, and efficiency?

These questions indicate a significant research gap in the evaluation and management of challenges in smart transportation.

1.2 Contributions and Novelty

Smart transport helps cities achieve sustainability goals while improving quality of life and making cities more attractive and competitive. This paper presents a hybrid MCDM framework based on grey system theory to address issues in smart transportation. The primary contributions of the suggested decision-making paradigm are summarized as follows:

- This study presents a methodological framework for solving the problem of assessment of challenges for decision-makers in the transport field.
- This study presents the use of the subjective weighting method G-DEMATEL as a model for the assessment of challenges in the field of transport. This is the first time the G-DEMATEL method has been applied in the literature on smart transport.
- The proposed evaluation framework can help solve other MCDM problems in the transport domain and apply them to other domains.

1.3 Organization of the Study

The study is organized as follows. Section 2 provides a complete literature assessment and highlights research needs. Section 3 describes the MCDM framework utilized in the analysis. The fourth section introduces the data and displays the outcomes of the suggested methodology. Section 5 focuses on the practical and managerial ramifications. Finally, Sect. 6 examines the results, limits, and future directions.

2 Literature Review

The literary portion is split into two subsections. The first sub-section summarizes transport studies that used MCDM techniques to give some background information. The second sub-section discusses the literature on research that used the G-DEMATEL technique.

2.1 MCDM Studies in the Field of Transportation

Although several studies have used MCDM approaches in smart transportation, this article provides a quick overview of a few of them.

Khan et al. [5] planned to apply weight identification and ranking algorithms to uncover, analyze, and prioritize essential transport features to create smart transportation systems. They employed the VIKOR technique to determine the top-ranked smart transport systems, emphasizing transportation effectiveness, safety, efficiency, and sustainability. Alamoodi et al. [6] used an Entropy weighting strategy and a 2-tuple linguistic T-spherical fuzzy choice by opinion score method to pick electric buses. Elomiya et al. [7] used fuzzy AHP, GIS, and SWARA methods to select an electric vehicle charging station location. Sekmen et al. [8] evaluated urban transport preferences with the AHP method. Zagorskas and Turskis [9] evaluated alternative road connections with the G-ARAS method. Sujanto et al. [10] used fuzzy Delphi and BWM methods to understand green logistics in Taiwan's maritime transport sector.

Golui et al. [11] proposed a Fermatean fuzzy TOPSIS method for electric vehicle selection.

In addition, a summary of some of the studies in the transport field evaluated with G-DEMAEL is also given.

Sahu et al. [12] suggested an integrated G-DEMATEL, ANP, and G-DANP model for identifying and prioritizing issues influencing India's public transportation system. Durmaz and Budak [13] used G-DEMATEL-ANP's comprehensive method to identify barriers to implementing sustainable transport systems in urban logistics and develop strategic initiatives by ranking these barriers in order of significance. Karuppiah et al. [14] utilized G-DEMATEL to evaluate the facilitators of sustainable transportation in the Indian leather industry.

In general, studies in the field of transport have been evaluated with different MCDM methods [15–19].

2.2 *Studies Used the G-DEMATEL Technique*

Some recent studies applied the G-DEMATEL, a subjective weighting procedure. These studies can be summarized as follows.

Psychogiou and Tsoulfas [20] used a G-DEMATEL technique to investigate 14 elements influencing confidence in the wine supply chain. Chen et al. [21] utilized the G-DEMATEL approach in conjunction with the grey ISM to construct a hierarchical framework for investigating the links between the causes of urban issues. Zaman et al. [22] used a G-DEMATEL approach to visualize the prioritization and interactions of obstacles and facilitators to green procurement of environmentally responsible items in the pharmaceutical business. Govindan [23] employed two distinct MCDM techniques, BWM and G-DEMATEL, to identify the best and most appropriate circular economy strategies and assess widely gathered success metrics. Xing et al. [24] utilized the G-DEMATEL approach to quantify the degree of effect of factors influencing removal quality and visualize their causality, as well as the ISM to create a hierarchical structure to investigate the matching connection between these elements. They used the MICMAC approach to determine the function and status of components. Nazeri et al. [25] created a cause-and-effect graph using the G-DEMATEL approach better to comprehend the link between sustainable supply chain risks. Sohrabi [26] used G-DEMATEL and the AHP approach to examine the components influencing the cold's agility. Menon and Ravi [27] proposed the G-DEMATEL method to assess the impediments influencing sustainable supply chain implementations for the electronics sector.

2.3 Research Gaps

According to an evaluation of the past research in the literature, investigations conducted with MCDM techniques in the transport industry are primarily centred in different branches utilizing various weighting and ranking methodologies. Therefore, the limited number of studies conducted in the smart transport branch, which is an indispensable part of the transport sector, constituted the first motivation of this study. In addition, when the literature is examined, a generally applied and accepted criterion for evaluating difficulties in the transport sector has yet to be determined. In addition, considering the existing studies, there is no commonly acknowledged decision support tool or mathematical model for evaluating alternatives in the assessment process for real-world decision-makers and managers in the transportation industry.

Grey system theory (GST) is one of the valuable tools widely used in the MCDM literature to address uncertainties and incomplete information about the decision-making environment [4]. Compared to fuzzy set and rough set theories, grey theory has some critical advantages:

- It can improve the decision-making process by providing less computational complexity [28, 29].
- It allows decision-makers to minimize inconsistencies arising from uncertainties and complex situations related to the decision problem by using intrinsic numbers [30, 31].
- This provides the opportunity to obtain more robust, stable, and reliable results in the case of small, limited, and uncertain data [32, 33].

Although there are other approaches for weighing assessment criteria in the MCDM literature, the G-DEMATEL method is employed in this study due to the following benefits. The benefits of the G-DEMAYEL technique are outlined as follows:

- It integrates classic DEMATEL with grey theory [34].
- Experts may communicate their thoughts using language phrases based on their expertise and experience, addressing uncertainty and inaccurate assessments.
- It effectively analyses causal linkages and interactions between components.
- Integrating subjective evaluations improves problem-solving efficiency.

3 Methodology for Research

This section describes the basic algorithm for the proposed integrated technique. G-DEMATEL calculates the weights of the difficulties in smart transport based on the stages in the suggested MCDM methodology.

3.1 Grey System Theory

Deng [35] introduced the grey system theory to address uncertainties in circumstances with discrete data and imperfect knowledge. Grey system analysis is carried out with grey numbers, grey equations and/or grey matrices [36]. When it comes to grey numbers, neither black nor white concepts may come to mind once in a while, their values cannot be clearly expressed in numbers, but they are known to take values in which range. Grey number is $(\otimes x)$, $\otimes x \in \left[\underline{\otimes x}, \overline{\otimes x}\right] = \left[x\prime \in x \middle| \underline{\otimes x} \leq x\prime \leq \overline{\otimes x}\right]$ with x being a real number where, $\otimes x$ and $\overline{\otimes x}$ refer to the lower and upper limits of the number of $\otimes x$ greys, respectively. $\otimes x_1$ and $\otimes x_2$ are used for basic mathematical operations that can be done with grey numbers.

$$\otimes x_1 + \otimes x_2 = \left[\underline{x_1} + \underline{x_2}, \overline{x_1} + \overline{x_2}\right] \tag{1}$$

$$\otimes x_1 - \otimes x_2 = \left[\underline{x_1} - \overline{x_2}, \overline{x_1} - \underline{x_2}\right] \tag{2}$$

$$\otimes x_1 x \otimes x_2 = \left[\min\left(\underline{x_1}.\underline{x_2}, \underline{x_1}.\overline{x_2}, \overline{x_1}.\underline{x_2}, \overline{x_1}.\overline{x_2}\right), max\left(\underline{x_1}.\underline{x_2}, \underline{x_1}.\overline{x_2}, \overline{x_1}.\underline{x_2}, \overline{x_1}.\overline{x_2}\right)\right] \tag{3}$$

$$\otimes x_1 : \otimes x_2 = \left[\underline{x_1}, \overline{x_1}\right].\left[\frac{1}{\underline{x_2}}, \frac{1}{\overline{x_2}}\right] \tag{4}$$

$$k \in R^+, k. \otimes x = \left[k.\underline{x}, k.\overline{x}\right] \tag{5}$$

3.2 G-DEMATEL Method

The application steps of the G-DEMATEL approach, which integrates the grey numbers and the DEMATEL method, are given below [37, 38]:

Step 1. Determining the evaluation scale of grey relationships.

Let "n" be the number of criteria used in the study and "k" as the number of decision-makers selected for the study. Each decision maker is tasked with evaluating the direct influence of criterion "i" on criterion "j" using an integer scale ranging from 0 to 4. This scale is defined by descriptors that depict the level of impact, namely, “no influence”, “low influence”, “medium influence”, “high influence”, and “very high influence”. The evaluation is conducted across a set of "n" pre-identified criteria. Table 1 presents the linguistic terms and the corresponding Grey numbers [39].

Table 1 Grey linguistic expression scale

Crisp values	Linguistic variables	Interval grey number
0	No influence	[0, 0]
1	Low influence	[0, 1]
2	Medium influence	[1, 2]
3	High influence	[2, 3]
4	Very high influence	[3, 4]

Consequently, a comprehensive set of "k" initial relationship matrices was developed using the evaluations of the decision-maker's influences.

Step 2. Creating the appropriate Grey matrix for each initial relationship matrix.

Using the values obtained in Table 1 and Step 1, the matching Grey matrices Eq. (6) are used to calculate an upper and lower range of values [40].

$$\otimes G_{ij}^{k} = \left(\underline{\otimes} G_{ij}^{k}, \overline{\otimes} G_{ij}^{k}\right) \tag{6}$$

where $1 \leq k \leq l; 1 \leq i \leq n; 1 \leq j \leq n$.

Step 3. Finding the average of grey relation matrices.

Using the "k" Grey relation matrices, the average grey relational matrix $\left[\otimes G_{ij}\right]$ is obtained by Eq. (7).

$$\left[\otimes G_{ij}\right] = \left(\frac{\sum \underline{\otimes} G_{ij}^{k}}{l}, \frac{\sum \overline{\otimes} G_{ij}^{k}}{l}\right) \tag{7}$$

Step 4. Calculation of crisp matrices using average Grey matrices.

Crips matrices are obtained using the three-step procedure in Eqs. (8–13) [40, 41].

Step 4.1. Obtaining lower and upper normalized values.

$$\underline{\otimes} \dot{G}_{ij} = \left(\underline{\otimes} G_{ij} - j^{min} \underline{\otimes} G_{ij}\right) / \Delta_{min}^{max} \tag{8}$$

where $\underline{\otimes} \dot{G}_{ij}$ represents the normalized lower limit value of the grey number $\underline{\otimes} \breve{G}_{ij}$.

$$\overline{\otimes} \dot{G}_{ij} = \left(\overline{\otimes} G_{ij} - j^{min} \overline{\otimes} G_{ij}\right) / \Delta_{min}^{max} \tag{9}$$

where $\overline{\otimes} \dot{G}_{ij}$ represents the normalized upper limit value of the grey number $\overline{\otimes} \breve{G}_{ij}$.

$$\Delta_{min}^{max} = j^{max} \overline{\otimes} G_{ij} - j^{min} \underline{\otimes} G_{ij} \tag{10}$$

Step 4.2. Calculation of the total normalized crips value.

$$X_{ij} = \left(\frac{\underset{-}{\otimes}\dot{G}_{ij}\left(1 - \underset{-}{\otimes}\dot{G}_{ij}\right) + \left(\overline{\otimes}\dot{G}_{ij} * \overline{\otimes}\dot{G}_{ij}\right)}{\left(1 - \underset{-}{\otimes}\dot{G}_{ij} + \overline{\otimes}\dot{G}_{ij}\right)} \right) \quad (11)$$

Step 4.3. Calculation of final net values.

$$X^*_{ij} = \left(min \underset{-}{\otimes} \dot{G}_{ij} + \left(X_{ij} * \Delta^{max}_{min}\right) \right) \quad (12)$$

$$X = \left[X^*_{ij}\right] \quad (13)$$

Step 5. Obtaining normalized direct-relation matrix.

Using Eqs. (14) and (15), the normalized direct relationship matrix "N" is produced. This matrix's entries are all between 0 and 1.

$$L = \frac{1}{1 \leq i \leq n^{max} \sum_{j=1}^{n} X^*_{ij}} \quad (14)$$

$$N = L * X \quad (15)$$

where the initial crisp relationship matrix is called X, the normalizing factor is called L, and the normalized direct relation matrix is denoted by N.

Step 6. Determining the total relationship matrix "S" using Eq. (16)

$$S = N(I - N)^{-1} \quad (16)$$

where I represents a unit matrix.

Step 7. Obtaining causal parameters.

The symbols R and C represent the total of rows and columns, respectively. It may be calculated using Eqs. (17) and (18) as follows:

$$R_i = \sum_{j=1}^{n} s_{ij} \quad (17)$$

$$C_j = \sum_{i=1}^{n} s_{ij} \quad (18)$$

Step 8. Calculation of criterion weights.

Table 2 Profile of experts

DMs	Duty	Graduation degree	Experience	Age
Expert-1	Academician	Civil engineering	25	58
Expert-2	City and regional planner	City and regional planning	16	44
Expert-3	Transport engineer	Transport engineering	18	51

We use Eqs. (19) and (20) to get the weight coefficients of the criterion.

$$\omega_i = \sqrt{(R_i + C_i)^2 + (R_i - C_i)^2} \tag{19}$$

$$w_i = \frac{\omega_i}{\sum_{i=1}^{n} \omega_i} \tag{20}$$

Step 9. Generating the causal diagram.

A causal diagram is created using the values derived from Eqs. (19) and (20).

4 Empirical Analysis and Results

4.1 Data

To get more reasonable and realistic outcomes, an expert committee made up of three individuals with at least fifteen years of experience in the transportation business was constituted before the use of the G-DEMATEL algorithm. Table 2 provides comprehensive details on the three experts chosen to serve on the assessment committee. In addition, the evaluation criteria proposed by these three experts and their detailed explanations used in analysing the challenges faced in smart transportation are given in Table 3.

4.2 Application Findings of the G-DEMATEL Approach

By utilizing the individual perspectives of three distinct experts, Grey direct relationship matrices are created for employing Eq. (6) with the Grey numbers in Table 1. These matrices are given in Table 4.

Table 5 displays the combined Grey direct relationship matrix that was produced by averaging the expert-created Grey direct relationship matrices using Eq. (7).

A final relationship matrix, X, was generated by using Eqs. (8–13). A normalized direct-relation matrix was produced using Eq. (14), with a normalization factor of

Table 3 Evaluation criteria

Criteria	Code	Description
Data security and confidentiality	C1	The security and user privacy of data collected and processed in smart transport systems should be ensured
Environmental factors	C2	Climate change, natural disasters and environmental impacts may affect the design and operation of smart transport systems
Inadequate infrastructure and resources	C3	The existing infrastructure is not suitable for new technologies and insufficient resources may hinder the development of smart transport
Integration challenges	C4	Difficulties may arise in the integration of different transport systems, data sharing and interoperability
Social acceptance and security concerns	C5	It is important that new technologies are accepted by society and safety concerns are addressed. In particular, the safety and ethical issues of technologies such as autonomous vehicles fall into this category
Costs	C6	High investment costs and operating expenses may limit the expansion of smart transport projects
Technological transformation	C7	Rapid technological change may make it difficult for existing infrastructure and systems to adapt to this change
Legal and regulatory barriers	C8	Inadequacy or incompatibility of existing legal regulations may hinder the development and implementation of smart transport projects

0.0640. For the row and column total, using Eqs. (17–18). Table 6 also provides the weights for the criterion and their priority ranking, which were determined using Eqs. (19–20).

As can be seen in Table 6, the group that affects the criterion includes C3, C5, C6, and C8 with positive (R–C) values. The group affected includes C1, C2, C4, and C7 criteria with negative (R–C) values. The effect diagram in Fig. 1 is used to assist in determining the interactions between the criteria in the last phase of the approach, which involves using the (R + C) and (R–C) values.

The relative relevance of each criterion is shown by (R + C) values. It can be observed that the C5 criterion has the greatest value when looking at the (R + C) values. The criteria C6 and C3 come after this one.

When the influencing criteria are taken into account, it can be shown that C3 is the criterion with the greatest (R–C) value. The C6 and C8 criteria come after this criterion, in that order. The three criteria with the highest (R–C) values may thus be said to have the most effect on the other criteria.

Table 4 Grey direct relationship matrix for expert-1–2-3

Criteria	C1	C2	C3	C4	C5	C6	C7	C8
C1	[0,0],[0,0],[0,0]	[1,2],[1,2],[1,2]	[1,2],[1,2],[3,4]	[1,2],[1,2],[1,2]	[2,3],[3,4],[0,1]	[0,1],[2,3],[1,2]	[0,1],[0,1],[1,2]	[0,1],[0,1],[2,3]
C2	[0,1],[1,2],[1,2]	[0,0],[0,0],[0,0]	[0,1],[0,1],[1,2]	[2,3],[2,3],[2,3]	[2,3],[1,2],[1,2]	[2,3],[3,4],[0,1]	[1,2],[0,1],[1,2]	[0,1],[0,1],[2,3]
C3	[3,4],[3,4],[3,4]	[2,3],[2,3],[2,3]	[0,0],[0,0],[0,0]	[2,3],[2,3],[3,4]	[3,4],[2,3],[2,3]	[2,3],[0,1],[2,3]	[1,2],[0,1],[3,4]	[1,2],[0,1],[1,2]
C4	[1,2],[0,1],[0,1]	[2,3],[2,3],[0,1]	[2,3],[2,3],[0,1]	[0,0],[0,0],[0,0]	[2,3],[1,2],[1,2]	[1,2],[1,2],[1,2]	[0,1],[1,2],[1,2]	[0,1],[0,1],[2,3]
C5	[2,3],[1,2],[2,3]	[3,4],[3,4],[2,3]	[0,1],[0,1],[1,2]	[2,3],[2,3],[1,2]	[0,0],[0,0],[0,0]	[3,4],[3,4],[1,2]	[3,4],[1,2],[3,4]	[0,1],[2,3],[2,3]
C6	[2,3],[0,1],[2,3]	[0,1],[0,1],[1,2]	[1,2],[1,2],[1,2]	[2,3],[2,3],[1,2]	[3,4],[3,4],[1,2]	[0,0],[0,0],[0,0]	[3,4],[2,3],[3,4]	[3,4],[0,1],[3,4]
C7	[0,1],[1,2],[2,3]	[0,1],[0,1],[2,3]	[1,2],[1,2],[0,1]	[0,1],[0,1],[1,2]	[0,1],[1,2],[3,4]	[0,1],[0,1],[1,2]	[0,0],[0,0],[0,0]	[2,3],[0,1],[0,1]
C8			[0,1],[1,2],[3,4]	[0,1],[0,1],[1,2]	[0,1],[0,1],[0,1] [3,4],[3,4],[1,2]	[0,1],[1,2],[1,2]	[0,1],[2,3],[2,3] [3,4],[2,3],[2,3]	[0,0],[0,0],[0,0]

Table 5 Combined Grey direct relationship matrix

Criteria	C1	C2	C3	C4	C5	C6	C7	C8
C1	[0.000,0.000]	[0.667,1.667]	[2.000,3.000]	[0.667,1.667]	[1.667,2.667]	[1.000,2.000]	[0.333,1.333]	[0.667,1.667]
C2	[0.667,1.667]	[0.000,0.000]	[0.667,1.667]	[1.667,2.667]	[1.333,2.333]	[1.667,2.667]	[0.667,1.667]	[0.667,1.667]
C3	[3.000.4.000]	[2.000,3.000]	[0.000,0.000]	[1.667,2.667]	[2.333,3.333]	[1.333,2.333]	[1.333,2.333]	[0.667,1.667]
C4	[0.333,1.333]	[1.000,2.000]	[1.667,2.667]	[0.000,0.000]	[1.333,2.333]	[1.000,2.000]	[0.667,1.667]	[0.667,1.667]
C5	[1.667,2.667]	[2.333,3.333]	[0.333,1.333]	[1.333,2.667]	[0.000,0.000]	[2.333,3.333]	[2.333,3.333]	[1.333,2.333]
C6	[1.333,2.333]	[0.333,1.333]	[1.333,2.333]	[1.333,2.333]	[2.333,3.333]	[0.000,0.000]	[2.667,3.667]	[2.000,3.000]
C7	[1.000,2.000]	[0.667,1.667]	[0.333,1.333]	[0.667,1.667]	[1.333,2.333]	[0.333,1.333]	[0.000,0.000]	[0.667,1.667]
C8	[1.333,2.333]	[1.000,2.000]	[0.000,1.000]	[1.667,2.667]	[0.667,1.667]	[1.333,2.333]	[2.333,3.333]	[0.000,0.000]

Table 6 Row and column totals of the total relationship matrix

Criteria	R	C	R + C	R–C	Weights	Rank
C1	0,6633	0,8060	1,4693	–0,1428	0,1147	5
C2	0,6869	0,7332	1,4201	–0,0463	0,1104	7
C3	1,0732	0,6184	1,6916	0,4549	0,1362	3
C4	0,6292	0,8710	1,5002	–0,2419	0,1181	4
C5	1,0400	0,9694	2,0094	0,0707	0,1563	1
C6	1,0000	0,8119	1,8119	0,1881	0,1416	2
C7	0,5012	0,8986	1,3999	–0,3974	0,1131	6
C8	0,7597	0,6450	1,4047	0,1147	0,1096	8

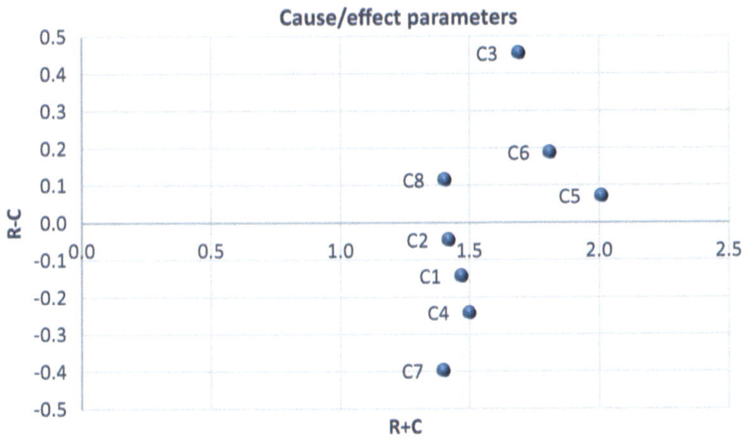

Fig. 1 Effect diagram

The C7 criterion with the lowest (R–C) value is the one that is most impacted by the other criteria when the affected criteria are taken into account. The criteria C4 and C1 come after this one.

5 Discussion, Practical and Managerial Implications

Smart transport projects offer a significant opportunity to solve the transport challenges of modern cities. However, the success of these projects is significantly influenced by the interaction and management of several factors. Identifying and prioritizing these factors is vital for the effective management of projects.

Some practical implications of our study, in which experts analyzed the challenges in smart transport, are as follows:

- The main practical implication is to present a hybrid framework for assessing challenges in smart transport. The evaluation framework covering the grey theory-based decision-making procedure (G-DEMATEL) is presented.
- It methodologically presents an evaluation framework for solving decision-making problems in smart transport. Furthermore, a choice issue is solved using this study's framework for the first time.
- It may be integrated into the resolution of additional decision-making issues related to smart transportation and can be used to solve other problems in the field of transport.
- This study can help decision-makers in the transport field to make correct and reasonable decisions by providing a flexible, easy-to-use, and modifiable evaluation framework.

The managerial implications of our paper can be summarized as follows:

- Systematic assessment of smart city challenges provides strategic information to municipalities, policy-makers and senior managers. Accordingly, the findings of the model proposed in this study can primarily guide the management team in improving the challenges of smart cities.
- The proposed evaluation framework provides necessary decision support to managers who want to provide smart transport services to identify the most critical challenge in smart transport.
- Our findings help municipalities and policymakers, the two most essential parts of the transport service network, to make long-term strategic decisions to address the most critical challenge for their citizens.
- The methodology introduced simplifies assessing challenges in smart transport and facilitates benchmarking analysis for transport sector stakeholders.

6 Conclusions, Limitations, and Future Directions

The main objective of this study is to identify, prioritize, and address the challenges encountered in smart transportation. Considering the economic and social role of municipal services, challenges in the transportation sector must be periodically measured and compared. Assessing challenges in smart transport is crucial for enhancing the quality of life for citizens and ensuring sustainable cities.

According to the weighting results, the most critical challenge is social acceptance and safety concerns (C5 = 0.1563). Other challenges are ranked in order of importance as follows: costs (C6 = 0.1416) > lack of infrastructure and resources (C3 = 0.1362) > integration challenges (C4 = 0.1181) > data security and privacy (C1 = 0.1147) > technological transformation (C7 = 0.1131) > environmental factors (C2 = 0.1104) > legal and regulatory barriers (C8 = 0.1096). These results indicate that social acceptance and safety concerns have been identified as the most critical challenge. This underscores the vital importance of societal acceptance and ensuring safety for the success of a project or initiative. Lack of social acceptance or security concerns can significantly impact the reputation of a project or initiative, as well as user trust, thereby affecting its chances of success. On the other hand, costs, lack of infrastructure and resources, integration challenges, data security and privacy, technological transformation, environmental factors, and legal and regulatory barriers are ranked in order of importance. These results guide in identifying key challenges that can affect the success of projects or initiatives. Particularly, it is observed that costs, lack of infrastructure, and integration challenges are the most important factors after social acceptance and safety concerns.

According to the given information, lack of infrastructure and resources, social acceptance and safety concerns, costs, and legal and regulatory barriers are identified as causal criteria. On the other hand, data security and privacy, environmental factors, integration challenges, and technological transformation are considered as outcome criteria. This distinction allows us to understand the root causes and effects of the problems more clearly. While causal criteria generally represent the fundamental reasons behind a project or initiative, outcome criteria typically focus on the effects or consequences of these reasons. For example, causal criteria such as lack of infrastructure and resources and costs may be factors that negatively impact the success of a project, while outcome criteria such as data security and privacy and technological transformation focus on the consequences or results of project failure. This differentiation can help us make better decisions about which areas to focus on when developing solution strategies. For example, a high ranking in costs may indicate that solutions aimed at reducing costs should be prioritized, while a high ranking in social acceptance and safety concerns highlights the need to address the concerns of society.

The central findings of this study are the identification and prioritization of key factors to be considered in smart transportation projects. Our numerical data show that social acceptance, infrastructure improvements, cost reduction, technological transformation, and environmental factors are critically important for the sustainability and success of projects. The significant findings of this study provide valuable guidance for strategic decision-makers and practitioners in the future of smart transportation. The success of smart transportation is crucial for ensuring sustainable development and meeting the needs of society. Therefore, in the planning and management of smart transportation projects, the identified factors should be considered and addressed in a balanced manner. The contribution of this study is to provide a framework for understanding the complexity of smart transportation and developing more effective strategies in this field. This study can help decision-makers and industry experts better understand and address the challenges faced by smart transportation, thereby contributing to a more sustainable and efficient transportation system.

This paper presents a decision support framework within grey system theory to address the problem of ranking the challenges that can be faced in smart transport. In future studies, the scope of the research can be extended by adding other challenges (e.g., social and governance challenges). There are several limitations to this study. Firstly, prioritizing the identified challenges may vary depending on the characteristics and context of implementing the smart city. Furthermore, the assessment and management of these challenges may differ in different geographical regions and cultural contexts. The findings of this study can be used as a general guide. Furthermore, a variety of MCDM approaches based on distinct fuzzy sets, including Fermatean fuzzy, hesitant fuzzy sets, intuitionistic fuzzy, neutrosophic fuzzy, image fuzzy, Pythagorean fuzzy, q-rung orthopair fuzzy, and global fuzzy sets, can be used to examine the subject issue in future studies.

References

1. Bibri, S.E., Krogstie, J.: Smart sustainable cities of the future: An extensive interdisciplinary literature review. Sustain. Cities Soc. **31**, 183–212 (2017). https://doi.org/10.1016/j.scs.2017.02.016
2. Evans, J., Karvonen, A., Raven, R.: The experimental city. Routledge (2017)
3. Hollands, R.G., Andrae, A.S.: Smart cities: Big data, civic hackers, and the quest for a new utopia. WW Norton & Company (2015)
4. Shaju, K., Babu, S., Thomas, B.: Analysing effectiveness of grey theory-based feature selection for meteorological estimation models. Eng. Appl. Artif. Intell.Artif. Intell. **123**, 106243 (2023). https://doi.org/10.1016/j.engappai.2023.106243
5. Khan H.U., Ali, F., Sohail, M., Nazir, S., Arif, M.: Decision making for selection of smart vehicle transportation system using VIKOR approach. Int. J. Data Sci. Anal. 1–15 (2024). https://doi.org/10.1007/s41060-024-00537-6
6. Alamoodi, A.H., Albahri, O.S., Deveci, M., Yüksel, S., Mohamad Sharaf, I., et al.: Selection of electric bus models using 2-tuple linguistic T-spherical fuzzy-based decision-making model. Expert. Syst. Appl. **249**, 123498 (2024). https://doi.org/10.1016/j.eswa.2024.123498
7. Elomiya, A., Křupka, J., Jovčić, S., Švadlenka, L., Pamucar, D., et al.: A hybrid suitability mapping model integrating GIS, machine learning, and multi-criteria decision analytics for optimizing service quality of electric vehicle charging stations. Sustain. Cities Soc. **106**, 105397 (2024). https://doi.org/10.1016/j.scs.2024.105397
8. Sekmen, F., Demirkol, I., Gökırmak, H.: Evaluation of urban transportation preferences with analytical hierarchy process method. Qual. Quant. **58**(3), 2087–2101 (2024). https://doi.org/10.1007/s11135-023-01731-7
9. Zagorskas, J., Turskis, Z.: Enhancing sustainable mobility: Evaluating new bicycle and pedestrian links to car-oriented industrial parks with ARAS-G MCDM approach. Sustainability (Switzerland) **16**(7), 2994 (2024). https://doi.org/10.3390/su16072994
10. Sujanto, R.Y., Kao, S.L., Yang, M.F.: Multicriteria Assessment of Green Logistics in Taiwan's Maritime Freight Transport: Green Packaging and Green Transportation as Driving Aspects. Trans. Nav. **18**(1), 35–44 (2024). https://doi.org/10.12716/1001.18.01.02
11. Golui, S., Mahapatra, B.S., Mahapatra, G.S.: A new correlation-based measure on Fermatean fuzzy applied on multi-criteria decision making for electric vehicle selection. Expert Syst. Appl. **237**, 121605 (2024). https://doi.org/10.1016/j.eswa.2023.121605
12. Sahu, S., Shanker, S., Kamat, A., Barve, A.: India's public transportation system: the repercussions of COVID-19. Public Transport **15**(2), 435–478 (2023). https://doi.org/10.1007/s12469-023-00320-z
13. Durmaz, N., Budak, A.: Evaluation of the barriers to the use of sustainable transportation systems in city logistics with an integrated grey DEMATEL-ANP approach. In: Handbook of Research on Promoting Sustainable Public Transportation Strategies in Urban Environments, pp. 293–312 (2023). https://doi.org/10.4018/978-1-6684-5996-6.ch014
14. Karuppiah, K., Sankaranarayanan, B., Ali, S.M.: Exploring key enablers of sustainable transportation in small-and medium-sized manufacturing enterprises. Kybernetes **51**(11), 3394–3418 (2022). https://doi.org/10.1108/K-05-2021-0416

15. Demir, G.: Evaluation of sustainable green building indicators by fuzzy multi-criteria decision making. In: Sahoo, L., Senapati, T., Yager, R.R. (eds). Real Life Applications of Multiple Criteria Decision-Making Techniques in Fuzzy Domain. Studies in Fuzziness and Soft Computing, vol. 420. Springer, Singapore (2023). https://doi.org/10.1007/978-981-19-4929-6_16
16. Hussain, A., Ullah, K., Senapati, T., Moslem, S.: Complex spherical fuzzy Aczel Alsina aggregation operators and their application in assessment of electric cars. Heliyon **9**(7) (2023). https://doi.org/10.1016/j.heliyon.2023.e18100
17. Kakati, P., Senapati, T., Moslem, S., Pilla, F.: Fermatean fuzzy Archimedean Heronian Mean-Based Model for estimating sustainable urban transport solutions. Eng. Appl. Artif. Intell. **127**(Part B), 107349 (2024). https://doi.org/10.1016/j.engappai.2023.107349
18. Moslem, S.: A novel parsimonious spherical fuzzy analytic hierarchy process for sustainable urban transport solutions. Eng. Appl. Artif. Intell.Artif. Intell. **128**, 107447 (2024). https://doi.org/10.1016/j.engappai.2023.107447
19. Sarkar, A., Moslem, S., Esztergár-Kiss, D., Akram, M., Jin, LeSheng, Senapati, T.: A hybrid approach based on dual hesitant q-rung orthopair fuzzy Frank power partitioned Heronian mean aggregation operators for estimating sustainable urban transport solutions. Eng. Appl. Artif. Intell.Artif. Intell. **124**, 106505 (2023). https://doi.org/10.1016/j.engappai.2023.106505
20. Psychogiou, E.-E., Tsoulfas, G.T.: Critical factors affecting trust in the wine supply chain in Greece: A grey DEMATEL spproach. Logistics **8**(1), 24 (2024). https://doi.org/10.3390/logistics8010024
21. Chen, W., Tian, Y., Wang, Y., Yan, H., Wang, Y.: Cause identification and coupling relationship analysis of urban problems: A case study of poor parking convenience. Buildings **14**(2), 516 (2024). https://doi.org/10.3390/buildings14020516
22. Zaman, S.I., Ali, M.R., Khan, S.A.: Exploring interrelationships among barriers and enablers of green procurement for a sustainable supply chain. Int. J. Procure. Manag. **19**(2), 226–252 (2024). https://doi.org/10.1504/IJPM.2024.136051
23. Govindan, K.: How digitalization transforms the traditional circular economy to a smart circular economy for achieving SDGs and net zero. Transp. Res. Part E: Logist. Transp. Rev. **177**, 103147 (2023). https://doi.org/10.1016/j.tre.2023.103147
24. Xing, M., Luo, X., Zan, Y., Jin, H., Luo, J, et al.: Identification of factors affecting hoisting quality of large offshore structures and analysis of their coupling relationship based on grey-DEMATEL-ISM-MICMAC. Ocean. Eng. **280**, 114805 (2023). https://doi.org/10.1016/j.oceaneng.2023.114805
25. Nazeri, A., Farahani, M., Kazemi, S.M.M., Safaie, N.: A combined approach to risk assessment in the sustainable supply chain. Int. J. Logist. Syst. Manag. **44**(3), 300–353 (2023). https://doi.org/10.1504/IJLSM.2023.129364
26. Sohrabi, R.: Agility factors in cold supply chain using GREY-DEMATEL and AHP. Int. J. Ind. Eng. **33**(2), 1–14 (2022). https://doi.org/10.22068/ijiepr.33.2.10
27. Menon, R.R., Ravi, V.: An analysis of barriers affecting implementation of sustainable supply chain management in electronics industry: a Grey-DEMATEL approach. J. Model. Manag. **17**(4), 1319–1350 (2022). https://doi.org/10.1108/JM2-02-2021-0042
28. Aslani, B., Rabiee, M., Tavana, M.: An integrated information fusion and grey multi-criteria decision-making framework for sustainable supplier selection. Int. J. Syst. Sci.: Oper. & Logist. **8**(4), 348–370 (2021). https://doi.org/10.1080/23302674.2020.1776414
29. Ulutaş, A., Balo, F., Sua, L., Demir, E., Topal, A., Jakovljević, V.: A new integrated grey MCDM model: Case of warehouse location selection. Facta Univ., Ser.: Mech. Eng. **19**(3), 515–535 (2021). https://doi.org/10.22190/FUME210424060U
30. Adali, E.A., Öztaş, G.Z., Öztaş, T., Tuş, A.: Assessment of European cities from a smartness perspective: An integrated grey MCDM approach. Sustain. Cities Soc. **84**, 104021 (2022). https://doi.org/10.1016/j.scs.2022.104021
31. Torkayesh, A.E., Zolfani, S.H., Kahvand, M., Khazaelpour, P.: Landfill location selection for healthcare waste of urban areas using hybrid BWM-grey MARCOS model based on GIS. Sustain. Cities Soc. **67**, 102712 (2021). https://doi.org/10.1016/j.scs.2021.102712

32. Bai, C., Sarkis, J.: Integrating sustainability into supplier selection with grey system and rough set methodologies. Int. J. Prod. Econ. **124**(1), 252–264 (2010). https://doi.org/10.1016/j.ijpe.2009.11.023
33. Xia, X., Govindan, K., Zhu, Q.: Analyzing internal barriers for automotive parts remanufacturers in China using grey-DEMATEL approach. J. Clean. Prod. **87**, 811–825 (2015). https://doi.org/10.1016/j.jclepro.2014.09.044
34. Bhatia, M.S., Srivastava, R.K.: Analysis of external barriers to remanufacturing using the grey-DEMATEL approach: An Indian perspective. Resour. Conserv. Recycl.. Conserv. Recycl. **136**, 79–87 (2018). https://doi.org/10.1016/j.resconrec.2018.03.021
35. Deng, J.L.: Control problems of grey systems. Syst. Control Lett. **1**(5), 288–294 (1982). https://doi.org/10.1016/S0167-6911(82)80025-X
36. Deng, J.L.: Introduction to grey system theory. J. Grey Syst. **1**(1), 1–24 (1989)
37. Bai, C., Sarkis, J.: Grey-based DEMATEL model for evaluating business process management critical success factors. Int. J. Prod. Econ. **146**(1), 281–292 (2013). https://doi.org/10.1016/j.ijpe.2013.07.011
38. Gupta, H., Barua, M.K.: A grey DEMATEL-based approach for modeling enablers of green innovation in manufacturing organizations. Environ. Sci. Pollut. Res.Pollut. Res. **25**, 9556–9578 (2018). https://doi.org/10.1007/s11356-018-1261-6
39. Tseng, M.L.: A causal and effect decision making model of service quality expectation using grey-fuzzy DEMATEL approach. Expert Syst. Appl. **36**, 7738–7748 (2009). https://doi.org/10.1016/j.eswa.2008.09.011
40. Rajesh, R., Ravi, V.: Modeling enablers of supply chain risk mitigation in electronic supply chains: a Grey–DEMATEL approach. Comput. Ind. Eng.. Ind. Eng. **87**, 126–139 (2015). https://doi.org/10.1016/j.cie.2015.04.028
41. Rajesh, R., Ravi, V., Venkata, R.R.: Selection of risk mitigation strategy in electronic supply chains using grey theory and digraph-matrix approaches. Int. J. Prod. Res. **53**(1), 238–257 (2015). https://doi.org/10.1080/00207543.2014.948579

T-spherical Fuzzy Group Decision-Making Using Subjective and Objective Weights of Experts and Copula Aggregation Operators

Lavanya Golipally, Usha Rani Naathi, Bishnupada Debnath, and Abhijit Saha

Abstract T-spherical fuzzy (T-SF) sets deal with degrees of belongingness, abstinence, and non-belongingness which make them superior in comparison to fuzzy sets and intuitionistic fuzzy sets. In any group decision-making process, the experts' judgments include ambiguity and uncertainty, particularly in selecting alternatives based on a given collection of criteria. The existing research on T-SF sets do not focus on both the subjective and objective weights. Consequently, the outcomes get distorted. To tackle this situation, in this work, we utilize the concepts of consistency and similarity between the decision experts so as to determine the decision experts' subjective and objective weights, respectively, under the T-SF environment. For the purpose of aggregation, T-SF weighted Copula aggregation operators are developed to avoid loss of information. We discuss the elegant properties of these proposed aggregation operators. We provide a case study regarding open-source software LMS selection to focus on the practicability and usefulness of the proposed approach.

Keywords T-spherical fuzzy sets · Copula weighted aggregation · Group decision-making · Open-source software LMS selection

---

L. Golipally · U. R. Naathi
Department of Artificial Intelligence and Data Science, B V Raju Institute of Technology, Narsapur, Telangana 502313, India

B. Debnath
Department of Mathematics, Rajiv Gandhi University, Doimukh, Papumpare, Arunachal Pradesh 791112, India
e-mail: bishnupada.debnath@rgu.ac.in

A. Saha (✉)
Department of Computing Technologies, SRM Institute of Science and Technology (SRMIST), Kattankulathur, Tamil Nadu 603203, India
e-mail: abhijit84.math@gmail.com

L. Sahoo et al. (eds.), *Decision Making Under Uncertainty Via Optimization, Modelling, and Analysis*, Studies in Systems, Decision and Control 558,
https://doi.org/10.1007/978-981-96-0085-4_28

1 Introduction

The concept of fuzzy set (FS) (Zadeh, 1965) originated because of the inclusion of vague human appraisals when addressing real-world issues. Intuitionistic fuzzy set (IFS) [1], an extension of FS, contains belongingness and non-belongingness degrees (MD & NBD) for handling imprecision. Consequently, IFSs are more potent and competent in solving uncertain MCDM issues than FSs. Decision-makers are constrained by Atanassov's IFS in two ways. Firstly, it limits the total of MD & NBDup to a unit interval, making it impossible for decision-makers to award these grades on their own. Secondly, it does not work well for issues when opinions have both a degree of abstinence and a degree of rejection rather than being of a yes or no type. Cuong [2] addressed this problem by introducing the picture fuzzy set (PFS), a generalized model in FS theory that describes four features of an imprecise event. Human opinion can be expressed in decision-making as favor, abstinence, disfavor, and refusal degree, which are represented by four PFS membership functions. Similar to Atanassov's IFS, Cuong's PFS limits the decision maker from establishing the degrees by their own interest in some way by stating that the sum of the three degrees cannot exceed 1. Mahmood et al. (2018) established the framework of the spherical fuzzy set (SFS) by extending PFS and presented a remedy for the limiting structure of PFS. The notion of SFS is expanded to encompass the T-spherical fuzzy set (T-SFS) by incorporating a parameter T, which provides decision-makers the flexibility to select membership grade values from any point within the interval [0, 1]. The advanced tool that can accurately and flexibly express human opinion regarding any imprecise occurrence is a T-SFS with no limitations. Ullah et al. [3] defined T-SF weighted aggregation operators (AOs). A few T-SF-improved interactive AOs were proposed by Garg et al. [4]. Liu et al. [5] addressed an MCDM issue by defining the T-SF-Muirhead mean AOs. In MCDM problems, the applicability of various AOs is discussed. For T-SFS, the Hamacher AOs were defined by Ullah et al. [6]. Einstein hybrid AOs based on T-SFSs were employed by Munir et al. [7] to solve difficulties in MCDM problems. Ullah et al. [6] used the T-SF Hamacher AOs to accomplish search and rescue tasks using robots. The T-SF power AOs were defined by Garg et al. [8] and used to solve MCDM problems. Garg et al. [9] introduced an approach based on the T-SF Maclaurin symmetric mean operator. Weighted spatial T-SF methods (c-means algorithm) were proposed by Xian et al [10] for image segmentation. Some information measures with T-SFSs were introduced by Shen et al. [11] and used in pattern recognition. Mahmood et al. [12] combined Aczel-Alsina AOs, bipolar fuzzy sets, and complex fuzzy sets to deal with multiple attribute decision-making issues. Jamil and Riaz [13] developed cubic bipolar fuzzy AOs with priority degrees. Alghazzawi et al. [14] utilized bipolar fuzzy AOs for emerging technology selection.

1.1 Research Gaps and Our Motivation

In the existing information aggregation methods [4–9] based on T-SFSs, experts' assessments were carried out independently making it difficult to draw a consistent conclusion. Specifically, these existing methods [4–9] fail to depict the uncertainty of expressing information with the interaction among experts. Accordingly, the assessment results get distorted.

To combine all the input data into a single entity, aggregation operators (AOs) are very useful and so they have been efficaciously employed for data processing, specifically decision-making, shape understanding, data analytics, and neural networks. Although aggregation operators like algebraic AOs (Mahmmod et al., 2018), improved interactive AOs [4], Einstein AOs [7], Hamachar AOs [6], power AOs [8] etc. were utilized for aggregating T-SF information, a combination of Copula [15], [16, 17] and T-SFSs hasn't yet been used for aggregation.

1.2 Contributions

This study contributes significantly by addressing these gaps in the literature through the following key contributions.

- The subjective weights of experts are calculated based on the concept of consistency harmonious weight index (CHWI), and the objective weights of the experts are calculated using the assessment similarity of experts.
- T-SF Copula weighted averaging (T-SFCWA) and T-SF Copula weighted geometric (T-SFCWG) AOs are developed.
- A case study of the selection of open-source software LMS packages is provided.

1.3 Organization of the Work

The structure of this paper is as follows: A summary of basic and relevant concepts is given in Sect. 2. The definitions and characteristics of Copula operations between T-SF elements are covered in Sect. 3, which also introduces the T-SFCWA and T-SFCWG AOs. The T-SF group decision-making scheme is furnished in Sect. 4. Section 5 describes the case study under investigation along with the solution. In Sect. 6, the conclusions are given.

2 Basic Concepts

Definition 1 (Mahmood et al., 2018): A T-Spherical fuzzy (T-SF) set Υ on a given set U is described by $\Upsilon = \{\langle u_i, a(u_i), b(u_i), c(u_i)\rangle | u_i \in U\}$, where $a, b, c : U \to [0, 1]$ are the belongingness degree (BD), abstinence degree (AD), and non-belongingness degree (NBD) respectively of $u_i \in U$, satisfying $0 \leq (a(u_i))^T + (b(u_i))^T + (c(u_i))^T \leq 1$ $(T \geq 1)$. An SVN set Υ with only one object converts to a T-SF number (T-SFN), and we write $\Upsilon = (a, b, c)$ or $\Upsilon =< a, b, c >$ for $a, b, c \in [0, 1]$ and $0 \leq a^T + b^T + c^T \leq 1$.

Definition 2 [9]: The accuracy value and score value of a T-SFN $\Upsilon =< a, b, c >$ are defined as

$$p(\Upsilon) = a^T + b^T + c^T \tag{1}$$

$$q(\Upsilon) = 0.5 \times (1 + a - b - c) \tag{2}$$

Clearly, $p(\Upsilon), q(\Upsilon) \in [0, 1]$. For the T-SFNs $\Upsilon_1 =< a_1, b_1, c_1 >$ and $\Upsilon_2 =< a_2, b_2, c_2 >$, a ranking rule is:
$\Upsilon_1 \succ \Upsilon_2$ if $q(\Upsilon_1) > q(\Upsilon_2)$; and $\Upsilon_1 \succ \Upsilon_2$ if $q(\Upsilon_1) = q(\Upsilon_2),\ p(\Upsilon_1) > p(\Upsilon_2)$.

Definition 3 Nelsen [15], Bagical et al., 2015; [16, 17]**:** A copula is a mapping $F : [0, 1] \times [0, 1] \to [0, 1]$ satisfying the following conditions:

i. $F(x, 0) = F(0, x) = 0,\ F(x, 1) = F(1, x) = x\ \forall x \in [0, 1]$
ii. $F(x_1, y_1) + F(x_2, y_2) - F(x_2, y_1) - F(x_1, y_2) \geq 0$, for $x_1, y_1, x_2, y_2 \in [0, 1]$ with $x_1 \leq x_2$ and $y_1 \leq y_2$.

Definition 4 Nelsen [15], Bagical et al., 2015; [16, 17]**:** An Archimedean copula is a mapping $F : [0, 1] \times [0, 1] \to [0, 1]$ defined by $F(x, y) = B(A(x) + A(y))$ where $A : [0, 1] \to [0, \infty)$ is a strictly decreasing function and $B : [0, \infty) \to [0, 1]$ is given as: $B(x) = \begin{cases} A^{-1}(x), & x \in [0, A(0)] \\ 0, & x \in [A(0), \infty] \end{cases}$

An Archimedean copula is termed as a strict Archimedean copula if F is strictly increasing on $[0, 1] \times [0, 1]$ and B becomes identical with A. In such a scenario $F(x, y) = A^{-1}(A(x) + A(y))$.

3 *T*-SF Copula Operations and Associated Aggregation Operators

Here, we present Copula operations on T-SF elements as well as T-SF Copula weighted averaging (T-SFCWA) and T-SF Copula weighted geometric (T-SFCWG) aggregation operators (AOs). Throughout this section, we shall assume that $\Upsilon_m =< a_m, b_m, c_m >$ $(m = 1, 2, ..., K)$ is a collection of T-SFNs.

Definition 5: For $\theta, \theta_1, \theta_2 > 0$, the Copula (Archimedean Copula) operations among *T*-SF numbers are

1. $$\Upsilon_1 \oplus \Upsilon_2 = \left\langle \sqrt[T]{1 - A^{-1}(A(1-a_1^T) + A(1-a_2^T))}, \sqrt[T]{A^{-1}(A(b_1^T) + A(b_2^T))}, \sqrt[T]{A^{-1}(A(c_1^T) + A(c_2^T))} \right\rangle.$$
2. $$\Upsilon_1 \otimes \Upsilon_2 = \left\langle \sqrt[T]{A^{-1}(A(a_1^T) + A(a_2^T))}, \sqrt[T]{1 - A^{-1}(A(1-b_1^T) + A(1-b_2^T))}, \sqrt[T]{1 - A^{-1}(A(1-c_1^T) + A(1-c_2^T))} \right\rangle.$$
3. $$\theta\Upsilon_1 = \left\langle \sqrt[T]{1 - A^{-1}(\theta A(1-a_1^T))}, \sqrt[T]{A^{-1}(\theta A(b_1^T))}, \sqrt[T]{A^{-1}(\theta A(c_1^T))} \right\rangle.$$
4. $$\Upsilon_1^\theta = \left\langle \sqrt[T]{A^{-1}(\theta A(a_1^T))}, \sqrt[T]{1 - A^{-1}(\theta A(1-b_1^T))}, \sqrt[T]{1 - A^{-1}(\theta A(1-c_1^T))} \right\rangle.$$

Theorem 1: For $\theta, \theta_1, \theta_2 > 0$,

1. $\Upsilon_1 \oplus \Upsilon_2 = \Upsilon_2 \oplus \Upsilon_1$
2. $\Upsilon_1 \otimes \Upsilon_2 = \Upsilon_2 \otimes \Upsilon_1$
3. $\theta(\Upsilon_1 \oplus \Upsilon_2) = (\theta\Upsilon_1) \oplus (\theta\Upsilon_2)$
4. $(\Upsilon_1 \otimes \Upsilon_2)^\theta = (\Upsilon_1^\theta) \otimes (\Upsilon_2^\theta)$
5. $(\theta_1 + \theta_2)\Upsilon_1 = (\theta_1\Upsilon_1) \oplus (\theta_2\Upsilon_1)$
6. $\Upsilon_1^{\theta_1+\theta_2} = (\Upsilon_1^{\theta_1}) \oplus (\Upsilon_1^{\theta_2})$

Proof: Follows from Definition 5.

Definition 6: Suppose β_m is the weights of Υ_m with the condition $\beta_m \in [0,\ 1]$ and $\sum_{m=1}^{K} \beta_m = 1$. Then we define the *T*-SFCWA operator as

$$T - SFCWA(\Upsilon_1, \Upsilon_2,\ \ldots,\ \Upsilon_K) = \mathop{\oplus}_{m=1}^{K} (\beta_m \Upsilon_m) \tag{3}$$

Theorem 2: The aggregated value $T - SFCWA(\Upsilon_1, \Upsilon_2,\ \ldots,\ \Upsilon_K)$ is also an*T*-SFN. In addition, we get

$$T - SFCWA(\Upsilon_1, \Upsilon_2,\ \ldots,\ \Upsilon_K) = \left\langle \sqrt[T]{1 - A^{-1}(\sum_{m=1}^{K} \theta_m A(1-a_m^T))}, \sqrt[T]{A^{-1}(\sum_{m=1}^{K} \theta_m A(b_m^T))}, \sqrt[T]{A^{-1}(\sum_{m=1}^{K} \theta_m A(c_m^T))} \right\rangle \tag{4}$$

Proof: Follows from Definition 5 and Definition 6.

Below we furnish a few properties of the *T*-SFCWA operator.

Theorem 3: If $\Upsilon_0 =< a_0, b_0, c_0 >= \Upsilon_m \forall m$, then $T - SFCWA(\Upsilon_1, \Upsilon_2,\ \ldots,\ \Upsilon_K) = \Upsilon_0$.

Theorem 4: $\Upsilon^- \prec T - SFCWA(\Upsilon_1, \Upsilon_2,\ \ldots,\ \Upsilon_K) \prec \Upsilon^+$ where $\Upsilon^- = \left\langle \min_m a_m, \max_m b_m, \max_m c_m \right\rangle$ and $\Upsilon^+ = \left\langle \max_m a_m, \min_m b_m, \min_m c_m \right\rangle$.

Theorem 5: If $\Upsilon'_m =< a'_m, b'_m, c'_m > \ (m = 1, 2, ..., K)$ be another collection of T-SFNs such that $a_m \leq a'_m, b_m \geq b'_m, c_m \geq c'_m \forall m$, then $T - SFCWA(\Upsilon_1, \Upsilon_2, ..., \Upsilon_K) \prec T - SFCWA(\Upsilon'_1, \Upsilon'_2, ..., \Upsilon'_K)$.

Definition 7: Suppose β_m is the weights of Υ_m with the condition $\beta_m \in [0, 1]$ and $\sum_{m=1}^{K} \beta_m = 1$. Then we define the T-SFCWA operator as

$$T - SFCWA(\Upsilon_1, \Upsilon_2, ..., \Upsilon_K) = \bigoplus_{m=1}^{K} (\beta_m \Upsilon_m) \tag{5}$$

Theorem 6: The aggregated value $T - SFCWA(\Upsilon_1, \Upsilon_2, ..., \Upsilon_K)$ is also a T-SFN. In addition, we get

$$T - SFCWG(\Upsilon_1, \Upsilon_2, ..., \Upsilon_K) = \left\langle \sqrt[T]{A^{-1}(\sum_{m=1}^{K} \theta_m A(a_m^T))}, \sqrt[T]{1 - A^{-1}(\sum_{m=1}^{K} \theta_m A(1 - b_m^T))}, \sqrt[T]{1 - A^{-1}(\sum_{m=1}^{K} \theta_m A(1 - c_m^T))} \right\rangle \tag{6}$$

Proof: Follows from Definition 5 and Definition 7.

Below we furnish a few properties of the T-SFCWG operator.

Theorem 7: If $\Upsilon_0 =< a_0, b_0, c_0 >= \Upsilon_m \forall m$, then $T - SFCWG(\Upsilon_1, \Upsilon_2, ..., \Upsilon_K) = \Upsilon_0$.

Theorem 8: $\Upsilon^- \prec T - SFCWG(\Upsilon_1, \Upsilon_2, ..., \Upsilon_K) \prec \Upsilon^+$ where $\Upsilon^- = \left\langle \min_m a_m, \max_m b_m, \max_m c_m \right\rangle$ and $\Upsilon^+ = \left\langle \max_m a_m, \min_m b_m, \min_m c_m \right\rangle$.

Theorem 9: If $\Upsilon'_m =< a'_m, b'_m, c'_m > \ (m = 1, 2, ..., K)$ be another collection of T-SFNs such that $a_m \leq a'_m, b_m \geq b'_m, c_m \geq c'_m \forall m$, then $T - SFCWG(\Upsilon_1, \Upsilon_2, ..., \Upsilon_K) \prec T - SFCWG(\Upsilon'_1, \Upsilon'_2, ..., \Upsilon'_K)$.

4 Group decision-making methodology

The steps of the proposed methodology are as follows:

Step 1: Construct the primary assessment matrices whose elements are T-SFNs.

Assume that the alternatives X_i $(i = 1, 2, ..., P)$ are under evaluation of the experts E_r $(r = 1, 2, ..., L)$ where each alternative is assessed against the criteria V_j $(j = 1, 2, ..., Q)$ under the T-SF environment. Suppose the primary assessment matrices are: $M_r = \left[\Upsilon_{rij}\right]_{P \times Q} = [< a_{rij}, b_{rij}, c_{rij} >]_{P \times Q}$.

Step 2: Compute experts' weights.

Suppose $W_r =$ weight of E_r,

$W_r' =$ subjective weight of E_r,
$W_r'' =$ objective weight of E_r.

The consistency of $M_r = \left[\Upsilon_{rij}\right]_{P\times Q}$ as well as the resemblance between the ambiguous experts' E_r and E_s $(s \neq r)$ are expressed by W_r. W_r' reflects the consistency harmonious weight index (CHWI) $(\varphi(E_r))$, and is displayed as

$$\varphi(E_r) = \sum_{j=1}^{Q} Q \bigg/ \left(\sum_{i=1}^{p}\sum_{t=1}^{p}(q(\Upsilon_{rij}) \times q(\Upsilon_{rtj}))\right) \tag{7}$$

Using the consistency [9] of M_r, W_r' is represented by

$$W_r' = \varphi(E_r) \bigg/ \sum_{r=1}^{L}\varphi(E_r)\ (r = 1, 2, ..., L) \tag{8}$$

In the event that the closeness between M_r and M_s $(r \neq s)$ is high, the impact of M_r is more important.

Let η_{rs} be the angle between M_r and M_s. Then ρ_{rs} can be computed by

$$\eta_{rs} = \frac{\langle\delta(M_r),\ \delta(M_s)\rangle}{\|\delta(M_r)\|\ \times\ \|\delta(M_s)\|} \tag{9}$$

where $\delta(M_r) =$ derived vector of $M_r = \left(\Upsilon_{r11}, \Upsilon_{r21}, ..., \Upsilon_{rP1}, \Upsilon_{r21}, \Upsilon_{r22}, ..., \Upsilon_{rP2}, ..., \Upsilon_{r1Q}, \Upsilon_{r2Q}, ..., \Upsilon_{rPQ}\right)$ and $\langle\delta(M_r),\ \delta(M_s)\rangle = \sum_{i=1}^{P}\sum_{j=1}^{Q}(q(\Upsilon_{rij}) + q(\Upsilon_{sij}))(\ \|\,.\,\|$ being the positive square root of $\langle.\,,\ .\rangle$.

Obviously, $0 \leq \eta_{rs} \leq 1\ \forall\, r, s$. η_{rs} expresses two types of similarities, namely (i) similarity between M_r and M_s and (ii) similarity between $\delta(M_r)$ *and* $\delta(M_s)$.

$$\text{Suppose } \eta_r = \sum_{s=1, s\neq r}^{l} \eta_{rs} \tag{10}$$

The similarity among M_r and M_s is expressed by η_r. Then η_r is normalized (Eq. (11)) to obtain the objective weight of each criterion.

$$W_r'' = \frac{\eta_r}{\sum_{r=1}^{l}\eta_r} \tag{11}$$

Using a linear combination of ϖ_r' and W_r'', weights of experts are computed as:

$$W_r = \gamma W_r' + (1-\gamma)W_r'' \tag{12}$$

γ determines which weight is dominant in Eq. (12). If γ is high, experts tend to favor the subjective weights; experts like the objective weights when is γ is small, experts prefer the objective weights. The risk attitude of specialists is represented by the parameter γÎ [0, 1].

Step 3: Obtain the aggregated T-SF decision matrix (A-T-SFDM).

The A-T-SFDM is $\left[\Upsilon_{ij}\right]_{P\times Q} = [< a_{ij}, b_{ij}, c_{ij} >]_{P\times Q}$ where:

$$\begin{aligned} \Upsilon_{ij} &= T - SFCWA(\Upsilon_{1ij}, \Upsilon_{2ij}, \ldots, \Upsilon_{Kij}) = \mathop{\oplus}_{r=1}^{L} (W_r \Upsilon_{rij}) \\ \text{or } \Upsilon_{ij} &= T - SFCWG(\Upsilon_{1ij}, \Upsilon_{2ij}, \ldots, \Upsilon_{Kij}) = \mathop{\otimes}_{r=1}^{L} (\Upsilon_{rij})^{W_r} \end{aligned} \quad (13)$$

Step 4: Normalize the A-T-SFDM.

Suppose that the normalized A-T-SFDM is $\left[\tilde{\Upsilon}_{ij}\right]_{P\times Q} = [< \tilde{a}_{ij}, \tilde{b}_{ij}, \tilde{c}_{ij} >]_{P\times Q}$, where

$$\tilde{\Upsilon}_{ij} = \begin{cases} < a_{ij}, b_{ij}, c_{ij} >, & \text{if } C_j \text{ is beneficial} \\ < c_{ij}, 1 - b_{ij}, a_{ij} >, & \text{if } C_j \text{ is non - beneficial} \end{cases} \quad (14)$$

Step 5: Obtain the final A-T-SFDM.

The final A-T-SFDM is $[\Upsilon_i]_{P\times Q} = [< a_i, b_i, c_i >]_{P\times Q}$ where

$$\begin{aligned} \Upsilon_i &= T - SFCWA(\tilde{\Upsilon}_{i1}, \tilde{\Upsilon}_{i2}, \ldots, \tilde{\Upsilon}_{iQ}) = \mathop{\oplus}_{j=1}^{Q} (w_j \tilde{\Upsilon}_{ij}) \\ \text{or } \Upsilon_i &= T - SFCWG(\tilde{\Upsilon}_{i1}, \tilde{\Upsilon}_{i2}, \ldots, \tilde{\Upsilon}_{iQ}) = \mathop{\otimes}_{j=1}^{Q} (\tilde{\Upsilon}_{ij})^{w_j} \end{aligned} \quad (15)$$

Step 6: Generate the ranking order of alternatives.

The ranking order can be generated based on the scores $q(\Upsilon_i)$ $(i = 1, 2, \ldots, P)$.

5 Case Study and Solution

The performance of an organization is negatively affected if the open-source software learning management system (OSS-LMS) packages are not properly assessed before using them. A tough task for a client is to pick the best one among the OSS-LMS packages that are found in the market. A selection of the best OSS-LMS packages can be regarded as a group decision-making problem. Suppose, an educational institute wants to choose an efficient OSS-LMS package out of six OSS-LMS packages, namely- Moodle (X_1), Sakai (X_2), eFront (X_3), ATutor (X_4), Canvus (X_5) and Dokeos (X_6). For the purpose of identifying and evaluating agile and sustainability criteria, a committee of three experts was constituted. Through an extensive examination of the literature using the Scopus and Web of Science databases, the committee has selected 6 criteria, namely: security and privacy (V_1), e-learning standards (V_2), reliability (V_3), usability (V_4), functionality (V_5), and maintainability (V_6). The initial assessments (in terms of the linguistic terms *VVH*: Very very high ($<0.9, 0.4, 0.5>$), *VH*: Very high ($<0.8, 0.5, 0.5>$), *H*: High ($<0.6, 0.2, 0.4>$), *M*: Medium ($<0.5, 0.3, 0.4>$), *L*: Low ($<0.3, 0.2, 0.6>$)) of the three experts are shown in Table1.

Using Eqs. (7) and (8), we calculate the expert's subjective weights. We calculate the experts' objective weights using Eqs. (9), (10) and (11). Equation (12) determines the total weight (for $\gamma = 0.5$) of each expert. All these are summarized in Table 2. Now, utilizing Eq. (13) we obtain aggregated T-SF matrix (Table 3) (using $T-SFCWA$ operator).

Since all attributes are of benefit type, normalization is not applicable here. Suppose that the criteria weights are 0.25, 0.20, 0.15, 0.14, 0.12, and 0.14, respectively. Then the final aggregated T-SFNs obtained by utilizing Eq. (15) ($T-SFCWA$ operator) are as follows:

$\Upsilon_1 = <0.7628, 0.2443, 0.4362>$, $\Upsilon_2 = <0.7752, 0.2848, 0.4440>$, $\Upsilon_3 = <0.7518, 0.2530, 0.4412>$,

$\Upsilon_4 = <0.7295, 0.2665, 0.4311>$, $\Upsilon_5 = <0.7332, 0.2434, 0.4141>$, $\Upsilon_6 = <0.7818, 0.2776, 0.4436>$.

The scores of the final aggregated T-SFNs (obtained by using $T-SFCWA$ operator) are, respectively, 0.5412, 0.5232, 0.5288, 0.5160, 0.5378, and 0.5303. Therefore, the ranking order is: $X_1 \succ X_5 \succ X_6 \succ X_3 \succ X_2 \succ X_4$ and so "Moddle" is the most suitable alternative.

6 Conclusions

In this work, we have defined Copula operations between T-SF numbers based on which T-SF Copula weighted aggregation operators have been developed to tackle the information loss due to aggregation. All essential features of these proposed operators have been presented. Then, the T-SF group decision-making scheme was furnished where the concepts of consistency and similarity between the decision experts so as to determine the decision experts' subjective and objective weights, respectively, under the T-SF environment. Next, a case study of open-source software LMS selection has been introduced. Outcomes demonstrate that the developed model can effectively address challenges arising in scenarios involving uncertainty based on T-SF sets.

The drawback of the proposed operators lies in their inability to manage dependencies among multi-inputs. In future applications, a potential enhancement involves

Table 1 Primary assessment results

	E1						E2						E3					
	V1	V2	V3	V4	V5	V6	V1	V2	V3	V4	V5	V6	V1	V2	V3	V4	V5	V6
X1	VVH	H	VH	M	H	H	H	VH	L	H	M	VH	VH	VVH	H	H	M	H
X2	VH	H	L	M	VVH	M	VVH	M	H	VH	M	H	VH	M	VH	VVH	M	VH
X3	H	M	VH	VVH	M	VH	VVH	M	L	VH	M	M	H	H	L	M	L	VVH
X4	VH	M	L	VH	H	M	M	H	VVH	M	H	VH	M	VH	M	H	H	VVH
X5	M	VVH	H	M	H	H	H	M	M	H	H	VVH	M	H	VVH	H	VH	M
X6	M	H	VH	VH	M	VVH	VH	VVH	L	M	H	VH	VVH	M	VH	H	M	H

Table 2 Weights of experts

	Subjective	Objective	Final
E1	0.3345	0.3335	0.334
E2	0.3402	0.3327	0.336
E3	0.3253	0.3338	0.330

Table 3 Aggregated T-SF matrix

	V1	V2	V3	V4	V5	V6
X1	< 0.8302, 0.2717, 0.4558 >	< 0.8297, 0.2722, 0.4560 >	< 0.6773, 0.2266, 0.4741 >	< 0.5726, 0.2187, 0.4000 >	< 0.5402, 0.2469, 0.4000 >	< 0.7039, 0.2268, 0.4246 >
X2	< 0.8506, 0.4558, 0.5000 >	< 0.5402, 0.2469, 0.4000 >	< 0.6764, 0.2262, 0.4732 >	< 0.8257, 0.3671, 0.4560 >	< 0.7929, 0.3222, 0.4244 >	< 0.6891, 0.2598, 0.4241 >
X3	< 0.8046, 0.2246, 0.4246 >	< 0.5398, 0.2474, 0.4000 >	< 0.6459, 0.2266, 0.5580 >	< 0.8266, 0.3676, 0.4564 >	< 0.4562, 0.2474, 0.4368 >	< 0.8255, 0.3667, 0.4558 >
X4	< 0.6755, 0.3320, 0.4244 >	< 0.6891, 0.2598, 0.4241 >	< 0.7883, 0.2565, 0.4740 >	< 0.6900, 0.2608, 0.4244 >	< 0.6000, 0.2000, 0.4000 >	< 0.8257, 0.3671, 0.4560 >
X5	< 0.5404, 0.2467, 0.4000 >	< 0.7985, 0.2570, 0.4244 >	< 0.7973, 0.2564, 0.4241 >	< 0.5726, 0.2187, 0.4000 >	< 0.7025, 0.2262, 0.4241 >	< 0.7992, 0.2565, 0.4246 >
X6	< 0.8257, 0.3671, 0.4560 >	< 0.7992, 0.2565, 0.4246 >	< 0.7452, 0.2765, 0.5261 >	< 0.6900, 0.2608, 0.4244 >	< 0.5404, 0.2467, 0.4000 >	< 0.8306, 0.2730, 0.4564 >

merging the Copula operator with operators like Maclaurin symmetric mean, Heronian mean, Hamy mean, or Bonferroni mean to address this limitation. Additionally, a dispersion measure or maximum deviation measure based optimization model can be developed to generate criteria weights.

References

1. Atanassov, K.T.: Intuitionistic fuzzy sets. Fuzzy Sets and Syst **20**, 87–96 (1986)
2. Cuong, B.C.: Picture fuzzy sets. JComput Sci Cybern **30**, 409 (2014)
3. Ullah, K., Mahmood, T., Jan, N.: Some averaging aggregation operators for T-spherical fuzzy sets and their applications in multi-attribute decision making. In: Proceedings of the International Conference on Soft Computing and Machine Learning (ICSCML), Wuhan, China, 26–28 (2019)
4. Garg, H., MunirM, U.K., Mahmood, T., Jan, N.: Algorithm for T-spherical fuzzy multi-attribute decision making based on improved interactive aggregation operators. Symmetry **10**, 670 (2018)

5. Liu, P., Khan, Q., Mahmood, T., Hassan, N.: T-spherical fuzzy power Muirhead mean operator based on novel operational laws and their applications in multi-attribute group decision making. IEEE Access **7**, 22613–22632 (2019)
6. Ullah, K., Mahmood, T., Garg, H.: Evaluation of the performance of search and rescue robots using T-spherical fuzzyHamacher aggregation operators. Int J of Fuzzy Syst **22**(2), 570–582 (2020)
7. Munir, M., Kalsoom, H., Ullah, K., Mahmood, T., Chu, Y.M.: T-spherical fuzzy einstein hybrid aggregation operators and their applications in multi-attribute decision making problems. Symmetry **12**, 365 (2020). https://doi.org/10.3390/sym12030365
8. Garg, H., Ullah, K., Mahmood, T., Hassan, N., Jan, N. (2021) T-spherical fuzzy power aggregation operators and their applications in multi-attribute decision making. J. Ambi. Intelland. Humani. Comput. 12:9067–9080
9. Garg, H., Ahmad, A., Ullah, K., Mahmood, T., Ali, Z.: Algorithm for multiple attribute decision-making using T-spherical fuzzy Maclaurin symmetric mean operator. Iranian J Fuzzy Syst **19**(6), 111–124 (2022)
10. Xian, S., Cheng, Y., Chen, K.: A novel weighted spatial T-spherical fuzzy C-means algorithms with bias correction for image segmentation. Int J of Intell Syst **37**(2), 1239–1272 (2022)
11. Shen, X., Sakhi, S., Ullah, K., Abid, M.N., Jin, Y.: Information measures based on T-spherical fuzzy sets and their applications in decision making and pattern recognition. Axioms **11**(7), 302 (2022)
12. Mahmood, T., Rehman, U., Ali, Z.: Analysis and application of Aczel-Alsina aggregation operators based on bipolar complex fuzzy information in multiple attribute decision making. Inf. Sci. **619**, 817–833 (2023)
13. Jamil, N., Riaz, M.: Cubic bipolar fuzzy aggregation operator with priority degree with multi-criteria decision-making. Scientia Iranica **31**(5), 388–416 (2024)
14. Alghazzawi, D., Abbas, S., Alolaiyan, H., Khalifa, H.A.W., Alburaikan, A., Xin, Q., Razaq, A.: Dynamic bipolar fuzzy aggregation operators: A novel approach for emerging technology selection in enterprise integration. AIMS Mathematics **9**(3), 5407–5430 (2024)
15. Nelsen RB. An introduction to copula. Springer Science and Business Media (2013)
16. Han, B., Tao, Z., Chen, H., Zhou, L., Liu, J.: A new computational model based on Archimedean copula for probabilistic unbalanced linguistic term set and its application to multi-attribute group decision-making **140**, 106264 (2020)
17. Qiyas, M., Abdullah, S., Khan, F., Naeem, M.: Banzhaf-Choquet Copula based aggregation operators for managing fractional orthotriple fuzzy information. Alex. Eng. J. **61**(6), 4659–4677 (2022)
18. Bacigal, T., Mesiar, R., Najjari, V.: Generators of copulas and aggregation. Inf. Sci. **306**, 81–87 (2015)
19. Baidya, J., Garg, H., Saha, A., Mishra, A.R., Rani, P., Dutta, D.: Selection of third party reverses logistic providers: an approach of BCF-CRITIC-MULTIMOORA using Archimedean power aggregation operators. Complex and Intell Syst (2021). https://doi.org/10.1007/s40747-021-00413-x
20. Zadeh, L.A.: Fuzzy sets. Inf. Cont. **8**(3), 338–353 (1985)

Effect of Internal Delay on the Dynamics of a Mean-Field Diffusive Coupled Oscillating System

Saumendra Sankar De Sarkar and Saumen Chakraborty

Abstract Coupled oscillators are being studied extensively to explore different dynamical phenomena in nature and society. Time delays, namely processing delay and propagation delay play a vital role in influencing the response of a coupled oscillating system. Delay, if not negligible, is to be considered in modelling a system to replicate the system dynamics as obtained in the real world. It has a significant contribution to collective behaviour if the number of interacting components is large in the network. This chapter deals with the dynamics of a mean-field diffusive coupled Van der Pol oscillator-based system with an inherent delay in each oscillator unit. The component oscillators become unstable through inhomogeneous limit cycle (IHLC) and quasi-periodic (QP) oscillations. Analytical studies and numerical simulation have been presented to explore the coupled system response with different values of delay, diffusive coupling strength as well as number of oscillators. With the increase in delay time, coupling strength or the number of oscillators in the coupled system, the complexity of the system response is increased and the system gradually becomes unstable.

Keywords Van der Pol oscillator · Coupled oscillator · Oscillation quenching

1 Introduction

A broad category of systems in natural science, engineering and social science are well described by interacting oscillatory processes. In this respect, studies on the collective behaviour of coupled oscillators have emerged as a very important approach for modelling and predicting the behaviour of such systems. Extensive

S. S. De Sarkar (✉)
Department of Physics, Raniganj Girls' College, Searsole Rajbari, West Bengal 713358, India
e-mail: ssdesarkar@gmail.com

S. Chakraborty
Department of Physics, Bidhan Chandra College, Asansol, West Bengal 713304, India
e-mail: saumenbcc@gmail.com

L. Sahoo et al. (eds.), *Decision Making Under Uncertainty Via Optimization, Modelling, and Analysis*, Studies in Systems, Decision and Control 558,
https://doi.org/10.1007/978-981-96-0085-4_29

studies on the dynamics of a network of oscillators with various oscillatory dynamics (oscillator type) and coupling schemes are found in the literature [1–12]. From the practical point of view, a time delay between the interacting units, comprising the coupled system, arises due to the propagation of signals with a limited speed. This is known as propagation delay or transmission delay. On the other hand, a finite response time is required by the internal subsystem of each unit to sense the received information and then react accordingly. This is recognized as a processing delay. Time delays typically mean a combination of these two delays.

Processing time delay is a significant factor in biological systems, such as the Mackey–Glass equation for blood cell production [13] and protein synthesis due to sequential production [14] to name a few. The transmission of information between the neurons in a neural network is not also instantaneous [15]. Time delays are also evident in population dynamics [16], enzyme-catalyzed reactions [17], laser physics [18], climate systems [19], human behaviour [20] etc. They also influence the economic and financial systems controlling asset prices [21] and market stability [22]. Engineering studies also show physical delays in the mixing processes of liquid or gaseous fluids and chemical reactors [23], highlighting the importance of considering this delay time in system modelling. Therefore, the time delay is crucial in system modelling to describe real-world systems.

In this work, we try to address the possible effect of an inherent delay in each component oscillator as well as the number of oscillators on the evolution of the network dynamics. Delays may be neglected if they have no appreciable effects on the design and analysis of a dynamical system. Otherwise, delays have to be incorporated explicitly in the system model, according to the underlying mechanism of the intermediate steps, towards a better description of the system dynamics. The effects of time delay in network dynamics are still a subject of immense interest to researchers [24–33]. For the present study, we choose oscillatory systems governed by Van der Pol (VdP) equations and coupled by mean field diffusion (MFD). The delay term has been introduced in the mathematical equation representing a VdP oscillator (VdPO) to achieve our goal. The choice of the VdP equation is obvious as it is considered one of the most important models for several oscillatory processes in nature and society. VdPOs are used to model various natural and engineering systems such as the conduction function of the human heart [34], biological [35] and chemical oscillations [36], action potentials of neurons [37] and semiconductor lasers [38]. It has been studied in detail in the context of chaos [39, 40] also. Coupled VdPOs attract the interest of the researchers to the plethora of complex collective behaviours like Phase locking, synchronization, quenching of oscillation and chimera [41–51]. Among several coupling schemes, MFD coupling is one of the very popular schemes in studying such coupled systems. A wide variety of complex and rich behaviour in MFD-coupled systems are well reported [52–60]. Motivated by these practical aspects, we have introduced delay in each oscillator model in a network of MFD-coupled VdPOs and studied its effects on the system dynamics.

Earlier, the effect of an inherent delay due to processing the damping term in a class of an electrical circuit model of VdPO and its effect on the system behaviour in isolated conditions was studied in [61]. The response time of the circuit components may introduce such a delay while processing the damping term. The magnitude of this processing delay may be small for a single oscillator; however, it may have a significant contribution to the collective behaviour if the network consists of a large number of interacting units. This chapter deals with the response of the coupled system with the variation of this intrinsic delay through analytical analysis and numerical simulation of the system equations. We present the system dynamics considering the processing delay only, not the delay due to propagation through the component oscillators due to the coupling to reduce the complexity of the problem. Moreover, in many cases, processing delay has much more influence than propagation delay [15]. The present study describes that an MFD-coupled VdPO network shows oscillation quenching states (QS) with proper values of coupling strength. These QS are of two types: Amplitude Death (AD) and Oscillation Death (OD). AD refers to the case where the individual oscillators of the coupled system converge at a common homogeneous steady state (HSS), whereas, in OD, the oscillators populate different inhomogeneous steady states (IHSS) [62, 63] as a result of symmetry breaking bifurcation. Further investigation shows that the quenched state (QS) region decreases with the increase of internal delay. The coupled system becomes unstable faster with increasing delay. Moreover, it is observed that the system shows inhomogeneous limit cycle (IHLC) and quasi-periodic (QP) oscillations before entering the unstable region. Extended investigation with the variation of the number of oscillators in the network reveals the creation of multiple non-symmetric IHSS with an increase in the number of oscillators.

The next part of this chapter is organized as follows. Section 2 comprises the system equation formulation and the analytical analysis of the stability of the equilibrium points of the system. A verification using XPPAUT [64] has also been included in Sect. 2. Section 3 describes the numerically obtained results in detail. A systematic study of the coupled systems' behaviour with a large number of oscillators is also included in this section. Finally concluding remarks on the present work have been summarized in Sect. 4.

2 System Equation Formulation: MFD-Coupled VdPO with Internal Delay

The following first-order equations describe a conventional VdPO (CVdPO):

$$\frac{dx}{dt} = y \tag{1a}$$

$$\frac{dy}{dt} = \epsilon\left(1 - x^2\right)y - x \tag{1b}$$

Here the first term on the right-hand side of Eq. (1b) represents the nonlinear damping term and ϵ denotes the strength of the damping.

The Eq. (1a) can easily be modelled in an electronics laboratory using off-the-shelf components. Implementation of the damping term in the equation requires adders and multipliers. In these chips, the functional elements are operational amplifiers (OPAMP) which may introduce a little bit of delay throughout its operation. Again, the integrators, required to model the oscillator circuit, are also implemented by OPAMPs. Hence, a delay will also come into play between its output and input. So, an overall processing delay may be assumed between the two inputs (in opposite phases) at the adder. Again, a propagation delay, whatever small it may be, will have to be considered due to the transmission of signals through different paths between the interacting blocks or elements. We have mentioned, in the introduction section, that processing delay has more contribution than the propagation delay in most of the practical cases. Hence, we assume that there is an inherent delay in the electronic model of such VdPO due to the processing of the damping term. We represent the resultant delay by a delay block in Fig. 1 placed between the multiplier and the adder [61].

To write the system equations of the proposed delayed system we consider three state variables x, y and z, where, x and y are integrator outputs and z is the delayed version of the damping term, i.e.,

$$z = e^{-p\tau}\left[\left(1 - x^2\right)y\right] \tag{2}$$

where τ represents the delay and p represents the operator $\frac{d}{dt}$. By using the first-order Pade approximation [65],

$$e^{-p\tau} = \frac{1 - \frac{p\tau}{2}}{1 + \frac{p\tau}{2}} \tag{3}$$

In Eq. (2) and Eq. (3) one can get

$$\tau\frac{dz}{dt} = 2\left[\left(1 - x^2\right)y - z\right] + \tau\left[2xy^2 - \left(1 - x^2\right)(\epsilon z - x)\right] \tag{4}$$

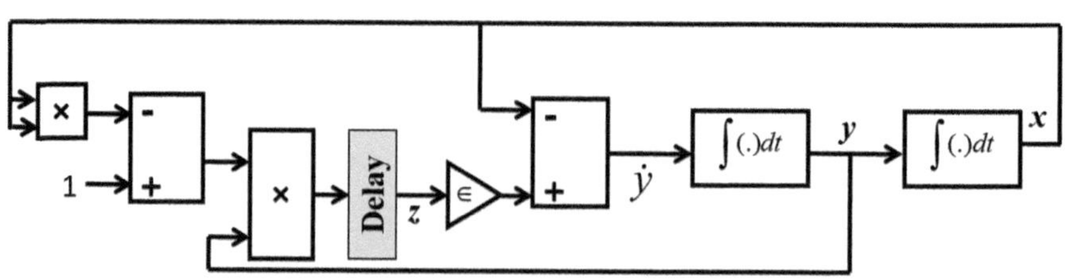

Fig. 1 Block diagram of a VdPO with inherent delay

The equation set that describes the dynamics of the proposed delayed system is given as follows (for details see ref [61]):

$$\frac{dx}{dt} = y \tag{5a}$$

$$\frac{dy}{dt} = \epsilon z - x \tag{5b}$$

$$\tau \frac{dz}{dt} = 2\left[\left(1 - x^2\right)y - z\right] + \tau\left[2xy^2 - \left(1 - x^2\right)(\epsilon z - x)\right] \tag{5c}$$

where τ represents the delay. It can be easily seen that for $\tau = 0$ the above equations reduce to Eq. (1a,b).

Considering these delayed expressions, a network system of N number of VdPOs with an internal delay interacting through MFD coupling can now be modelled mathematically as

$$\frac{dx_i}{dt} = y_i + d_s(m_s\overline{x} - x_i) \tag{6a}$$

$$\frac{dy_i}{dt} = \epsilon z_i - x_i \tag{6b}$$

$$\tau \frac{dz_i}{dt} = 2\left[\left(1 - x_i^2\right)y_i - z_i\right] + \tau\left[2x_i y_i^2 - \left(1 - x_i^2\right)(\epsilon z_i - x_i)\right] \tag{6c}$$

Here $i = 1 \cdots N$ and the mean-field term of the coupled system is $\overline{x} = \frac{1}{N}\sum_{i=1}^{N} x_i$. The diffusive coupling strength is given by d_s; m_s is the strength of the mean-field density ($0 < m_s < 1$).

To understand the dynamics and behaviour of the modified system analytically we consider the limiting case of two identical VdP oscillators with inherent delay. The steady-state values of the coupled system as obtained from Eq. (6a–c) are as follows:

(a) a trivial fixed point at the origin (0, 0, 0, 0, 0, 0) and
(b) two more coupling-dependent nontrivial fixed points:
(i) $(x^\dagger, y^\dagger, z^\dagger, x^\dagger, y^\dagger, z^\dagger)$ where $x^\dagger = \frac{y^\dagger}{d_s(1-m_s)}$, $y^\dagger = \pm\sqrt{\frac{d_s(1-m_s)\{\epsilon d_s(1-m_s)-1\}}{\epsilon\{1-\tau d_s(1-m_s)\}}}$ and $z^\dagger = \frac{x^\dagger}{\epsilon}$, and
(ii) $\left(x^\ddagger, y^\ddagger, z^\ddagger, -x^\ddagger, -y^\ddagger, -z^\ddagger\right)$ where $x^\ddagger = \frac{y^*}{d_s}$, $y^\ddagger = \pm\sqrt{\frac{d_s^2 - \frac{d_s}{\epsilon}}{1-\tau d_s}}$ and $z^\ddagger = \frac{x^*}{\epsilon}$.

Nontrivial fixed points are dependent on delay parameter τ. But for $\tau = 0$, nontrivial fixed points converge to the conventional one.

The stability of the trivial fixed point (0,0,0,0,0,0) is investigated by constructing the linear transformation Jacobian matrix of the system at the origin. The characteristic equation is a cubic equation of the eigenvalue λ and it is as given below:

$$\tau\lambda^3 + (2 + \in\tau + d_s(1-m_s)\tau)\lambda^2 - \left\{\frac{2\in}{\tau} - 1 - d_s(1-m_s)\left(\frac{2}{\tau} + \in\right)\right\}\lambda + \frac{2}{\tau}\{1 - \in d_s(1-m_s)\} = 0 \tag{7}$$

The condition of stability is examined by applying the Routh-Hurwitz array technique to the above equation. The limiting relations are given by

$$\tau > 0 \tag{8a}$$

$$d_s > \frac{2\epsilon - \tau}{(2+\epsilon\tau)(1-m_s)} \tag{8b}$$

$$d_s < \frac{1}{\epsilon(1-m_s)} \tag{8c}$$

The limit of d_s can be obtained from this equation for a given set of other design parameters.

From the nontrivial steady state expression, we derive two Pitchfork bifurcation (PB) points PB1 and PB2, at

$$(d_s)_{PB1} = \frac{1}{\epsilon} \tag{9a}$$

$$(d_s)_{PB2} = \frac{1}{\epsilon(1-m_s)} \tag{9b}$$

Thus, the Pitchfork bifurcation points remain unaffected by the internal delay. From Eq. (8b), we get two Hopf bifurcation (HB) points HB1 and HB2 at

$$(d_s)_{HB1} = \frac{2\epsilon - \tau}{(2+\epsilon\tau)(1-m_s)} \tag{10a}$$

$$(d_s)_{HB2} = \left(\frac{2\epsilon - \tau}{2+\epsilon\tau}\right) \tag{10b}$$

Here we see that the Hopf bifurcation points depend on the parameter τ. Thus, the internal delay affects the quenching and revival of oscillation of the system which happens through Hopf bifurcation only.

To find the stability of the non-trivial fixed point, at first, we construct the linear transformation Jacobian matrix of the system at $(x^\dagger, y^\dagger, z^\dagger, x^\dagger, y^\dagger, z^\dagger)$ with state Eqs. (6a–6c). The characteristic equation is a cubic equation of the eigenvalue λ and it is as given:

$$\lambda^3 + (C + d_s(1 - m_s))\lambda^2 + (Cd_s(1 - m_s) - B \in +1)\lambda + (C - A \in -d_s(1 - m_s)B \in) = 0 \tag{11}$$

Similarly, constructing the linear transformation Jacobian matrix of the system at $(x^{\ddagger}, y^{\ddagger}, z^{\ddagger}, -x^{\ddagger}, -y^{\ddagger}, -z^{\ddagger})$, the characteristic equation becomes

$$\lambda^3 + (C + d_s)\lambda^2 + (Cd_s - B\epsilon + 1)\lambda + (C - A\epsilon - d_s B\epsilon) = 0 \tag{12}$$

where

$$A = 2y^{\dagger/\ddagger^2} - 3x^{\dagger/\ddagger^2} + 2x^{\dagger/\ddagger}\left(\in z^{\dagger/\ddagger} - \frac{2}{\tau}y^{\dagger/\ddagger}\right) + 1 \tag{13a}$$

$$B = \frac{2}{\tau}\left(1 - x^{\dagger/\ddagger^2}\right) + 4x^{\dagger/\ddagger}y^{\dagger/\ddagger} \tag{13b}$$

$$C = \frac{2}{\tau} + \in \left(1 - x^{\dagger/\ddagger^2}\right) \tag{13c}$$

The condition of stability may be obtained by applying the Routh-Hurwitz array technique to Eq. (11). The limiting relations are given by

$$d_s > \frac{B\epsilon - 1}{C(1 - m_s)} \tag{14a}$$

$$d_s > -\frac{C}{(1 - m_s)} \tag{14b}$$

$$d_s < \frac{C - a\epsilon}{B\epsilon(1 - m_s)} \tag{14c}$$

$$d_s > \frac{1}{(1 - m_s)}[\sqrt{\frac{1}{4}\left(1 + \frac{1}{C}\right)^2 - \left(\frac{A}{C} - B\right)\epsilon} - \frac{1}{2}\left(1 + \frac{1}{C}\right)] \tag{14d}$$

Similarly, the stability analysis using Eq. (12) by applying the Routh-Hurwitz array technique gives the following limiting relations:

$$d_s > \frac{B\epsilon - 1}{C} \tag{15a}$$

$$d_s > -C \tag{15b}$$

$$d_s < \frac{C - a\epsilon}{B\epsilon} \tag{15c}$$

$$d_s > [\sqrt{\frac{1}{4}\left(1+\frac{1}{C}\right)^2 - \left(\frac{A}{C}-B\right)\epsilon} - \frac{1}{2}\left(1+\frac{1}{C}\right)] \tag{15d}$$

Solving the transcendental Eqs. (14a–d) and (15a–d), graphically, we obtain the possible bifurcation points as shown in Fig. 2 for $\tau = 0.1$.

The values and nature of different bifurcation points have also been found using XPPAUT [64]. Figure 3a and b show the bifurcation diagrams with d_s as a control parameter for $\tau = 0.02$ and $\tau = 0.1$, respectively. In both cases, we consider $m_s = 0.4$ and $\epsilon = 0.4$. From Fig. 3a it is observed that the stable fixed point gives birth to two stable fixed points through PB1 at $d_s = 2.5$ (red line to red line), thus it is supercritical, but at $d_s = 4.16$ (i.e.PB2) and at $d_s = 5.59$ (i.e. PB3) an unstable fixed point gives birth of two unstable fixed point (black line to black line) so, those are subcritical. The bifurcation points PB1 and PB2 can easily be verified from (9a) and (9b). It is seen that, for $\tau = 0.02$ the stable fixed points in Fig. 3a enter into the stable limit cycle through HB1 at $d_s = 0.6613$ (supercritical) and to unstable limit cycle through HB2 at $d_s = 0.3968$ (subcritical), respectively. From Fig. 3b, for $\tau = 0.1$, it is to be noted that PB1 and PB2, exist at the same values of d_s as obtained for $\tau = 0.02$. It is obvious because PB1 and PB2 do not depend on the parameter τ, as evident from Eq. 9a,b. But PB3 occurs at a different value $d_s = 5.86$, for $\tau = 0.1$,

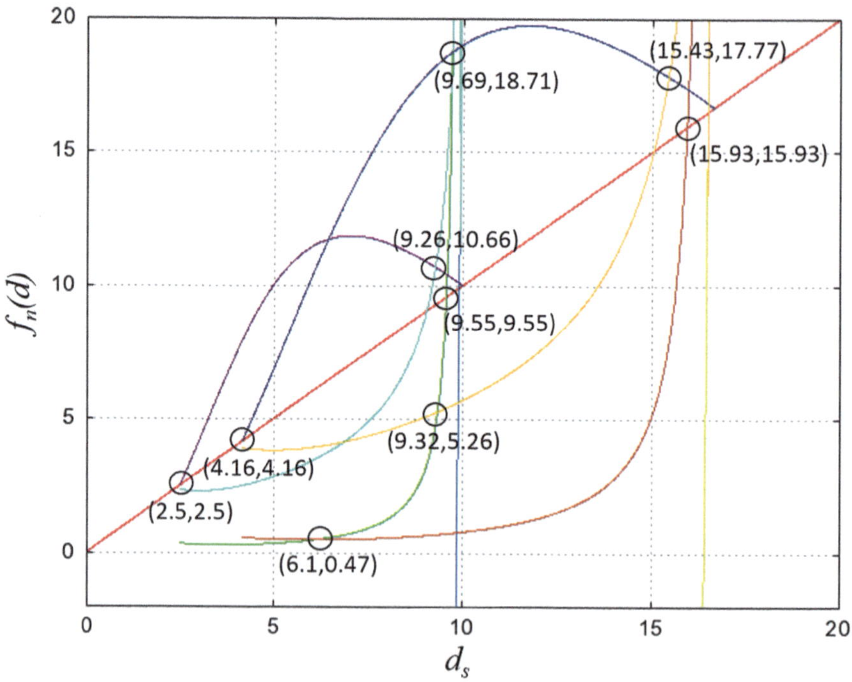

Fig. 2 Graphically obtained possible bifurcation points of the proposed system with $\tau = 0.1$ and possible bifurcation points which are also obtained by XPPAUT are shown by a black small circle

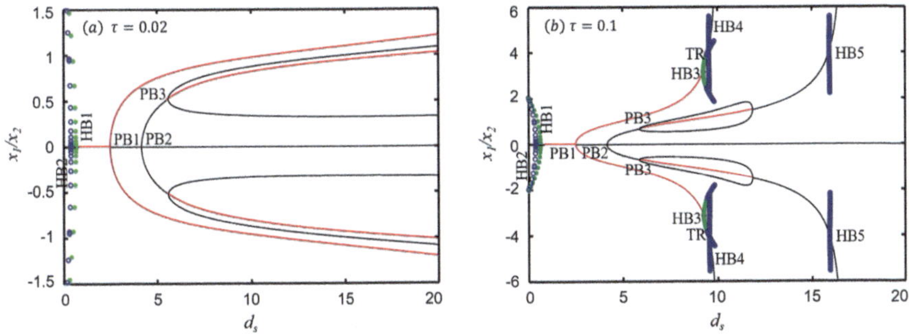

Fig. 3 Bifurcation diagram (using XPPAUT) for two MFD coupled (m_s= 0.4, ϵ = 0.4) VdPOs having internal delay (a) τ = 0.02, (b) τ = 0.1 with d_s as control parameter [Eqs. (6a–6c)]. The red (grey) line is for the stable fixed point, the black lines are for unstable fixed points, the green solid circle represents the amplitude of the stable limit cycle that emerges through Hopf bifurcation and the blue open circle represents that of an unstable limit cycle

indicating that PB3 is τ dependent. Moreover, the supercritical and subcritical Hopf bifurcations occur at d_s = 0.6358 (HB1) and d_s = 0.3831 (HB2), respectively. These bifurcation points are in good agreement with (10a) and (10b), which shows the dependence on the parameter τ. In addition to the above, for τ = 0.1, we find Hopf bifurcation points HB3 at d_s = 9.257 (Fig. 3b), where the red line changes into a solid green circle, which means stable fixed point gives birth of stable limit cycle and HB4 at d_s = 9.552 through which the whole system becomes unstable, respectively. In between HB3 and HB4 we find a TR at d_s = 9.432 where a solid green circle changes into a blue open circle. This means a torus bifurcation at this point. These observations are supported by our numerical simulation, carried out in the next part.

3 Numerical Simulation Results

To illustrate the influence of time delay on the system, the dynamics of the present system have been studied concerning different control parameters. The 4th-order Runge–Kutta technique is used to study the dynamics of the proposed system. The dynamics have been studied in the steady behaviour region, i.e., the initial transients are carefully eliminated. Initially, we examine the dynamics with τ = 0.02 for which the proposed system exhibits behaviour like a conventional one. The observation is presented through a bifurcation diagram in Fig. 4a. We choose ϵ = 0.4, m_s = 0.4 and then observe the system behaviour for different d_s values. It is observed that with increasing d_s, the coupled system moves from the oscillating state to the oscillation cessation state. At first, both the system converges to trivial steady state (HSS, AD state) from oscillating state (OS) at d_s = 0.6613 (HB2) and for further increment of d_s such AD state splits into two different steady-state values through symmetry breaking at $d_s = 2.5$ (PB1). These states are parameter dependent, i.e., non-trivial (IHSS, OD states).

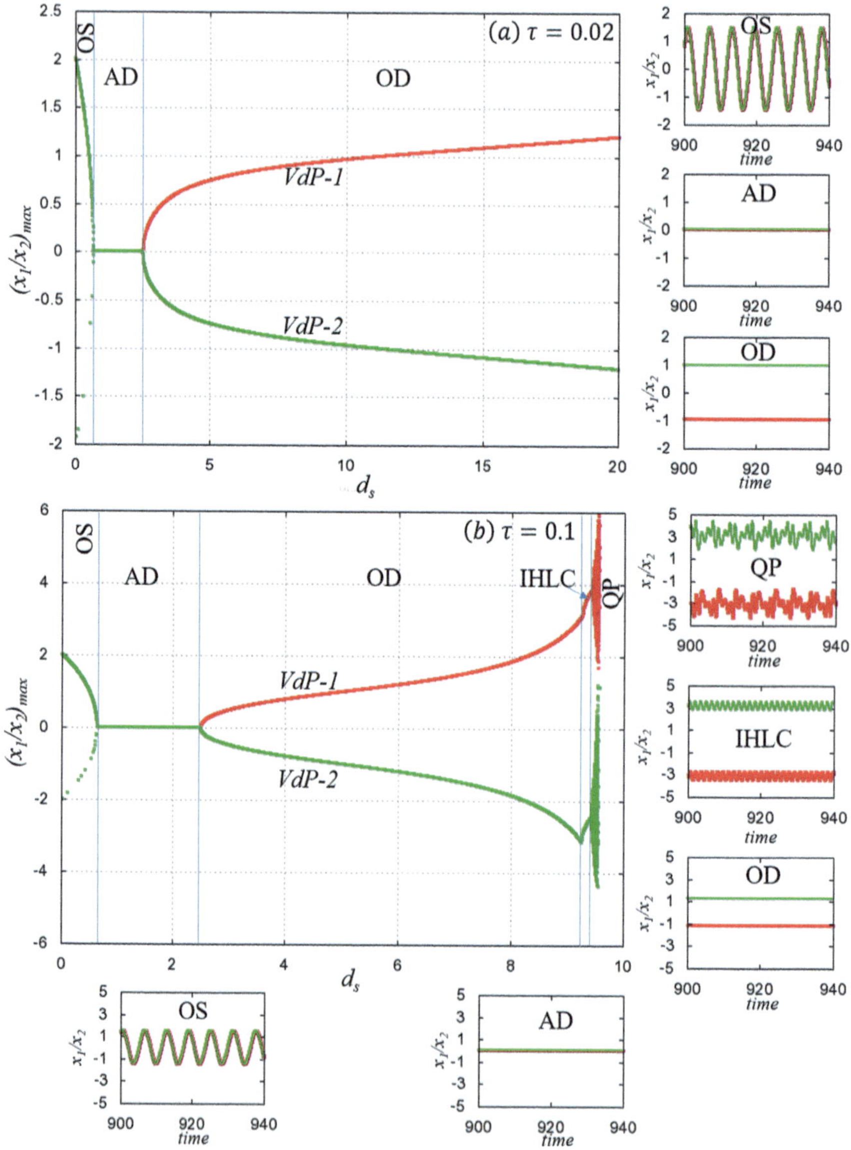

Fig. 4 Numerically obtained bifurcation diagram and time series plot of x_1 and x_2 with d_s as the control parameter for **a** $\tau = 0.02$: system converges from OS to AD, then AD to OD **b** $\tau = 0.1$: OS to AD, AD to OD and then OS through IHLC, finally the coupled system becomes unstable through quasi-periodicity (QP). The values of other parameters are $\epsilon = 0.4$, $m_s = 0.4$

Then we increase the delay value. For $\tau = 0.1$, we find HB2 occurs at $d_s = 0.646$ and PB1 occurs at $d_s = 2.5$ as predicted in the analytical studies. In addition to this we find the OD state of the system enters into the oscillating state, i.e., regeneration of oscillation occurs at $d_s = 9.254$. Moreover, in this region each oscillator enters into non-identical oscillation from its OD state i.e., IHLC is generated (Fig. 4b). For higher d_s value ($d_s = 9.421$) the proposed system enters into the quasi-periodic (QP) behavior from IHLC and creates a torus-like attractor. This was also obtained by XPPAUT analysis (Fig. 3b). Finally, the coupled system became unstable at $d_s = 9.554$.

In detail observation, as shown in Fig. 5, we find there is a slight increase in the AD region as we increase the value of τ but the OD region decreases faster and the whole system becomes unstable faster through IHLC and quasi-periodicity. The quasi-periodicity has been quantified by a return map. Figure 6a shows the torus nature of the attractor and 6(b) shows the return map of local maxima of the co-ordinate $x_1[M_{n+1}(x_1) = f(M_n(x_1))]$. The one-dimensional closed loop map justifies the occurrence of quasi-periodicity in the system under study.

We, now, extend our study to a network of VdPOs ($N_{max} = 100$). We increase the number of oscillators in the network step by step and record the appearance of the quenched states. We have carried out the numerical simulation of the system Eq. (6a–c) 30 times consecutively with 30 random initial conditions. We choose the design parameters values as $\epsilon = 0.4$, $m_s = 0.4$, $\tau = 0.1$ and $d_s = 5$ so that the system is in the OD region. We see that only OD is found in each simulation for a two-oscillator network, but as we increase the oscillator number the OD becomes less probable and at the same time the initial value-dependent QS are born. The number of such QS also increases with the increment in number of oscillators. For example, when the number of oscillators is four, OD states appear 18 times in a trial of 30 simulations with random initial conditions. In the rest of the cases, the oscillators populate in QS (+1.123438, −0.830254). Similarly, if we consider a six-oscillator network, OD occurs 06 times only. The other possible outcomes are the asymmetric QS (+1.086701, −0.894153) and a single QS (−0.753602). We continue the observations for N = 8, 10, 20 and 30. During investigation we find that the occurrence of symmetry-breaking OD states gradually decreases as we increase the oscillator number in the network and more initial value-dependent quenched states appear in the system dynamics. This is demonstrated in Table 1.

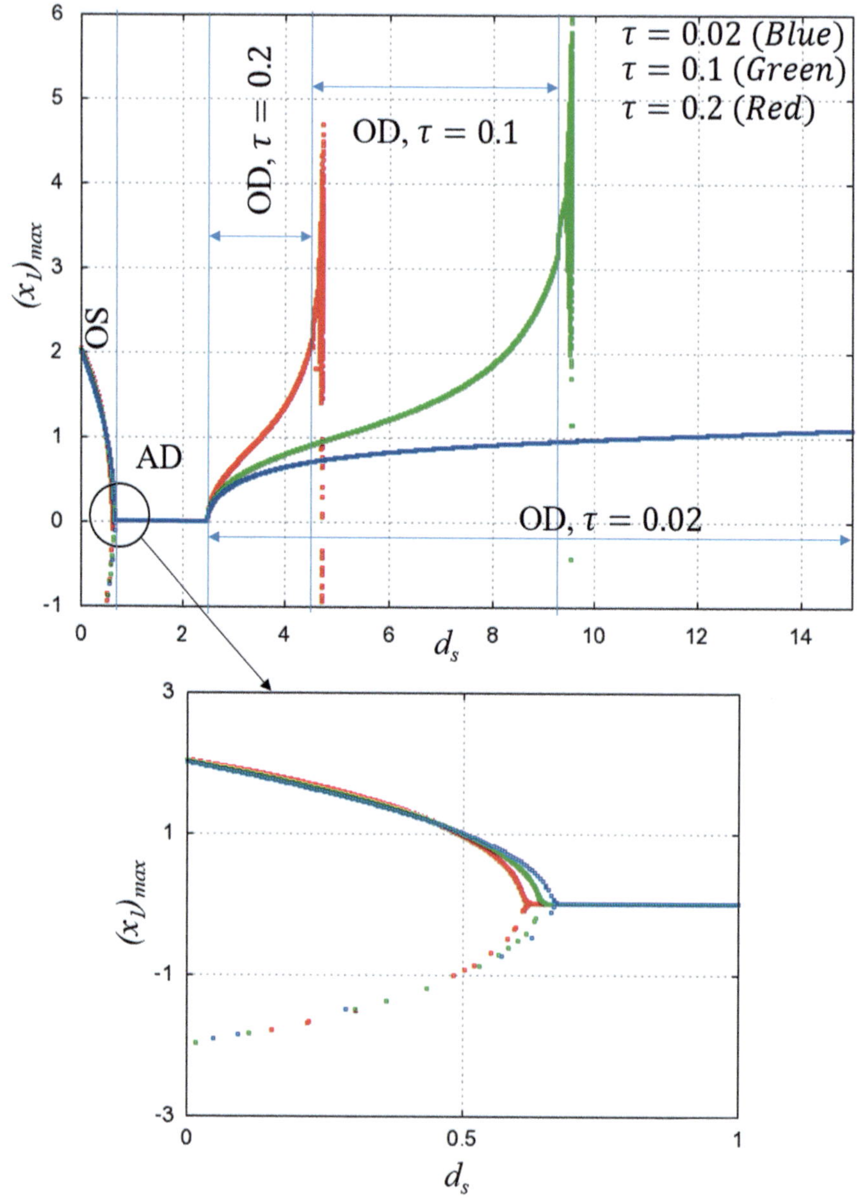

Fig. 5 Numerical study of shifting of HB points and other bifurcation points with the variation of τ

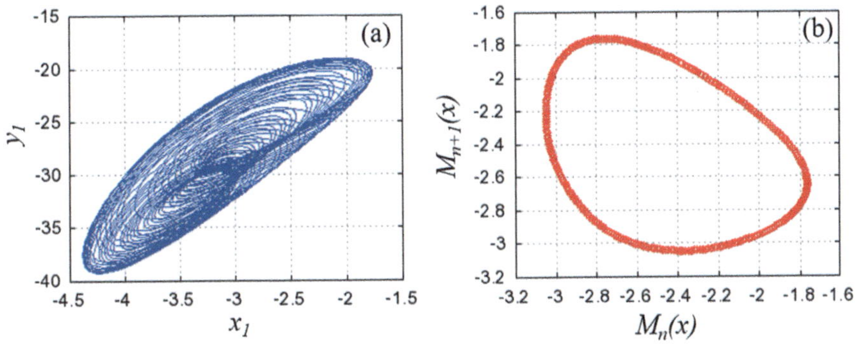

Fig. 6 **a** The torus nature of the attractor and **b** First-return map of the maxima of the coordinate x. The one-dimensional closed loop map justifies the occurrence of quasi-periodicity in the system under study. ($\epsilon = 0.4$, $m_s = 0.4$, $\tau = 0.1$ and $d_s = 9.45$)

Table 1 Numerical simulation results with $d_s = 5$, $\epsilon = 0.4$, $m_s = 0.4$ and $\tau = 0.1$

Number of oscillators	Oscillation death (OD)		Other quenched states (QS)
	OD value	No. of times appeared	
2	(±1.000000)	30	Not found
4		18	(+1.123438, −0.830254)
6		06	(+1.086701, −0.894153), (−0.753602)
8		04	(+1.123438, −0.830254), (+1.066599, −0.922886)
10		03	(+1.101985, −0.869752), (+1.142798, −0.786264), (+1.054032, −0.939312)
20		02	(+1.123438, −0.830254), (+1.101985, −0.869752), (+1.142798, −0.786264), (+1.054032, −0.939312), (+1.027778, −0.970582), (+1.078789, −0.905864)
30		Not Found	(+1.086701, −0.894153), (+1.054032, −0.939312), (+1.101985, −0.869752), (+1.070705, −0.917282), (+1.036695, −0.960378), (+1.116498, −0.843854)

It is further mentioned that in another trial of 30 simulations with random initial values, the cases of the appearance of OD state may be different and the QS obtained are also not the same as shown in Table 1. However, with the increase of several oscillators, the appearance of OD states becomes less probable and the number of initial value-dependent quenched states gradually increases. In other words, the dynamics obtained in each trial will not be the same as described in Table I but will be similar.

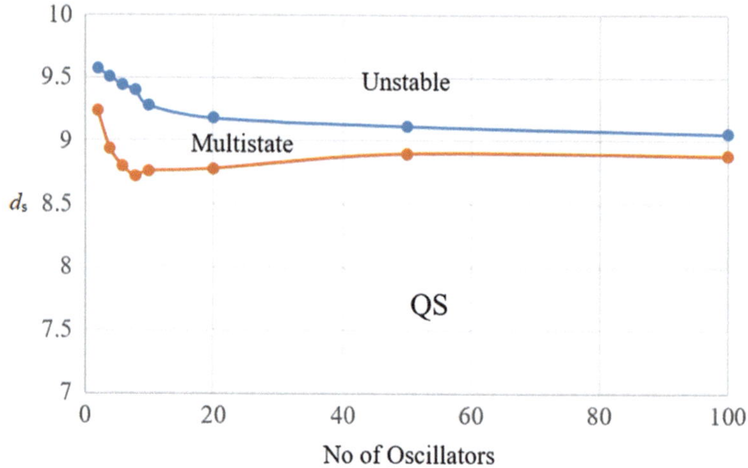

Fig. 7 The limiting value of d_s beyond which the system becomes unstable with the variation of oscillator number and the transition region where multiple states (like QS, OS, etc.) are observed for $\epsilon = 0.4$, $m_s = 0.4$ and $\tau = 0.1$

Further investigation has been made on the system dynamics for the variation of the diffusive strength d_s for a fixed oscillator number N. Figure 7 depicts a schematic diagram of two limiting values of d_s, up to which the network remains stable and beyond which it becomes unstable for a particular value of N. In between these two values, the network shows initial value-dependent multiple states (like QS, OS, unstable state, etc.). We have simulated the system Eq. (6a–c) 30 times consecutively with 30 random initial conditions for fixed values of d_s and N. In these simulations, we consider that value of d_s as the limiting value for a stable (unstable) state for which we get a stable (unstable) state above 90% times. The orange line indicates the region up to which the system remains stable and beyond the blue line the system becomes unstable. For d_s values in the region between these two lines, multiple states (QS, OS) are observed. It is evident from the diagram that the instability of the whole system becomes faster as the oscillator number is increased. We depict the existence of multiple states (Fig. 8), at the transition region (for d_s =9) between the stable and unstable zones obtained by random variation of initial conditions $d_s$in a network of 100 VdPOs with $\epsilon = 0.4$, $m_s = 0.4$ and $\tau = 0.1$.

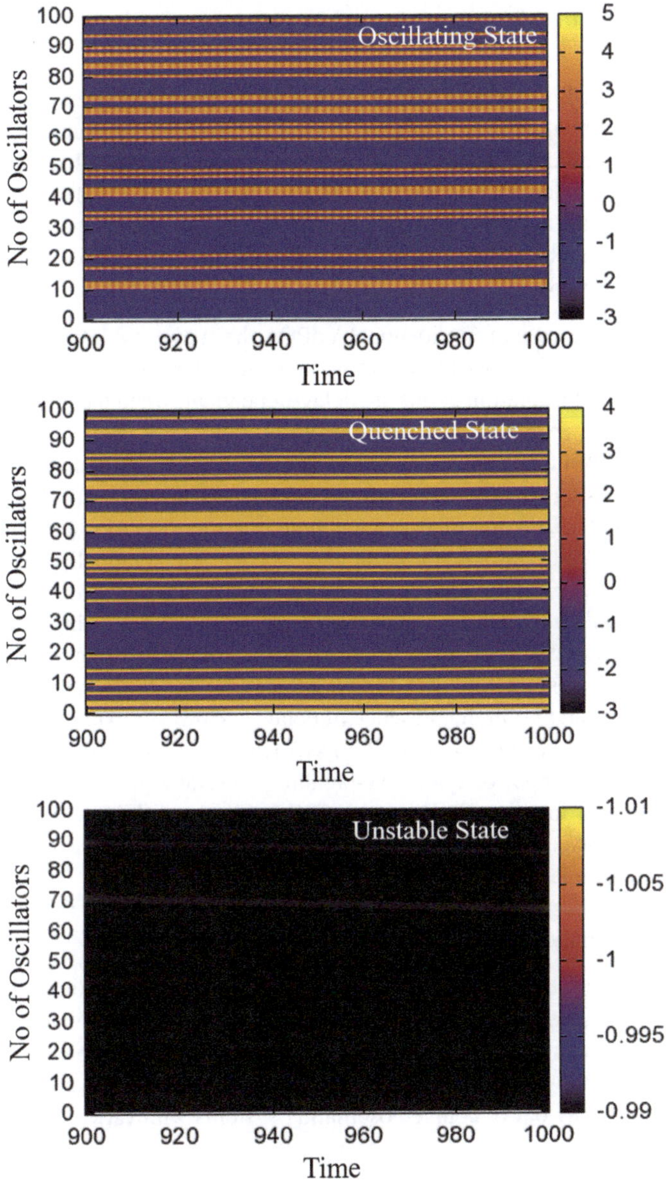

Fig. 8 The existence of multiple states in the network of 100 VdPOs for $d_s = 9, \epsilon = 0.4, m_s = 0.4$ and $\tau = 0.1$

4 Conclusion

The dynamics of MFD-coupled VdPOs with internal delay have been explored in this chapter. This internal delay is attributed to the processing of the damping term in an electrical circuit model of a VdPO. The response of the system with different values of the processing delay (τ) is reported considering the variation of diffusive coupling strength (d_s) and the number of component oscillators (N) in the network. The observations noted through extensive theoretical explanation and numerical simulation are summarized as follows:

(i) In the limiting case of two identical VdPOs, the studies with the variation of d_s reveal that the Pitchfork bifurcation points are independent of the parameter τ but the Hopf bifurcation points are delay-dependent, when the mean field diffusion strength (m_s) is kept fixed. The system converges from oscillating states (OS) to amplitude death states (AD), then AD to oscillation death states (OD) with increasing d_s. The OD state then enters OS through an inhomogeneous limit cycle (IHLC) and finally, the coupled system becomes unstable through quasi-periodicity (QP). As the parameter τ is increased gradually, the AD region is observed for a little wider range of d_s, but the range of d_s becomes narrow for observation of OD. It is found that the presence of processing delay reduces the oscillation cessation region and makes the coupled system unstable faster.

(ii) Only OD states are found in the case of a two-oscillator network for fixed values of d_s and m_s in the presence of delay τ. However extended studies with the increase in the number of oscillators (N) in the network explore that the probability of occurrence of symmetry-breaking OD states gradually decreases and generation of initial value-dependent QS is observed. Initial value-dependent multiple states (QS, OS) become larger in number with the increase in N with fixed d_s, m_s and τ. The system dynamics show deterioration, i.e., transition to the unstable region for smaller d_s (with fixed m_s and τ) with the increment of oscillators in the network.

As delay is very common in any natural system, the present study may find its importance from a design point of view of coupled oscillators as well as in exploring various complex natures of different physical, chemical and biological systems that are analogous to networks of oscillators. This study may also be extended to explore the dynamical behaviour of coupled oscillating systems with various other coupling schemes.

Appendix

Abbreviations

VdPO	Van der Pol Oscillator
CVdPO	Conventional Van der Pol Oscillator
MFD	Mean field diffusion
QS	Quenching states
OS	Oscillating states
HSS	Homogeneous steady states
IHSS	Inhomogeneous steady states
IHLC	Homogeneous limit cycle
AD	Amplitude death
OD	Oscillation death
QP	Quasi-periodic
PB	Pitchfork bifurcation
HB	Hopf bifurcation

References

1. Strogatz, S.H., Mirollo, R.E.: Stability of incoherence in a population of coupled oscillators", J. Stat. Phys. **63**, 613 (1991)
2. Strogatz, S.H.: From Kuramoto to Crawford: Exploring the onset of synchronization in populations of coupled oscillators. Physica D **143**, 1 (2000)
3. Daido, H., Nakanishi, K.: Aging transition and universal scaling in oscillator networks. Phys. Rev. Lett. **93**,104101 (2004)
4. Hong, H., Strogatz, S.H.: Mean-field behavior in coupled oscillators with attractive and repulsive interactions. Phys. Rev. **E 85**, 056210 (2012)
5. Guo, Y., Jiang, W., Niu, B.: Multiple scales and normal forms in a ring of delay coupled oscillators with application to chaotic Hindmarsh-Rose neurons. Nonlinear Dyn. **71**, 515–529 (2013)
6. Pikovsky, A., Rosenblum, M.: Dynamics of globally coupled oscillators: Progress and perspectives. Chaos **25**, 097616 (2015)
7. Zou, W., Senthilkumar, D.V., Nagao, R., et al.: Restoration of rhythmicity in diffusively coupled dynamical networks. Nat. Commun. **6**, 7709 (2015)
8. Woo, J.H., Honey, C.J., Moon, J.Y.: Phase and amplitude dynamics of coupled oscillator systems on complex networks. Chaos **30**, 121102 (2020). https://doi.org/10.1063/5.0031031
9. Kitsunai, S., Cho, W., Sano, C., Saetia, S., Qin, Z., Koike, Y., Frasca, M., Yoshimura, N., Minati, L.: Generation of diverse insect-like gait patterns using networks of coupled Rössler systems. Chaos **30**, 123132 (2020)
10. Xu, C., Wang, X., Zheng, Z., Cai, Z.: Stability and bifurcation of collective dynamics in phase oscillator populations with general coupling. Phys. Rev. E **103**, 032307 (2021)
11. Burylko, O., Martens, E.A., Bick, C.: Symmetry breaking yields chimeras in two small populations of Kuramoto-type oscillators. Chaos **32**(9), 093109, (2022). https://doi.org/10.1063/5.0088465

12. Gogoi, P.B., Kumarasamy, S., Prasad, A., Ramaswamy, R.: Phase slips in coupled oscillator systems. Phys. Rev. E **108**, 014209 (2023)
13. Mackey, M., Glass, L.: Oscillation and chaos in physiological control systems. Science **197**, 287–289 (1977) https://doi.org/10.1126/science.267326
14. Mather, W., Bennett, M.R., Hasty, J., Tsimring, L.S.: Delay-induced degrade-and-fire oscillations in small genetic circuits. Phys. Rev. Lett. **102**, 068105 (2009)
15. Milton, J.G.: Time delays and the control of biological systems: An overview. IFAC-Papers On Line **48**(12), 87–92(2015). https://doi.org/10.1016/j.ifacol.2015.09.358
16. Kuang, Y.: Delay differential equations with applications to population dynamics, (Academic Press, New York, 1993)
17. Hinch, R., Schnell, S.: Mechanism equivalence in enzyme-substrate reactions: Distributed differential delay in enzyme kinetics. J. Math. Chem. **35**(3), 253–264 (2004)
18. Soriano, M.C., García-Ojalvo, J., Mirasso, C.R., Fischer, I.: Complex photonics: Dynamics and applications of delay-coupled semiconductors lasers. Rev. Modern Phys. **85**(1), 421 (2013)
19. Andersson, L.S., Lundberg, P.A.: Delayed albedo effects in a zero-dimensional climate model. J. Atmos. Sci. **45**, 2294–2305 (1988)
20. Bielczyk, N., Foryś, U., Platkowski, T.: Dynamical models of Dyadic interactions with delay. J. Math. Sociol. **37**(4) (2013)
21. Dibeh, G.: Speculative dynamics in a time-delay model of asset prices. Physica A **355**(1), 199–208 (2005)
22. Li, J.C., Mei, D.C.: The influences of delay time on the stability of a market model with stochastic volatility. Phys. A: Stat. Mech. Its Appl. **392**(4), 763–772 (2013). https://doi.org/10.1016/j.physa.2012.10.028
23. Emmanuel, W., Fridman, E., Sename, O., Dugard, L.: Recent results on time-delay systems: Analysis ad Control (Springer International, Switzerland, 2016)
24. Kouda, A., Mori, S.: Analysis of a ring of mutually coupled van der Pol oscillators with coupling delay. IEEE Trans. Circuits Syst. **28**(3), 247–253(1981). https://doi.org/10.1109/tcs.1981.1084964
25. Reddy, D.V.R., Sen, A., Johnston, G.L.: Experimental evidence of time-delay-induced death in coupled limit-cycle oscillators. Phys. Rev. Lett. **85**(16), 3381–3384(2000). https://doi.org/10.1103/physrevlett.85.3381
26. Earl, M.G., Strogatz, S.H.: Synchronization in oscillator networks with delayed coupling: A stability criterion. Physical Review E **67**(3) (2003). https://doi.org/10.1103/physreve.67.036204
27. Yao, C., Ma, J., Li, C., He, Z.: The effect of process delay on dynamical behaviors in a self-feedback nonlinear oscillator. Commun. Nonlinear Sci. Numer. Simul. **39**, 99–107 (2016)
28. Zou, W., Zhan, M. Kurths, J.: The impact of propagation and processing delays on amplitude and oscillation deaths in the presence of symmetry-breaking coupling. Chaos **27**(11), 114303 (2017)
29. Premraj, D., Suresh, K., Thamilmaran, K.: Effect of processing delay on bifurcation delay in a network of slow-fast oscillators. Chaos **29**(12), 123127 (2019). https://doi.org/10.1063/1.5123417
30. Yao, C., He, Z., Zou, W.: On Oscillation behavior driven by processing delay in diffusively coupled inactive systems: Cluster synchronization and multistability. Chaos **30**, 123137 (2020). https://doi.org/10.1063/5.0025958
31. Ibrahim, M.M., Kamran, M.A., Mannan, M.M.N., Jung, H., Kim, S.: Lag synchronization of coupled time-delayed FitzHugh–Nagumo neural networks via feedback control. Sci. Rep. **11**, 3884 (2021). https://doi.org/10.1038/s41598-021-82886-x
32. Hui, N., Biswas, D., Banerjee, T., Kurths, J.: Effects of propagation delay in coupled oscillators under direct–indirect coupling: Theory and experiment. Chaos **31**(7), 073115 (2021). https://doi.org/10.1063/5.0057311
33. Medvedev, G.S., Mizuhara, M.S., Phillips, A.: A global bifurcation organizing rhythmic activity in a coupled network. Chaos **32**(8), 083116 (2022). https://doi.org/10.1063/5.0089946

34. Ladeira, G., Lima, G., Balthazar, J.M., Tusset, A.M., Bueno, A.M.: P and T Waves Heart Modeling with Van der Pol Oscillator, 24th ABCM International Congress of Mechanical Engineering, (December 3–8, 2017, Curitiba, PR, Brazil, 2017)
35. Winfree, A.T.: Biological rhythms and the behavior of populations of coupled oscillators. J. Theor. Biol. **16**, 15–42 (1967). https://doi.org/10.1016/0022-5193(67)90051-3
36. Kuramoto, Y: Chemical oscillations, waves, and turbulence, (Springer-Verlag, Berlin, 1984). https://doi.org/10.1007/978-3-642-69689-3
37. FitzHugh, R.: Impulses and physiological states in theoretical models of nerve membranes. Biophysics J **1**, 445466 (1961)
38. Vahala, K., Yariv, A.: Semiclassical theory of noise in semiconductor lasers - Part I. IEEE J. Quantum Electron. **19**(6), 1096–1101 (1983). https://doi.org/10.1109/jqe.1983.1071986
39. Ge, Z.M., Hsu, M.Y.: Chaos in a generalized van der Pol system and in its fractional order system. Chaos Solitons Fractals **33**, 17111745 (2007)
40. De Sarkar, S.S., Sharma, A.K., Chakraborty, S.: Chaos, antimonotonicity and coexisting attractors in Van der Pol oscillator based electronic circuit. Analog Integr. Circ. Sig. Process. **110**, 211–229 (2022). https://doi.org/10.1007/s10470-021-01934-8
41. Pinto, S. E de, S., Lopes, S.R., Viana, R.L.: Collective behavior in a chain of van der Pol oscillators with power-law coupling. Phys. A: Stat. Mech. Its Appl. **303**(3–4), 339–356 (2002). https://doi.org/10.1016/S0378-4371(01)00549-0
42. Li, X., Ji, J.C., Hansen, C.H.: Dynamics of two delay coupled van der Pol oscillators. Mech. Res. Commun. **33**(5), 614–627 (2006). https://doi.org/10.1016/j.mechrescom.2005.09.009
43. Suchorsky , M.K., Rand, R.H.: A pair of van der Pol oscillators coupled by fractional derivatives. Nonlinear Dyn. **69**, 313–324 (2012). https://doi.org/10.1007/s11071-011-0266-1
44. Kengne, J., Chedjou, J.C., Kom, M., Kyamakya, K., Tamba, V.K.: Regular oscillations, chaos, and multi-stability in a system of two coupled van der Pol oscillators: numerical and experimental studies. Nonlinear Dyn. **76**, 11191132 (2014)
45. Banerjee, T., Biswas, D., Ghosh, D., Bandyopadhyay, B., Kurths, J.: Transition from homogeneous to inhomogeneous limit cycles: Effect of local filtering in coupled oscillators. Phys. Rev. E Phys. Rev. E **97**(4), 042218 (2018)
46. Ulonska, S., Omelchenko, I., Zakharova, A., Schöll, E.: Chimera states in networks of VanderPol oscillators with hierarchical connectivities. Chaos **26**, 094825 (2016). https://doi.org/10.1063/1.4962913
47. Zhao, N., Sun, Z., Yang, X., Xu, W.: Explosive death of conjugate coupled Van der Pol oscillators on networks. Phys. Rev. E **97**, 062203 (2018). https://doi.org/10.1103/PhysRevE.97.062203
48. Dixit, S., Sharma, A., Shrimali, M.D.: The dynamics of two coupled Van der Pol oscillators with attractive and repulsive coupling. Phys. Lett. A **383**(32), 125930 (2019). https://doi.org/10.1016/j.physleta.2019.125930
49. Tegnitsap, J.V.N., Fotsin, H.B., Ngouonkadi, E.B.M.: Magnetic coupling based control of a chaotic circuit: Case of the van der Pol oscillator coupled to a linear circuit. Chaos, Solitons & Fractals **152**, 111319 (2021). https://doi.org/10.1016/j.chaos.2021.111319
50. Telem, A.N.K., Rajagopal, K., Fonzin, T.F., Ramakrishnan, B.: On the dynamics of a system of two coupled van der Pol oscillators subjected to a constant excitation force: effects of broken symmetry. Eur. Phys. J. Spec. Top. **230**, 3551–3564 (2021). https://doi.org/10.1140/epjs/s11734-021-00232-8
51. Tegnitsap, J.V.N., Tsefack, M.B.S., Ngouonkadi, E.B.M., Fotsin, H.B.: On the modeling of some triodes-based nonlinear oscillators with complex dynamics: case of the Van der Pol oscillator. Physica Scripta **96**(12), 125269 (2021)
52. Banerjee, T., Ghosh, D.: Experimental observation of a transition from amplitude to oscillation death in coupled oscillators. Phys. Rev. E **89**, 062902 (2014)
53. Banerjee, T.: Mean-field-diffusion–induced chimera death state. Eur. Lett. **110**(6), 60003 (2015)
54. Chakraborty, S., Dandapathak, M., Sarkar, B.C.: Oscillation quenching in third order phase locked loop coupled by mean field diffusive coupling. Chaos **26**, 113106 (2016). https://doi.org/10.1063/1.4967389

55. Verma, U.K., Sharma, A., Kamal, N.K., Kurths, J., Shrimali, M.D.: Explosive death induced by mean–field diffusion in identical oscillators. Sci. Rep. **7**, 7936 (2017). https://doi.org/10.1038/s41598-017-07926-x
56. Kumar, K., Biswas, D., Banerjee, T., Zou, W., Kurths, J., Senthilkumar, D.V.: Revival and death of oscillation under mean-field coupling: Interplay of intrinsic and extrinsic filtering. Phys. Rev. E **100**, 052212 (2019)
57. Guo, Q., Yang, X., Liu, J., Xu, W.: Death behavior of mean-field coupled van der pol oscillators induced by time-delayed feedback. Int. J. Bifurc. Chaos **30**(3), 2050040 (2020). https://doi.org/10.1142/S0218127420500406
58. Chakraborty, S., Dandapathak, M., De Sarkar, S.S.: Effect of self feedback on mean-field coupled oscillators: Revival and quenching of oscillations. Int. J. Bifurc. Chaos **31**(5), 2150078 (2021). https://doi.org/10.1142/S0218127421500784
59. Dixit, S., Chowdhury, S.N., Ghosh, D., Shrimali , M.D.: Dynamic interaction induced explosive death. Eur. Lett. **133**(4), 40003 (2021)
60. Liu, S., Sun, Z., Zhao, N., Xu, W.: Explosive death induced by environmental coupling. Commun. Nonlinear Sci. Numer. Simul. **98**, 105774 (2021). https://doi.org/10.1016/j.cnsns.2021.105774
61. Chakraborty, S., De Sarkar, S.S.: Dynamical behavior of a van der pol oscillator circuit with internal delay. Manager's J. Circuits & Syst. **8**(2), 16–21 (2020). https://doi.org/10.26634/jcir.8.2.18161
62. Koseska, A., Volkov, E., Kurths, J.: Parameter mismatches and oscillation death in coupled oscillators. Chaos **20**, 023132 (2010)
63. Koseska, A., Volkov, E., Kurths, J.: Oscillation quenching mechanisms: Amplitude vs. oscillation death. Phys. Rep. **531**, 173–199 (2013)
64. Ermentrout, B.: Simulating, analyzing, and animating dynamical systems: A guide to Xppaut for researchers and students (Software, Environments, Tools, SIAM Press, 2002)
65. Franklin, G.F., Powell, J.D., Emami-Naeini, A: Feedback control of dynamic systems (3rd ed., MA: Addison Wiley, 1994)

Markovian Brand Switching Model for Long-Term Steady-State Market Shares: A Study on Toothpaste Market

Ranjan Kumar Gupta, Debdip Khan, and Sudatta Banerjee

Abstract The objectives of this chapter are to determine the long-run/steady-state market shares of different popular brands of a product by initially constructing a one-step transition probability matrix to illustrate brand interplay, inform brands about primary competitors and recommend promotional strategies, and determine long-term steady-state market shares by calculating steady-state transition probabilities. The commonly visualized fluctuations in the demands of different popular brands of different products around us have been the main motivators in doing the study. The product toothpaste has been selected and the methodology of Markovian analysis has been implemented. Data about the current toothpaste brands purchased by customers in the Purba Bardhaman district of West Bengal, India, as well as potential future purchases, were gathered through a questionnaire-based survey.

Keywords Brand switching · Markov chain · Toothpaste · Market share · Steady state

1 Introduction

It is commonly observed that the demands of various brands of different FMCGs around us fluctuate with time. This makes us curious to know or predict the market shares of different brands in the long run/steady state. In this chapter, an effort has been made to fulfil our objective of determining the steady-state market share of toothpaste brands. For the said purpose, the methodology of Markovian Analysis has been utilized.

In the field of marketing, which encapsulates entrepreneurial pursuits [1], the act of crafting a unique name, symbol, or design to distinguish a product from others is

R. K. Gupta (✉)
Department of Management and Marketing, West Bengal State University, Barasat, WB, India
e-mail: rangupt@rediffmail.com

D. Khan · S. Banerjee
Department of Business Administration, Burdwan Raj College, Burdwan, WB, India

L. Sahoo et al. (eds.), *Decision Making Under Uncertainty Via Optimization, Modelling, and Analysis*, Studies in Systems, Decision and Control 558,
https://doi.org/10.1007/978-981-96-0085-4_30

known as branding. Brand switching occurs when a loyal customer shifts allegiance to a competitor, indicating a deliberate change in purchasing behaviour.

Among the various models employed in marketing to analyze brand switching, the Markovian Brand Switching Model stands out. Grounded in Markov chains, this method is predominantly used to study Brand Loyalty within marketing contexts.

Markov, a Russian mathematician, introduced Markov chains in the early 1900s as a mathematical model for analyzing random phenomena. While the model's origins date back to this era, its significant applications in brand loyalty studies began in the latter half of the twentieth century.

Pioneering work on brand loyalty was undertaken by Brown [2], followed by Osgood and Tannenbaum [3] who proposed principles of congruity in predicting attitude changes. Anderson and Goodman [4] focused on statistical inference of Markov chains.

Subsequent researchers like [5] explored the dynamics of brand loyalty, while [6] assessed Markov brand switching models. This era saw experimentation with the integration of brand switching with established mathematical models such as stochastic models [7], linear learning models [8], and Markov models for switching regressions [9].

Research expanded into various areas in subsequent years, including multivariate Markov switching factor models [10], non-stationary Markov analysis [11], and Markov-switching time series models [12], among others.

Aypar and Tuncay [13] analyzed brand loyalty using Markov chains, while Rundle and Mackay [14] and [15] evaluated brand loyalty performance. Boas [16] examined consumer learning, brand loyalty, and competition, while [17] tested different brand-switching models and [18] worked on production–transportation problems.

In 2010, Puligadda and Ross explored branding effects on variety perception, and Shaker and Basem focused on relationship marketing and organizational performance. In more recent years, [19] predicted customer churn and retention rates in Nigeria's mobile telecommunication industry using Markov chain analysis, and [20] emphasized inventory control and manufacturing firm performance. Bishop et al. [21] provided a unique solution to quantum stochastic differential inclusions. In 2020. Gupta et al. worked on customer perception and brand preference in the health drinks market. Very recently [22] developed a production optimization model based on multi-criteria decision analysis for the maintenance of environmental sustainability.

Apart from these, the past few decades have witnessed several research works based on the application of established mathematical tools done by researchers like [23, 24], 25, 26, 27, Paramanik et al. [28], Sahoo et al. [29]. Recent studies from the last three years have continued to explore and expand the applications of the Markovian Brand Switching Model, focusing on various industries and incorporating advancements in data analysis and computational methods. Recent research has applied the Markovian Brand Switching Model across different markets to gain insights into consumer behaviour. For instance, [30] employed the model to study brand-switching behaviour in the smartphone industry. By analyzing customer data from multiple brands, they were able to identify key factors that influence brand loyalty and predict future market shares with greater accuracy. In the automotive sector, a study by Singh and Kaur [31] utilized Markovian analysis to understand brand-switching patterns among car buyers in India. Their findings highlighted the importance of after-sales service and customer satisfaction in retaining customers, providing actionable insights for car manufacturers to improve their retention strategies.

The integration of the Markovian Brand Switching Model with advanced data analytics and machine learning techniques has been a significant trend in recent studies. Wang et al. [32] demonstrated the effectiveness of combining Markov chains with machine learning algorithms to predict brand switching in the fast-moving consumer goods (FMCG) sector. Their hybrid model outperformed traditional methods, offering more accurate predictions and deeper insights into consumer preferences. Similarly, [33] explored the use of big data analytics in conjunction with Markovian models to analyze brand-switching behaviour in the online retail industry. By leveraging large datasets and sophisticated algorithms, they were able to uncover complex switching patterns and provide more nuanced recommendations for marketing strategies.

The digital transformation of markets has also influenced the application of the Markovian Brand Switching Model. In a study by Chen and Zhang [34], the impact of digital marketing efforts on brand-switching behaviour in the fashion industry was examined. Their research showed that online marketing campaigns and social media interactions significantly affect consumer loyalty and switching decisions, underscoring the need for brands to adapt their strategies in the digital age. Moreover, [35] investigated the role of e-commerce platforms in brand-switching behaviour for electronic products. They found that factors such as user interface design, customer reviews, and personalized recommendations play a crucial role in influencing consumer choices, thereby affecting brand loyalty and switching rates.

With the growing emphasis on sustainability, recent studies have also explored the intersection of sustainable consumer behaviour and brand switching. Johnson et al. [36] applied the Markovian Brand Switching Model to understand how sustainability initiatives by brands influence customer loyalty in the beverage industry. Their findings indicated that brands with strong sustainability credentials are more likely to retain customers and attract switchers from less sustainable competitors. In the context of the cosmetics industry, a study by Park and Lee [37] examined the impact of eco-friendly product lines on brand-switching behaviour. They found that consumers are increasingly favouring brands that prioritize environmental sustainability, leading to higher retention rates and lower switching probabilities.

However, to our knowledge based on the literature review, no studies have applied Markov analysis to toothpaste brand switching in the Purba Bardhaman district of West Bengal.

Toothpaste is an essential product with a significant customer base globally, including in India, particularly in states like West Bengal. This study conducted a sample survey among toothpaste consumers in the Bardhaman district of West Bengal, employing a questionnaire method. The aim was to assess the current market shares of various toothpaste brands, transition probabilities between brands, and long-term steady-state market shares.

Calculations of long-term steady-state market shares were based on the common Markovian assumption that during the study period, no radical changes in marketing strategies were employed by any toothpaste brands, or if changes occurred, they were similar across all brands.

Additionally, efforts were made to inform each brand of their potential competitors—brands from which they may lose market share and those from which they could win back customers—allowing brands to tailor their marketing strategies accordingly.

2 Objectives

This research paper aims to achieve the following objectives:

(a) Assessing the current market shares of various toothpaste brands in the Purba Bardhaman district of West Bengal, India.
(b) Constructing a one-step transition probability matrix that illustrates the interplay between different toothpaste brands.
(c) Informing various brands about their primary potential competitors and offering recommendations to devise effective promotional and advertising strategies against rival brands.
(d) Determining the long-term steady-state market shares of different toothpaste brands by computing the steady-state transition probabilities among them.

3 Methodology

In the Purba Bardhaman district of West Bengal, India, a comprehensive study was conducted to understand consumer behaviour concerning toothpaste brand preferences. A sample of 500 consumers was selected for this study, with the primary objective of examining brand loyalty and predicting future market trends through the application of Markovian analysis.

Sampling Methodology.

The sampling method employed was primarily convenience sampling due to its practicality and ease of implementation. Convenience sampling involves selecting participants who are readily available and willing to participate. However, to enhance the representativeness of the sample and reduce bias, efforts were made to introduce randomness wherever feasible. This was achieved by approaching consumers at various locations such as markets, shopping centres, and residential areas at different times of the day.

Each of the 500 participants was provided with a detailed questionnaire designed to capture critical information relevant to the study. The questionnaire was structured into three main sections:

(a) Demographic Profile: This section included questions about the respondent's age, gender, income level, educational background, and family size. This demographic information is crucial as it helps in segmenting the market and understanding the influence of various demographic factors on brand preferences.
(b) Current Toothpaste Brand Preference and Past Usage: Participants were asked to indicate their current toothpaste brand, the brands they had used in the past, and the reasons for any brand switches. This section aimed to capture the historical brand usage patterns and identify the factors influencing brand loyalty and switching behaviour.
(c) Anticipated Toothpaste Brand for Next Purchase: This section solicited information about the respondents' future purchase intentions. Participants were asked to predict which brand they would likely choose for their next purchase and the reasons behind their choice. This data is vital for forecasting future market trends and understanding the potential shifts in brand loyalty.

Markovian Analysis

The data gathered from the respondents underwent a thorough Markovian analysis. This analysis involved several critical steps:

Computation of a One-Step Transition Probability Matrix: The one-step transition probability matrix represents the likelihood of a consumer switching from one brand to another in a single period. This matrix was computed by analyzing the respondents' past and anticipated brand usage data. Each element of the matrix indicates the probability of a switch from one brand to another.

Formulation of Steady-State Equations: Steady-state equations were formulated based on the transition probability matrix. These equations represent the long-term behaviour of the system, where the probabilities of switching between brands reach an equilibrium. The steady-state equations help in predicting the market share of each brand in the long run.

Calculation of Steady-State Probabilities: Using the steady-state equations, the steady-state probabilities for each brand were calculated. These probabilities indicate the long-term market share of each brand, assuming that the switching probabilities remain constant over time.

The results of the Markovian analysis provided valuable insights into the toothpaste market in the Purba Bardhaman district. The steady-state probabilities revealed the expected long-term market shares of the different brands. Brands with higher steady-state probabilities are likely to dominate the market in the future, while those with lower probabilities may need to adopt new strategies to retain their market share.

The transition probability matrix highlighted the brands that are most vulnerable to losing customers and those that are likely to gain new customers. This information is crucial for brands to understand their competitive position and to devise effective marketing strategies.

4 Markov Chain and Its Properties

One special type of discrete-time stochastic process is called a Markov Chain.

Discrete-time stochastic process.

If the value of a system characteristic at time t (t can be any of the discrete points in time that may be labelled as 0, 1, 2, 3, …) is denoted by X_t, then in most of the situations, before time t, X_t is not known with certainty and is a random variable. A description of the relation between the random variables X_0, X_1, X_2, X_3... is known as a discrete-time stochastic process.

Definition of Markov Chain

With the assumption that at any time a discrete-time stochastic process can be in one of a finite number of states labelled 1,2,…,s, that discrete-time stochastic process qualifies as a Markov chain if, for $t = 0, 1, 2, 3,\ldots$ and all states

$$P(X_{t+1} = i_{t+1} | X_t = i_t,\ X_{t-1} = i_{t-1}, \ldots,\ X_1 = i_1,\ X_0 = i_0) = P(X_{t+1} = i_{t+1} | X_t = i_t)$$

Further, the stationary assumption of Markov chain is that for all states i and j and all t,

$P(X_{t+1} = j | X_t = i) = p_{ij}$ is independent of t; where p_{ij} is the probability that given the system is in state i at time t, the system will be in state j at time $(t + 1)$. p_{ij} is the one-step transition probability.

Thus, the probability distribution of the state at a time $(t + 1)$ depends on the state at time t and does not depend on the states the chain passes through on the way to i, at time t.

Transition Probability Matrix

The initial probability distribution for a Markov chain may be given by the vector $q = [q1\ q2 \ldots qs]$, where $qi = P(X0 = i)$. The transition probability matrix P is given by

$$P = \begin{bmatrix} p_{11} & p_{12} & \cdots & p_{1s} \\ p_{21} & p_{22} & \cdots & p_{2s} \\ . & . & \cdots & . \\ . & . & \cdots & . \\ . & . & \cdots & . \\ p_{s1} & p_{s2} & \cdots & p_{ss} \end{bmatrix} \quad \textit{such that}, \ \sum_{j=1}^{s} p_{ij} = 1, \ \textit{for each } i.$$

If a stationary Markov chain is in state i at time m, then the probability that n periods later it will be in state j is given by

$P(X_{m+n} = j|X_m = i) = P(X_n = j|X_0 = i) = p_{ij}(n)$, where $p_{ij}(n)$ is known as the $n-$ step transition probability and independent of m.

It is known that for $n > 1$, $p_{ij}(n) = (i, j)$th element of the matrix P^n. Also, the probability of being in state j at time n is given by $\sum_{i=1}^{s} q_i p_{ij}(n) = q(j\text{ th column of } P^n)$.

Steady-State Probabilities

For an s-state ergodic Markov chain, there exists a vector $\pi = [\pi_1\ \pi_2\ \pi_3 \ldots \pi_s]$, such that

$$\lim_{n\to\infty} P^n = \begin{bmatrix} \pi_1 & \pi_2 & \cdots & \pi_s \\ \pi_1 & \pi_2 & \cdots & \pi_s \\ \cdot & \cdot & \cdots & \cdot \\ \cdot & \cdot & \cdots & \cdot \\ \cdot & \cdot & \cdots & \cdot \\ \pi_1 & \pi_2 & \cdots & \pi_s \end{bmatrix} \quad ie. \ \lim_{n\to\infty} p_{ij}(n) = \pi_j$$

Thus, for large n, P^n approaches a Matrix with identical rows.

This vector $\pi = [\pi_1\ \pi_2\ \pi_3 \ldots \pi_s]$ is known as the steady-state distribution or equilibrium distribution and $\pi_1, \pi_2, \pi_3, \ldots \pi_s$ are the steady-state probabilities. Now, at the steady state, $p_{ij}(n + 1) = p_{ij}(n) = \pi_j$. The steady-state probability distribution can be obtained by solving the following equations:

$$\pi = \pi P \tag{1}$$

$$\sum_{i=1}^{s} \pi_j = 1 \tag{2}$$

5 Analysis and Findings

Respondents were queried regarding their current toothpaste brand preference and their intended brand for their next purchase. Based on these responses, we computed the current market shares for each toothpaste brand and constructed a one-step transition probability matrix.

Current Market Shares

From the responses obtained, regarding the toothpaste brands currently used by the respondents of our sample, the present market shares of the different toothpaste

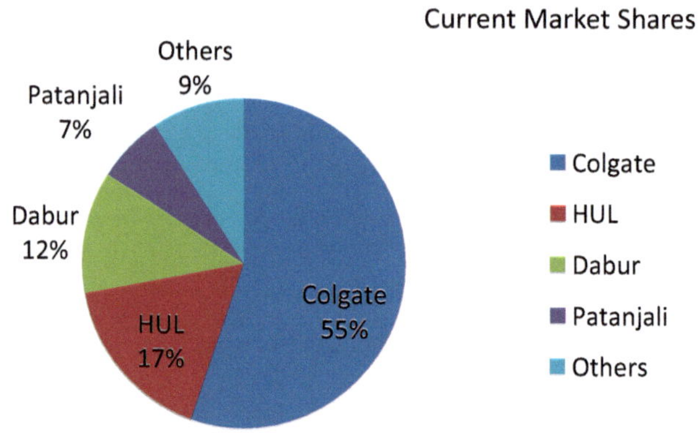

Fig. 1 Estimated Market Shares of Toothpaste Brands Among Respondents in Purba Bardhaman

brands in the district of Purba Bardhaman are estimated to be as follows (as depicted in Fig. 1 and also in Table 1): Colgate–55%, HUL-17%, Dabur-12%, Patanjali-7% and others -9%.

From the responses obtained from each of the 500 respondents regarding the toothpaste brands that they intended to purchase in their next purchase, the following data comprising of the frequency of respondents switching their brands and also the frequency of customers retaining their old brands have been obtained (As depicted in Table 1).

Taking each row in the above table, by dividing each cell frequency by the total frequency of the corresponding row the one-step transition probability matrix is obtained and displayed below in Table 2.

One-Step Transition Probability Matrix

One-step Transition probability matrix of the toothpaste brands taking the views of respondents constituting the sample is as follows:

The transition matrix provides significant insights into consumer behaviour. Colgate demonstrates substantial brand loyalty, with 58% of its customers remaining

Table 1 Consumer responses on retaining or switching toothpaste brands in their next purchase

Currently Purchased Brands	Frequency of customers brands purchased in the next purchase						Current market shares (T/GT)
	Colgate	HUL	Dabur	Patanjali	Others	Total (T)	
Colgate	161	55	35	13	11	275	0.55
HUL	27	27	9	2	20	85	0.17
Dabur	17	7	29	7	0	60	0.12
Patanjali	9	0	7	5	14	35	0.07
Others	2	7	4	0	32	45	0.09
Total (GT)						500	1.00

Table 2 Transition probability matrix of consumer movement between toothpaste brands

Currently purchased brands	Transition probability, $P_{ij}(1)$ brands purchased in the next purchase				
	Colgate	HUL	Dabur	Patanjali	Others
Colgate	0.586	0.2	0.127	0.047	0.04
HUL	0.317	0.317	0.106	0.024	0.236
Dabur	0.283	0.117	0.483	0.117	0
Patanjali	0.257	0	0.2	0.143	0.4
Others	0.044	0.156	0.089	0	0.711

loyal to the brand. However, there's a noteworthy challenge posed by the combined influence of HUL and Dabur, which accounts for the defection of approximately 32% of Colgate's customer base. Similarly, HUL shows a brand loyalty of 31%, but an equal percentage of its customers are poised to switch to Colgate. While Colgate currently enjoys a dominant position in the toothpaste market, the potential emergence of HUL and Dabur as a unified force poses a future threat to Colgate's market share.

The data analysis emphasizes the critical need for Colgate to strategically direct its promotional and advertising endeavours towards mitigating the influence of competitors like HUL and Dabur. This imperative arises from the discernible pattern in consumer behaviour, wherein Colgate emerges as the frontrunner in brand loyalty, with an impressive 58% of its customer base exhibiting unwavering allegiance. Close behind, Dabur demonstrates a substantial 48% brand loyalty, underscoring its position as a formidable competitor. Meanwhile, HUL maintains a respectable 31% brand loyalty among its customers.

An intriguing insight surfaces regarding Patanjali customers, who notably exhibit a marked reluctance to switch to HUL for their future purchases. This observation suggests a unique consumer sentiment among Patanjali loyalists, potentially influenced by factors such as brand perception or product attributes.

Furthermore, an intriguing revelation from the data table pertains to consumers who currently do not purchase Colgate, HUL, Dabur or Patanjali products. Remarkably, this segment expresses a notable disinterest in these brands for their future purchases, with an impressive retention rate of 71% for alternative brands. This finding underscores the entrenched consumer preferences and the challenges associated with penetrating this segment of the market.

The formulation of steady-state Eqs. (3–8) based on the derived one-step transition probability matrix is a critical step in predicting long-term market dynamics. These equations, developed with the assistance of Eqs. (1) and (2), are fundamental in projecting the future market shares of toothpaste brands. It is essential to recognize that this analytical process operates under the common Markovian assumption, which posits that the marketing strategies employed by toothpaste brands remained relatively stable throughout the study period. This assumption ensures consistency in the analysis and facilitates reliable projections of market dynamics.

In summary, the analysis underscores the significance of brand loyalty and consumer behaviour dynamics in shaping the competitive landscape of the toothpaste market. By leveraging these insights and adhering to the Markovian framework, companies like Colgate can devise effective strategies to maintain their market position and counteract potential threats posed by rival brands.

Now, equation (1) can be written as

$$[\pi_1\pi_2\pi_3\pi_4\pi_5]\begin{bmatrix} 0.5860.2000.1270.0470.040 \\ 0.3170.3170.1060.0240.236 \\ 0.2830.1170.4830.1170.000 \\ 0.2570.0000.2000.1430.400 \\ 0.0440.1560.0890.0000.711 \end{bmatrix} = [\pi_1\pi_2\pi_3\pi_4\pi_5]$$

Or,

$$0.586\pi_1 + 0.317\pi_2 + 0.283\pi_3 + 0.257\pi_4 + 0.044\pi_5 = \pi_1 \tag{3}$$

$$0.210\pi_1 + 0.317\pi_2 + 0.117\pi_3 + 0.000\pi_4 + 0.156\pi_5 = \pi_2 \tag{4}$$

$$0.127\pi_1 + 0.106\pi_2 + 0.483\pi_3 + 0.200\pi_4 + 0.089\pi_5 = \pi_3 \tag{5}$$

$$0.047\pi_1 + 0.024\pi_2 + 0.117\pi_3 + 0.143\pi_4 + 0.000\pi_5 = \pi_4 \tag{6}$$

$$0.040\pi_1 + 0.236\pi_2 + 0.000\pi_3 + 0.400\pi_4 + 0.711\pi_5 = \pi_5 \tag{7}$$

And equation (2) can be written as

$$\pi_1 + \pi_2 + \pi_3 + \pi_4 + \pi_5 = 1 \tag{8}$$

Solving the above set of steady-state equations using the software Mathemetica 7.0, the following solution has been obtained. These values give the long-run predicted market share of each brand.

$$\pi_1 = 0.3260(\text{refers to the steady} - \text{state market share of Colgate})$$

$$\pi_2 = 0.1896(\text{refers to the steady} - \text{state market share of HUL})$$

$$\pi_3 = 0.1814(\text{refers to the steady} - \text{state market share of Dabur})$$

$$\pi_4 = 0.0480(\text{refers to the steady} - \text{state market share of Patanjali})$$

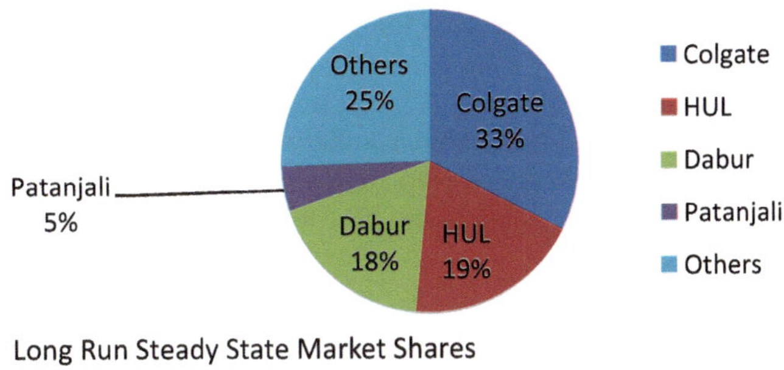

Fig. 2 Projected Long-Term Market Shares of Toothpaste Brands

$$\pi_5 = 0.2550(\text{refers to the steady} - \text{state market share of Others})$$

Thus, the long-run market shares of the different Toothpaste Brands will be as follows (as depicted in Fig. 2):

Colgate—32.60%.
Hul—18.96%.
Dabur—18.14%.
Patanjali—4.80%.
Others—25.50%.

From the values of current market shares and long-run steady-state market shares, it is evident that there is a gross difference between the two.

The transition matrix, as depicted in Table 2, unveils an intriguing trend in brand switching among toothpaste consumers, particularly the movement from Colgate to HUL, and subsequently from HUL to other brands. This pattern suggests a latent phenomenon influenced by consumers' underlying psychological preferences, potentially driven by their inclination towards seeking medicated products. The data indicates that while the "Others" brand does not directly diminish Colgate's market share, it indirectly erodes it through the conduit of HUL, indicating a significant fraction of Colgate's clientele shifting allegiance due to perceived medicinal benefits.

Conversely, customers transitioning from Colgate to Dabur exhibit a remarkable sense of loyalty towards Dabur, refraining from switching to alternative brands in their subsequent purchases. This loyalty underscores Dabur's entrenched perception as a provider of medicinal toothpaste, cultivated over an extended period. Consequently, competing brands lacking strong medical associations struggle to sway customers away from Dabur, highlighting the pivotal role of medicinal properties in consumer decision-making processes.

Furthermore, the transition matrix underscores a shift in consumer priorities, with contemporary toothpaste consumers placing greater emphasis on the health benefits offered by toothpaste formulations rather than mere taste, flavour or freshness. This trend underscores the pivotal role played by medicinal properties in shaping consumer preferences and influencing brand loyalty within the toothpaste market.

In light of these findings, it becomes imperative for Colgate to adapt its product offerings and marketing strategies to align with evolving consumer preferences. To safeguard and enhance its market share, Colgate must prioritize the development and promotion of toothpaste formulations enriched with medicinal properties. By emphasizing the health benefits of its products, Colgate can effectively resonate with the preferences of modern consumers and potentially capture a larger share of the market.

Thus, the transition matrix analysis unveils valuable insights into the dynamics of brand switching and consumer behaviour within the toothpaste market. The prevalence of brand loyalty, particularly towards medicated toothpaste brands like Dabur, underscores the significance of medicinal properties in influencing consumer choices. By leveraging these insights and adapting its strategies accordingly, Colgate can position itself competitively in the market and secure its foothold against evolving consumer preferences.

6 Conclusion

This chapter analyzed toothpaste brand preferences among consumers in the Purba Bardhaman district of West Bengal, employing a Markovian approach to predict long-term market shares. By surveying 500 consumers and analyzing their current and intended brand choices, we constructed a one-step transition probability matrix to model brand-switching behaviour.

Key Findings.

Current Market Shares:

Colgate dominates with a 55% market share. HUL holds 17%, Dabur 12%, Patanjali 7%, and other brands collectively 9%.

Brand Loyalty and Switching: Colgate retains 58% of its customers, but faces significant competition from HUL and Dabur, losing about 32% to these competitors. HUL and Dabur also exhibit strong brand loyalty, retaining 31% and 48% of their customers, respectively. Patanjali and "Others" show varied loyalty, with unique switching patterns indicating customer preferences for specific product attributes.

Steady-State Market Shares: Long-term market predictions show Colgate at 32.60%, HUL at 18.96%, Dabur at 18.14%, Patanjali at 4.80%, and Others at 25.50%. The steady-state analysis indicates a potential shift in market dynamics, with Colgate's dominance challenged by a combination of HUL, Dabur, and other brands.

Implications for Marketing Strategies: The transition probability matrix and steady-state market shares provide critical insights for brand strategy development. Colgate needs to focus on reinforcing its medicinal properties to counter the competitive threat from HUL and Dabur. Marketing campaigns should highlight health benefits to retain and attract health-conscious customers. HUL and Dabur should capitalize on their strong brand loyalty by emphasizing their unique product attributes and differentiating them from Colgate. Patanjali and Others should target niche markets and leverage specific product features that appeal to distinct consumer segments.

Consumer Behaviour Insights: The analysis reveals a significant trend where consumers prioritize health benefits over other product attributes such as taste or freshness. This trend suggests that brands focusing on medicinal properties and health-related marketing can potentially gain a competitive edge.

Recommendations:

Targeted Advertising: Brands should tailor their promotional strategies based on the identified switching behaviours. Colgate, in particular, should direct its advertising efforts against HUL and Dabur.

Product Innovation: To maintain market share, brands need to continuously innovate and enhance their product offerings, particularly emphasizing health benefits.

Consumer Education: Educating consumers about the health benefits of their products can help brands build stronger loyalty and reduce switching.

In conclusion, the Markovian analysis provides a robust framework for understanding consumer behaviour and predicting long-term market trends. Brands can leverage these insights to devise effective marketing strategies and maintain a competitive advantage in the dynamic toothpaste market of Purba Bardhaman.

References

1. Ansoff, H.I.: Corporate Strategy, Penguin Books Ltd., McGraw Hill (1987)
2. Brown, G.: Brand loyalty: Fact or fiction. Advert. Age **24**, 28–35 (1953)
3. Osgood, C., Tannenbaum, P.: The principles of congruity in the prediction of attitude change. Psychol. Rev. **62**(1), 42–55 (1955)
4. Anderson, T.W., Goodman, L.A.: Statistical inference about Markov chains. Ann. Math. Stat. **28**, 89–110 (1957)
5. Harary, F., Lipstein, B.: The dynamics of brand loyalty: a marketing approach. Oper. Res. **Jan-Feb** 78–82 (1962)
6. Ehrenberg, A.S.C.: An appraisal of markov brand switching models. J. Mark. Res. **2**(3) 47–62 (1965)
7. Farley, J., Kuehn, A.: Stochastic models of brand switching. In: Science in Marketing, Schwartz, G (ed), New York: John Wiley & Sons Inc (1965)
8. Carman J.M.: Brand switching and linear learning models. J. Advert. Res. **6**, 23–31 (1966)
9. Goldfeld, S.M., Quandt, R.E.: A Markov model for switching regressions. J. Econ. **1**, 3–16 (1973)
10. Kim, M.-J., Yoo, J.-S.: New index of coincident indicators: A multivariate Markov switching factor model approach. J. Monet. Econ. **36**, 607–630 (1995)
11. Zepeda, L.: Technical change and the structure of production: A non-stationary Markov Analysis. Eur Rev Agric Econ **22**, 41–60 (1995)
12. Hamilton, J.D.: Specification testing in Markov-switching time series models. J. Econ. **70**, 127–157 (1996)
13. Aypar, U., Tuncay, C.: Analysis of brand loyalty with Markov chains (2001)
14. Rundle-Thiele, S., Mackay, M.: Accessing the performance of brand loyalty measures. J. Serv. Mark. **15**(7), 529–546 (2001)
15. Ehrenberg, A.S.C., Uncles, M.D., Goodhardt, G.J.: Understanding brand performance measures: Using Dirichlet benchmarks. J. Bus. Res. **57**(1) 291–307 (2002)

16. Boas, V.: Consumer learning, brand loyalty, and competition. Mark. Sci. **23**(1), 134–145 (2004)
17. Roy, D., Lahiri, I.: Some tests for suitability of brand switching model. Eur. J. Mark. **38**(5/6), 524–536 (2004)
18. Gupta, R.K., Bhunia, A.K.: An application of real coded genetic algorithm (RCGA) for Integer Linear programming in production-transportation problems with flexible transportation cost. Adv. Model. Optim. **8**, 73–85 (2006)
19. Adebiyi, .O., Oyatoye, E.O., Mojekwu, J.N.: Predicting customer churn and retention rates in Nigeria's mobile telecommunication industry using markov chain. Modelling. .ActaUniv.Sapientiae. Econ. Bus. **3**(1), 67–80 (2015)
20. Adeyeye, J.O., Ogunnaike, O., Amaihian, A., Olokundun, M, Inelo, F.: Inventory control and performance of manufacturing firms. J. Eng. Appl. Sci. **11**(2), 199–203 (2016)
21. Bishop, S.A., Agarana, M.C., Okagbue, H. Oghonyon, J. G.: On unique solution of quantum stochastic differential inclusions. In: Proceedings of the World Congress on Engineering and Computer Science 1(Oct. 2017). San Francisco, USA (2017)
22. Khan, D., Gupta, R.K.: Production optimization with the maintenance of environmental sustainability based on multi-criteria decision analysis. Environ., Dev. Sustain. (2023). https://doi.org/10.1007/s10668-023-03316-8.
23. Gupta, R., Khan, D.: Heuristic solutions for interval valued games. Iran. J. Numer. Anal. Optim. **12**(1), 187–200 (2022). https://doi.org/10.22067/ijnao.2021.71321.1048
24. Gupta, R.K., Khan, D.: A technique to solve mixed strategy non-cooperative zero sum games with more than two players. Int. J. Oper. Res. **49**(3), 385–402 (2024)
25. Roy, D., Khan, D.: Integrated model for line balancing with workstation inventory management. Int. J. Ind. Eng. Comput. **1**, 139–146 (2010)
26. Roy, D., Khan, D.: Designing of an assembly line based on reliability approach. An Int. J. Optim. Control.: Theor. Appl. (IJOCTA) **1**(1), 45–52 (2011)
27. Deb, I, Gupta, R.K.: A genetic algorithm based heuristic optimization technique for solving balanced allocation problem involving overall shipping cost minimization with restriction to the number of serving units as well as customer hubs. Results Control. Optim. **11**, 100227 (2023)
28. Paramanik, R., Mahato, S.K., Kumar, N., Bhattacharyee, N., Gupta, R.K.: Optimization of system reliability for multi- level RAPs in intuitionistic fuzzy atmosphere using genetic algorithm. Results Control. Optim. **9**. 100175 (2022)
29. Sahoo, L., Senapati, T., Yager, R.R. (Eds.): Real life applications of multiple criteria decision making techniques in Fuzzy domain, studies in fuzziness and soft computing (STUDFUZZ, vol. 420), Springer Nature (2023)
30. Yuan, J., Chen, L. and Zhao, Y.: Predicting smartphone brand switching behavior using Markov chains. J. Mark. Anal. **9**(2), 115–128 (2021)
31. Singh, R., Kaur, G.: Understanding brand switching in the Indian automotive market: A Markovian approach. J. Bus. Res. **143**, 462–477
32. Wang, X., Liu, H., Zhang, T.: Combining Markov chains and machine learning for FMCG brand switching prediction. J. Data Sci. Anal. **11**(4), 305–319 (2022)
33. Li, P., Wang, R. and Sun, Q.: Big data analytics and markovian models for online retail brand switching. Electron. Commer. Res. Appl. **56**, 101124 (2023)
34. Chen, Y., Zhang, Z.: Digital marketing and brand switching in the fashion industry: A Markovian analysis. Fash. Mark. Manag. **25**(3), 354–370 (2021)
35. Ahmed, S., Rahman, M.: The influence of e-commerce on brand switching for electronic products: A Markov chain approach. J. Retail. Consum. Serv. **64**, 102738 (2022)
36. Johnson, L., Smith, K., Brown, M.: Sustainability and brand loyalty in the beverage industry: A Markovian perspective. Sustain. Mark. **7**(1), 98–112 (2023)

37. Park, J., Lee, H.: Eco-friendly products and brand switching in the cosmetics industry: a markovian approach. J. Sustain. Mark. **10**(2), 149–165 (2024)
38. Gupta, R.K., Khan, D., Ghosh, P.: A study om the customers' perception of different children's health drinks. South Asian J. Mark. Manag. Res. **10**, 29–39 (2020)
39. Puligadda, S., Ross, W. Jr.: The unique effects of branding on variety perception. J. Brand. Manag. **18**, 134–149 (2010)

Root Hair Algorithm: A Swarm Intelligence Algorithm

Nabajyoti Bhattacharjee, Nabendu Sen, and Laxminarayan Sahoo

Abstract In the field of computational mathematics, direct search algorithm has a greater role in solving optimization and decision-making problems. In this direction, many algorithms have been developed to solve the optimization problems, and these algorithms are classified into two major categories: Evolutionary Algorithms (Genetic Algorithm, Differential Evolution, etc.) and Swarm Intelligence (Particle Swarm Algorithm, Ant Colony Algorithm, etc.). In Swarm Intelligence Algorithms, the behaviour of a population in a group of particles or individuals is observed and the pattern of transferring the information from the particle of the previous generation to the new generation is obtained through mathematical expressions and logical progression of the algorithm. In the present paper, an attempt has been made to develop a swarm intelligence algorithm by following the behaviour of root hair searching underground water for the plant. The algorithm is coded in SciLab 6.1.0 and 15 experimental problems are selected from Jamil et al. (2013) to test the algorithm. A fair comparison has been made with the MEAN ± SD for the selected problem adopted from (Wahab et al. in PLoS ONE 10, 2015) by standard algorithms such as Particle Swarm Optimization, Ant Colony Algorithm, Genetic Algorithm and Differential Evolution.

Keywords Swarm intelligence · Particle swarm optimization · Genetic algorithm · Ant colony algorithm · Differential evolution · Root hair algorithm

---

N. Bhattacharjee
Department of Mathematics, Pandit Deendayal Upadhyaya Govt. Model College, Katlicherra 788162, India

N. Sen (✉)
Department of Mathematics, Assam University, Silchar 788011, India
e-mail: nsen08@yahoo.com

L. Sahoo
Department of Computer and Information Science, Raiganj University, Raiganj 733134, India

L. Sahoo et al. (eds.), *Decision Making Under Uncertainty Via Optimization, Modelling, and Analysis*, Studies in Systems, Decision and Control 558,
https://doi.org/10.1007/978-981-96-0085-4_31

1 Introduction

In the field of operation research, engineering and econometrics we come across certain situations, systems and environments that need to be optimized to achieve certain goals. During the study of real-world problems, the objective for the problem is usually nonlinear and in most cases with too many variables. In such situations, we apply evolutionary algorithms to obtain a suitable solution for our problem. There are many evolutionary algorithms now available in the field of computational mathematics and many more will join in the future. Among all the algorithms the swarm intelligence algorithm holds a special place for its flexibility and ability to perform under large dimensions. In our present paper, we propose a swarm intelligence algorithm motivated form the water suction strategy of root hair. There is much research that has been done to develop, modify and test the efficiency of evolutionary algorithms [1], making a comprehensive review of swarm optimization algorithms. The experimental problems are necessary to examine the algorithms and several benchmark problems in this direction [2] provide a detailed survey on benchmark experimental problems. In the research of swarm intelligence, many contributions have been made in this direction like the Particle Swarm Optimization Algorithm (PSO) (*Particle Swarm Optimization—an Overview|ScienceDirect Topics*, n.d.), Honey Bee Optimization Algorithm (HBO) (*Honey Bees Inspired Optimization Method: The Bees Algorithm—PMC*, n.d.), Wasp Swarm Algorithm (WS) [3], Ant Colony Optimization (ACO) [4], Firefly Algorithm (FA) (*Firefly Algorithm—an Overview|ScienceDirect Topics*, n.d.), etc. Among all these swarm algorithms, the particle swarm algorithm is exploited most to develop several modified versions of PSO such as weighted PSO, constriction PSO, Hamdan & Zawawi (2009) and MPSO. The ongoing research on optimization algorithms, various other algorithms (see Table 1) were developed based on evolutionary theories and physical processes.

2 Description of Algorithm

In a plant there are two parts, the shoot and the root, the purpose of the root is to hold the plant on the ground and to suck water from the soil. It is observed that root hair tends to grow in a direction in which the probability of the amount of water

Table 1 List of various algorithms

Evolutionary algorithm	Genetic algorithm [5]
	Genetic programming [6]
	Evolution strategies [7]
Swarm algorithm	Particle swarm optimisation [8]
	Firefly algorithm [9]
	Monkey search [10]

Fig. 1 Pictorial representation of root hair algorithm in nature

suction improves every time it moves forward. A root hair multiplies its number in the direction in which the root hair can suck maximum hair (see Fig. 1).

3 Root Hair Algorithm (Swarm Intelligence)

In our present optimization algorithm, we consider the strategy of root hair. A set of root hair tends to grow in a direction to suck water as much as possible. So we follow the same policy to find the minimum value of a function. Let F be the amount of amount of water pulled by the root. Let H be the head of a single root tip, Q be the capacity of the water suction of each root tip and N rate at which the water is pulled by a root hair. Let v be the specific speed of water suction then we have

$$v = \frac{N\sqrt{Q}}{H^{0.75}}$$

Let the rate at which the water is pulled by a root hair be defined as

$$N = Bestsuctionrate - Currentsuctionrate$$

Let $\{x_{ij}\}$ be the initial set of roots hair suction rate, where x_{ij} represents ith root hair in jth position. Let g_j represent the global best suction rate at jth position.

When we find the global best water suction rate at jth position and v_jth be the specific speed of water suction of ith root at jth position. Then the specific velocity equation of ith root hair at $(j+1)$th position is given by

$$v_{j+1}^{i} = v_{j}^{i} + \frac{(g_j - x_{ij})r\sqrt{Q}}{H^{0.75}}$$

where r is a random number uniformly distributed between 0 and 1. Further, we associate an inertial weight Δ to the specific velocity to stabilize the solution. Motivated by weighted PSO we adopt the definition of weight in [] that is

$$\Delta = \Delta_{max} - \frac{(\Delta_{max} - \Delta_{min})}{Maxite} j$$

where Δ_{max} and Δ_{min} are the upper and lower limit of inertial weights and *Maxite* is the maximum iteration we consider for our computation purpose. Then the resulting specific speed is

$$v_{j+1}^{i} \rightarrow \Delta v_{j+1}^{i}$$

The new set of root hair suction rate in the direction of the best suction rate is defined as;

$$x_{i(j+1)} = x_{ij} + v_{j+1}^{i}$$

Pictorial representation of Behaviors of RHA has been depicted in Fig. 2. The computational algorithm of the proposed procedure is as follows:

Computational Algorithm

Step 1: Let us initialize n root hair.
Step 2: Set the global best as the root hair for which the water suction is maximum from the initial set of root hair.
Step 3: Set the initial velocity of water suction be zero.
Step 4: Compute the amount of water that can be sucked by each root hair.
Step 5: Find the root hair G for which the water suction is maximum.
Step 6: Initialize n new root hair in the direction of G.
Step 7: Repeat Step IV.
Step 8: Update global best.
Step 9: Stopping criterion.

Figure 3, depicting the decision-making process of the computational algorithm.

4 Experimental Problems

The experimental problems are designed to test the performance of algorithms to optimize a function. There are five categories in which test functions can be classified; Continuity, Differentiability, Separable, Scalable and Unimodal or Multimodal. For our present paper, we classify the function into two categories Separable Unimodal; Non-separable Unimodal; Separable Multimodal and Non-Separable Multimodal. The following benchmark problems (see Table 2) are selected to perform the test.

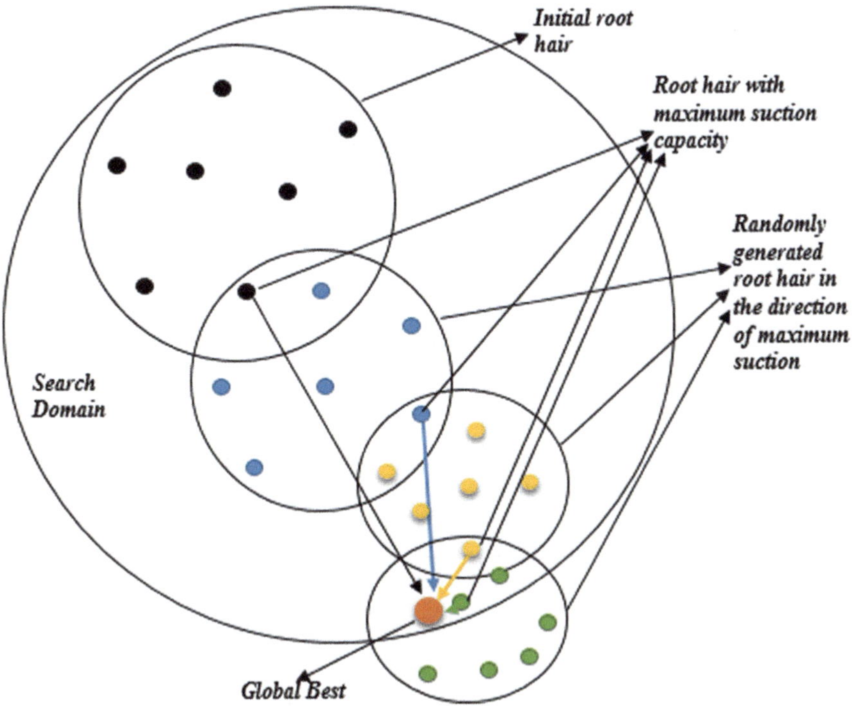

Fig. 2 Pictorial representation of behaviors of RHA

5 Performance Evaluation of Experimental Results

In this, the results obtained in the above problems are analysed with other calculated test results obtained by using GA, ACO, PSO and DE and are adopted from [1]. All the test problems are coded in SciLab 6.1.0 in the system with 12 GB RAM Intel i3-processor and in Windows 10 environment. The comparison is made on the Mean Error (**Mean** $\pm$ **SD**). The result is considered **0.0** if the obtained value is $a \times 10^{-10}$ and 30 independent runs are performed to evaluate both Mean and Standard Deviation.

The value of the parameters is given by
Number of Hair $n = 100$
Capacity $Q = 20$
Head $H = 5$
Maximum weight $w_{Max} = 0.9$
Minimum weight $w_{Min} = 0.4$
Maximum iteration $Max_{ite} = 100$
Maximum run $Max_{Run} = 30$

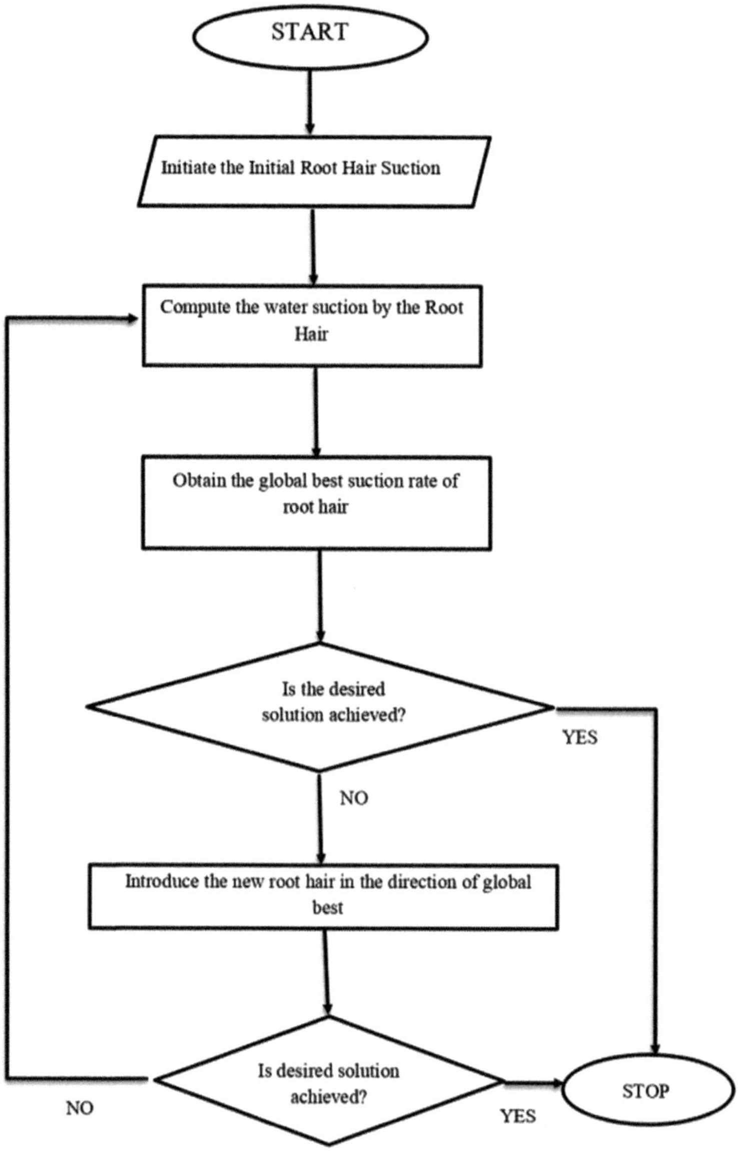

Fig. 3 Flowchart

To compare the values obtained on solving the benchmark problems by RHA with the standard algorithms stated above it is wise to analyse the solution of the problems taking 5 at a time (see Table 3).

Table 2 Benchmark problems

Sl. No	Function	Optimal value	Dim	Range	Property
1	Ackley 1	0	30	[−35,35]	Non-separable multimodal
2	Beale	0	2	[−4.5,4.5]	Non-separable unimodal
3	Bohachevsky 1	0	2	[−100,100]	Separable multimodal
4	Bohachevsky 2	0	2	[−100,100]	Non-separable multimodal
5	Bohachevsky 3	0	2	[−100,100]	Non-separable multimodal
6	Booth	0	2	[−10,10]	Non-separable unimodal
7	Branin	0.3978873	2	[−5,10;0,15]	Non-separable multimodal
8	Colville	0	4	[−10,10]	Non-separable multimodal
9	McCormick	−1.9133	2	[−1.5,4;−3,3]	Non-separable multimodal
10	Easom	−1	2	[−100,100]	Non-separable multimodal
11	Dixon-Price	0	24	[−10,10]	Non-separable unimodal
12	Rosenbrock	0	2	[−100,100]	Non-separable unimodal
13	Sum Square	0	30	[−10,10]	Separable unimodal
14	Sphere	0	30	[0,10]	Separable multimodal
15	Trid 6	−50	6	[−36,36]	Non-separable multimodal

The results obtained for the Ackley 1 indicate that RHA performs well as compared to other algorithms, however, the time of evaluation is better for PSO. All the algorithms provide the best results for Bohachevsky i, $i = 1, 2, 3$ except for the case when RHA stuck to the locally at 0.18 for $i = 3$, though it took less time of evaluation as compared to other algorithms but the does not show impressive results for $i = 3$. For Beale, RHA took the maximum time to evaluate the result but it produced an exact solution with zero error, along with RHA, the result obtained with PSO and DE contains zero error and takes almost half the time of evaluation for DE and least for PSO. Table 4 displayed the performance evaluation of experimental problem 6–10 with other algorithms.

Table 3 Performance evaluation of experimental problems 1–5 with other algorithms

Sl. No	Experimental Problem	GA	DE	PSO	ACO	RHA
1	**Ackley 1**	17.194 ± 0.79083 (4.0344 s)	12.795 ± 0.8414 (11.5375 s)	16.004 ± 0.53105 (**2.9844 s**)	15.884 ± 1.221 (8.9734)	2.998421 ± 0.87365 (8.672825)
2	**Beale**	0.70313 ± 0.0(3.107 s)	0.0 ± 0.0 (4.9531 s)	0.0 ± 0.0 (**1.9094 s**)	7.0313 ± 0.0 (2.8938)	0.0 ± 0.0 (7.072584)
3	**Bohachevsky 1**	0.0 ± 0.0 (6.351 s)	0.0 ± 0.0 (5.1844s)	0.0 ± 0.0 (1.93 2 s)	0.0 ± 0.0 (1.8641 s)	0.0 ± 0.0 (**1.34010**)
4	**Bohachevsky 2**	0.0 ± 0.0 (5.7203 s)	0.0 ± 0.0 (5.0547 s)	0.0 ± 0.0 (2.0359 s)	0.0 ± 0.0 (1.8547 s)	0.0 ± 0.0 (**1.6453**)
5	**Bohachevsky 3**	0.0 ± 0.0 (5.5688 s)	0.0 ± 0.0 (5.0203 s)	0.0 ± 0.0 (1.9438 s)	0.0 ± 0.0 (1.8547 s)	0.18 ± 0.0 (0.670047)

Table 4 Performance evaluation of experimental problem 6–10 with other algorithms

Sl. No	Experimental Problems	GA	DE	PSO	ACO	RHA
6	**Booth**	0.0 ± 0.0(2.5125 s)	0.0 ± 0.0(4.7984 s)	0.0 ± 0.0(1.8234 s)	0.0 ± 0.0(1.8719 s)	0.0 ± 0.0(**0.61859 s)**
7	**Branin**	0.39789 ± 0.0(5.9844 s)	0.39789 ± 0.0(4.9469 s)	0.39789 ± 0.0(1.9719 s)	0.39789 ± 0.0(**1.8563 s)**	0.37887 ± 0.0(0.76167 s)
8	**Colville**	0.0 ± 0.0(2.6875 s)	1.4017 ± 2.1101(4.7859 s)	0.0 ± 0.0(**1.892 s)**	66.160 ± 37.94(2.2703 s)	0.0024538 ± 9.65842(6.324979 s)
9	**McCormick**	−1.9134 ± 0.0(2.7969 s)	−1.9134 ± 0.0(2.3266 s)	−1.9134 ± 0.0(**1.8844 s)**	−1.9134 ± 0.0(5.0672 s)	−1.9132 ± 0.0(4.58641 s)
10	**Easom**	−1 ± 0.0(6.0281 s)	−1 ± 0.0(4.8734 s)	−1 ± 0.0(**1.887 s)**	−1 ± 0.0(2.0016 s)	−1 ± 0.0(2.30356 s)

The performance of all the algorithms is significant for the Booth function and as per the speed is concerned RHA produces results faster than others. For the Branin function, RHA produces the best result both in terms of optimal value and speed. GA and PSO performed well and produced the best results for the Colville function and PSO outruns all the other algorithms in terms of speed. RHA performed well for the Colville function but the time of execution was slow as compared to other algorithms. All the algorithms performed well for the Easom function; however, the time of execution is the least in PSO. Table 5, displayed the performance evaluation of experimental problem 11–15 with other algorithms.

The performance of RHA in Dixon-Price, Rosenbrock and Sphere functions is better than another algorithm. For the Sum-Square and Trid 6 problem the solution is not satisfactory rather mean diverges from the actual optimal value. However, the actual solution can be achieved with RHA when Max_{ite} Increases (see Table 6).

6 Sensitivity Analysis

In this section, the parameters involved in the algorithm are analysed subject to the benchmark problem Rosenbrock function. The change in the parameters is given by $N \to N * (1 + r)$ where, $r \in \{-0.5, -0.25, 0.25, 0.5\}$.Analysis results have been displayed in Table 7.

Interpretation of Sensitivity Analysis

- The solution is closer to the optimal value of 0 if the number of hairs increases.
- The solution improves with the increase in capacity.
- The solution is closer to the optimal value when the value of Head is near 5.
- The weights w_{Max} and w_{Min} show better results when it is closer to the values 0.9 and 0.4, respectively.

7 Conclusion

In this paper, an attempt has been made to develop an optimization algorithm inspired by the strategy of water suction of root hair and weighted particle swarm optimization. The algorithm is tested with 15 experimental problems selected randomly from [2], which are divided into 3 different categories based on the number of variables. Parameters of the algorithm are set and executed in SciLab 6.1.0 to get the Mean $\pm$ SD which is then compared with the existing solution of [1]. The number of iterations and maximum run are kept the same as per [1] so the total data for mean and standard deviation remains the same and a fair comparison can be made. The result obtained suggests that, the root hair algorithm performed better than GA, DE, PSO and ACO. However, the performance of RHA in solving separable problems is not satisfying, though the algorithm performed significantly well in the case of

Table 5 Performance evaluation of experimental problem 11–15 with other algorithms

Sl. No	Experimental Problems	GA	DE	PSO	ACO	RHA
11	**Dixon-Price**	110290 ± 34184(4.1063 s)	1314500 ± 6382.2(11.3109 s)	4963300 ± 3431700(**3.1 s**)	1870800 ± 434440(6.7469 s)	45202.43 ± 14049.422(8.314 s)
12	**Rosenbrock**	12493000 ± 8672500(4.0797 s)	3890100 ± 2241700(11.0344 s)	702890000 ± 489370000(2.9797 s)	110510000 ± 21694000(6.9359 s)	12041 ± 28574.895(**0.96347 s**)
13	**Sum Square**	17.376 ± 3.7449×10^{-15}(3.8938 s)	7.7637 ± 1.4868(11.4047 s)	3.7357 ± 0.18203(**3.1422 s**)	5.6363 ± 0.40719(6.9031 s)	447.25 ± 2086.9261(8.99323 s)
14	**Sphere**	644.15 ± 168.76(**4.3531 s**)	5594.2 ± 1509.1(10.9984 s)	104540 ± 1860.3(7.3219 s)	17596 ± 1.8603(7.3219 s)	589.428 ± 136.16(8.6775994 s)
15	**Trid 6**	−0.25 ± 12.293(3.1516 s)	−47.697 ± 5.0327(5.8891 s)	−5 ± 0.0(2.0859 s)	−24.3 ± 9.3339(2.7422 s)	4.068965 ± 10.39893(**2.5115 s**)

Table 6 Performance breakdown of all the algorithms

Category	Number of functions	GA	DE	PSO	ACO	RHA
Best performing method	15	6	7	7	5	**11**
Unimodal separable	1	0	0	**1**	0	0
Unimodal inseparable	4	1	2	1	1	**4**
Multimodal separable	2	1	**2**	1	1	1
Multimodal inseparable	8	4	4	5	3	**6**
Unimodal	5	1	2	2	1	**4**
Multimodal	10	5	6	6	4	**7**
Separable	3	1	**2**	**2**	1	1
Inseparable	12	5	6	6	4	**10**

Table 7 Sensitivity Analysis of the Parameters of RHA

Parameters	% changed	Changed value	x_1	x_2	Optimal value	Time elapse
$n = 100$	−50	50	4.4777	20.0521	12.0949	0.4369
	−25	75	1.00008	1.000159	6.283×10^{-9}	0.6593
	25	125	1	1	4.981×10^{-13}	1.09085
	50	150	1	1	5.116×10^{-15}	1.30723
$Q = 20$	−50	10	0.9984852	0.9969656	0.0000023	0.8757315
	−25	15	1.0000807	1.00001618	6.53×10^{-9}	0.8719258
	25	25	0.9999989	9,999,979	1.118×10^{-12}	0.8740488
	50	30	0.9999997	0.9999994	1.254×10^{-13}	0.879733
$H = 5$	−50	2.5	1.0048781	1.009771	0.0000238	0.8784215
	−25	3.75	1.000005	1.0000101	2.528×10^{-11}	0.8730672
	25	6.25	1.0000002	1.0000003	2.794×10^{-14}	0.8748685
	50	7.5	1.0000008	1.0000016	6.318×10^{-13}	0.8743909
$w_{Max} = 0.9$	−50	0.45	0.994391	0.9888006	0.0000315	0.8807585
	−25	0.675	0.9999152	0.9998301	7.204×10^{-9}	0.8845122
	25	1.125	1.0000147	1.0000271	1.797×10^{-10}	0.876673
	50	1.35	0.9991651	0.998588	0.0000073	0.8789276
$w_{Min} = 0.4$	−50	0.2	0.9999724	0.9999447	7.614×10^{-10}	0.8763335
	−25	0.3	0.9999999	0.99999998	9.012×10^{-15}	0.8735013
	25	0.5	0.9999986	0.99999968	2.328×10^{-11}	0.8773104
	50	0.6	1.0000895	1.0001849	1.143×10^{-8}	0.8810503

inseparable problems. Sensitivity analysis of the parameters of RHA is performed and the result suggests that increasing the number of hair (population) and capacity parameters can improve the global best. Further, the solution reaches to optimal value if the head $H = 5$.

This is the first time a numerical optimization algorithm has been developed inspired by the water suction strategy of root hair as far as the author's best knowledge. The obtained result suggests that the algorithm performed well or even better in some cases than the existing algorithms. The proposed algorithm can be extended to get a more improved solution, furthermore, experimental problems can be selected and solved to test the efficiency and robustness of RHA.

References

1. Wahab, M.N.A., Nefti-Meziani, S., Atyabi, A.: A comprehensive review of swarm optimization algorithms. PLOS ONE **10**(5), e0122827 (2015). https://doi.org/10.1371/journal.pone.0122827
2. Jamil, M., Yang, X.-S.: A Literature Survey of Benchmark Functions For Global Optimization Problems. Int. J. Math. Model. Numer. Optim. **4**(2), 150 (2013). https://doi.org/10.1504/IJMMNO.2013.055204
3. Pinto, P., Runkler, T.A., Sousa, J.M.: Wasp swarm optimization of logistic systems. In: B. Ribeiro, R.F. Albrecht, A. Dobnikar, D.W. Pearson, N.C. Steele (Eds.) Adaptive and Natural Computing Algorithms (pp. 264–267). Springer (2005). https://doi.org/10.1007/3-211-27389-1_63
4. Selvi, V., Umarani, Dr. R.: Comparative analysis of ant colony and particle swarm optimization techniques. Int. J. Comput. Appl. **5**(4), 1–6 (2010). https://doi.org/10.5120/908-1286
5. Devooght, R.: Multi-objective genetic algorithm (2011). https://www.semanticscholar.org/paper/Multi-objective-genetic-algorithm-Devooght/bfeef4cf230d4db51ef332237dae8530f2b5f613
6. Koza, J.R., Poli, R.: Genetic Programming. In: E.K. Burke, G. Kendall (Eds.), Search methodologies: Introductory tutorials in optimization and decision support techniques (pp. 127–164). Springer US (2005). https://doi.org/10.1007/0-387-28356-0_5
7. Beyer, H.-G., Schwefel, H.-P.: Evolution strategies—A comprehensive introduction. Nat. Comput. **1**, 3–52 (2002). https://doi.org/10.1023/A:1015059928466
8. Pereira, G.: Particle swarm optimization (2011)
9. Kouziokas, G.: Firefly Algorithm, Harmony Search and Cat Swarm Algorithm (pp. 111–130) (2023). https://doi.org/10.1201/9781003247746-6
10. Kuliev, E., Kureichik, V.: Monkey search algorithm for ECE components partitioning. J. Phys.: Conf. Ser. **1015**, 042026 (2018). https://doi.org/10.1088/1742-6596/1015/4/042026
11. Firefly Algorithm—An overview|ScienceDirect Topics (n.d.). Retrieved July 15, 2023, from https://www.sciencedirect.com/topics/engineering/firefly-algorithm
12. Honey Bees Inspired Optimization Method: The Bees Algorithm—PMC (n.d.) Retrieved July 15, 2023, from https://www.ncbi.nlm.nih.gov/pmc/articles/PMC4553508/
13. Particle Swarm Optimization—An overview|ScienceDirect Topics. (n.d.). Retrieved July 15, 2023, from https://www.sciencedirect.com/topics/engineering/particle-swarm-optimization

Pythagorean Fuzzy Ordered Weighted Averaging Aggregation Operator Based on Appropriate Score Function and Their Application to Multi-criteria Decision-Making in IT Project Management

Jogjiban Chakraborty, Sathi Mukherjee, and Laxminarayan Sahoo

Abstract Multi-Criteria Decision-Making (MCDM) plays a pivotal role in addressing complex decision problems with multiple conflicting criteria. Within the context of Pythagorean Fuzzy Sets (PFSs), the Pythagorean fuzzy ordered weighted averaging aggregation operator (PFOWA operator) serves as a mathematical tool for combining data from multiple sources. Here, we have also proposed Einstein-Ordered Weighted Averaging aggregation operator (EOWA operator) to solve the MCDM problem incorporating appropriate score function measure of Pythagorean Fuzzy Sets. With the use of Pythagorean Fuzzy Sets along with PFOWA and EOWA operators incorporating an appropriate score function, this research investigates the application of multi-criteria decision-making within IT project management. The proposed operators aim to address the inherent uncertainties and complexities associated with IT project environments by leveraging Pythagorean Fuzzy Numbers. Through the integration of a suitable score function, the PFOWA and EOWA operators allow for a nuanced consideration of the importance and ranking of criteria, contributing to a more refined decision-making process. The study explores the practical application of these two operators in IT project management, providing insights into their effectiveness in handling diverse and conflicting criteria. The findings contribute to the advancement of decision support systems in IT project management, offering a promising avenue for enhancing the precision and adaptability of decision-making processes in this dynamic field. Here, we have determined the more suitable team based on a few criteria in the current competitive environment in the

J. Chakraborty
Department of Mathematics, Bankura University, Bankura 722133, India

S. Mukherjee
Department of Mathematics, Gobinda Prasad Mahavidyalaya, Bankura 722133, India

L. Sahoo (✉)
Department of Computer and Information Science, Raiganj University, Raiganj 733134, India
e-mail: lxsahoo@gmail.com

L. Sahoo et al. (eds.), *Decision Making Under Uncertainty Via Optimization, Modelling, and Analysis*, Studies in Systems, Decision and Control 558,
https://doi.org/10.1007/978-981-96-0085-4_32

IT industry with the help of appropriate scoring measures for Pythagorean fuzzy sets using the aggregation approach. Subsequently, a numerical example has been considered to assess the performance of the scoring function that was enriched with Pythagorean Fuzzy Set Measures across a range of multi-criteria decision scenarios. Finally, the experimental results of the study have been presented.

Keywords Pythagorean fuzzy sets · Multi-criteria decision-making · PFOWA operator · EOWA operator · Uncertainty · Decision-makers · IT project management

Abbreviations

AHP	Analytic Hierarchy Process
ANP	Analytic network process
DM	Decision matrix
DMs	Decision-makers
ELECTRE	Elimination and Choice Translating Reality
EOWA	Einstein Ordered Weighted Averaging aggregation
FS	Fuzzy Set
IFS	Intuitionistic fuzzy set
IS	Information System
IT	Information technology
MCDM	Multi-criteria decision-making
MINLP	Mixed integer non-linear programming
NLP	Non-linear programming
PFN	Pythagorean fuzzy number
PFOWA	Pythagorean fuzzy ordered weighted averaging
PFS	Pythagorean fuzzy set
PM	Project Management
PS	Project selection
R&D	Research and Development
TOPSIS	Technique for Order of Preference by Similarity to Ideal Solution
VIKOR	VIekriterijumsko KOmpromisno Rangiranje

1 Introduction

To effectively complete a project, a set of related activities need to be scheduled and carried out in a particular order to produce a unique output (product or service) within a certain period. This process is known as project management (PM) [1]. Given the perspective of the multi-criteria decision-making (MCDM) technique, the specifics of a research project called information technology (IT) management projects are explored. The processes of establishing and operating IT according to the needs for information accessibility and procurement have grown along with the economic necessity and advantages of information. The techniques, abilities, and ideas of project management are foremost among these fields. Organisations understand the importance of project management in terms of efficiently organising and managing work to transform resources into new IT products, creating IT services for customers inside and outside the organisation, planning and constructing information systems (IS) and bringing about organisational transformation. The flowchart for the entire IT project Management System is depicted in Fig. 1. As an outcome, IT grew to prominence as a setting in which project management emerged, leading to a large amount of cross-disciplinary study at the interface of projects and IT [2, 3]. Applications for software development must be updated and changed often to support enterprise systems. Numerous software development activities are involved in these ongoing IS/IT modifications. Undertaken as projects, these endeavours are typically sophisticated and occur in a constantly changing environment where commercial circumstances and technological advancements occur on a regular basis. Software development projects have an elevated rate of failure, which has resulted in catastrophic effects on organisations. Many researchers have attempted to identify the variables influencing the development of IS/IT projects and have explored different strategies for controlling it [4].

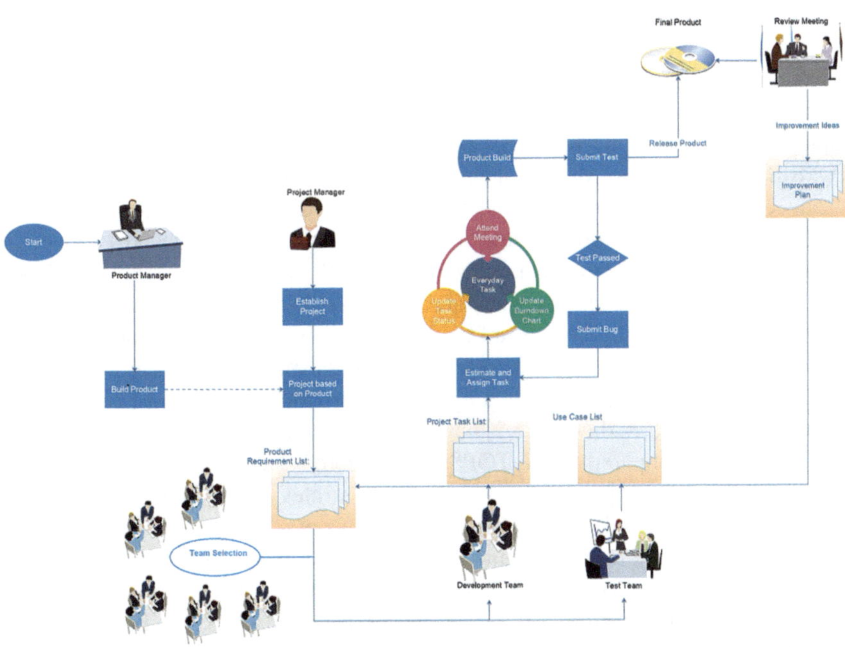

Fig. 1 Flowchart for the entire IT project management system

The lack of comprehensive knowledge on the organisational phenomena that Argote [5] is studying is a broad definition of uncertainty. Requirements uncertainty, according to Nidumolu [6], is the discrepancy between the quantity of information held by the developers and the information required to establish user requirements. In the rapidly evolving landscape of Information Technology (IT) project management, the complexity and multifaceted nature of decision-making processes pose significant challenges to project managers. As IT projects involve numerous criteria, often with varying degrees of importance and uncertainty, effective decision support systems become paramount for successful project outcomes. The integration of fuzzy set theory into decision-making models has emerged as a promising approach to address the ambiguity inherent in IT project environments. The problem of project selection (PS) holds significant importance in managerial decision-making. The decision-maker, or decision makers, have a broad range of selection criteria available to them when it comes to projects that have contradictory or inaccessible characteristics. Comparing different criteria and using pairwise comparison to determine their relative relevance is a difficult task for decision-makers (DMs). Thus, selecting a project is a multi-criteria decision-making (MCDM) problem [7]. Furthermore, the objectives and parameters of project selection models are typically taken to be deterministic or crisp. However, due to insufficient and unavailable data, real-world project selection problems are associated with a variety of imprecise/fuzzy sources. As such, applying probability theory or fuzzy theory may be a useful strategy. Different fields use different strategies for choosing projects. Most well-done works frequently result in complex mathematical programming, such as non-linear programming (NLP) or mixed-integer nonlinear programming (MINLP) [8–13]. On the other hand, a variety of multicriteria decision-making (MCDM) techniques have been established recently to address PS problems. For the project selection problem, Mohanty [14] employed The Technique for Order of Preference by Similarity to Ideal Solution (TOPSIS) method as an MCDM technique. In prior research, Mohanty et al. [15] introduced an approach to R&D project selection based on a fuzzy analytical network process (ANP). For the project selection problem, Enea & Piazza [16] suggested a restricted fuzzy analytical hierarch process (AHP) technique. Mahmood Zadeh et al. [17] used the TOPSIS approach and fuzzy AHP to choose projects. AHP and fuzzy TOPSIS algorithms were used by Amiri [18] for project selection in oil-field development. A fuzzy ELECTRE technique for choosing the best project using the TOPSIS method has been presented by Daneshvar Rouyendegh and Erol [19]. On the other hand, some studies [20–22] have demonstrated that VIKOR yields superior outcomes when compared to the TOPSIS approach. To solve the project selection problem, this research provides a hybrid fuzzy strategy that combines AHP and VIKOR techniques. Zadeh [23] introduced the fuzzy set (FS) concept to address uncertainty in a more efficient manner. The situation between "exists" and "does not exist" is presented by a fuzzy set by giving elements the membership degree $\mu(x)$ about a set. The notion of the intuitionistic fuzzy set (IFS) has been proposed by Atanassov [24, 25] and contained non-membership degree $\gamma(x)$ along with $\mu(x)$ satisfying $0 \leq \mu(x) + \gamma(x) \leq 1$ and $\mu(x) + \gamma(x) + \pi(x) = 1$, where $\pi(x)$ is the hesitant function. Considering this, it has been found that IFSs perform better than

FS at simulating human expressions like perception, knowledge and behaviour. The Pythagorean Fuzzy Sets (PFS) approach, developed by Yager [26, 27], is predicated on the idea that the squared sum of the degrees of membership and non-membership is less than one. The use of membership degree, non-membership degree and the degree of hesitation to express complex information are the main advantages of such extended FSs. In past studies, several score functions were constructed and examined to quantify PFSs [28–35]. Several applications have efficiently made use of these measures. The Pythagorean fuzzy set (PFS) is a useful tool for managing ambiguity. Numerous PFS scores are a well-known study with real-world applications in a variety of domains, such as medical diagnosis, clustering, decision-making and pattern analysis. When it comes to multi-criteria decision-making (MCDM), when complex decisions require an analysis of several competing criteria, the precise evaluation and prioritisation of options are critical. In real-world decision scenarios, uncertainties and imprecisions are inherent, and traditional approaches frequently find it difficult to account for them. Fuzzy set theory has been used to address vagueness and ambiguity in decision-making processes as a solution to this problem. Conventional fuzzy sets, however, do not always capture the complexity of uncertainty well enough. As an expansion of the conventional fuzzy sets, Pythagorean Fuzzy Sets (PFS) have come to light in recent years, providing a more thorough framework for more accurately representing ambiguous data. This work explores the transformative effects of scoring functions when strengthened with Pythagorean Fuzzy Set Measures, focusing on the relationship between MCDM and Pythagorean Fuzzy Set theory. To overcome the limitations of existing techniques and open the door to more reliable and accurate decision-making processes, this study combines the strength of scoring function with the sophisticated representation capabilities of Pythagorean Fuzzy Sets. This research focuses on the development and application of a Pythagorean Fuzzy Ordered Weighted Averaging (PFOWA) aggregation operator [36], augmented by an appropriate score function, to facilitate multi-criteria decision-making in the realm of IT project management. The Pythagorean Fuzzy Numbers (PFN), an extension of classical fuzzy numbers, enable a more nuanced representation of uncertainty by incorporating a membership degree, non-membership degree, and a hesitation degree. This inherent flexibility aligns well with the intricacies of IT project management where decision criteria often exhibit degrees of imprecision. The PFOWA aggregation operator extends the traditional Ordered Weighted Averaging (OWA) approach by incorporating Pythagorean Fuzzy Numbers. This extension allows project managers to not only aggregate criteria but also consider the relationships among them, offering a more comprehensive and contextually relevant decision-making framework. Integral to the proposed PFOWA aggregation operator is the incorporation of an appropriate score function. This function plays a pivotal role in assigning weights to criteria based on their relative significance, reflecting the preferences and priorities of decision-makers. The utilization of a tailored score function enhances the precision and adaptability of the PFOWA operator, ensuring a more accurate representation of decision elements in the IT project management context. The research aims to bridge the gap between theoretical advancements in fuzzy set theory and their practical applications in IT project management. To solve the MCDM problem in the context of IT project management, we have also proposed the Einstein ordered weighted averaging (EOWA) aggregation operator. By exploring the effectiveness of the PFOWA and EOWA aggregation operator with an appropriate score function, the study seeks to provide project managers with a powerful tool for navigating the intricate decision landscapes inherent in IT projects. As IT projects involve diverse stakeholders, conflicting objectives, and evolving technological landscapes, the developed model holds the potential to significantly enhance the decision-making capabilities of project managers, ultimately contributing to the success of IT projects in an increasingly competitive and dynamic industry. To choose the best team, the aggregations based multi-criteria decision-making method is used combined with a novel Pythagorean fuzzy scoring measure strategy. First, a relevant score measure from previously conducted studies has been taken into consideration here. To deal with ambiguous and conflicting information, the suggested work makes use of four criteria and five distinct alternatives/options in a fuzzy environment. The findings demonstrate that, in IT project management, the recommended technique is a more workable approach to choosing the most suitable development teams for software products. Next, a numerical example has been examined to evaluate the effectiveness of an aggregation's techniques combined with suitable scoring function enhanced with Pythagorean Fuzzy Set Measures in a variety of multi-criteria decision situations. Finally, the result of the outcomes has been presented.

The remaining work is displayed as follows. In Sect. 2, the basic fuzzy mathematics concepts that are employed throughout the work are explained. Section 3 gives specifics on how the Entropy Method is used to calculate weights for criteria. Section 4 provides an explanation of how to arrange performances in MCDM using PFOWA and EOWA operators. Section 5 includes a case study, a numerical experiment, and remarks. Section 6 of the study contains concluding remarks.

2 Preliminaries

We have defined a few fundamental mathematical terms in this section that are utilised throughout the entire manuscript.

Definition 2.1 (Yager, [26]).

Let U be a universe of discourse. A Pythagorean fuzzy set (PFS) P in U is given by $P = \{< u, \mu_P(u), \gamma_P(u) >: u \in U\}$, where $\mu_P : U \to [0, 1]$ denotes the degree of membership and $\gamma_P : U \to [0, 1]$ [0, 1] denotes the degree of non-membership of the element $u \in U$ to the set P respectively and the condition that $0 \leq \mu_P^2(u) + \gamma_P^2(u) \leq 1,\ u \in U$.

The degree of hesitation that the element $u \in U$ belongs to the PFS P is denoted by $\pi_P(u)$ and is defined by $\pi_P(u) = \sqrt{1 - \mu_P^2(u) - \gamma_P^2(u)}$.

Now, according to Li et al. [51], we may denote $P = (\mu_P, \gamma_P)$ as Pythagorean fuzzy number (PFN).

Let $P = (\mu_P, \gamma_P)$ be the PFN then hesitation of the PFN $P = (\mu_P, \gamma_P)$ is $\pi_P = \sqrt{1 - \mu_P^2 - \gamma_P^2}$.

Definition 2.2 (Zhang & Xu [37]).

Let $P = (\mu_P, \gamma_P)$, $P_1 = (\mu_{P_1}, \gamma_{P_1})$ and $P_2 = (\mu_{P_2}, \gamma_{P_2})$ are three PFNs and a scalar $\lambda > 0$; then

$$P_1 + P_2 = (\sqrt{\mu_{P_1}^2 + \mu_{P_2}^2 - \mu_{P_1}^2 \mu_{P_2}^2}, \gamma_{P_1} \gamma_{P_2})$$

$$P_1 P_2 = (\mu_{P_1} \mu_{P_2}, \sqrt{\gamma_{P_1}^2 + \gamma_{P_2}^2 - \gamma_{P_1}^2 \gamma_{P_2}^2})$$

$$\lambda P = (\sqrt{1 - (1 - \mu_p^2)^\lambda}, \gamma_P^\lambda)$$

$$P^\lambda = (\mu_P^\lambda, \sqrt{1 - (1 - \gamma_p^2)^\lambda})$$

Definition 2.3 (Zhang and Xu [37]).

Let $P = (\mu_P, \gamma_P)$ be the PFN, then the accuracy degree of $P = (\mu_P, \gamma_P)$ can be defined as follows:

$$H(P) = \mu_P^2 + \gamma_P^2$$

Definition 2.4 Zhang and Xu [37]).

Let $P = (\mu_P, \gamma_P)$ be the PFN, then the score function of $P = (\mu_P, \gamma_P)$ can be defined as follows:

$$S(P) = \mu_P^2 - \gamma_P^2$$

Definition 2.5 (Zhang, [30]).

Let $P = (\mu_P, \gamma_P)$ be the PFN, then the closeness index of $P = (\mu_P, \gamma_P)$ can be defined as follows:

$$R(P) = \frac{1 - \gamma_P^2}{2 - \mu_P^2 - \gamma_P^2}$$

It is to be noted that $H(P) \in [0, 1]$, whereas $S(P) \in [-1, 1]$. Here, it is to be mentioned that a function $F(.)$ is positive if $F(.) \in [0, 1]$ and negative if $F(.) \in [-1, 0)$. In the purpose of ranking of PFNs, most of the researchers have considered score function lies in the interval -1 and 1. Also, some of the researchers have proposed score function lies in the interval 0 and 1. For more details, one may refer to the work of Sahoo (2020). So, to maintain the same, in this chapter, we have considered the score function the same as the closeness index defined by Zhang [30].

So, we have considered $S(P) = \frac{1-\gamma_P^2}{2-\mu_P^2-\gamma_P^2}$ as a score function.

Theorem 2.1 The score function $S(P) = \frac{1-\gamma_P^2}{2-\mu_P^2-\gamma_P^2}$ is positive, i.e., $S(P) \in [0, 1]$.

Proof: For any PFN $P = (\mu_P, \gamma_P), 0 \le \mu_P^2 + \gamma_P^2 \le 1$ and $0 \le \mu_P \le 1; 0 \le \gamma_P \le 1$ Consequently $\gamma_P^2 \le 1 \Rightarrow 0 \le 1 - \gamma_P^2 \le 1$. Also, $2 - \mu_P^2 - \gamma_P^2 > 0$. So, we may write $0 \le \frac{1-\gamma_P^2}{2-\mu_P^2-\mu_P^2} \le \frac{1}{2-\mu_P^2-\mu_P^2} \le 1$. Therefore, $0 \le \frac{1-\gamma_P^2}{2-\mu_P^2-\mu_P^2} \le 1 \Rightarrow S(P) \in [0, 1]$.

Definition 2.6 Let $P_1 = (\mu_{P_1}, \gamma_{P_1})$ and $P_2 = (\mu_{P_2}, \gamma_{P_2})$ be two PFNs. Then ranking or order relation between P_1 and P_2 are as follows:

(i) $P_1 \succ P_2$ if and only if either $S(P_1) > S(P_2)$ or $S(P_1) = S(P_2)$ and $H(P_1) > H(P_2)$.

(ii) $P_1 \prec P_2$ if and only if either $S(P_1) < S(P_2)$ or $S(P_1) = S(P_2)$ and $H(P_1) < H(P_2)$.

(iii) $P_1 \approx P_2$ iff $S(P_1) = S(P_2)$ and $H(P_1) = H(P_2)$.

Definition 2.7 (Yager and Abbasov [38]).

Let $P_j = (\mu_{P_j}, \gamma_{P_j}), j = 1, 2, ..., n$ be the collection of PFNs; then Pythagorean fuzzy ordered weighted averaging aggregation operator can be defined as follows:

$PFOWA_w(P_1, P_2, ..., P_n) = w_1 P_{\sigma(1)} + w_2 P_{\sigma(2)} + ... + w_n P_{\sigma(n)}$, where $w = (w_1, w_2, ..., w_n)^T$ is the weighted vector of $P_j = (\mu_{P_j}, \gamma_{P_j}), j = 1, 2, ..., n$ such that $0 \le w_j \le 1, j = 1, 2, ..., n$ and $\sum_{j=1}^{n} w_j = 1$.

Here, $(\sigma_{(1)}, \sigma_{(2)}, ..., \sigma_{(n)})$ is a permutation of $(1, 2, ..., n)$ such that $P_{\sigma(j)} \le P_{\sigma(j-)}$, for all $j = 1, 2, ..., n$.

Theorem 2.2 (Rahman et al., [36]).

Let $P_j = (\mu_{P_j}, \gamma_{P_j}), j = 1, 2, ..., n$ be the collection of PFNs; then their aggregated value by using the PFOWA operator is also a PFN, and:

$PFOWA_w(P_1, P_2, ..., P_n) = \left(\sqrt{1 - \prod_{j=1}^{n} (1 - \mu_{\sigma(j)}^2)^{w_j}}, \prod_{j=1}^{n} \gamma_{\sigma(j)}^{w_j} \right)$, where $w = (w_1, w_2, ..., w_n)^T$ is the weighted vector of $P_j = (\mu_{P_j}, \gamma_{P_j}), j = 1, 2, ..., n$ such that $0 \le w_j \le 1, j = 1, 2, ..., n$ and $\sum_{j=1}^{n} w_j = 1$.

Here, $(\sigma_{(1)}, \sigma_{(2)}, ..., \sigma_{(n)})$ is a permutation of $(1, 2, ..., n)$ such that $P_{\sigma(j)} \prec P_{\sigma(j-1)}$, for all $j = 1, 2, ..., n$.

Definition 2.8 Let $R_j, j = 1, 2, ..., n$ be the collection of real numbers; then Einstein-ordered weighted averaging aggregation operator can be defined as follows:

$EOWA_w(R_1, R_2, ..., R_n) = w_1 \otimes R_{\sigma(1)} \oplus w_2 \otimes R_{\sigma(2)} \oplus ... \oplus w_n \otimes R_{\sigma(n)}$, where $w = (w_1, w_2, ..., w_n)^T$ is the weighted vector of $R_j, j = 1, 2, ..., n$ such that $0 \le w_j \le 1$, $j = 1, 2, ..., n$ and $\sum_{j=1}^{n} w_j = 1$.

Here, $(\sigma_{(1)}, \sigma_{(2)}, ..., \sigma_{(n)})$ is a permutation of $(1, 2, ..., n)$ such that $R_{\sigma(j)} \le R_{\sigma(j-1)}$, for all $j = 1, 2, ..., n$. It is to be noted that for two real numbers x and y, we have used Einstein t-norm for product $x \otimes y$ and t-conorm for sum $x \oplus y$. i.e., $T(x, y) = x \otimes y = \frac{xy}{1+(1-x)(1-y)}$ and $S(x, y) = x \oplus y = \frac{x+y}{1+xy}$.

3 Weight Calculation for Criteria Using the Entropy Method

This section has provided a step-by-step computing approach for entropy-based criteria weight calculation. For further information, refer to Shannon's work [39]. If $D = (S(P_{ij}))_{m\times n}$ be the decision matrix and $w = (w_1, w_2, ..., w_n)^T$ is the weight such that $0 \le w_j \le 1$ along with $\sum_{j=1}^{n} w_j = 1$. Following are the different steps to calculate the criteria weight w_j, $j = 1, 2, ..., n$.

Step 1: For $j = 1, 2, ..., n$ calculate $E_j = -\frac{1}{\log(n)} \sum_{i=1}^{m} p_{ij} \log(p_{ij})$ where $p_{ij} = \frac{S(P_{ij})}{\sum_{j=1}^{n} S(P_{ij})}$. It is to be worth mentioning that $\lim_{p_{ij}\to 0.} p_{ij} \log p_{ij} \to 0$.

Step 2: For $j = 1, 2, ..., n$ calculate $w_j = \frac{1-E_j}{\sum_{j=1}^{n} 1-E_j}$.

4 Method to Use PFOWA and EOWA Operators to Arrange Performances in MCDM

This section explains the usage of the PFOWA and EOWA operators for solving an MCDM problem when the weights of the criteria are unknown and can be determined by using Shannon entropy. Let $D = (P_{ij})_{m\times n}$ be the decision matrix. Here, each element of the decision matrix be expressed in terms of PFNs as $P_{ij} = (\mu_{ij}, \gamma_{ij})$. Now, using the entropy method mentioned in Sect. 2.3 we have calculated the criteria weights $w = (w_1, w_2, ..., w_m)^T$. After that, we employed $PFOWA_w(P_1, P_2, ..., P_n)$ and $EOWA_w(S(P_1), S(P_2), ..., S(P_n))$ operators and derived attribute values of each alternative T_i, $i = 1, 2, ..., m$. We first computed the attribute values using the PFOWA and EOWA operators, and then we computed the corresponding scoring function values. Based on this, the choice with the greatest function value score is the most effective selection.

4.1 Computational Procedure

Step1: Form the decision matrix.

Let there be m alternatives $T_1, T_2, ..., T_m$ and n criteria $C_1, C_2, ..., C_n$ and $P_{ij} = (\mu_{ij}, \gamma_{ij})$, $i = 1, 2, ..m$; $j = 1, 2, ..., n$ is either benefit or cost for the decision makers' point of view. Then the decision matrix (DM) is as follows:

$$D = (P_{ij})_{m\times n} = \begin{matrix} T_1 \\ T_2 \\ \nu \\ T_m \end{matrix} \begin{pmatrix} C_1 & C_2 & \dots & C_n \\ (\mu_{11}, \gamma_{11}) & (\mu_{12}, \gamma_{12}) & \dots & (\mu_{1n}, \gamma_{1n}) \\ (\mu_{21}, \gamma_{21}) & (\mu_{22}, \gamma_{22}) & \dots & (\mu_{2n}, \gamma_{2n}) \\ \text{A} & \text{A} & \dots & \text{A} \\ (\mu_{m1}, \gamma_{m1}) & (\mu_{m2}, \gamma_{m2}) & \dots & (\mu_{mn}, \gamma_{mn}) \end{pmatrix}$$

Step 2: Calculate the score function value of each P_{ij} then normalized DM can be written as follows:

$$S((P_{ij})_{m\times n}) = \begin{matrix} T_1 \\ T_2 \\ \nu \\ T_m \end{matrix} \begin{pmatrix} C_1 & C_2 & \dots & C_n \\ S(P_{11}) & S(P_{12}) & \dots & S(P_{1n}) \\ S(P_{21}) & S(P_{22}) & \dots & S(P_{2n}) \\ \text{A} & \text{A} & \dots & \text{A} \\ S(P_{m1}) & S(P_{m2}) & \dots & S(P_{mn}) \end{pmatrix}$$

Here, $S(P_{ij}) = S((\mu_{ij}, \gamma_{ij}))$ be the score function value for the Pythagorean fuzzy set which is either benefit or cost in view of decision makers under uncertainty.

Step 3: Calculate the weight vector $w = (w_1, w_2, ..., w_n)^T$ using the entropy method.

Step 4: Calculate permutation matrix D_σ, where $(\sigma_{(1)}, \sigma_{(2)}, ..., \sigma_{(n)})$ is a permutation of $(1, 2, ..., n)$ such that $P_{\sigma(j)} \prec P_{\sigma(j-1)}$, for all $j = 1, 2, ..., n$.

Step 5: Employed $PFOWA_w(P_1, P_2, ..., P_n)$ and $EOWA_w(S(P_1), S(P_2), ..., S(P_n))$ operators and derived attribute values of each alternative T_i, $i = 1, 2, ..., m$ and computed the corresponding scoring function values.

Step 7: The optimal choice may then be determined by ranking the attribute values based on the score function, with the alternative with the highest values being the one that is most desirable.

5 Numerical Example (A Case Study)

ABC Tech Solutions, a leading IT company, was tasked with selecting a suitable project management methodology for an upcoming software development project. The decision-makers needed to consider multiple criteria, including cost, project duration, team expertise and client preferences. Recognizing the need for a comprehensive decision-making approach, the project manager decided to employ MCDM techniques with Pythagorean fuzzy sets. This case study delves into the utilization of Pythagorean Fuzzy Sets to facilitate project management based on four distinct criteria: (i) Team Expertise: Skills and experience of the project team. (C_1), (ii) Client Preferences: Specific requirements and expectations (C_2) (iii) Project Duration: Time constraints and deadlines (C_3) and (iv) Cost: Budget constraints and resource allocation (C_4). By leveraging Pythagorean Fuzzy Sets, we aim to quantify the multi-faceted attributes of each app while accommodating the inherent uncertainties in decision-making.

5.1 *Problem Statement*

This case study considers five teams viz.T_1, T_2, T_3, T_4 and T_5 as alternatives. These teams are evaluated based on four criteria: benefits (C_1 & C_2) and Time/costs (C_3 & C_4). Pythagorean Fuzzy Numbers are employed to express the attributes' varying degrees of uncertainty.

5.2 *Methodology*

(I) Criteria Definition:

Benefits: (a Team Expertise: Skills and experience of the project team. (C_1): On a scale of 0 to 1. (b) Client Preferences: Specific requirements and expectations (C_2): On a scale of 0 to 1.

Costs: (a) Project Duration: Time constraints and deadlines (C_3): On a scale of 0 to 1. (b) Cost: Budget constraints and resource allocation (C_4): On a scale of 0 to 1.

(i) *Fuzzy Number Representation:* Each criterion's value is represented as a Pythagorean Fuzzy Number $P = (\mu_P, \gamma_P)$, where μ_P is the degree of membership γ_P is the degree of non-membership of the Pythagorean Fuzzy set.
(ii) *Scoring Functions:* We have considered the following scoring function to aggregate the different criteria.

$$S(P) = \frac{1 - \gamma_P^2}{2 - \mu_P^2 - \gamma_P^2}$$

(iii) *Solution Procedure:* Score function based Multi-Criteria decision-making based on Pythagorean Fuzzy Aggregation Operators viz. PFOWA & EOWA Operators to arrange Performances in MCDM.

For the purpose of this article, we have chosen five top-rated software developer teams viz.T_1, T_2, T_3, T_4 and T_5 as alternatives and four criteria viz., (i) Team Expertise: Skills and experience of the project team. (C_1), (ii) Client Preferences: Specific requirements and expectations (C_2) (iii) Project Duration: Time constraints and deadlines (C_3) and (iv) Cost: Budget constraints and resource allocation (C_4).

Table 1 shows the decision matrix in terms of Pythagorean fuzzy number rating values in decision-makers point of view.

Now, we have normalized the decision matrix. For the cost criteria, we have interchanged a membership degree with a non-membership degree. However, benefit criteria remain unchanged. The converted decision matrix has been presented in Table 2.

For the decision matrix displayed in Table 2 now, we have employed a score function measure. The converted decision matrix has been presented in Table 3.

Using the entropy method mentioned here, we have calculated the weight vector $w = (0.25, 0.24, 0.26, 0.25)^T$. Now, we have obtained the permutations based on score function values and the corresponding permutation matrix based on score function has been displayed in Table 4. A permutation decision matrix based on score function values has also been presented in Table 5.

Using $PFOWA_w(P_{\sigma(1)}, P_{\sigma(2)}, P_{\sigma(3)}, P_{\sigma(4)})$ and decision matrix presented in Table 4, we have obtained the following:

Table 1 Decision matrix for the problem

	C_1	C_2	C_3	C_4
T_1	(0.79, 0.45)	(0.37,0.62)	(0.65,0.30)	(0.73, 0.53)
T_2	(0.61,0.70)	(0.28,0.73)	(0.27,0.90)	(0.54, 0.82)
T_3	(0.31,0.60)	(0.32,0.51)	(0.45,0.54)	(0.36, 0.62)
T_4	(0.49, 0.58)	(0.36,0.63)	(0.48,0.33)	(0.48,0.57)
T_5	(0.30,0.40)	(0.85,0.32)	(0.63, 0.71)	(0.78,0.49)

Table 2 Decision matrix (benefit and cost criteria)

	C_1	C_2	C_3	C_4
T_1	(0.79, 0.45)	(0.37,0.62)	(0.30, 0.65)	(0.53, 0.73)
T_2	(0.61,0.70)	(0.28,0.73)	(0.90, 0.27)	(0.82, 0.54)
T_3	(0.31,0.60)	(0.32,0.51)	(0.54, 0.45)	(0.62, 0.36)
T_4	(0.49, 0.58)	(0.36,0.63)	(0.33, 0.48)	(0.57, 0.48)
T_5	(0.30,0.40)	(0.85,0.32)	(0.71, 0.63)	(0.49, 0.78)

Table 3 Converted decision matrix based on score function values

	C_1	C_2	C_3	C_4
T_1	0.680	0.416	0.388	0.394
T_2	0.448	0.336	0.830	0.684
T_3	0.415	0.452	0.530	0.586
T_4	0.466	0.409	0.463	0.533
T_5	0.480	0.764	0.549	0.340

Table 4 Permutation decision matrix based on score function

	$P_{\sigma(1)}$	$P_{\sigma(2)}$	$P_{\sigma(3)}$	$P_{\sigma(4)}$
T_1	(0.79, 0.45)	(0.37,0.62)	(0.53, 0.73)	(0.30, 0.65)
T_2	(0.90, 0.27)	(0.82, 0.54)	(0.61,0.70)	(0.28,0.73)
T_3	(0.62, 0.36)	(0.54, 0.45)	(0.32,0.51)	(0.31,0.60)
T_4	(0.57, 0.48)	(0.49, 0.58)	(0.33, 0.48)	(0.36,0.63)
T_5	(0.85,0.32)	(0.71, 0.63)	(0.30,0.40)	(0.49, 0.78)

Table 5 Permutation decision matrix based on score function values

	$S(P)_{\sigma(1)}$	$S(P)_{\sigma(2)}$	$S(P)_{\sigma(3)}$	$S(P)_{\sigma(4)}$
T_1	0.680	0.416	0.394	0.388
T_2	0.830	0.684	0.448	0.336
T_3	0.586	0.530	0.452	0.415
T_4	0.533	0.466	0.463	0.409
T_5	0.764	0.549	0.480	0.340

$PFOWA_w(T_1) = (0.57, 0.60)$ and $S(PFOWA_w(T_1)) = 0.48$.
$PFOWA_w(T_2) = (0.75, 0.52)$ and $S(PFOWA_w(T_2)) = 0.62$.
$PFOWA_w(T_3) = (0.48, 0.47)$ and $S(PFOWA_w(T_3)) = 0.50$.
$PFOWA_w(T_4) = (0.45, 0.54)$ and $S(PFOWA_w(T_4)) = 0.47$.
$PFOWA_w(T_5) = (0.66, 0.50)$ and $S(PFOWA_w(T_5)) = 0.57$.

Using $EOWA_w(S(P)_{\sigma(1)}, S(P)_{\sigma(2)}, S(P)_{\sigma(3)}, S(P)_{\sigma(4)})$ and decision matrix presented in Table 5, we have obtained the following:

$$EOWA_w(T_1) = 0.33$$

$$EOWA_w(T_2) = 0.43$$

$$EOWA_w(T_3) = 0.35$$

$$EOWA_w(T_4) = 0.32$$

$$EOWA_w(T_5) = 0.39$$

The best team based on PFOWA and EOWA Operators have been found using our suggested approach. The detailed results have been presented in Table 6.

Using the PFOWA and EOWA operators, the best team can be selected based on determined values. If the top team T_2 is not selected to complete the project on time, team T_5 will take over. A similar assignment will be made for the rest of the

Table 6 Results of the numerical experiment

Operator used	Rank of the alternatives	Best team
PFOWA	$T_2 \succ T_5 \succ T_3 \succ T_1 \succ T_4$	T_2
EOWA	$T_2 \succ T_5 \succ T_3 \succ T_1 \succ T_4$	T_2

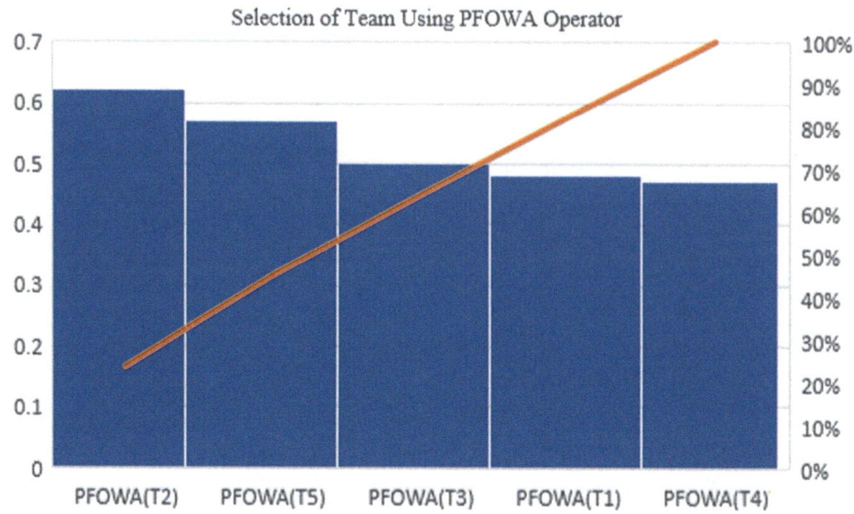

Fig. 2 Suitable team selection as per PFOWA operator

team. Therefore, based on the merit of the computed values, the project manager might assign the work. The team selection can be visualised in Figs. 2 and 3. It needs to be mentioned that the PFOWA and EOWA operators yield the same outcomes. Therefore, we asserted that the two operators are equivalent. It is also important to note that when we use our scoring function in association with the aggregation approach to address decision-making problems, it performs well for both operators.

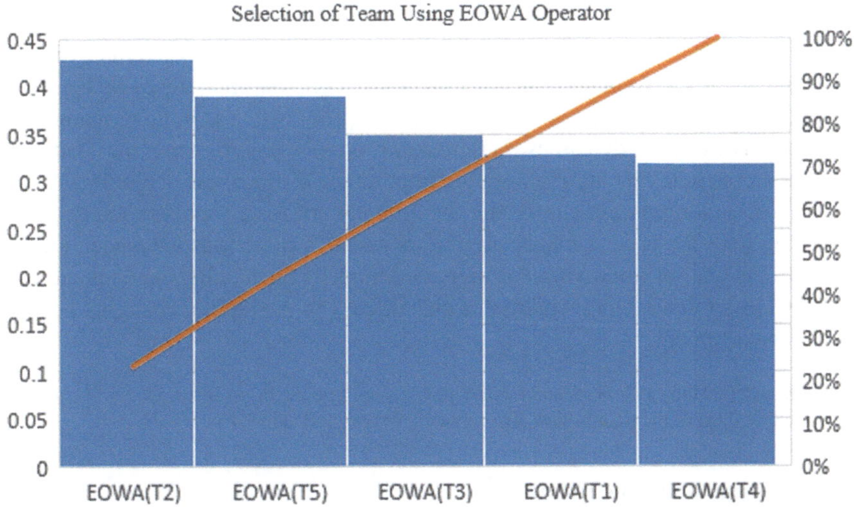

Fig. 3 Suitable team selection as per EOWA operator

6 Conclusion

In conclusion, the research conducted to select the best team by the project manager using Multiple Criteria Decision-Making (MCDM) techniques with the incorporation of Pythagorean Fuzzy Numbers and the aggregation operators PFOWA (Pythagorean Fuzzy Ordered Weighted Averaging) and EOWA (Einstein Ordered Weighted Average) has provided valuable insights into handling the varying degrees of uncertainty associated with project team selection. By using Pythagorean Fuzzy Numbers, the ambiguity and imprecision present in the decision-making process could be represented more realistically. A thorough evaluation of the criteria has been made possible by the PFOWA and EOWA aggregation operators, which considered the significance and the ordered structure of the decision variables. This method improved the process of making decisions by empowering project managers to choose their teams wisely by considering a variety of factors in a comprehensive manner. The findings of the study add to the expanding amount of knowledge in MCDM, especially when it comes to project management, where making decisions is influenced by a variety of competing factors and unknowns. To help decision-makers handle complex choice environments, Pythagorean Fuzzy Numbers, and the aggregation operators PFOWA and EOWA are applied. This eventually improves the efficacy and efficiency of team selection processes. As for the future scope of research, several avenues present themselves for further exploration and refinement. Firstly, the extension of the proposed model to accommodate a wider range of criteria and additional decision-making scenarios would enhance its applicability in diverse project environments. Furthermore, the incorporation of realistic case studies and the practical implementation of the devised framework may furnish empirical confirmation and elucidate the model's efficacy in factual project environments. Additionally, hybrid strategies that incorporate machine learning algorithms for predictive analytics or Pythagorean Fuzzy Numbers with other methods of decision-making could be investigated. This would facilitate a decision-making process that is more dynamic and adaptive, taking team dynamics and changing project requirements into account. The study concludes by highlighting the importance of fuzziness and ambiguity in the team selection process and laying the groundwork for a sophisticated decision support system for project managers. Future research directions have been identified with the goal of improving the suggested model's practicality, generalizability, and flexibility to maintain its relevance and efficacy in the rapidly changing field of project management.

Author Contribution J.C., S.M. and L.S. formulated and studied the problem. J.C., S.M. and L.S. wrote the first draft of the manuscript. All authors have read and agreed to the published version of the manuscript.

Compliance with Ethical Standards

Conflict of Interest The authors confirm that they do not have any competing interests.

Ethical Approval None of the authors of this article have conducted any experiments using humans or animals.

Informed Consent All individuals who took part in the study provided their informed permission.

References

1. Huljenic, D., Desic, S., Matijasevic, M.: Project management in research projects. In: Proceedings of the 8th International Conference on Telecommunications, 2005. ConTEL 2005. (Vol. 2, pp. 663–669). IEEE (2005)
2. Wu, X., Tsai, J. C. A., Lei, Y.: Information technology project management research: A review of works by influential pioneers. Proj. Manag. J. 87569728231171056 (2023)
3. Tesch, D., Ireland, L. R., Liu, J. Y. C.: Project management: IS/IT research challenges. J. Int. Technol. Inf. Manag. **17**(1), 4 (2008)
4. Chen, H.G., Jiang, J.J., Chen, J.C., Shim, J.T.: The impacts of conflicts on requirements uncertainty and project performance. J. Int. Technol. Inf. Manag. **13**(3), 2 (2004)
5. Argote, L.: Input uncertainty and organizational coordination in hospital emergency units. Adm. Sci. Q. 420–434 (1982)
6. Nidumolu, S.: The effect of coordination and uncertainty on software project performance: residual performance risk as an intervening variable. Inf. Syst. Res. **6**(3), 191–219 (1995)
7. Salehi, K.: A hybrid fuzzy MCDM approach for project selection problem. Decis. Sci. Lett. **4**(1), 109–116 (2015)
8. Badri, M.A., Davis, D., Davis, D.: A comprehensive 0–1 goal programming model for project selection. Int. J. Proj. Manag. **19**(4), 243–252 (2001)
9. Benli, O.S., Yavuz, S.: Making project selection decisions: a multi-period capital budgeting problem. Int. J. Ind. Eng.-Theory Appl. Pract. **9**(3), 301–310 (2002)
10. Gabriel, S.A., Kumar, S., Ordonez, J., Nasserian, A.: A multi-objective optimization model for project selection with probabilistic considerations. Socio-Econ. Plan. Sci. **40**(4), 297–313 (2006)

11. Huang, X.: Optimal project selection with random fuzzy parameters. Int. J. Prod. Econ. **106**(2), 513–522 (2007)
12. Carlsson, C., Fullér, R., Heikkilä, M., Majlender, P.: A fuzzy approach to R&D project portfolio selection. Int. J. Approx. Reason. **44**(2), 93–105 (2007)
13. Rabbani, M., Aramoon Bajestani, M., Baharian Khoshkhou, G.: A multi-objective particle swarm optimization for project selection problem. Expert. Syst. Appl. **37**(1), 315321 (2010)
14. Mohanty, R.P.: Project selection by a multiple-criteria decision-making method: an example from a developing country. Int. J. Proj. Manag. **10**(1), 31–38 (1992)
15. Mohanty, R.P., Agarwal, R., Choudhury, A.K., Tiwari, M.K.: A fuzzy ANP-based approach to R&D project selection: a case study. Int. J. Prod. Res. **43**(24), 5199–5216 (2005)
16. Enea, M., Piazza, T.: Project selection by constrained fuzzy AHP. Fuzzy Optim. Decis. Mak. **3**(1), 39–62 (2004)
17. Mahmoodzadeh, S., Shahrabi, J., Pariazar, M., Zaeri, M.S.: Project selection by using fuzzy AHP and TOPSIS technique. world academy of science. Eng. Technol. **30**, 333–338 (2007)
18. Amiri, M.P.: Project selection for oil-fields development by using the AHP and fuzzy TOPSIS methods. Expert. Syst. Appl. **37**(9), 6218–6224 (2010)
19. Daneshvar Rouyendegh, B., Erol, S.: Selecting the best project using the Fuzzy ELECTRE Method. Math. Probl. Eng. (2012)
20. Chu, M.T., Shyu, J., Tzeng, G.H., Khosla, R.: Comparison among three analytical methods for knowledge communities' group-decision analysis. Expert. Syst. Appl. **33**(4), 1011–1024 (2007)
21. Opricovic, S., Tzeng, G.H.: Compromise solution by MCDM methods: A comparative analysis of VIKOR and TOPSIS. Eur. J. Oper. Res. **156**(2), 445–455 (2004)
22. Raei, R., Jahromi, M.: Portfolio optimization using a hybrid of fuzzy ANP, VIKOR and TOPSIS. Manag. Sci. Lett. **2**(7), 2473–2484 (2012)
23. Zadeh, L.A., Klir, G.J., Yuan, B.: Fuzzy sets, fuzzy logic, and fuzzy systems: selected chapters (Vol. 6). World scientific (1996)
24. Atanassov, K.T., Atanassov, K.T.: Intuitionistic fuzzy sets (pp. 1–137). Physica-Verlag HD (1999)
25. Atanassov, K.T.: More on intuitionistic fuzzy sets. Fuzzy Sets Syst. **33**(1), 37–45 (1989)
26. Yager, R.R.: Pythagorean fuzzy subsets. In 2013 joint IFSA world congress and NAFIPS annual meeting (IFSA/NAFIPS) (pp. 57–61). IEEE (2013)
27. Yager, R.R.: Pythagorean membership grades in multicriteria decision making. IEEE Trans. Fuzzy Syst. **22**(4), 958–965 (2013)
28. Wan, S. P., Li, D.F.: Atanassov's intuitionistic fuzzy programming method for heterogeneous multiattribute group decision making with Atanassov's intuitionistic fuzzy truth degrees. IEEE Trans. Fuzzy Syst. **22**(2), 300–312 (2013)
29. Chen, S.M., Cheng, S.H., Lan, T.C.: Multicriteria decision making based on the TOPSIS method and similarity measures between intuitionistic fuzzy values. Inf. Sci. **367**, 279–295 (2016)
30. Zhang, X.: Multicriteria Pythagorean fuzzy decision analysis: A hierarchical QUALIFLEX approach with the closeness index-based ranking methods. Inf. Sci. **330**, 104–124 (2016)
31. Peng, X., Dai, J.: Approaches to Pythagorean fuzzy stochastic multi-criteria decision making based on prospect theory and regret theory with new distance measure and score function. Int. J. Intell. Syst. **32**(11), 1187–1214 (2017)
32. Peng, X.: Algorithm for Pythagorean fuzzy multi-criteria decision making based on WDBA with new score function. Fundam. Inform. **165**(2), 99–137 (2019)
33. Huang, C., Lin, M., Xu, Z.: Pythagorean fuzzy MULTIMOORA method based on distance measure and score function: its application in multicriteria decision making process. Knowl. Inf. Syst. **62**, 4373–4406 (2020)
34. Li, D., Zeng, W.: Distance measure of Pythagorean fuzzy sets. Int. J. Intell. Syst. **33**(2), 348–361 (2018)
35. Li, Y., Sun, G., Li, X.: A novel unified ranking method of Pythagorean fuzzy numbers characterized by the residual sector area (2022)

36. Rahman, K., Abdullah, S., Ali, A., Amin, F.: Pythagorean fuzzy ordered weighted averaging aggregation operator and their application to multiple attribute group decision-making. EURO J. Decis. Process. **8**, 61–77 (2020)
37. Zhang, X., Xu, Z.: Extension of TOPSIS to multiple criteria decision making with Pythagorean fuzzy sets. Int. J. Intell. Syst. **29**(12), 1061–1078 (2014)
38. Yager, R.R., Abbasov, A.M.: Pythagorean membership grades, complex numbers, and decision making. Int. J. Intell. Syst. **28**(5), 436–452 (2013)
39. Shannon, C.E.: A mathematical theory of communication. Bell Syst. Tech. J. **27**(3), 379–423 (1948)

Author Index

L. Sahoo et al. (eds.), *Decision Making Under Uncertainty Via Optimization, Modelling, and Analysis*, Studies in Systems, Decision and Control 558,
https://doi.org/10.1007/978-981-96-0085-4

GPSR Compliance

The European Union's (EU) General Product Safety Regulation (GPSR) is a set of rules that requires consumer products to be safe and our obligations to ensure this.

If you have any concerns about our products, you can contact us on ProductSafety@springernature.com

In case Publisher is established outside the EU, the EU authorized representative is:

Springer Nature Customer Service Center GmbH
Europaplatz 3
69115 Heidelberg, Germany

Batch number: 08164694

Printed by Printforce, the Netherlands